DYNAMIC ASPECTS OF CONFORMATION CHANGES IN BIOLOGICAL MACROMOLECULES

DYNAMIC ASPECTS OF CONFORMATION CHANGES IN BIOLOGICAL MACROMOLECULES

PROCEEDINGS OF THE 23RD ANNUAL MEETING
OF THE SOCIÉTÉ DE CHIMIE PHYSIQUE
ORLÉANS, 19–22 SEPTEMBER 1972

Edited by

CHARLES SADRON

Professeur au Muséum National d'Histoire Naturelle
Directeur du Centre de Biophysique Moléculaire
(C.N.R.S.)

D. REIDEL PUBLISHING COMPANY

DORDRECHT-HOLLAND / BOSTON-U.S.A.

First printing: December 1973

Library of Congress Catalog Card Number 72–97962

ISBN-13: 978-94-010-2581-2 e-ISBN-13: 978-94-010-2579-9
DOI: 10.1007/978-94-010-2579-9

Published by D. Reidel Publishing Company,
P.O. Box 17, Dordrecht, Holland

Sold and distributed in the U.S.A., Canada, and Mexico
by D. Reidel Publishing Company, Inc.
306 Dartmouth Street, Boston,
Mass. 02116, U.S.A.

Softcover reprint of the hardcover 1st edition 1973

TABLE OF CONTENTS

PREFACE VII

ORGANIZING COMMITTEE IX

ACKNOWLEDGMENTS IX

LIST OF PARTICIPANTS XI

B. PULLMAN / Quantum-Mechanical Studies on the Conformation of Biomolecules 1

M. PTAK / Analyse conformationnelle des biopolymères par les méthodes semi-empiriques 23

M. GOLDBERG / l'État natif est-il l'état fondamental? 57

E. M. BRADBURY, P. D. CARY, C. CRANE-ROBINSON, and H. W. E. RATTLE / Conformations and Interactions of Histones and Their Role in Chromosome Structure 67

C. TONIOLO / Conformational Equilibria and Stability of Oligopeptides 87

A. HVIDT / Isotopic Hydrogen Exchange in Solutions of Biological Macromolecules 103

P. COLSON and J. P. GASPARD / Determination of the Relaxation Spectrum of Macromolecules in Solution by the Analysis of the Electric Birefringence 117

PH. WAHL and J. C. BROCHON / Resolution of the Fluorescence Spectra of Proteins Using Decay Measurements 129

L. F. TEN EYCK and P. D. PULSINELLI / Tertiary and Quaternary Structural States of Hemoglobin 137

K. WÜTHRICH, R. M. KELLER, and R. BAUMANN / Proton and Carbon-13 Nuclear Magnetic Resonances in Hemes and Hemoproteins: New Aspects for the Investigation of the Molecular Conformations 151

R. G. SHULMAN, D. R. KEARNS, B. R. REID, and Y. P. WONG / A High Resolution NMR Study of Hydrogen Bonded Protons in tRNA 165

B. ALPERT, R. BANERJEE et L. LINDQVIST / Cinétique des changements de conformation tertiaires de l'hémoglobine étudiée par photolyse-laser 171

A. BAUDRAS / Transitions réversibles d'activité et de structure quaternaire de la L(+) lactate: cytochrome *c* oxydoréductase (cytochrome b_2) de la levure *Hansenula anomala* 181

J. WYMAN / On Allosteric Models 207

H. BUC, M. H. BUC, M. MORANGE, L. C. OUDIN, and H. WINKLER / Conformational Changes in Glycogen Phosphorylase *b* from Rabbit Skeletal Muscle. Equilibrium and Kinetic Studies 225

R. CERF / Quelques problèmes de cinétique conformationnelle de systèmes coopératifs 247
D. THUSIUS, G. FOUCAULT, and F. GUILLAIN / The Analysis of Chemical Relaxation Amplitudes and Some Applications to Reactions Involving Macromolecules 271
M. LAZDUNSKI, C. LAZDUNSKI, C. PETITCLERC, D. CHAPPELET, M. FOSSET, C. GACHE, et M. DELAAGE / Mécanismes enzymatiques de type flip-flop 285
J. P. EBEL / Structure tridimensionnelle des acides ribonucléiques de transfert 301
J. J. HOPFIELD, S. OGAWA, and R. G. SHULMAN / The Rate of Carbon Monoxide Binding to Hemoglobin Kansas 327
J. RAMSTEIN, J. F. HOGREL, M. DOURLENT, M. LENG, et C. HÉLÈNE / Interactions de la proflavine avec différents ADN et polynucléotides de synthèse 333
C. HÉLÈNE, H. BORAZAN, J. L. DIMICOLI, J. C. MAURIZOT, M. DURAND, et J. J. TOULME / Interactions des acides aminés aromatiques avec les acides nucléiques 349
D. GENEST and PH. WAHL / Energy Transfers Study in the DNA–Ethidium Bromide Complex by means of Anisotropy Decay 367
M. LENG, V. DEUBEL, et J. C. THRIERR / Étude des interactions entre l'acide polyribo-uridylique ou l'acide polyribobromo-uridylique et des polycations 381
F. TRAVERS, A. M. MICHELSON, et P. DOUZOU / Acide poly-uridylique à basse température 393
W. GUSCHLBAUER, D. THIELE, M.-T. SAROCCHI, et CH. MARCK / Phénomènes d'hystérèse dans les polynucléotides 401
D. VASILESCU, M. TEBOUL, H. KRANCK, et B. CAMOUS / Étude de la transconformation thermique du Na–DNA en solution par la spectrographie de bruit 411
SH. LIFSON / Recent Developments in the Consistent Force Field Calculations 421
M. IWATSUBO, J. M. JALLON, et A. DI FRANCO / Étude cinétique des changements conformationnels de la glutamate déshydrogénase par substrats et effecteurs 431
J. M. YON et J.-P. TENU / Mouvements moléculaires consécutifs à la fixation de métaux dans la β-galactosidase 447
B. LABOUESSE / Mouvements intramoléculaires dans la chymotrypsine 459
R. D. KORNBERG / The Diffusion of Phospholipids in Membranes 477
PH. DEVAUX, C. J. SCANDELLA, et H. M. MCCONNELL / Étude par marquage de spin de la diffusion latérale dans les membranes biologiques 487
S. FERMANDJIAN, D. GREFF, et P. FROMAGEOT / Angiotensine II et analogues: Études conformationnelles comparatives 493
J. P. CHANGEUX / Caractérisation, isolement et purification de la protéine réceptrice de l'acétylcholine à partir de l'organe électrique du gymnote 511

PREFACE

On the day after the 1959 Cambridge Congress, during which the International Union of Pure and Applied Biophysics was founded, a biophysics section was formed within the Society of Physical Chemistry (Société de Chimie Physique).

Since then, three of the Society's annual meetings (the 11th, 17th, and 23rd) were devoted exclusively to the physico-chemical study of biological systems.

The first of these was held in June 1961 at a hotel in Col de Voza, at the foot of an alpine glacier above Chamonix.

The second, in May 1967, took place in the more learned setting of the venerable rooms of the National Museum of Natural History in Paris.

The third – the one dealt with in the present volume – was recently held at Orléans-La Source in the newly built lecture theatres of the young University, which is near the great Institutes of the National Centre for Scientific Research (CNRS), on the Sologne plateau.

These three stages are milestones of an evolution which characterises (at least schematically) the explosive evolution of biological physico-chemistry.

The first colloquium, with the title 'Deoxyribonucleic Acid: Structure, Synthesis and Functions', actually marks the first contact of the physical chemist with one of the then most prestigious biological macromolecules, the structure of which had just been discovered, and in this way celebrated one of the first and most striking successes of molecular biology.

The second, less biologically orientated, was devoted to the general study of helicoidal macromolecules in solution, thus marking a return to the fundamental preoccupations of the physical chemistry of macromolecules.

The third, however, represents a considerable leap forward because it not only examines critically the problems of the determination of a variety of structures but also of the dynamics of the structural changes these molecules undergo under the influence of physical as well as chemical agents, thus showing the way towards a general study of the relationship between structure and function.

It should perhaps be emphasised that, in spite of everything, we are still confined to the study of the macromolecular compounds in isolation from their native structural unit, and still far removed from the latter's complexities.

Nevertheless we would like to point out that the last session of the congress was devoted to membranes and their properties, thus providing an important starting point leading to the physico-chemical study of the structures within the organic entity of the living cell.

Without resorting to the daring extrapolations sometimes indulged in by the physical chemist when he allows himself to be too easily seduced by the wonderful world of biological facts, we do believe that this colloquium – no doubt one of many others equally as significant – represents a step forward on the long road towards combining in one unified theory both inanimate and living nature. We have been profoundly and sincerely inspired by all the eminent scientists, from abroad and from France, who honoured us with their presence and to whom we extend our warmest feelings of gratitude.

C. SADRON
Chairman of the Organizing Committee

ORGANIZING COMMITTEE

C. Sadron, *Chairman*
C. Troyanowsky, *General Secretary*
R. Banerjee
P. Bothorel
H. Buc
J. Coursaget
M. Daune
L. Monnerie
J. Monod
M. Ptak
B. Pullman
A. Rassat
Miss J. Yon
Mrs. O. Yovanovitch

ACKNOWLEDGMENTS

The 23rd Annual Meeting of the Société de chimie physique was organized with the financial help of the following organizations: Délégation générale à la recherche scientifique et technique; Ministère de l'éducation nationale: Direction de la coopération universitaire internationale; Direction des recherches et moyens d'essais; Commissariat à l'énergie atomique; and the following firms and industrial groups: Compagnie française de raffinage; Laboratoires Sandoz; Société l'Air Liquide; Société Kodak Pathé; Union des industries chimiques. We should like to renew our thanks to all the above.

LIST OF PARTICIPANTS

Albiser, G., Laboratoire de biophysique, Nancy
Alfsen, Mme, Faculté de Medecine, Laboratoire des Etats liés moléculaires, Paris
Alpert, B., Institut de biologie physico-chimique, Paris
Arrio, B., Institut de biochimie, Orsay
Aslanian, Mme, Physique des solides, Ecole Normale Supérieure, Paris
Aufauvre, F., Centre Universitaire des Sciences et Techniques, Clermont-Ferrand
Banerjee, R., Institut de biologie physico-chimique, Paris
Baudras, A., Institut de physiologie, Toulouse
Bechet, J. J., Enzymologie physico-chimique et moléculaire, Orsay
Berliner, C., Chimie générale II, Université de Bruxelles (Belg.)
Berteaud, A. J., Laboratoire du CNRS, Bellevue
Blandin, Mme, Centre de génétique moléculaire, Gif s/Yvette
Bothorel, P., Centre de recherches Paul Pascal, 33-Talence
Bottreau, A. M., Optique ultra-hertzienne, 33-Talence
Bouhet, J. C., Service de biochimie, CEN, Saclay
Bourgarit, J. J., Unité de physique biologique, Paris VII
Bourgoin, D., Institut Pasteur, Service de biophysique, Paris
Bradbury, E. M., Biophysics Laboratories, Portsmouth Polytechnic (G.B.)
Brochon, J. C., Centre de biophysique moléculaire, Orléans
Buc, H., Biochimie cellulaire, Institut Pasteur, Paris
Buc, M. H., Mme, Biochimie cellulaire, Institut Pasteur, Paris
Caloin, M., Acoustique moléculaire, Strasbourg
Cambrai, M., Laboratoire de recherches Unilever, La Plaine St Denis
Cardinaud, R., Département de biologie, CEN, Gif s/Yvette
Castaing, Mme, Laboratoire de biochimie physique, Orsay
Cerf, Acoustique moléculaire, Université B. Pascal, Strasbourg
Chaillot, B., Biologie physico-chimique, IBANA, Dijon
Chambron, J., Institut de physique biologique, Strasbourg
Changeux, J. P., Neurobiologie moléculaire, Institut Pasteur, Paris
Charlier, M., Centre de biophysique moléculaire, Orléans
Charmasson, R., Laboratoire de physique des liquides, Université de Provence, Marseille
Checcucci, A., Laboratorio per lo studio delle proprieta fisiche di bimolecule e cellule, Pisa (Ital.)
Clar, Mme, Laboratoire de recherches Unilever, La Plaine St Denis
Claverie, J. M., Unité de physique biologique, Université Paris VII
Clement, R., Groupe de biophysique du CRM, Strasbourg

Colonna-Cesari, F., Laboratoire de biophysique, Nancy
Colson, P., Service de chimie physique, Université de Liège (Belg.)
Colson-Guastalla, Mme, Laboratoire de biologie physico-chimique, Université de Montpellier
Cotrait, M., Laboratoire de cristallographie et de physique cristalline, Talence
Coursaget, J., Département de biologie du CEA, Gif s/Yvette
Daune, M., Université Blaise Pascal, CRM, Strasbourg
De Coen, J. L., Département de biologie moléculaire, Université de Bruxelles (Belg.)
Delbos, G., Optique ultra-hertzienne, 33-Talence
Devaux, P., Stanford University, California (U.S.A.)
Di Franco, A., Mlle, Enzymologie physico-chimique et moléculaire, Orsay
Donzel, B., Centre de biophysique moléculaire, Orléans
Dornand, J., ENSCM, Montpellier
Douzou, P., Institut de biologie physico-chimique, Paris
Dunand, J. J., Société CAMECA, Courbevoie
Ebel, J. P., Laboratoire de chimie biologique, Université L. Pasteur, Strasbourg
Ellenberger, M., Département de biologie, CEN, Saclay
Engelborghs, Y., Département de biologie moléculaire, Université de Bruxelles (Belg.)
Fermandjian, S., Service de biochimie, CEN Saclay., Gif s/Yvette
Filitti-Wurmser, S., Mme, Institut de biologie physico-chimique, Paris
Finer, E. G., Laboratoire de recherches Unilever, Welwyn (G.B.)
Fougnot, C., Mlle, Centre scientifique et polytechnique, Saint-Denis
Frédéricq, E., Laboratoire de chimie physique, Université de Liège (Belg.)
Freymann, R., Laboratoire de Physique moléculaire expérimentale, Université de Paris VI
Fuchs, R., Laboratoire de biophysique du CRM, Strasbourg
Gache, C., Centre de biochimie et de biologie moléculaire, Marseille
Garnier, J., Laboratoire de biochimie physique de l'INRA, Orsay
Garrigou-Lagrange, C., Mlle, Centre de recherches P. Pascal, 33-Talence
Gaspard, J. P., Institut Laue-Langevin, 38-Grenoble
Genest, D., Centre de biophysique moléculaire, Orléans
Gentou, C., Service de biochimie médicale, Faculté de médecine, Clermont-Ferrand
Gérard, D., Laboratoire de physique, UER de Sciences, Strasbourg
Gingold, Service de biophysique, CEN Saclay, Gif s/Yvette
Goldberg, M., Biochimie cellulaire, Institut Pasteur, Paris
Grosjean, H., Laboratoire de chimie biologique, Université libre de Bruxelles (Belg.)
Grouselle, M. J., Institut de biochimie, Orsay
Guinand, S., Mlle, Laboratoire de biologie physico-chimique, Orsay
Guschlbauer, W., Service de biochimie du CEN Saclay, Gif s/Yvette
Hanss, M., U.E.R. de médecine et biologie, Bobigny
Hélène, C., Centre de biophysique moléculaire, Orléans
Hellemans, L., Biol. Dynamics, Université de Louvain (Belg.)
Her, C., Laboratoire de recherches Unilever, La Plaine St-Denis

Holguin-Hueso, J., Département de biologie du CEA, Gif s/Yvette
Horn, P., Laboratoire de biophysique, Centre 1er cycle, Nancy
Huet, Mme, Unité de physique biologique, Paris VII
Hvidt, A., Mme, Laboratoire de chimie III, Université de Copenhague (P.B.)
Iwatsubo, M., Centre de génétique moléculaire, Gif s/Yvette
Joly, M., Service de biophysique, Institut Pasteur, Paris
Jozefonvicz, J., Mme, Centre scientifique et polytechnique, Saint-Denis
Kamenka, Mme, Laboratoire des Interactions moléculaires, Université de Montpellier
Karplus, M., Harvard University – Institut de biologie physico-chimique, Paris
Kellershorn, Enzymologie physico-chimique et moléculaire, Orsay
Kornberg, R., Stanford Univ., USA – MRC, Molecular biology, Cambridge, U.K.
Labeyrie, F., Mme, Centre de génétique moléculaire, Gif s/Yvette
Labouesse, B., Laboratoire de biochimie, Université de Bordeaux, 33-Talence
Lam, T. H., Département de biologie du CEN Saclay, Gif s/Yvette
Lami, H., Laboratoire de physique, UER Pharmacie, Strasbourg
Lang, M., Centre de recherches sur les macromolécules, Strasbourg
Lawrence, J. J., Laboratoire de biologie cellulaire, CEN Grenoble
Lazdunski, M., Centre de biologie moléculaire, Marseille
Lecompte, Mlle, Centre scientifique universitaire, Saint-Denis
Lefevre, J. F., Centre de recherches sur les macromolécules, Strasbourg
Leicknam, J. P., SRIRMA, CEN Saclay, Gif s/Yvette
Lenfant, J. P., Laboratoire de biochimie physique, INRA, Orsay
Leng, M., Centre de biophysique moléculaire, Orléans
Levilliers, Mlle, Institut de biochimie, Orsay – Hopital Cochin, Paris
Lifson, S., Weizmann Institute, Rehovot, (Isr.)
Loze, C. de, Mlle, Service de chimie physique, CNRS, Thiais
Lussan, C., Centre de recherches Paul Pascal, 33-Talence
Magat, M., Physico-chimie des rayonnements, Orsay
Maigret, B., Institut de biologie physico-chimique, Paris
Majoube, Département de biologie, CEN Saclay, Gif s/Yvette
Mani, J. C., ENSCM, Montpellier
Marche, P., Département de biologie, CEN Saclay, Gif s/Yvette
Mareschi, M., Centre d'études cryogéniques de l'Air Liquide, 38-Sassenage
Marzat, C., Optique ultra-hertzienne, Université de Bordeaux
Matricon, J., Laboratoire de physique des solides, Université de Paris VII
Maurizot, J. C., Centre de biophysique moléculaire, Orléans
Melki, G., Laboratoire de biophysique, Centre anticancéreux, Rennes
Menèzes, L. C., Institut de biochimie, Orsay
Mestdagh, Mlle, Université de Louvain (Belg.)
Metzger, G., Centre de recherche sur les macromolécules, Strasbourg
Michels, B., Acoustique moléculaire, Université L. Pasteur, Strasbourg
Monsigny, M., Centre de biophysique moléculaire, Orléans

Morange, M., Biologie moléculaire, Institut Pasteur, Paris
Morel, J. E., Département de biologie du CEN Saclay, Gif s/Yvette
Morgat, J. L., Service de biochimie du CEA, Gif s/Yvette
Morin, P., Laboratoire de biologie physico-chimique, Université de Montpellier
Muller, D., Centre scientifique universitaire de Saint-Denis
Murcia, X. de, Laboratoire de biophysique, Université L. Pasteur, Strasbourg
Nabredick, E., Mme, Département de biologie, CEN Saclay, Gif s/Yvette
Pairault, X., Hôpital Cochin, Paris
Pantaloni, D., Laboratoire d'enzymologie du CNRS, Gif s/Yvette
Pernollet, J. C., Laboratoire de biochimie physique, INRA, Orsay
Persoons, A., Université de Louvain (Belg.)
Pinset, I., Mme, Département de biologie, CEN Saclay, Gif s/Yvette
Popot, J. L., Unité de neurobiologie moléculaire, Institut Pasteur, Paris
Pouradier, J., Société Kodak Pathé, Vincennes
Pouyet, J., Centre de recherches sur les macromolécules, Strasbourg
Premillat, S., Laboratoire de biophysique, centre de 1er cycle, Nancy
Ptak, M., Centre de biophysique moléculaire, Orléans
Pudles, J., Institut de biochimie, Orsay
Pullman, B., Institut de biologie physico-chimique, Paris
Ramstein, J., Centre de biophysique moléculaire, Orléans
Rassat, A., Centre d'Etudes Nucléaires de Grenoble
Reiss, C., Centre de génétique moléculaire, Gif s/Yvette
Rinaudo, M., Mme, CERMAV, Grenoble
Rodriguez, J. M.
Ruysschaert, J. M., Chimie générale II, Université libre de Bruxelles (Belg.)
Sadron, C., Centre de biophysique moléculaire, Orléans
Sanson, A., Centre de biophysique moléculaire, Orléans
Schmitt, B., Unité de physique biologique, Université Paris VII
Schott, B., Groupe de physique des solides, Université Paris VII
Schreiber, J. P., Centre de recherches sur les macromolécules, Strasbourg
Schulz, J. A., Laboratoire d'acoustique moléculaire, Strasbourg
Schweitz, H., Centre de biochimie moléculaire, Marseille
Shulman, R. G., Bell Laboratories, Murray Hill (U.S.A.)
Sturm, Centre de recherches sur les macromolecules, Strasbourg
Tassin, J. J., Société CAMECA, Courbevoie
Tchuempe-Tchuente, G., Laboratoire de physique médicale, Rennes
Teissie, J., Chimie macromoléculaire, EPCI, Paris
Ten Eyck, L. F., MRC Laboratories of Molecular Biology, Cambridge (G.B.)
Tenu, J. P., Laboratoire d'enzymologie physico-chimique et moléculaire, Orsay
Thiele, Mme, Département de biologie, CEN Saclay, Gif s/Yvette
Thierry, C., Mme, Département de biologie, CEN Saclay, Gif s/Yvette
Thiery, J., Département de biologie, CEN Saclay, Gif s/Yvette
Thiriet, Y., Acoustique moléculaire, Strasbourg

Thusius, D., Laboratoire d'enzymologie physico-chimique et moléculaire, Orsay
Toniolo, C., Institut de chimie organique, Université de Padoue (Ital.)
Travers, F., Institut de biologie physico-chimique, Paris
Troyanowsky, C., Laboratoire de chimie physique, EPCI, Paris
Vasilescu, D., Laboratoire de biophysique, Université de Nice
Venard, R., Mlle, Laboratoire de biologie physico-chimique, Orsay
Viallet, P., Laboratoire de chimie physique, Centre Universitaire de Perpignan
Vigny, P., Laboratoire Curie, Institut du Radium, Paris
Vincent, J. P., Centre de biologie moléculaire du CNRS, Marseille
Weill, G., Centre de recherches sur les macromolécules, Strasbourg
Weintraub, Mme, Laboratoire des Hormones, Hôpital du Kremlin-Bicêtre
Wurmser, R., Institut de biologie physico-chimique, Paris
Wüthrich, K., Institut de biologie moléculaire et biophysique, Zürich (Suisse)
Yon, J., Mlle, Laboratoire d'enzymologie physico-chimique et moléculaire, Orsay
Zana, R., Centre de recherches sur les macromolécules, Strasbourg

QUANTUM-MECHANICAL STUDIES ON THE CONFORMATION OF BIOMOLECULES

BERNARD PULLMAN

Institut de Biologie Physico-Chimique, Laboratoire de Biochimie Théorique, associé au CNRS, 13, rue Pierre et Marie Curie, Paris 5, France

Summary. In contrast to 'empirical' computations which divide the potential energy of a system into several discrete contributions and evaluate each of them separately, quantum-mechanical methods enable the evaluation of the total molecular energy of a molecular system corresponding to any given configuration of the constituent atoms. The author describes the results of a quantum-mechanical calculation on the conformational properties of the amino-acid residues of proteins and of the constituents of the nucleic acids. The theoretical results are compared with the conformations of the compounds and the corresponding biopolymers. Comparison indicates the superiority of the quantum-mechanical PCILO method over the empirical and simpler quantum-mechanical procedures such as EHT or CNDO/2.

Résumé. A la différence des calculs 'empiriques' qui décomposent l'énergie potentielle d'un système en composantes qu'ils évaluent individuellement, les calculs quantiques permettent une évaluation globale de l'énergie interne d'un système moléculaire. L'auteur présente les résultats des calculs quantiques effectués pour les résidus acides aminés des protéines et pour les constituants des acides nucléiques. Une comparaison est faite avec les propriétés conformationnelles de ces entités dans les composés modèles et dans les biopolymères. On démontre la supériorité des calculs effectués par la méthode quantique PCILO sur les calculs empiriques et sur les calculs effectués par des méthodes quantiques plus simples, EHT ou CNDO/2.

1. Introduction

Because of its great complexity and the serious difficulties encountered in its study, the problem of the conformational properties of biological molecules and polymers is an excellent field for possible fruitful collaboration between theory and experiment. In fact, parallel to the accumulation of experimental data, there has been during the last few years an extremely prominent development of theoretical investigations in this field.

The bulk of these studies, although theoretical, are generally referred to as 'empirical' because of their mode of dealing with the problem. They consist of *partitioning the potential energy* of the system into several discrete contributions, such as nonbonded interactions, electrostatic interactions, barriers to internal rotations, hydrogen-bonding, etc., which are then evaluated with the help of empirical formulas deduced from studies on model compounds of small molecular weight. In the simplest approximation of these procedures (the 'hard sphere' approximation), introduced by Ramachandran and his collaborators [1] and which practically inaugurated this area of research, the problem is even limited to the sole evaluation of allowed or forbidden contacts, with the help of Van der Waals (or similar) radii.

Praiseworthy as such attempts are, they suffer from two obvious drawbacks. In the first place, whatever the practical justification for the partitioning of the total potential energy into a series of components, the procedure involves necessarily an

C. Sadron (ed.), Dynamic Aspects of Conformation Changes in Biological Macromolecules. 1–22. *All Rights Reserved.*

element of arbitrariness and possibly incompleteness. In fact, most authors take into consideration different selected components only and frequently different sets of them in different publications. Secondly, the fundamental formulas and parameters used to define the various components are far from being well established and differ, often appreciably, from one author to another.

A more rigorous theoretical investigation, starting from a more fundamental point of view and taking less cognizance of empirical data, seems therefore highly desirable. Such an investigation can be made through a direct quantum-mechanical treatment of the problem. In fact, this is the more desirable as at the comparatively small interatomic distances involved in this problem the various electronic interactions involved cannot be treated satisfactorily by the classical procedures.

Such a quantum-mechanical treatment has now, at least in principle become possible due essentially to the recent development of methods which, operating always within the general scheme of the molecular orbital method, deal simultaneously with all valence, σ and π, or even all (including inner shell) electrons. Such methods, the most prominent among which are the Extended Hückel Method [2], the CNDO/2 method [3], the PCILO method [4, 5, 6] and the so-called *ab initio* procedure [7] are able to evaluate the *total molecular energy* corresponding to any given configuration of the constituent atoms and are thus able to choose the preferred ones. In this way they avoid the artificial decomposition of the energy into constituent terms and the empiricism of their estimation. The precision of the quantum-mechanical results depends naturally on the refinement of the quantum-mechanical method utilized but they all follow the common principle of a *global* evaluation of molecular energy.

Among the quantum-mechanical methods used preferentially in our laboratory is the PCILO method. It is also the method which has been the most extensively applied to biochemical conformational problems and which has produced the greatest amount of results. We shall describe here some of its essential contributions in the field of fundamental biological molecules and compare them with the results of other methods and with the available experimental data.

2. Proteins and Their Constituents

As is well known the theoretical investigations in this field have been in their overwhelming majority based on the so-called 'dipeptide' approximation, the definitions and conventions of which are illustrated in Figure 1 [8].

The fundamental observation (see e.g. [9]) which is at the basis of this approximation is that *because of the planarity of the peptide unit, the flexibility of the main backbone of the polypeptide chain originates essentially from the possibilities of rotation about the* $N{-}C^{\alpha}$ and $C^{\alpha}{-}C'$ *single bonds adjacent to the* α*-carbons*. Consequently the conformation of, say, a pair of peptide units can be specified by giving the values of the two dihedral angles Φ and Ψ around the C^{α} carbon joining these units. The conformation of the backbone of the whole polypeptide chain may be described by indicating the sequence of these angles along the chain.

The second fundamental observation indicates that while the values assigned to one angle of a pair, Φ_i, depend markedly on the values assigned to the other angle of the same pair, Ψ_i, *the interactions associated with rotations of one such pair are largely independent of the angles assumed by the neighbouring pairs* Φ_{i-1}, Ψ_{i-1} and Φ_{i+1}, Ψ_{i+1}. In other words *short-range interactions* of an amino acid residue in a polypeptide chain essentially involve the two peptide units directly connected with the residue.

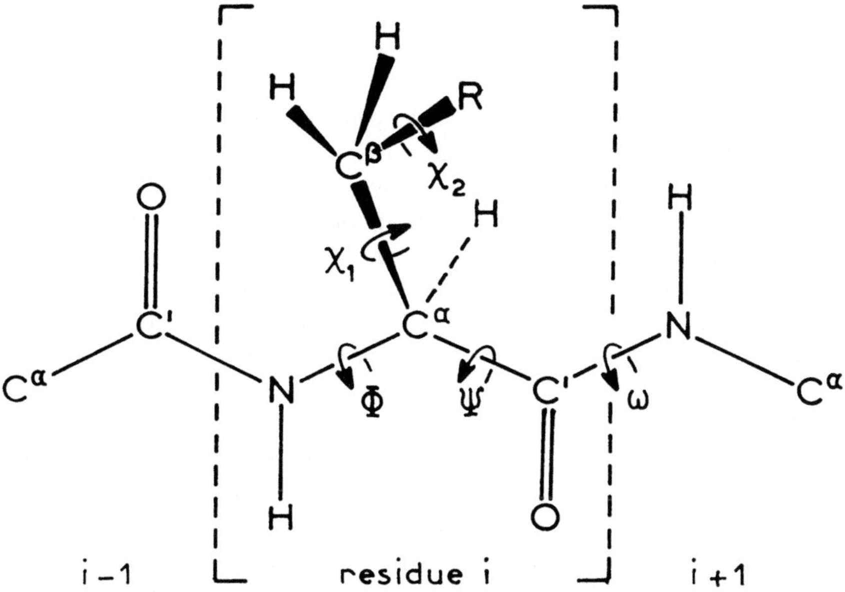

Fig. 1. Standard conventions for studying the conformaton of polypeptides (brackets indicate limits of a residue).

due. As a result of this state of affairs each pair of related rotational angles Φ and Ψ can, at least at first approximation, be treated separately from the others of the chain.

It is for these reasons that both the empirical and the quantum-mechanical computations have as an essential *first objective* the study of the conformational properties of α-amino acid residues as manifested in 'dipeptides' which are blocks corresponding to the fragment of a polypeptide chain represented in Figure 1. More specifically the calculations refer generally to structures I:

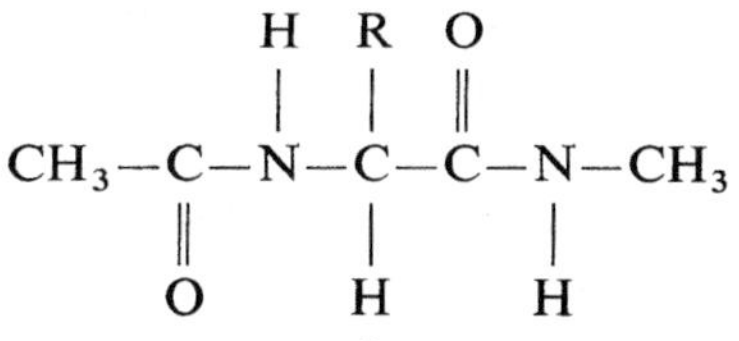

I

It is expected that the study of these model compounds will provide information on their conformational characteristics which should help us to understand the behaviour of the residues in larger units, polypeptides and proteins and thus to elucidate the origin of the conformational properties of these polymers. It should, however, be fully realized from the very beginning that because we are dealing essentially with a model and whatever the *a priori* justification for the utilization of the particular model adopted, we are necessarily faced with a double problem:

(1) the validity of the calculations for the very model compound itself;

(2) the validity, or better the significance of the calculations for the system which the model is supposed to represent, i.e., in our case for the conformation of the amino acid residues *in* proteins.

The attraction of the second aspect of the problem, by far of course the more important in the long term, was so overwhelming that it induced most authors working in the field to neglect the first aspect. This was regretable because it prevented the most straightforward evaluation of the validity of the different types of computation We shall see the significance of this remark shortly.

Let us consider in the first place two fundamental and most studied examples: those of the glycyl and alanyl residues and look, to start with, at the results from the viewpoint of their significance for the model dipeptide compound.

A. GLY RESIDUE

The PCILO conformational energy map for the ground state of this residue as obtained by the study of I with $R=H$ is available in Figure 2 [10]. The most stable conformation is predicted for the combination $\Phi=90°$, $\Psi=240°$ and the symmetric position $\Phi=270°$, $\Psi=120°$. It corresponds to a seven-membered ring (Figure 3a) called C_7 stabilized by an intramolecular hydrogen bond between $H^2 \ldots O^1$.

Secondary, local minima appear 2 kcal/mole above the global preceding minimum in the regions $\Phi=0°$, $\Psi=0°$ and $\Phi=180°$, $\Psi=90°$ (and in the symmetrical area). The first of these secondary minima corresponds to the fully extended (FE) form of the molecule stabilized by a weak intramolecular hydrogen bond formed between $O^2 \ldots H^1$ and leading to a cyclic pentagonal structure called C_5 (Figure 3b). The other secondary minimum (called conformation M) corresponds to a minimum of the repulsive part of the energy along the line $\Phi=180°$ (Figure 3c).

How do other methods behave with respect to these specific predictions? This is indicated in Table I, in which for reasons which will shortly become obvious we have limited our attention to the C_7 and C_5 conformations. It can be immediately seen that none of the other procedures predicts energy minima for *both* these conformations. Empirical computations generally show no energy minima for any of them, although some include the C_7 region in the allowed zone. Only one such recent calculation by Popov *et al.* [11] explicitly predicts C_7 as the most stable conformer. The EHT calculations, especially the most detailed of them of Momany *et al.* [12], predict a stable C_5 form, but ignore the C_7 form.

The last column in Table I indicates the experimental results on the preferred con-

TABLE I

The preferred conformations of the GLY dipeptide (parentheses indicate allowed conformation which do not correspond, however, to energy minima)

Conformations \ Methods		Hard sphere		+ H bond	Empirical potential functions				Quantum-mechanical					Exp. (l)
									EHT				PCILO (j), (k)	
		(a)	(b)	(c)	(d)	(e)	(f)	(g)	(h)	(i)	(m)	(n)		
C_5	$\Phi°$	–	–	–	–	–	–	–	?	0	45	90	0	0
	$\Psi°$								0	0	0	0	0	0
C^7	$\Phi°$	–	–		(90)	(90)	(100)	113	–	–	–	–	90	105
	$\Psi°$				(250)	(270)	(270)	249					240	230
Order of decreasing stability ↓					R_α, L_α (C_7)								C_7 C_5	C_7 C_5

References to Table I

(a) Ramakrishnan, C. and Ramachandran, G. N.: 1965, *Biophys. J.* **5**, 911.
(b) Leach, S. J., Nemethy, G., and Scheraga, H. A.: 1966, *Biopolymers* **4**, 369.
(c) Ramachandran, G. N., Venkatachalam, C., and Krimm, S.: 1966, *Biophys. J.* **6**, 849.
(d) Scott, R. A. and Scheraga, H. A.: 1966, *J. Chem. Phys.* **45**, 2091.
(e) Brant, D. A., Miller, W. G., and Flory, P. J.: 1967, *J. Mol. Biol.* **23**, 47.
(f) Ponnuswamy, P. K.: 1970, Thesis, University of Madras, India.
(g) Popov, E. M., Lipkind, G. M., Arkhipova, S. F., and Dashevskii, V. G.: 1968, *Mol. Biol.* **2**, 622.
(h) Hoffman, R. and Imamura, A.: 1969, *Biopolymers* **7**, 207.
(i) Momany, F. A., McGuire, R. F., Yan, J. F., and Scheraga, H. A.: 1971, *J. Phys. Chem.* **75**, 2286.
(j) Maigret, B., Pullman, B., and Dreyfus, M.: 1970, *J. Theoret. Biol.* **26**, 321.
(k) Maigret, B.: 1971, Thesis, University of Paris VI.
(l) Marraud, M.: 1971, Thesis, University of Nancy.
(m) Hopfinger, A. J.: 1970, Thesis, Case Western Reserve University.
(n) Govil, G.: 1970, *J. Chem. Soc.* A 2464.

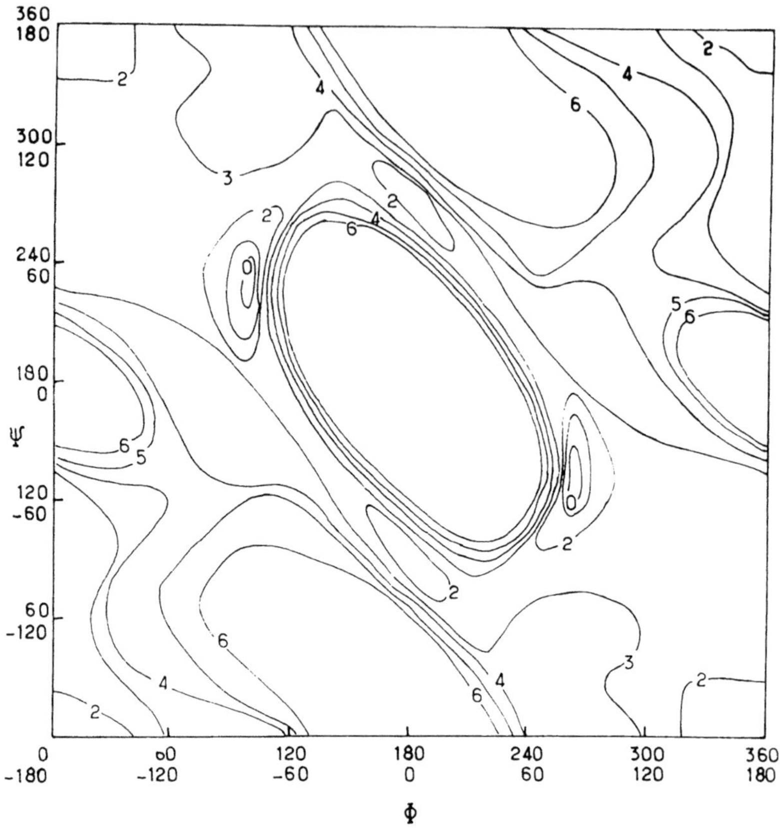

Fig. 2. PCILO conformational energy map for the GLY residue. Isoenergy curves in kcal/mole with respect to the global minimum taken as energy zero.

formations of the model compound obtained by NMR and infrared techniques by Neel, Lascombe and co-workers [13–16]. The experiments of these authors, carried out in inert solvents (CCl_4), correspond to conditions not too far removed from the isolated molecule used in the computations. The results indicate the existence of the model compound in the two preferred conformations *predicted by the PCILO calculations*: the seven membered hydrogen-bonded ring (C_7) with angles $\Phi \approx 105°$, $\Psi \approx 230°$ and the fully extended form, involving a five-membered hydrogen-bonded ring (C_5), with angles $\Phi = \Psi = 0°$. A refinement of the experimental technique of quantitative analysis of the dipeptide [17] leads to an approximate evaluation of the proportion of the two forms. The agreement with the theoretical predictions is again satisfactory, the C_7 form being the predominant one. There is no indication as to the existence of the M form.

B. ALA RESIDUE

The PCILO conformational energy map is presented in Figure 4 [1]. It indicates the

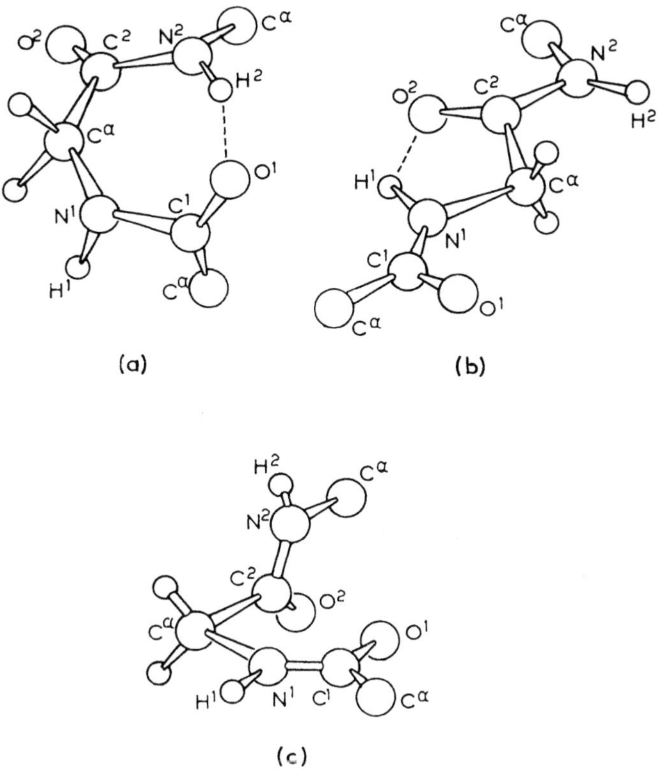

Fig. 3. The preferred conformations of the GLY dipeptide. (a) C_7; (b) C_5; (c) M.

existence of a nearly degenerate (two-fold) global minimum associated with the coordinates $(\Phi, \Psi)=(270°, 150°)$ and $(90°, 240°)$. The first of these is the most stable by about 0.5 kcal/mole. Both correspond to a folded, seven-membered ring conformation with a stabilizing intramolecular hydrogen bond between $H^2 \dots O^1$, similar to the preferred C_7 conformation found for the glycyl residue. The two conformations differ only in the orientation of the β-CH_3 group with respect to the heptagonal ring: the arrangement is axial for the conformation $(\Phi, \Psi)=(270°, 150°)$ (conformation C_7^{ax}) and equatorial for the conformation $(\Phi, \Psi)=(90°, 240°)$ (conformation C_7^{eq}).

A secondary minimum, about 3 kcal/mole above the global one is seen for $(\Phi, \Psi)= =(0°, 0°)$ and corresponds to the totally extended form, involving a weak hydrogen bond between $O^1 \dots H^2$ (conformation C_5).

Because of the closeness of the energies of the conformations C_7^{ax} and C_7^{eq} obtained in this first calculation, a number of refinements have been carried out in order to distinguish between the two. The most significant results were obtained by allowing flexibility for χ^1 and minimizing simultaneously Φ, Ψ and χ^1 and through the construction of a probability map (Figure 5). The first of the refinements displaces the global minimum to C_7^{eq} and the second shows that the probability of C_7^{eq} is greater than that of C_7^{ax} [18].

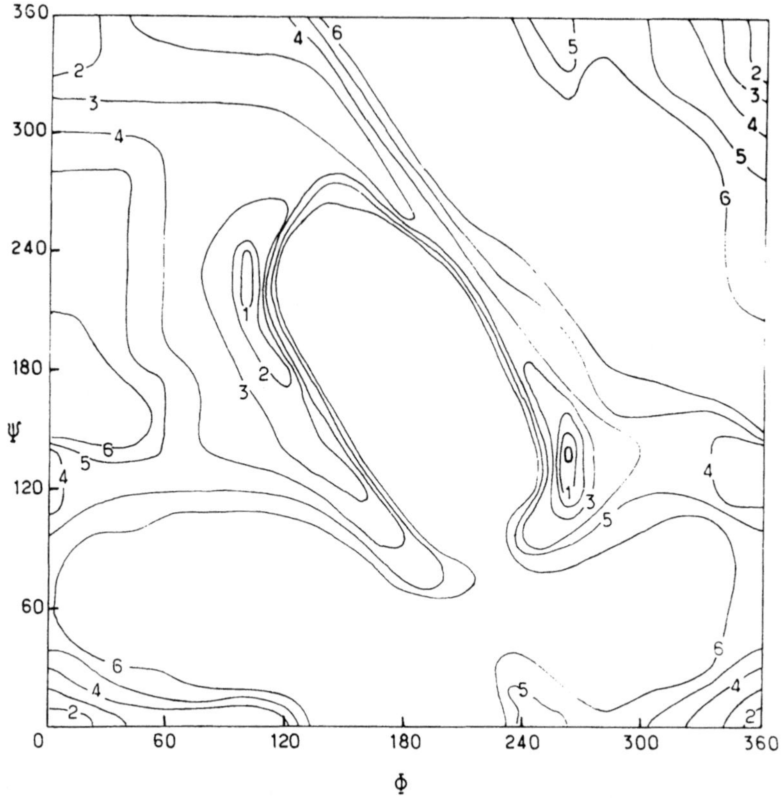

Fig. 4. PCILO conformational energy map for the ALA residue. Isoenergy curves in kcal/mole with respect to the global minimum taken as energy zero.

We may now again confront these PCILO predictions concerning the most stable form(s) of the model compound with: (a) the predictions obtained by other theoretical methods; (b) the experimental results on this very compound. This is done in Table II. The conclusions which may be drawn from it are practically the same as those obtained in the case of the GLY residue. We observe the general insufficiency of the empirical procedures, with the exception of a recent refined calculation by the deflation technique [19], to account for the preferential stability of the C_7 *and* the C_5 conformations and a similar inadequacy in this respect of the EHT method. The CNDO/2 results [12] deserve particular mention. At first sight the CNDO/2 conformational energy map (Figure 6) resembles the PCILO map. There are, however, in fact, important differences in detail. Thus CNDO/2 fixes the global minimum at $\Phi = 170°$, $\Psi = 280°$, the M region, which is only a secondary minimum in the PCILO calculations. It predicts the C_7^{ax} form to be the next most stable conformation, more stable than C_7^{eq}. It predicts also a local minimum for the FE conformation but about 4.1 kcal/mole above the global minimum. It is also useful to stress that altogether the stability contours of the CNDO calculations are, for a given level above the global minimum, much more restricted

TABLE II

The preferred conformations of the ALA dipeptide (parentheses indicate allowed conformations which do not correspond, however, to energy minima)

Methods / Conformations		Hard sphere (a)	(b)	+ H bond (c)	Empirical potential functions (d)	(e)	(f)	(g)	Flexibility (p)	Minimization (p)	Quantum-mechanicel EHT (h)	EHT (i)	CNDO[a] (i)	PCILO (j)	PCILO (r)	Exp. (l)
C_5	$\Phi°$		–		–	–	–	–	–	22		50	10	0	9	20
	$\Psi°$									345		340	350	0	344	340–360
C_7^{eq}	$\Phi°$	–	–		(95)	–	(100)	108	–	105	–	–	110	90	102	105
	$\Psi°$				(250)		(270)	250		252			235	240	220	230
C_7^{ax}	$\Psi°$	–	–	–	–	–	–	241	(260)	250	–	–	250	270	255	–
	$\Psi°$							113	(100)	111			130	150	140	
Order of decreasing stability ↓								C_7^{ax} C_7^{eq}		C^{eq} C_5 C_7^{ax}			M C_7^{ax} M′ C_7^{eq} C_5	C^{7ax} C_7^{eq} C_5	C^{7eq} C_5 C_7^{ax} R_α M	C_7^{eq} C_5

[a] M: $\Phi = 170°$, $\Psi = 280°$.
M′: $\Phi = 175°$, $\Psi = 105°$.

References to Table II

(a)–(n) Same as in Table I.
(o) Gibson, K. D. and Scheraga, H. A.: 1966, *Biopolymers* **3**, 709.
(p) Crippen, G. M. and Scheraga, H. A.: 1969, *Proc. Nat. Aacad. Sci. US* **64**, 42.
(r) Maigret, B., Pullman, B., and Perahia, D.: 1971, *J. Theoret. Biol.* **31**, 269.

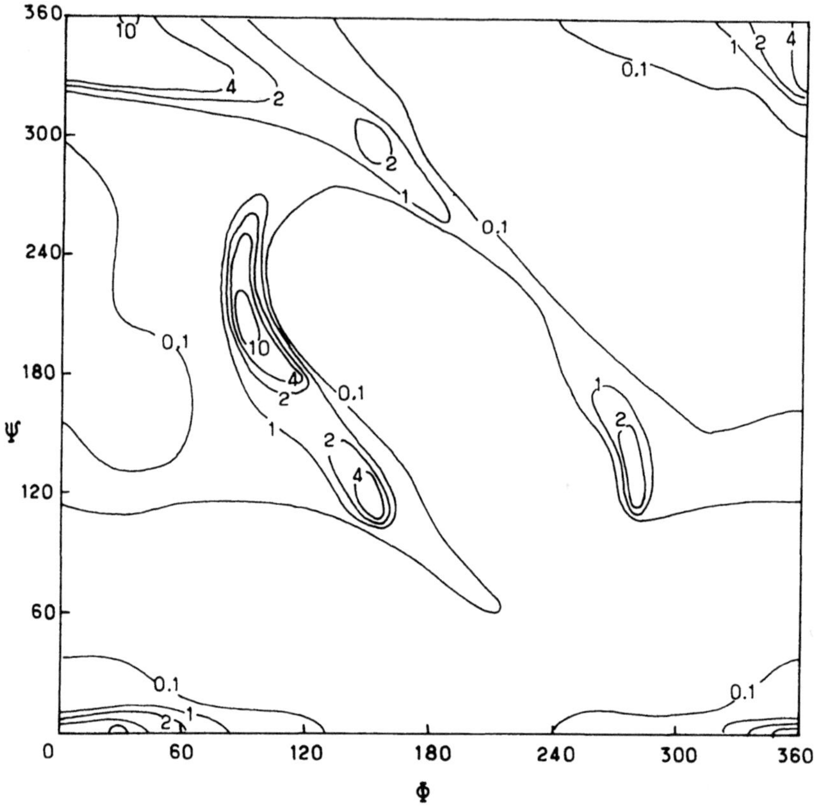

Fig. 5. The probability map for the ALA residue.

than the PCILO contours. Thus, e.g., the 5 kcal/mole CNDO contour is completely centered around C_7 and M minima. One must go as far as the 8 kcal/mole isoenergy contour in order to englobe the regions enclosed in the 5 kcal/mole contour of the PCILO computations.

The experimental results on the conformation of the model dipeptide in solution as obtained by Neel and Marraud [14–16] are indicated in the last column of Table II and they are in agreement with the PCILO results in particular the probability ones. On the other hand, it must be mentioned that there is a division among the experimentalists as to the preeminence of the C_7^{eq} or C_7^{ax} form [16, 20, 21], and some even question the preeminence of the C_7 form over the C_5 one [21].

The conclusion to be drawn from these examples seems to us straight-forward. They indicate both the precision of the PCILO calculations with respect to the preferred conformations of the model dipeptides and the probable superiority from that point of view of these calculations with respect to the majority of the other procedures.

By now these calculations have been extended to all the amino acid residues of proteins. As indicated in Figure 7, the model dipeptide should assume the C_7^{eq} and C_5 ones as the most stable conformations for the majority of them. Recent investiga-

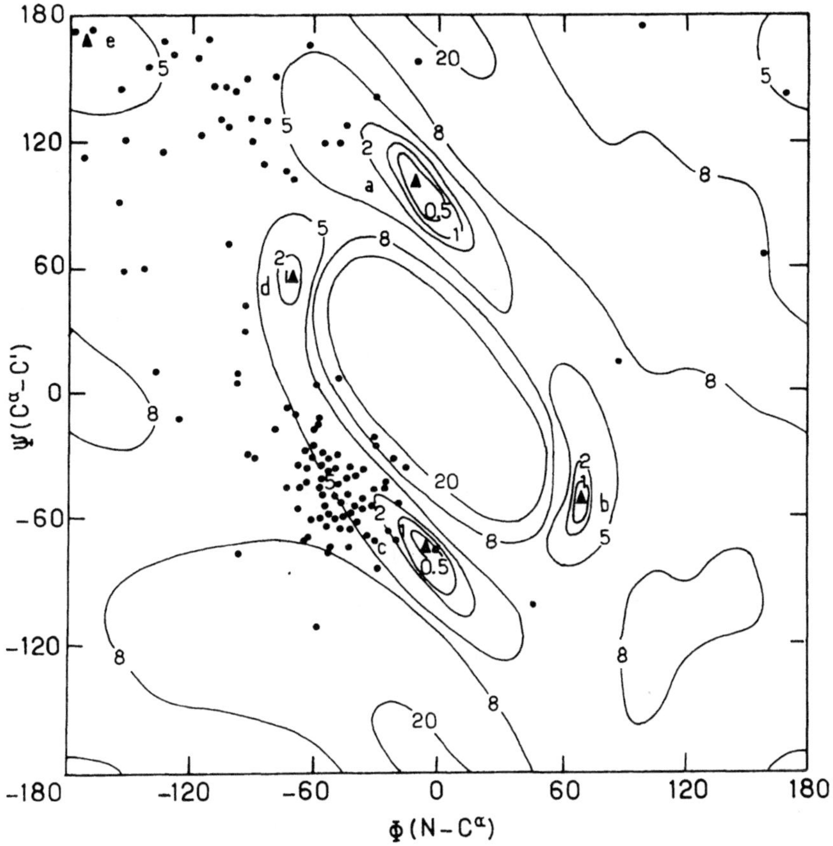

Fig. 6. CNDO conformational energy map for the ALA residue, after [12]. Note that the angles are counted following the new convention; ● experimental conformations of this residue in globular proteins (added by us).

tions by Neel and Marraud confirm this prediction for all residues which these authors have investigated.

We now turn to the second aspect of the problem, namely the significance of the theoretical results for the conformation of the residues in proteins.

Let us again consider first the two fundamental examples of the GLY and ALA residues.

C. GLY RESIDUE

Figure 8 presents:

(a) the contours of the PCILO computations limited at the value of 6 kcal/mole above the deepest minimum;

(b) the limits of a typical 'allowed' conformational space as obtained in the 'hard sphere' approximation of the empirical methods and which is also typical of the results obtained in a number of more developed empirical calculations using partionited potential energy functions, at least the older ones;

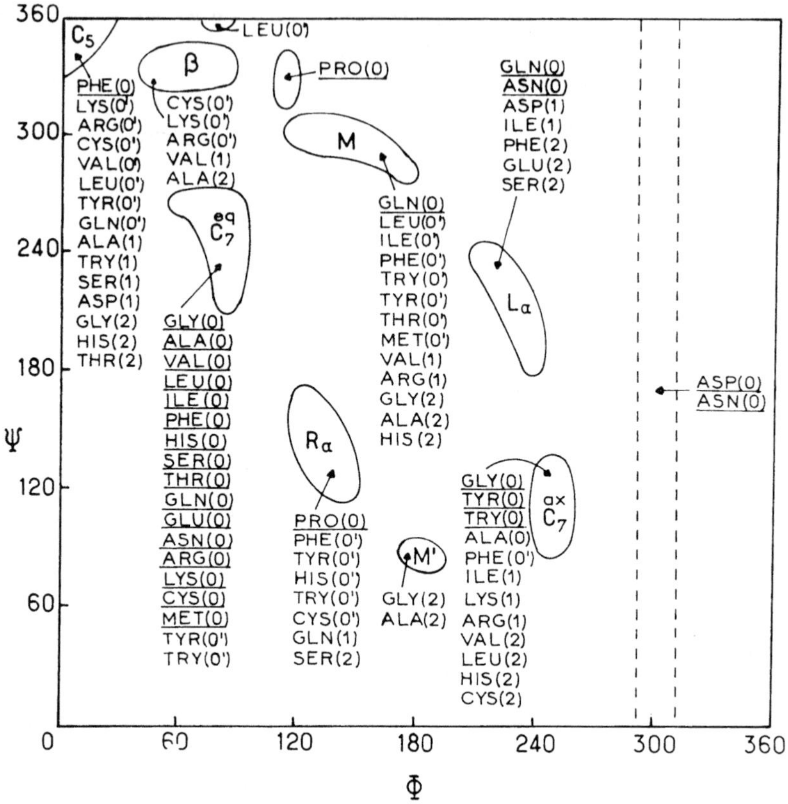

Fig. 7. The stable conformations predicted by the PCILO method for the α-amino acid residues of proteins. The underlined conformation is the most stable one. The numbers in brackets after the names of the residues indicate in kcal/mole the energy of the corresponding conformation with respect to the global minimum.

(c) the experimental conformations of the glycyl residues in globular proteins as determined by high resolution X-ray studies.

The examination of this figure shows that:

(a) The theoretical predictions of the PCILO method are extremely satisfactory as regards *the overall zone of conformational stability* of the glycyl residue. Practically all experimental conformations fall within the limits indicated by the traced contour. It is also evident that this agreement is better than that obtained with the hard sphere or the early partitioned potential energy calculations. A large number of representative points are located outside the limits of these empirical approximations which allow less conformational freedom than the PCILO ones. It must nevertheless be mentioned that the situation with the more recent empirical and the EHT computations is quite comparable, from this point of view to that corresponding to the PCILO calculation.

(b) On the other hand the experimental points do not especially cluster around the energy minima of the PCILO results, in particular around the global one, but are widely scattered. In view of the discussion of the preceding section it is extremely

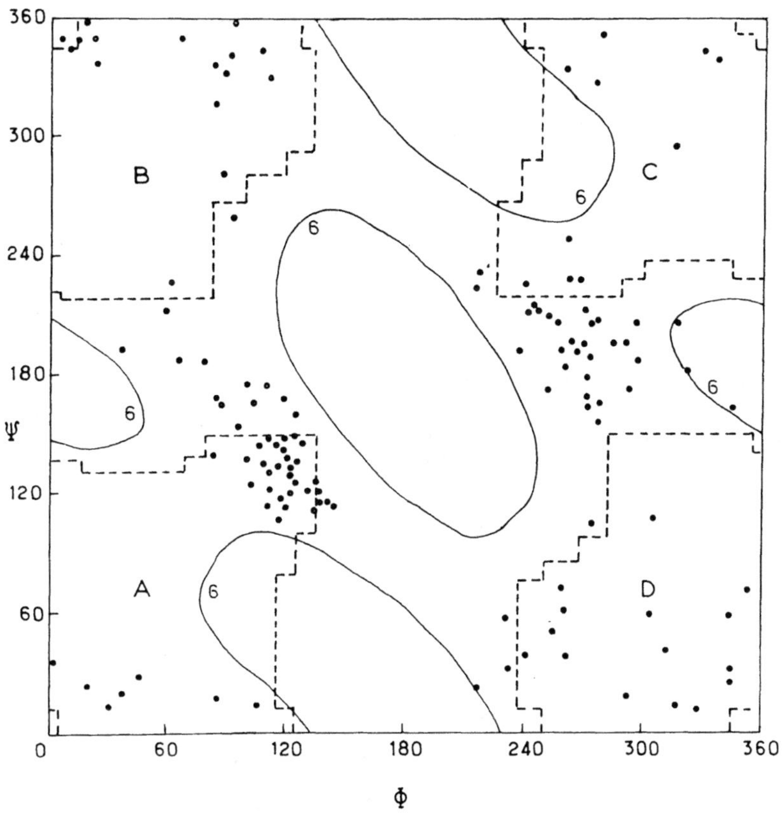

Fig. 8. Conformations of the GLY residues in globular proteins; —— limits of the PCILO stable zone of the *GLY* residue at 6 kcal/mole above the global minimum; --- limits of the stable zone in typical empirical computations.

reasonable that it should be like this, the situation being due to the *high specificity of the calculations for the model compound* for which they have been carried out. The intra-dipeptide hydrogen bonds typical of these specific conformations have only a restricted probability and are only very seldom found in proteins, where the majority of hydrogen bonds observed correspond to interactions between residues further separated in the chain, which are not taken into account in the calculations based on the dipeptide model.

It is, however, significant to observe that the experimental conformations of the GLY residues in proteins do not cluster around the energy minima on the refined empirical map of Figure 8 or the EHT map either, although these maps do not have the 'excuse' of specificity with respect to the model dipeptide. The results are thus very similar in this respect to the PCILO ones.

D. ALA RESIDUE

The results for the ALA residue essentially substantiate the conclusions drawn form

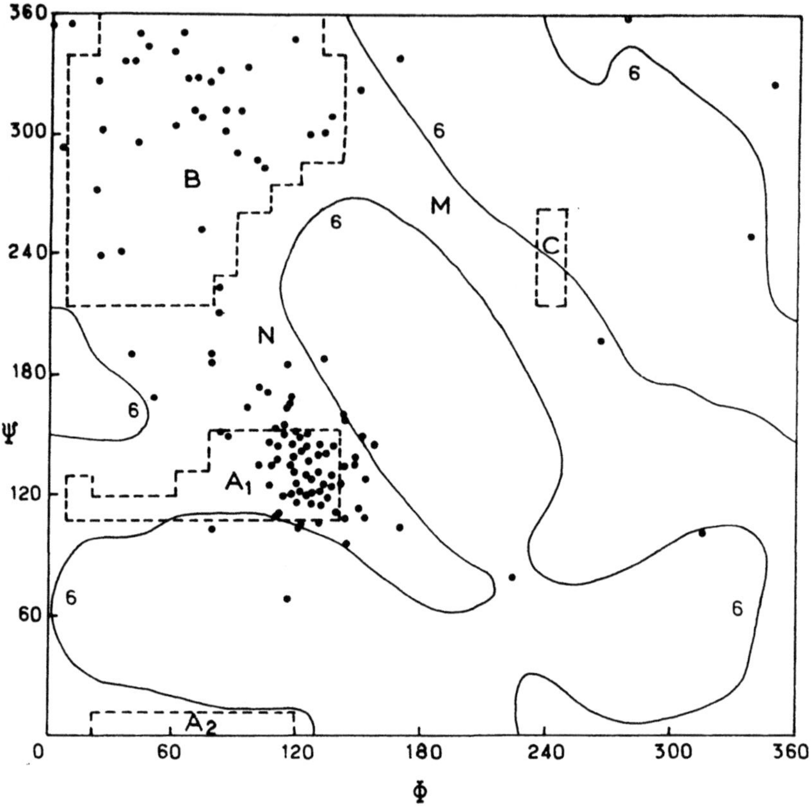

Fig. 9. Conformations of the ALA residues in globular proteins; —— limits of the PCILO stable zone of the *ALA* residue at 6 kcal/mole above the global minimum; --- limits of the stable zone in typical empirical computations.

the study of the GLY residue, as can be judged from the data of Figure 9. As in the case of the glycyl residue it is obvious that the PCILO and the experimental data agree satisfactorily in so far *as the general contours of the stability zones* are concerned. The significance of the probability map which has the advantage of eliminating from the stability region a large part of the conformations corresponding to $\Phi > 180°$ may be underlined. The agreement appears somewhat better with these quantum-mechanical calculations than with the older empirical ones. In this case again more recent empirical computations are close to the PCILO results and so are also the EHT results. The results of the CNDO computations are more disputable. As already mentioned before, these computations when considered within the usual limit of, say, 5 kcal/mole above the global minimum give an extremely restricted conformational stability zone, centered essentially around the C_7 and M conformational regions. Within these limits they are unable to account for the conformational distribution of the residue in proteins. With a limit of 8 kcal/mole the situation is more comparable to PCILO (Figure 6).

On the other hand the detailed distribution of the experimental points does not strictly relate to the computed importances of the different zones in any of these calculations. Thus, in particular, the density of the presence of the alanyl residues in the R_α zones appears greater than the calculated probability for that region. This is, however, a normal situation as this density originates from the helical portions of the proteins, the importance of which cannot be expected to be properly reproduced in calculations on dipeptides.

As said above, we are now in possession of the conformational energy maps for all the amino acid residues of proteins. We cannot reproduce here all of them. We can only say that their examination confirms the general conclusions deduced from the presentation of the cases of the GLY and ALA residues.

3. Nucleic Acids and Their Constituents

The conformational properties of polynucleotide chains depend on a series of torsion angles (Figure 10) which may be divided into three groups:

(1) the glycosidic torsion angle χ defining the relative orientation of the purine and pyrimidine bases with respect to the sugar;

(2) the torsion angles of the backbone Φ', ω', ω, Φ, Ψ, Ψ';

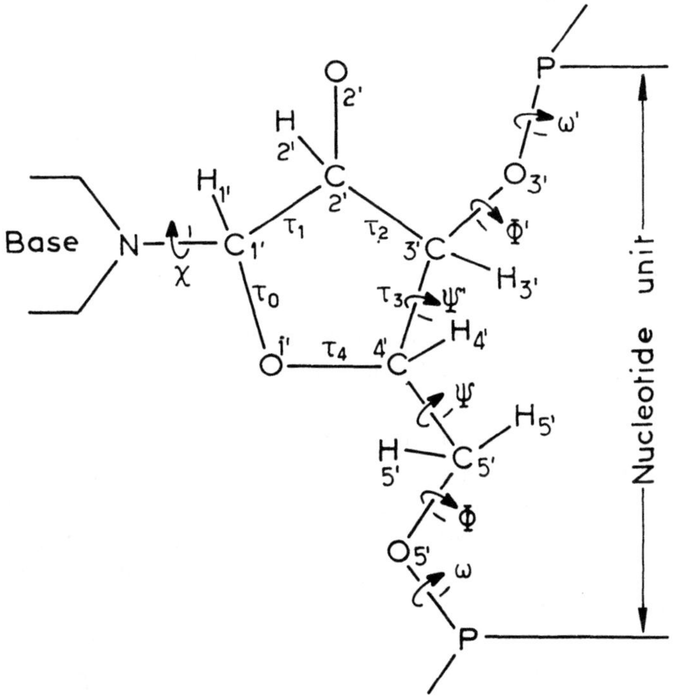

Fig. 10. The principal torsion angles in polynucleotides. Notations following Sundaralingam, M.: 1969, *Biopolymers* 7, 821.

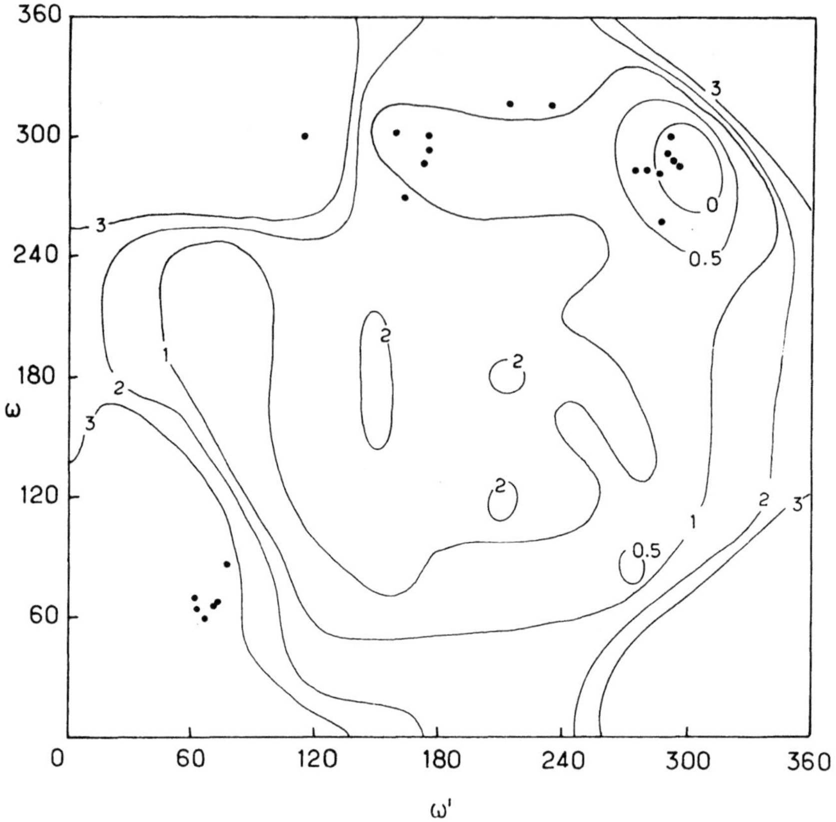

Fig. 11. (ω'–ω) conformational energy map. Isoenergy curves in kcal/mole with respect to the global minimum taken as energy zero. Constructed with $\Phi' = 240°$, $\Phi = 180°$ and $\Psi = 60°$; ● experimental results.

(3) the torsion angles about the bonds of the sugars: τ_0–τ_4, defining the pucker of this constituent.

In the study of nucleosides special attention has frequently been devoted to the orientation of the exocyclic CH_2OH group of the sugar. This is defined by the torsion angles Φ_{CO} and Φ_{OO}.

This represents a large number of degrees of freedom. Even if we put aside the torsions of the bonds of the sugars and adopt the usual division of the puckering of the sugar into the classical four principal types: C(3′)-endo, C(2′)-endo, C(3′)-exo and C(2′)-exo, we are still left with six essential torsion angles in the mononucleotide unit. If we admit a priori that each of these torsions can adopt three preferred values (a reasonable a priori estimation), the number of possible combinations is $3^6 = 729$. Taking the four principal puckering of the sugars into account, a dinucleoside phosphate may have about 3000 acceptable conformations. One of the goals of the theoretical work is to operate a selection of the most stable and most probable among these possible conformations so as to reduce this number.

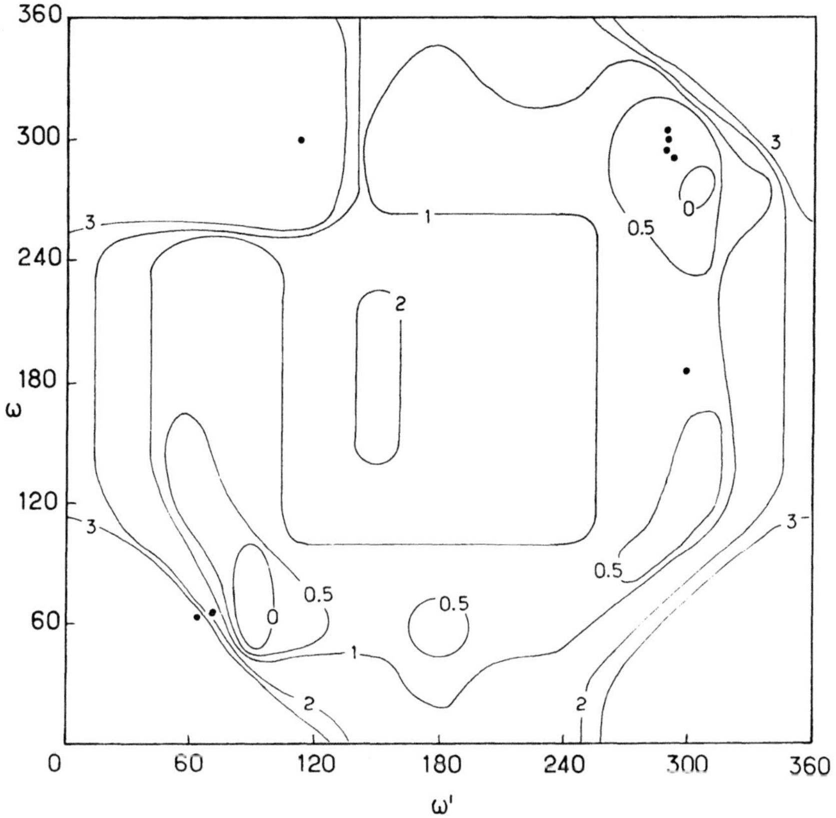

Fig. 12. (ω′–ω) conformational energy map. Isoenergy curves in kcal/mole with respect to the global minimum taken as energy zero. Constructed with $\Phi' = 240°$, $\Phi = 180°$ and $\Psi = 180°$; ● experimental results.

The work has been highly successful [22–24]. We can only illustrate here by selected examples the large amount of results obtained and we shall do so for some of the torsion angles of the backbone.

Four types of conformational energy maps have been constructed corresponding to the four possible combinations of two consecutive such angles: (ω′–ω), (Φ′–ω′), (ω–Φ), (Φ–Ψ). For the construction of these maps, preselected values have been adopted for the torsions of the angles not involved in the particular map under consideration, based both upon the conclusions of Sundaralingam's compilation and upon the indications of the potential energy and Extended Hückel calculations. These preselected values are: 300° (preferred), 60° and 180° for ω and ω′, 60° (preferred), 180° and 300° for Ψ, 180° for Φ and 240° for Φ′. The conformational energy maps have been established for each pair of adjacent rotation angles with the appropriate combinations of these different preselected values for the remaining torsion angles.

As an example let us consider *the (ω′–ω) conformational energy map.*

It is presented in Figure 11. In this map, the most stable of this category, the

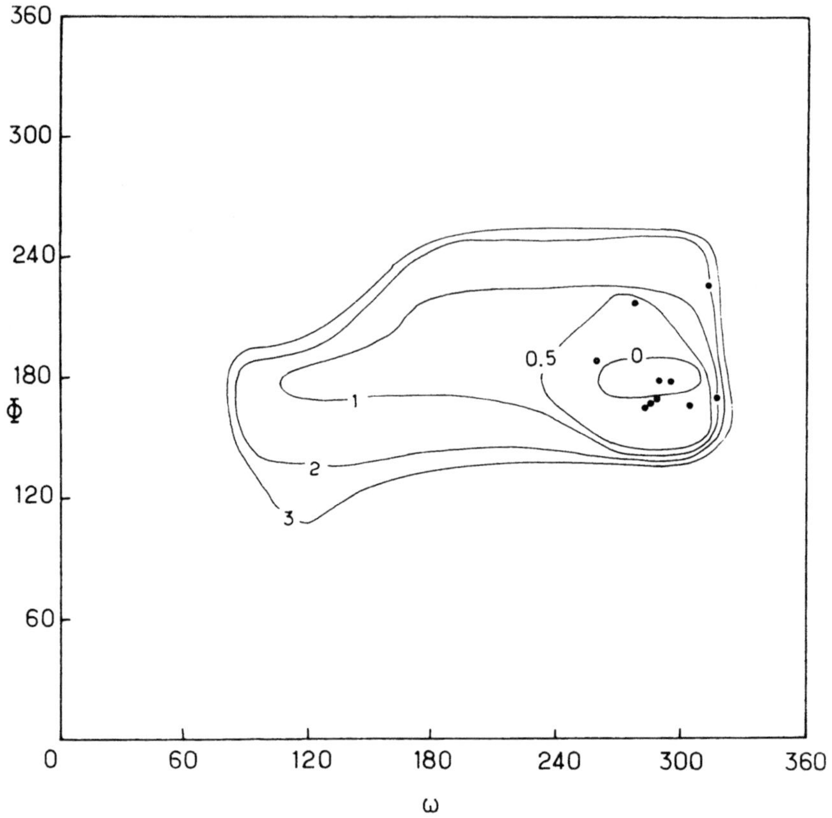

Fig. 13. (ω–Φ) conformational energy map. Isoenergy curves in kcal/mole with respect to the global minimum taken as energy zero. Constructed with $\Phi' = 240°$, $\omega' = 300°$ and $\Psi = 60°$; ● experimental results.

preselected values of the remaining angles are $\Phi' = 240°$, $\Phi = 180°$, $\Psi = 60°$. The global minimum is centered around $\omega' = 300°$, $\omega = 300°$. There is a secondary minimum around $(\omega'–\omega) = (270°, 90°)$ at about 0.5 kcal/mole above the global one.

It may be observed that there is a large flat area of conformational stability included within the 1 or 2 kcal/mole isoenergy curves. The experimental conformations plotted on this map seem to cluster around three combinations of (ω', ω). There is in the first place a number of representative points around the global minimum. These points represent essentially nucleic acids and polynucleotides. Then, vitamins B_{12} cluster around $(\omega', \omega) = (180°, 300°)$ in a region devoid of a local minimum but within the 1 kcal/mole contour. Finally there are experimental points which cluster around $(\omega', \omega) = (60°, 60°)$ in a region of relatively higher energy. The explanation that can be suggested is that these compounds are not true nucleotides and have ambiguous torsion angles.

Figure 12 shows the (ω', ω) map obtained with the preselected value of $\Psi = 180°$ instead of 60°, the other preselected angles remaining the same. This time there are two practically equivalent global minima, centered around $(\omega', \omega)=(300°, 270°)$ and $(90°, 90°)$. These minima are about 2.2 kcal/mole above the global minimum of the preceding map and are both surrounded by a relatively large 0.5 kcal/mole contour. The corresponding experimental points cluster around the two minima.

We may compare our results with those obtained by the Extended Hückel Theory (EHT) and by the empirical computations. EHT results [25] predict the existence of *seven equivalent* regions of energy minima located around $(\omega', \omega)=(60°, 60°)$, $(60°, 180°)$ $(180°, 60°)$, $(180°, 180°)$, $(180°, 300°)$, $(300°, 180°)$ and $(300°, 300°)$. This means that although this procedure indicates minima in the actually stable regions it is unselective with respect to these minima and places at the same level regions which are of high energy in the PCILO computations and which, in fact, are unoccuped by experimental conformations. Obviously the EHT procedure is less precise than the PCILO one. It may be added that the EHT map was obtained with Ψ fixed at 180° which, as we have seen from the PCILO results, is not the most stable arrangement.

In the empirical computations [20], the (ω', ω) map was constructed for dimethyl phosphate as reference compound. The results are very similar to those of the EHT procedure and indicate seven energy minima located in the same regions as those of EHT. Following the details of the computational procedure, some or others of these regions represent global minima: thus, when only nonbonded and torsional interactions are considered, the global minima are located at $(\omega', \omega)=(60°, 60°)$ and $(300°, 300°)$; when electrostatic interactions are included in the computations with the dielectric constant $\varepsilon=4$, the global minimum shifts to $(180°, 180°)$; when $\varepsilon=10$, however, the global minima move again to $(60°, 60°)$ and $(300°, 300°)$. These results suffer thus from the existence of too great a number of local minima, some in regions of apparently little significance, and also from the difficulty in unambiguously choosing the preferred ones.

Let us consider yet another example – that of the $(\omega$–$\Phi)$ *conformational energy map*.

The most stable map (Figure 13) corresponds to the preselected values $\Phi'=240°$; $\omega'=330°$, $\Psi=60°$. It presents a global minimum at $\omega=270°$–$300°$ and $\Phi=180°$. The experimental conformations of all the compounds whose other torsion angles are close to the preselected values cluster around this minimum.

The EHT computations [25], carried out with the hypothesis of $\Phi'=240°$, $\omega'=300°$ and $\Psi=180°$, which does not correspond to the most stable arrangement as evaluated by PCILO, predict three equivalent global minima centered around $(\omega, \Phi)=(180°, 120°)$ $(180°, 180°)$ and $(300°, 180°)$. In contrast to the PCILO results they do not select the $(300°, 180°)$ minimum as the deepest one, and place it thus at the same level as the $(180°, 120°)$ minimum which does not appear in the PCILO computations. The experimental data, as have been seen, favor the PCILO results.

As to the potential energy computations [26] they indicate nine regions of predicted stability centered around $(\omega, \Phi)=(60°, 60°)$, $(60°, 180°)$, $(60°, 300°)$ $(180°, 60°)$, $(180°, 180°)$, $(180°, 300°)$, $(300°, 60°)$, $(300°, 180°)$ and $(300°, 300°)$. They place the global minimum in the $(180°, 180°)$ zone. These results, very different from the quantum-mechanical ones, whether EHT of PCILO, do not reflect the experimental situation.

Very similar results are obtained for the other couples of the torsion angles. Altogether they indicate, with the exception of a few isolated cases, that overall agreement between the PCILO predictions and the experimental data is satisfactory, and is much better than that obtained with the empirical or the EHT method. It may be added that very recently Olson and Flory [27] devoted a large study to an attempt aimed at refining the previous empirical computations of Lakshminarayanan and Sasisekharan through the evaluation of the *statistical weights* of the low energy domains of the conformational energy maps. Their efforts lead to what can only be considered as very unsatisfactory results. In cases in which the empirical computations of the Indian authors indicated a number of equivalent local energy minima, the calculations of Olson and Flory select the most probable ones. This leads in general to a pronounced disagreement with the available experimental information. In particular one may note the occurrence of the highest statistical weights on the ω–ω' (Ψ'–Ψ'' in their notations), ω–Φ (Ψ''–Φ') and Φ–Ψ (Φ'–Φ'') maps around the 180°–180° values of the corresponding torsion angles (tt conformation) for which the conformations are practically never observed.

The PCILO calculations indicate also that the most stable arrangements of two consecutive torsion angles consist generally of the combination of the two individually preferred values with, sometimes, a small deformation. Thus the most stable combination of ω, ω' correspond to 300° with 300° and 90° with 90° (large stable zones); of Φ', ω' to 180–240° with 270° (large stable zone) followed by 180°–210° with 90°; of ω, Φ to 270°–300° with 180°, followed by 60° with 180°; of Φ–Ψ to 180° with 60° followed by 180° with 300° or with 180°.

PCILO computations have by now been completed for the remaining degrees of freedom appearing in a polynucleotide chain. Detailed studies have been devoted in particular, to the torsion about the glycosydic bond (the problem of the *syn-anti* conformations), to the orientation of the exocyclic CH_2OH group of the sugar in nucleotides (the problem of the *gg*, *gt* or *tg* conformations) and to the puckering of the ribose or deoxyribose rings.

4. Conclusions

The two cases discussed illustrate the success of the quantum-mechanical approach to the conformational porperties and problems of biomolecules and its superiority over the classical 'empirical' type of computations provided, however, that the quantum-mechanical method is of a relatively high level of refinement. The particular value of the PCILO procedure which combines such a refinement with high efficiency due to the rapidity of its computations is especially worth underlining. Of course, this does not mean that the 'empirical' methods must necessarily always be less satisfactory than the quantum-mechanical ones. By their very nature they may certainly be adapted so as to reproduce the results of the refined quantum computations, even of PCILO. This adaptation remains, however, to be done.

The two cases discussed above illustrate also the two most current possibilities that occur in treatments in which model compounds are used for the study of conformational properties of large polymeric systems. In the case of 'dipeptides' the high specificity of the model, the large zones of stable values predicted for the essential torsion angles, the specific effects appearing only in the polymers (e.g., long distance hydrogen bonds) make the model only partially appropriate. On the other hand, in the case of the 'mononucleotide unit', its principle conformational properties seem much more satisfactorily representative of those of the corresponging polymers.

Quantum-mechanical calculations have also been carried out for a large number of other biomolecules: disaccharides [28], steroids [29], retinals [30] and more recently for a large series of compounds of pharmacological interest (for a general review see [31].

References

[1] See e.g. Ramachandran, G. N. and Sasisekharan, V.: 1968 *Adv. Protein Chem.* **23**, 283.
[2] Hoffmann, R.: 1963, *J. Chem. Phys.* **39**, 1397; 1964, **40**, 2745.
[3] Pople, J. A. and Segal, G. A.: 1965, *J. Chem. Phys.* **43**, S136; 1966, **44**, 3289.
[4] Diner, S. Malrieu, J. P., and Claverie, P.: 1969, *Theoret. Chim. Acta (Berlin)* **13**, 1.
[5] Malrieu, J. P., Claverie, P., Diner, S.: 1969, *Theoret. Chim. Acta (Berlin)* **13**, 18.
[6] Jordan, F., Gilbert, M., Malrieu, J. P., and Pincelli, U.: 1969, *Theoret. Chim. Acta (Berlin)* **15**, 211.
[7] Clementi, E.: *Chem. Rev.* **68**, 341.
[8] Edsall, J. T., Flory, P. J., Kendrew, J. C., Liquori, A. M., Nemethy, G., Ramachandran, G. N., and Scheraga, H. A.: 1966, *Biopolymers* **4**, 121.
[9] Flory, P.: 1969, *Statistical Mechanics of Chain Molecules*, Wiley-Interscience, New York.
[10] Maigret, B., Pullman, B., and Dreyfus, M.: 1970, *J. Theoret. Biol.* **26**, 321.
[11] Popov, E. M., Lipkind, G. M., Arkhipova, J. F., and Dashevskii, U. G.: 1968, *Mol. Biol.***2**, 622.
[12] Momany, F. A., McGuire, R. F., Yan, J. F., and Scheraga, H. A.: 1971, *J. Phys. Chem.* **75**, 2286.
[13] Avignon, M., Huang, P. V., Lascombe, J., Marraud, M., and Neel, J.: 1969, *Biopolymers* **8**, 69.
[14] Maraud, M., Neel, J., Avignon, M., and Huang, P. V.: 1970, *J. Chim. Phys.* **67**, 959.
[15] Thong, C. M., Canet, D., Granger, P., Marraud, M., and Neel, J.: 1969, *C.R. Acad. Sci. Paris* **269**, 580.
[16] Marraud, M.: 1971, Thesis, University of Nancy.
[17] Avignon, M. and Huang, P. V.: 1970, *Biopolymers* **9**, 427.
[18] Maigret, B., Pullman, B., and Perahia, D.: 1971, *J. Theoret. Biol.* **31**, 269.
[19] Crippen, G. M. and Scheraga, H. A.: 1969, *Proc. Nat. Acad. Sci. US* **64**, 42.
[20] Bystrov, V. F., Portnova, S. L., Tsetlin, V. I., Ivanov, V. T., and Ovchinnikov, Yu. A.: 1969, *Tetrahedron* **25**, 493.
[21] Smolikova, J., Vitek, A., and Blaha, K.: 1971, *Collection Czechoslov. Chem. Commun.* **36**, 2474.
[22] Berthod, H. and Pullman, B.: 1971, *Biochim. Biophys. Acta* **232**, 595.
[23] Berthod, H. and Pullman, B.: 1971, *Biochim. Biophys. Acta* **246**, 359.
[24] Pullman, B., Perahia, D., and Saran, A.: 1972, *Biochim. Biophys. Acta* **1**, 269.
[25] Saran, A. and Govil, G.: 1971, *J. Theoret. Biol.* **33**, 407
[26] Sasisekharan, V. and Lakshminarayanan, A. V.: 1969, *Biopolymers* **8**, 505.
[27] Olson, W. K. and Flory, P. J.: 1972, *Biopolymers* **11**, 25, 57.
[28] Giacomini, M., Pullman, B., and Maigret, B.: 1970, *Theoret. Chim. Acta* **19**, 347.
[29] Caillet, J. and Pullman, B.: 1970, *Theoret. Chim. Acta* **17**, 377.
[30] Langlet, J., Pullman, B., and Berthod, H.: 1970, *J. Chim. Phys.* **67**, 480.
[31] Pullman, B. and Courrière, Ph.: 1973, *Conformations of Biological Molecules and Polymers* (ed. by B. D. Bergmann and B. Pullman), Academic Press, New York, p. 547.

DISCUSSION

De Coen: J'ai été assez étonné de vous entendre dire que seuls les calculs quantiques dans l'approximation PCILO permettent de prévoir l'existence des formes C_7 et C_5 des dipeptides mises en évidence expérimentalement. En effet, un calcul semi-empirique de l'énergie conformationnelle de l'unité dipeptidique comprenant un terme de Van der Waals décrit par n'importe quelle série de paramètres proposés par différents auteurs (Scheraga, Flory, Liquori, etc.) et un terme électrostatique traduisant l'existence d'un moment dipolaire associé à la liaison peptidique montre qu'effectivement les formes C_5 et C_7 sont les plus stables. Je me permets par ailleurs de souligner qu'un avantage net des calculs empiriques sur les calculs quantiques est le temps de calcul sur ordinateur beaucoup plus faible qu'ils nécessitent.

Pullman: Je suis désolé mais je n'ai aucun souvenir d'une publication de votre part d'il y a cinq ou six ans dans laquelle vous avez trouvé par le calcul un minimum énergétique dans la zone C_7^{eq}. Je vous serais reconnaissant de m'en donner la référence. Il est tout à fait inexact que l'introduction dans les calculs empiriques de la composante électrostatique conduit nécessairement à l'existence des formes C_7. Cela dépend entièrement de la valeur de la constante diélectrique choisie (ce choix étant dans les conditions actuelles tout à fait arbitraire). Il suffit pour cela de regarder les résultats de Popov qui a montré qu'en faisant varier cette constante on peut passer d'un minimum global pour une forme ouverte à un minimum pour une forme C_7.

Quoi qu'il en soit, même si la zone C_7 apparaît comme permise dans certains calculs empiriques, il est certain que sa signification a totalement échappé aux empiristes jusqu'aux calculs de Popov et les nôtres et l'apparition des résultats expérimentaux pour les dipeptides.

Bourgoin: Les énergies calculées sont-elles des énergies internes ou des énergies libres? Pour quelle température sont faits ces calculs?

Pullman: Les calculs sont représentatifs des molécules isolées, dans le vide. La température n'intervient pas dans les calculs.

Bothorel: Je veux simplement signaler que les travaux d'Avignon ont été effectués au Centre Paul Pascal et ont été dirigés conjointement par C. Garrigou-Lagrange, J. Lascombe et P. Bothorel. Les techniques utilisées étaient la spectrométrie de vibration et la diffusion Rayleigh dépolarisée, technique développée dans notre centre.

Hélène: Quelle est, a l'heure actuelle, la taille maximale d'un oligopeptide dont on peut calculer la conformation par la méthode PCILO?

Pullman: Le programme fonctionne pour 100 atomes lourds. Il peut être agrandi.

ANALYSE CONFORMATIONNELLE DES BIOPOLYMÈRES PAR LES MÉTHODES SEMI-EMPIRIQUES

MARIUS PTAK

Centre de Biophysique Moléculaire du CNRS, 45045 Orléans Cédex, France

Résumé. Le calcul des conformations des molécules biologiques a d'abord été fait à l'aide de méthodes dites empiriques. Le principe de ces méthodes est maintenant bien connu. Il a néanmoins semblé utile d'en faire une mise au point pour montrer notamment comment on peut justifier la décomposition de l'énergie potentielle d'interaction moléculaire en plusieurs termes. Il est intéressant de constater qu'aux minimums de l'énergie d'interaction de Van der Waals correspondent des conformations stables que l'on retrouve en minimisant l'énergie totale. Une comparaison des différentes fonctions susceptibles de représenter la liaison hydrogène est également faite. Enfin, un résumé des principaux résultats obtenus avec les peptides permet notamment de comparer les prédictions théoriques aux conformations déterminées par différentes méthodes expérimentales.

Summary. Conformations of biological molecules have been first calculated by empirical (or semi-empirical) methods. The basis of these methods are yet well known. Nevertheless, it remains some interest in a review of these methods in order to show specially how it is possible to justify the decomposition of the potential of molecular interaction into several terms. It is interesting to note that the minimums of Van der Waals interactions correspond to stable conformation which are not very different of those found by minimized total energy. A comparison of the different functions, used to represent the hydrogen bond, is also presented. Finally, a summary of theoretical predictions concerning peptides allow comparison with different experimental studies of the conformation of these molecules.

1. Généralités sur la conformation des biopolymères

La spécificité de macromolécules biologiques telles que les acides nucléiques et les protéines est étroitement liée à leur conformation. Dans des conditions voisines de celles du milieu biologique (solvant, pH, force ionique, température) ces polymères ont une structure tridimensionnelle stable essentiellement déterminée par (a) les interactions intramoléculaires, (b) les interactions avec les ions et le solvant constituant le milieu extérieur.

Certaines conformations ont une symétrie de type hélicoïdal. C'est le cas des protéines de structure telles que les collagènes, les soies, etc., dont les chaînes polypeptidiques de composition chimique simple sont sous forme d'hélice α ou β. C'est aussi le cas des acides désoxyribonucléiques supports du code génétique, qui sont constitués à partir de quatre bases nucléiques fixées dans un ordre donné le long d'une chaîne organophosphorée. Deux chaînes polynucléotidiques sont assemblées par liaisons hydrogène pour former la double hélice de Watson et Crick [1].

D'autres conformations, au contraire, ont une géométrie beaucoup plus complexe déterminant des sites de contrôle, des sites de réaction, des zones d'interaction spécifique entre macromolécules, etc. C'est le cas des acides ribonucléiques et surtout des enzymes. Dans ce dernier cas, vingt acides aminés différents sont utilisés (forme L) pour édifier les chaînes polypeptidiques dont la conformation dépend, en partie, de

C. Sadron (ed.), Dynamic Aspects of Conformation Changes in Biological Macromolecules. 23–55.

l'ordre d'enchaînement des constituants. Le cas de ces polymères est donc particulièrement intéressant pour tenter d'expliquer ou de prévoir par voie théorique, la conformation d'une séquence peptidique à partir des interactions existant entre les atomes de la chaîne et, si possible, entre ces atomes et l'environnement. Les premières descriptions des conformations des polypeptides par Pauling et Corey [2] sont essentiellement basées sur la notion de rayon de Van der Waals limitant les contacts interatomiques et sur la formation du nombre maximum de liaisons hydrogène —CO...HN— intra- ou intermoléculaires. En 1963, Ramachandran *et al.* [3] utilisant les approximations très simplificatrices de la méthode des 'sphères dures' étudient d'une façon détaillée les régions stériquement permises pour une unité 'dipeptidique' dont la conformation dépend de deux paramètres ϕ et ψ qui sont respectivement les angles de rotation autour des liaisons $N—C^{\alpha}$ et $C^{\alpha}—C'$. Leach *et al.* [4] développent un traitement sur ordinateur de cette manipulation des sphères dures ou sphères de Van der Waals. En 1965, Liquori *et al.* [5] proposent d'utiliser des fonctions potentiel plus élaborées, incluant notamment les interactions de Van der Waals, pour calculer la conformation de polypeptides tels que: poly-L-alanine et polyglycine. Ce procédé, très rapidement développé par plusieurs groupes (Scheraga *et al.* [6]; Brant et Flory [7]; Ramachandran *et al.* [8]; Popov *et al.* [9]), pour calculer la conformation des peptides, est à la base de ce que l'on convient d'appeler les méthodes semi-empiriques (ou empiriques selon les auteurs).

Il faut dès le départ, bien en souligner les modalités d'application.

(a) L'énergie d'interaction n'est pas calculée globalement mais décomposée en un certain nombre de termes indépendants. Ces termes sont justifiés par des calculs quantiques et ajustés à partir de résultats expérimentaux (cristallographie, thermodynamique).

(b) En dehors de cas particuliers, il n'est pas tenu compte des interactions avec le solvant, mais seulement des interactions intramoléculaires.

(c) Compte tenu du volume et de la complexité des calculs, l'application à une macromolécule est seulement possible pour un polymère stéréorégulier.

Le développement récent des méthodes quantiques [11–13] a pu faire penser que les méthodes semi-empiriques allaient être rapidement abandonnées. En fait, ces dernières méthodes permettent, dans de nombreux cas, de prévoir très valablement les conformations et sont d'un usage beaucoup plus commode pour le traitement sur ordinateur. Elles continuent donc d'étre très largement employées d'autant plus que la définition des fonctions potentiel s'améliore rapidement d'une part et qu'il semble possible maintenant de tenir compte des interactions avec le solvant ou avec d'autres molécules d'autre part.

Dans tous les cas, un calcul conformationnel comporte nécessairement une part importante d'empirisme. Ce n'est qu'un moyen pour mieux comprendre l'élaboration d'une conformation macromoléculaire et pour obtenir différentes structures possibles entre lesquelles l'expérience (RMN, IR, DC, etc.) permettra éventuellement de choisir. Le but ultime restant de saisir la signification physique et surtout biologique des résultats ainsi obtenus.

2. Principes du calcul conformationnel

On part du principe que les conformations stables d'une molécule correspondent aux minimums de l'énergie totale d'interaction entre les atomes constitutifs. Avant de décrire les méthodes de calcul de cette énergie, il convient de définir le cadre dans lequel elles s'appliquent. Il est commode pour cela de prendre l'exemple des protéines qui donnent les cas les plus complexes de relation entre la structure primaire et les structures secondaire et tertiaire.

La conformation d'une protéine à l'état 'natif' en solution résulte de la combinaison d'un certain nombre de facteurs schématiquement décrits ainsi:

2.1. Structure primaire

Un certain nombre d'acides aminés est strictement nécessaire pour maintenir une conformation donnée d'une chaîne polypeptidique. Les autres peuvent être échangés selon des lois qui sont en fait les bases de l'évolution moléculaire [14]. De l'ordre d'enchaînement dépendent surtout les interactions à courte distance déterminant des structures secondaires souvent voisines de structures type telles que: hélice α, repliement de chaîne (β-turn), alternance α–β, etc. Les interactions avec le solvant peuvent également déterminer des conformations dépendant de cet ordre. C'est ainsi qu'un polymère dans lequel alternent des séquences de polarités différentes prend, dans un solvant donné, une conformation exposant vers l'extérieur les groupements inter-agissant préférentiellement avec le solvant. C'est aussi un résultat bien connu que dans l'hémoglobine les résidus hydrophobes sont tous dirigés vers l'intérieur de la protéine. Un calcul d'interaction intramoléculaire ne tient pas compte de ce facteur.

2.2. Interactions intramoléculaires

Elles contribuent de façon essentielle à l'élaboration de la conformation. Il est en effet extrêmement remarquable de constater qu'un très grand nombre de résidus de protéines de structure connue (myoglobine, hémoglobine, lyzozyme, RNase, etc.) se trouve dans des zones voisines des minimums d'énergie calculés pour l'unité peptidique conformationnelle dans laquelle les interactions diminuent très vite pour $r > 7$ Å. Une première approche consiste donc, pour élaborer la conformation d'une chaîne, à combiner une suite de conformations élémentaires des résidus. C'est l'idée du code stéréochimique de Liquori [15] développée aussi par De Coen [16]. La conformation réelle nécessite des déformations de ces conformations élémentaires pour rendre compte des interactions entre parties éloignées de la chaîne, de la fermeture de boucles par des ponts S—S et des interactions avec le solvant. L'importance attachée au calcul des conformations des peptides est justifiée; ce calcul peut fournir la conformation la plus probable d'une petite molécule, et les éléments de description d'une structure tertiaire et aussi d'une pelote statistique.

2.3. Interaction avec le milieu extérieur

(a) L'interaction avec le solvant et le cas échéant des ions ne peut actuellement être représentée dans une théorie microscopique, notamment dans le cas d'un polymère non stéréorégulier. Il lui correspond en fait un terme entropique dans l'énergie du système protéine–solvant. Pour les protéines en solution aqueuse, la conformation conduit à des interactions particulières entre les groupements non polaires souvent appelées 'interactions hydrophobes'.

(b) Nous rappellerons pour mémoire que la conformation d'une enzyme peut dépendre de l'interaction avec un effecteur. Les méthodes empiriques ne permettent guère d'aborder ce problème pour le moment. Néanmoins, la prise en compte d'interactions entre molécules différentes est appelée à des développements importants, c'est ainsi que Scheraga *et al.* [10] ont calculé récemment l'interaction entre un substrat et un site spécifique d'enzyme.

(c) Enfin, on peut se poser le problème de savoir si la conformation d'une protéine résulte d'un 'folding' progressif au cours de la biosynthèse avec le cas échéant, intervertion du support ribosomal ou au contraire, d'une transition pelote statistique → structure stable n'intervenant qu'à la libération de la chaîne complète. La nature de l'état natif sera discutée par Goldberg. Rappelons seulement que la renaturation d'une protéine ne conduit pas forcément à l'état d'énergie minimum, mais à l'état favorisé par la cinétique de transition. Le calcul conformationnel permet de trouver les minimums de l'énergie (dont les valeurs absolues n'ont guère d'intérêt). L'exploration de la surface d'énergie permet de donner une loi de répartition statistique des différents états conformationnels correspondants.

2.4. Conclusions

(1) La première étape est de calculer les conformations possibles de l'unité conformationnelle d'un polymère à partir de l'énergie d'interaction intramoléculaire.

(2) On peut étendre les résultats au cas d'un polymère stéréorégulier en utilisant plusieurs dizaines ou même une centaine de paramètres.

(3) Les résultats peuvent également servir de base à un code stéréochimique permettant des calculs sur les oligomères.

(4) On peut également les utiliser pour déterminer certains paramètres de pelote statistique.

(5) Enfin, on peut envisager de faire un calcul global de l'énergie d'une molécule ayant une dizaine de constituants. Les difficultés de la minimisation et le volume des calculs donnent alors les limites de la méthode.

3. Conventions – géométrie – minimisation

Nous continuerons à prendre les peptides comme exemple.

L'unité conformationnelle utilisée dans la plupart des calculs semi-empiriques est représentée sur la Figure 1. En première approximation la conformation de cette unité est indépendante de celle de ses plus proches voisines. On remarque qu'il ne s'agit pas de l'unité peptidique conventionnelle, ni du résidu, mais de l'ensemble des atomes dont les interactions ne dépendent que des deux paramètres ϕ et ψ (pour le

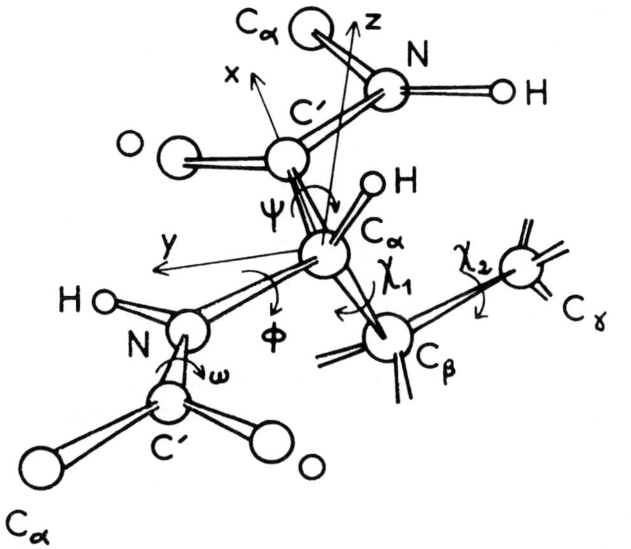

Fig. 1. Représentation de l'unité peptidique conformationnelle avec un système d'axes de référence possible dont l'origine est le C^{α} central.

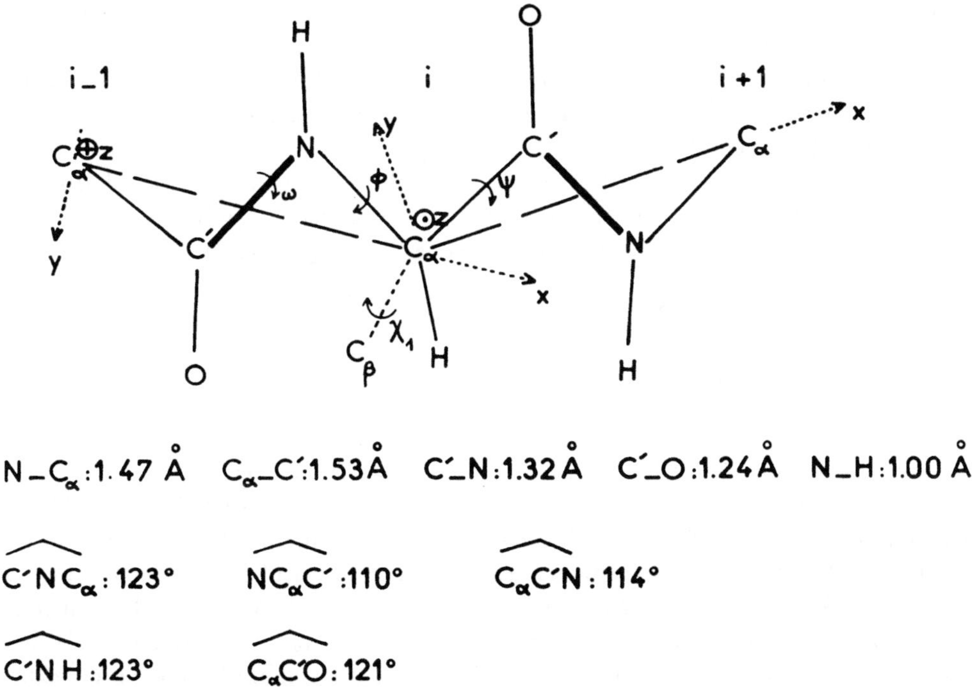

Fig. 2. Géométrie du squelette peptidique. Les coordonnées atomiques peuvent être rapportées à des référentiels ayant pour origines les C^{α} successifs (Brant et Flory).

squelette). La présence de deux liaisons CO—NH a conduit à utiliser improprement le terme de dipeptide. Unité peptidique conformationnelle ou unité peptidique indépendante (UPI de De Coen) semblent plus appropriées. Les deux liaisons peptidiques sont *trans*: $\omega=0$ et la conformation du squelette dépend des deux paramètres ϕ et ψ. La conformation de la chaîne latérale dépend des angles de rotation: χ autour des liaisons C^{α}—C^{β}, C^{β}—C^{γ}, etc. Les définitions et conventions de numération et de sens ont été récemment précisées par l'IUPAC–IUB [17]. Dans la plupart des calculs, les angles de valence et les longueurs des liaisons sont fixes (Figure 2). Dès que la chaîne s'allonge, de faibles variations de ces valeurs donnent des variations importantes de la géométrie qui est déterminante dans le cas de fermeture de boucle [18], ou de la comparaison avec des données cristallographiques.

Les coordonnées cartésiennes des atomes sont calculées dans un référentiel fixe lié à un atome C^{α} ou N, choisi comme origine. La liaison peptidique étant considérée comme *trans* est plane, Brant et Flory utilisent un système de liaisons virtuelles C_i^{α}—C_{i+1}^{α} (Figure 2). Les transformations de coordonnées sont effectuées à l'aide de matrices de translation et de rotation définies pour chaque passage d'un atome à un autre atome en remontant la chaîne (squelette ou chaîne latérale) en direction de l'origine.

$$\mathbf{r}_{ji}^{(k)} = \mathbf{p}_i^{(k)} + T_{i\to k}\mathbf{r}_{ji}^{(i)}.$$

$\mathbf{r}_{ji}^{(k)}$ est la position de l'atome ji dans le référentiel k, $\mathbf{r}_{ji}^{(i)}$ est la position de cet atome dans le référentiel i, $\mathbf{p}_i^{(k)}$ est la position de l'origine de i dans k, et $T_{i\to k}$ est la matrice 3×3 de rotation $i\to k$.

Les coordonnées finales d'un atome dans le référentiel fixe sont obtenues à partir d'un produit de matrices dépendant de la longueur de la chaîne et des systèmes de coordonnées intermédiaires choisis.

Un cas particulier intéressant est celui des conformations hélicoïdales. Parmi les différentes méthodes possibles pour générer une hélice, celle de Miyazawa *et al.* [19] permet de calculer simplement les paramètres de l'hélice (Figure 3).

$$\begin{aligned}
&\cos t = (\mathbf{C}\cdot\mathbf{C}')/\mathbf{C}^2\\
&\varrho_1(1-\cos t) = C/2\\
&h^2 + 2\varrho_1^2(1-\cos t) = B^2\\
&h\sin t = \mathbf{B}\cdot(\mathbf{C}\times\mathbf{C}')/\mathbf{C}^2 .
\end{aligned}$$

Dans le cas d'une hélice binaire $t=\pi$, $\varrho_1=C/4$, $h=(B^2-C^2/4)^{1/2}$.

La recherche du minimum global et des minimums secondaires de l'énergie est faite par minimisation d'une fonction dépendant de 2 à n paramètres. La plupart des méthodes sont basées sur la connaissance des énergies et le calcul de gradients. n est en général limité à une dizaine et on emploie la méthode de Davidson [20] ou mieux, la méthode dite de 'steepest descent' [21]. Pour des oligomères construits à partir d'éléments de conformations connues on recherche souvent les minimums de la molécule au voisinage des minimums des constituants en prenant le risque d'omettre des conformations nouvelles spécifiques de la séquence.

4. Énergie d'interactions intermoléculaires*

La méthode semi-empirique est basée sur la possibilité de décomposer l'énergie d'interaction en plusieurs termes:

(a) barrières de rotation autour des liaisons simples,
(b) interactions électrostatiques,
(c) interactions de Van der Waals,
(d) liaisons hydrogène,
(e) déformation des angles et des liaisons de valence.

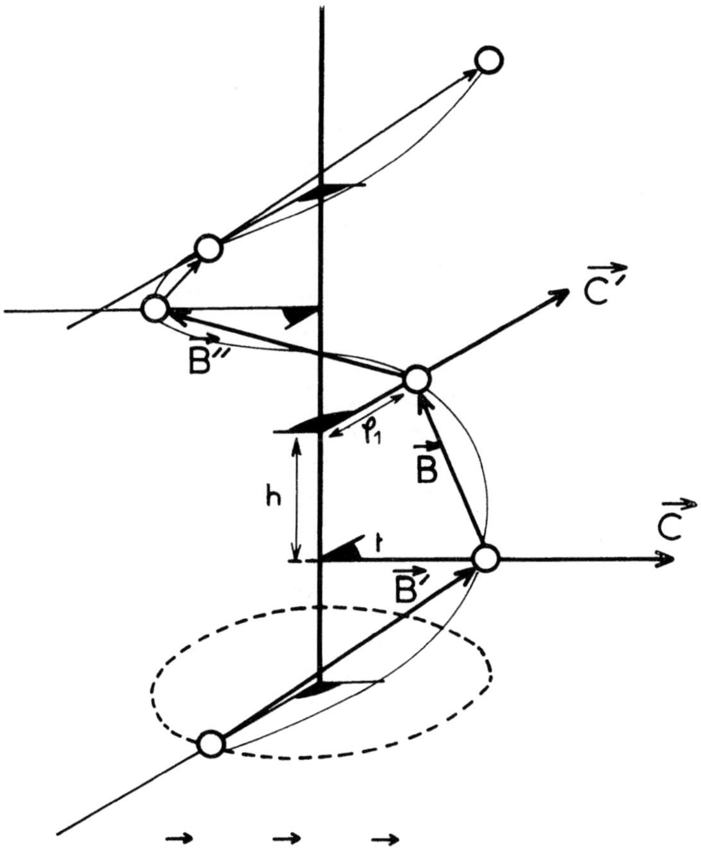

$$\vec{C} = \vec{B'} - \vec{B}$$

$$\vec{C'} = \vec{B} - \vec{B''}$$

Fig. 3. Paramètres d'une hélice.

* 'Intermoléculaires' est le terme consacré pour désigner les interactions entre atomes voisins qui constituent, dans notre cas, une même molécule. Il n'y a pas lieu de confondre avec l'interaction entre deux molécules différentes.

Dans une gamme donnée de distances interatomiques*, on cherche les fonctions potentiel les plus simples rendant le mieux compte des données expérimentales qui sont les configurations d'équilibre dans des cristaux, les barrières de rotation moléculaire, les coefficients de viriel, etc.

Cette méthode est justifiée par le théorème d'Hellmann et Feynman [22]. En négligeant les forces magnétiques et les effets relativistes ces auteurs ont montré que les forces exercées sur les noyaux déterminées à partir d'une surface d'énergie potentielle définie par la mécanique quantique sont très exactement celles que l'on obtient dans un calcul électrostatique classique compte tenu de la densité de probabilité de présence électronique obtenue à partir de l'équation de Schrödinger [22]. (On se place dans l'approximation de Born–Oppenheimer.)

On se bornera à calculer les interactions électriques entre atomes 'non liés', c'est-à-dire distants de plus de deux liaisons, dans les différents cas:

(a) interactions purement électrostatiques,
(b) interactions dues à l'induction,
(c) interactions dues aux fluctuations.

En fait, le terme (b) est pratiquement toujours négligé, les deux autres termes jouissent d'une propriété d'additivité. On rajoute un terme d'échange pour traduire l'existence de barrières de rotation. Le cas de la liaison hydrogène sera traité à part. N'entrent pas en ligne de compte les interactions 'fortes', c'est-à-dire les liaisons chimiques les effets de résonance et de transferts de charge.

Remarque. La décomposition précédente n'est donc pas totalement arbitraire. Elle conduit à prélever dans l'énergie les termes principaux et à ajuster leur paramètres en fonction des données de l'expérience. Pour des distances interatomiques faibles ($\simeq$somme des rayons de Van der Waals) et pour les interactions d'échange, les méthodes quantiques semblent cependant préférables.

4.1. Barrières de rotation autour de liaisons simples

La spectroscopie hertzienne montre que la rotation autour d'une liaison C—C n'est pas libre. Il existe une énergie potentielle de torsion qui est de l'ordre de 2,7 à 3 kcal mole^{-1} pour des molécules telles que l'éthane, et dont la symétrie est d'ordre 3. Cette valeur ne peut être expliquée seulement à l'aide des interactions électriques entre atomes non liés, c'est-à-dire entre protons. Il faut ajouter un terme de torsion qu'on attribue en général aux effets d'échange entre les électrons des liaisons C—H et qui sera un fonction périodique $E_T = f(\theta)$ où θ est l'angle de rotation, et la période est $2\pi/3$ ou $2\pi/6$.

Dans un peptide, des termes de torsion interviennent pour les rotation ϕ et ψ dans le squelette et les rotations χ dans les chaînes latérales.

L'évaluation de l'énergie de torsion est difficile, notamment pour la liaison N—C$^{\alpha}$

* Le long d'une chaîne polymérique, les interactions à courte distance désignent les interactions entre plus proches voisins ($r < 10$ Å par exemple) les interactions à longue distance s'exercent entre parties éloignées de la chaîne. Dans la théorie des forces intermoléculaires, les interactions à courte distance désignent les interactions entre atomes unis par une liaison chimique, les interactions à longue distance s'exercent entre atomes non liés.

pour laquelle on manque de données expérimentales sur des molécules modèles. (On prend habituellement le nitrométhane.) On convient de prendre pour le squelette:

$$E_T = \tfrac{1}{2}E(\phi)(1 + \cos 3\phi) + \tfrac{1}{2}E(\psi)(1 - \cos 3\psi).$$

La hauteur des barrières $E(\phi)$ et $E(\psi)$ varient, selon les auteurs, de 0 à 1,5 kcal mole^{-1} [23]. Brant et Flory [7] ont montré que pour la pelote statistique $\langle r^2 \rangle$, carré moyen de la distance entre extrémités de chaîne, était très peu sensible à ces valeurs. De Santis *et al.* [5] négligent le plus souvent ces termes qui ne sont pas prépondérants dans le cas de conformation du type hélice α ou structure β. Pour les chaînes latérales, il faut ajuster les paramètres en fonction de la nature de ces chaînes: chaînes aliphatiques (Ala, Val, Leu, etc.), chaînes polaires (Asp, Glu, Lys, Arg, etc.), chaînes à groupements aromatiques (Phe, Tyr, Try), chaînes à fonction esters (benzylglutamate, benzylaspartate, etc.). Un cas particulier intéressant est celui du pont disulfure. Pullman *et al.* [24] ont montré que le minimum d'énergie de rotation autour de la liaison S—S correspond à un angle dihedral $=90°$. Dans l'énergie totale, le terme E_T est certainement celui qui est défini avec le plus d'empirisme. Le développement de la spectroscopie de rotation et la recherche théorique de nouvelles fonctions (Lifson) laissent espérer une meilleure définition de cette énergie torsionnelle.

4.2. Interactions électrostatiques

La liaison peptidique —C(=O)—N(H)— possède un moment dipolaire permanent $=3,7$ *D*.

Brant et Flory [7] ont démontré l'importance des interactions dipolaires pour déterminer les conformations du squelette peptidique. Ces interactions interviennent également lorsque les chaînes latérales ont un moment non nul. Pour calculer l'énergie électrostatique d'interaction E_E, on préfère utiliser l'approximation des monopoles à celle des dipoles qui présente des difficultés [25]. Le premier termed 'interaction entre des charges ponctuelles est:

$$E_E = \sum \frac{1}{4\pi\varepsilon} \frac{q_i q_j}{r_{ij}}.$$

($\sum$: somme sur toutes les charges partielles, ε: permittivité $\varepsilon = \varepsilon_0 \varepsilon_r$, ε_0: permittivité du vide.)

Il faut déterminer une distribution de charges partielles compatible avec les moments dipolaires des liaisons, le moment dipolaire total et la charge globale de la molécule. L'évaluation des charges σ est habituellement faite par la méthode de Del Ré [26], qui prend en compte tous les atomes de la molécules portant chacun des charges partielles. Les charges π importantes pour les liaisons peptidiques, les groupements carbonyl, les cycles aromatiques sont évaluées par la méthode de Berthod et

Pullman [27]. Des méthodes quantiques plus élaborées telles que la méthode EHT (Extended Hückel Theory) [28] ou mieux CNDO/2 de Pople et Segal [29], ont été également utilisées pour déterminer la distribution des charges partielles permettant de rendre compte des interactions. Il faut bien insister sur le fait que dans tous les cas, il s'agit de procédés pour aboutir à une expression finale de l'énergie suffisamment satisfaisante et non d'une description de charges réelles expérimentalement mesurables. L'approximation des monopoles a été récemment contestée par Rein [30] qui propose que pour des distances <5 Å, l'interaction entre deux molécules, c'est-à-dire deux distributions de charges, comporte des termes multipolaires dominants (quadripolaires) et que même au-delà de 20 Å, les termes du second ordre du développement en série de l'énergie ne sont pas négligeables.

Il est difficile de préciser la valeur de la permittivité ε. Il s'agit en effet d'une constante microscopique qui ne reflète en rien les propriétés diélectriques macroscopiques du solvant. Il faudrait distinguer différents cas:

(1) deux atomes entre lesquels il n'y a pas de molécules de solvant,

(2) deux atomes séparés par une molécule de solvant de position et d'orientation connue,

(3) deux atomes séparés par deux ou plusieurs molécules de solvant.

Les interactions dépendent de la polarisabilité des atomes, de leur action sur l'environnement, de la nature du solvant, etc. Suivant les auteurs, ε varie de 1 à 4, valeurs surtout valables pour le cas (1) et donnant des énergies de l'ordre du kcal mole^{-1}. A longue distance, l'énergie diminue très vite surtout dans l'eau et si ε reste constant, les interactions électrostatiques sont surévaluées. Enfin, nous rappellerons que si ε varie avec la distance, sa valeur dépend de la conformation de la molécule, et n'est pas la même pour toutes les paires d'atomes en interaction.

4.3. Interactions de van der Waals

Les effets de polarisation (ou d'induction), c'est-à-dire les interactions entre charges permanentes et dipoles induits sont en général faibles ($\simeq 10^{-3}$ kcal mole^{-1}) et donc négligeables. On peut en tenir compte dans le cas d'interactions entre charges et groupements aromatiques hautement polarisables (interactions entre COO^- et un résidu tryptophane dans une protéine, par exemple).

Par contre, les interactions entre charges fluctuantes ou encore interactions de Van der Waals–London ou interactions de dispersion sont du même ordre de grandeur que les interactions électrostatiques et interviennent de façon importante et souvent prépondérante. Ces interactions attractives introduites dans la théorie des gaz réels, assurent la cohésion des gaz rares à l'état liquide et ont fait l'objet de déterminations théoriques et expérimentales diverses [31]. Dans le cas simple de deux molécules A et B contenant plusieurs électrons, de symétrie sphérique et dans l'état fondamental l'énergie d'interaction due aux fluctuations est au deuxième ordre (modèle de Drude)

$$E_{\mathrm{VW}} \simeq -\frac{1}{(4\pi\varepsilon_0)^2}\frac{\alpha_{\mathrm{a}}\alpha_{\mathrm{b}}}{n^4}\frac{1}{r^6}\frac{I_{\mathrm{a}}I_{\mathrm{b}}}{I_{\mathrm{a}}+I_{\mathrm{b}}};$$

α_a, α_b sont les polarisabilités des molécules, n l'indice de réfraction du milieu, I_a et I_b les énergies d'ionisation des molécules A et B.

Une approche plus exacte est basée sur un traitement quantique du problème à l'aide d'une méthode de perturbation par exemple, qui donne une expression du type:

$$E_{VW} \simeq - C/r^6 .$$

Le coefficient C est donné par l'équation de Slater–Kirkwood dans laquelle Pitzer [32] introduit les nombres 'éffectifs' d'électrons:

$$C = \frac{\frac{3}{2}e(\hbar/m^{1/2})\,\alpha_a\alpha_b}{(\alpha_a N_a)^{1/2} + (\alpha_b N_b)^{1/2}} .$$

BUCKINGHAM

$$E_{ij} = \frac{a_{ij}}{\exp^{b_{ij} r_{ij}}} - \frac{c_{ij}}{r_{ij}^6}$$

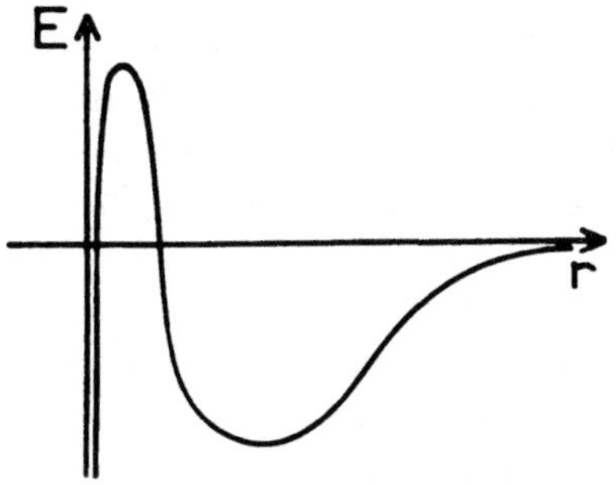

LENNARD-JONES

$$E_{ij} = \frac{d_{ij}}{r_{ij}^{12}} - \frac{c_{ij}}{r_{ij}^6}$$

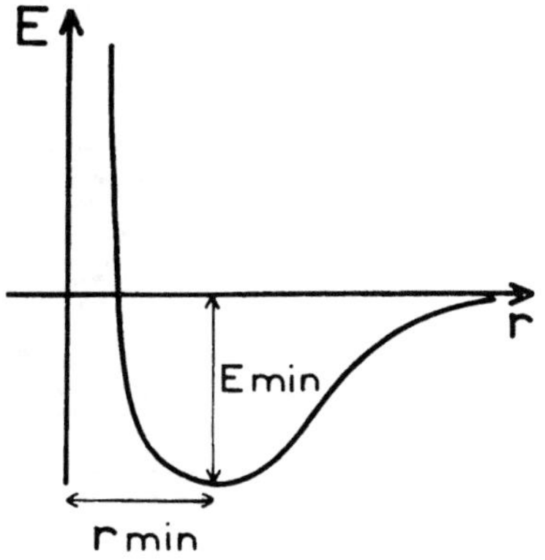

$$E_{ij} = E\min_{ij} \left[\frac{r\min_{ij}}{r_{ij}}\right]^6 \left[\left[\frac{r\min_{ij}}{r_{ij}}\right]^6 - 2.0\right]$$

Fig. 4. Représentation graphique des fonctions de Buckingham et de Lennard–Jones.

Jehle [33] a récemment proposé une technique de 'reaction field' particulièrement intéressante pour le traitement des fluctuations.

On groupe en général le terme attractif de Van der Waals et un terme répulsif apparaissant lorsque la distance entre atomes est de l'ordre de grandeur de la somme des rayons de Van der Waals. Parmi les nombreuses fonctions proposées [34], les plus utilisées sont alors (Figure 4):

$$E_{ij} = \frac{a_{ij}}{\exp(b_{ij} r_{ij})} - \frac{c_{ij}}{r_{ij}^6} \quad \text{Buckingham}$$

$$E_{ij} = \frac{d_{ij}}{r_{ij}^{12}} - \frac{c_{ij}}{r_{ij}^6}, \quad \text{Lennard–Jones}$$

où r_{ij} est la distance entre les atomes i, j. Les coefficients a_{ij}, b_{ij}, c_{ij}, d_{ij}, indépendants de r_{ij} sont, dans la pratique, ajustés d'après les résultats de spectrométrie hertzienne et de cristallographie. La fonction de Lennard–Jones n'a pas l'inconvénient de tendre vers $-\infty$ quand $r \to 0$. E_{ij} décroît très vite au-delà de 7 Å.

Pour le peptide on calcule $E_{\mathrm{VW}} = \sum_{ij} E_{ij}$, $\sum_{ij}$: somme sur toutes les paires d'atomes 'non liés' c'est-à-dire séparés par plus de deux liaisons.

4.4. Liaisons hydrogène

L'énergie de ces liaisons comprise entre 3 et 8 kcal mole^{-1} est nettement supérieure à E_{VW} et doit donc être incluse dans le calcul de l'énergie d'interaction avec la plus grande précision possible. Dans les peptides, il s'agit essentiellement de liaisons NH...CO qui sont linéaires dans les hélices et telles que la distance N— ...O soit

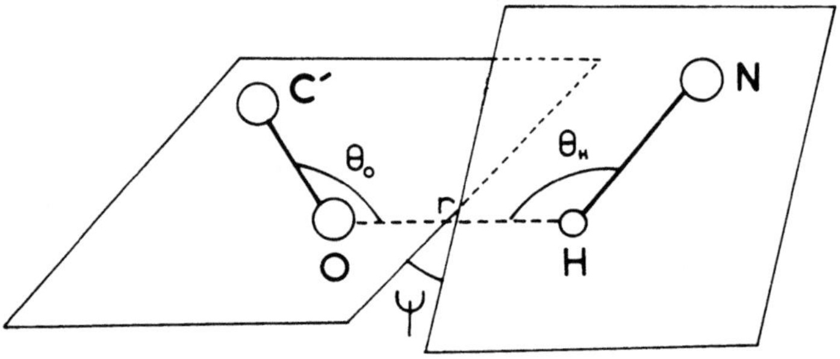

$$E_H = 4\Sigma \left[\left[\frac{\sigma}{r}\right]^{12} - \left[\frac{\sigma}{r}\right]^{6} \right] - \frac{\mu_O \mu_H}{r^3} g\left[\theta_O, \theta_H, \Psi\right]$$

$$g\left[\theta_O, \theta_H, \Psi\right] = 2 \cos\theta_O \cos\theta_H - \sin\theta_O \sin\theta_H \cos\Psi$$

Fig. 5. Énergie de la liaison hydrogène selon la relation de Stockmayer utilisée par A. M. Liquori [36].

égale à 2,85 Å. De très nombreuses fonctions ont été proposées pour représenter la liaison hydrogène. Elles doivent rendre compte de l'énergie et de la géométrie de la liaison. Il semble particulièrement difficile de décrire la directivité de la liaison souvent reliée à l'hybridation sp^2 de l'oxygène.

(a) *Relation de Stockmayer* [35]

Surtout utilisée par le groupe de Liquori [36], elle est constituée d'une fonction de Lennard–Jones et d'un terme d'interaction dipolaire introduisant une dépendance angulaire (Figure 5).

$$E_H = 4\varepsilon\left[\left(\frac{\sigma}{r}\right)^{12} - \left(\frac{\sigma}{r}\right)^{6}\right] - \frac{\mu_O\mu_H}{r^3}\, g\,(\theta_O, \theta_H, \psi)$$

$$g\,(\theta_O, \theta_H, \psi) = 2\cos\theta_O \cos\theta_H - \sin\theta_O \sin\theta_H \cos\psi\,.$$

μ_O, μ_H sont des moments dipolaires centrés sur les atomes O et H et respectivement dirigés suivant les liaisons C=O et H—N.

(b) *Relation de Lippincott et Schroeder* [37]

Cette relation a été souvent utilisée avec différentes modifications destinées à rendre compte des effets directionnels. Pour une liaison hydrogène linéaire X—H...Y, ces auteurs ont proposé d'abord une fonction comprenant quatre termes:

(a) deux termes représentant les interactions X—H et H...Y,

(b) un terme répulsif et un terme attractif.

$$E_H = \overbrace{D\{1 - \exp[-n(r - r_0)^2/2r]\}}^{\text{X—H}} - \overbrace{D^*\{1 - \exp[-n^*(R - r - r_0^*)^2/2(R - r)]\}}^{\text{H}\ldots\text{Y}} + {}$$
$$+ \underbrace{A\exp(-bR)}_{\text{rép.}} - \underbrace{B/R^m}_{\text{attr.}}.$$

r_0 est la distance X—H à l'équilibre et r cette distance lorsque la liaison H est formée; r^* est la distance H...Y à l'équilibre, R est la distance X...Y, D, n et D^*, n^* sont des paramètres reliés aux énergies des liaisons X—H et H...Y.

Les constantes A, b du terme répulsif, et B, m (souvent $m \simeq 6$) sont ajustées d'après les données expérimentales. Cette relation a été appliquée au cas de la liaison H dans les peptides.

Moulton et Kromhout [38] ont introduit une dépendance angulaire en multipliant l'exponentielle du premier terme par $\cos^2\theta$ et l'exponentielle du second par $\cos^2\theta'$ (θ et θ' sont les angles entre les liaisons N—H et H...O et les directions 'normales' de liaison des atomes N et O). Scott et Scheraga [39] ont supposé la liaison X—H de longueur fixe, ce qui supprime le premier terme de E_H. Pour tenir compte de l'existence de deux doublets sur l'oxygène, ils ont pris deux termes pour la liaison H...Y et proposent une expression telle que:

$$E_H = -D^* \cos^2\theta_1 \exp[-n^*(R-r-r_0^*)^2/2(R-r)] -$$
$$- D^* \cos^2\theta_2 \exp[-n^*(R-r-r_0^*/2(R-r)] +$$
$$+ A\exp(-bR) - \tfrac{1}{2}A(R_0/R)^m \exp(-bR_0),$$

R_0 étant la distance minimum N— ...O $\simeq$2,85 Å; θ_1, θ_2 étant les angles entre la liaison H ... O et les directions des doublets libres de O.

La forme de la dépendance angulaire fut ensuite modifiée par Ooi *et al.* [40.] Balusubramanian *et al.* [41] ont proposé d'autres modifications de la relation de Moulton et Kromhout, qui reste néanmoins d'une utilisation laborieuse, compte tenu du nombre élevé de paramères. Ils montrent alors la possibilité d'approximer la fonction E_H de la manière suivante:

$$E_H = E(R) + \exp[p_3\Delta(q_1\theta^2 + q_2\theta^3)] \text{ avec } E(R) = E_{min} + p_1\Delta^2 + p_2\Delta^3.$$

R= distance N— ... O; $\theta = \widehat{HNO}$, $\Delta = R - R_{min}$; p_1, p_2, p_3, q_1, q_2 étant cinq constantes ajustées d'après l'expérience.

(c) *Relations empiriques de Scheraga*

La difficulté d'utilisation des relations de Lippincott et Schroeder a conduit Scheraga à rechercher des relations empiriques représentant au mieux les données des rayons X, de la diffraction des neutrons, de la thermodynamique etc. L'idée de base est que les effets directionnels doivent être contenus dans un calcul global de l'interaction totale entre deux molécules:

La première relation proposée [42] est:

$$R_1AH...BR_2.$$
$$E_{tot} = \left\{\begin{matrix}\text{Buckingham}\\ \text{or}\\ \text{Lennard–Jones}\end{matrix}\right\} + S(r),$$

où $S(r)$ est l'énergie d'interaction entre tous les atomes autres que H et B (énergie essentiellement électrostatique, E_{VW} étant petit).

La seconde relation proposée [43] basée sur un calcul d'orbitales moléculaires du type CNDO/2 est de la forme

$$E_{tot} = \sum_{ij} E_{min}\left(\frac{r_{ij}^{min}}{r_{ij}}\right)^6 \left\{\left(\frac{r_{ij}^{min}}{r_{ij}}\right)^6 - 2{,}0\right\} + \frac{A_k}{R_{O...H}^{12}} - \frac{B_k}{R_{O...H}^{10}} + 166{,}0 \sum_{ij} \frac{q_i q_j}{r_{ij}}.$$

Les fonctions de Buckingham ou de Lennard–Jones sont remplacées par une fonction en R^{-12}, R^{-10} appelée potentiel GHB (General Hydrogen Bond). Les coefficients A_k, B_k sont ajustés à partir des valeurs trouvées pour les liaisons H dans des dimères de petites molécules modèles.

L'usage de fonctions empiriques a l'avantage de simplifier le traitement sur ordinateur. Les fonctions actuellement utilisées donnent (au moins dans les cas classique) une bonne représentation de la liaison H. Une interprétation plus fine nécessite évidemment des méthodes quantiques qui évaluent globalement les interactions en tenant compte des termes coulombiens et d'échange et, dans un calcul SCF, des termes de polarisation et de transfert de charge.

4.5. Déformations des angles et des liaisons de valence

On utilise en général des fonctions du type:

$$E_{\Delta r} = \tfrac{1}{2}K_{\Delta r}(\Delta r)^2 \qquad E_{\Delta\theta} = \tfrac{1}{2}K_{\Delta\theta}(\Delta\theta)^2$$

pour décrire les variations d'énergie associées à des variations Δr des longueurs de liaisons et $\Delta\theta$ des angles de valence. K_r, K_θ sont évalués d'après les résultats de l'infrarouge et conduisent à des valeurs telles que:

$$E_{\Delta r} \simeq 0{,}5 \text{ kcal mole}^{-1} \text{ pour } \Delta r = 0.01 \text{ Å}$$
$$E_{\Delta\theta} \simeq 0{,}3 \text{ kcal mole}^{-1} \text{ pour } \Delta\theta = 5^\circ .$$

Ces deux termes sont ajoutés aux termes d'interactions dépendant de ϕ et ψ et font donc intervenir deux paramètres supplémentaires pour la minimisation.

4.6. Conclusion

L'énergie totale sera donc une somme de termes tels que:

$$E = E_{\mathrm{T}} + E_{\mathrm{E}} + E_{\mathrm{VW}} + E_{\mathrm{H}} + E_{\Delta r} + E_{\Delta\theta} .$$

On recherche les minimums de cette fonction auxquels correspondent les conformations stables de la molécule.

5. Quelques exemples d'application

5.1. Peptides. Polypeptides

La Figure 6 rappelle les résultats obtenus par Ramachandran à l'aide de la méthode des sphères dures, c'est-à-dire en tenant compte seulement des encombrements stériques pour un 'dipeptide' improprement appelé glycyl-L-alanine et qui correspond en fait à l'unité peptidique conformationnelle précédemment décrite. La conformation qui ne dépend que des deux paramètres ϕ et ψ est représentée sur une carte conformationnelle à deux dimensions. Sur cette carte, sont portés les points représentatifs d'unités de conformations régulières type: hélice α droite: α_{R}, hélice α gauche: α_{L}, structures β// et anti// et structure de la polyglycine II [7]. Cette méthode basée sur la notion de rayon de Van der Waals est évidemment approximative et permet essentiellement de déterminer les zones interdites en fonction de la nature du résidu peptidique considéré. La méthode peut être étendue à des oligopeptides [5] et à des homopolypeptides [44]. La détermination des minimums d'énergie permet de raffiner considérablement ces résultats.

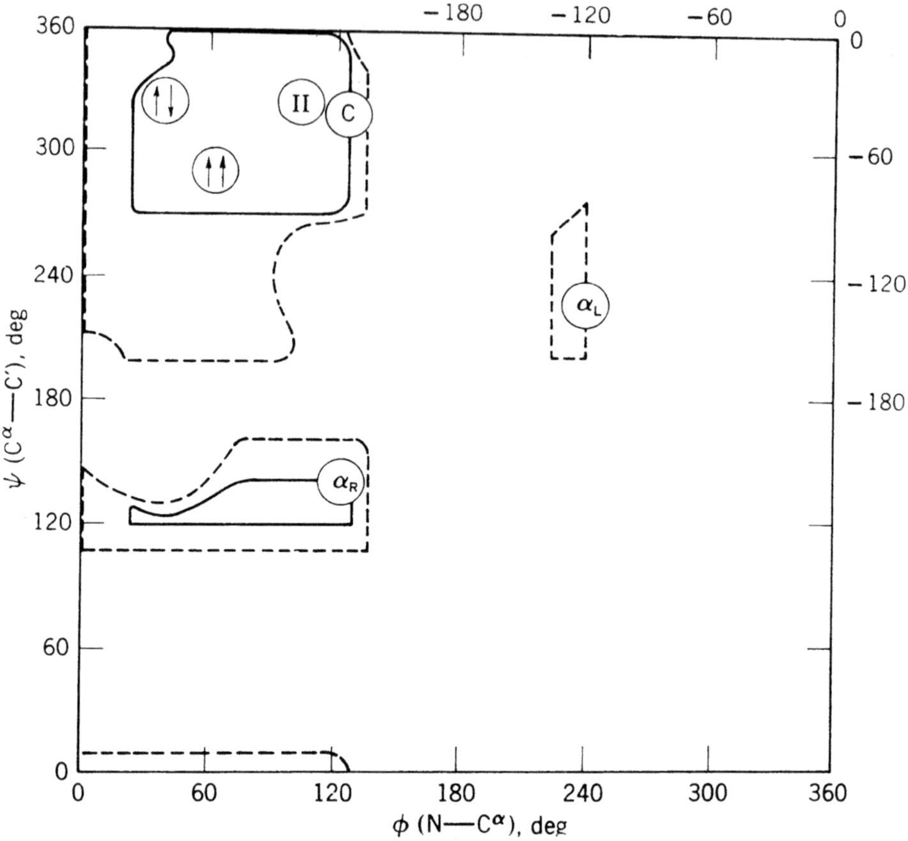

Fig. 6. Carte stérique de Ramachandran obtenue par la méthode des sphères dures pour le résidu L-alanyl. Les contours en trait plein correspondent aux conformations normalement permises (les traits pointillés correspondent aux 'limites extérieures'). (d'après [7])

5.1.1. *Unité peptidique conformationnelle*

Sur la Figure 7, Scott et Scheraga [39] ont supposé les contours isoénergétiques définissant des minimums d'énergie, aux contours des zones stériquement permises. Les minimums correspondant aux conformations α_R, α_L, β sont, bien entendu, situés dans ces zones. Une telle carte conformationnelle énergétique donne déjà une meilleure idée de la stabilité d'une conformation et des stabilités relatives des différentes formes.

D'une façon générale, la forme générale des contours isoénergétiques et les valeurs des énergies des minimums dépendent beaucoup du choix des coefficients des différentes fonctions potentiel, notamment de la valeur et de la grandeur des déformations des liaisons et des angles de valence. Les valeurs absolues de l'énergie n'ont guère de signification. Pour une méthode de calcul donnée, on peut néanmoins valablement comparer, sur une carte conformationnelle, les stabilités relatives de différentes conformations et avoir un ordre de grandeur des barrières d'énergie les séparant.

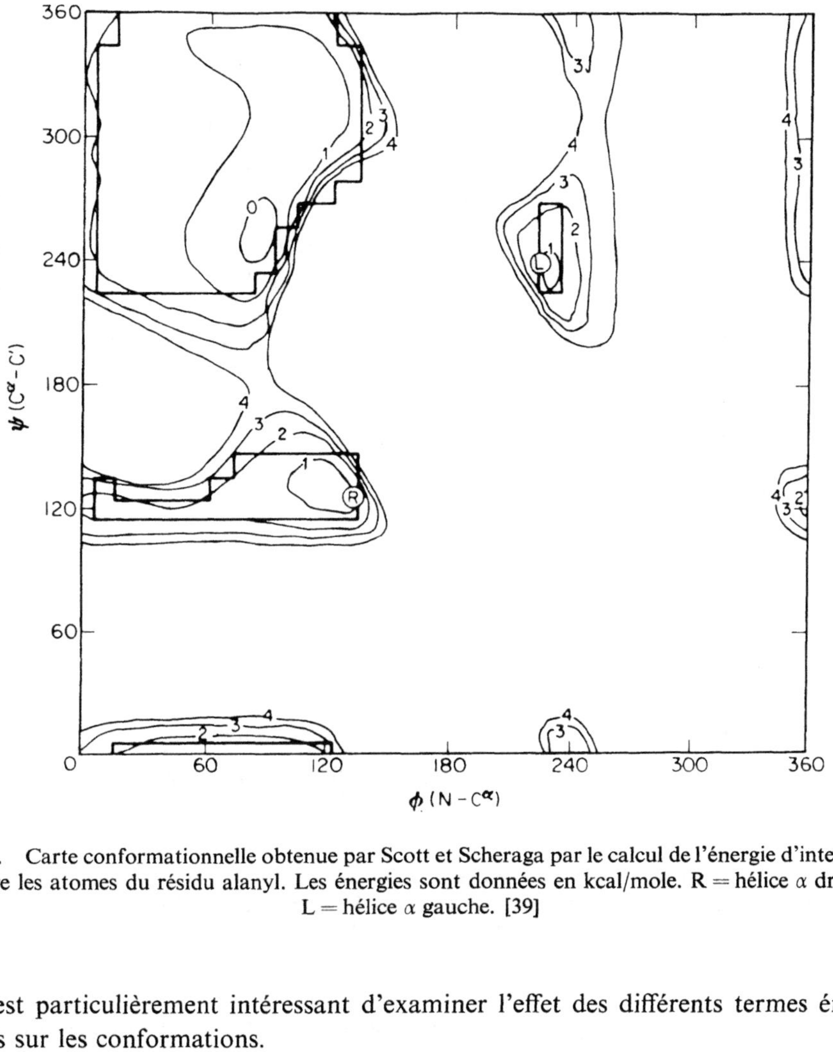

Fig. 7. Carte conformationnelle obtenue par Scott et Scheraga par le calcul de l'énergie d'interaction entre les atomes du résidu alanyl. Les énergies sont données en kcal/mole. R = hélice α droite; L = hélice α gauche. [39]

Il est particulièrement intéressant d'examiner l'effet des différents termes énergétiques sur les conformations.

(a) Pour un résidu alanyl en prenant seulement E_{VW} on obtient des contours qui reproduisent remarquablement bien les zones stériquement permises et dont les minimums se trouvent au voisinage des points définissant les hélices α_R et α_L. C'est là un résultat remarquable, au moins pour les homopolypeptides: les forces de dispersion (attraction en $1/r^6$) seules, préfigurent déjà des structures stéréorégulières compactes du type hélice α. On trouve également une zone de faible énergie au voisinage des structures β (hélice binaire). Ceci confirme que pour des polymères en chaîne, les structures de symétrie hélicoïdale sont des structures stables pour des interactions intramoléculaires attractives.

(b) Si on ajoute à E_{VW} les termes E_E et E_T, la forme générale des contours ne varie pas d'une façon déterminante. Les cartes données par Brant et Flory [45] illustrent ce point. Il faut remarquer que les minimums obtenus par ces auteurs sont beaucoup

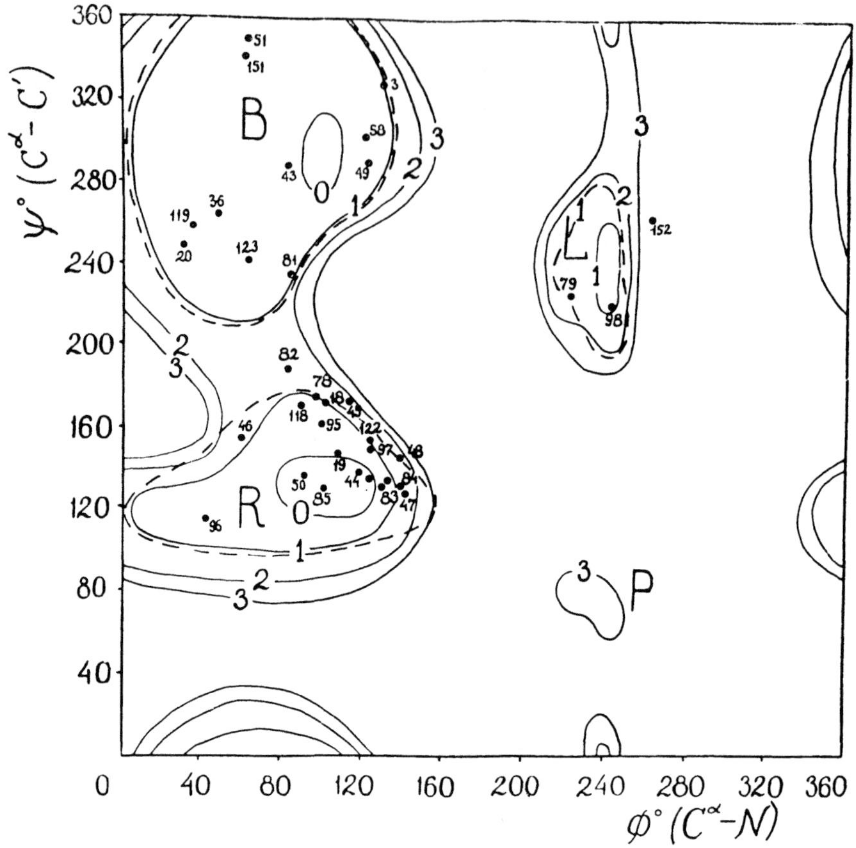

Fig. 8. Carte conformationnelle obtenue par Popov pour le résidu alanyl. Les contours pointillés sont obtenus par minimisation en tenant compte de la déformation des angles de valence. Les points correspondent aux ıésidus alanyl de la myoglobine.

plus aigus que ceux trouvés par Scott et Scheraga. Un autre fait est que si l'on inclue E_E, le minimum absolu se trouve dans la zone des structures β et non plus au voisinage de l'hélice α_R. Brant et Flory attribuent ce déplacement aux interactions dipolaires entre liaisons peptidiques qui sont répulsives dans l'hélice α et attractives dans l'hélice binaire.

(c) La distortion des angles de valence conduit à un élargissement important des contours énergétiques et à une diminution notable des barrières séparant les zones de stabilité. La carte conformationnelle obtenue n'est pas sans rappeler celle obtenue par les méthodes quantiques où les régions permises dessinent un 'croissant' reliant les zones β, α_R, α_L[46, 47] (Figure 8).

(d) Sur la Figure 9 est représentée la carte conformationnelle du résidu alanyl donnée par Liquori [36] et à partir de laquelle il élabore son hypothèse du code stéréochimique.

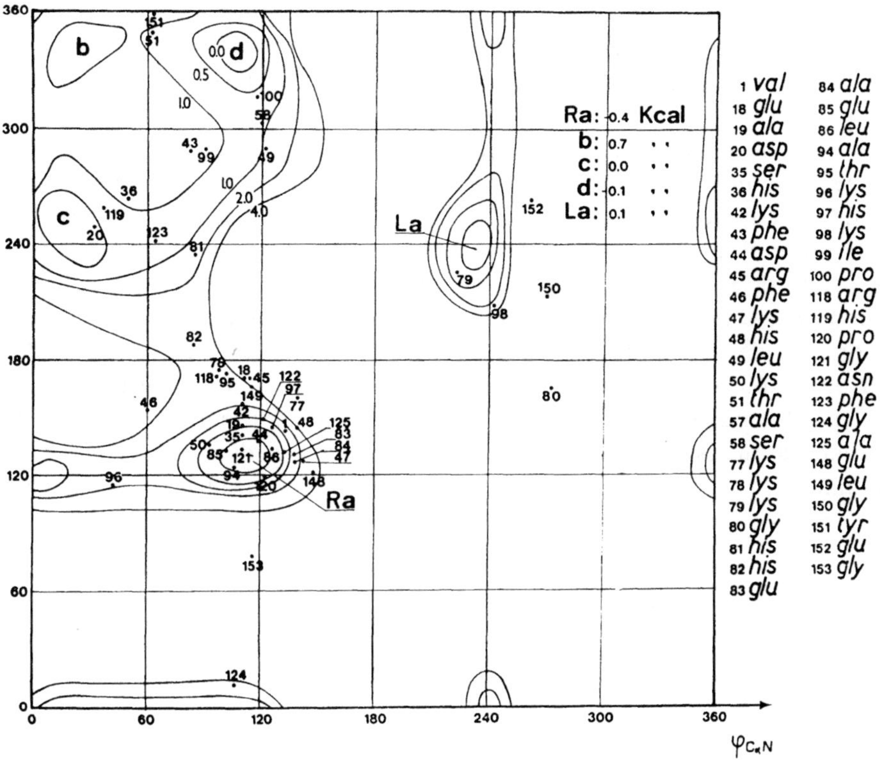

Fig. 9. Carte conformationnelle de Liquori. Les points représentent les résidus de la myoglobine. [36]

L'énergie calculée incluant E_{VW} et E_T présente cinq minimums principaux:

R_a: qui permet de générer l'hélice α_R,
b: qui permet de générer l'hélice β,
c: qui permet de générer l'hélice γ (hélice ternaire),
d: qui permet de générer l'hélice S (hélice trouvée pour la polyproline),
L_a: qui permet de générer l'hélice α_L.

Pour la proline ($\phi = -60° =$ constante) seuls R_a et d sont autorisés. Pour la glycine qui a un C^α symétrique et un proton pour chaîne latérale les zones d'énergie minimum sont beaucoup plus larges.

L'introduction de E_E permet à De Coen [16] de rajouter deux minimums:

e: $\phi = -80°$, $\psi = 80°$.
e^*: $\phi = 60°$, $\psi = -80°$,

qui correspondent à des conformations avec liaison hydrogène intramoléculaire NH…OC, $3 \rightarrow 1$ fermant un cycle à 7 atomes (C^7) (ces conformations ont été bien étudiées par Neel *et al.* [48] et Pullman [13]).

Partant de la remarque que nous avons faite, c'est-à-dire qu'un grand nombre de résidus dans les protéines connues sont situés sur la carte conformationnelle au voisinage des minimums précédents, Liquori propose de construire les chaînes polypeptidiques à partir de ces conformations élémentaires ou 'codons stéréochimiques'.

La comparaison avec l'expérience dans le cas de la myoglobine montre que ce concept peut être utile pour décrire la structure tertiaire des protéines. De Coen a introduit des améliorations notables en examinant le rôle des chaînes latérales. On sait que la position des minimums d'énergie potentielle du squelette sont peu sensibles à la nature de la chaîne latérale, au moins lorsque celle-ci est aliphatique et peu volumineuse. Dans une séquence, les interactions entre chaînes latérales et le squelette (importantes dans le cas de chaînes latérales polaires) peuvent néanmoins modifier la probabilité pour qu'un résidu adopte l'une des conformations de base. Il est également possible que des interactions entre chaînes latérales jouent un rôle de discrémination parmi les conformations possibles. Ce sont évidemment ces facteurs qui déterminent en partie la spécificité conformationnelle d'une séquence, en tenant compte du volume et du caractère hydrophobe ou hydrophile des chaînes latérales.

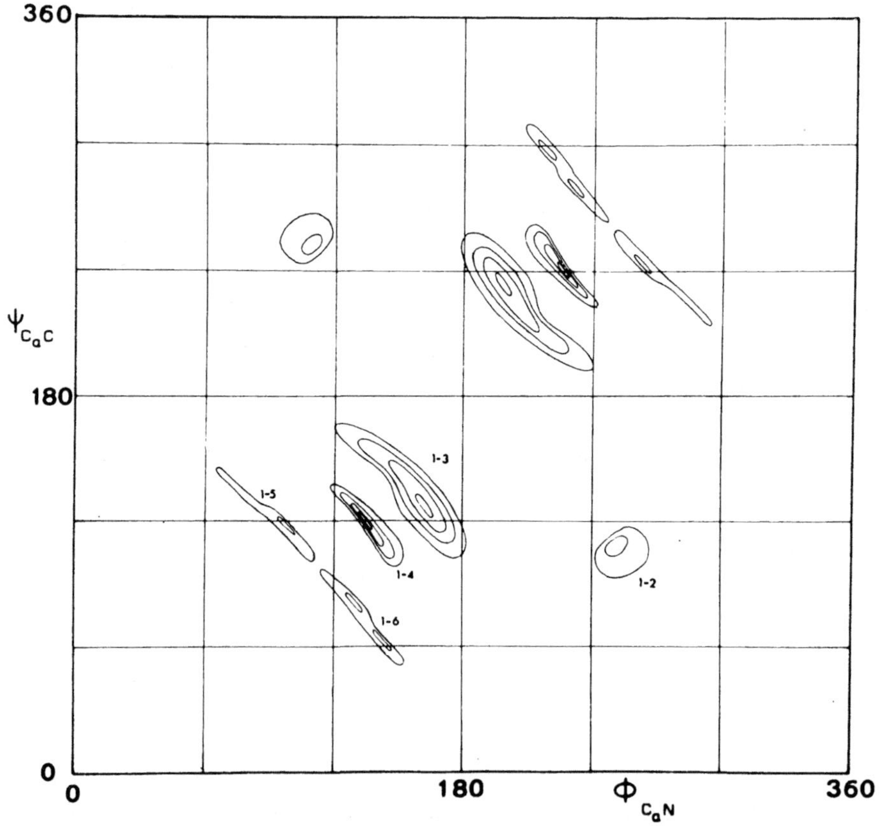

Fig. 10. Minimums d'énergie pour la liaison hydrogène correspondant à différents types d'hélice. [36]

5.1.2. *Liaisons hydrogène. Polypeptides*

Très souvent les liaisons hydrogène interviennent pour stabiliser des conformations correspondant aux minimums de l'énergie potentielle comprenant tous les termes sauf E_H ou seulement $(E_{VW}+E_T)$ par exemple. Les liaisons intramoléculaires sont du type $3\rightarrow 1$, $4\rightarrow 1$, $5\rightarrow 1$, $6\rightarrow 1$, $7\rightarrow 1$, etc. (cette notation respecte la nomenclature standard, elle correspond aux liaisons 1—2, 1—3, 1—4, 1—5, 1—6, désignées par Liquori.) Une interaction NH...O=C dans le même résidu au voisinage de la conformation étendue est proposée par Neel stabilisant une conformation en C^5, mais la nature exacte de cette liaison reste discutée. La liaison $4\rightarrow 1$ existe dans l'hélice 3_{10} et le 'β-turn' [49], la liaison $5\rightarrow 1$ est celle de l'hélice α.

Il est très remarquable de constater qu'en prenant une fonction de Stockmayer on obtienne des minimums d'énergie de la liaison H au voisinage des minimums trouvés pour les autres termes de l'énergie potentielle (Figure 10).

On voit notamment que les conformations en C^7 sont trouvées, c'est-à-dire qu'en prenant tous les termes dans l'énergie: E_{VW}, E_E, E_T, E_H, on obtient une carte conformationnelle qui n'est pas foncièrement différente de celle obtenue par les méthodes quantiques.

Utilisant cette énergie totale, Scott et Scheraga [39] ont calculé la conformation d'un certain nombre de polypeptides dont la poly-L-alanine. Les minimums trouvés pour l'unité ne sont que faiblement modifiés et ces auteurs ont trouvé que l'hélice α_R est légèrement (0,01 kcal mole^{-1}/residu) plus stable que l'hélice α_L.

5.1.3. *Pelote statistique*

Les conformations déterminées pour l'unité peptidique peuvent servir pour déterminer certaines caractéristiques de la pelote statistique.

(a) *Carré moyen de la distance entre extrémités de chaîne.* Suivant la procédure proposée par Brant et Flory [7], on définit:

$$\langle r^2\rangle_0/n_p l_p^2\,,$$

où $\langle r^2\rangle_0$: carré moyen de la distance entre extrémités; l_p: longueur de la liaison virtuelle C^α—C^α (3,8 Å); n_p: nombre de liaisons virtuelles.

Lorsque $n_p\rightarrow\infty$

$$\langle r^2\rangle_0/n_p l_p^2 \rightarrow (E+\langle\tilde{T}\rangle)\,(E-\langle\tilde{T}\rangle)^{-1}\,.$$

E: matrice identité d'ordre 3. $\langle\tilde{T}\rangle$ est une matrice moyennée de la façon suivante:

$$\langle\tilde{T}\rangle = Z^{-1}\sum_{\phi}\sum_{\psi}\tilde{T}\exp(-E/RT) \qquad Z=\sum_{\phi}\sum_{\psi}\exp(-E/RT)\,.$$

Cette moyenne peut être faite pour chaque terme E de l'énergie totale E_{tot}. $\tilde{T}$ est la matrice qui fait passer d'un système de coordonnées associé à une liaison virtuelle au système associé à la liaison virtuelle $i-1$.

(b) *Carré moyen du moment dipolaire.* $\langle\mu^2\rangle_0$ peut être calculé par une méthode de moyenne ou à partir des moments individuels de 3,7 D de la liaison peptidique, et en tenant compte des groupements ionisés de bouts de chaîne.

(c) *Valeur moyenne de la constante de couplage* $^3J_{NH-_\alpha CH}$. En résonance protonique, on sait que la constante $^3J_{NH-_\alpha CH}$ peut être relieé par une relation de type Karplus à l'angle dièdre θ: H–N–αC–H [50]:

$$J = A\cos^2\theta - B\cos\theta + C\sin^2\theta \quad \text{avec} \quad \theta = |\phi - 60| \quad \text{pour un résidu L}.$$

Les valeurs exactes des coefficients sont discutées [51]. Les plus récentes ont été proposées par Ramachandran *et al.* [52].

$$A = 7{,}9 \qquad B = 1{,}55 \qquad C = 1{,}35.$$

La constante ^3J peutêtre moyennée de la même façon que les précédentes grandeurs:

$$\langle J\rangle = \frac{\sum\limits_{\phi}\sum\limits_{\psi} J\exp(-E/RT)}{\sum\limits_{\phi}\sum\limits_{\psi} \exp(-R/RT)}.$$

Les valeurs de $\langle r^2\rangle_0/n_p l_p^2$ et $\langle J\rangle$ peuvent être déterminées expérimentalement par des mesures hydrodynamiques d'une part et par RMN d'autre part. Les valeurs de $\langle\mu^2\rangle_0$ ont été mesurées pour des oligopeptides zwitterioniques [53] et concordent relativement bien avec les valeurs calculées par les méthodes empiriques et aussi avec celles calculées par les méthodes quantiques. L'accord entre l'expérience et les théories empiriques reste excellent pour $\langle J\rangle$ et $\langle r^2\rangle_0/n_p l_p^2$. Pour cette dernière grandeur, la méthode PCILO donne par contre des valeurs très inférieures aux valeurs expérimentales. En fait, cette méthode a été surtout appliquée à l'étude des conformations en C_7 des dipeptides modèles faisant intervenir des interactions à courte distance qui sont peut-être surestimées [53].

5.1.4. *Peptides cycliques*

Ces molécules de dimensions limitées (nombre de résidus allant de 5 à 12 pour la plupart) offrent un excellent moyen de comparer les résultats expérimentaux aux prédictions du calcul conformationnel. En effet, les conformations en solution peuvent être étudiées et déterminées au moins en partie par différentes méthodes dont le dichroïsme circulaire et surtout la RMN sont les plus efficaces. Par ailleurs, pour une dizaine de résidus, les méthodes semi-empiriques permettent de calculer un certain nombre de conformations possibles de la molécule qui sont fortement dépendantes des interactions intramoléculaires.

En fait, la RMN et le calcul conformationnel sont de plus en plus souvent utilisés en parallèle. Pour une valeur de ^3J, la fonction $J=f(\theta)$ peut donner de 2 à 4 valeurs de θ, donc de ϕ. Certaines sont éliminées par suite d'encombrement stérique. Le calcul conformationnel peut permettre de choisir parmi les autres, celles dont la probabilité est maximum. Les premiers calculs faits sur la gramicidine S, l'oxytocine et la vasopressine [23] n'ont pas été bien confirmés par l'expérience. La gramicidine S est un exemple typique: la conformation du squelette de ce décapeptide dépend de 20 para-

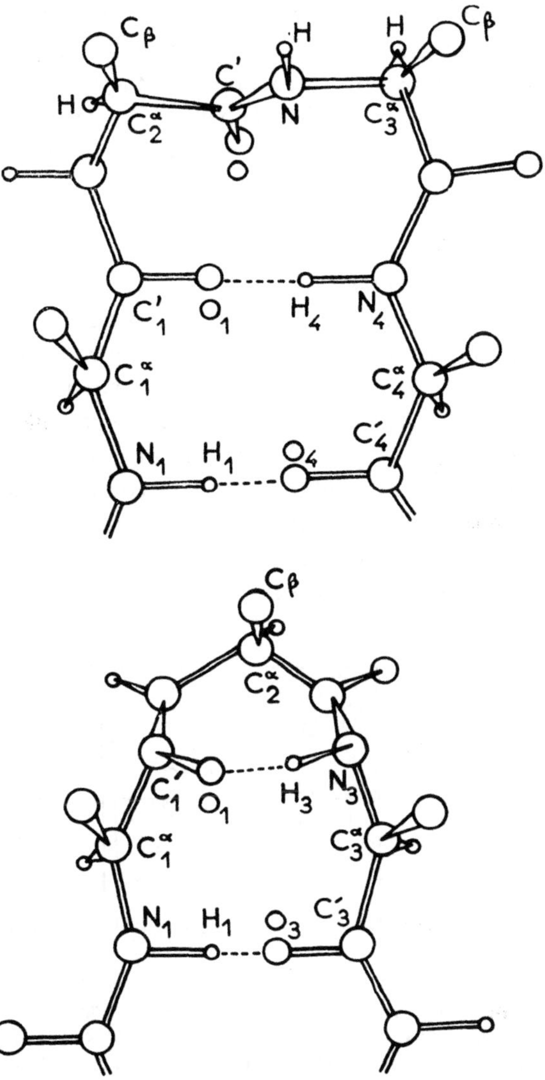

Fig. 11. Modes de repliement possibles d'une chaîne polypeptidique: (a) en haut: formation d'une boucle β; (b) en bas: formation d'un cycle à 7 atomes.

mètres rotationnels. Il faut ajouter à cela les 16 angles de rotation des chaînes latérales. Il est clair que le volume des calculs correspondants est considérable et on s'est contenté au départ d'évaluations partielles de l'énergie conformationnelle. Actuellement, on a bien mis en évidence l'existence de structures type comme le 'β-turn' permettant la formation de liaisons H intramoléculaires (Figure 11), qui aident beaucoup à la compréhension des conformations des cycles peptidiques. De Santis [54] a calculé la conformation de la gramicidine S en construisant une séquence des conformations locales les plus stables et en raffinant la conformation ainsi obtenue d'après les résultats expérimentaux d'une part, par minimisation de l'énergie totale d'autre part. La conformation obtenue semble en bon agrément avec les résultats de diffraction des rayons X actuellement connus sur la structure de cet antibiotique.

En utilisant les conformations locales, Tonelli [55] a récemment proposé des modèles pour l'évolidine dont un, au moins, est en bon accord avec les résultats de l'étude RMN faite par Kopple [56].

5.1.5. *Protéines*

Même avec des fonctions potentiel très simples, il est actuellement hors de question de calculer la conformation d'une chaîne polypeptidique qui peut comporter plusieurs centaines de résidus. L'utilisation du code stéréochimique et le raffinement progressif de la conformation d'une chaîne en cours de croissance peut néanmoins conduire à des résultats intéressants pour comprendre le phénomène de nucléation [57]. On peut aborder le problème d'un autre côté en partant des conformations connues de protéines que l'on compare aux résultats du calcul conformationnel. C'est l'idée de Pohl [58] qui a reporté sur la carte conformationnelle (ϕ, ψ) les valeurs de toutes les paires d'angles mesurées sur les cristaux de myoglobine, lyzozyme, α-chymotrypsine, carboxypeptidase A, RNase-S (Figure 12).

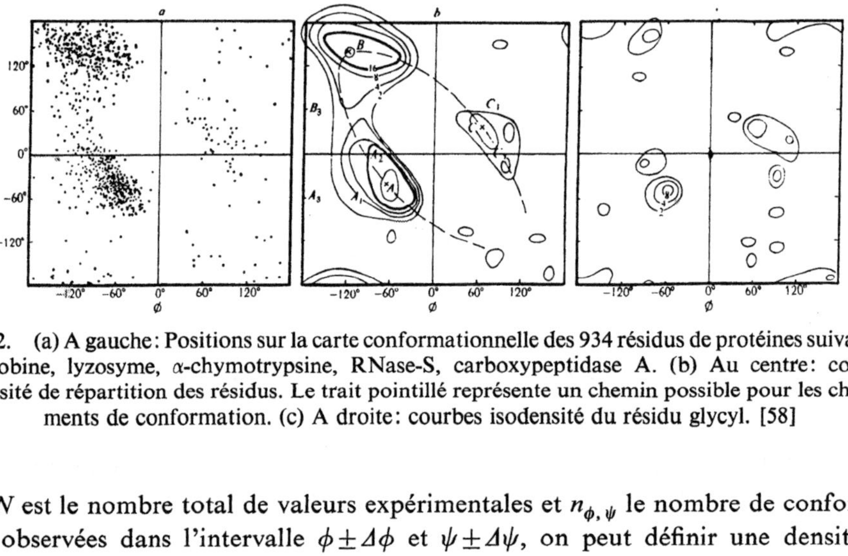

Fig. 12. (a) A gauche: Positions sur la carte conformationnelle des 934 résidus de protéines suivantes: myoglobine, lyzosyme, α-chymotrypsine, RNase-S, carboxypeptidase A. (b) Au centre: courbes isodensité de répartition des résidus. Le trait pointillé représente un chemin possible pour les changements de conformation. (c) A droite: courbes isodensité du résidu glycyl. [58]

Si N est le nombre total de valeurs expérimentales et $n_{\phi,\psi}$ le nombre de conformations observées dans l'intervalle $\phi \pm \Delta\phi$ et $\psi \pm \Delta\psi$, on peut définir une densité de probabilité

$$q_{\phi,\psi} = \frac{n_{\phi,\psi}}{4\pi N} \Delta\phi \cdot \Delta\psi ,$$

que l'on peut relier à l'énergie conformationnelle par une relation telle que:

$$q_{\phi,\psi} = Z^{-1} \exp(-E_{\phi,\psi}/RT),$$

où Z serait l'équivalent d'une fonction de partition.

Il est très frappant de constater que sur les 934 valeurs reportées sur la carte conformationnelle, près de 90% des points se trouvent dans une zone comprenant la région α_R, la région β et la région qui les relie. Les contours isodensité calculés pour des cellules de $\Delta\phi = \pm 10°$ et $\Delta\psi = \pm 10°$ avec $n_{\psi,\phi} = 2$, 4, 8, 16, 32 (pour $n_{\psi,\phi}$ on prend $E = 9$ kcal mole^{-1} au-dessus du minimum le plus profond α_R) prennent une analogie remarquable avec les contours isoénergétiques.

Ceci confirme l'idée qu'une structure protéique doit pouvoir être construite en partant des conformations élémentaires que l'on déforme plus ou moins. Ces déformations rendent compte des effets spécifiques de chaînes latérales et des effets de solvant.

On pourrait également définir un chemin conformationnel de repliement de la chaîne polypeptidique qui passerait, sur la carte conformationnelle, par les zones α_R, β, α_L et aussi C_7 en empruntant les passages de plus basse énergie. Un objectif particulièrement important du calcul conformationnel est donc de prévoir le mode de passage d'un type de conformation à l'autre et la stabilité relative des structures stéréo-régulières et des structures obtenues par combinaison de 'codons stéréochimiques' (Figure 13).

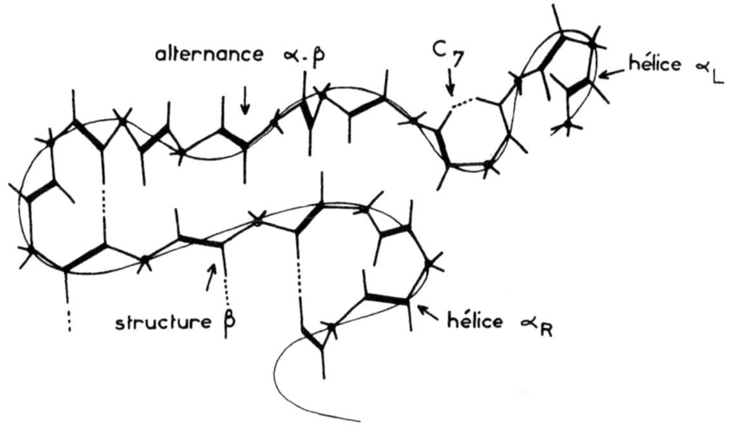

Fig. 13. Possibilités de repliement d'une chaîne polypeptidique basées sur la succession de tronçons de conformation stables.

5.2. Autres applications

Les progrès faits dans le calcul des conformations de peptides s'expliquent par le fait que ces conformations dépendent d'un nombre relativement limité de paramètres, que les moments dipolaires sont connus et qu'on a pu calculer la répartition des charges partielles par différentes méthodes. Il n'en va pas de même pour d'autres molécules biologiques telles nucléotides, oses, phospholipides, et c'est plus récemment que des calculs d'énergie ont pu être effectués pour ces molécules [10].

5.2.1. *Nucléosides – nucléotides*

La conformation du squelette polynucléotidique dépend de quatre paramètres princi-

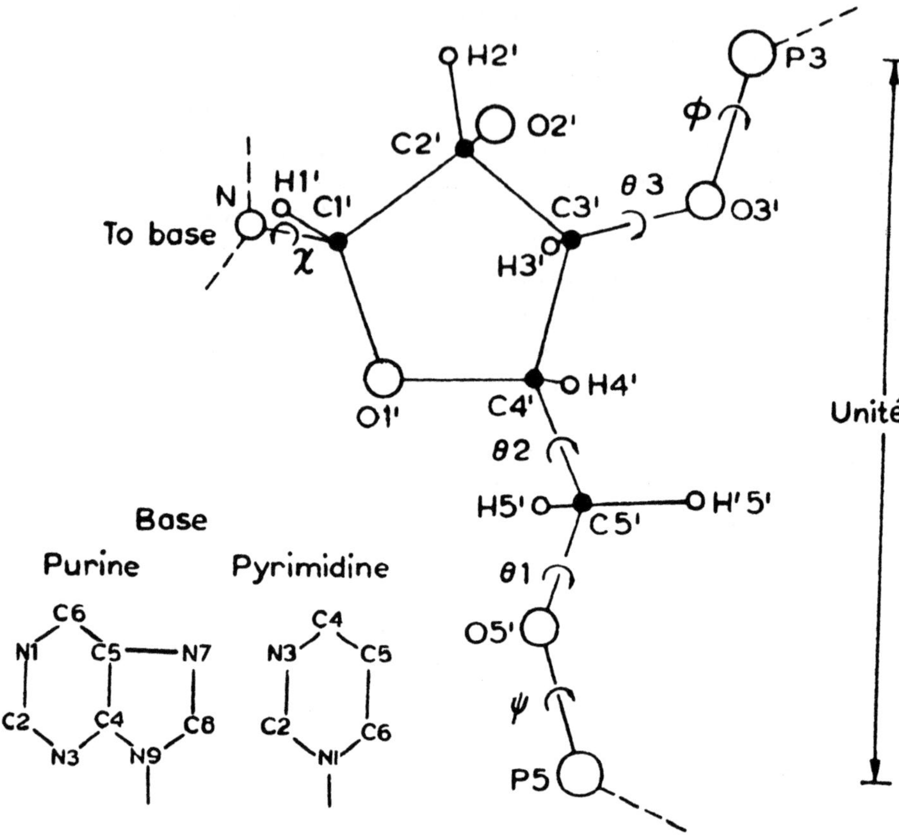

Fig. 14. Unité nucléotidique conformationnelle.

paux notés θ_1, θ_2, θ_3 et χ selon Sasisekharan (Figure 14) (Une uniformisation des notations a été étudiée au cours du 5ème symposium de Jerusalem 1972.) On peut également faire intervenir les angles de torsion autour des cinq liaisons du cycle ribose, ce qui permet notamment de distinguer les formes endo- et exo- du sucre.

Les premiers résultats ont été obtenus en utilisant la méthode des sphères dures [59]. Ils mettent en évidence un nombre limité de conformations autour des valeurs $\theta_1 = 180°$, $\theta_2 = \pm 60°$, $\theta_3 = 210–270°$, et $\chi = 30°$, 220°, conformations qui ne sont pas trop éloignées de celles observées dans les cristaux. Le calcul d'une distribution possible des charges partielles de type σ par la méthode de Del Ré et de type π par la méthode de Hückel [60] a permis à Sasisekharan *et al.* d'aborder un calcul d'énergie par les méthodes semi-empiriques [61]. Les termes pris en compte sont E_E, E_T, et E_{VW}. Dans une première étape, ils étudient l'effet de la rotation χ autour de la liaison sucre-base et du 'puckering' endo- ou exo- du sucre et montrent l'importance des forces de Van der Waals dans la stabilisation des rotamères. Les conformations stables sont principalement dans la région anti- ($\chi \simeq 0°$) et dans la région syn- ($\chi \simeq 100°$).

La forme anti semble plus stable pour les pyrimidines, notamment lorsque le sucre est 3′-endo. Les purines peuvent adopter les deux conformations syn- et anti- sans effet réellement déterminant de la conformation du sucre. Dans une deuxième étape, la conformation de la chaîne phosphate–sucre est déterminée. L'énergie présente des minimums pour $\theta_1 \simeq 180°$ ($\pm 80°$), $\theta_2 = 60°$, 300° et $\theta_3 = 180°$, 280°.

Pour une molécule modèle tel que le diméthylphosphate, les conformations les plus stables sont gauche–gauche (gg), par rotation autour des deux liaisons phosphodiester.

De nombreux calculs ont été faits récemment à l'aide des méthodes quantiques EHT, CNDO et PCILO [10], qui permettent surtout de préciser les résultats obtenus par les méthodes semi-empiriques. Il convient de rappeler qu'il est particulièrement difficile de déterminer les conformations en solution des nucléosides et des nucléotides par RMN du fait de l'imprécision relative des relations empiriques du type 'Karplus' reliant les angles de rotation aux constantes de couplage proton–proton et de l'existence d'équilibres conformationnels [62, 63]. Des études théoriques et expérimentales, il ressort que la conformation préférentielle des nucléosides est syn- et celle des nucléotides est anti-.

5.2.2. *Saccharides*

L'unité constitutive des polysaccharides est l'α-D-glucose pour l'amylose et le β-D-glucose pour la cellulose. Les résidus sont liés en 1 → 4′. En calculant l'énergie d'interaction des atomes non liés Rao *et al.* [64] ont montré que la forme chaise C_1 de Reeves [65] de α-D-glucose était la plus stable parmi les 8 formes possibles. Le disaccharide est formé de deux résidus α-D-glucose reliés en 1–4′ par l'intermédiaire d'un atome d'oxygène. Les deux résidus étant sous forme C_1, la conformation de l'ensemble ne

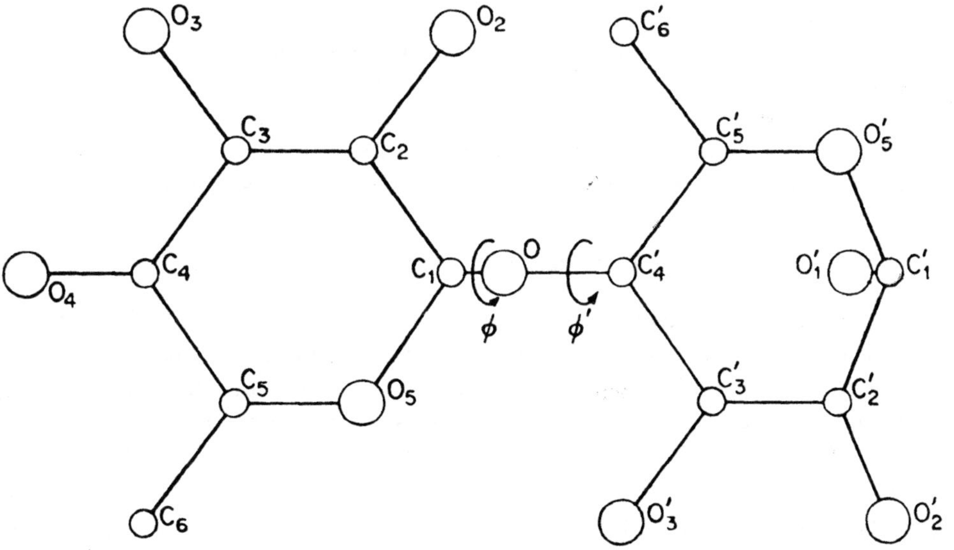

Fig. 15. Unité disaccharidique conformationnelle.

dépend que des angles de rotation ϕ et ϕ' (Figure 15) et peut être représentée sur une carte conformationnelle à deux dimensions. On trouve un minimum relativement large, centré vers $\phi \simeq -30°$, $\phi' \simeq 170°$. A partir des conformations correspondant à cette zone d'énergie minimum on peut construire des hélices à 6 résidus par tour ($n=6$, $h=1,33$ Å) dont la forme gauche a la plus basse énergie et n'est guère différente de la forme V de l'amylose observée par diffraction des rayons X.

Pour la forme β des calculs d'énergie faisant intervenir les termes E_T, E_{VW}, E_H [66, 67] ont suivi les études préliminaires par la méthode des sphères dures [68]. L'effet de la liaison $1 \rightarrow 2$, $1 \rightarrow 3$ ou $1 \rightarrow 4$ a été mis en évidence. Dans ce dernier cas, qui est celui de la cellulose, la région d'énergie minimum correspond assez bien à une hélice $n=2$, $h=5,15$ Å. Cela n'est cependant pas une démonstration de l'existence effective de formes régulières hélicoïdales dans la cellulose native.

5.2.3. *Autres molécules*

Il est clair que les méthodes semi-empiriques peuvent être utilisées pour calculer les conformations de beaucoup d'autres molécules et macromolécules biologiques. Liquori *et al.* [69] ont calculé les conformations de composés pharmacologiques tels que la muscarine et l'acétylcholine. La liste de ces composés s'est considérablement allongée avec l'introduction des méthodes quantiques [13] qui s'appliquent plus aisément à des molécules relativement petites.

Dans un autre domaine, il faut citer la tentative de Vanderkooi [10], qui a cherché à déterminer les conformations stables des têtes polaires de phospholipides. La conformation de ces têtes polaires dépend beaucoup de leur état d'hydratation [70]. Parmi les quatre conformations proposées, la forme allongée 4 et une forme repliée 2 ou 3 (hydratée) semblent bien correspondrent aux résultats obtenus par Wilkins [71] qui a établi les profils de densité électronique des têtes polaires par diffraction des rayons X.

Citons encore les calculs de Cornibert et Marchessault sur le poly-β-hydroxybutyrate [72] et ceux de Cleland sur les polysaccharides ioniques [73].

Dans la plupart de ces cas, les calculs permettent d'éclairer d'une façon particulièrement utile des résultats expérimentaux (rayons X, mesures morphologiques, RMN) d'une précision relative et dont l'interprétation est quelquefois complexe.

6. Conclusions – perspectives

A. (1) Les calculs semi-empiriques (comme les calculs quantiques d'ailleurs) s'appliquent à une molécule isolée plongée dans un diélectrique isotrope et homogène. C'est une situation évidemment très éloignée de la réalité puisqu'on ne tient pas compte des interactions avec d'autres molécules et avec le solvant. Des tentatives sont actuellement faites pour introduire ces interactions. Il est cependant remarquable de constater que les conformations trouvées expérimentalement se placent pratiquement toujours parmi celles qui sont prévues par voie théorique. Ceci conduit à penser que, au moins pour les petites molécules et les homopolymères, les interactions avec l'environnement (solvant et autres molécules) 'sélectionnent' parmi les différentes conformations possibles celle qui est compatible avec les données géométriques et qui conduit à un minimum de l'énergie totale du système. Un exemple bien connu est celui des structures β. L'hélice binaire est intrinsèquement une conformation stable. A l'état solide, cette structure est compatible avec la formation d'un réseau régulier de liaisons hydrogène entre les chaînes assurant la stabilité macroscopique du système. La structure β est donc un état favorable pour un homopolypeptide à l'état solide. Nous avons également relevé que dans les protéines, un très fort pourcentage des résidus se trouvent dans les zones d'énergie minimum. La sélection entre les formes stables de l'unité peptidique conformationnelle faisant intervenir:

(a) Des interactions spécifiques entre les chaînes latérales et le squelette peptidique. Ceci justifie la réalisation d'un tableau comparatif complet et détaillé des conformations de vingt amino-acides naturels [74].

(b) Des interactions entre résidus voisins. Ceci nécessite le calcul complet de la conformation de dipeptides réels (à deux résidus) pour déterminer d'éventuels effets de 'doublet', dans une séquence [75].

(c) Des interactions à longue distance permettent la formation de zones hydrophobes, de ponts S—S ou l'assemblage de sous-unités.

On met donc en évidence le rôle primordial des interactions à courte distance et des phénomènes de 'nucléation' d'une chaîne polymérique. Il est cependant prématuré d'en déduire un schéma des modes de repliement d'un polypeptide ou de conclure que la conformation d'une protéine ne correspond pas au minimum global d'énergie.

(2) Les calculs d'énergie donnent une image statique des conformations. Il est évidemment nécessaire d'introduire le poids statistique des différentes conformations pour donner une image plus proche de la réalité d'une molécule en solution.

(3) La décomposition de l'énergie en plusieurs termes a l'avantage de montrer le rôle primordial des interactions de Van der Waals et des liaisons hydrogène dans la stabilisation d'une conformation. Ceci permet aux cristallographes et spectroscopistes une première approche souvent fructueuse des conformations étudiées expérimentalement. L'influence des interactions électrostatiques se fait surtout sentir aux extrémités de chaîne ou dans le cas de chaînes latérales très polaires. En aucun cas, le calcul n'a pour but de donner les coordonnées des atomes dans une conformation donnée avec la précision de méthodes telles que la diffraction des rayons X et dans certains cas la RMN.

(4) Il est évidemment possible d'améliorer encore la définition des fonctions potentiel représentant les différents termes de l'énergie d'interaction. Ceci est particulièrement vrai pour l'énergie torsionnelle E_T et aussi pour E_H. Des calculs quantiques raffinés (calculs *ab initio* de Pople par exemple) doivent permettre de mieux comprendre les interactions au sein de petites molécules modèles. L'accumulation des résultats expérimentaux sur la conformation de ces molécules (diffraction des rayons X, spectroscopie de rotation, RMN) et sur la formation des liaisons hydrogène doit permettre de définir avec plus de précision les coefficients des fonctions empiriques utilisées pour représenter ces interactions.

Il faut par ailleurs attirer l'attention sur la nécessité de tenir compte des possibilités de déformation des liaisons et surtout des angles de valence, qui se traduisent par un abaissement notable des barrières d'énergie entre différentes conformations.

B. L'introduction des interactions entre molécules et des interactions avec le solvant conduit à une augmentation importante du nombre de paramètres dans le calcul des minimums d'énergie. Ceci est un argument supplémentaire pour l'utilisation des méthodes semi-empiriques.

(1) Une méthode d'étude du 'packing' d'un cristal a été développée par McGuire *et al.* [76] et appliquée au cas d'homopolypeptides. Pour la poly-L-alanine sous forme β, le 'packing' des chaînes est dû essentiellement à la formation de liaisons hydrogène intermoléculaires et aux interactions entre chaînes latérales. Les interactions intramoléculaires sont peu affectées et les structures calculées avec les structures déterminées par diffraction des rayons X. Ces résultats s'appliquent également au poly-L-aspartate de benzyle sous forme d'hélice ω pour lequel les interactions intramoléculaires jouent un rôle plus important, les interactions intermoléculaires s'établissant essentiellement entre les cycles benzéniques.

(2) Krimm et Venkatachalam [77] ont pour la première fois introduit des molécules de solvant dans le calcul de l'énergie. Ils ont choisi pour cela la poly-L-proline sous forme hélicoïdale et une molécule d'eau fixée sur le groupe —C=O, et ont calculé $E_{VW} + E_E$. Ils ont de plus introduit un facteur entropique pour la fixation des molécules d'eau et peuvent de ce fait obtenir l'énergie libre pour chaque conformation de polymère.

$$F = E_{min} - TS,$$

où E_{min} est le minimum de l'énergie potentielle d'interaction.

La fixation de l'eau modifie sensiblement les conformations de la chaîne peptidique et favorise notamment la forme Pro_n (II) par rapport à la forme Pro_n (I).

C. Ces deux contributions constituent certainement les développements les plus importants pour l'avenir des méthodes semi-empiriques de calcul des conformations. Elles doivent permettre de comprendre de façon de plus en plus détaillée la façon dont s'élabore une conformation macromoléculaire, c'est-à-dire de fournir des moyens de la contrôler et de la reproduire pour des molécules préparées par synthèse.

Références

[1] Watson, J. D. et Crick, F. H. C.: 1953, *Nature* **171**, 737.

[2] Pauling, L. et Corey, R. B.: 1950, *J. Am. Chem. Soc.* **72**, 5349. (voir également série 1951, *Proc. Nat. Acad. Sci. US* **37**, 235–729).

[3] Ramachandran, G. N., Ramakrishnan, C., et Sasisekharan, V.: 1963, *J. Mol. Biol.* **7**, 95.

[4] Leach, S. J., Nemethy, G., et Scheraga, H. A.: 1966, *Biopolymers* **4**, 369.

[5] De Santis, P., Giglio, E., Liquori, A. M., et Ripamonti, A.: 1965, *Nature* **206**, 455.

[6] Scheraga, H. A., Scott, R. A., Vanderkooi, G., Leach, S. J., Gibson, K. D., Ooi, T., et Nemethy, G.: 1967, dans *Conformation of Biopolymers* (G. N. Ramachandran, éd.), Vol. 1, Academic Press, London, p. 43.

[7] Flory, P. J.: 1969, *Statistical Mechanics of Chain Molecules*, Interscience New-York.
[8] Ramachandran, G. N. et Sasisekharan, V.: 1968, dans *Advances in Protein Chemistry*, Vol. 23, Academic Press, Londres, p. 283.
[9] Popov, E. M., Dashevskii, V. G., Lipkind, G. M., et Azkhipova, S. F.: 1968, *Mol. Biol. (URSS)* **2**, 612.
Popov, E. M., Lipkind, G. M., Azkhipova, S. F., et Dashevskii, V. G.: 1968, *Mol. Biol. (URSS)* **2**, 622.
[10] Vème Symposium de Jérusalem sur *Conformation of Biological Molecules and Polymers* (E. B. Bergmann et B. Pullman, éds.), à paraître.
[11] Hoffmann, R. et Imamura, A.: 1969, *Biopolymers* **7**, 207.
[12] Yan, J. F., Momany, F. A., Hoffmann, R., et Scheraga, H. A.: 1970, *J. Phys. Chem.* **74**, 420,
[13] Pullman, B.: 1971, dans *Aspects de la chimie quantique contemporaine* (R. Daudel et B. Pullman. éds.), C.N.R.S., Paris.
[14] Jukes, T. H.: 1966, *Molecules and Evolutions*, Columbia University Press, New York-Londres.
[15] Liquori, A. M.: 1969, *Quart. Rev. Biophys.* **2**, 1.
[16] De Coen, J. L.: 1970, *J. Mol. Biol.* **49**, 405.
[17] Voir: 1971, *Biochim. Biophys. Acta* **229**, 1.
[18] Go, N. et Scheraga, H. A.: 1970, *Macromolecules* **3**, 178.
[19] Sugeta, H. et Miyazawa, T.: 1967, *Biopolymers* **5**, 673.
[20] Davidson, W. C.: 1959, AFC Research and Development Report, ANL 5990.
Fletcher, R. and Powell, M. J. D.: 1963, *Computer J.* **6**, 163.
[21] Fletcher, R. et Reeves, C. M.: 1964, *Computer J.* **7**, 149.
[22] Hirschfelder, J. O., Curtis, C. F., et Bird, R. B.: 1967, *Molecular Theory of Gases and Liquids*, Wiley, New York.
[23] Scheraga, H. A.: 1968, *Advances in Physical Organic Chemistry* (V. Gold, éd.), Academic Press, Londres, p. 103.
[24] Perahia, D. et Pullman, B.: 1971, *Biochem. Biophys. Res. Commun.* **43**, 65.
[25] Bradley, D. F., Lifson, S., et Honig, B.: 1964, dans *Electronic Aspects of Biochemistry* (B. Pullman, éd.), Academic Press, New York, p. 77.
[26] Del Re, G.: 1964, dans *Electronic Aspects of Biochemistry* (B. Pullman, éd.), Academic Press, New York, p. 221.
[27] Berthod, H. et Pullman, A.: 1965, *J. Chim. Phys.* **62**, 942.
[28] Hoffmann, R.: 1963, *J. Chem. Phys.* **39**, 1397; 1964, **40**, 2474, 2480, 2745.
[29] Pople, J. A. et Segal, G. A.: 1966, *J. Chem. Phys.* **44**, 3289.
[30] Rein, R., Rabinowitz, J. R., et Swissler, T. J.: 1972, *J. Theor. Biol.* **34**, 215.
[31] Ewing, G. E.: 1972, *Angew. Chem., Intern. Ed.* **11**, 486.
[32] Pitzer, K. S.: 1959, *Adv. Chem. Phys.* **2**, 59.
[33] Jehle, H., Parke, W. C., et Salyers, A.: 1964, dans *Electronic Aspects of Biochemistry* (B. Pullman, éd.), Academic Press, New York, p. 313.
[34] Venkatachalan, C. M. et Ramachandran, G. N.: 1967, dans *Conformation of Biopolymers* (G. N. Ramachandran, éd.), Vol. 1, Academic Press, Londres, p. 83.
[35] Stockmayer, W. H.: 1941, *J. Chem. Phys.* **9**, 398.
[36] Liquori, A. M.: 1969 dans *Symmetry and Function of Biological Systems at the Macromolecular Level*, (A. Engström et B. Strandberg, éds.), A. Linqvist et Wiksell, Stockholm, p. 101.
[37] Lippincott, E. R. et Schroeder, R.: 1955, *J. Chem. Phys.* **23**, 1099.
Schroeder, R. et Lippincott, E. R.: 1957, *J. Phys. Chem.* **61**, 291.
[38] Moulton, W. G. et Kromhout, R. A.: 1956, *J. Chem. Phys.* **25**, 34.
[39] Scott, R. A. et Scheraga, H. A.: 1966, *J. Chem. Phys.* **45**, 2091.
[40] Ooi, T., Scott, R. A., Vanderkooi, G., et Scheraga, H. A.: 1967, *J. Chem. Phys.* **46**, 4410.
[41] Balasubramanian, R., Chidambaram, R., et Ramachandran, G. N.: 1970, *Biochim. Biophys. Acta* **221**, 182, 196.
[42] Poland, D. et Scheraga, H. A.: 1967, *Biochemistry* **6**, 3791.
[43] McGuire, R. F., Momany, F. A., et Scheraga, H. A.: 1972, *J. Phys. Chem.* **76**, 375.
[44] Nemethy, G. et Scheraga, H. A.: 1965, *Biopolymers* **3**, 155.
[45] Brant, D. A. et Flory, P. J.: 1965, *J. Am. Chem. Soc.* **87**, 663; **87**, 2791.
[46] Gibson, K. D. et Scheraga, H. A.: 1966, *Biopolymers* **4**, 709.
[47] Lipkind, G. M., Azkhipova, S. F., et Popov, E.: 1970, *Izv. SSSR*, Ser. Khim. **2**, 315.

[48] Avignon, M., Huong, P. V., Lascombe, J., Marraud, M., et Neel, J.: 1969, *Biopolymers* **8**, 69. Marraud, M., Neel, J., Avignon, M., et Huong, P. V.: 1970, *J. Chim. Phys.* **67**, 959.
[49] Venkatachalam, C. M.: 1968, *Biopolymers* **6**, 1425.
[50] Bystrov, V. F., Portnova, S. L., Tsetlin, V. I., Ivanov, V. T. et Ovchinnikov, Yu. A.: 1969, *Tetrahedron* **25**, 493.
[51] Thong, C. M., Canet, D., Granger, P., Marraud, M., et Neel, J.: 1969, *C. R. Acad. Sci. Paris*, Ser. C **269**, 580.
[52] Ramachandran, G. N., Chandrasekaran, R., et Kopple, K. D.: 1971, *Biopolymers* **10**, 2113.
[53] Tonelli, A. E.: 1971, *Macromolecules* **4**, 618.
[54] De Santis, P. et Liquori, A. M.: 1971, *Biopolymers* **10**, 699.
[55] Tonelli, A. E.: 1972, *Macromolecules* **5**, 286.
[56] Kopple, K. D.: 1971, *Biopolymers* **10**, 1139.
[57] De Coen, J. L.: voir [10].
[58] Pohl, F. M.: 1971, *Nature New Biol.* **234**, 277.
[59] Sasisekharan, V., Lakshminarayanan, A. V., et Ramachandran, G. N.: 1967, dans *Conformation of Biopolymers*, Vol. 2, Academic Press, Londres-New York, p. 641. Lakshminarayanan, A. V. et Sasisekharan, V.: 1970, *Biochim. Biophys. Acta* **204**, 49.
[60] Renugopalakrishnan, V. Lakshminarayanan, A. V., et Sasisekharan, V.: 1971, *Biopolymers* **10**, 1159.
[61] Lakshminarayanan, A. V. et Sasisekharan, V.: 1969, *Biopolymers* **8**, 475; **8**, 489; **8**, 505.
[62] Schweizer, M. P., Witkowski, J. T., et Robins, R. K.: 1971, *J. Am. Chem. Soc.* **93**, 277.
[63] Hruska, F. E., Smith, A. A., et Dalton, J. G.: 1971, *J. Am. Chem. Soc.* **93**, 4334.
[64] Rao, V. S. R., Sundararajan, P. R., Ramakrischnan, C., et Ramachandran, G. N.: 1967, dans *Conformation of Biopolymers*, Vol. 2, Academic Press, Londres-New York, p. 721.
[65] Reeves, R. E.: 1950, *J. Am. Chem. Soc.* **72**, 1499.
[66] Rees, D. A. et Skerrett, R. J.: 1968, *Carbohyd. Res.* **7**, 334.
[67] Sathyanarayana, B. K. et Rao, V.S.R.: 1971, *Biopolymers* **10**, 1605.
[68] Ramachandran, G. N.: 1968, dans *Structural Chemistry and Molecular Biology* (A. Rich and N. Davidson, éds.), W. H. Freeman, San Francisco, p. 77.
[69] Liquori, A. M., Damiani, A., et De Coen, J. L.: 1968, *J. Mol. Biol.* **33**, 439. Liquori, A. M., Damiani, A., et Elefante, G.: 1968, *J. Mol. Biol.* **33**, 445.
[70] M. Ptak: à paraître.
[71] Wilkins, M. H. F.: 1972, dans *Membrane Structure and Its Biological Application*, (Conference NY Acad. Sci, June 2–4 1971 (D. E. Green et J. F. Danielli, éds), in press.
[72] Cornibert, J. et Marchessault, R. H.: communication personnelle.
[73] Cleland, R. L.: 1971, *Biopolymers* **10**, 1925.
[74] Sasisekharan, V. et Ponnuswamy, P. K.: 1971, *Biopolymers* **10**, 583; et réf. citées.
[75] Popov, E. M., Pletnev, V. Z., Lipkind, G. M., et Archipova, C. F.: 1970, *Khimia PS* **2**, 191.
[76] McGuire, R. F., Vanderkooi, G., Momany, F. A., Ingwall, R. T., Crippen, G. M., Lotan, N., Tuttle, R. W., Kashuba, K. L., et Scheraga, H. A.: 1971, *Macromolecules* **4**, 112.
[77] Krimm, S. et Venkatachalam, C. M.: 1971, *Proc. Nat. Acad. Sci. US* **68**, 2468.

DISCUSSION

Pullman: Je voudrais faire remarquer que c'est la zone C_7^{ax} beaucoup moins importante et non pas la zone C_7^{eq} qui apparaît comme un minimum local dans les calculs de Scheraga dans lesquels il admet la flexibilité du modèle. Aucun minimum n'apparaît dans ces calculs pour la forme C_7^{eq}.

De Coen: La carte d'énergie conformationnelle du dipeptide calculée à l'aide des paramètres des fonctions de Van der Waals proposés par Scheraga en 1966 incluant une contribution électrostatique ($t = 4{,}0$) montre effectivement un minimum au voisinage de la forme C_7. La même carte d'énergie calculée par Flory en 1965 ne possède pas de minimum dans cette région mais cela est dû uniquement au fait que cet auteur a considéré la chaîne latérale de l'alanine comme un seul groupe méthyl au lieu d'introduire séparément l'atome de carbone et les trois hydrogènes du groupe CH_3.

Ptak: Les minimums d'énergie correspondant aux conformations C_7 sont trouvés par des méthodes semi-empiriques (De Coen, Popov). Il suffit d'introduire des interactions électrostatiques. Si l'on introduit une liaison H, ce minimum est renforcé. Ces conformations C_7 peuvent être importantes pour de petits peptides en solution. Il faudrait certainement en faire une étude plus complète en RMN dans différents solvants. Leur importance dans le cas des conformations de protéines mérite d'être étudiée avec détail.

H. Buc: (1) Quelle est la précision obtenue aujourd'hui par les méthodes cristallographiques sur les angles ψ et ϕ et sur les angles caractérisant les chaînes latérales? (2) Quelle précision obtient-on par les méthodes spectroscopiques en solution? (3) Dans le cas oú les points expérimentaix approchent en diagramme $<\phi, \psi<$ des valeurs énergétiques hautes ($\simeq$ 6 kcal/mole) ne peut-on penser que d'autres mécanismes dont on ne tient pas compte dans le calcul, comme la distorsion de la liaison 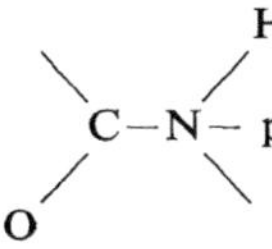 peuvent intervenir pour restabiliser cette conformation?

Ptak: (1) En aucune façon un calcul conformationnel n'a pour but de remplacer la diffraction des rayons X pour déterminer des conformations. Il donne des valeurs indicatives sur les conformations stables correspondant à des minimums d'énergie plus ou moins profonds. Il peut aider à comprendre des résultats expérimentaux tels que de RMN et de DC par exemple. (2) Pour les molécules en solution, la RMN, qui est certainement la méthode la mieux adaptée, utilise une relation semi-empirique du type Karplus, qui conduit à des valeurs possibles de ϕ à $\pm 10°$ au moins (pour des molécules de petite masse). C'est l'utilisation parallèle des deux méthodes théorique et expérimentale qui permet de proposer une conformation possible (représentée au mieux par un modèle).

Pullman: Les expérimentateurs considèrent souvent que les valeurs des angles ϕ et ψ sont valables avec $\pm 10°$. Vous avez mal compris la signification de la limite de 6 kcal/mole sur nos diagrammes d'iso-énergie. Il n'y a pas de limites supérieures dans les calculs eux-mêmes. L'energie est calculée pour chaque point de la carte. C'est dans la présentation des résultats que, pour plus de clarté, nous nous arrêtons à la courbe de 6 kcal/mole. Les courbes supérieures allant d'ailleurs à cette limite. Cela n'exclut pas l'existence possible des minimums d'énergie vers des énergies plus élevées.

D'autre part, nous constatons, c'est une observation de faits, que pratiquement toutes les conformations observées pour les résidus α-aminés des protéines, en solutions ou dans les cristaux, tombent à l'intérieur de la zone de stabilité conformationnelle théorique indiquée avec cette limite. Cela indique que, contrairement aux conclusions que l'on aurait pu tirer des résultats des anciens calculs empiriques, ni les protéines ni l'environnement n'imposent à ces résidus de conformations 'extraordinaires' de hautes énergies. Elles choisissent uniquement entre les conformations de basse énergie permises.

A propos de la comparaison de Tonelli, que vous avez citée entre les résultats quantiques et l'expérience, Tonelli a commis des erreurs numériques grossières dans la présentation des résultats quantiques. Je les ai rectifiées dans ma communication au Colloque sur les Conformations des molécules et polymères biologiques, Jérusalem, avril 1972 (Academic Press, sous presse), auquel vous avez participé. Lorsque ces erreurs numériques sont corrigées, il y a un excellent accord entre la théorie et l'expérience en ce qui concerne le moment dipolaire moyen des diastéréoisomères zwitterioniques des oligo-alanines. La valeur de la distance moyenne reste trop faible mais ceci est naturel compte tenu de la spécificité géométrique du dipeptide modèle qui ne se prête pas à une telle extrapolation, que nous n'avons d'ailleurs, pour notre part, jamais essayé de faire.

L'ÉTAT NATIF EST-IL L'ÉTAT FONDAMENTAL?

MICHEL GOLDBERG

Chef de Laboratoire à l'Institut Pasteur, Professeur à l'Université Paris VII, France

Résumé. Après avoir défini l'état natif d'une protéine, nous rappellerons les arguments classiques en faveur de l'identité des états natif et fondamental. Nous donnerons quelques exemples de l'influence des caractéristiques cinétiques du repliement de la chaîne polypeptidique sur l'état atteint après 'renaturation'. Ces exemples, ainsi que l'effet de divers cofacteurs sur la restructuration de protéines dénaturées, indiquent que les propriétés cinétiques du 'folding' peuvent parfois l'emporter sur les propriétés énergétiques. Nous montrerons enfin comment les associations quaternaires peuvent modifier les propriétés structurales et énergétiques des protomères et étendrons ces observations au cas des chaînes peptidiques structurées en plusieurs 'globules'.

Summary. This conference is aimed at showing how some classical molecular-biology experiments can raise some questions about the widely accepted physical-chemical conclusions concerning the identity of the native and fundamental states of a protein.

It is not intended to be either a general conference about this topics, or a review of the vast amount a work related to this subject.

To begin with, an operational definition of the 'native state' of a protein is given and, from this mere definition, it appears that the native and fundamental states of a protein can hardly be compared.

The main experimental arguments suggesting the identity of the two states (i.e. spontaneous refolding of denatured proteins) are discussed and experiments proving the existence of a 'kinetic trough' during the refolding of urea-denatured β-D-galactosidase are reported.

By studying the kinetics of renaturation of denatured β-D-galactosidase with the kinetics of ω-complementation between incomplete fragments of the galactosidase protomer, intra- and inter-chain stereospecific associations are compared and the influence of the kinetic properties of this association is discussed.

Finally, the influence of quaternary association on the structure of a protomer is discussed and illustrated by the structural and energetical coupling between the tertiary and quaternary structure of *E. coli* tryptophanase.

Even though none of the experiments reported conclusively show that the native state is structurally very different from the fundamental state, it is suggested that the classical experiments do not conclusively demonstrate the identity of the two states. The main reason for that appears to be the inadequacy of the definitions of the two states, one being essentially operational while the other is solely thermodynamic.

1. Préambule

Lorsque les organisateurs de cette réunion de la Société de Chimie Physique m'ont soumis le sujet de cette conférence, je me proposais de faire un exposé général sur l'état de la question, fondé sur une étude bibliographique.

L'examen du programme de cette réunion m'a montré que la quasi totalité des conférences portait au contraire sur le travail spécifique d'un laboratoire. Pour rester dans la note générale, j'ai donc choisi d'orienter mon exposé sur les recherches de notre groupe et, en particulier de Madame Jacqueline London, Monsieur Olivier Raibaud et Mademoiselle Cécile Skrzynia. Je tiens à préciser que c'est là la seule raison pour laquelle il ne sera pas fait mention des très nombreux travaux relatifs à ce sujet et dont nos résultats ne représentent qu'une parcelle très modeste.

Il est d'autre part souligné que le but de cet exposé est de montrer combien la

C. Sadron (ed.), Dynamic Aspects of Conformation Changes in Biological Macromolecules. 57–65.

notion d'‘état natif’ définie ci-dessous d'une manière opérationnelle, et la notion d'‘état fondamental’ dont la définition en termes physiques est parfaitement précise et rigoureuse, sont difficilement comparables. Cette remarque n'est sans doute pas inutile dans le cadre d'une réunion groupant physico-chimistes et biologistes dont le langage n'est pas toujours parfaitement harmonisé.

2. Introduction

Il est très généralement admis aujourd'hui que toute l'information nécessaire à la formation de la structure tridimensionnelle fonctionnelle (ou native) d'une protéine est contenue dans la séquence des amino-acides de cette protéine. Nous nous rangerons à cette hypothèse, solidement fondée sur diverses preuves expérimentales, en formulant toutefois une remarque fondamentale: la nature de l'environnement influe sans aucun doute sur la structure de la protéine, que ce soit par la nature du solvant ou par l'existence d'interactions avec divers ligands spécifiques, de petite taille ou macromoléculaires.

La question se pose alors de savoir quel principe fondamental permet la traduction d'une information ‘à une dimension’ (la structure primaire) en une information ‘à trois dimensions’ (la structure tridimensionnelle de la protéine). Deux écoles de pensée s'opposent à ce sujet; d'une part divers groupes (Lifson, Liquori, Ramachandran, Scheraga, etc.) tentent de prédire par des calculs énergétiques la conformation native d'une chaîne polypeptidique de séquence donnée en l'identifiant par hypothèse à la conformation thermodynamiquement la plus stable (état fondamental). D'autres travaux, moins physico-chimiques certes, semblent suggérer que cette hypothèse pourrait être mise en doute, et que des facteurs strictement cinétiques peuvent conduire la chaîne peptidique à un état métastable. Nous allons discuter quelques-uns des arguments expérimentaux à l'appui de ces deux modèles.

3. Définitions

Afin d'alléger la discussion, deux termes fondamentaux vont être dès à présent définis:

(a) *l'état natif*: cet état correspond à la conformation, ou à l'ensemble de conformations en équilibre ou non, qui confèrent à la protéine ses propriétés fonctionnelles.

(b) *le repliement*: c'est l'ensemble des changements conformationnels qui accompagnent le passage d'une forme dénaturée à l'état natif de la protéine. L'équivalent anglo-saxon de ce terme est ‘folding’.

4. Les arguments expérimentaux

4.1. L'ÉTAT NATIF EST L'ÉTAT FONDAMENTAL

A l'appui de cette affirmation vient un ensemble impressionnant d'expériences conduisant, à partir d'une protéine totalement dénaturée par perturbation de l'environnement (pH, urée, température, etc.), à la réapparition de l'activité enzymatique lors de l'élimination de l'agent perturbant. Plusieurs dizaines de protéines ont ainsi été dénaturées et renaturées, avec des rendements de renaturation de l'ordre de 100%.

Ce type d'expériences montre que les protéines ainsi dénaturées se replient spontanément vers l'état natif, sans l'aide de quelque autre composant cellulaire que ce soit puisque la renaturation est réalisée sur une préparation enzymatique parfaitement purifiée.

Bien que de tels résultats semblent indiquer que l'état natif corresponde à l'état fondamental, ils appellent un certain nombre de réserves qui restreignent sérieusement la généralité, voir le bien-fondé, de cette conclusion. En premier lieu, il convient de remarquer que seules les tentatives positives de renaturation sont rapportées dans la littérature; ainsi, personne n'est censé savoir que divers groupes de chercheurs (E. Fischer, H. Buc, nous-mêmes) ont sans succès essayé de renaturer la phosphorylase b du muscle de lapin dénaturée par l'urée. Compte tenu de cette remarque, il est clair qu'on ne connaîtra sans doute jamais la liste des protéines qui n'ont pu être renaturées. En second lieu, nous rappellerons que la recherche des conditions de renaturation est une entreprise longue et pénible au cours de laquelle il faut faire varier de très nombreux paramètres (température, pH, force ionique, nature du tampon, ions divalents, vitesse d'élimination de l'agent perturbant, concentration de protéines, etc.); bien souvent, les conditions qui donnent une bonne renaturation sont très différentes des conditions de stabilité maximum de la structure native. Ainsi, la β-D-galactosidase d'*E. coli* est considérablement plus résistante à l'inactivation thermique, à la dénaturation par l'urée, à la dénaturation par le pH, en présence qu'en l'absence d'ions Mg^{++} [1]. Pourtant, le rendement de la renaturation est bien meilleur lorsque celle-ci est réalisée en l'absence d'ions Mg^{++} [2].

Ce dernier argument indique clairement que, en dehors de l'aspect strictement énergétique, il faut rechercher d'autres facteurs pouvant influencer le repliement des protéines globulaires.

4.2. L'ÉTAT NATIF PEUT NE PAS ÊTRE L'ÉTAT FONDAMENTAL

Nous allons rapporter quelques expériences illustrant l'importance des paramètres cinétiques du repliement.

4.2.1. *Rôle des paramètres cinétiques dans le repliement*

La β-D-galactosidase d'*Escherichia coli* peut être dénaturée en urée 8 M et renaturée en éliminant l'urée par dialyse [3, 4]. Nous avons obtenu d'excellents rendements de renaturation en éliminant l'urée par dilution, ce qui a permis une étude cinétique du repliement. Les résultats de cette étude peuvent être résumés de la manière suivante:

(1) La renaturation par dilution est un processus lent (demie vie de l'ordre de 10 minutes), dont la vitesse ne dépend pas de la concentration de protéine. Le phénomène cinétiquement limitant est donc probablement le repliement des monomères et non pas l'association des protomères.

(2) Lorsque l'élimination de l'urée est faite en deux temps (une première dilution ramène la concentration d'urée à X molaire, X étant une valeur comprise entre 8 et 1;

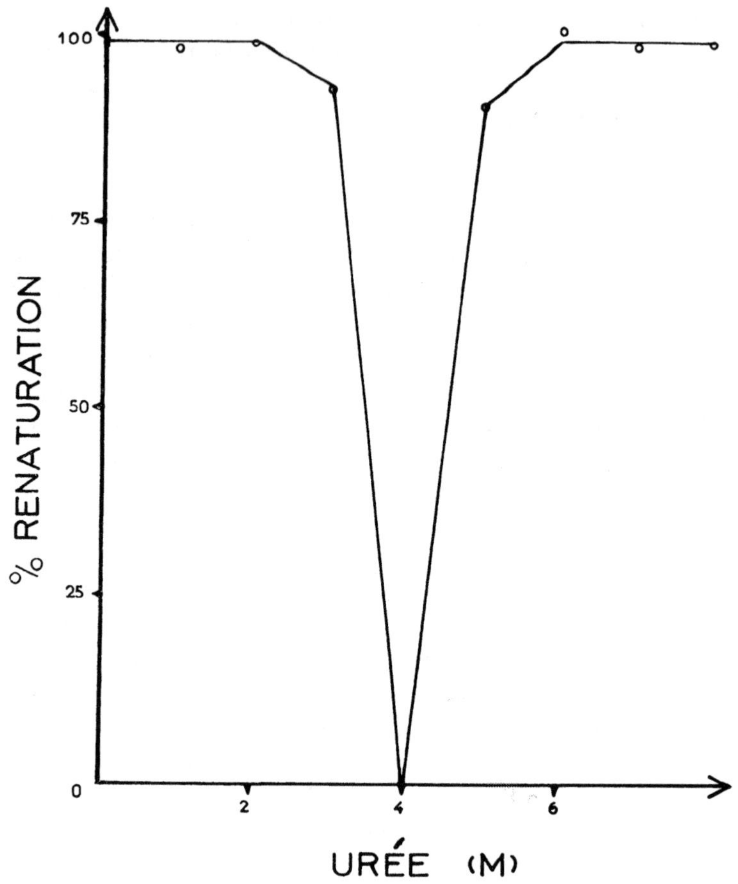

Fig. 1.

la protéine séjourne une heure à *X* M urée; une seconde dilution élimine l'urée) la cinétique et le rendement de la renaturation dépendent de la valeur de *X* (voir Figure 1).

Ainsi, si la protéine a séjourné une heure en urée 7, 6, 5 ou 3 M, les résultats sont identiques à ceux observés par dilution directe. (cinétique monomoléculaire – rendement élevé).

Si la protéine a séjourné en urée 2 M, le rendement demeure élevé, mais la cinétique devient d'ordre compris entre 2 et 3. D'autre part, une étude des propriétés de sédimentation de la préparation obtenue par dilution de l'urée 8 M à 2 M indique que la protéine reste monomérique. Lors de l'élimination de l'urée par la seconde dilution, la réaction cinétiquement limitante est donc la réassociation des monomères.

Enfin, si la protéine a séjourné en urée 4 M, le rendement de la renaturation devient nul ($<0{,}01\%$). Nous avons observé un phénomène analogue avec la tryptophanase d'*Escherichia coli*, la concentration d'urée intermédiaire critique étant 3 M. L'existence, dans ces deux cas au moins, d'une concentration d'urée empêchant la renaturation indique clairement qu'au moins deux voies distinctes peuvent conduire à un état replié:

$$\begin{array}{lccccl} & & & X \to \to & I & \text{(inactif)} \\ & & \nearrow & & & \\ & D \to \to \to D' & & & & \\ & & \searrow & & & \\ & & & \to \to & N & \text{(natif)} \\ \text{urée} & 8\,\mathrm{M} & & X\,\mathrm{M} & 0\,\mathrm{M} & \end{array}$$

Lorsque la protéine ne séjourne que très peu à la concentration critique X M, la voie D′—N est cinétiquement favorisée. Au contraire, un séjour prolongé à la concentration critique permet à la transconformation D′—X de se produire. L'état obtenu est un état stable, la transconformation de I vers N ne se faisant pas.

Il est donc clair que, suivant la voie cinétiquement favorisée par les conditions expérimentales, le repliement peut aboutir à deux états stables distincts dont on ne peut dire lequel est le plus stable.

Ce type de résultats permet dans doute d'interpréter les échecs dans certaines tentatives de renaturation.

4.2.2. *Rôle des liaisons covalentes dans le repliement des protéines*

L'étude de la 'complémentation du type ω' entre mutants d'*Escherichia coli* sur la β-D-galactosidase [5], celle des propriétés physico-chimiques de l'enzyme complémenté [6] et celle des propriétés immunologiques des fragments complémentants [7] ont permis d'aboutir à un modèle selon lequel au moins deux structures globulaires, à repliement indépendant, constituent le protomère de la galactosidase [6].

L'étude cinétique de la complémentation entre ces globules [8] montre que leur association est beaucoup plus lente quand ils proviennent de peptides indépendants (complémentation) que lorsqu'ils sont portés par la même chaîne peptidique (renaturation de l'enzyme sauvage). L'explication de cette observation pourrait être que la complémentation entre globules indépendants est une réaction intermoléculaire, donc concentration–dépendante; au contraire, dans le cas de la renaturation de l'enzyme sauvage, cette association intramoléculaire ne dépend pas de la concentration de protéines et peut se faire rapidement même à faible concentration de protéine. Le rôle des liaisons covalentes qui, dans l'enzyme sauvage, unissent les globules serait donc de favoriser, cinétiquement, l'association stéréospécifique non covalente entre globules.

Le rôle fondamental des paramètres cinétiques du repliement est confirmé par l'effet catalytique du thio-phényl-éthyl-β-D-galactoside (analogues du substrat) dans la complémentation ω [8].

Cet aspect strictement cinétique de l'influence des liaisons covalentes mérite d'être souligné, indépendamment des interactions entre globules, à propos du repliement des protéines globulaires en général. En effet, la stabilité d'une conformation particulière de la protéine est assurée par un nombre élevé d'interactions non covalentes, de faible énergie, mais dont la présence *simultanée* au sein de la protéine apporte une contribution énergétique importante à l'édifice moléculaire. Dans l'hypothèse où quelques interruptions seraient introduites dans la continuité covalente de la chaîne peptidique, l'existence simultanée des interactions faibles entre résidus provenant de quelque 4 ou 5 peptides serait très peu probable au cours du repliement de la protéine. Cette condition ne peut avoir une probabilité importante d'être satisfaite qu'au cours d'un processus monomoléculaire. C'est peut-être autant dans cette direction que dans une explication strictement énergétique, qu'il faut chercher l'interprétation de l'impossibilité de réactiver après dénaturation des zymogènes activés par coupure de la chaîne peptidique.

4.2.3. *Rôle des contraintes quaternaires sur la structure des protomères*

Au moins dans certaines conditions expérimentales dont nous convenons bien sûr qu'elles sont artificielles, nous avons vu que le repliement de structures globulaires monocaténaires peut aboutir à un état conformationnel stable à partir duquel peuvent se construire des états oligomériques par suite d'interactions stéréospécifiques entre chaînes peptidiques.

Diverses études suggèrent pourtant que la formation de la structure quaternaire peut s'accompagner de modifications énergétiques et conformationnelles de l'état des chaînes peptidiques. C'est le cas, en premier lieu, des protéines auxquelles s'applique le modèle allostérique; toutefois, le fait même que ces protéines existent sous au moins deux conformations en équilibre rapide et spontané nous écarte du propos de cette conférence car, d'après notre définition, ces deux conformations ne représentent qu'un seul état.

Nous allons donc donner une illustration expérimentale d'un tel couplage énergétique et structural entre niveaux tertiaire et quaternaire dans le cas d'une protéine ne présentant aucun caractère allostérique, la tryptophanase d'*Escherichia coli*.

Cet enzyme est constitué de quatre chaînes peptidiques identiques portant chacune une molécule de pyridoxal-5-phosphate (PLP) [9]. L'apo-enzyme se distingue de l'holo-enzyme par deux propriétés remarquables:

(a) Le coefficient de sédimentation, à 20 °C, de l'apo-TPase est proche de 9,5 S alors que celui l'holo-TPase est proche de 10,5 S [10].

(b) A 4° C, l'apo-enzyme est dimérique alors que l'holo-enzyme reste tétramérique. L'apo-dimère peut être réassocié en tétramère soit par addition de PLP, soit par élévation de température [10].

Nous avons préparé des molécules 'hybrides', c'est-à-dire des tétramères ayant n molécules de PLP liées de manière covalente à la protéine (par réduction de la base de Schiff entre la protéine et le PLP) et $4-n$ sites PLP libres. Ces hybrides ont donc certains protomères à l'état 'apo' et d'autres à l'état 'holo-réduit'. (Nous remarquerons que l'holo-enzyme réduit présente les mêmes propriétés de stabilité à froid et de sédimentation que l'holo-enzyme.) La stabilité thermique de l'apo-enzyme, de l'holo-enzyme et des hybrides a été étudiée en mesurant la perte d'activité en fonction du temps de chauffage à des températures appropriées. Les résultats obtenus indiquent que l'holo-enzyme est considérablement plus stable que l'apo-enzyme; il suffit de lier deux PLP à un tétramère pour lui conférer la stabilité de l'holo-enzyme complet. Le PLP présente donc une grande coopérativité quant à la stabilisation de l'ensemble des protomères de l'enzyme.

Nous sommes arrivés à la même conclusion par l'étude de la stabilité de la forme tétramérique à 4 °C.

Le PLP agit, selon toute vraisemblance, en stabilisant la conformation tertiaire du protomère auquel il est lié. Or, la fixation du PLP entraîne un changement conformationnel important de la protéine [10]. On peut donc en déduire que la stabilisation d'un protomère par fixation de PLP s'accompagne d'un changement de structure de ce protomère, lequel impose une transconformation similaire aux autres protomères par l'intermédiaire des contraintes quaternaires.

Ces études ont donc permis de mettre en évidence, dans le cas d'une protéine parfaitement michaélienne, l'existence d'un couplage énergétique et structural entre les structures tertiaires et quaternaires.

Par analogie avec la mise en évidence d'états métastables et même en admettant qu'un monomère natif est dans son état fondamental, on peut se demander si l'état oligomérique obtenu par l'association d'un nombre N rigoureusement défini de monomères correspond encore à l'état fondamental de ces N chaînes peptidiques considérées non plus chacune isolément, mais comme un système de N éléments.

Nous allons montrer pourquoi le caractère stéréospécifique des interactions quaternaires élimine la possibilité que la conformation des protomères soit considérablement modifiée lors de la formation de l'oligomère, même si l'état obtenu ainsi n'est pas métastable. En effet, les contraintes quaternaires s'exercent sur les protomères lorsqu'ils sont associés. Or cette association, stéréospécifique par essence, ne peut s'exercer que si les monomères se sont, au préalable, correctement repliés. Si les interactions quaternaires devaient bouleverser la structure des protomères, ceux-ci perdraient leurs aires d'association spécifiques, donc les contraintes quaternaires disparaitraient. Il est donc clair que ces interactions entre protomères ne peuvent modifier trop considérablement l'état conformationnel des monomères.

Il apparaît donc que, malgré les modifications structurales et énergétiques introduites par la formation des associations quaternaires, chaque protomère ne peut être suffisamment perturbé pour atteindre un minimum énergétique qui, même s'il était considérablement plus stable que l'état 'natif' des protomères isolés, en serait structuralement très différent.

L'expérience déjà mentionnée de la renaturation de la galactosidase à partir de l'urée 8 M en passant per l'urée 2 M confirme que le repliement du protomère précède la formation de l'oligomère, et donne tout son poids à l'argumentation développée ci-dessus.

5. Discussion

Nous avons donné un certain nombre d'arguments théoriques et expérimentaux qui suggèrent que l'état natif peut ne pas être l'état fondamental. Il convient toutefois de signaler qu'aucun de ces arguments n'est une preuve de la non-identité de ces deux états. Au cours de cette discussion, nous avons avant tout voulu montrer la prudence extrême dont il faut faire preuve dans l'étude complexe des mécanismes de repliement des protéines.

En effet nous avons vu que la structure quaternaire influe sur la structure tertiaire, et vice versa. Que dans le repliement de grandes chaînes polypeptidiques, un phénomène voisin de la formation des structures quaternaires peut avoir lieu lors de l'interaction entre globules. Que la formation de ces globules peut être gouvernée par un processus cinétique autant que strictement énergétique.

C'est peut-étre dans ce dernier point que réside l'explication d'un fait expérimental troublant: les études cinétiques réalisées sur la renaturation de petites protéines monomériques telles que la chaîne α de la tryptophane synthétase [11], la ribonucléase [12], le chymotrypsinogène [13], etc. montrent que leur repliement est très rapide, de l'ordre de quelques millisecondes à quelques secondes. Au contraire, la renaturation de grandes protéines est considérablement plus lente, de l'ordre de plusieurs minutes au moins. Il semble donc bien qu'un processus spécifique aux grandes chaînes peptidiques soit à l'origine de ce ralentissement de leur repliement.

Que ce processus soit l'association spécifique entre globules qui exige une transconformation d'un ou plusieurs globules est une hypothèse séduisante; elle reste cependant à démontrer.

Références

[1] Goldberg, M.: résultats non publiés.
[2] Ullmann, A. et Monod, J.: 1969, *Biochem. Biophys. Res. Commun.* **35**, 35.
[3] Zipser, D.: 1963, *J. Mol. Biol.* **7**, 113.
[4] Perrin, D. et Monod, J.: 1963, *Biochem. Biophys. Res. Commun.* **12**, 425.
[5] Ullmann, A., Perrin, D., Jacob, F., et Monod, J.: 1965, *J. Mol. Biol.* **12**, 918.
[6] Goldberg, M. E.: 1970, *J. Mol. Biol.* **46**, 441.
[7] Celada, F.: communication personnelle.
[8] Ullmann, A. et Monod, J.: 1970, dans *The Lactose Operon* (Beckwith et Zipser, éds.), Cold Spring Harbor Laboratory, p. 265.
[9] London, J. et Goldberg, M. E.: 1972, *J. Biol. Chem.* **247**, 1566.
[10] Morino, Y. et Snell, E. E.: 1967, *J. Biol. Chem.* **242**, 5591.
[11] Hardman, J. K.: communication personnelle.
Goldberg, M. E.: résultats non publiés.
[12] Tsong, T. Y., Baldwin, R. L., et Elson, E. L.: 1972, *Proc. Nat. Acad. Sci. US* **69**, 1809.
[13] Tsong, T. Y. et Baldwin, R. L.: 1972, *J. Mol. Biol.* **69**, 145.

DISCUSSION

Sadron: Vous avez considéré le cas général d'une macromolécule (par ex. une protéine) qui se trouve dans un état donné, par exemple au sein d'un solvant donné à une température donnée, et dont vous modifiez les propriétés en changeant progressivement les conditions extérieures pour ramener finalement celles-ci à leurs caractéristiques initiales. Vous vous êtes alors préoccupé de savoir dans quelle mesure la macromolécule elle-même, après ce cycle fermé de transformations, était ou non revenue à son état conformationnel initial. Pour cela vous avez utilisé comme témoin une propriété particulière de la macromolécule en expérience, par exemple son activité enzymatique. Ne pensez-vous pas qu'il puisse être à craindre que l'activité en question puisse rester inchangée alors que pourtant certaines modifications de conformation aient pu subsister? Pensez-vous que peut-être il y aurait intérêt, dans le même état d'esprit, de voir dans quelle mesure certaines modifications de conformation (observées par une autre méthode) peuvent rester sans effet sur telle ou telle propriété fonctionnelle de la macromolécule?

Goldberg: Bien sûr, l'activité ne représente qu'un critère; mais c'est le plus sensible pour tester l'état natif (propriété fonctionnelle). D'autre part, dans de nombreux cas, d'autres critères (propriétés optiques, stabilité thermique, sédimentation) ont permis de conclure à l'identité de la protéine 'initiale' et de la protéine dénaturée puis renaturée. Une seule exception à ma connaissance: le ribonucléase dont certaines propriétés sont légèrement modifiées.

Karplus: Your very interesting results certainly demonstrate that it is possible to 'trap' a protein molecule in a nonnative conformation by introducing special conditions. Since each of the experiments can be explained *without* assuming that the native state is not the minimum free energy state under normal conditions, the results do not provide evidence against the concept that the native state is that of minimum energy.

Goldberg: My aim was not to decide which conformation is the most stable. I just wanted to show that a 'kinetic trough' can exist, and hence that the native state might be kinetically more than energetically favoured.

Engelborchs: Ne faut-il pas ajouter à la liste d'informations contenues dans la conformation native, une contribution d'information qui provient des conditions (au sens large) de la synthèse; non seulement la direction de la synthèse, mais aussi les conditions de solvant, proximité des ribosomes, etc.?

Goldberg: Sans aucun doute et c'est ce que j'ai voulu dire en parlant du solvant et de l'environnement, y compris les interactions avec des ligands spécifiques, comme pouvant jouer sur la conformation atteinte par une protéine.

Ptak: La représentation des états d'énergie minimum de l'unité conformationnelle sur une carte donne une description statique de la molécule. Pour la molécule réelle en solution, composée de une ou plusieurs unités, il est essentiel de rappeler que la conformation résulte d'une distribution statistique des états d'énergie selon une loi de Boltzmann par exemple.

La cinétique du processus de 'folding' est tout à fait compatible avec l'existence des différentes conformations de l'unité. Une image simple serait que la cinétique de réaction sélectionne les conformations des unités nécessaires pour former la structure tertiaire. L'état final pourrait être très différent du minimum absolu de l'énergie. La représentation des conformations des protéines de structure comme sur une carte (ϕ, ψ) en donne une idée très claire.

De Coen: En ce qui concerne le problème du reploiement de la chaîne polypeptidique, je pense que les expériences du groupe de Goodman notamment ont clairement démontré qu'il n'est pas nécessaire de disposer d'une très longue chaîne pour avoir l'apparition d'un phénomène de structuration. La présence d'une dizaine de résidus semble suffisante pour assurer la formation d'un noyau conformationnel dans un solvant organique ou un mélange de solvant organique et d'eau. Dès lors, il paraît très raisonnable d'accepter un schéma séquentiel du reploiement d'un polypeptide. Dans cette hypothèse, diverses régions de la chaîne peptidique subiraient en premier lieu un arrangement spatial local contrôlé par des forces à courtes distances. Ensuite ces premiers noyaux structurés s'organiseraient au mieux entre eux pour donner naissance à la structure tertiaire de la molécule. Si un tel schéma, il me paraît très peu probable que la conformation réalisée de cette manière corresponde effectivement à l'état d'énergie le plus bas sauf dans quelques cas exceptionnels. Dans la majorité des cas, il me semble plus réaliste de penser qu'un état métastable à durée de vie plus ou moins longue est réalisé.

CONFORMATIONS AND INTERACTIONS OF HISTONES AND THEIR ROLE IN CHROMOSOME STRUCTURE

E. M. BRADBURY, P. D. CARY, C. CRANE-ROBINSON and H. W. E. RATTLE

Biophysics Laboratories, Physics Department, Portsmouth Polytechnic,
Gun House, Hampshire Terrace, Portsmouh, Great Britain

Abstract. High resolution nuclear magnetic resonance spectroscopy has been used to locate segments of the histone polypeptide chain which are involved in (i) histone–histone interactions (ii) histone–DNA interactions and (iii) interactions of F1 in chromatin. It is shown that the basic end of chain segments are the primary sites of interaction with DNA while central apolar segments are involved in interactions between histones. The interactions of histone F1 have been followed in chromatin gel and from this a scheme is proposed describing the role of F1 molecules in the chromatin condensation process.

Résumé. La spectroscopie de RMN à haute résolution a servi à localiser les segments de la chaîne polypeptidique de l'histone qui participent respectivement: (1) aux interactions histone–histone; (2) aux interactions histone–ADN; (3) aux interactions de F1 dans la chromatine. On établit que les sites primaires des interactions avec l'ADN sont les extrémités basiques de segments de chaîne tandis que les interactions entre histones mettent en jeu des segments centraux apolaires. Les interactions de l'histone F1 ont été suivies dans un gel de chromatine et conduisent à proposer un schéma qui décrit le rôle des molécules F1dans le processus de condensation de la chromatine.

1. Introduction

The close association of histones with DNA in the nuclei of cells of higher organisms resulted in the hypothesis of Stedman and Stedman [1] that histones are involved in the genetic regulation of these systems. Although many studies have been reported in the last decade purporting to demonstrate this role for histones there are many contradictions and clear, unambiguous evidence has yet to be produced. A second proposal concerning the role, or roles, of histones is that they are involved in controlling the conformational changes which chromosomes undergo during mitosis. It should be emphasized that because of the very large size of the DNA molecules in mammalian cells these conformational changes are probably among the most complex to be found in any biological system. Further, since the function of a biological macromolecule is determined by its precise conformation, factors which control the conformations of chromatin are probably those involved in controlling its activity. In this way histones may be considered to be involved in some control mechanism although the level at which they act is unknown. A distinction must be made between the gross repression of chromatin activity at a particular stage of the cell cycle and the precise control of the activity of specific genes. In the latter control mechanisms there is as yet no strong evidence to implicate histones.

In this contribution we should like to review briefly the conformational studies which have been made on histones. More detailed reviews of this subject have recently been presented [2, 3].

C. Sadron (ed.), Dynamic Aspects of Conformation Changes in Biological Macromolecules. 67–85.

2. Histones

It has long been thought that the distributions of amino acids along the polypeptide chains of histones are highly irregular. This was first suggested by Phillips [4] from tryptic digests studies of the histones; some histones yielded insoluble 'cores' containing a very high proportion of apolar amino acids and the amino acid compositions of the peptides obtained showed that the average separation between basic residues varied from 0 to 8 residues. The extent of this irregularity has now been fully revealed by the amino acid sequences of the histones F2A1 [5], F2B [6], F2A2 [7, 8], F3[9] and the partial sequence of F1 [10, 11]. All these histone sequences contain regions at the ends of the polypeptide chains, which are very rich in basic and helix destabilizing residues while other regions have low basicity and contain a high proportion of apolar residues, aromatic and other active residues such as serine, threonine and acidics. These features have led to suggestions that regions rich in basic residues are the primary sites for interaction with DNA while the apolar regions contain a high potential for the formation of secondary conformation and are available for histone–histone interactions. Because of this highly nonuniform distribution of residues in these proteins it has proved possible to use high resolution NMR spectroscopy to locate those regions of the polypeptide chains in which the mobilities of residues have been restricted by inclusion in helical segments and in interactions with other histones or with DNA. Residues which are located specifically in different regions of the polypeptide chain are thus used to report on the mobilities of that part of the chain.

There are two basic assumptions involved in these studies, the first is that increasing the ionic strength of aqueous solutions of histones induces conformational changes and interhistone interactions similar to those found in the ionic environment of DNA. This assumption can be verified only by studying the interactions of histones in chromatin. Recently it has been found that the interactions of the very lysine rich histone F1 can be followed in chromatin on varying the ionic strength. As will be shown later the segment of the F1 molecules involved in interchain interactions in the chromatin gel is the same as found for the isolated F1 molecules in aqueous salt solutions. The second assumption is that marked losses in apparent areas of only certain resonance peaks (when compared to other peaks in the same NMR spectrum) are a result of dipolar broadening due to restricted motion of specific regions of the polypeptide chain for one or other of the reasons given above. Since histone molecules consist of single polypeptide chains and paramagnetic ions are excluded from the system this assumption is most probably valid.

A. NMR STUDIES OF HISTONE FRACTIONS

Earlier high resolution NMR studies were carried out largely at 100 MHz with additional studies at 220 MHz. These showed that histones F1 [12], F2A1 [12] and F2B [13] in aqueous solution gave well defined high resolution NMR spectra which agreed closely with the spectra expected from random coil polypeptide chains. Optical spectroscopic studies [14] had shown that histones were largely in the random coil form in water and with increase in ionic strength all of the histone fractions, except F1, were found to undergo a partial coil → helix transition to 25–30% helix content. Increase in ionic strength also had a marked effect on the high resolution NMR spectra of histones with only certain of the resonance peaks showing a marked apparent loss in peak area. Based on a visual comparison of spectral changes it was suggested that the segments involved in salt-induced helix formation and histone–histone interactions contained a high proportion of apolar residues and were low in basic residues; 50 to 102 for histone F2A1 [12], 51 to 100 for histone F1 [12] and 60 to 102 for histone F2B [13]. In addition preliminary studies have been reported on the interactions of histones F2B and F1 with DNA [15]. These led to the tentative conclusion that whereas the whole of the F1 molecule was involved in the interaction in the case of F2B the regions rich in basic residues were found to be the primary sites of interaction with pure DNA.

As the conclusions drawn in these earlier studies were obtained from a visual inspection of the spectral changes induced in the histones much of the information contained in the spectra is not fully utilised, i.e. only prominent spectral changes were analysed whereas there are interesting changes in the shapes of the envelopes of overlapping resonance peaks which contain additional information. To make a more detailed analysis of these changes computer programmes have been written for the simulation of histone spectra and of the induced spectral changes [16]. A standard peak shape is used which is obtained by modifying a Gaussian shape by trial fittings to observed NMR line shapes. The resonance of each chemical group in the protein is then specified on a punched card by its position on a frequency scale, width in Hz and area representing the number of protons contributing to the resonance. Peaks representing all groups of protons in the protein are added by the computer and plotted to give a simulated spectrum. A close fit between the observed and simulated spectra was obtained for the random coil form of all the histones. Changes in the simulated spectra could then be introduced by broadening out the resonance peaks corresponding to a given section of the histone polypeptide chain and comparing in detail the resultant simulated envelope with that observed for the modified histone spectrum. It was found that the modified simulated spectra approached the observed spectra in $NaCl/D_2O$ solutions only when resonance peaks from residues in closely defined sections of the molecule are broadened. For example [16], the segments of histone F2B defined in this way were similar, but not identical, to those proposed by a visual comparison of the spectral changes. Thus whereas the visual comparison led to the conclusion that the segment 60–102 was involved in conformational changes and interactions between histones the computer comparison of the spectral changes gave the closest fit when peaks from residues 31 to 102 were broadened. The computer simulation therefore allows a more precise comparison of the NMR spectral changes than is possible with visual inspection and more detailed conclusions can be drawn.

B. HISTONE F2A1

Historically bovine F2A1 was the first histone to be fully sequenced [5]. In addition

DeLange *et al.* sequenced F2A1 from pea seedling [17] and showed that except for two conservative replacements the sequences of F2A1 from these widely differing tissues were identical. This rigid conservation of sequence is unique and must reflect the stringent requirements imposed on each residue of the polypeptide chain by its function. DeLange *et al.* [5] also pointed out the highly assymmetric nature of the distribution of residues along this polypeptide chain. Thus it can be shown that where as the amino half of the molecule has a ratio of basic to acidic residues of 15.6:1, for the

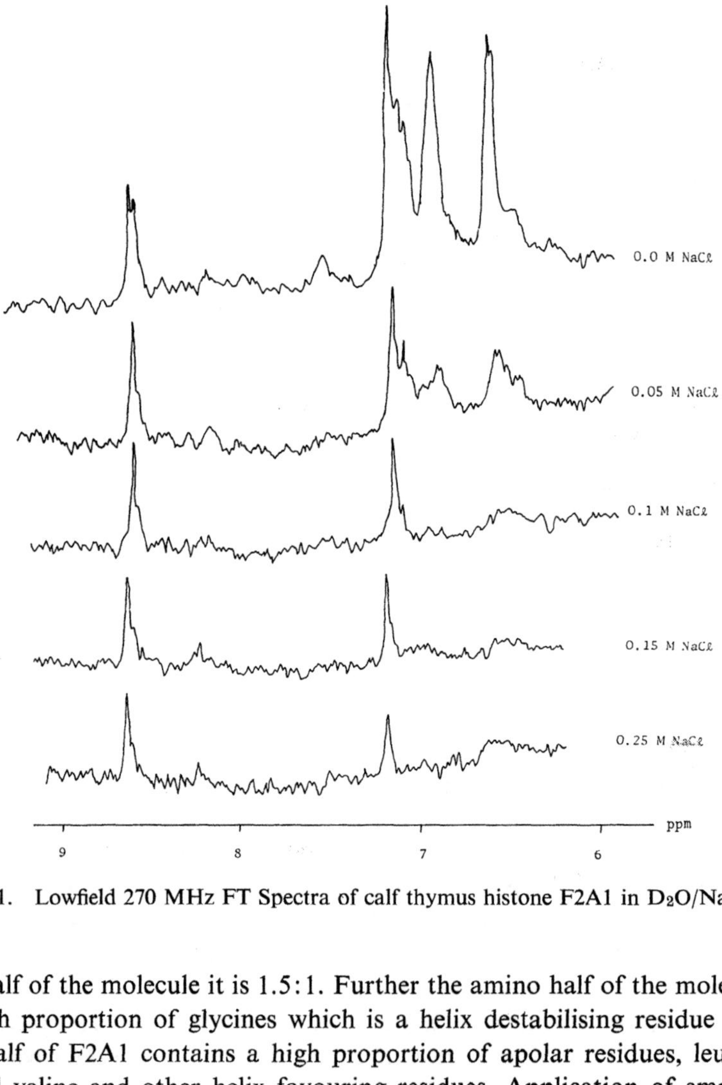

Fig. 1. Lowfield 270 MHz FT Spectra of calf thymus histone F2A1 in D_2O/NaCl.

carboxyl half of the molecule it is 1.5:1. Further the amino half of the molecule contains a high proportion of glycines which is a helix destabilising residue while the carboxyl half of F2A1 contains a high proportion of apolar residues, leucine, isoleucine and valine and other helix favouring residues. Application of empirical sequence–conformation rules all indicate that the sequences with potential for helix formation lie largely in the nonbasic apolar segments. Aromatic residues are also distributed in an irregular manner. In the amino half of the molecule there is one, aromatic His 18, while in the carboxyl half there are Tyr 51, Phe 61, Tyr 72, His 75 Tyr 88, Tyr 98 and Phe 100. Figure 1 shows the effect of increase in ionic strength on the aromatic spectrum. In water the aromatic resonances are well developed as expected for a random coil protein. As can be seen a small increase in ionic strength to 0.05 M NaCl causes an appreciable loss in apparent peak area of the tyrosine and the phenylalanine resonances. By 0.1 M NaCl these resonances have broadened to such an extent as to become almost unobservable while the sharp resonances remaining are the C_2 and C_4 protons of a histidine residue. These resonances must be assigned to histidine 18 since the only other histidine (75) is among the other aromatics whose resonances have broadened. Clearly the conformational changes and interactions which cause the loss of mobility and thus peak broadening involves at least the segment from 51 to 102. That the restriction of molecular mobility goes right to the end of the F2A1 polypeptide chain is illustrated by the complete loss of peak area from the phenylalanine at 100. Marked spectral changes are also observed in the upfield region of the spectrum as shown in Figure 2. With increase in ionic strength the peak showing the largest effect is that at 0.93 ppm due to the methyl groups of valine, leucine and isoleucine. Included in this diagram are the computer simulated spectra for the random coil polypeptide 2D (which is to be compared with the spectrum of F2A1 in water 2A) and that obtained by broadening resonances from segment 33 to 102. The latter computed spectrum was found to give the best fit with the

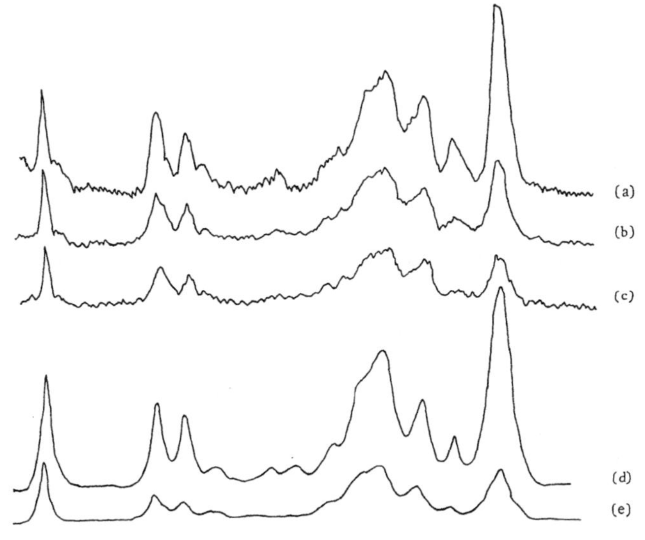

Fig. 2. Observed 220 MHz upfield spectra of histone F2A1 in (a) D_2O, (b) 0.1 M NaCl/D_2O and (c) 0.2 M NaCl/D_2O. Simulated spectra of (d) the random coil form of F2A1 and (e) that obtained by broadening resonances from segment 33 to 102.

observed spectrum of F2A1 in 0.2 M NaCl. Therefore it is suggested that a small increase in the ionic strength of an aqueous solution of F2A1 causes a conformational change in the segment 33 to 102, the resulting structure acting as a site for interchain interactions between F2A1 molecules. Hydrodynamic studies [18] have shown that increases in ionic strength causes the formation of large aggregates. NMR studies show that the sites of aggregation are located in a specific segment of the polypeptide chain. The region not involved in these interactions is the basic segment of the chain from 1 to 32. Addition of DNA to an F2A1 solution results in a loss in peak area of the resonances from the sidechains of the basic residues and a gain in area of the upfield peak at 0.93 ppm due to the methyl resonances of leucine, isoleucine and valine. This latter peak becomes the most prominent in the spectrum of the F2A1/DNA complex [19]. It appears from this that the segment of F2A1 which is the primary site of interaction with DNA is complementary to the segment which is involved in salt-induced intrachain conformational changes and protein interchain interactions.

C. HISTONE F2B

The effect of increase in salt molarity and change of pH on the conformations and interactions of histone F2B has been studied by optical and high resolution NMR spectroscopy. Optical rotatory dispersion studies show that F2B in water is in the random coil conformation and increase in salt molarity or pH [14] causes an increase in the ORD parameters characteristic of a helical conformation to give about 30% helix at 1.0 M NaCl. The absence of the extended β-conformation can be demonstrated by infrared spectroscopy. NMR studies of F2B [13] show that only specific resonance peaks are broadened by the salt -and pH-induced conformational changes and interactions. These are the apolar residues, valine, leucine and isoleucine, the arginine residues, acidic and aromatic residues. A computer simulation [16] of the random coil F2B NMR spectrum and the salt modified spectra show that the best fit between observed and calculated spectra is obtained when the resonances from residues in the segment 31 to 102 are broadened. It is suggested therefore that this segment is involved in the F2B conformational changes and interchain interactions.

The above conclusions were largely confirmed by cleaving F2B at methionine 59 and 62 with cyanogen bromide and studying the salt-dependent conformational behaviour and interactions of each half of the molecule [20]. Striking differences in behaviour were observed. ORD results showed that the amino half of F2B (1 to 58) remained fully in the random coil form, from 0 to 2.5 M NaCl. The NMR changes showed however, that interchain interactions could be induced at very high salt between the half molecules at segment 31 to 50. In contrast the carboxyl half of F2B was found to show considerable changes both in the ORD parameters and in the NMR spectra at much lower salt molarities. The ORD results showed that the carboxyl half of F2B was in the random coil form in water but increase of salt molarity caused a rapid conformational change to about 60% helix content at 0.7 M NaCl. The total helix content of the whole of F2B is thus contained in the carboxyl half of the molecule. Marked changes in the NMR spectra were observed on increasing ionic strength and at 0.7 M NaCl the segment which has its mobility restricted by the conformational changes and interchain interactions was found to be 66 to 102. These results showed therefore that the same regions of the half molecules were involved in conformational changes and interactions as found for the whole molecule. For the whole molecule the interacting segment was found to be 31 to 102, whilst for the half molecules it was 31 to 50 and 66 to 102. Not unexpectedly some of the residues adjacent to the cleavage sites are mobile in the half molecules at high salt. In general, however, the averages of the ORD parameters and of the NMR spectra of the half molecules are close to those observed for the whole molecule. It follows that as regards salt- induced conformational changes and interactions the behaviour of each half of the molecule is largely independent of the other half. This finding supports the general proposal that regions of histones with different character, as regards their amino acid composition, have distinct functions.

D. F2B/DNA INTERACTIONS

100 MHz results for F2B-DNA interactions [15] show that as the salt is progressively dialysed from solution the resonance peaks from the basic regions of the polypeptide chain show a marked loss in peak area while the resonance peaks from the nonbasic regions become more pronounced. The results indicate that as the DNA histone interactions became stronger they overcome the interchain histone–histone interactions and the segments involved in these latter interactions become more mobile; the basic regions of the histones are those involved in the histone DNA interactions. From computer simulations of the NMR spectra of F2B when complexed with DNA the best fit between observed and calculated spectra was obtained when resonances from residues in the segments 1 to 30 and 103 to 125 were broadened [16]. It is proposed that these segments are the primary sites for interaction with DNA and it is of considerable interest that they are exactly complementary to the segment 31 to 102 which was found to be involved in salt- and pH-induced interactions of F2B molecules.

E. F2A2

The recent sequence determination [7, 8] of this protein has allowed analysis of the NMR spectral changes on increase in ionic strength. Similar to F2B, ORD studies [14] showed that F2A2 was largely in the random coil in water and increase of the salt molarity to 1 M NaCl resulted in an increase in helicity to 30%. The NMR spectra over the same range of salt molarities show broadening and loss in apparent area of specific peaks only. As can be seen in Figure 3, the peaks markedly affected are the 0.93 ppm from the methyl groups of valines, leucines and isoleucines and the CH_2 resonances of glutamic acid residues. More dramatic effects are seen for the resonances from aromatic residues shown in the spectra in Figure 4. In water solution the resonances of the 4 histidines at positions 31, 82 123 and 124, the single phenylalanine at 25 and 3 tyrosines at 39, 50 and 57 are well developed. Increasing the salt molarity to 0.2 M NaCl results in the broadening and apparent loss in area of the tyrosines. Clearly the segment involved in the initial conformation and interchain interactions

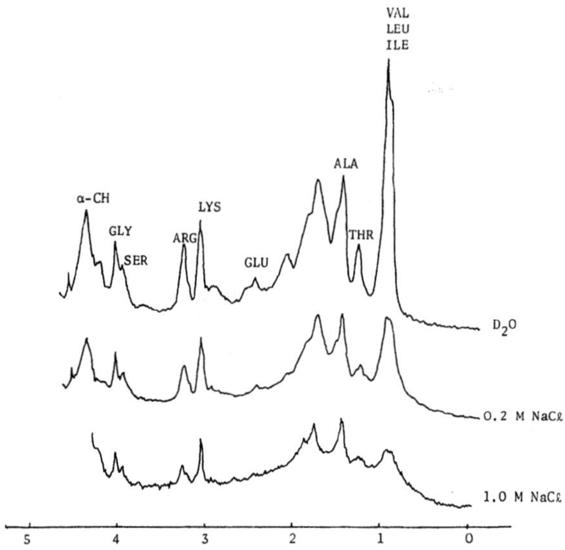

Fig. 3. Highfield 270 MHz spectra of calf thymus histone F2A2 in D_2O/NaCl.

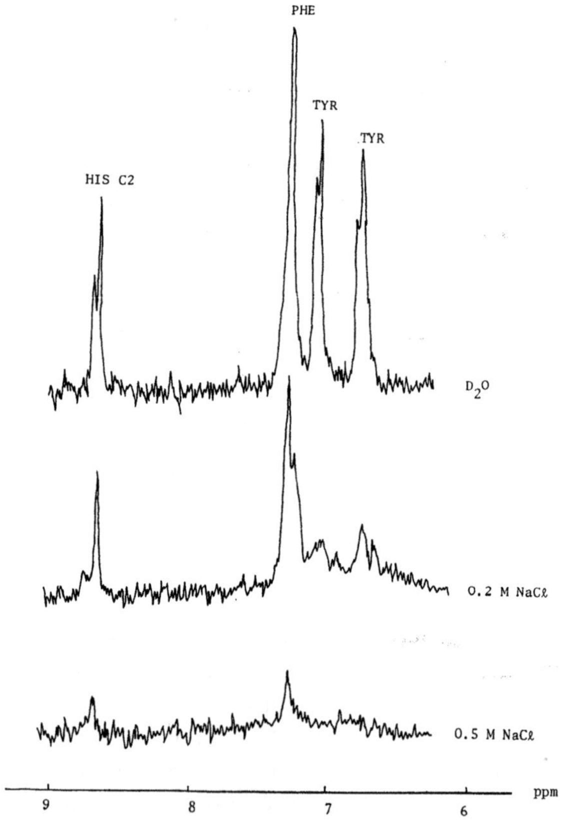

Fig. 4. Lowfield 270 MHz spectra of calf thymus histone F2A2 in D_2O/NaCl.

is that which includes the tyrosine at 39, 50 and 57. By 0.5 M NaCl the resonance from phenylalanine 25 is broadened and resonances from some of the histidines are also affected. The segment involved in the conformational changes and interactions must therefore be extended to include histidine 31 and phenylalanine 25. Although not complete, computer simulations suggest that the segment from 25 to at least residue 92 is involved in the conformational changes and interactions. As can be seen from Figure 3 this segment is rich in apolar, aromatic and acidic residues.

The interaction of F2A2 with DNA results in a similar NMR behaviour to that observed for the interaction of F2A1 and F2B with DNA, i.e. the peak from methyl groups of apolar residues becomes the most prominent while the areas of peaks from the sidechain groups of the lysines and arginines are considerably reduced. It follows that the basic segments of F2A2 are the primary sites of interaction with DNA while the apolar central region has mobility and is available for interaction other than with DNA.

Similar studies have been performed on histone F3 with similar conclusions.

F. LYSINE-RICH HISTONES

There are many features which distinguish the lysine-rich histones from the other histones. They are the most basic of the histones and are approximately twice as big, having a molecular weight of 21000 compared to the range of 11000 to 15000 found for the others. In spite of the high basic charge they are the first histones to be dissociated from the chromatin complex on increase of ionic strength with simple salts. Of all the histones the lysine-rich histones show the most variability. Thus Cole and his coworkers [21] have shown the presence of microheterogeneity in rabbit thymus F1 histones and have suggested that differences in patterns of microheterogeneity are related to tissue specificity. The most striking example of tissue specificity, however, is to be found for the chicken erythrocyte lysine rich histone F2C which largely, though not completely replaces the F1 histones found in other tissues. F2C is more basic than F1 and contains an appreciably higher proportion of arginine. Palau *et al.* [22] has found that the lysine-rich histone from sea urchin sperm $\phi 1$ also contains a higher proportion of arginine and suggests that the more metabolically inactive tissues contain lysine rich histones of higher basicity.

We have studied the interactions of calf thymus F1 and the F1 fragment 1–72 and 73–216, chicken erythrocyte F2C and sea urchin sperm $\phi 1$. Partial sequence data is available for F1 [10, 11] and F2C [23].

G. F1 AND F1 FRAGMENTS

Earlier 100 and 220 MHz NMR studies [12] of the effect of increasing ionic strength on the interactions of F1 showed that resonances from the apolar residues leucines, valines and isoleucines and the aromatic residues were broadened with a marked loss in apparent area at high salt molarity. From the composition of peptides [24] it was suggested that the segment 51 to 100 was involved in these interactions. More recently the sequence of the first 107 residues has become available [10, 11] and this partial

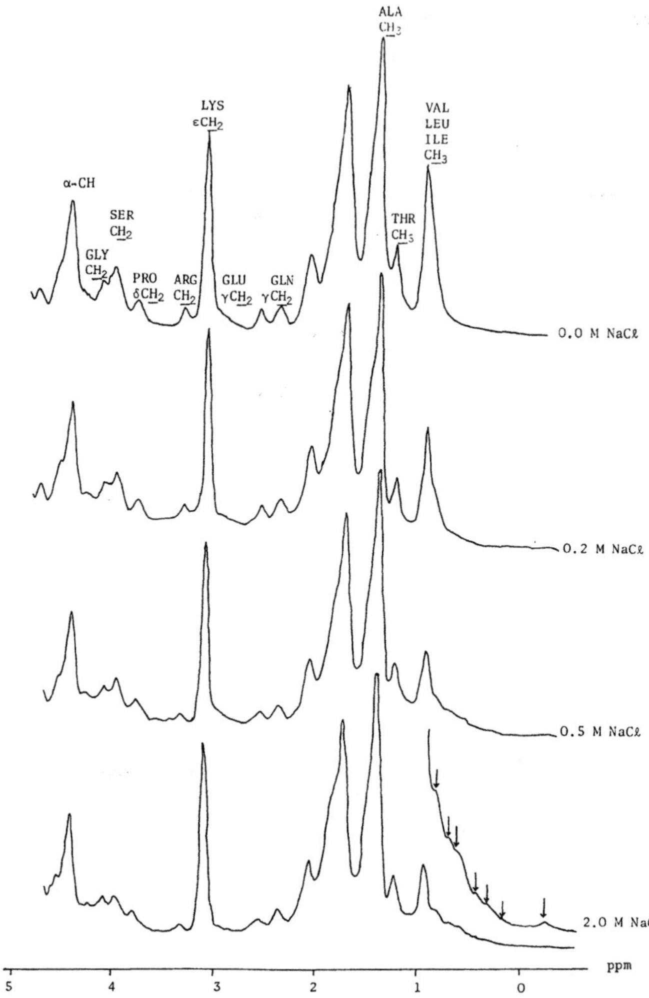

Fig. 5. 270 MHz FT spectra of thymus F1, 50 mg/ml in D_2O/NaCl(pH 3), 2*K* pulses.

sequence shows several very interesting features. The first 40 residues contain 14 alanines, 9 lysines, 7 prolines and 5 other random coil-favouring residues (3 glycines and 2 serines) to give a total of 35, or 87% of these residues. Although the sequence of the carboxyl half 107 to 216 is unknown its amino acid composition shows that it contains 88% of the residues listed above (42 lysines, 33 alanines, 15 prolines, 5 glycines and 2 serines). The high basic charge of these end segments and their similarity of composition has led Cole and his coworkers to suggest that they are the primary sites for interaction with DNA. The intermediate segment 40 to 106 is considerably different in character from the end segments and contains 16 isoleucine and valine residues out of a total of 20 for the whole molecular and the only two aromatic residues tyrosine 72 and phenylalanine 106. This segment is also appreciably less basic than

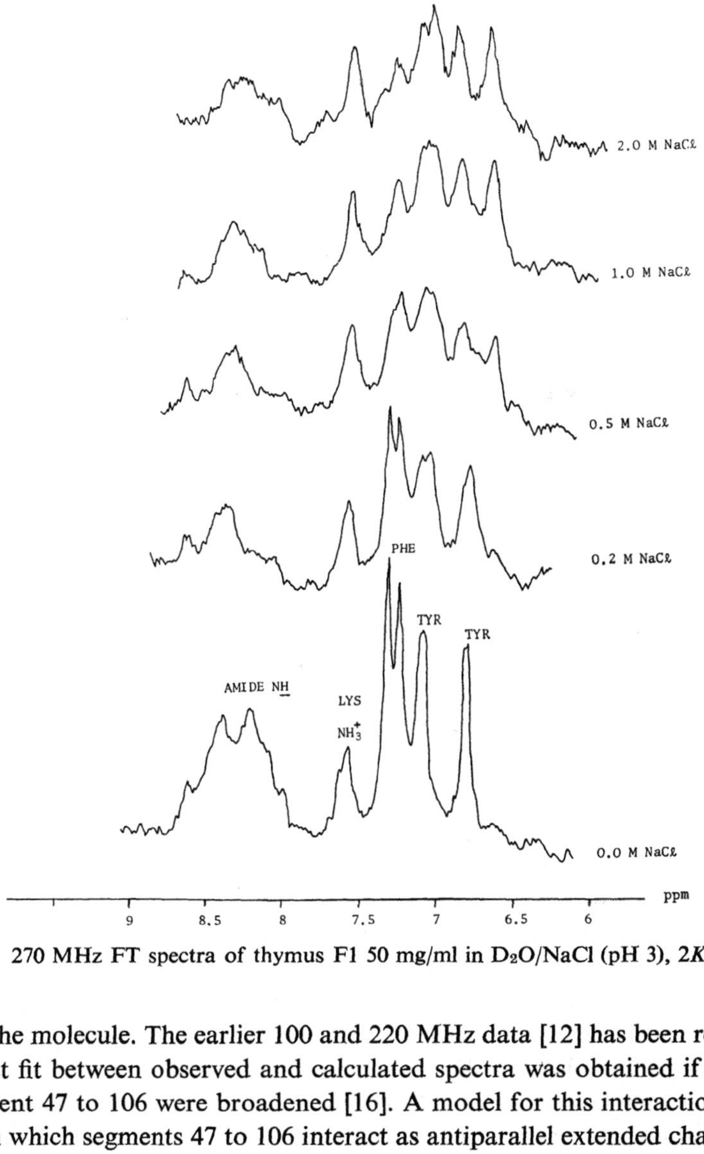

Fig. 6. 270 MHz FT spectra of thymus F1 50 mg/ml in D_2O/NaCl (pH 3), 2*K* pulses.

the rest of the molecule. The earlier 100 and 220 MHz data [12] has been re-evaluated and the best fit between observed and calculated spectra was obtained if resonances in the segment 47 to 106 were broadened [16]. A model for this interaction has been proposed in which segments 47 to 106 interact as antiparallel extended chains [16]. In this scheme there are many favourable sidechain interactions between antiparallel chains.

Recently we have re-examined calf thymus F1 at 270 MHz using the Bruker 270 FT spectrometer of the Oxford Enzyme Group. The upfield spectra (Figure 5) show, with increase in salt molarity, that as the area of the peak at 0.93 ppm decreases weak resonances appear upfield of this peak with one resonance peak as high as −0.24 ppm. These upfield shifted resonances result from the perturbations of the resonances of the apolar residues leucine, isoleucine and valine by particular magnetic environments in the interacted form of the F1 molecules. Some of these upfield peaks result from ring-current shifted resonances of apolar residues in the vicinity of aromatic groups and it is interesting that in the model [16] of the F1 interaction tyrosine 72 is aligned with leucine 83 and phenylalanine 106 with valine 89. The spectrum of the aromatic residues (Figure 6) also shows additional complexity with peaks shifting as the salt molarity is increased. These differences can be attributed to particular environments for the aromatic residues in the interacted state of F1.

Fortunately, the only tyrosine residue in the F1 molecule is located almost exactly in the middle of the interacting segment at position 72. In collaboration with J. Palau F1 has been cleaved with N-bromosuccinamide to give the fragments 1–71 and 73–216. The upfield and aromatic spectra of the C-terminal 73–216 were found to be largely unchanged even by 2.0 M NaCl. This segment therefore remains fully mobile over the complete salt range and there is no evidence of any salt-induced interactions. Similarly for the N-terminal 1–72 segment (except for a slight broadening of the methyl peak at 0.93 ppm) the spectrum also remains unchanged up to a salt molarity of 2.0 M NaCl. These results contrast with those obtained for the halves of the F2B molecules for which it was shown that both the NMR and ORD data averaged out to the behaviour of the whole molecule. It was concluded that the interactions of the half molecules were the same as those present in the whole molecule. The interactions are presumably between parallel F2B molecules so that the interactions sites remain on cleaving the molecule. In the case of F1 the absence of interactions between the fragments of molecules is consistent with the proposed model in which interactions occur between antiparallel chains. In the F1 interaction model segment 48 to 71 interacts with segment 84 to 106 and cleavage at tyrosine 72 isolates these interaction sites.

H. F2C

F2C is the unique lysine rich histone fraction found in mature erythrocytes. Its composition shows that it is more basic than the F1 histones and moreover contains a high proportion of arginine residues. The tentative sequence has been determined for the first 78 residues [25] and this shows that unlike F1, there is no transition in the character of the sequence at residue 40. In the partial sequence of F2C the apolar residues, leucines, isoleucines and valines are uniformly distributed as are most other residues. The composition data [23] show, however, that the amino third of the molecule is appreciably less basic than the carboxyl two thirds and contains most of the aromatic residues (2 tyrosines and 2 histidines) with 1 tyrosine and 1 phenylalanine in the carboxyl two thirds of the molecule.

This molecule provided the first indication that highly specific interactions and conformation existed for histones [26]. Figure 7 shows the 270 MHz FT spectrum of F2C in D_2O and 1 M NaCl/D_2O. Two striking features were observed on increase of ionic strength. Firstly the marked increase in complexity of the aromatic region and secondly the appearance of ring-current shifted resonances upfield of the methyl peak at 0.93 ppm. The perturbed spectrum is generated (i) on increasing the concentrations of F2C in the absence of salt, e.g. from 10 to 50 mg at 20 °C (ii) on reducing the temperature of a 50 mg/ml solution from 38° to 20 °C in pure water and (iii) on increasing

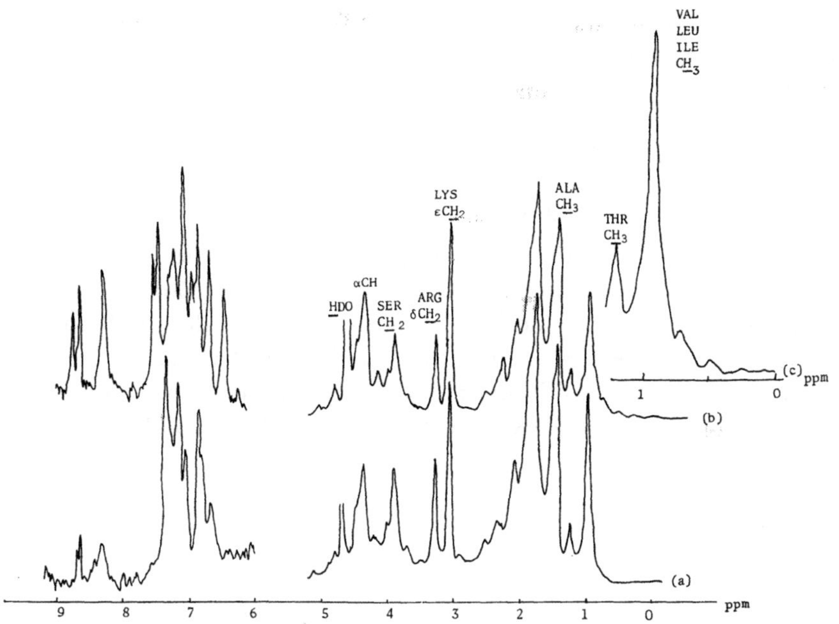

Fig. 7. 270 MHz Fourier transform spectra of chicken erythrocyte histone F2C at 38°C, protein concentration 50 mg/ml; (a) spectrum in D_2O (2*K* pulses); (b) spectrum in D_2O/1 M NaCl (2*K* pulses); (c) expansion of 0 to 1 ppm region of spectrum (b).

the ionic strength to 1 M NaCl of a 10 mg/ml solution .The spectral changes shown in Figure 7 exactly parallel those observed when an enzyme is denatured, e.g. lysozyme and it must be concluded that intra and intermolecular interactions must occur for F2C which are as specific as those found in the native form of enzymes.

I. F1–DNA INTERACTIONS

Early 100 MHz results [12] suggested that, in contrast to the behaviour of the other histones, the whole of the F1 molecule became bound to the DNA. In these studies no differential behaviour was observed in the NMR spectrum on dialysing from high to low salt, the whole NMR spectrum of F1 simply becoming weaker. These observations have now been repeated twice with the same results. The binding of the two fragments 1 to 71 and 73 to 216 to DNA has also been studied and in a similar manner the whole spectrum for each fragment weakens as the salt molarity is reduced. It appears from these studies that the whole of F1 and of fragments of F1 becomes stabilized by the interaction with DNA at low salt. It may well be, however, that the interaction sites available to F1 molecules in the nucleoprotein complex are those provided by the complex of DNA with the other histones and the interactions of F1 with pure DNA may not be a realistic model. As mentioned earlier there are several reasons why the lysine-rich histones should be considered as different to the other histones. One of these was the observation that although they are the most basic of the histones they are the first to be dissociated from the nucleoprotein complex on increasing the salt molarity. By 0.6 M NaCl the lysine-rich histones are completely dissociated while the other histones remain bound in the complex. This observation prompted us to record NMR spectra of nucleoprotein gel over the range of salt molarities from water to 0.6 M NaCl.

J. F1 INTERACTIONS IN CHROMATIN GEL

Chromatin in water is a diffuse fibrous gel. Increasing the salt molarity to 0.1 to 0.2 M NaCl causes a contraction of the gel to about one-tenth of its original volume [27]. As the molarity is increased above 0.3 M NaCl the gel disperses and by 0.6 M NaCl the F1 molecules are dissociated and the gel has become fully dispersed. NMR spectra recorded over this range of salt molarities are shown in Figure 8. Unlike the interaction of F1 with DNA these spectra show differential broadening effects.

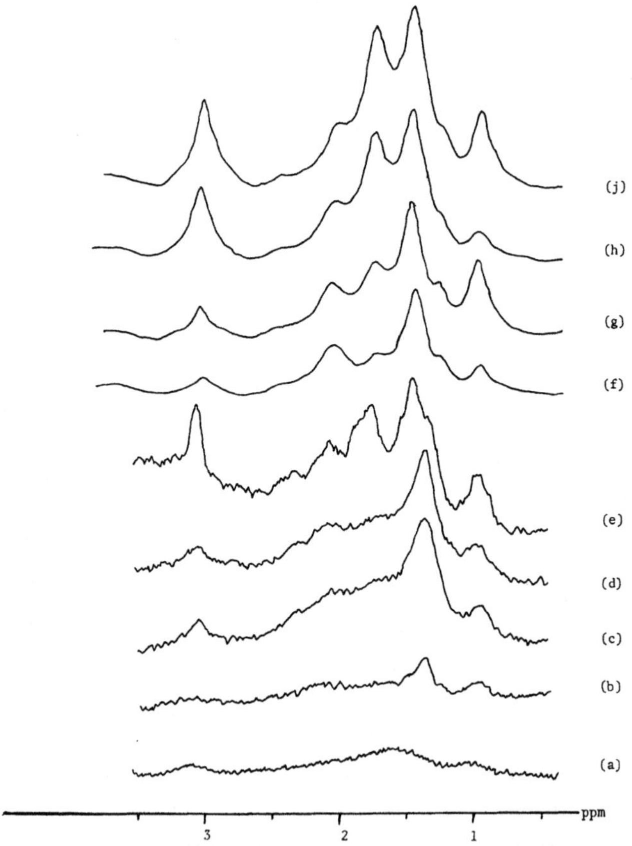

Fig. 8. High-field parts of (a) 100 MHz NMR spectrum of F1-depleted DNP; and 220 MHz spectra of total DNP in (b) D_2O, (c) 0.1 M NaCl, (d) 0.25 M NaCl, and (e) 0.5 M NaCl; simulated 220 MHz spectra of histone fraction F1, (j) in random-coil form, (h) with signals from residues 47–106 broadened, (f) and (g) and (h) and (j) but with the resonances of 40 lysine residues removed.

Figure 8a is the control and shows the signals from F1 depleted nucleoprotein in 0.6 M NaCl. Only weak broadened resonances are observed and it is assumed therefore that in the range of salt molarities 0 to 0.6 M NaCl the NMR signals from the gel will come only from the F1 molecules. The almost complete absence of signal in Figure 8b indicates that in the noncondensed gel in D_2O, virtually all the histones are bound firmly to the complex. The addition of 0.1 M NaCl (8c) which is accompanied by gel condensation, causes a large change in the spectrum; peaks appear which correspond mainly to alanine, proline and a few lysine residues, with very little other observable change. Increasing salt to 0.25 M (8d) produced very little change either in the spectrum, or in the state of the gel condensation, but a further increase to 0.5M

(a)

(b)

(c)

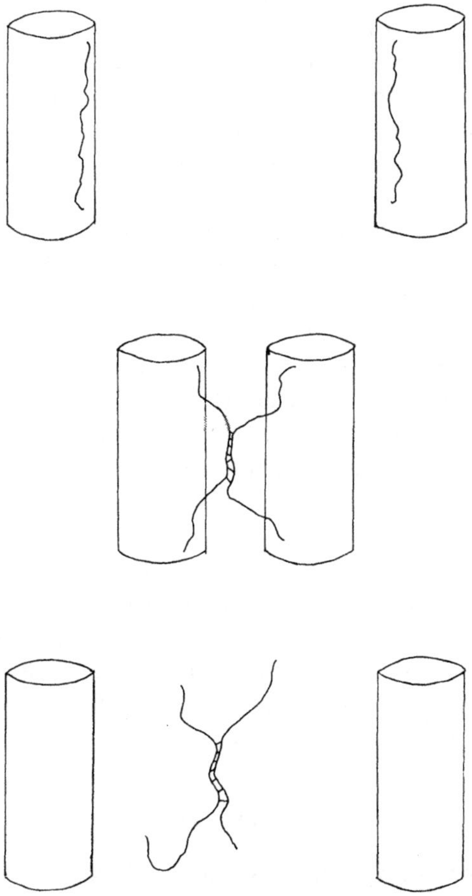

Fig. 9. Proposed mechanism for shrinkage of DNP gel. The cylinders represent all of the DNP except F1, the single lines F1 molecules; (a) in H_2O, with gel expanded and F1 bound to the rest of the complex; (b) in 0.1–0.3 M NaCl, region 47–106 of F1 freed from complex; F1–F1 antiparallel interactions cause gel shrinkage, as lysine-rich ends of F1 are still bound to the complex; (c) in 0.5 M NaCl. The F1 is now freed completely from the complex, leaving it free to expand to its original volume.

(8e), which corresponds both to complete re-expansion of the gel and to complete removal of the F1, yields a spectrum which is identical to that of F1 in 0.5 M NaCl, with the sole difference that all the linewidths are broader by a factor of approximately 1.8.

The most remarkable change in the spectrum between 0.25 and 0.5 M NaCl is the sudden appearance of the lysine peaks at an ionic strength which corresponds with the release of F1 from the DNP complex. Figure 8f shows that the spectrum observed for DNP at 0.1 or 0.25 M NaCl may be closely simulated by removing about 40 of the 60 lysine residues. It is important at this stage also to note that the simulations of the F1 spectrum at 0.1 M and 0.25 M NaCl (8c, d) also include the changes brought about by the association of the section 47–106 of the F1 molecules.

We may therefore draw the following conclusions from the spectra of Figure 8:

(a) Under the salt conditions used, no spectrum appears which is not directly attributable to the molecules of F1 in the complex. Removal of F1 also removes almost the entire observable spectrum (Figure 8a) even at a sodium chloride concentration at which a spectrum is very apparent in total DNP. The implication of this result is that all the remaining components of the system (the other histone fractions, DNA and acidic proteins) are too rigid throughout the experiment to give a high-resolution spectrum. A corollary of this conclusion is that the spectrum of F1 or parts of F1 must only be observable as long as those parts are not bound to the rigid complex.

(b) Under conditions of low ionic strength, where there is no condensation of the gel, all the F1 is quite firmly bound to the complex.

(c) Under conditions corresponding to *gel condensation*, most of the lysines of the F1 are firmly bound, and the spectrum which is visible corresponds to that obtained when an F1–F1 interaction of residues 47–106 is taking place. (Compare Figures 8c, f.)

(d) Complete removal of the F1 reduces re-expansion of the gel and a freeing of the lysine residues. The spectrum obtained still includes the interaction of residues 47–106.

An important conclusion of this study is that the segment of F1 47–106 which interacts in the chromatin gel when condensation of the gel takes place between 0.1 and 0.2 M NaCl, is the same segment of F1 which is involved in the salt-induced interactions of the pure histone in aqueous solutions. The results described above lead to the model shown in Figure 9 for the interactions of F1 molecules in the condensation of the chromatin gel at 0.1 to 0.2 M NaCl.

3. Conclusions

The above studies convincingly demonstrate the power of the technique of high resolution NMR in probing the conformations and interactions of histones (i) with themselves, (ii) with DNA and (iii) in chromatin gel. The application of this technique is a consequence of the asymmetric distribution of residues in the histone molecules and uses these residues to probe the mobilities of the segments of the polypeptide chain in which they are located. A general picture is emerging in which segments of histones rich in basic residues are the primary sites of interaction with DNA while the complementary segments which contain a high proportion of apolar aromatic, acidic and other active residues are the sites of histone–histone interactions. Through such interactions it is possible to envisage schemes whereby the conformation of DNA could be controlled in the interphase and metaphase chromosome. The highly condensed state of metaphase chromosomes with dimensions of the order of microns, contains a DNA double helix of overall length in the range of centimetres. The process of chromosomal coiling is reversible and the high degree of folding required to pack the DNA molecule into the condensed chromosome would demand considerable specificity of interaction both of the histones with DNA and of the histones with each other.

Large changes in the state of chromatin gel can be induced by small changes in salt molarity of 0.1 to 0.2 M NaCl. The interactions of the F1 molecules have been followed in the chromatin gel through these salt changes and this has led to the first detailed proposal concerning a structural role for the lysine rich histones. It is also significant that the self-interactions of other histones can be induced at relatively low salt molarities and it may be speculated that changes in the ionic strength of the cell nucleus play a role in the process of chromosome coiling.

Acknowledgements

This work is supported by the Science Research Council.

Acknowledgement is made to the Editor of the *Eur. J. Biochem.* for permission to reproduce Figure 2 and to the Editor of *Nature* for permission to reproduce Figures 7–9.

References

[1] Stedman, E. and Stedman, E.: 1950, *Nature* **166**, 780.
[2] Bradbury, E. M. and Crane-Robinson, C.: 1971, in *Histones and Nucleohistones* (ed. by D. M. P. Phillips, Plenum Press, London.
[3] Frédéricq, E.: 1971, in *Histones and Nucleohistones* (ed. by D. M. P. Phillips), Plenum Press, London.
[4] Phillips, D. M. P.: 1964, in *The Nucleohistones* (ed. by J. Bonner and P. O. P. T'so), Holden-Day, San Francisco, Calif.
[5] DeLange, R. J., Fambrough, D. M., Smith, E. L., and Bonner, J. J.: 1969, *J. Biol. Chem.* **244**, 319.
[6] Iwai, K., Ishikawa, K., and Hayashi, H.: 1970, *Nature* **226**, 1056.
[7] Sugano, H., Olson, M. O. J., Yeoman, L. C., Johnson, B. R., Taylor, C. W., Starbuck, W. C., and Busch, H. J.: 1972, *J. Biol. Chem.* **247** 3589.
[8a] Olson, M. O. J., Sugano, H., Yeoman, L. C., Johnson, B. R., and Starbuck, W. C.: 1971, *Fed. Proc.* **30**, 1295.
[8b] Olson, M. O. J., Sugano, H., Yeoman, L. C., Johnson, B. R., Jordan, J. J., Starbuck, W. C., and Busch, H.: 1972, *Physiol. Chem. Phys.* **4**, 10.
[9] DeLange, R. J., Hooper, J. A., and Smith, E. L.: 1972, *PNAS* **69**, 882.
[10] Bustin, M. and Cole, R. D.: 1970, *J. Biol. Chem.* **245**, 1458.
[11] Rall, S. C. and Cole, R. D.: 1971, *J. Biol. Chem.* **246**, 7175.
[12] Boublik, M., Bradbury, E. M., and Crane-Robinson, C.: 1970, *Eur. J. Biochem.* **14**, 486.
[13] Boublik, M., Bradbury, E. M., Crane-Robinson, C., and Johns, E. W.: 1970, *Eur. J. Biochem.* **17**, 151.

[14] Bradbury, E. M., Crane-Robinson, C., Goldman, H., Rattle, H. W. E., and Stephens, R. M.: 1967, *J. Mol. Biol.* **29**, 507.
[15] Boublik, M., Bradbury, E. M., Crane-Robinson, C., and Rattle, H. W. E.: 1971, *Nature* **229**, 149.
[16] Bradbury, E. M. and Rattle, H. W. E.: 1972, *Eur. J. Biochem.* **27**, 270.
[17] DeLange, R. J., Fambrough, D. M., Smith E. L., and Bonner, J.: 1969, *J. Biol. Chem.* **244**, 5069.
[18a] Edwards, P. A. and Shooter, K. V.: 1969, *Biochem. J.* **114**, 227.
[18b] Diggle, J. H. and Peacocke, A. R.: 1971, *FEBS Letters* **18**, 138.
[19] Bradbury, E. M., Cary, P. D., and Crane-Robinson, C.: unpublished data.
[20] Bradbury, E. M., Cary, P. D., Crane-Robinson, C., Riches, P. L., and Johns, E. W.: 1972, *Eur. J. Biochem.* **26**, 482.
[21] Kinkade, J. M. and Cole, R. D.: 1966, *J. Biol. Chem.* **241**, 5798.
[22] Palau, J., Ruiz Carrillo, A., and Subirana, J. A.: 1969, *Eur. J. Biochem.* **7**, 209.
[23] Greenaway, P. J. and Murray, K.: 1971, *Nature New Biol.* **229**, 233.
[24] Bustin, M., Rall, S. C., Stellwagen, R. H., and Cole, R. D.: 1969, *Science* **163**, 391.
[25] Greenaway, P. J.: 1970, Ph.D. Thesis, Edinburgh University.
[26] Bradbury, E. M., Crane-Robinson, C., and Johns, E. W.: 1972, *Nature* **238**, 262.
[27] Bradbury, E. M., Carpenter, B. G., and Rattle, H. W. E.: 1973, *Nature* **241**, 123.

DISCUSSION

Hélène: Have you any idea about the binding constants of your histone fractions to DNA and about the exchange rate between bound and free histone? Would you be able to see any change in chemical shifts of either aliphatic or aromatic protons upon binding histones to DNA?

Bradbury: We have not measured the binding constants of histones to DNA. On the binding of histones to DNA there is a marked loss in apparent peak area of certain peaks only. No chemical shift difference is observed, which leads us to the conclusion that we are in the slow exchange region.

Frédéricq: I would like to raise the question of the specificity of DNA–histone interaction. As you know, attempts to establish such specificity in systems containing isolated DNA and isolated histone fractions have been unsuccessful. In my laboratory, J. Dedek is trying an approach under milder conditions: he uses native nucleohistone dissociated at various salt concentrations, the same reassociated after dialysis, with or without excess histone addition. In all systems he found the same amout and the same fractions of histone combined to DNA. This indicates at least a partial specificity of interaction. Would it be possible to make similar experiments using the NMR method to check similarities in dissociated and reassociated DNA-histone complexes?

Bradbury: In general the spectra of the different histone fractions show many similarities. It would therefore be difficult to determine from the NMR spectrum of a mixture of histones in the presence of DNA exactly which histones were present and their amounts. Technically the problem could be solved if specifically labelled histone fractions were used. This, however, would be very time consuming and other approaches could probably be more profitable.

Mme Hvidt: What are the rates of the conformational charges of the histones induced by changes in salt concentrations?

E. M. Bradbury: Only a qualitative answer can be given to this question; that the rate of histone conformational change is rapid compared to the time necessary to record the NMR spectrum which is of the order of minutes. Lk, Wickett, Craig and Isenberg (*Biopolymers* **11** (1972) 375) have looked at the conformational changes in histone F2A1 by the intrinsic fluorescence of tyrosine residues and circular dichroism. They find on the addition of phosphate there is a rapid conformational change, i.e. rapid compared to the time of measurement, followed by a slow change of the order of minutes to hours. The former change they attribute to the rapid formation of α-helical segments while the latter is attributed to the formation of interchain β-structures.

Rassat: When you fit the different NMR spectra of your protein as salt concentration changes, you introduce a line-width parameter. What is this parameter? How many parameters do you introduce and how sensitive is your mobility or nonmobility assignment for a given segment to the variation of these parameters? Can you change the length of a segment? Finally, what is the precision of the correlation with the biochemical experiment using NAD?

Bradbury: The model which we use for our spectrum situations is a very simple one. Position width and area data obtained from amino-acid and polypeptide spectra is combined by the computer to give a simulated spectrum of the random coil form of the histone. When it is required to check whether a given segment of the molecule is restricted in motion by an interaction, all the resonances due to the residues in that segment are broadened to 1000 Hz. No allowance is made for the fact that, presumably, residues near the ends of the interacted segment have different mobilities to those in the centre. The length of the segment may be chosen to be any value.

For experiments involving interactions between a histone fraction and DNA, it is not likely that the precision of results found using the simulation method is at present better than $\pm$4–5 residues at each end of the segment. It is hoped that improvements in the computer programme, possibly allowing for mobility differences, will shortly improve this figure.

CONFORMATIONAL EQUILIBRIA AND STABILITY OF OLIGOPEPTIDES

CLAUDIO TONIOLO

Institute of Organic Chemistry, University of Padova, 35100 Padova, Italy

Summary. The analysis of the stabilization forces maintaining the multitudinous conformations found in proteins has largely profited by the study of model compounds, in which the number of conformations and hence the variety of vicinal interactions are very greatly reduced.

In this communication we wish to report some results obtained with oligo-γ-ethyl-L-glutamates, oligo-L-isoleucines, and oligo-L-phenylalanines.

A combination of modern techniques and instrumentations as well as the recent discovery of the fluoroalcohols, which are versatile solvents for a variety of peptide molecules and satisfactorily transparent in the far UV, enabled us to undertake a detailed conformational study of the abovementioned oligopeptides.

Both the average state of the molecules and their local conformational preferences irrespective of the overall secondary structure have been explored as a function of several parameters, such as the number of residues in the peptide chain, nature of the solvent, concentration and temperature.

Résumé. L'analyse des forces de stabilisation qui maintiennent les multiples conformations possibles dans les protéines a abondamment profité de l'étude de modèles dans lesquels le nombre des conformations, et par conséquent celui des interactions vicinales, sont largement réduits. On rend compte ici de quelques résultats obtenus avec des oligo-γ-ethyl-L-glutamates, oligo-L-isoleucines and oligo-L-phénylalanines.

L'association des techniques et appareillages modernes et de l'emploi des alcools fluorés, qui sont raisonnablement transparents dans l'ultraviolet lointain et dissolvent bien divers peptides, nous a permis d'entreprendre une étude conformationnelle détaillée de ces produits.

L'état moyen de la molécule et les conformations locales préférées, indépendamment de la structure secondaire d'ensemble, ont été examinés en fonction de divers paramètres tels que le nombre de résidus dans la chaîne peptidique, la nature du solvant, la concentration et la température.

1. Introduction

The basic conformations of the secondary structures of proteins have been theoretically predicted and experimentally found to be the α-helix, the β-pleated sheet, and the random chain or nonperiodic conformations. The last, of course, includes the many varied folds which help differentiate one protein from the other. The analysis of the stabilization forces maintaining these conformations in proteins has represented the matter of a number of investigations and has largely profited by the study of model compounds (Table I), in which the variety of conformations and hence of interactions is very greatly reduced. In this article we wish to report some results which we have recently obtained with three linear homo-oligopeptide series (entry B.b of Table I) derived from γ-ethyl-L-glutamate, L-isoleucine, and L-phenylalanine residues, respectively. The peculiar type of information which one might gain from the investigation of linear homo-oligopeptides concerns the combined effect of chain length and nature of side-chain on the stability of various structures, allowing one to evaluate the contribution of particular bonds within a particular conformation and, conse-

C. Sadron (ed.), Dynamic Aspects of Conformation Changes in Biological Macromolecules. 87–101.

TABLE I

Classification of model compounds of biologically active peptide molecules

A. Polypeptides[a]	a. Copolypeptides	I. Random copolypeptides
		II. Block copolypeptides
		III. Sequential copolypeptides
	b. Homo-polypeptides	
B. Oligopeptides[b]	a. Linear co-oligopeptides	I. Without a repeating sequence
		II. With a repeating sequence
	b. Linear homo-oligopeptides	
	c. Cyclic co-oligopeptides	I. Without a repeating sequence
		II. With a repeating sequence
	d. Cyclic homo-oligopeptides	
C. Amides	a. Linear amides	
	b. Monocyclic amides	
	c. Polycyclic amides	

[a] From α-amino acids; chain length $n \geqslant$ ca. 20.
[b] From α-amino acids; chain length c. $2 \leqslant n \leqslant 20$.

quently, to shed light on the nature of short-range interactions, which are of major importance with regard to the problem of the mechanism of protein folding. A combination of modern techniques and instrumentations (UV absorption and CD to 185 nm, and high resolution NMR) and the discovery of 2, 2, 2-trifluoroethanol (TFE), which is a versatile solvent for peptide molecules and satisfactorily transparent in the far UV, enabled us to undertake a detailed conformational study of the above-mentioned linear homo-oligopeptides. Both the average state of the molecules and their local conformational preferences have been explored as a function of several parameters, such as number of residues in peptide chain, nature of side-chain, concentration and temperature.

2. Synthetic Aspects

The problem of the selection of the synthetic approach is of paramount importance, since strictly monodisperse and optically pure oligopeptides are required for these physicochemical investigations. A careful search in the literature have revealed that the preparation of linear homo-oligopeptide series has been attempted by four different approaches: (a) controlled chemical or enzymatic hydrolysis of homo-polypeptides [1–4]; (b) controlled polymerization of amino acyl N-carboxy anhydrides, at low anhydride-to-initiator molar ratios [5]; (c) synthesis with the aid of an insoluble polymeric support (Merrifield's solid phase synthesis) [6, 7–9]; and (d) conventional synthesis in homogeneous phase [10, 11] (stepwise and/or fragment condensation methods).

The first two approaches give highly heterogeneous materials containing homologous series of homopeptides. The heterogeneity of these low molecular weight preparations makes their use of questionable validity for the determination of the relationship between size of oligomer and secondary and tertiary conformations. Nevertheless, it has been shown that in favourable cases such series can be resolved into their components by ion-exchange column chromatography [1–5]. The isolated peptide proved to be analytically pure and the optical configuration was fully retained. However, the technique employed for the separation of the individual members drastically limits the use of these two synthetic approaches to the few oligopeptide series which derive from amino acids carrying ionizable groups in their side-chains (i.e., homo-oligopeptides derived from glutamic acid, lysine, etc.).

On the other hand, the main objection to the use of Merrifield's solid phase synthetic approach [6] arises from the observation that failure sequences occur in which one or more amino acids are left out. This drawback depends on the fact that quantitative yields, which are essential in every coupling step in order to obtain homogeneous material, are not usually achieved. The number of failure sequences grows binomially whereas the amount grows linearly with increasing chain length [12]. In the synthesis of homo-oligopeptides the number of failure sequences is greatly reduced, since truncated sequences are identical to failure sequences with deletions. However, even in this case the use of ion-exchange chromatography is essential for the separation of the various components (Figure 1).

Because of these considerations, we resolved to prepare the homo-oligopeptide series by the conventional approach in homogeneous phase without underestimating the difficulties arising from the number of steps necessary to perform such synthesis. This choice proved to be satisfactory, since we have obtained the desidered chemically and optically pure compounds in adequate quantity. To this purpose we have em-

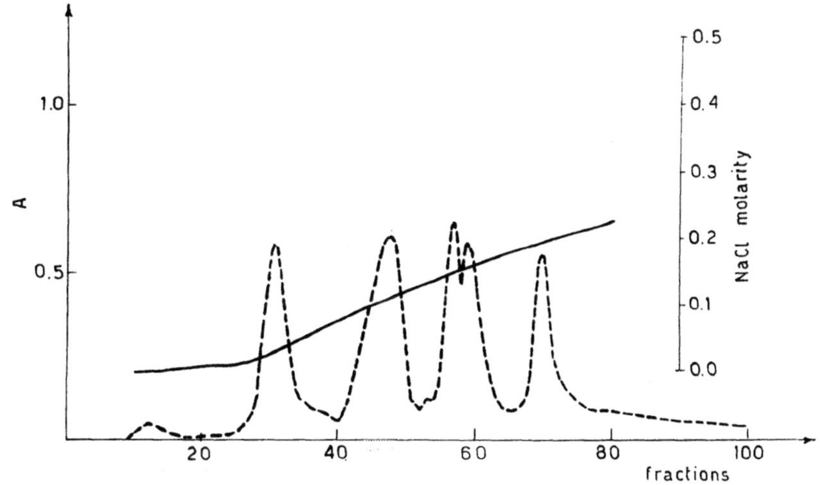

Fig. 1. Gradient elution chromatogram on a DEAE-cellulose resin of a sample containing the sequence (Glu–Glu–Glu–Glu)-prepared by the Merrifield's solid phase method (Toniolo, unpublished results).

ployed urethane groups for the protection of the amino function, the dicyclohexylcarbodiimide [13] and Rudinger's acylazide [14] methods for the formation of the peptide bonds, and alkyl esters for the protection of the carboxyl functions.

3. γ-Ethyl-L-Glutamate Studies

In 1960, Blout *et al.* [15] suggested that the conformation of synthetic homo-polypeptides in the solid state depend on their amino acid composition. In his classification the glutamic acid esters (i.e., γ-ethyl-L-glutamate) tend to form α-helical structures. Moreover, high molecular weight poly-γ-ethyl-L-glutamate exists in a right-handed α-helical conformation in a TFE solution, as ORD studies have clearly shown [16].

In this report we wish to describe an application of far-UV CD and 220 MHz NMR to the problem of the critical size for helix formation in the oligopeptide series derived from γ-ethyl-L-glutamate. These measurements have provided a more direct assessment of conformation than was possible by studying far-UV absorption and ORD spectra [16]. Our results [17, 18] demonstrate the existence of a critical size for helix formation in TFE, above which CD bands change dramatically and are substantially

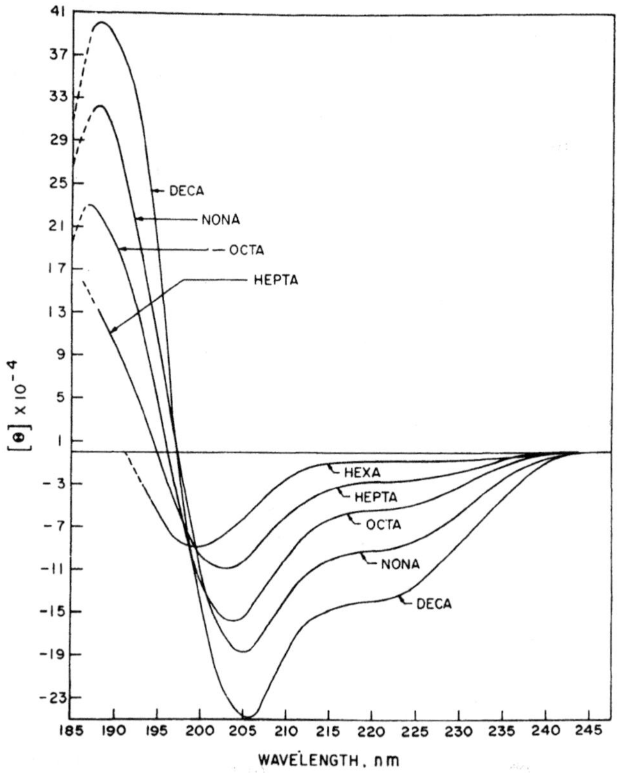

Fig. 2. Circular dichroism spectra of N-carbobenzoxy-γ-ethyl-L-glutamate oligomers in trifluoroethanol. Values for the total molar ellipticity of each oligomer are recorded.

like spectra of right-handed α-helices (Figure 2). The first negative band at 222 nm is therefore assigned to the $n \rightarrow \pi^*$ transition of the amide chromophere, while the negative band near 205 nm and the large positive band below 200 nm result from the exciton splitting of the $\pi \rightarrow \pi^*$ transition of the peptide groups in a helical array. A sharp increase in molar ellipticity occurs with the onset of helicity for the $n \rightarrow \pi^*$ transition. A plot of molar ellipticity at 222 nm versus number of residues in the peptide chain (n) is shown in Figure 3. The increase in slope provides obvious support for the onset of helicity. Moreover, a plot of wavelengths for the negative maximum in the 200 nm region versus n has revealed that in going from pentamer to hexamer to heptamer to octamer the positions of the bands redshift rather sharply. This is additional evidence for the splitting of this band and the critical size for helix formation.

However, it is difficult to make unambiguous secondary structural assignments based only on CD patterns. Many low molecular weight linear and cyclic peptides exhibit CD patterns reminiscent of those shown by highly helical synthetic polypeptides. It is one of the main theses of this report that similarity of CD patterns alone for oligopeptides and high molecular weight peptides are not sufficient to deduce

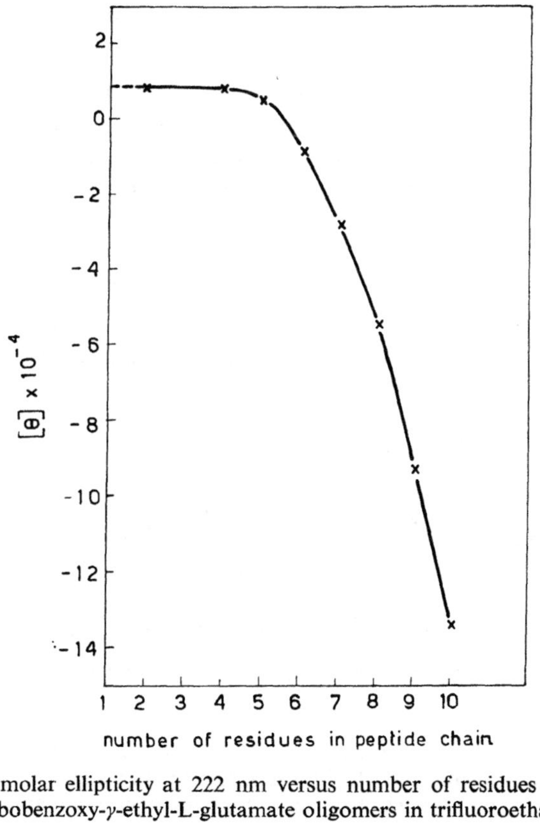

Fig. 3. Plot of total molar ellipticity at 222 nm versus number of residues in peptide chain for N-carbobenzoxy-γ-ethyl-L-glutamate oligomers in trifluoroethanol.

secondary structure. It must be recognized that end effects, unimportant in high polymer chains, play a dominant role in short peptides. End residues contribute intrinsically different values to optical properties. End effects lead to different conformational equilibria and stability from those seen with high polymers since solvation of terminal residues can differ substantially from internal residues.

To place our stereochemical deductions on a more firm footing we have also examined the N-carbobenzoxy-γ-ethyl-L-glutamate oligomers with the aid of 220 MHz NMR. The positions of resonances of amide N–H protons sensitively reflect the presence or absence of helical character. When these protons are involved in C=O...H–N bonds in helical regions of polypeptides, they are subjected to anisotropy fields and other environmental influences that profoundly affect their shielding. In TFE as a solvent the N–H proton resonances are well resolved and N–H chemical shifts are sensitive functions of oligomer length. For example, the hexamer exhibits peaks for each of the NH's (Figure 4) starting with the urethane-like N-terminal residue at highest field. The resonance at 7.36 ppm arises from the phenyl protons of the amino-blocking group and furnishes a convenient internal standard. All N–H proton resonances for the lower oligomers (dimer through hexamer) are resolved. For higher members the internal NHs appear to overlap. If we plot the position of the N-terminal and C-terminal NHs as a function of *n*, we see that both NH protons shift progressively to lower field in going from dimer to hexamer. We interpret this as a change from an extended conformation for the dimer to mixture of extended and folded forms for the trimer to hexamer. For the dimer, conformational alterations do not materially affect the geometric relationship of the two amide groups. From the trimer to the hexamer alteration of the internal rotation angles most dramatically alter the proximity of the N- and C-terminal amide groups. As a result, a substantial change

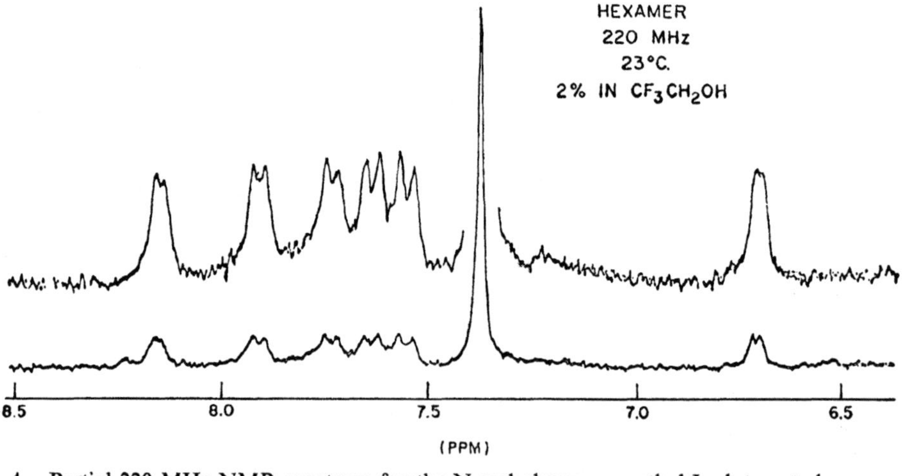

Fig. 4. Partial 220 MHz NMR spectrum for the N-carbobenzoxy-γ-ethyl-L-glutamate hexamer as a 2 % solution in trifluoroethanol at 23 °C. This spectrum shows the amide and phenyl protons. The upper spectrum shows the resonances of the lower spectrum at higher gain.

in environment is felt by these terminal groups. The equilibrium among all these conformations must be rapidly established on a NMR time scale, since we observed only a single, averaged one resonance for each amide proton. Commencing with the heptamer, stable helical structures exist which cause specific internal rotation angles to be cooperatively preferred. The terminal groups are conformationally frozen with respect to each other. The factor which changes with oligomer length is the intramolecular hydrogen bonding contacts.

These results suggest that NMR offers a tool to study conformational preferences of peptides and proteins irrespective of the overall helicity of the structure. Our explanation for the NMR data appears to be correct, since no change in chemical shift has been observed for the oligomers if the peptides are solvated to the extent that they do not form folded and stable helical forms.

In conclusion, we obtained a picture of critical size for helicity from CD and NMR spectra for the γ-ethyl-L-glutamate oligopeptides. Both techniques clearly indicate the onset of helicity at about the heptamer for these peptides in TFE; moreover, NMR technique indicates folding of the peptide chain prior to helix formation.†

4. Isoleucine Studies

Since it is a β, β-dialkyl-substituted amino acid, isoleucine has been classified by Blout *et al.* [15] as a non α-helix forming amino acid for steric reasons. In fact, poly-L-isoleucine in the solid state has been shown to exist in β-form by IR studies [20] and the ORD spectrum of a film in this conformation has been also reported [20]; however, because of its extremely low solubility, no solution studies on this polymer have been carried out up to now. In addition, it has been demonstrated by IR studies that the protected *t*-butyloxycarbonyl (BOC) oligomers ($1 \leqslant n \leqslant 5$) exhibit a disordered structure in solution [21, 22] and the BOC-hexamer an antiparallel β-conformation in the solid state [22].

In order to get some clues as to which conformation(s) long sequences of L-isoleucine residues might assume in solution, we have investigated the UV absorption and CD spectra†† in TFE of the first members of the homologous series with the general formula BOC–(L-Ile)$_n$–OMe ($2 \leqslant n \leqslant 8$) [23–25]. The far-UV absorption spectra of polypeptides in solution have been found to be conformationally dependent [26]. In this spectral region an absorption maximum at about 190 nm is associated with the $\pi \rightarrow \pi^*$ transition of the peptide chromophore, the intensity of which is significantly less for helical polypeptides than for the same polymers in the random coil conformation. Helical structure, in addition, generally show a slight blue shift of the 190 nm band and an appearance of a shoulder at about 205 nm. On the other hand,

† Although it is always difficult to compare CD and NMR results because of the different concentration ranges employed in the two types of experiments, it has been demonstrated that in the γ-ethyl-L-glutamate oligopeptide series associations do not occur in TFE even in the fairly concentrated solutions used for NMR studies [19].

†† Solubility problems prevented systematic NMR conformational studies.

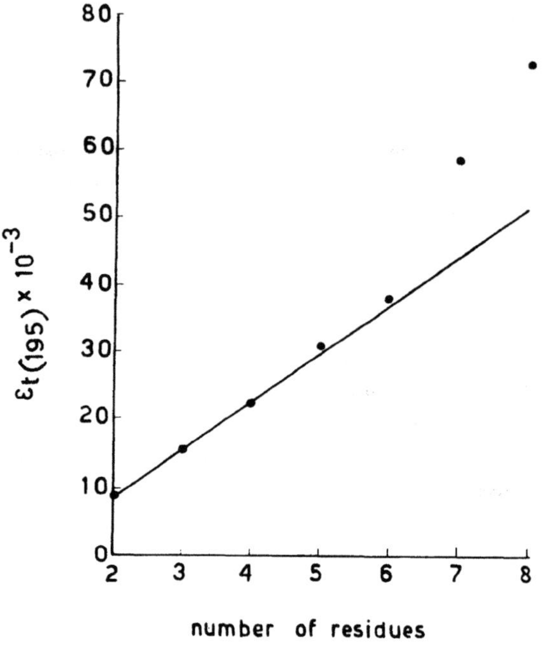

Fig. 5. Plot of total molar extinction coefficients of the N-*t*-butyloxycarbonyl-L-isoleucine oligomers in trifluoroethanol at 25 °C.

hyperchromism and a 5 nm bathochromic shift indicate the presence of a β-conformation. The λ_{max} of the L-isoleucine oligomers shift from below 189 nm for the dimer to 195.5 nm for the hepta- and octapeptides. There is a 4 nm redshift from the hexamer to the heptamer which can be used to support the existence of β-conformations for these oligopeptides. A more convincing proof, however, came from an analysis of absorption intensities. Figure 5 depicts a plot of total molar extinction coefficients at 195 nm versus n. The hyperchromic effect commences at the heptamer, indicating the onset of the β-form at this stage. Lastly, the UV spectrum of the heptamer has been examined at various temperatures in TFE. A decrease of the hyperchromic effect and the blueshift of the peptide $\pi \rightarrow \pi^*$ transition were observed as the temperature was increase from 25 °C to 65 °C. This effect has been shown to be completely reversible. The most likely explanation of this behaviour is that the increase of temperature causes the dissociation of the aggregates.

In order to confirm this preliminary conformational analysis we have examined the L-isoleucine oligopeptides in TFE using CD. Results of these studies are presented in Figure 6. The CD spectra of the tri-, tetra- and pentapeptide show a negative dichroic band at 195 nm, and a negative dichroic band in this spectral region is typical of random polypeptide chains [27]. The CD spectrum of the dipeptide is rather different, but here the end groups may play a predominant role. The above results seem to indicate that the lower oligomers ($n<6$) are present in an unordered conformation

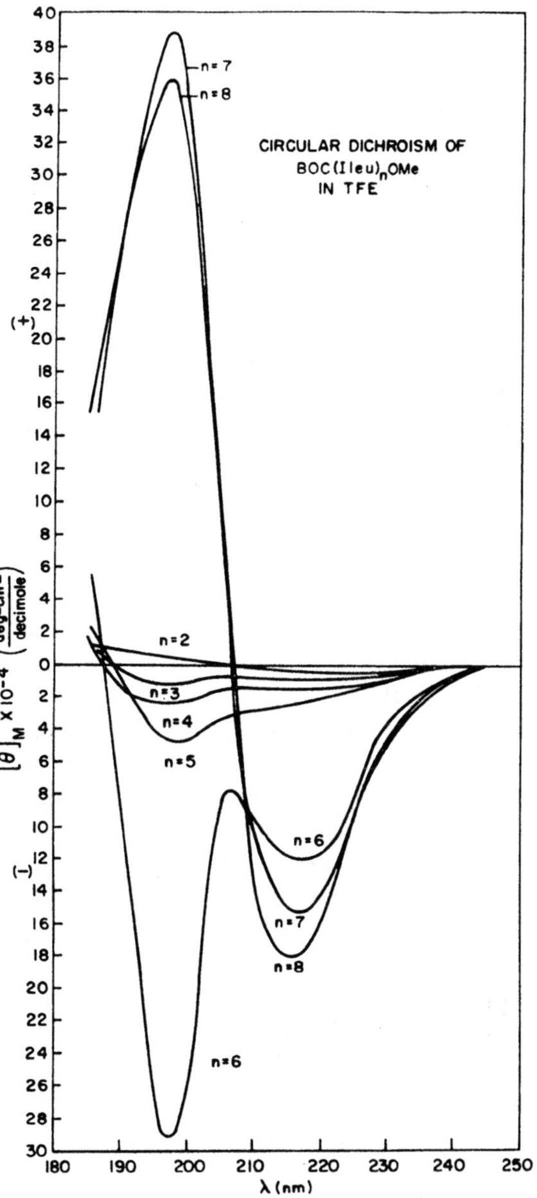

Fig. 6. Circular dichroism spectra of the N-*t*-butyloxycarbonyl-L-isoleucine oligomers in trifluoroethanol (concentration about 2 mg/ml).

in TFE. At the hexamer, however, an abrupt increase in molar ellipticity at both 197 nm and 216 nm is observed. Subsequent investigation of the hexamer at various concentrations showed that the CD patterns are concentration dependent indicating associated structures. The spectral patterns of the hexamer do not conform to those observed for a β-conformation. It is particularly difficult to explain the large ellipticity at 197 nm. At present, we can only conclude that at about two mg/ml the hexamer exists in an associated structure which does not appear to be a typical β-conformation. The spectral patterns of the heptamer and octamer are completely different from those of the lower members of the series. The CD of these oligomers is characterized by a negative Cotton effect centered at 216 nm and a positive spectral band centered at 197.5 nm. The overall shape of the CD spectra is very similar to that reported for polypeptides in a β-conformation [28]. We conclude that both the heptamer and octamer are present in β-forms in TFE. Furthermore, concentration studies indicate that for the heptamer the β-conformation is intermolecular, whereas the octamer forms a stable intramolecular β-conformation. We feel that the additional stability of the octamer cannot be explained by the effect of one added residue on an intermolecular β-structure. In contrast, at these short chain lengths, the extra residue might be expected to lead to an intramolecular species, which is appreciably more stable than that formed by the heptamer. The additional fact that the homopolypeptide derived from L-valine, which is sterically similar to L-isoleucine, forms a cross-β-structure [29] gives further evidence that our oligomers could also have this tendency.

In conclusion, our CD analysis of the L-isoleucine oligomers suggests that the higher oligomers form β-conformations in TFE. These findings confirm preliminary conformational information obtained using UV absorption spectroscopy. It is interesting to note that although the hexamer associates in TFE it does not form a β-structure. It is possible that the formation of a β-conformation may require a critical chain length analogous to that observed for oligomers which form helices. In the case of the isoleucine oligomers this critical chain length appears to be seven residues. A comparison of the β-structures formed by the isoleucine oligopeptides with those formed by oligo-alanines [30] indicates that the former species are considerably more stable. These findings agree fully with our expectations. The isoleucine peptides have a much higher hydrocarbon content than their alanine counterparts. In a fairly polar medium such as TFE, therefore, they would be expected to have a greater tendency to associate. Such lyophobic interactions are known to be enthalpy-driven [31]. In addition, the fact that steric interactions destabilize helical forms in the isoleucine oligomers adds to the relative stability of their β-conformations.

5. Phenylalanine Studies

In 1966 Auer and Doty [32] suggested that the class of α-helix forming amino acids as proposed by Blout *et al.* [15] should be further divided into two groups: a group with a single aliphatic substituent on the β-carbon (i.e., alanine) which forms unhindered helices and a group which forms helices destabilized by steric interaction between bulky side chains and backbone (i.e., phenylalanine). In the latter case the rotation of the β-carbon of the benzylic side chain about the bond joining it to the α-carbon is severely restricted, and the rotation of the phenyl group in turn is also critically diminished. The correctness of such sub-classification has been proved for polypeptides [32–35]. However, quite recently, Goodman *et al.* [30] demonstrated that the homo-oligopeptides derived from L-alanine do not form helices in pure TFE; instead, the onset of a β-form has been shown at the level of the heptamer in the N-carbobenzoxy series. In the present report we wish to discuss some preliminary UV and CD results obtained in TFE by examining the first members of the homologous series with the general formula BOC–(L-Phe)$_n$–OMe ($1 \leqslant n \leqslant 6$).* A study of the chirospectroscopic properties of oligopeptides with aromatic side chain appears to be of interest because the CD contribution of sidechain chromophores both in the near and far-UV region and its dependence on conformation is not yet sufficiently understood. For this reason, the establishment of the conformation in which aromatic oligopeptides exist in solution is difficult [36]. An additional point of interest of this study concerns the possibility of comparing the vibrational fine structure in the near-UV CD and absorption spectra of the oligophenylalanines [37–39].

Absorption spectra of the BOC-oligophenylalanines have been measured between 280 and 200 nm. In this region, bands corresponding to ${}^1A_{1g} \rightarrow {}^1B_{2u}$ and ${}^1A_{1g} \rightarrow {}^1B_{1u}$ benzene transitions are observed. The higher energy transition appears as a band in the monomer whereas as a pronounced shoulder in the higher homologs (Figure 7).

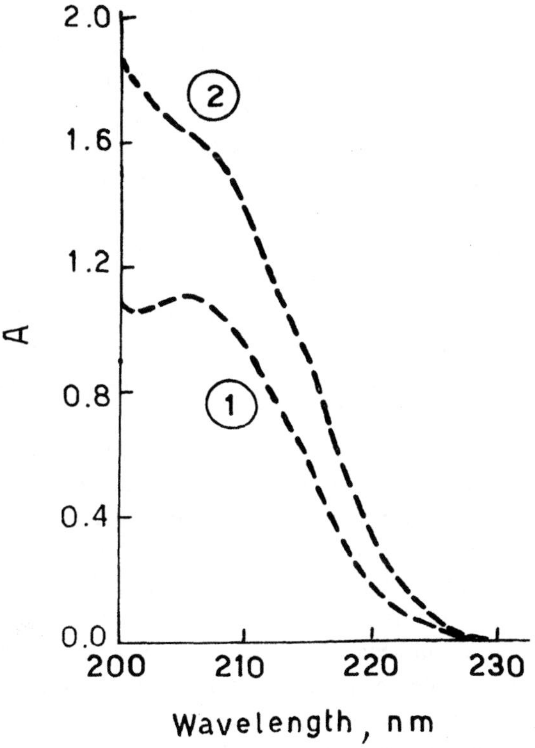

Fig. 7. Far-UV absorption spectra of BOC–L–Phe–OMe (1) and BOC–(L-Phe)$_2$–OMe (2) in trifluoroethanol; 0.5 mm cell; concentration 0.65 mg/ml.

* The peptides with chain length higher than 6 are not soluble in TFE.

This finding seems to be related to a substantial increase in intensity of an absorption band located below 200 nm. Conversely, the vibrational structure of the 257 nm band of the monomer does not change in going to dimer and higher oligomers: in fact one can note in each case the series of vibrational bands characteristic of the substituted benzene nucleus [40]. In addition, at room temperature the absorption bands are much better resolved than at 60 °C [39]. Finally, a plot of total molar extinction coefficients at 257 nm versus *n* does not show any evidence of hyper- or hypochromicity [41].

In evaluating the near-UV transitions we have considered phenylalanine to be a derivative of toluene, as was first suggested by Sponer [42]. Our study extends Sponer's earlier investigation in two relevant respects. First, our analysis correlates both absorption and CD spectra, whereas Sponer had access only to the absorption spectrum. Secondly, we take toluene to have a local C_{2v} symmetry [39] instead of the C_s originally assigned by Sponer. The manuscript including the detailed analysis of the vibrational structure of the 257 nm band of the oligophenylalanines is in preparation [43].

The CD spectra of oligophenylalanines with 1 to 6 phenylalanine residues present a single intense and positive Cotton effect at 214–218 nm in the 230–205 nm region in TFE. From Figure 7 it appears that in the same wavelength region the phenylalanine peptides show an absorption band (shoulder) which has its maximum at 206–207 nm and a shoulder near 215 nm. Cotton effects in the 215 nm region of phenylalanine and several of its derivatives have been reported [34, 38, 44]. From these investigations it has been concluded that both the $^1A_{1g} \rightarrow {}^1B_{1u}$ transition of the phenyl ring and the $\pi \rightarrow \pi^*$ transition of the peptide chromophore contribute to optical activity. Below 205 nm the CD patterns become uncertain because of the highly unfavourable signal to noise ratios.

The near-UV CD Cotton effects of L-phenylalanine were first described by Moscowitz *et al.* [45] on the basis of ORD measurements. Recent improvements in CD instrumentation have permitted a detailed investigation of the optically active absorption bands in phenylalanine compounds [37–39]. For a discussion on the nature of the transitions involved the reader is referred to [44, 46]. Figure 8 illustrates the UV and CD spectra of BOC–$(\text{L-Phe})_2$–OMe in the 280–230 nm region at room temperature in TFE. The UV absorption bands (shoulders) appear at 266.5, 263, 260, 257, 252, 247 and 241 nm. We have already seen that the vibronic structure of the 257 nm band is preserved identical in all the oligophenylalanine series ($1 \leqslant n \leqslant 6$). However, the CD pattern of the dimer (weak negative Cotton effects at 267, 260.5, 254.5 and 247 nm) changes with increasing chain length [47]. The CD band at 254.5 nm appears as a shoulder in the spectra of pentamer and hexamer, and the CD band at 247 nm appears as a shoulder at the trimer and tetramer and finally disappears at the pentamer and hexamer. These findings seem to be related to the substantial enhancement in intensity of the negative dichroic band located at 239–236 nm. When one compares each vibrational component in the UV and CD spectra above 255 nm, the components appears to change sign of rotational strength, i.e., the bands at 266.5 and 260 nm have corresponding negative CD maxima, and the bands at 263 and

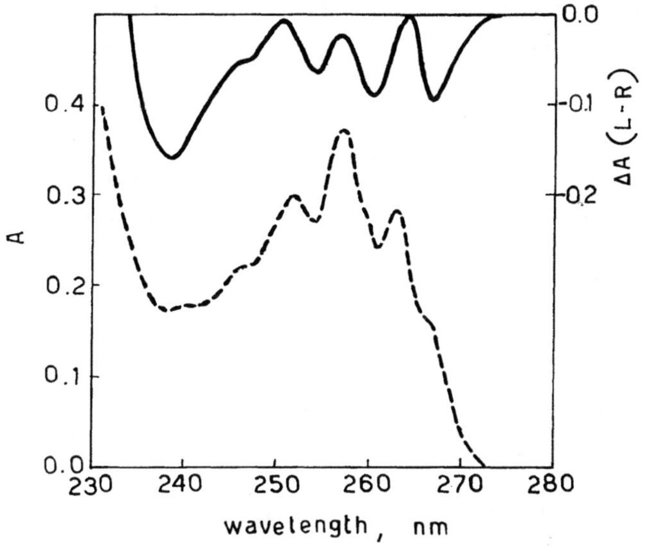

Fig. 8. Near-UV absorption (---) and circular dichroism (——) spectra of BOC–(L-Phe)$_2$–OMe in trifluoroethanol; absorption: 1 cm cell, concentration 0.42 mg/ml; circular dichroism: 1 cm cell, concentration 0.66 mg/ml, sensitivity 4×10^{-4}.

257 nm correspond to extrema which are directed in a positive manner. A similar situation at room temperature is seen in N-acetyl-L-phenylalaninamide [38, 39], L-phenylalanyl-L-phenylalanine [37] and poly-L-phenylalanine [37]. Conversely, BOC–L-Phe–OMe, in addition to the CD bands above reported for the dimer, exhibits two positive CD maxima at 264 and 257.5 nm. This patterns parallels those reported for N-acetyl-L-phenylalanine [39] and its alkyl esters [39]. The highest energy CD band reported in Figure 8 shifts from 239 nm for the dimer to 236 nm for the hexamer. Its peculiar interest resides on the fact that no apparent absorption counterpart is present in the UV absorption spectrum. We ascribe this band to the residual difference between the overlapping tails of the narrower positive band at 214–218 nm and a broader negative amide (ester) band extending above 250 nm. A preliminary conformational analysis of the oligophenylalanines in TFE indicates that their CD spectra below 245 nm do not resemble any of the spectra characteristic of peptide compounds in periodic or nonperiodic conformations. This finding is certainly due to overlapping bands resulting from aromatic transitions and main-chain amide chromophores. The total molar ellipticities at 220 and 240 nm increase gradually from dimer to pentamer. This result along with the similarity of the CD patterns beginning with BOC–L-Phe–OMe, for which structure formation is highly unlikely, suggest that these oligomers do not form ordered structures in TFE, at least at the concentrations examined (0.5 mg/ml). At the hexamer, however, an abrupt increase in molar ellipticity per residue at 240 nm, which parallels a net decrease in molar ellipticity per residue at 220 nm, seems to indicate the onset of an ordered conformation. Nevertheless, we

believe that further CD investigations at different concentrations and temperatures are required to establish the nature of such conformation and of the forces which stabilize it.

Acknowledgements

Much of work summarized here resulted from collaborative effort with the group directed by Prof. M. Goodman (University of California, San Diego, La Jolla, Calif., U.S.A.) to whom the author is personally indebted for his initial stimulation, generous hospitality and uninterrupted encouragement and assistance.

References

[1] Stewart, J. W. and Stahmann, M. A.: 1962, *Polyamino Acids, Polypeptides and Proteins* (ed. by M. A. Stahmann), The University of Wisconsin Press, Madison, p. 95.
[2] Sober, H. A.: *ibid.*, p. 105.
[3] Miller, W. G.: *ibid.*, p. 291.
[4] Yaron, A. Otey, M. C., Sober, H. A., Katchalski, E., Ehrlich-Rogozinski, S., and Berger, A.: 1972, *Biopolymers* **11**, 607.
[5] Ikeda, S. and Imae, T.: 1972, *Biopolymers* **11**, 493.
[6] Merrifield, R. B.: 1963, *J. Am. Chem. Soc.* **85**, 2149.
[7] Grahl-Nielsen, O. and Tritsch, G. L.: 1969, *Biochemistry* **8**, 187.
[8] Krumdieck, C. L. and Baugh, C. M.: 1969, *Biochemistry* **8**, 1568.
[9] Nitecki, D. E. and Goodman, J. W.: 1970, in *Peptides: Chemistry and Biochemistry* (ed. by B. Weinstein), M. Dekker, New York, p. 435.
[10] Goodman, M. and Boardman, F.: 1963, *J. Am. Chem. Soc.* **85**, 2483.
[11] Goodman, M. and Langsam, M.: 1966, *Biopolymers* **4**, 275.
[12] Bayer, E., Hagenmaier, H., Jung, G., Parr, W., Eckstein, H., Hunziker, P., and Sievers, R. E.: 1971, in *Peptides 1969* (ed. by E. Scoffone), North-Holland Publ. Co., Amsterdam, p. 65.
[13] Sheehan, J. C. and Hess, G. P.: 1955, *J. Am. Chem. Soc.* **77**, 1067.
[14] Rudinger, J. and Honzl, J.: 1961, *Collection Czech. Chem. Commun.* **26**, 2333.
[15] Blout, E. R., de Lozé, C., Bloom, S. M., and Fasman, G. D.: 1960, *J. Am. Chem. Soc.* **82**, 3787.
[16] Goodman, M. and Rosen, I.: 1964, *Biopolymers* **2**, 537.
[17] Goodman, M., Verdini, A. S., Toniolo, C., Phillips, W. D., and Bovey, F. A.: 1969, *Proc. Nat. Acad. Sci. US* **64**, 444.
[18] Goodman, M., Toniolo, C., and Verdini, A. S.: 1971, in *Peptides 1969* (ed. by E. Scoffone), North-Holland Publ. Co., Amsterdam, p. 207.
[19] Temussi, P. A. and Goodman, M.: 1971, *Proc. Nat. Acad. Sci. US* **68**, 1767.
[20] Blout, E. R. and Shechter, E.: 1963, *Biopolymers* **1**, 565.
[21] Shields, J. E., McDowell, S. T., Pavlos, J., and Ray, G. R.: 1968, *J. Am. Chem. Soc.* **90**, 3549.
[22] Widmer, U. and Lorenzi, G. P.: 1971, *Chimia* **25**, 236.
[23] Toniolo, C.: 1970, *Chim. Ind. (Milan)* **52**, 1039.
[24] Toniolo, C.: 1971, *Biopolymers* **10**, 1707.
[25] Goodman, M., Naider, F., and Toniolo, C.: 1971, *Biopolymers* **10**, 1719.
[26] Gratzer, W. B.: 1967, in *Poly-α-Amino Acids* (ed. by G. D. Fasman), Vol. 1, M. Dekker, New York, p. 177.
[27] Beychok, S.: *ibid.*, p. 293.
[28] Townend, R., Kumosinski, T. F., Timasheff, S. N., Fasman, G. D., and Davidson, B.: 1966, *Biochem. Biophys. Res. Commun.* **23**, 163.
[29] Fraser, R. D. B., Harrap, B. S., Mac Rae, T. P., Stewart, F. H., and Suzuki, E.: 1965, *J. Mol. Biol.* **12**, 482.
[30] Goodman, M., Naider, F., and Rupp, R.: 1971, *Bioorganic Chem.* **1**, 310.
[31] Nemethy, G.: 1967, *Angew. Chem. Intern. Edit.* **6**, 195.
[32] Auer, H. E. and Doty, P.: 1966, *Biochemistry* **5**, 1716.

[33] Fasman, G. D.: 1967, *Poly-α-Amino Acids* (ed. by G. D. Fasman), Vol. 1, M. Dekker, New York, p. 499, and references therein.
[34] Peggion, E., Verdini, A. S., Cosani, A., and Scoffone, E.: 1969, *Macromolecules* **2**, 170.
[35] Sage, H. J. and Fasman, G. D.: 1966, *Biochemistry* **5**, 286.
[36] Engel, J., Liehl, E., and Sorg, C.: 1971, *Eur. J. Biochem.* **21**, 22.
[37] Urry, D. W.: 1968, *Ann. Rev. Phys. Chem.* **19**, 477.
[38] Simmons, N. S., Barel, A. O., and Glazer, A. N.: 1969, *Biopolymers* **7**, 275.
[39] Horwitz, J., Strickland, E. H., and Billups, C.: 1969, *J. Am. Chem. Soc.* **91**, 184.
[40] Bayley, C. R., Ingold, C. K., Poole, H. G., and Wilson, C. L.: 1946, *J. Chem. Soc.* **A**, 222.
[41] Lapp, C. and Laustriat, G.: 1971, *J. Chim. Phys.* (*France*) **68**, 1333.
[42] Sponer, H.: 1942, *J. Chem. Phys.* **10**, 672.
[43] Toniolo, C. and Bonora, G. M.: in preparation.
[44] Goodman, M., Davis, G. W., and Benedetti, E.: 1968, *Accts. Chem. Res.* **1**, 275, and references therein.
[45] Moscowitz, A., Rosenberg, A., and Hansen, A. E.: 1965, *J. Am. Chem. Soc.* **87**, 1813.
[46] Goodman, M. and Toniolo, C.: 1968, *Biopolymers* **6**, 1673.
[47] Toniolo, C., Peggion, E., Bonora, G. M., and Palumbo, M.: *Bioorganic Chem.*, submitted for publication.

DISCUSSION

Shulman: In your first peptide series why do you say that the change of NMR shifts with chain length comes from a different average between two states, rather than a dependence of the shift upon the chain length in one conformation?

Toniolo: We have interpreted the progressive shifts to lower field of both N-terminal and C-terminal N–H resonances as a function of the number of amino acid residues in the γ-ethyl-L-glutamate oligopeptide chain as arising from the interconversion of folded and extended forms for the oligopeptides. The possibility of different interpretations of this experimental finding and some evidences against them have been discussed in the paper of Temussi and Goodman (*Proc. Nat. Acad. Sci. US* **68** (1971) 1767).

ISOTOPIC HYDROGEN EXCHANGE IN SOLUTIONS OF BIOLOGICAL MACROMOLECULES

AASE HVIDT

Chemistry Laboratory III, H. C. Ørsted Institute, Universitetsparken 5, DK 2100 Copenhagen Ø, Denmark

Summary. In the isotope exchange experiments discussed the substance to be studied is dissolved in water with a different isotopic composition of the hydrogen atoms, and the approach of the solution towards isotopic equilibrium (the 'relaxation') is observed. The measured relaxation times of the isotopic exchange of peptide hydrogen atoms in protein solutions are much longer than the corresponding relaxation times in solutions of randomly coiled polypeptides. The interpretation of the kinetics of the slow exchange in protein solutions is discussed in terms of conformational properties of the proteins.

Résumé. Dans les expériences d'échange isotopique discutées la substance à étudier est dissoute dans une eau dont l'hydrogène a une composition isotopique différente, et l'on suit l'évolution de la solution vers l'équilibre isotopique ('relaxation'). Pour les atomes d'hydrogène peptidiques de solutions protéiques, les temps de relaxation mesurés pour l'échange isotopique sont très supérieurs à ceux des solutions de polypeptides en pelote statistique. L'interprétation de la cinétique de cet échange lent en solution protéique est discutée en fonction des propriétés conformationnelles des protéines.

1. Introduction

Systematic investigations of the kinetics of the exchange of hydrogen isotopes in aqueous protein solutions were initiated in the early fifties by Linderstrøm-Lang [1, 2], and during the past two decades this field of research has acquired an additional, and steadily increasing interest. Extensive studies have confirmed Linderstrøm-Lang's original observation that a part of the hydrogen atoms in protein molecules exchange with hydrogen from the solvent more slowly than the corresponding hydrogen atoms in low molecular weight compounds or in randomly coiled macromolecules. The generally accepted explanation of this phenomenon is that the slowly exchanging hydrogen atoms are located in regions of the protein molecules which are not freely exposed to the solvent. Measurements of the exchange rates, using isotopes, can thus be used as a tool to test out the accessibility of the interior of protein molecules.

Several experimental methods have been developed for the measurement of the exchange of hydrogen isotopes in aqueous solution. The general principles underlying these methods are described in the following. For details of the various experimental techniques the reader is referred to available review articles [3, 4, 5] and to the references therein.

2. The Hydrogen Isotope Exchange Reaction

The hydrogen exchange experiments discussed here are initiated by the dissolution of a substance having a certain isotopic composition of hydrogen atoms in water

C. Sadron (ed.), Dynamic Aspects of Conformation Changes in Biological Macromolecules. 103–115.

with a different isctopic composition. The approach of the system to isotopic equilibrium between the solute and the solvent is then observed.

The isotopic exchange of the hydrogen atom of a solvent-exposed group P may be described by the following reaction scheme

$$\mathrm{PH} + \mathrm{SH}^* \underset{k^-}{\overset{k^+}{\rightleftharpoons}} \mathrm{PH}^* + \mathrm{SH}\,, \tag{1}$$

where S is the solvent, H and H* are different hydrogen isotopes (protium, deuterium or tritium), and k_+ and k_- are second-order rate constants.

The time dependence of the concentrations of the reactants in reaction scheme (1) is conveniently expressed in terms of the deviation, Δ, of the actual concentrations from those at equilibrium:

$$\begin{aligned}
[\mathrm{PH}] &= [\mathrm{PH}]_{\mathrm{eq}} + \Delta \\
[\mathrm{SH}^*] &= [\mathrm{SH}^*]_{\mathrm{eq}} + \Delta \\
[\mathrm{PH}^*] &= [\mathrm{PH}^*]_{\mathrm{eq}} - \Delta \\
[\mathrm{SH}] &= [\mathrm{SH}]_{\mathrm{eq}} - \Delta
\end{aligned} \tag{2}$$

$$\begin{aligned}
\mathrm{d}\Delta/\mathrm{d}t &= - k_+ [\mathrm{PH}]\,[\mathrm{SH}^*] + k_- [\mathrm{PH}^*]\,[\mathrm{SH}] = \\
&= - \{k_+ ([\mathrm{PH}]_{\mathrm{eq}} + [\mathrm{SH}^*]_{\mathrm{eq}}) + k_- ([\mathrm{PH}^*]_{\mathrm{eq}} + [\mathrm{SH}]_{\mathrm{eq}})\}\, \Delta - \\
&\quad - (k_+ - k_-)\, \Delta^2 .
\end{aligned}$$

For *dilute* solutions we have $[\mathrm{PH}]+[\mathrm{PH}^*] \ll [\mathrm{SH}]+[\mathrm{SH}^*]$, and $\Delta \ll [\mathrm{SH}^*]+[\mathrm{SH}]$, so

$$\begin{aligned}
\mathrm{d}\Delta/\mathrm{d}t &\cong - \{k_+ [\mathrm{SH}^*]_{\mathrm{eq}} + k_- [\mathrm{SH}]_{\mathrm{eq}}\}\, \Delta = \\
&= - k\Delta \\
\Delta &\cong \Delta_0 \exp(-kt) = \Delta_0 \exp(-t/\tau) .
\end{aligned} \tag{3}$$

In Equation (3) Δ_0 is the value of Δ at time zero, and

$$k = k_+ [\mathrm{SH}^*]_{\mathrm{eq}} + k_- [\mathrm{SH}]_{\mathrm{eq}}$$

is a pseudo first-order rate constant. In most cases one of the hydrogen isotopes is present in great excess, so that one of the terms in the expression for k is vanishingly small.

For dilute solutions, in which all groups of the solute are exposed to the solvent, the equilibration is thus a 'relaxation' process, characterized by a spectrum of first-order rate constants k, or of time constants ('relaxation times') $\tau = k^{-1}$.

It is now assumed that the groups of a solute capable of exchange can be classified into one of two classes, depending on whether or not they are, at the time in question, freely exposed to the solvent. The exchange takes place in the solvent exposed states exclusively, and it follows first order kinetics.

The range of relaxation times which can be measured by the isotope exchange methods depends on the time of observation of the experiments, and on the nature of the substances studied. In all cases it falls in the range 10^2–10^5 sec. This restricts the discussion almost exclusively to hydrogen atoms located in peptide groups [3].

3. Hydrogen Exchange in Solutions of Polypeptides and Proteins

3.1. Randomly coiled, freely drained polypeptides

Investigations of solutions of low molecular weight peptides [6–9], and of randomly coiled polypeptides [9–13] have demonstrated that the exchange of hydrogen atoms of peptide groups is catalyzed by protons and by hydroxyl ions [14]. The exchange rates depend on the chemical nature and the electric charge of the substituents near the peptide groups, but for peptide groups in polypeptide chains this influence is found to be relatively weak. To a good approximation the exchange of the presumably solvent-exposed peptide hydrogen atoms in solutions of randomly coiled polypeptide chains follows simple first-order kinetics:

$$\Delta = \Delta_0 \exp(-k_0 t).$$

In dependence of pH the rate constant k_0 at θ°C is expressed by the following formula [3, 13, 14]:

$$k_0 \cong (10^{-\mathrm{pH}} + 10^{\mathrm{pH}-6})\, 10^{(\theta-25)\,0.05}\ \mathrm{sec}^{-1}. \tag{4}$$

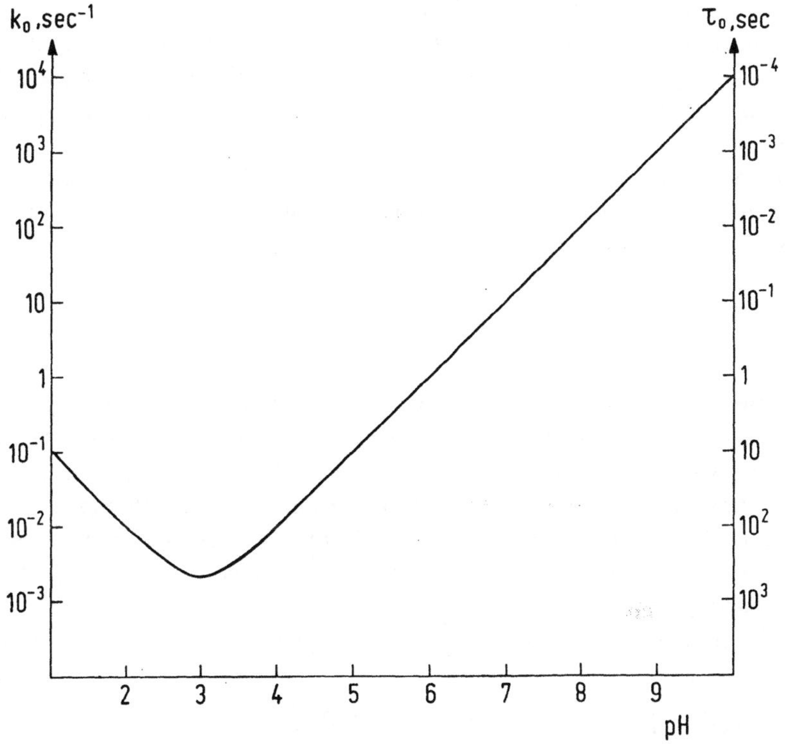

Fig. 1. pH dependence of the rate constant k_0, or the relaxation time τ_0, of the exchange of solvent-exposed peptide groups at 25°C, plotted according to Equation (4).

A graphical representation of k_0, or of $\tau_0 = k_0^{-1}$, at 25°C, as a function of pH, is shown on Figure 1.

The hydrogen exchange methods here discussed permit only estimates of k_0 at values of pH near pH 3, where the exchange rates are minimal. Formula (4) is based on the assumption, supported by NMR investigations [15], that the catalytic effect of protons and hydroxyl ions on the exchange reaction persists at values of pH beyond the range in which the isotopic exchange rates have been directly measured.

3.2. Folded Polypeptides and Proteins

A. *Theoretical*

The exchange rates of peptide hydrogen atoms in aqueous solutions of folded polypeptides or proteins are lower than the exchange rates in solutions of randomly coiled polypeptides. The slow exchange indicates that some of the peptide groups in these solutions are protected from exposure to the solvent, and, thereby, hindered in exchanging their hydrogen atoms.

In order to give an interpretation at the molecular level of the kinetics of the slow exchange reactions it is assumed that macromolecules in solution fluctuate spontaneously between different conformations. In some of these conformations a given peptide group is protected by other parts of its molecule from exposure to the solvent, and therefore prevented from exchange of its hydrogen atom. In other conformations the peptide group considered is freely exposed to the solvent, and the probability of the exchange of its hydrogen atom expressed by the rate constant k_0, given by Equation (4).

The exchange mechanism outlined is described by the following reaction scheme for the isotopic exchange of the hydrogen atom of a given, the ith, peptide group of a polypeptide or protein:

$$\mathrm{PH}_\mathrm{u}^{(i)} \underset{k_-^{(i)}}{\overset{k_+^{(i)}}{\rightleftharpoons}} \mathrm{PH}_\mathrm{e}^{(i)} \underset{\text{exchange}}{\overset{k_0}{\rightharpoonup}} \mathrm{PH}^{*(i)}. \qquad (5)$$

$\mathrm{PH}_\mathrm{u}^{(i)}$ denotes polypeptide or protein molecules in which the ith peptide group is unexposed to the solvent, and has not yet exchanged its hydrogen isotope. $\mathrm{PH}_\mathrm{e}^{(i)}$ denotes molecules in which this same group is exposed to the solvent, its hydrogen isotope still being unexchanged. $\mathrm{PH}^{*(i)}$ denotes molecules in which the ith peptide group has exchanged its hydrogen isotope. $k_+^{(i)}$ and $k_-^{(i)}$ are first-order rate constants, typical of the ith peptide group, and k_0 is the exchange rate constant of solvent exposed peptide groups, given by Equation (4).

The differential equations expressing the change with time of the concentration of the reactants in reaction scheme (5)

$$\begin{aligned}
\mathrm{d}[\mathrm{PH}_\mathrm{u}^{(i)}]/\mathrm{d}t &= -k_+^{(i)}[\mathrm{PH}_\mathrm{u}^{(i)}] + k_-^{(i)}[\mathrm{PH}_\mathrm{e}^{(i)}]\\
\mathrm{d}[\mathrm{PH}_\mathrm{e}^{(i)}]/\mathrm{d}t &= k_+^{(i)}[\mathrm{PH}_\mathrm{u}^{(i)}] - (k_-^{(i)} + k_0)[\mathrm{PH}_\mathrm{e}^{(i)}]\\
\mathrm{d}[\mathrm{PH}^{*(i)}]/\mathrm{d}t &= k_0\,[\mathrm{PH}_\mathrm{e}^{(i)}]
\end{aligned}$$

lead to the following expression for the time dependence of the fraction, $x^{(i)}$, of molecules, which have not exchanged the hydrogen isotope of the ith peptide group (14):

$$x^{(i)} = x_1^{(i)} \exp(-\beta_1^{(i)} t) + x_2^{(i)} \exp(-\beta_2^{(i)} t), \tag{6}$$

where

$$\begin{aligned}
\beta_1^{(i)} &= \tfrac{1}{2}\{k_+^{(i)} + k_-^{(i)} + k_0 + [(k_+^{(i)} + k_-^{(i)} + k_0)^2 - 4k_+^{(i)}k_0]^{1/2}\}\\
\beta_2^{(i)} &= \tfrac{1}{2}\{k_+^{(i)} + k_-^{(i)} + k_0 - [(k_+^{(i)} + k_-^{(i)} + k_0)^2 - 4k_+^{(i)}k_0]^{1/2}\}\\
x_1^{(i)} &= \frac{k_0([\mathrm{PH}_e^{(i)}(0)]/[\mathrm{P}]) - \beta_2^{(i)}}{\beta_1^{(i)} - \beta_2^{(i)}}\\
x_2^{(i)} &= \frac{\beta_1^{(i)} - k_0([\mathrm{PH}_e^{(i)}(0)]/[\mathrm{P}])}{\beta_1^{(i)} - \beta_2^{(i)}}.
\end{aligned} \tag{7}$$

In deriving expression (7), [P] is the total polypeptide or protein concentration, and the following initial conditions are assumed: $[\mathrm{PH}_e^{(i)}] = [\mathrm{PH}_e^{(i)}(0)]$, and $[\mathrm{PH}^{*(i)}] = 0$, at $t = 0$. If there is equilibrium at time zero between $\mathrm{PH}_e^{(i)}$ and $\mathrm{PH}_u^{(i)}$ conformations, we have $[\mathrm{PH}_e^{(i)}(0)]/[\mathrm{P}] = k_+^{(i)}/(k_+^{(i)} + k_-^{(i)})$, so that the time course of $x^{(i)}$ (Equation (6)) is completely specified by the rate constants $k_+^{(i)}$, $k_-^{(i)}$ and k_0.

The total exchange of peptide hydrogen atoms in a solution is considered to be a sum over i of terms like (6). If n is the number of peptide groups per macromolecule, the fraction of unexchanged peptide hydrogen atoms, at time t, is expressed by

$$x = n^{-1} \sum_i [x_1^{(i)} \exp(-\beta_1^{(i)} t) + x_2^{(i)} \exp(-\beta_2^{(i)} t)]. \tag{8}$$

Two extreme cases are of immediate interest:

$$k_+^{(i)} + k_-^{(i)} \ll k_0, \tag{9a}$$

$$k_+^{(i)} + k_-^{(i)} \gg k_0, \tag{9b}$$

for which the expressions for $x_1^{(i)}$, $x_2^{(i)}$, $\beta_1^{(i)}$, and $\beta_2^{(i)}$, in terms of the parameters $k_+^{(i)}$, $k_-^{(i)}$ and k_0, are summarized in Table I.

(a) In this case the conformational motility of the macromolecules, here expressed

TABLE I

xpressions for the parameters (7), calculated for the cases (9a, b). The expressions given in parentheses apply when conformational equilibrium exists at time zero (see text)

	$x_1^{(i)}$	$x_2^{(i)}$	$\beta_1^{(i)}$	$\beta_2^{(i)}$
.) $k_+^{(i)} + k_-^{(i)} \ll k_0$	$[\mathrm{PH}_e^{(i)}(0)]/[\mathrm{P}]$ $(= k_+^{(i)}/(k_+^{(i)} + k_-^{(i)}))$	$1 - [\mathrm{PH}_e^{(i)}(0)]/[\mathrm{P}]$ $(= k_-^{(i)}/(k_+^{(i)} + k_-^{(i)}))$	k_0	$k_+^{(i)}$
) $k_+^{(i)} + k_-^{(i)} \gg k_0$	0	1	$k_+^{(i)} + k_-^{(i)}$	$k_0 k_+^{(i)}/(k_+^{(i)} + k_-^{(i)})$

by $k_+^{(i)} + k_-^{(i)}$, is low when compared with the rate constant of exchange of solvent-exposed peptide groups. Any peptide group which, at time zero or subsequently, becomes exposed to the solvent, completes its isotopic exchange before it possibly – by a subsequent conformational transition of the macromolecule – gets buried in the interior of the conformation. The rate of exchange of peptide groups, which at time zero are unexposed to the solvent, is determined by the rate constants $k_+^{(i)}$ of the conformational transitions in which these groups become exposed to the solvent. The exchange rate will be *independent of pH* if the $k_+^{(i)}$'s are independent of pH, i.e., if no change of the macromolecular conformation occurs as a function of pH.

(b) In this case the rate of the conformational transitions of the macromolecule is high as compared with k_0. The exchange of the ith peptide group is characterized by the overall rate constant $\beta^{(i)} = k_0 k_+^{(i)} / (k_+^{(i)} + k_-^{(i)})$, which is proportional to k_0, and to the fraction $\varrho^{(i)} = k_+^{(i)} / (k_+^{(i)} + k_-^{(i)})$ of the molecules, whose ith peptide group, at a given time, is exposed to the solvent. $\beta^{(i)}$ *depends on pH* through k_0, and at pH>4 the exchange rates increase by a factor of 10 per pH unit in cases where the conformational properties of the macromolecules do not change with pH (see Equation (4)).

Measurements of x as a function of time can always be analyzed, in accordance with Equation (8), as a sum of first-order reactions, leading to estimates of exchange rate constants for all or part of the peptide groups of the macromolecules, and it is of crucial importance for the interpretation of the rate constants estimated to be able to distinguish between cases (a) and (b). In case (a) the exchange rate constants reflect *dynamic* aspects of the macromolecules, namely the frequencies, $k_+^{(i)}$, of the conformational transitions by which the interior regions become exposed to the solvent. In case (b) the exchange rate constants, or rather the ratios $\beta^{(i)}/k_0$, measure *thermodynamic* features of the macromolecules, expressed by the equilibrium probability, $\varrho^{(i)}$, of finding the peptide groups in solvent-exposed states

$$\varrho^{(i)} = k_+^{(i)} / (k_+^{(i)} + k_-^{(i)}),$$

by the equilibrium constant $K^{(i)}$

$$K^{(i)} = k_+^{(i)} / k_-^{(i)} = \varrho^{(i)} / (1 - \varrho^{(i)}),$$

or by the change in standard free energy, $\Delta G^{0\,(i)}$, of the conformational transitions by which the peptide groups become exposed to the solvent

$$\Delta G^{0(i)} = -RT \ln K^{(i)}.$$

Studies of the pH dependence of the exchange rates should, in principle, enable us to distinguish between the cases (a) and (b). The sensitivity of most polypeptide and protein conformations to changes in pH complicates the situation, but existing data present overwhelming evidence [5] that the exchange is generally catalyzed by protons and by hydroxyl ions, so that case (b), the so-called EX_2 mechanism [3], prevails in solutions of polypeptides or proteins. Other reaction mechanisms have been suggested [16, 17, 18] and discussed [2, 10, 19, 21, 22], but the pH dependence of the exchange rates bears strong evidence in favour of the EX_2 mechanism, and there are no experimental results known to the author which are incompatible with this mechanism. In the following discussion of available experimental data on the exchange in solutions of polypeptides and proteins it is assumed that the exchange occurs in accordance with case (9b), the EX_2 mechanism.

It is a consequence of the EX_2 mechanism (see Table I, and (9b)) that

$$k_+^{(i)} \gg \beta^{(i)} \quad (> 10^{-3}\,\mathrm{sec}^{-1}). \tag{10}$$

The *measurable* peptide groups, and that is the majority, if not all the peptide groups of polypeptide and protein molecules in solution, are exposed to the solvent in conformational transitions of the macromolecules occurring at frequencies much higher than $10^{-3}\,\mathrm{sec}^{-1}$. The equilibrium probability of the fluctuations, $\varrho^{(i)}$, can be calculated from the exchange rate constants measured

$$\varrho^{(i)} = \beta^{(i)}/k_0 .$$

Since it is the ratios $\beta^{(i)}/k_0$ rather than the absolute values of the constants $\beta^{(i)}$, which are characteristic for the polypeptides and proteins studied, it is preferable [5, 22] to present kinetic measurements of the exchange as plots of x (or $\ln x$) versus the dimensionless parameter $(k_0 t) = t/\tau_0$, i.e., to measure time relative to the hydrogen exchange relaxation time of solvent-exposed peptide groups:

$$x = n^{-1} \sum_i \exp(-\beta^{(i)} t) = n^{-1} \sum_i \exp(-\varrho^{(i)} (k_0 t)) \tag{11}$$

$$\mathrm{d}\ln x/\mathrm{d}(k_0 t) = -(xn)^{-1} \sum_i \varrho^{(i)} \exp(-\varrho^{(i)} (k_0 t)) = -\langle \varrho \rangle_x . \tag{12}$$

In a plot of $\ln x$ versus $k_0 t$ the slope of the rate curve, for any value of x, is the average value of ϱ, $\langle \varrho \rangle_x$, for the unexchanged peptide groups. In particular the initial slope (i.e., the slope for $x=1$, $t=0$) is the average value of ϱ for all peptide groups present in the macromolecule.

The change of independent variable from t to $(k_0 t)$ has the advantage that the rate curves obtained are described by parameters $\varrho^{(i)}$ typical of the substances studied, rather than of the exchange reaction, so that measurements of the exchange, made at different values of pH and temperature, become directly comparable in terms of conformational properties of the polypeptides and proteins studied.

B. *Experimental Results on Polypeptides*

The kinetics of the H-T exchange in aqueous solutions of the homopolymer polyglutamic acid (PG), and of copolymers of glutamic acid and alanine (PGA) was studied by Ikegami and co-workers [11, 12]. The exchange was measured at various values of pH in the range pH 3–7, where a helix-coil transition of the polymers is observed.

Results obtained for a sample of polyglutamic acid, with the molecular weight $M = 27000$, were interpreted in accordance with the reaction scheme (5), assuming case (9b) (the EX_2 mechanism) to apply [11]. This assumption is reasonable since the local conformation in which the peptide group hydrogen was found to be unexchangeable (the PH_u conformation) could be identified as the helix conformation, and it is known from temperature-jump studies [23] that the halftime of the helix-coil relaxation is below 10^{-5} sec, i.e., much smaller than the relaxation times measured of the exchange reaction. The helical content of the polypeptides, from measurements of optical rotatory dispersion and UV absorption, was found to agree satisfactorily with values of $(1-\langle\varrho\rangle)$, estimated according to Equation (12) from the initial slope of the exchange rate curves [11].

At low values of the helical content of the polypeptide the measured exchange rates follow simple first-order kinetics, but at high helical content deviations from first-order kinetics are observed as shown in Figure 2 by a plot on $\ln x$ versus $(k_0 t)$ for polyglutamic acid at pH 4.79, 20°C. $\langle\varrho\rangle$, determined from the initial slope of this plot, is of the order of 10^{-2} so that, under the conditions of the exchange experiment, the helical content of the polypeptide is practically 100%.

The observed deviations from simple first-order kinetics are in qualitative accordance with the results of a recent theoretical analysis by Miller [24] of the exchange rates of helical polypeptides. Due to the cooperative nature of the helix formation, the random coil probability, $\varrho^{(i)}$, of a peptide group of a helical polypeptide is a function of the chain position, i, and deviations from first-order kinetics are therefore to be expected, even for homopolypeptides [24].

More pronounced deviations from first-order kinetics, and slower exchange rates, are observed for the series of copolymers of L-glutamic acid and L-alanine, studied by Ikegami and Kono [12]. Figure 3 presents data on one of these copolymers, for

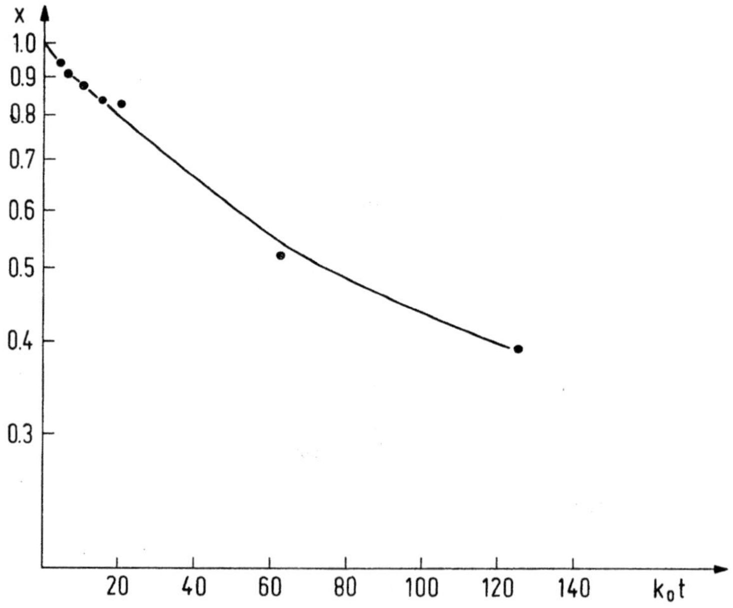

Fig. 2. The fraction, x, of unexchanged peptide hydrogen atoms in aqueous solutions of polyglutamic acid, pH 4.79, 20°C, as a function of $(k_0 t)$. The data are replotted from Figure 3 in [11].

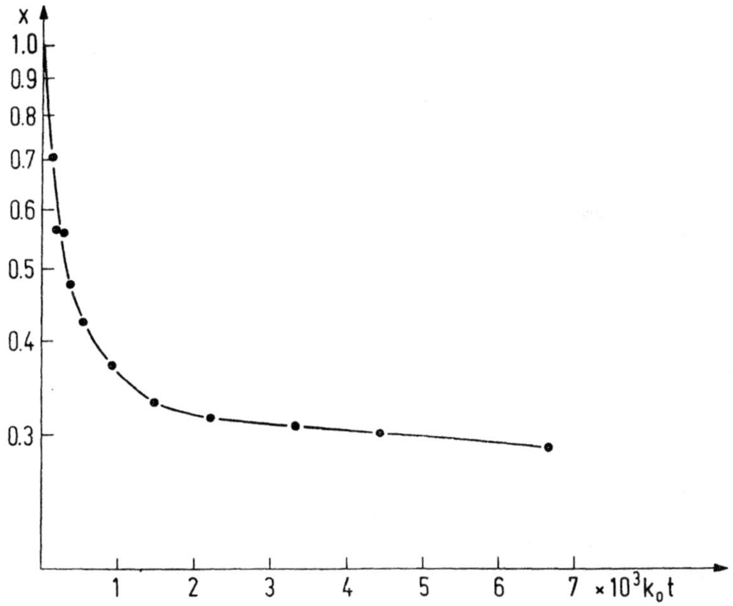

Fig. 3. The fraction, x, of unexchanged peptide hydrogen atoms in aqueous solutions of a copolymer of L-glutamic acid and L-alanine, pH 6.04, 20°C, as a function of (k_0t). The data are replotted from Figure 15 in [12].

which Glu:Ala = 30:70, M = 35000, at pH 6.04, 20°C. The initial slope of the rate curve corresponds to a ϱ value of the order of 10^{-3}, while the final slope indicates that for about one third of the peptide groups the value of ϱ is, on average, of the order of 10^{-5}. The most slowly exchanging hydrogen atoms are, most likely, located in the midpart of the helices, where the probability of exposure to the solvent is minimal, but superhelical conformations may contribute to the shielding of the peptide groups as well. The high content of nonpolar sidechains of the polypeptide suggests that conformations, similar to those of the globular proteins, may be formed, in which the nonpolar groups are located in interior regions of the polypeptide matrix.

C. *Experimental Results on Proteins*

The relaxation times of the exchange of peptide hydrogen atoms in aqueous protein solutions most often cover a range of eight or more decades [5]. The accessible range of relaxation times in a protein solution is hardly more than two decades, say, $5 \times 10^2 - 5 \times 10^4$ sec, so for given conditions of pH and temperature only a minor fraction of the peptide groups in a protein solution exchange their hydrogen atom at measurable rates. The exchange rates vary, however, with pH in accordance with the EX_2 reaction mechanism, case (9b). Peptide hydrogen atoms which remain unexchanged after a period of 20 hours at pH 3, may exchange at measurable rates at pH 6, and exchange at rates too fast to be measured at pH 9. By varying pH we may, so to speak, scan the peptide groups of a protein solution.

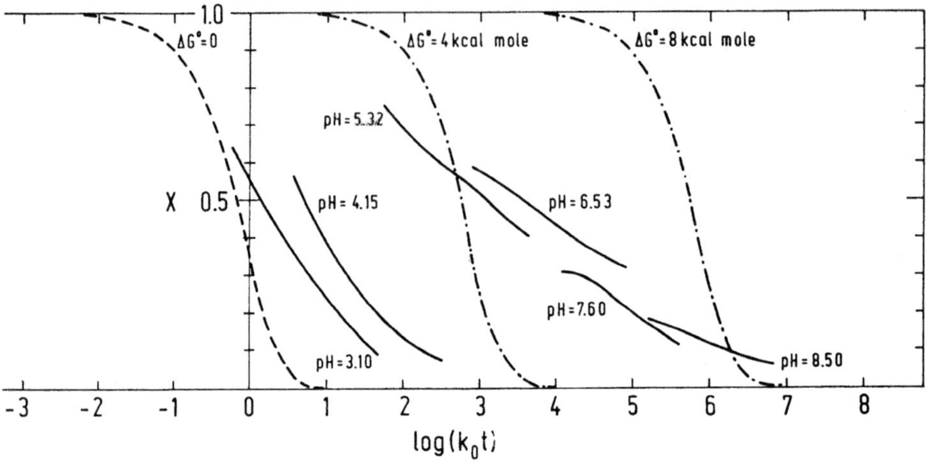

Fig. 4. The fraction, x, of unexchanged peptide hydrogen atoms in aqueous solutions of serum albumin, 25°C, as a function of $\log(k_0t)$. The broken and the dotted curves are calculated for hypothetical polypeptides, for which the change in standard free energy ΔG^0, following the exposure of the peptide groups to the solvent, is 0, 4, and 8 kcal/mole, respectively. The data are replotted from Figure 2b in [22].

Plots of x versus $\log(k_0t)$ are convenient for a comparison of data for proteins obtained under various conditions of pH and temperature [5, 22]. In this representation a simple first-order rate curve has the S-like shape shown by the broken curves in Figure 4. The corresponding value of the equilibrium probability, ϱ, of finding the peptide groups in a solvent-exposed state can be identified as the reciprocal of the value of (k_0t) for which the first-order rate curve has an inflexion point, i.e., where $x=e^{-1}$ [22].

When no shift of the conformational equilibrium of a protein occurs as a function of pH, the fragments of the rate curves, measured at a series of values of pH, can be pieced together in an x versus $\log(k_0t)$ plot as one single curve, a 'relaxation spectrum', characteristic of the protein. This has been the case, for example, for lysozyme at pH 3.2–8.7, 0°C, for carbonic anhydrase, pH 5–9.4, 4°C, for trypsin, pH 2.8–6.7, 23°C, and for chymotrypsinogen pD 2.1–11.2, 21°C, all summarized, by Willumsen, in a recent review article [5].

The rate curves of the various peptide groups present in protein solutions overlap to such an extent that a detailed analysis of the rate curve of a protein solution, in terms of the individual values of $\varrho^{(i)}$, is impossible. It has been shown, however [5, 22], that in plots of x versus $\log(k_0t)$ the area between the rate curve of a randomly coiled polypeptide (the broken curve, labelled $\Delta G^0=0$, in Figure 4) and that of a protein, equals $\langle \Delta G^0 \rangle/2.30RT$. $\langle G^0 \rangle$ is the average value of the change in standard free energy of the transitions by which the peptide groups are exposed to the solvent,

$$\langle \Delta G^0 \rangle = n^{-1} \sum_i \Delta G^{0(i)}.$$

The shape of the rate curves illustrates the distribution of the values of ΔG^0 of the individual peptide groups [5, 22].

Values of $\langle \Delta G^0 \rangle$ for a series of 16 proteins are reported in [5]. They vary from about 3 kcal/mole (for lysozyme at 0 °C) to about 7 kcal/mole (for metmyoglobin at 4 °C). Attempts to correlate the values of $\langle \Delta G^0 \rangle$ with the dimensions of the protein molecules have lead to the conclusion that no simple relationship seems to exist between the value of $\langle \Delta G^0 \rangle$ and the volume or the surface of a protein molecule [5].

The hydrogen exchange technique has been utilized recently [22] to study the thermodynamics of the conformational changes of serum albumin, known to occur as a function of pH. The exchange has been measured in aqueous solutions of this protein at a series of values of pH from pH 3 to 8.5, 25 °C. The fragments of the exchange-rate curves, obtained under different conditions of pH, show that the protein protects its peptide groups most efficiently at pH values near pH 6.5, as is demonstrated by the data reported in Figure 4. The broken curves on this figure are first-order rate curves calculated for hypothetical polypeptides for which $\Delta G^{0(i)} = \Delta G^0 = 0$, 4 and 8 kcal/mole, respectively. It is seen from Figure 4 that at pH 6.53 60% of the peptide groups of serum albumin are protected from the solvent by conformations stabilized by more than 4 kcal/mole. The unfolding of the protein in the acid range [25] is reflected by a displacement (with decreasing pH) of the rate curves to the left in Figure 4. At pH 3.1 less than 10% of the peptide groups of the protein are protected by conformations stabilized by more than 2 kcal/mole.

4. Concluding Remarks

The possibilities involved in studies of the kinetics of chemical reactions for characterization of macromolecules have been discussed by Morawetz [26]. We wish here to point out a few typical features of the exchange of hydrogen isotopes which form the basis of the choice of this particular reaction for studies of polypeptide and protein conformations.

The isotopic exchange distinguishes itself by causing at most only minor changes of the macromolecular conformations. While other, in principle similar reactions, such as the association/dissociation of protons, or the oxidation/reduction of sulphur groups [26] may bring about rather drastic conformational changes, the exchange of hydrogen isotopes can be studied while maintaining the conformation of the macromolecules practically unchanged. In the next place, the ubiquitous presence of peptide groups in all regions of a polypeptide or protein molecule, together with the suitable range of the exchange rate constant of the peptide hydrogen atom, open up the possibility of getting information about the stability of local conformations of the macromolecular matrix.

Measurements of properties like optical rotation or circular dichroism have lead to estimates of the average value of the helical content of polypeptides and proteins in solution, and X-ray crystallography has contributed the detailed structure of proteins in the crystalline state. It is the supplementary contribution to our knowledge about macromolecular conformations from hydrogen exchange studies, that local fluctuations of the conformations occur in solution by which the, on the average, interior regions of the molecules become exposed to the solvent. Experimental data on the exchange suggests that the frequencies of these fluctuations are much larger than 10^{-3} sec^{-1}, and that the probability of finding the peptide groups of protein molecules in solvent-exposed states vary from practically 1, for peptide groups located at the surface of the molecules, to 10^{-8} or less for the most deeply buried groups.

References

[1] Linderstrøm-Lang, K.: 1955, *Chem. Soc. (London)*, Spec. Publ. **2**, 1.
[2] Linderstrøm-Lang, K: 1958, in *Symposium on Protein Structure* (ed. by A. Neuberg), Methuen, London, p. 23.
[3] Hvidt, Aa. and Nielsen, S. O.: 1966, *Advan. Prot. Chem.* **21**, 287.
[4] Ottesen, M.: 1971, in *Methods of Biochemical Analysis* (ed. by D. Glick), Vol. 20, Wiley-Interscience, p. 135.
[5] Willumsen, L.: 1971, *C.R. Trav. Lab. Carlsberg* **38**, 14, 223.
[6] Nielsen, S. O., Bryan, W. P., and Mikkelsen, K.: 1960, *Biochim. Biophys. Acta* **42**, 550.
[7] Nielsen, S. O.: 1960, *Biochim. Biophys. Acta* **37**, 146.
[8] Klotz, I. M. and Frank, B. H.: 1965, *J. Am. Chem. Soc.* **87**, 2721.
[9] Molday, R. S., Englander, S. W., and Kallen, R. G.: 1972, *Biochemistry* **11**, 150.
[10] Bryan, W. P. and Nielsen, S. O.: 1960, *Biochim. Biophys. Acta* **42**, 552.
[11] Ikegami, A., Yamamoto, S., and Oosawa, F.: 1965, *Biopolymers* **3**, 555.
[12] Ikegami, A. and Kono, N.: 1967, *J. Mol. Biol.* **29**, 251.
[13] Englander, S. W. and Poulsen, A.: 1969, *Biopolymers* **7**, 379.
[14] Hvidt, Aa.: 1964, *C.R. Trav. Lab. Carlsberg* **34**, 299.
[15] Berger, A., Loewenstein, A., and Meiboom, S.: 1959, *J. Am. Chem. Soc.* **81**, 62.
[16] Rosenberg, A. and Chakravarti, K.: 1968, *J. Biol. Chem.* **243**, 5193.
[17] Woodward, C. K. and Rosenberg, A.: 1971, *J. Biol. Chem.* **246**, 4105.
[18] Woodward, C. K. and Rosenberg, A.: 1971, *J. Biol. Chem.* **246**, 4114.
[19] Benson, E. S. and Hallaway, E.: 1970, *J. Biol. Chem.* **245**, 4144.
[20] Bryan, W. P.: 1970, in *Recent Progress in Surface Science* (ed. by J. F. Danielli), Academic Press, New York, p. 101.
[21] Tanford, C.: 1970, *Advan. Prot. Chem.* **24**, 1.
[22] Hvidt, Aa. and Wallewik, K.: 1972, *J. Biol. Chem.* **247**, 1530.
[23] Lumry, R., Legare, R., and Miller, W. G.: 1964, *Biopolymers* **2**, 489.
[24] Miller, W. G.: 1970, *Biochemistry* **9**, 4921.
[25] Foster, J. F.: 1960, in *The Plasma Proteins* (ed. by F. W. Putnam), Vol. 1, Academic Press, New York, p. 179.
[26] Morawetz, H.: 1965, in *Macromolecules in Solution*, Interscience, New York, Ch. IX.

DISCUSSION

Goldberg: Could you briefly describe the technics involved in proton exchange measurement and the lowest time range observable with these technics?

Mme Hvidt: The H–D exchange in D_2O solutions can be measured by IR-spectroscopy. The CONH– group has a characteristic absorption band (the amide II band) at about 1550 cm^{-1}. D_2O is relatively transparent in this region, and the degree of exchange of the peptide groups can be measured by the disappearing amide II absorption. The exchange can also be traced by the use of T. In the T methods a sample of either the solvent or the solute is isolated at time intervals and the T content is measured by scintillation counting. The T content can be measured more accurately than the D content, but problems arise in isolating a sample of solvent or solute with an isotopic composition representative of that in the solution at a given time. The lower limit of the observable time range depends on the rate of dissolution of the substances studied. It is usually a few minutes.

Wüthrich: In NMR studies of peptides one often derives information about intramolecular shielding of the amide protons from exchange studies in mixed solvents. In a typical experiment one would add a small amount of D_2O to a solution of the peptide in DMSO, and then observe the rate of exchange of the amide protons. Do you have some comments on the interpretation of this type of measurement?

Mme Hvidt: No, but the exchange in mixed solvents may be different from the exchange in aqueous solutions.

H. Buc: Is it possible to use Raman spectroscopy for the same purpose looking at specific resonances (aromatic, SH residues) and at the same time at the overall 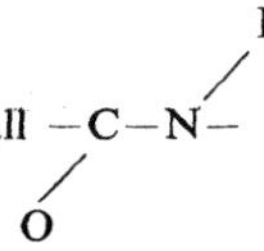 accessibility?

Mme Hvidt: Yes, but the concentration required for Raman spectroscopy may be so high that the simple kinetics of the dilute solutions is no longer obeyed.

Goldberg: The ΔG° between the 'folded' and 'unfolded' states of some 5 kcal/mole corresponds to a local or overall unfolding of the protein. If it is a local unfolding, this might answer Shulman's previous question: a total unfolding would be highly improbable.

Mme Hvidt: It corresponds to a local unfolding of the protein conformation.

Wüthrich: I would like to add a comment on some recent studies on the amide proton exchange in the basic trypsin inhibitor. This molecule consists of 58 amino acid residues and its molecular conformation in single crystals contains approximatively 3 turns of α-helix and an extended β-sheet region. We find that in a solution of this protein in D_2O at pO-values ranging from 1 to 7 the amide protons in 8 positions of the peptide chain are not noticeably exchanged after the solution had been standing for 200 hours at 30°C. Rapid exchange is achieved upon heating the solution to about 60°C, and at pD values of 9 or higher. In view of the small size of this molecule it is rather surprising that so many amide protons are so thoroughly shielded against exchange with the solvent.

Karplus: It may be of interest to mention related results found in a recent 250 MHz proton NMR study of the bovine pancreatic trypsin inhibitor by S. Karplus, G. Snyder and B. D. Sykes (*Biochemistry*, in press). This molecule consists of only 58 amino acids, has little helical content, and only a small amount of 'interior' according to the X-ray structure determination. It is, however, a very 'rigid' molecule as indicated by its retention of activity following exposure to high temperatures and pH extremes; this rigidity is probably in part due to the presence of three disulfide bridges and certainly plays an important role in its high association constants with trypsin. Examination of the inhibitor after *four months* in D_2O, at pH 7, showed eleven NH protons that had not exchanged. This seems like a most unusually slow rate of exchange relative to the times mentioned in the lecture. As the pH increases, so does the rate of exchange; at pH 13, all the protons have exchanged in two days, although the protein is not denatured as indicated by other resolved resonances.

Mme Hvidt: Hydrogen atoms of peptide groups for which $\varrho < 10^{-9}$ remain unexchanged after four months at room temperature, pH 7, and I will characterize a content of the protein of 20% of such peptide groups as unusually high. The protein is more reluctant to exchange its hydrogen atoms than is, for example, insulin, which has a comparable molecular weight, and a similar number of disulfide bridges. You mention that the exchange rates increase with increasing pH; it would be interesting to see if the rate constants are proportional to the hydroxyl ion concentration (i.e. if the time required to obtain a certain degree of exchange is inversely proportional to the hydroxyl ion concentration), or if the protein presents a case of measurable exchange rates for which the assumption of case (9b) (the EX_2 mechanism) is not fulfilled.

DETERMINATION OF THE RELAXATION SPECTRUM OF MACROMOLECULES IN SOLUTION BY THE ANALYSIS OF THE ELECTRIC BIREFRINGENCE

P. COLSON

Université de Liège, Institut de Chimie, Sart-Tilman, B-4000, Liège, Belgique

and

J. P. GASPARD

Institut Max von Laue – Paul Langevin, B.P. 156, 38042 Grenoble Cédex, France

Summary. A new method of moments is proposed for the study of the relaxation spectrum $f(\alpha)$* of the macromolecules in solution. This method gives maximal information on $f(\alpha)$ from the analysis of the electric birefringence of the solution and enables us to obtain exact error bounds on the integrated distribution. This method is applied to the solution of DNA and nucleohistone in the native and sonicated state. The molecular lengths are determined from the formula of Broersma. A simultaneous study of the sedimentation distributions enables the deduction of a relation between the sedimentation coefficient and the length of the particles in solution.

Résumé. Nous proposons une nouvelle méthode de moments pour l'étude du spectre de relaxation $f(\alpha)$* des macromolécules en solution. Cette méthode fournit une information maximale sur $f(\alpha)$ à partir de l'analyse de la biréfringence électrique de la solution et permet d'obtenir des bornes exactes sur la distribution intégrée. Cette méthode est appliquée aux solutions de DNA et nucléohistone à l'état natif et ultrasonné. Les longueurs moléculaires sont déterminées à partir des formules de Broersma. Par une étude simultanée des distributions de sédimentation, nous établissons une relation entre le coefficient de sédimentation et la longueur des particules en solution.

1. Introduction

The aim of this paper is to obtain the distribution of lengths of macromolecules in solution. The principle of the method [1] is to orientate the macromolecules by means of an electric field, then to switch off the field and to measure the disorientation as a function of time, using a property sensitive to the anisotropy of the solution, in our case the electric birefringence. From these measurements we approach the distribution of the relaxation times [2] and by the Burgers–Broersma relations we obtain the distribution of the lengths. First, we underline the difficulty of determining the distribution of relaxation times from birefringence measurements, especially in the case of a continuous distribution. Then, we present the advantages of the moment's method in this numerically ill-conditioned analysis. At the first stage it allows the characterization of the distribution (mean, width, skewness…) of relaxation time inverses in a very simple way. In the second stage, we show more precisely how to express the maximal information available in the form of exact error bounds on the integrated distribution and we present a method of approaching the actual distribution

* $\alpha = 1/\tau$ is the inverse of the relaxation time.
$\alpha = 1/\tau$ est l'inverse du temps de relaxation.

C. Sadron (ed.), Dynamic Aspects of Conformation Changes in Biological Macromolecules. 117–127.

of lengths by a continuous function. We apply the results to the DNA and nucleohistone (DNH) molecules, in the native and sonicated states. After sonication we find a strong displacement of the mean length towards the lower values and a narrowing of the distribution. A relation is deduced between the sedimentation coefficient and the length of a molecule, at various times of sonication which gives information about the flexibility of DNA and DNH.

2. Theory

2.1. Position of the Problem

In dilute solutions, i.e., when no interaction between macromolecules occurs, the disorientation is produced by thermal motion. At time t, the birefringence $\Delta n(t)$ – i.e., the difference in the refractive indices parallel and perpendicular to the electric field – is proportional to the number of oriented molecules. If the solution contains N species of macromolecules, assumed to have the shape of long rigid rods, each species having a relaxation time $\tau_i = 1/\alpha_i$ and a concentration A_i, then the normalized birefringence is given by [2, 4]:

$$y(t) = \frac{\Delta n(t)}{\Delta n(0)} = \sum_{i=1}^{N} A_i \exp(-\alpha_i t), \tag{1}$$

where $\Delta n(0)$ is the steady-state birefringence, i.e., the birefringence corresponding to the maximum orientation of the macromolecule at the considered field strength.

In the case of large macromolecules like DNA or DNH, the distribution of lengths – and thus of relaxation times – is continuous. Equation (1) becomes:

$$y(t) = \int_0^\infty f(\alpha) \exp(-\alpha t) \, d\alpha, \tag{2}$$

where $f(\alpha)$ is the density of molecules with the relaxation time $\tau = 1/\alpha$. The problem we are faced with consists in determining $f(\alpha)$ from the measure of the decaying curve $y(t)$.

We should first emphasize that, from the numerical point of view, this problem is strongly ill-conditioned. Since the function $y(t)$ is structureless, it is hopeless to extract a large amount of information from it. The following example shows in a straightforward way the difficulty of the problem: let us assume that the distribution $f_1(\alpha)$ has a double square shape (see Figure 1) and let $y_1(t)$, the corresponding birefringence, be given by (2) (see Figure 2). We show that, within an accuracy of less than 1%, it is possible to find a square function $f_2(\alpha)$ whose corresponding $y_2(t)$ fits $y_1(t)$ (see Figures 1 and 2). The Laplace transform of $f_1(\alpha)$ or any other integral transform, gives rise to the same behaviour (see Figure 3). This means that it is hopeless to deduce any detailed structure of $f(\alpha)$ from the $y(t)$ measurements which in practice are done with an accuracy of 1%. At the best, one will be able to determine the rough shape of

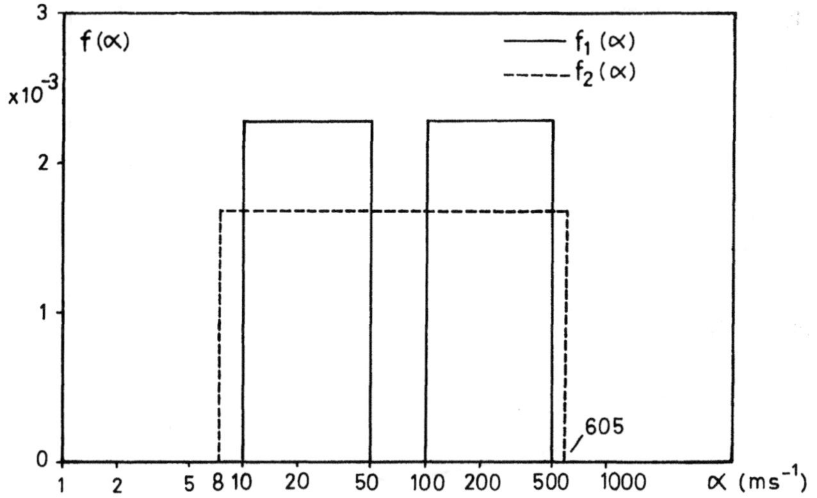

Fig. 1. Two generated differential distribution functions $f_1(\alpha)$ and $f_2(\alpha)$ with double and single square shape respectively.

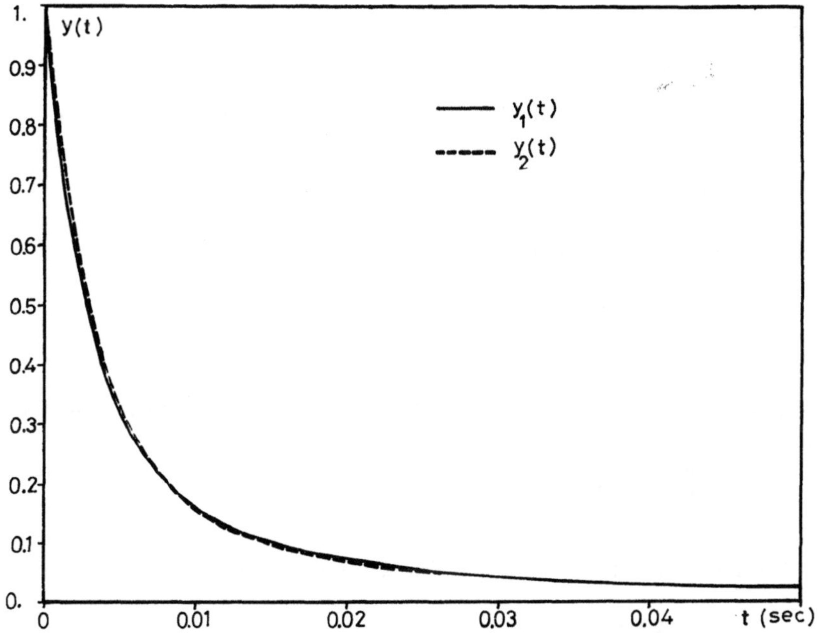

Fig. 2. Two birefringence curves $y_1(t)$ and $y_2(t)$ corresponding to $f_1(\alpha)$ and $f_2(\alpha)$ shown in Figure 1. $y_2(t)$ (dotted line) fits $y_1(t)$ within an accuracy of less than 1 %.

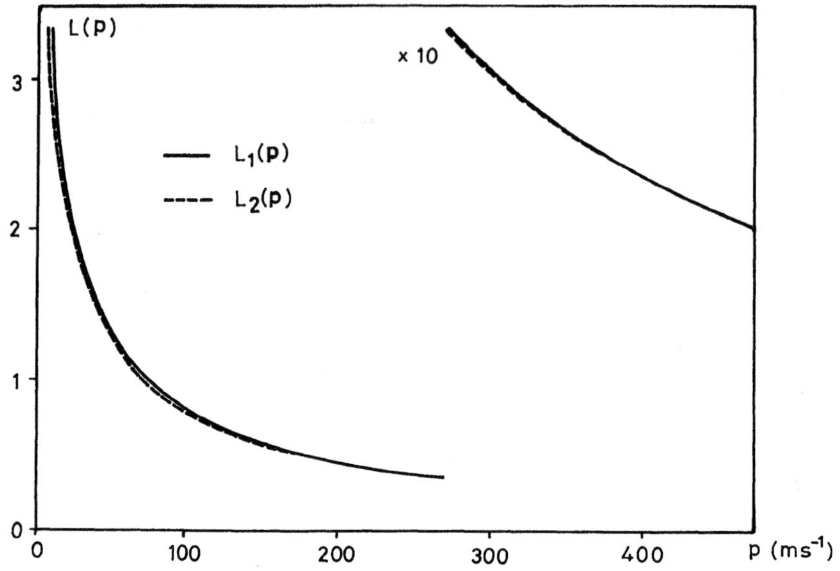

Fig. 3. The Laplace transforms $L_1(p)$ and $L_2(p)$ of the two generated birefringence curves $y_1(t)$ and $y_2(t)$. In the right side above are given the $L(p)$ transforms amplified ten times for the greatest values of p. The two curves are also strongly superimposed.

the distribution; the moment's method seems to be very adequate for this purpose. Indeed, the moments, as integral quantities, are only sensitive to the rough shape of $f(\alpha)$ and not to its fine structure.

2.2. Calculation of the moments of $f(\alpha)$

Let us consider the Laplace transform of $y(t)$ for the argument $p\,(p>0)$.

$$L(p) = \int_0^\infty \exp(-pt)\, y(t)\, \mathrm{d}t, \tag{3}$$

which is nothing else than the Hilbert transform of $f(\alpha)$

$$L(p) = \int_0^\infty \frac{f(\alpha)}{p+\alpha}\, \mathrm{d}\alpha. \tag{4}$$

By expanding the denominator of (4) in powers of α and using the definition of the moments

$$\mu_k = \int_0^\infty \alpha^k f(\alpha)\, \mathrm{d}\alpha, \tag{5}$$

we show directly that $L(p)$ expands in the moment's series

$$L(p) = \sum_{k=0}^{\infty} \frac{(-1)^k \mu_k}{p^{k+1}}. \tag{6}$$

From a numerical point of view, the Laplace transform has many advantages. It takes into account all the points of the curve $y(t)$ and so underemphasizes the random error on each measurement. Furthermore, as p is positive, the weighting factor $\exp(-pt)$ in the integral decreases, and so varies in the same way as the relative error of $y(t)$. Besides, it strongly diminishes the importance of the continuation of $y(t)$ from its largest t measurement (t_M) to infinity. In practice, we continue $y(t)$ by a single exponential

$$y(t) \sim A_M \exp(-\alpha_M t), \tag{7}$$

where A_M and α_M are determined from the far tail of $y(t)$ in a logarithmic plot ($\alpha_M = 1/\tau_M$, where τ_M is the relaxation time of the longest particles present in solution). From Equation (6) one can extract the N first moments by a linear least-square method after having noticed that the high-order momenta behave as

$$\mu_k \simeq A_M \alpha_M^k \quad k > N, \tag{8}$$

where α_M is the largest value of α referred to in Equation (7). Then the linear system to be solved is the following:

$$\sum_{k=0}^{N} \frac{(-1)}{p^{k+1}} \mu_k = L(p) - A_M \times \left(\frac{\alpha_M}{}\right)^{N+1} \frac{1}{p + \alpha_M}, \quad p > 0 \tag{9}$$

2.3. Semi-Quantitative Characterization of $f(\alpha)$ by its Moments

The first few moments of the function $f(\alpha)$ allow to give some parameters characteristic of the curve $f(\alpha)$ [5]. The moment μ_0 is simply the total concentration: with our normalization, it is equal to unity. The moment μ_1 gives the mean of α:

$$\bar{\alpha} = \mu_1 . \tag{10}$$

The square root of the second moment about the mean, $\mu_2' = \mu_2 - \mu_1^2$, measures the width of the distribution

$$\Delta\alpha = \sqrt{\mu_2'}, \tag{11}$$

while the third moment $\mu_3' = \mu_3 - 3\mu_2\mu_1 + 2\mu_1^3$ about the mean gives some indications on the skewness of the curve. In particular for symmetric curves, it vanishes. The even moments of large order are sensitive to the tail of the distribution.

2.4. Approximation of $f(\alpha)$ by a Continuous Function

As it appeared from the example of Section 2.1 it does not seem to be adequate to assume a certain functional form for $f(\alpha)$ with several parameters that are chosen so as to fit the birefringence measurements [6–8] unless one has physico-chemical evidence to assume such a form. The method we are developing does not assume any particular form for $f(\alpha)$. First, we express the maximal information available from the birefringence curves as exact error bonds on the integrated distribution. Then, we give an algorithm for the approximation of $f(\alpha)$.

From Equation (6) we can extract 5 to 9 significant moments, following the accuracy of the measurements. The moment's theory [9, 10] gives n exact bounds on the integrated distribution $F(\alpha)$.

$$F(\alpha) = \int_0^\alpha f(\alpha')\,\mathrm{d}\alpha', \tag{12}$$

if one knows $2n$ moments. By moving the position of these error bars, one gets two integrated curves $\bar{F}(\alpha)$ and $\underline{F}(\alpha)$, respectively upper and lower continuous bounds of the integrated distribution $F(\alpha)$ (Figure 4). Then we approximate the distribution by

$$f(\alpha) = \frac{1}{2}\frac{\mathrm{d}}{\mathrm{d}\alpha}\left[\bar{F}(\alpha) + \underline{F}(\alpha)\right]. \tag{13}$$

A detailed analysis of this algorithm and a comparison with the damped least-square method [11] will be discussed in a forthcoming paper [12].

2.5. Relation between Relaxation Time, Particle Length and Sedimentation Coefficient

The particle lengths are determined from the formula of Broersma [3] which expresses the relaxation time of a macromolecule as a function of its length and other physical

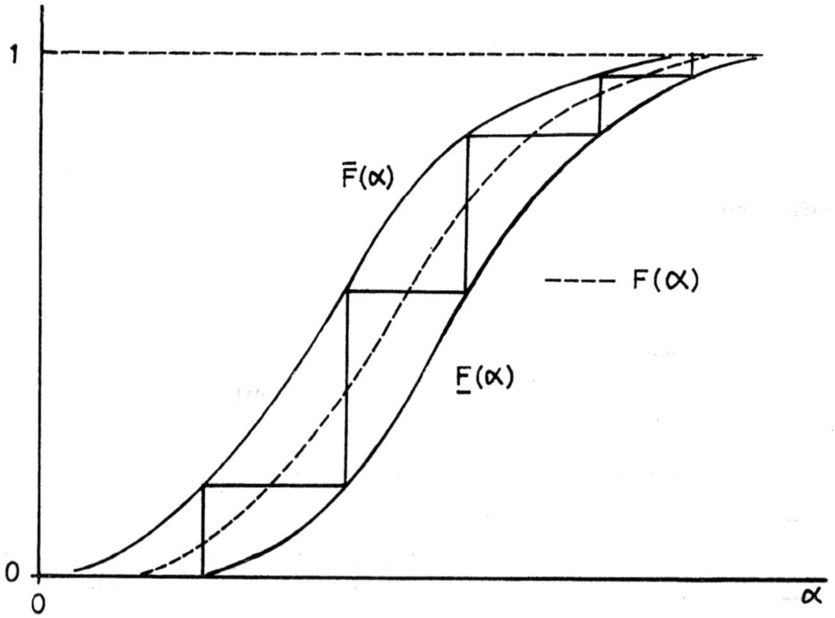

Fig. 4. Determination of the exact error bounds of the integrated distribution $F(\alpha)$ (dotted line). $\bar{F}(\alpha)$ and $\underline{F}(\alpha)$ are the upper and lower bounds of this integrated distribution.

parameters (solvent viscosity, temperature...). For this calculation, we may assume particle diameters of 20 and 30 Å for DNA and DNH respectively [13–14]:

$$l^3 = \frac{18kT\tau}{\pi\eta_0}\{\sigma - [1.57 - 7(1/\sigma - 0.28)^2]\}, \quad \sigma > 2 \tag{14}$$

where k is the Boltzmann constant, T the absolute temperature (K), η_0 the solvent viscosity and σ is equal to $\ln(2a/b)$, where a represents the half-length and b the radius of the cylindrical particle.

At the same time, Nicolaïeff and Litzler [15] have established for DNA a relation between the sedimentation coefficient and the particle length by calculating the cumulative distribution of the concentration versus the length, on the one hand, and versus the sedimentation coefficient, on the other hand. This relation is as follows:

$$s^\circ_{20,w} = 0.781\ l^{0.3}, \tag{15}$$

where $s^\circ_{20,w}$ is the sedimentation coefficient reduced to standard conditions and measured at 20 °C (expressed in Svedbergs) and l is the particle length (expressed in Ångströms) determined by electron microscopy.

These authors have verified this relation up to DNA molecular lengths of 8000 Å and values of its sedimentation coefficient lower than 11.5 s which corresponds to molecular weights lower than 1.4×10^6 according to the formula of Eigner and Doty [16].

Actually, the validity of Equation (15) has never been verified with longer molecules and particularly with soluble and/or sonicated DNH.

We think that the determination of the lengths by the birefringence decay curves is closer to the natural biological conditions than that made by electron microscopy.

3. Material and Methods

The calf thymus DNA and DNH were prepared by the method of Kay et al. [17] and of Frédéricq [18], respectively. Birefringence measurements were carried out with an apparatus described elsewhere by Houssier and Frédéricq [19]. However, a slight modification was made to lower the response of the detecting circuit by reducing the load resistance of the photomultiplier to 10 kΩ. The response time reached in this way was of the order of 1 μsec (determined by birefringence relaxation of nitrobenzene). This improvement enabled relaxation times to be measured on sonicated DNA and DNH. These samples were obtained by means of a Mullard 60 W ultrasonic generator driving an air-cooled magnetostrictive transducer at 20 kc/sec for various lengths of times. No hyperchromicity was observed after this treatment as verified by the melting profiles. We observed, however, that the melting temperature was slightly reduced by the ultrasonic treatment. Such a behaviour may be explained by the melting theories of DNA [20]. For electro-optical measurements, the solvents used were 1 mM NaCl (pH 6.5) and twice distilled water (pH 6.5) for DNA and DNH, respectively.

A Beckman model E ultracentrifuge with absorption optics was used for the measurements of the sedimentation distribution curves of DNA and DNH (25 °C). These distributions were calculated according to Schumaker and Schachman [21], and to Baldwin [22]. Corrections were made for concentration effects, whereas no diffusion effect could be detected. The molecular weight $\bar{M}_{s,\eta}$ was calculated from the Scheraga–Mandelkern equation [23]. The solvent used in the measurements of sedimentation and viscosity was 10 mM NaCl (pH 6.5); this eliminates charge effects and allows us to work at ionic strength values close to the birefringence measurements.

4. Results and Discussion

Table I gives the values of $\bar{\alpha}$, $\Delta\alpha$, $\Delta\tau$ and $\mu_3'/(\mu_2')^{3/2}$ (defined in Section 2.2) calculated both for native and sonicated DNH. We observe a strong displacement of the mean towards the shorter relaxation times and a sharpening of the distribution when the degradation increases, for both DNA and DNH. Unfortunately, the calculated moments μ_k are neither those of the distribution of the lengths nor those of the relaxation

TABLE I

Values of $\bar{\alpha}$, $\Delta\bar{\alpha}$, $\Delta\bar{\tau}$, as $=\mu'_3/(\mu'_2)^{3/2}$ for DNH native and sonicated during 15 min

	native DNH	DNH sonicated 15 min
α	8.5 ms^{-1}	91 ms^{-1}
$\Delta\bar{\alpha}$	15.3 ms^{-1}	93 ms^{-1}
$\Delta\tau$	0.4 ms	0.03 ms
as	2.5	300

times. Nevertheless, the knowledge of the two or three first moments of $f(\alpha)$ allows us to reject several incompatible types of functions. For instance, in our case, the logarithmic normal distribution, although used by some authors [4, 24, 25], has to be rejected because of its too large skewness.

To illustrate our calculation method, we are now treating an example applied to a DNH sample sonicated for 15 minutes under a nitrogen sweeping. Figure 5 gives the oscillogram of the electric birefringence of this sample measured at 12.3 kV/cm. As can be seen, the decay (or relaxation) curve is followed down to a few percent of the maximum value of the birefringence in order to minimize the error introduced by cutting off the data at t_M.

Figure 6 gives the differential distribution function $f(\alpha)$ approximated by a continuous function as illustrated in the case of DNH sonicated for 15 minutes ($E = 12.3$ kV/cm).

On the other hand, a bilogarithmic plot of sedimentation coefficient versus length for sonicated DNH and DNA enables us to deduce a relation between these two parameters of the macromolecules. So the relation deduced for DNH is as follows:

$$s^{\circ}_{25,w} = 0.065\, l^{0.85}, \tag{16}$$

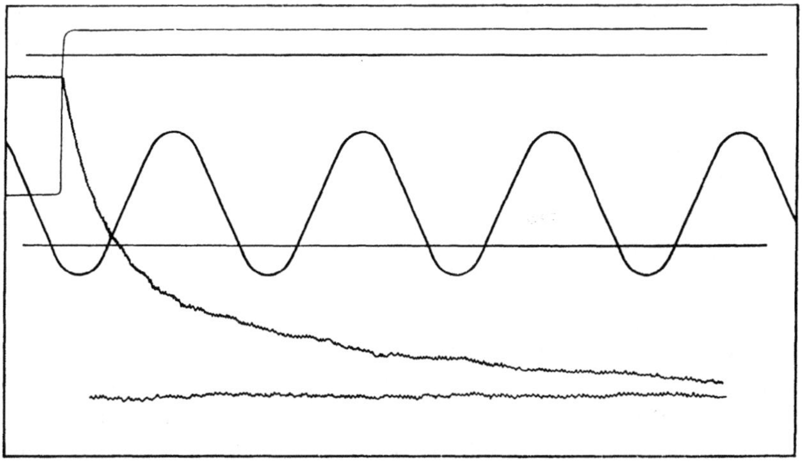

Fig. 5. Oscillogram of the electric birefringence relaxation of a DNH sample sonicated during 15 min. $E = 12.3$ kV/cm; water (pH 6.5); $\lambda = 550$ mμ. The upper trace of the oscillogram shows the electric pulse. The sine wave of standard frequency is also recorded (10 kHz). The relaxation is visualized by the decaying trace.

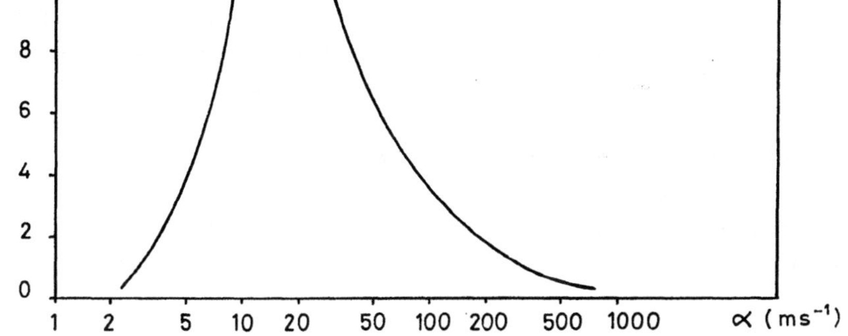

Fig. 6. Semi-logarithmic plot of the differential distribution function $f(\alpha)$ of a DNH sample sonicated during 15 min. The relaxation of birefringence from which the distribution function is determined, was measured at 12.3 kV/cm.

for a sonication time equal to or longer than 15 minutes (the minimum molecular weight obtained for DNH is of the order of 5.5×10^5 for 90 minutes of sonication). In the case of DNA, a unique relation cannot be deduced for the samples of native DNA and for the samples sonicated for various times. The slope of the straight line increases with the extent of degradation and takes a limit value of 0.5 for sonication times equal to or longer than 20 minutes. However, a relation whose slope is similar to that of Nicolaïeff and Litzler is deduced under the same conditions of sonication time (i.e., three minutes):

$$s^{\circ}_{25,w} = 1.34\, l^{0.31}. \tag{17}$$

We think that a relation between the sedimentation coefficient and the length of a molecule is only valid in a given range of s and l where the shape of the molecule does not change significantly. However, under experimental conditions of ionic strength, where the two relations (16) and (17) were deduced, DNH undergoes a smaller change in internal structure than DNA during degradation. Before the action of ultrasound, the latter is much more flexible than DNH and consequently, its flexibility and structural rigidity undergo by sonication a greater change than that of nucleohistone. Therefore, a unique relation cannot be established between these parameters from the native to the most degraded DNA. On the contrary, the same criterion justifies the applicability of relation (16) from DNH sonicated for 15 minutes to the most degraded samples of this macromolecule.

A more detailed treatment of the theoretical and practical aspects of this subject will be published elsewhere [12].

5. Conclusion

The distribution of the birefringence relaxation times of DNA and DNH is determined without introducing any hypothesis about this distribution function and by using the method of moments applied to the Laplace transform of the inverse of relaxation times.

The length distribution of native and sonicated DNA and DNH is estimated with regard to the extent of degradation; a sharpening of this distribution and a displacement of the median length to a lower value is observed when the degradation increases. This conclusion was also deduced by sedimentation measurements on these two macromolecules. In any case, a monodisperse system is never obtained even for the most degraded samples. These conclusions agree with those of Obrenovitch *et al.* [26], who make only sedimentation measurements on DNA.

A relation was deduced between sedimentation coefficient and length of the macromolecules. Such a relation gives evidence for a more pronounced change in the internal structure of DNA than that of DNH under the action of ultrasound.

Acknowledgements

The authors are grateful to Professors V. Desreux and E. Frédéricq and Dr C. Houssier, for many helpful discussions. One of us (P.C.) thanks the Fonds National de la Recherche Scientifique and the Fonds de la Recherche Fondamentale Collective for financial support.

References

[1] Frédéricq, E. and Houssier, C.: 1973, *Electric Birefringence and Electric Dichroism*, The Clarendon Press, Oxford.
[2] Benoit, H.: 1951, *Ann. Phys.* **6**, 561.
[3] Broersma, S.: 1960, *J. Chem. Phys.* **32** 1626.
[4] Paulson, C. M.: 1965, Ph.D. Thesis, University of California, Berkeley.
[5] Kendall, M. G. and Stuart, A.: 1963, *The Advanced Theory of Statistics*, Vol. 1, London.
[6] Watanabe, H.: 1966, Scientific Papers of the College of General Education, University of Tokyo **16**, 49.
[7] Matsumoto, M., Watanabe, H., and Yoshioka, K.: 1968, *Biopolymers* **6**, 929; 1970, **9**, 1307.
[8] Matsumoto, M., Watanabe, H., and Yoshioka, K.: 1972, *Koll. Z.Z.* **250**, 298.
[9] Gordon, R. G.: 1969, *Advan. Chem. Phys.* **15**, 79.
[10] Gaspard, J. P. and Cyrot-Lackmann, F.: 1973, *J. Phys. C.*, in press.
[11] Laiken, S. L. and Printz, H. P.: 1970, *Biochemistry* **9**, 1547.
[12] Colson, P. and Gaspard, J. P., to be published.
[13] Watson, J. D. and Crick, F. H. C.: 1953, *Nature* **171**, 737.
[14] Zubay, G. and Doty, P.: 1959, *J. Mol. Biol.* **1**, 1.
[15] Nicolaïeff, A. and Litzler, R.: 1969, *Biopolymers* **8**, 181.
[16] Eigner, J. and Doty, P.: 1965, *J. Mol. Biol.* **12**, 549.
[17] Kay, E. R. M., Simmons, N. S., and Dounce, A. L.: 1957, *J. Am. Chem. Soc.* **74**, 1724.
[18] Frédéricq, E.: 1962, *Biochim. Biophys. Acta* **55**, 300.
[19] Houssier, C. and Frédéricq, E.: 1966, *Biochim. Biophys. Acta* **120**, 113.
[20] Crothers, D. M., Kallenbach, N. R., and Zimm, B. H.: 1965, *J. Mol. Biol.* **11**, 802.
[21] Schumaker, V. N. and Schachman, H. K.: 1957, *Biochim. Biophys. Acta* **23**, 628.
[22] Baldwin, R. L.: 1954, *J. Phys. Chem.* **58**, 1081; *J. Am. Chem. Soc.* **76**, 402.
[23] Scheraga, H. A. and Mandelkern, L.: 1953, *J. Am. Chem. Soc.* **75**, 179.
[24] Espenscheid, W. F., Kerker, M., and Matijevic, E.: 1964, *J. Phys. Chem.* **68**, 3093.
[25] Takashima, S.: 1966, *Biopolymers* **4**, 663.
[26] Obrenovitch, A. and Aubel-Sadron, G.: 1971, *J. Chim. Phys.* **68**, 521.

RESOLUTION OF THE FLUORESCENCE SPECTRA OF PROTEINS USING DECAY MEASUREMENTS

PH. WAHL and J. C. BROCHON

Centre de Biophysique Moléculaire, 45-Orléans-02, France

Summary. It is shown that the determination of fluorescence decay at various emission wavelengths enables us to resolve the fluorescence spectra into several components.

The method is applied to the study of the endonuclease of *Staphylococcus aureus* fluorescence and of the human serum albumin emission. In the first case, one can decompose the spectrum into two components each of them being characterized by one decay time: the tyrosine emission with a decay time of 1.8 nsec and the tryptophan emission with a decay time of 5.5 nsec. In the second case, the spectrum is decomposed into three components: the tyrosine emission (1.5 nsec), the tryptophan emission (6.17 nsec) and a third component which has a weak contribution and which is ascribed to an other kind of tryptophan emission (decay time 12.08 nsec).

Résumé. On montre que la mesure des déclins de fluorescence à différentes longueurs d'onde d'émission peut permettre de résoudre les spectres de fluorescence en plusieurs composantes.

La méthode a été appliquée à l'émission de fluorescence de l'endonucléase du *Staphylococcus aureus* et de la sérum albumine humaine. Dans le premier cas, la méthode permet de décomposer le spectre en deux composantes caractérisées chacune par une constante de temps. Il s'agit de la tyrosine de constante de temps de 1,8 nsec et du tryptophane (5,5 nsec). Dans le second cas, le spectre est décomposé en trois composantes: la tyrosine (1,5 nsec), le tryptophane (6,17 nsec) et une troisième composante en faible proportion attribuée également au tryptophane (12,08 nsec).

1. Introduction

The protein fluorescence is essentially emitted by the tyrosyl and the tryptophanyl residues [1]. All proteins contain tyrosyl residues. One distinguishes two classes of proteins: the class A proteins which do not contain tryptophanyl residue and the class B proteins which contain tryptophanyl residues. Fluorescence spectra of class A proteins are similar to the tyrosine spectrum with little differences in the maximum position which occurs at 305 nm [1]. The fluorescence emission of class B proteins is dominated by the emission of tryptophanyl residues [1, 2]. The indole nucleus of tryptophan has a spectrum strongly dependent on the environment. As a result, the wavelength of the fluorescence spectrum maximum varies from one protein to another in the range of 325 nm to 355 nm [3]. It has been observed that the shapes and the widths of the fluorescence spectra of proteins of class B are very similar. However, solvent perturbation studies show that the emission of the different tryptophan residues, belonging to the same protein, may have different wavelength maxima [4].

Furthermore, it is possible to demonstrate the tyrosine contribution to the fluorescence of some class B proteins, by comparing the spectrum of the fluorescence excited at 280 nm where the tyrosines strongly absorb, with the fluorescence excited at 295 nm where the tyrosine absorption is weak [5]. Weber [2] has proposed a matrix method to detect the components of fluorescence spectra. Up to now no extensive use of this method has been made; this may be due to the low sensibility of the method.

C. Sadron (ed.), Dynamic Aspects of Conformation Changes in Biological Macromolecules. 129–136.

In the present work, we propose a new method, based on the analysis of the florescent decay at various wavelengths of emission. The method is applied to the study of the *Staphylococcus aureus* endonuclease and to the human serum albumin. Detailed accounts of this work will be given elsewhere [15, 19].

2. Fluorescence Decays Analysis

Let us assume that during the excited state no reaction occurs [11] and that no solvent reorientation with a relaxation time comparable to the fluorescence decay times occurs [9, 10]. Under these conditions, the decay time of each fluorescent species does not depend on the excitation and the emission wavelengths.

The fluorescent response of a solution to a very short exciting flash is a product of convolution [11] of the form:

$$s(t) = \int_0^t g(T)\, S(t - T)\, \mathrm{d}T, \tag{1}$$

where $g(T)$ is the response function which is obtained by placing a reflector instead of the cell in the apparatus. $S(T)$ is the density of probability of a fluorescent photon emission at the time t after the absorption of an excitation photon at time zero.

We assume that several fluorescent species are present in the solution, each of them characterized by its fluorescent decay and a fluorescent spectrum. For simplification, we assume in addition that each species has a single exponential decay. The more general case where the decays of the components are expressed by the sum of exponentials, may be also envisaged [15].

Let us consider the decay obtained at a given emission wavelength λ_e by an excitation at the wavelength λ_E. The normalized decay is then given by the following expression:

$$S(\lambda_E, \lambda_e, t) = \sum \frac{a_i}{\tau_i} (\lambda_e, \lambda_E) \exp(-t/\tau_i), \tag{2}$$

with $\sum a_i = 1$. τ_i is the time constant characteristic of species i, a_i represents the intensity contribution of species i, under the conditions of continuous excitation, coresponding to the ordinary spectral measurements [15].

Then the fluorescence intensity of species i is given by:

$$F_i(\lambda_E, \lambda_e) = a_i(\lambda_E, \lambda_e)\, F(\lambda_E, \lambda_e), \tag{3}$$

where $F(\lambda_E, \lambda_e)$ is the fluorescence intensity of the solution. The fluorescence spectrum of species i is obtained by plotting $F_i(\lambda_E, \lambda_e)$ in function of λ_e, for a fixed value of λ_E. Inversely for the excitation spectrum, λ_e is kept fixed and λ_E is varied.

It has been established [8, 17] that the moments of the resolution function $g(t)$ and the experimental decay $s(t)$ may be used to compute the quantities T', T'', $T^{(n)}$

defined by:

$$\begin{aligned} T' &= \sum a_i\tau_i \\ T'' &= \sum a_i\tau_i^2 \\ &\vdots \\ T^{(n)} &= \sum a_i\tau_i^n . \end{aligned} \tag{4}$$

Equations (4) may be used to determine the a_i's and the τ_i's, if T', T'', $T^{(n)}$ have been calculated by the moments.

Practically, the number of components directly determinable is limited, since the moments of high order are not known with a sufficient accuracy.

The resolution will be greatly facilitated if additional information is used. This is the case when only one species is emitting in a given wavelength range, so that the time constant of this species may be directly obtained.

The fluorescence decay measurement was made using a nanosecond flash apparatus with the single photoelectron method previously described [12–15]. The wavelength band is 15 nm for the excitation and 10 nm for the emission.

3. Results

3.1. Study of the Staphylococcus aureus endonuclease fluorescence

This enzyme contains one tryptophanyl residue and seven tyrosyl residues. The pH chosen for the measurements was neutral and the temperature was 21°C.

Results of fluorescence decay analysis are presented in Table I. Fluorescence decays are identical for all the emission wavelengths equal to or greater than 340 nm, with

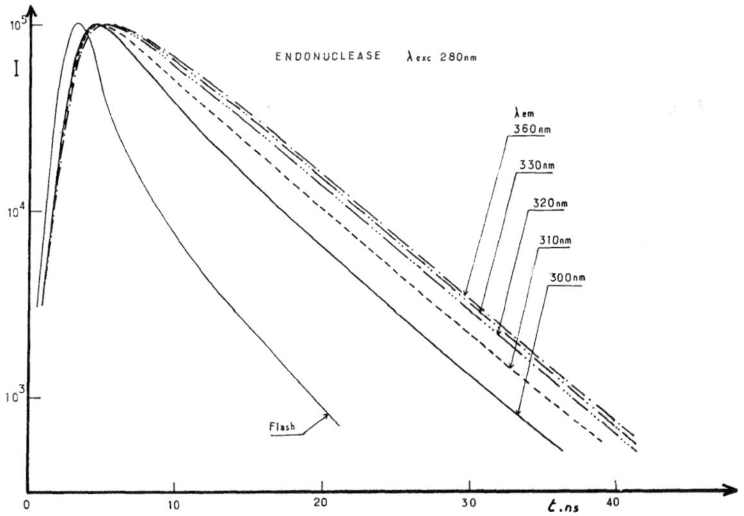

Fig. 1. The fluorescence decays of *Staphylococcus aureus* endonuclease at pH 7 for various emission wavelengths. All emissions with wavelength equal to or greater than 340 nm, have the same decay.

TABLE I

Results of decay analysis of *Staphylococcus aureus* endonuclease

Emission (nm)	Excitation at 280 nm				Excitation at 295 nm			
	τ (nsec)	τ_2 (nsec)	a_1	a_2	τ (nsec)	τ_2 (nsec)	a_1	a_2
300	1.8	5.5	0.552	0.448	–	–	–	–
310	1.8	5.5	0.278	0.722	–	–	–	–
320	1.8	5.5	0.098	0.912	1.8	5.5	0.06	0.94
330	1.8	5.5	0.039	0.961	1.8	5.5	0.04	0.96
340	–	5.5	–	1.0	–	5.5	–	1.0
350	–	5.5	–	1.0	–	5.5	–	1.0
360	–	5.5	–	1.0	–	5.5	–	1.0
380	–	5.5	–	1.0	–	5.5	–	1.0

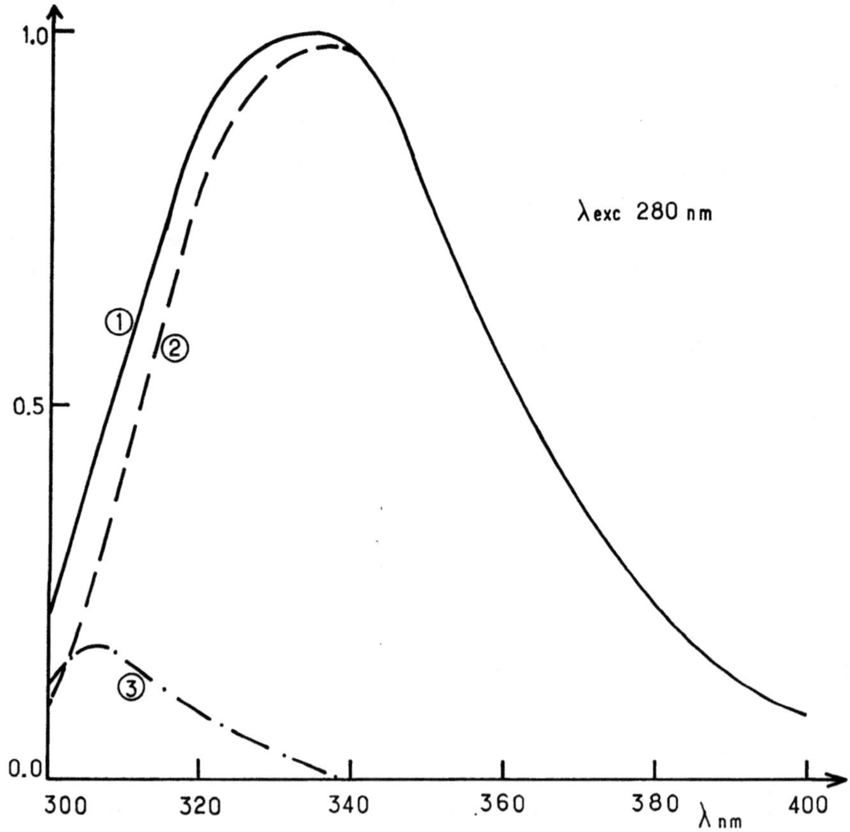

Fig. 2. Emission spectrum of *Staphyloccus aureus* endonuclease at pH 7. Decomposition into two components, the decay time of which is $\tau = 5.5$ nsec (2) and 1.8 nsec (3).

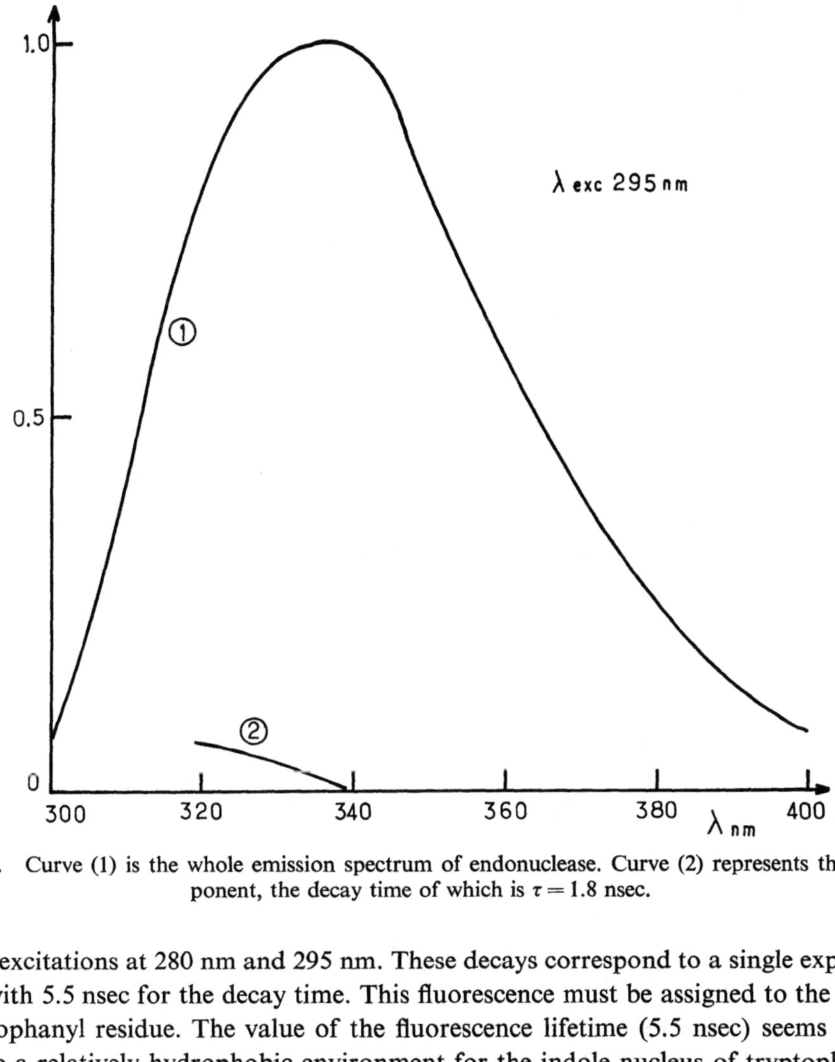

Fig. 3. Curve (1) is the whole emission spectrum of endonuclease. Curve (2) represents the component, the decay time of which is $\tau = 1.8$ nsec.

both excitations at 280 nm and 295 nm. These decays correspond to a single exponential with 5.5 nsec for the decay time. This fluorescence must be assigned to the single tryptophanyl residue. The value of the fluorescence lifetime (5.5 nsec) seems to indicate a relatively hydrophobic environment for the indole nucleus of tryptophan in this protein. Experimental decay curves for an excitation at 280 nm are shown in Figure 1. The decay is faster when the wavelength decreases from 330 nm to 300 nm. For all wavelengths under 340 nm no experimental decay curve corresponds to a single exponential We tried then to fit the experimental curve with two exponential decays. One of the time constant is chosen equal to 5.5 nsec and the other parameters are determined using the values T' and T'' computed by the moments. After a procedure of successive approximations the second constant time is found to be 1.8 nsec. The values of the coefficients a_1 and a_2 are given in Table I.

It can be seen that the contribution of the 1.8 nsec component increases when the emission wavelength decreases. The contribution is also greater for the 280 nm excitation than for the 295 nm excitation.

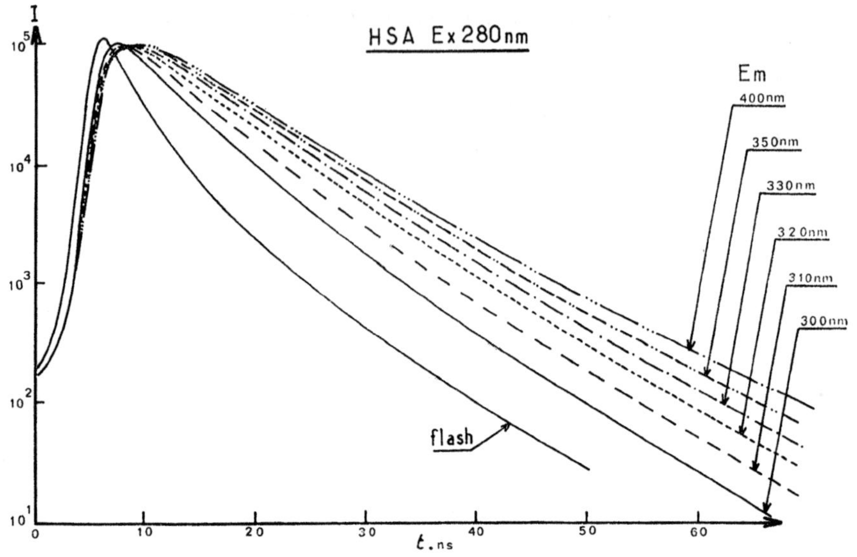

Fig. 4. Fluorescence decays of human serum albumin for various emission wavelengths. Reproduced from reference 15 with the permission of the publisher.

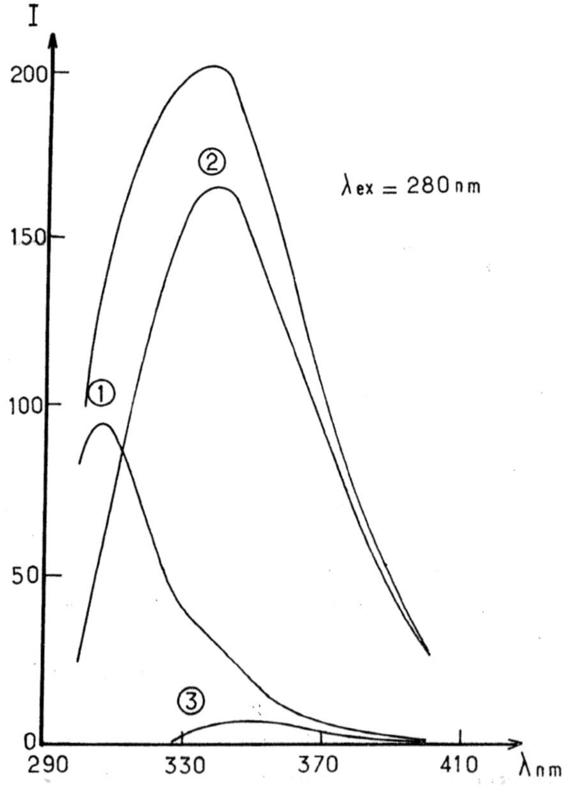

Fig. 5. Decomposition of the whole emission spectrum of human serum albumin in three components. Each component is characterized by its decay time: 1.5 nsec for component (1), 6.17 nsec for component (2), 12.08 nsec for component (3).
Reproduced from reference 15 with the permission of the publisher.

The application of the formula (3) allows to determine the spectrum of each component. These spectra are drawn on Figures 1 and 2. From these results the 1.8 nsec component must be assigned to the tyrosine fluorescence, while the 5.5 nsec component corresponds to the tryptophan.

3.2. Study of Human Serum Albumin

This protein contains 17 tyrosyl residues and one tryptophanyl residue. Decays measured at various wavelengths of emission with an excitation of 280 nm are shown in Figure 4.

Here one cannot fit the experimental curves with a sum of two exponentials. Then one is led to try three exponential terms. A detailed description of the mathematical method used in this decomposition is given elsewhere [15]. The time constants of the three components have been found to be 1.5 nsec, 6.17 nsec, 12.08 nsec. The corresponding fluorescence spectra are given in Figure 5.

The spectrum of the 1.5 nsec component obviously represents the tyrosine contribution. The 6.17 nsec component which is the major component, comes from tryptophan.

The presence of the third component was unexpected. Its spectral maximum is situated around 350 nm, its contribution is weak. Various purifications failed to eliminate this component, which led us to the conclusion that this emission comes from some tryptophan residues which are either chemically modified or rather have a configuration different from the other tryptophan residue.

4. Conclusion

In conclusion, we think that the method exposed here is able to resolve the fluorescence spectra of protein in several components. In the fluorescence of many proteins, the tyrosine contribution is thought to be negligible. This assumption is assumed to be generally true for the 295 nm excitation. However, using that excitation, we are able to detect a tyrosine emission in *Staphylococcus aureus* endonuclease as well as in the case of human serum albumin [15]. One must, however, recognize that the wavelength bands used here are rather wide allowing an excitation down to 288 nm. Anyway, we think that our method can be useful in displaying the tyrosine fluorescence in proteins where it has been assumed to be negligible. This method may well be useful also in order to differentiate the emission coming from different tryptophan residues, having different environments.

References

[1] Teale, F. W.: 1960, *Biochem. J.* **76**, 381.
[2] Weber, G.: 1961, *Nature* **190**, 27.
[3] Teale, F. W. J.: 1961, *Biochem. J.* **80**, 14P.
[4] Lehrer, S. S. and Fasman, G. D.: 1967, *J. Biol. Chem.* **242**, 4644.
[5] Vladimirov, Yu. A. and Ziminia, G. M.: 1965, *Biokim.* **30**, 1105.
[6] De Lauder, W. B. and Wahl, Ph.: 1970, *Biochemistry* **9**, 2750.
[7] Brochon, J. C. and Wahl, Ph.: 1972, *Eur. J. Biochem.* **25**, 20.

[8] Isenberg, I. and Dyson, R. D.: 1969, *Biophys. J.* **9**, 1337.
[9] Bakhshiev, N. G., Mazurenko, Yu. T., and Peterskaya, I. V.: 1966, *Opt. Spectrosk.* **21**, 5550.
[10] Brand, L. and Gohlke, J.: 1971, *J. Biol. Chem.* **246**, 2317.
[11] Birks, J. B. and Munro, I. H.: 1967, *Progr. React. Kinet.* **4**, 239.
[12] Koechlin, Y.: 1961, Thesis Paris.
[13] Laustriat, G., Pfeffer, G., Lami, H., and Coche, A.: 1969, Colloque International d'Électronique Nucléaire, Paris.
[14] Wahl, Ph.: 1969, *Biochim. Biophys. Acta* **175**, 55.
[15] Wahl, Ph. and Auchet, J. C.: 1972, *Biochim. Biophys. Acta* **285**, 99.
[16] Dyson, R. D. and Isenberg, I.: 1971, *Biochemistry* **10**, 3233.
[17] Wahl, Ph. and Lami, H.: 1967, *Biochim. Biophys. Acta* **133**, 233.
[18] Chen, R. F.: 1967, *J. Biol. Chem.* **242**, 173.
[19] Brochon, J. C., Wahl, Ph., and Auchet, J. C.: to be published.

DISCUSSION

Iwatsubo: Le spectre d'émission de HSA (no. 1 sur la figure $\tau = 1{,}5$ nsec) représente deux pics (ou un pic plus un épaulement). Pourriez-vous me donner une explication concernant ce deuxième pic?

Brochon: La constante de temps de déclin de 1,5 nsec obtenue pour l'émission de l'ensemble des 17 résidus tyrosines représente en fait une moyenne. On pourrait expliquer cet épaulement en admettant que le déclin de l'émission des tyrosines n'est pas simple et qu'il peut être décomposé lui-même en deux composantes.

Lami: Vous avez indiqué que la sérum albumine contient un seul résidu tryptophane et pourtant vous décelez deux composantes dans le déclin que vous attribuez à ce même résidu. Comment expliquez-vous ce résultat?

Brochon: La deuxième composante correspondant à $\tau = 12{,}08$ nsec est toujours présente après délipidation et désionisation du monomère. Dans la merceptoalbumine, qui a été isolée de ce même monomère, on retrouve également cette composante longue. En utilisant un tampon phosphate au lieu du tampon tris-HCl cette composante est encore présente. Il pourrait s'agir soit d'un résidu tryptophane modifié que les méthodes de purification ne peuvent éliminer, soit que cet unique résidu tryptophane de la HAS est placé dans un environnement différent caractérisé par une autre constante de temps de déclin.

TERTIARY AND QUATERNARY STRUCTURAL STATES OF HEMOGLOBIN

L. F. TEN EYCK and P. D. PULSINELLI
MRC Laboratory of Molecular Biology, Hills Road, Cambridge, Great Britain

Summary. X-ray diffraction studies have led to the solution of the structures of several conformational states of hemoglobin, including some partially oxygenated and partially oxidized forms. Comparison of these structures is beginning to show the conformational changes involved in the cooperative binding of oxygen to hemoglobin. This paper will attempt to summarize the data at hand and indicate where the missing structural information can be found.

Résumé. La diffraction des rayons X a permis d'élucider les structures de divers états conformationnels de l'hémoglobine, y compris ceux de certaines formes partiellement oxygénées ou oxydées. La comparaison de ces structures commence à mettre en évidence les changements conformationnels mis en jeu dans la liaison coopérative de l'oxygène à l'hémoglobine. On présente ici les données disponibles et indique dans quelle direction chercher les données structurales manquantes.

Hemoglobin is the classic example of a protein showing cooperative binding of ligands, a phenomenon of fundamental importance in biology. The affinity of oxyhemoglobin for oxygen is greater than the affinity of deoxyhemoglobin for oxygen. Monod, Wyman, and Changeux [1] have shown that such behaviour can be explained thermodynamically if there exist two forms of the protein in equilibrium with each other which have different affinities for ligand. The presence of the ligand then stabilizes the form with the higher affinity for ligand. This model appears to be basically correct for hemoglobin, although some modiffication of the pure allosteric model may be required to explain all of its properties.

The structures of both the oxidized and reduced forms of hemoglobin have been solved by X-ray diffraction [2–4]. The hemoglobin molecule is a tetramer composed of two pairs of identical subunits. Both the α-subunits and the β-subunits look remarkably like myoglobin, the oxygen storage protein of red muscle. The subunits have a very high alpha helix content, there being eight helical segments in the β-chaims (which contain 146 amino acids) and seven helical segments in the α-chaims (which contain 141 amino acids). Each subunit contains a heme group, which is the site of ligand binding. The heme group consists of a porphyrin ring with an iron atom in the center. Four of the six possible coordination positions of the iron atom are filled by the nitrogen atoms of the porphyrin ring; a fifth position is filled by a histidine imidazole group attached to the F helix. The sixth coordination position is vacant in deoxyhemoglobin (deoxy Hb) and filled by the ligand in all other normal forms of hemoglobin. The iron atom is ferrous in the physiological form of hemoglobin and will bind ligands such as oxygen and carbon monoxide. The iron is easily oxidized to the ferric state, in which case it will bind water or ionic ligands such as hydroxide, cyanide, azide, and flouride. There is an enzyme in the red blood cell, methemoglobin reductase, which reduces ferric hemoglobin (met Hb) to the normal ferrous form.

C. Sadron (ed.), Dynamic Aspects of Conformation Changes in Biological Macromolecules. 137–149.

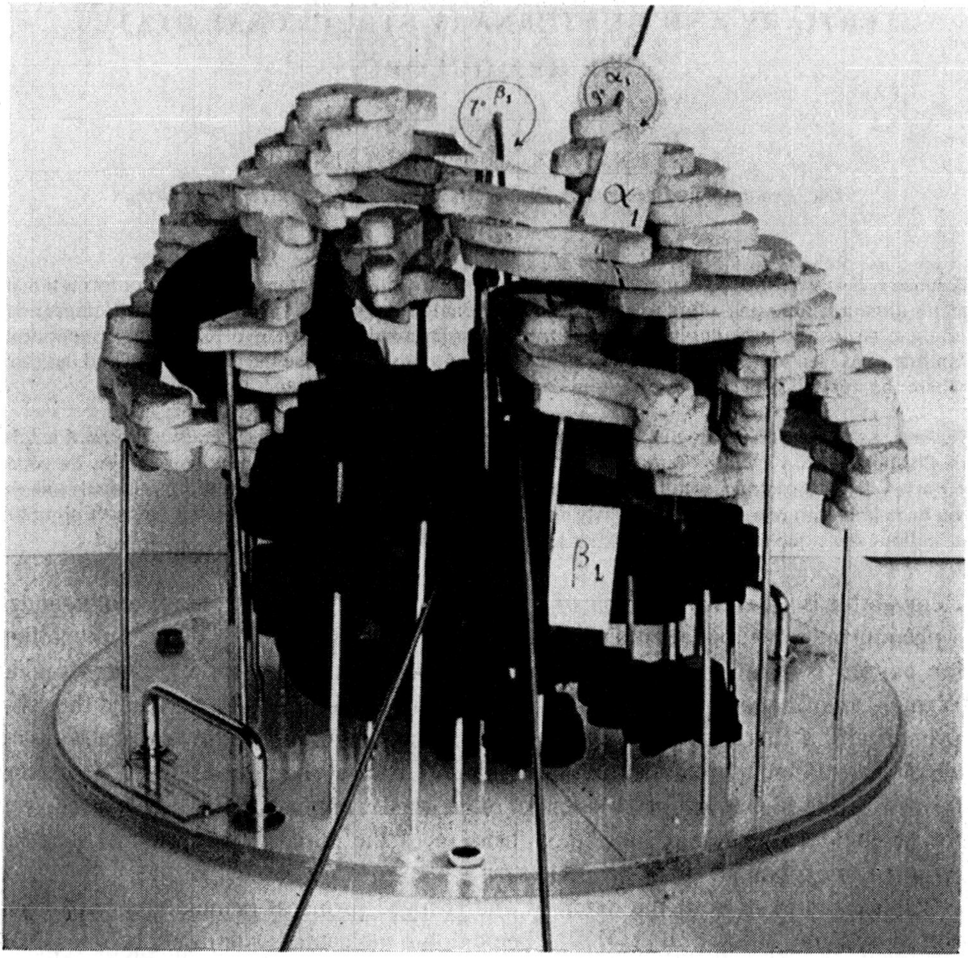

Fig. 1. A 5.5 Å resolution model of horse met Hb. The α-subunits are shown in white, the β-subunits are shown in black, and the disks represent the heme groups. On reduction to deoxy Hb the subunits rearrange themselves by rotation about the axes shown in the figure.

Figure 1 shows a low resolution model of horse met Hb. This is illustrative of all of the liganded structures, which are basically the same. On conversion to deoxy Hb there is a rearrangement of the subunits which is shown by the rods in the figure. The net result is that there is very little change at the interface between the α_1- and β_1-subunits but a shift of almost 6 Å at the interface between the α_1- and β_2 subunits. There is no form of hemoglobin known which exhibits cooperative ligand binding without also showing the rearrangement of the quaternary structure.

The allosteric model for cooperative binding predicts that any substance which binds preferentially to one of the two forms of the protein will have a regulatory effect on the binding of the ligand. Hemoglobin is subject to regulation by inorganic anions, several organic phosphates, and hydrogen ions. Of these allosteric effectors the most

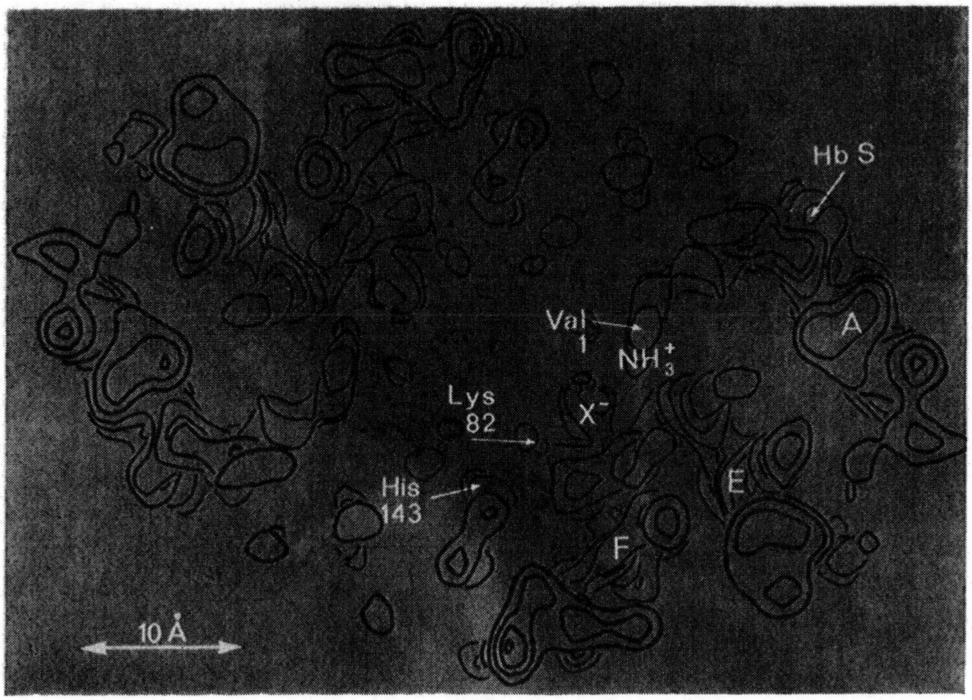

Fig. 2a.

Fig. 2. Binding of organic phosphates to deoxy Hb. (a) The native electron density at the bottom of the molecule between the two β-subunits. The cross marks the position of the molecular diad axis. Note that his site is a pocket surrounded by positive charges. (b) The same site with the difference map showing the binding of 2,3-diphosphoglycerate superimposed on the native map. The solid contours are positive differences. Binding of the strongly charged organic phosphate seems to significantly tighten the structure.

important seems to be one of the organic phosphates, 2,3-diphosphoglycerate (DPG). This compound is normally present in red blood cells and binds strongly to deoxy Hb in the ratio of one mole per mole of hemoglobin tetramer. Perutz [5] predicted that the binding site would be at the bottom of the molecule between the two β-chains on the basis of chemical information and model building experiments; this was confirmed by Arnone [6]. Figure 2a shows the native molecular electron density in this region; note that it is a pocket lined with positive charges. Figure 2b is the same region with the difference map showing the bound DPG. Figure 3 is the structural interpretation of the binding site.

Another organic phosphate, inositol hexaphosphate (IHP), is commonly used in experiments on hemoglobin. This compound does not occur physiologically in mammals but does have all of the effects on hemoglobin of DPG. In fact IHP is a much more potent allosteric effector than DPG, and is very cheaply available as phytic acid; hence its wide experimental use. Dr Arnone has recently confirmed that the binding site on deoxy Hb for IHP is the same as that for DPG. Adenosine triphhospate is also

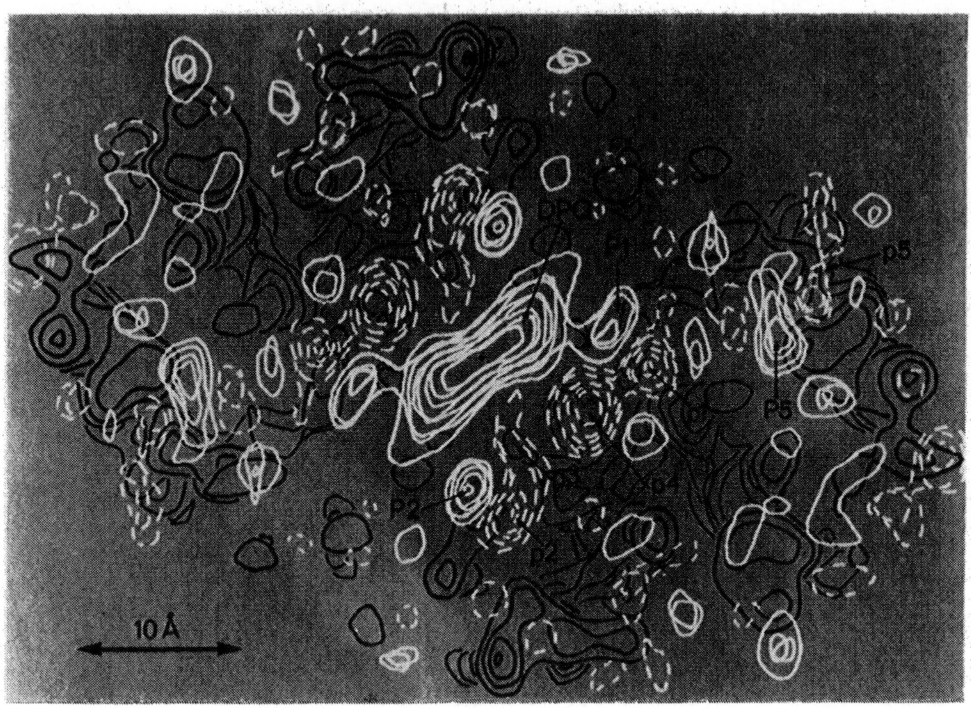

Fig. 2b.

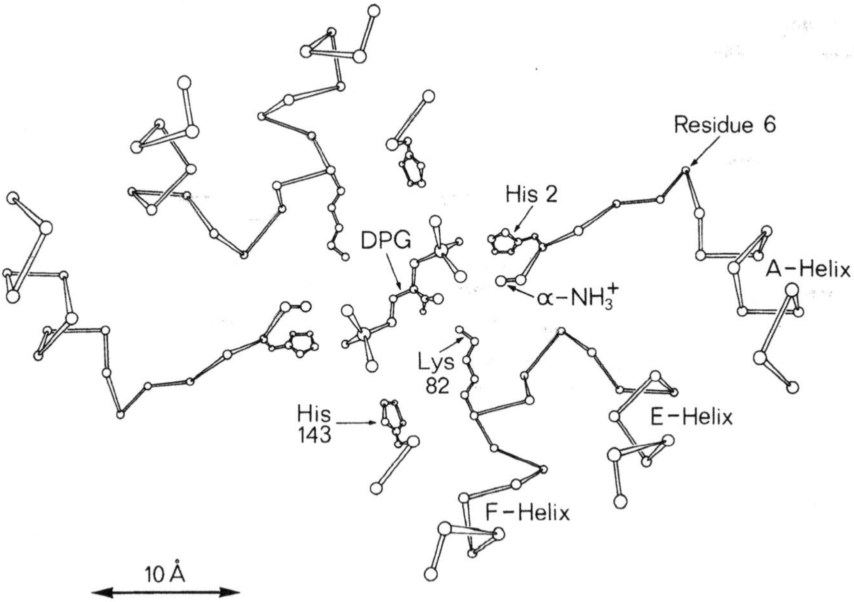

Fig. 3. A molecular interpretation of the binding of 2,3-DPG to human deoxy Hb.

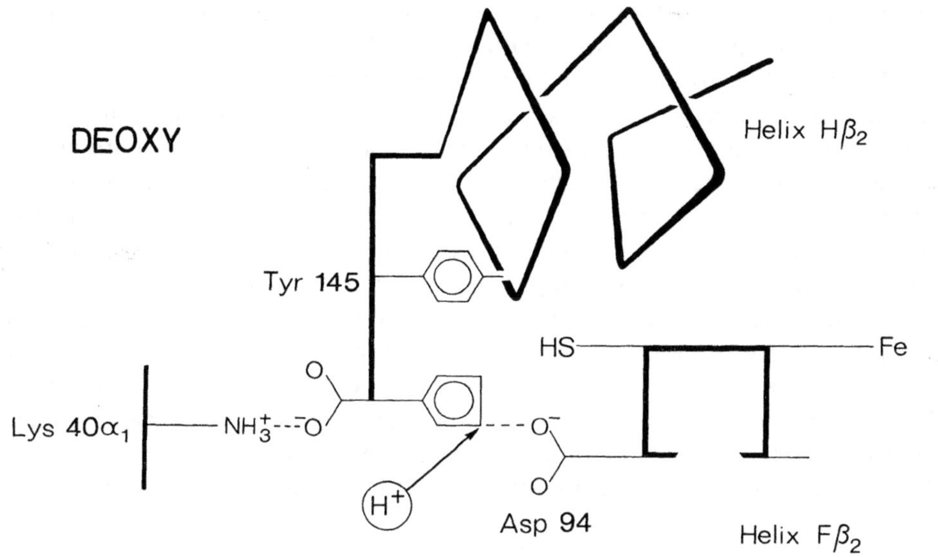

Fig. 4a.

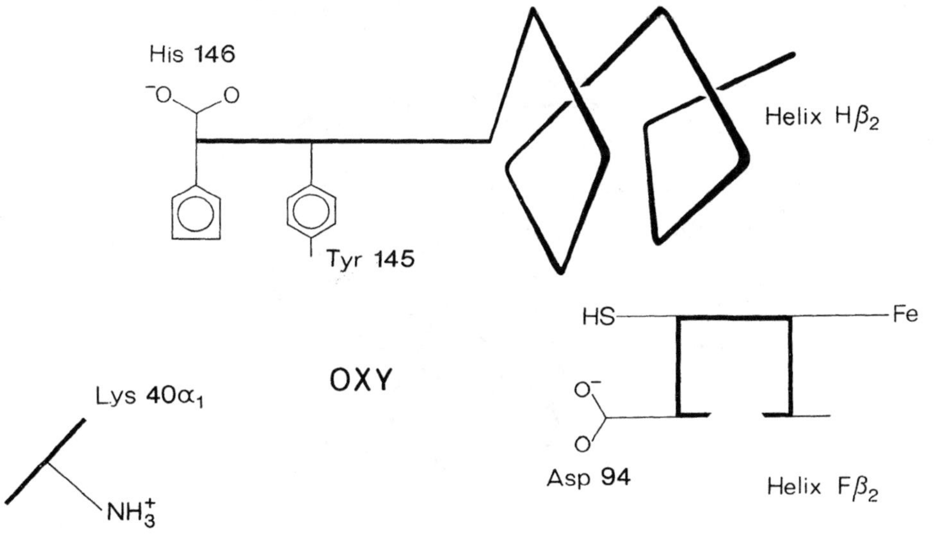

Fig. 4b.

Fig. 4. A comparison of the structure of the C-terminus of the β-subunits in the oxy and deoxy Hb structures. In the deoxy structure histidine 146 β participates in two salt bridges, one with a lysine from the opposite α-chain and one with an aspartate from the same β-chain. In the oxy structure the two C-terminal residues are not localized.

an allosteric effector for hemoglobin but the binding site has not yet been located. It probably binds at the same site as DPG and IHP.

The effect of hydrogen ions on the oxygen affinity of hemoglobin was finally understood when the structure of deoxy Hb was solved. The deoxyhemoglobin structure contains a number of salt bridges involving the C-terminal residues of both the α- and the β-subunits. Figures 4 and 5 illustrate these salt bridges schematically. Several of them involve charged groups which can be titrated near neutral pH. In the oxyhemoglobin structure the C-terminal residues are all delocalized and the salt bridges cannot possibly be formed. These salt bridges are very important for maintaining the structure of the deoxy tetramer. It is possible to selectively remove the various C-terminal residues by treatment with carboxypeptidase; all such derivatives have reduced cooperativity and increased oxygen affinity, which indicates that the relative stability of the oxyhemoglobin structure has been increased. Studies of the crystal structures [7] have shown that as long as one set of salt bridges is left intact the derivative can form a basically normal deoxyhemoglobin structure. When both sets of salt bridges are per-

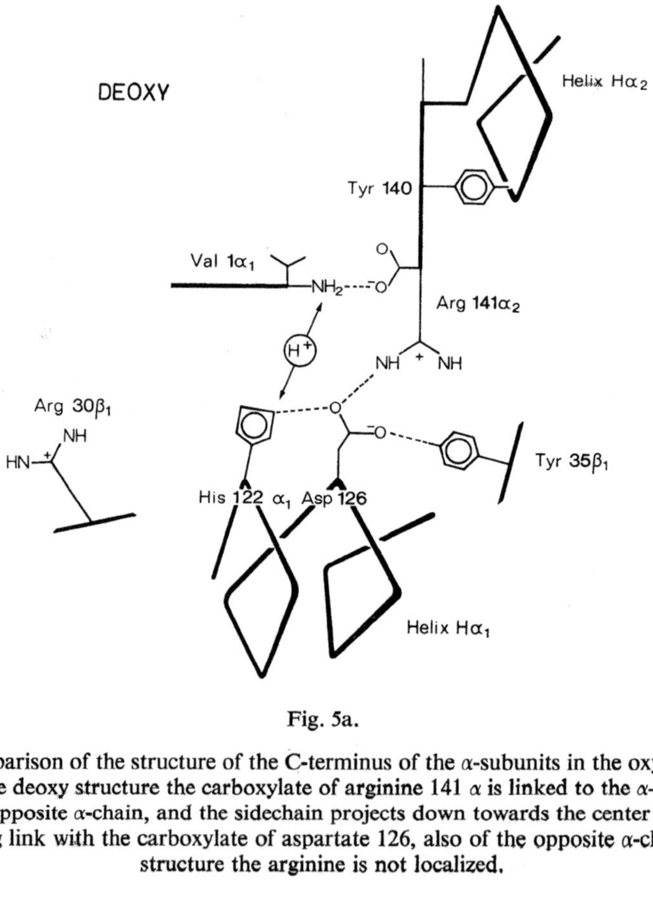

Fig. 5a.

Fig. 5. A comparison of the structure of the C-terminus of the α-subunits in the oxy and deoxy Hb structures. In the deoxy structure the carboxylate of arginine 141 α is linked to the α-amino group of valine 1 of the opposite α-chain, and the sidechain projects down towards the center of the molecule to form a strong link with the carboxylate of aspartate 126, also of the opposite α-chain. In the oxy structure the arginine is not localized.

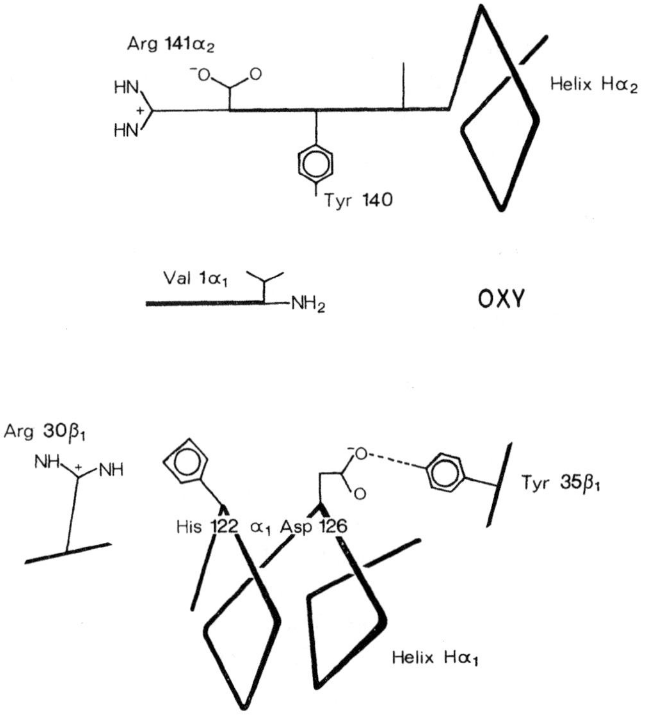

Fig. 5b.

turbed the resulting derivative can only form structures related to the oxyhemoglobin structure, even in the absence of any ligand.

The derivative in which the C-terminal arginine of the α-subunits has been removed is particularly interesting. When crystals of this derivative are grown in the absence of oxygen the human hemoglobin crystallizes isomorphously with native human deoxy-hemoglobin. The horse hemoglobin derivative crystallizes isomorphously with horse *oxyhemoglobin*. The reason for this peculiarity is that for human hemoglobin the deoxy structure is much less soluble than the oxy structure, while for horse hemo-globin the oxy structure is much less soluble than the deoxy structure. This in turn implies that there is, in this derivative, a substantial amount of the oxy structure even in the absence of oxygen. This is in agreement with the predictions of the allosteric model for hemoglobin.

The liganded forms of hemogoblin are all basically the same. They are characterized by a particular arrangement of the subunits, no C-terminal salt bridges, a large shift at the α_1–β_2 subunit interface relative to deoxyhemoglobin, and an iron atom which is in or near the plane of the heme group. The basic structure to which all of the others are referred is that of met Hb, in which the iron is ferric and the ligand is water. Oxyhemoglobin at acid pH is converted to this form in an X-ray beam, and therefore cannot be studied crystallographically. The structure of met Hb was solved to high resolution by Perutz *et al* [2]. It has recently been found that this form of hemoglobin is in a thermal equilibrium between high spin ($s=5/2$) and low spin ($s=1/2$) forms which can be shifted towards the high-spin form by addition of IHP. There is substantial evidence that the high-spin form in the presence of IHP has a quaternary structure similar to that of deoxy Hb, but it is not yet known just how far the similarity goes (Perutz, private communication).

Since oxyhemoglobin is converted to methemoglobin by X-rays, the structure of carbonmonoxyhemoglobin (CO-Hb) is being studied in our laboratory by Dr. Elizabeth Heidner. This work is still in progress but some preliminary results are available. There are no major differences between CO-Hb and met Hb, but there are small differences in some important parts of the molecule. There is a shift of the reactive sulphydryl group towards the heme group. There is some perturbation of the heme pocket because CO is a bulkier ligand than water, and there are some small shifts at the $\alpha_1-\beta_2$ subunit interface.

The stereochemical trigger for the transition between the two quaternary structures has been located, but the complete sequence of events is far from clear. Pauling and Coryell [8] observed that on oxygenation of deoxyhemoglobin the spin state of the ferrous iron changed from $s=2$ to $s=0$. Williams [9] pointed out that this corresponds to a marked reduction in the radius of the iron atom. Hoard [10] predicted on the basis of crystallographic studies of ferric porphyrin complexes that the low spin

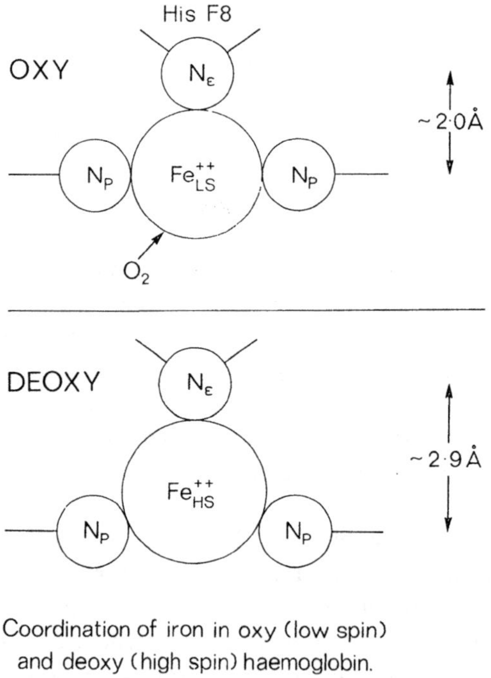

Fig. 6. The stereochemical trigger. Binding of ligand to the iron atom causes a change in the length of the iron–nitrogen bonds which is amplified by the geometry of the heme group into a large shift of the position of the iron atom.

form of the iron could fit into the center of the porphyrin ring, while in the high spin form the iron-nitrogen bonds would be sufficiently long that the iron atom would have to be substantially out of the plane of the porphyrin ring. A careful inspection of the electron density map of deoxyhemoglobin has shown that the iron atom is indeed substantially out of the plane of the porphyrin ring. Oxygenation should therefore pull the iron atom into the center of the porphyrin ring, shortening the distance between the plane of the heme group and helix F by nearly 1 Å. This is illustrated in Figure 6. The complete effect of this is unknown but it is believed to cause tertiary structure changes in the protein which directly weaken or break the salt bridges formed by the C-terminal residues.

In order to further understand the mechanism of ligation of hemoglobin we would like to study hybrid structures, that is, structures in which one pair of subunits is fully liganded and the other pair of subunits is fully reduced. There are several such systems available. It is possible to make hemoglobins in which one pair of subunits is ferric,

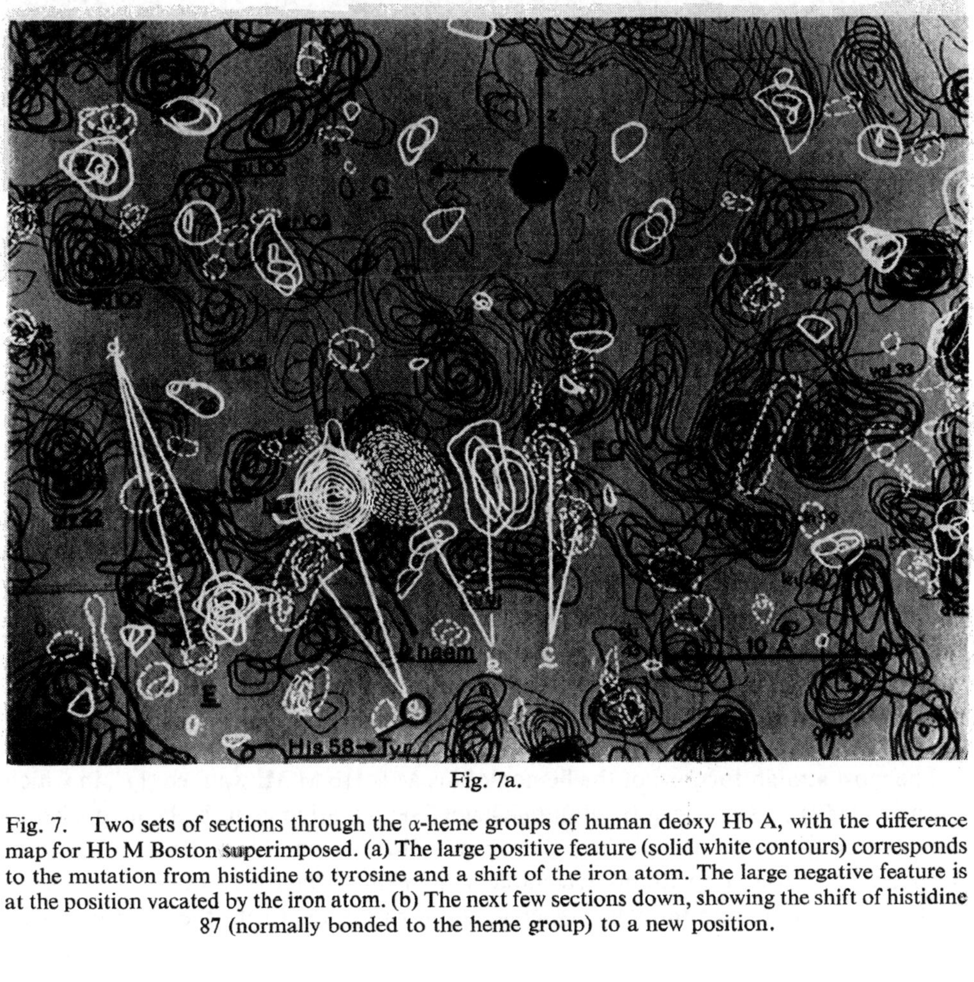

Fig. 7a.

Fig. 7. Two sets of sections through the α-heme groups of human deoxy Hb A, with the difference map for Hb M Boston superimposed. (a) The large positive feature (solid white contours) corresponds to the mutation from histidine to tyrosine and a shift of the iron atom. The large negative feature is at the position vacated by the iron atom. (b) The next few sections down, showing the shift of histidine 87 (normally bonded to the heme group) to a new position.

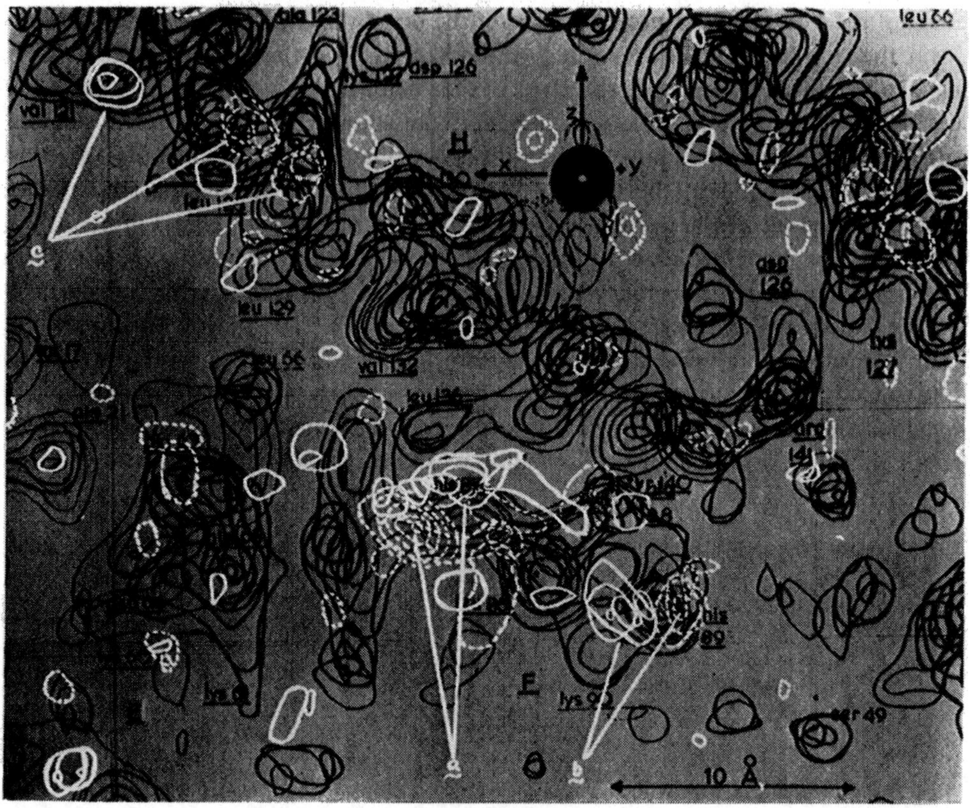

Fig. 7b.

with a cyanide ligand; these cyanomet hybrids should have the desired structures. We are currently attempting to grow crystals of these hybrids suitable for X-ray diffraction studies. Another approach which initially seemed quite promising was the study of a class of mutant hemoglobins known as the Hb M mutants. These mutants have one pair of chains ferric, and are not reduced by methemoglobin reductase. There are five known examples. Four of the five are replacements of the various heme-related histidines by tyrosines. The fifth is a replacement of a valine in the heme pocket by glutamic acid. A curious feature of these mutants is that they can all be crystallized in the deoxy quaternary structure (with the possible exception of Hb M Saskatoon, which has not yet been crystallized). The structures of four of the five hemoglobins M have now been studied, and it appears that although they may be useful for other purposes they are not good models of hybrid structures.

The most straightforward of the hemoglobins M is Hb M Milwaukee [11], in which valine E11 (67) β is replaced by glutamic acid. The mutation is in the heme pocket in the position normally occupied by the ligand. This hemoglobin forms a completely normal deoxy structure with no detectable movement of the iron atom into the plane of the heme group even though it is ferric and forms a bond with the glutamate side chain. If the α-chains are oxidized, Hb M Milwaukee crystallizes isomorphously with human met Hb, which shows that the β-subunits can accommodate themselves to the oxy structure. Spectroscopic studies indicate that the β-chains are capable of taking on the oxy form if the α-chains are reacted with CO. Thus there is probably a high spin-low spin equilibrium in this mutant like that mentioned previously for met Hb.

The structure of Hb M Boston has recently been solved by Pulsinelli (in preparation). This hemoglobin has the histidine in the heme pocket of the α-subunits (which does not interact with the iron) replaced by tyrosine. Figure 7 shows the difference electron density map superimposed on the native electron density map. It is very clear that the iron atom has moved drastically and that the normal bond to the proximal histidine is broken. The new structure is shown in Figure 8. The tyrosine forms a phenoxy bond with the iron, which is pulled *through* the heme group. The iron atom is now five coordinated ferric, and the connection between the home group and helix F is broken.

Hemoglobins M Hyde Park and M Iwate (also known as M Kankakee) have been studied by Greer [12]. Hb M Hyde Park has the heme-linked histidine of the β-subunits replaced by tyrosine and under deoxy conditions will crystallize isomorphously with normal human deoxy Hb. The difference map in the vicinity of the β-hemes is extremely complex and difficult to interpret. The deoxy crystals do not break up when

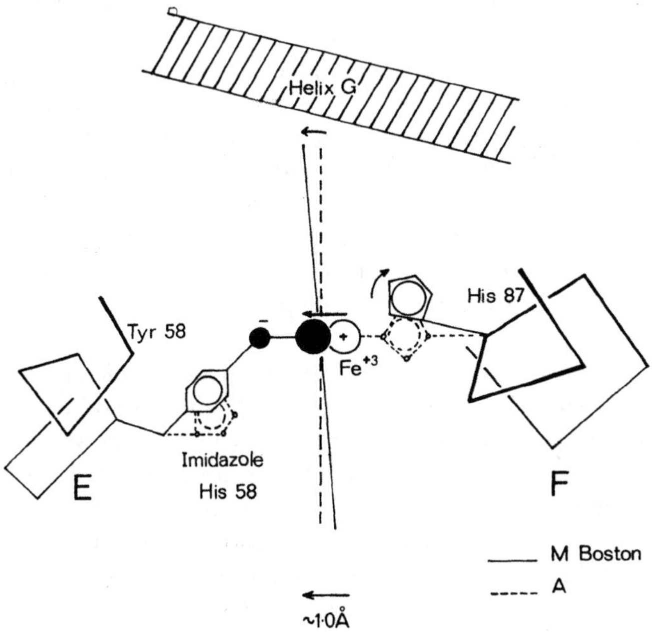

Haemoglobin M Boston

Fig. 8. The molecular interpretation of the Hb M Boston structure at the α-hemes.

exposed to air and Pulsinelli has collected a 3.5 Å set of data on the form with oxidized α-subunits. These data were extremely difficult to collect and the resulting electron density map is difficult to interpret because the β-subunits have a very abnormal structure due to the mutation. Nevertheless this is our best hybrid structure so far.

Hemoglobin M Iwate has the heme-linked histidines of the α-subunits replaced by tyrosine and crystallizes in a different space group from that of normal human deoxy Hb. Therefore it had to be solved as a new structure and only a 5.5 Å resolution map is available. The arrangement of the subunits is the same as that for normal human deoxy Hb and does not change when the normal β-chains are oxidized by exposure to air. The unreactive α-subunits are locked in a deoxy-like conformation because the extra length of the tyrosine compared with histidine compensates (and perhaps overcompensates) for the displacement of the iron atom into the plane of the heme. The distance between helix F and the heme group is similar to that of normal human deoxy Hb.

The above results seem to indicate that the hemoglobins M are extremely abnormal and should be treated with considerable caution. They certainly do not form good hybrid structures. Once the abnormalities of the structure have been determined by X-ray analysis, however, these mutants can be used to considerable advantage for such things as spectroscopic experiments, since many of the properties of the ferric subunits differ markedly from those of the normal subunits.

Another approach to the problem of determining intermediate structures in the ligation reaction of hemoglobin is to inhibit the change of quaternary structure in some fashion and then oxidize or reduce the crystals. The structures observed by such methods are almost certainly not kinetic intermediates (for that matter, neither are the hybrid structures) but should be very closely related to the kinetic intermediates. The first approach that was tried was the use of a chemical reagent, bis-maleido-methyl ether (BME). This reagent prevents the formation of the deoxy subunit arrangement; thus met Hb crystals may be reduced and the differences due to reduction studied [7, 13]. The difficulty with this particular approach is in finding a suitable reagent which does not perturb the normal structure. BME causes some significant changes from the normal met Hb structure which complicate analysis of the differences due to reduction.

Leigh Anderson has developed an alternative approach which is essentially to cast the molecules in plastic. He grows his crystals in the presence of acrylamide, which is polymerized by exposure to UV light after the crystals have grown. Crystals of deoxy Hb prepared in this manner can be oxidized by air slowly to met Hb without breaking up and are stable in the X-ray beam for periods varying from hours to days. On rapid oxidation or exposure to carbon monoxide the crystals break up. It is clear that the resulting map must be an average between the normal deoxy structure and molecules in various stages of oxidation. This procedure might therefore give some idea of the sequence of events in the oxidation process. At the moment the acrylamide procedure and the cyanomet hybrids seem to offer the most hope of determining the course of the ligation reactions of hemoglobin.

References

[1] Monod, J., Wyman, J., and Changeux, J. P.: 1965, *J. Mol. Biol.* **12**, 88.
[2] Perutz, M. F., Muirhead, H., Cox, J. M., and Goaman, L. G. C.: 1968, *Nature* **228**, 551.
[3] Bolton, W. and Perutz, M. F.: 1970, *Nature* **228**, 551.
[4] Muirhead, H. and Greer, J.: 1970, *Nature* **228**, 516.
[5] Perutz, M. F.: 1970, *Nature* **228**, 734.
[6] Arnone, A.: 1972, *Nature* **237**, 146.
[7] Perutz, M. F. and Ten Eyck, L. F.: 1971, in *Cold Spring Harbor Symposia on Quantitative Biology* **XXXVI**, 295.
[8] Pauling, L. and Coryell, C. D.: 1936, *Proc. Nat. Acad. Sci. U.S.* **22**, 210.
[9] Williams, R. J. P.: 1961, *Fed. Proc.* **20**, No. 3 (suppl. 10), 5.
[10] Hoard, J. L.: 1966, in *Hemes and Heme Proteins* (ed. by B. Chance, R. W. Estabrook, and T. Yonetani), Academic Press, New York, p. 9.
[11] Perutz, M. F., Pulsinelli, P. D., and Ranney, Helen M.: 1972, *Nature New Biol.* **237**, 259.
[12] Greer, J.: 1971, *J. Mol. Biol.* **59**, 107.
[13] Moffat, J. K.: 1971, *J. Mol. Biol.* **58**, 79.

DISCUSSION

Banerjee: I would like to hear you commenting, from a structural viewpoint, on the cases which have very small cooperativity in oxygenation reaction but rather large Bohr effects. Examples are hemoglobin valency hybrids $\alpha^{III}\beta_2^{II}$ and $\alpha_2^{II}\beta_2^{III}$. (Banerjee, Stetzkowsky and Henry, *J. Mol. Biol.*, in press.)

Ten Eyck: We are presently attempting to crystallize the cyanomet hybrids and can answer your questions in detail then. At present it seems certain that a quaternary structure change must release all protons involved in salt bridges. The question as to whether an oxidized subunit can maintain salt bridges in the deoxy quaternary structure remains open, but can perhaps be answered by studying partially oxidized forms.

Shulman: Banerjee raises the interesting question as to how the Bohr effect can be divided into parts. In the MWC model there are three contributions which must be evaluated to explain the hybrid results, as well as the results by the Rome group on the simultaneous release of protons and binding of oxygen. These three are the proton release per O_2 bound in the T-form, in the R-form and finally when the $T \rightarrow R$ switch occurs. When these contributions are evaluated, it should be possible to assign quantitative contributions to particular groups, probably those suggested by Perutz.

Wütrich: In view of what is generally known on the behaviour of iron (III) in coordination compounds it is rather surprising that the iron in the ferric subunits of hemoglobin M Boston is found to be penta-coordinated. I am tempted to guess that this observation in the single crystals of the hemoglobin does not necessarily mean that iron (III) will also be penta-coordinated in the molecule in solution.

Ten Eyck: The five-coordinated ferric state is certainly an unusual one possible only with very strong ligands such as phenoxy and methoxy ions. There is no question that Hb M Boston shows this state. When fully oxidized this hemoglobin crystallizes in an oxylike structure which may be six-coordinated. With respect to possible crystallization artifacts, the lattice energy is only 5 kcal/mole.

PROTON AND CARBON-13 NUCLEAR MAGNETIC RESONANCES IN HEMES AND HEMOPROTEINS: NEW ASPECTS FOR THE INVESTIGATION OF THE MOLECULAR CONFORMATIONS

KURT WÜTHRICH, REGULA M. KELLER, and RUDOLF BAUMANN

Institut für Molekularbiologie und Biophysik, Eidgenössische Technische Hochschule, 8049, Zürich, Switzerland

Summary. Applications of nuclear magnetic resonance spectroscopy to the investigation of the molecular conformations in paramagnetic proteins require a detailed knowledge of the electronic states in the paramagnetic centers. The data presented in this paper, which include the determination of the orientation of the electronic g-tensor in ferricytochrome b_5 in aqueous solution at ambient temperature, and measurements of the carbon-13 hyperfine shifts in several low spin iron (III) porphyrin complexes, should thus provide a basis for future conformational studies in low spin ferric hemoproteins. At the same time new insights are obtained into the spin density delocalization in these heme compounds.

Résumé. L'application de la spectroscopie de RMN à l'étude des conformations moléculaires dans les protéines paramagnétiques exige une connaissance détaillée des états électroniques des centres paramagnétiques. Les données présentées incluent la détermination de l'orientation du tenseur électronique g dans le ferricytochrome b_5 en solution aqueuse à température ambiante et la mesure des déplacements hyperfins du carbone-13 dans divers complexes ferriques à bas spin de la porphyrine. Elles devraient fournir la base d'études conformationnelles sur les hémoprotéines ferriques à bas spin. Elles fournissent également de nouveaux aperçus sur la délocalisation de la densité de spin dans ces composés hémiques.

1. Introduction

In the high resolution proton nuclear magnetic resonance (NMR) spectra of diamagnetic proteins the molecular conformations are mainly manifested by chemical shifts which arise from the local ring current fields in the aromatic amino acid residues and the prostethic groups [1, 2]. In paramagnetic proteins the nuclear resonances are further affected by the local magnetic fields of the unpaired electrons, and the resulting hyperfine shifts will in general also depend on the molecular conformations [2, 3]. Hence a detailed knowledge of the electronic states in the paramagnetic centers is a prerequisite for NMR studies of the molecular conformations in paramagnetic proteins.

For many hemoproteins the biological roles involve interconversions between diamagnetic and paramagnetic states of the heme iron, e.g., in myoglobin

$$Mb^{II}(Fe^{2+}, S=2) \overset{O_2}{\rightleftarrows} Mb^{II}O_2(Fe^{2+}, S=0) \tag{1}$$

or in cytochrome c

$$Cytc^{III}(Fe^{3+}, S=\tfrac{1}{2}) \overset{e^-}{\rightleftarrows} Cytc^{II}(Fe^{2+}, S=0). \tag{2}$$

Since the proton NMR spectra of hemoproteins contain in general a considerable

C. Sadron (ed.), Dynamic Aspects of Conformation Changes in Biological Macromolecules. 151–163.

number of resolved resonances [2], NMR is in principle a suitable technique for the investigation of the changes in the protein conformations which might occur in the course of these reactions. However, even though some approximative approaches appear to have been quite successful [4], the lack of sufficiently accurate data on the electronic structures in the heme groups has so far prevented rigorous investigations of this type. In the light of some recent developments it appears now as if these difficulties could be overcome for type (2) systems, which involve only the low spin ferric and ferrous states. The present paper describes the results of proton and carbon-13 NMR studies of low spin ferric heme compounds which we expect to contribute substantially towards this goal.

2. General Aspects

For the purpose of the present discussion the resonance position in ppm for each proton in a low spin ferric hemoprotein may be written as a sum of three terms

$$\Delta\nu = \Delta\nu_d + \Delta\nu_c + \Delta\nu_{pc}. \tag{3}$$

$\Delta\nu_d$ is the position which the resonance would have in the absence of the unpaired electron, and $\Delta\nu_c$ and $\Delta\nu_{pc}$ are the contact shift and pseudocontact shift which arise from the interactions with the unpaired electron. Depending on the relative importance of the three terms in Equation (3), the protons in the protein may be divided into three classes. The resonance positions of protons in certain locations relative to the heme will be determined entirely by $\Delta\nu_d$, and these lines can in favorable cases be identified on the basis of their behaviour at variable temperatures [2]. In the second class there are the protons of the heme and the axial ligands of the heme iron, which can experience sizeable contact and pseudocontact shifts, and finally there are all the other protons for which $\Delta\nu$ is determined by $\Delta\nu_d + \Delta\nu_{pc}$. In principle, if the atomic coordinates of the molecule and the electronic g-tensor are known, approximate values for $\Delta\nu_d$ and $\Delta\nu_{pc}$ [4, 5, 6] can be computed for every proton, and once a resonance line is assigned to specific protons its contact shift can then be determined with Equation (3).

Until recently the lack of data on the electronic g-tensor presented a major obstacle for the separation of contact and pseudocontact shifts in low spin ferric heme compounds. Recently the g-tensor was determined by electron paramagnetic resonance (EPR) spectroscopy in single crystals of cyanoferrimyoglobin [7] and ferricytochrome c [8]. In both proteins the g-tensor has orthorhombic symmetry. The pseudocontact shift for the ith proton in an orthorhombic system is given by Equations (4) and (15).

$$\Delta\nu'_{pc_i} = \frac{\beta^2 S(S+1)}{9kTr_i^3}\left(g_x^2(1 - 3\cos^2\omega_{xi}) + g_y^2(1 - 3\cos^2\omega_{yi}) + g_z^2(1 - 3\cos^2\omega_{zi})\right) \tag{4}$$

$$\Delta\nu_{pc_i} = F_i \Delta\nu'_{pc_i}. \tag{5}$$

β is the Bohr magneton, S the effective spin ($S = 1/2$ for low spin Fe^{3+}), k the Boltzmann constant, T the absolute temperature, r_i the distance between proton i and the iron, ω_{xi}, ω_{yi} and ω_{zi} the direction cosines of the position vector $\mathbf{r}_i$ relative to the principal axes of the g-tensor. $\Delta\nu'_{pc_i}$ is the pseudocontact shift for the situation where the electron spin would be located entirely on the iron, and only the ground state would be populated. F_i in Equation (5) then accounts for the influence of spin delocalization from the iron to the ligands, thermal mixing of electronic states, and Zeeman mixing with low lying excited states [6]. F_i will in general depend on the location of the nuclei relative to the heme group [9].

If the molecular geometry and the electronic g-tensor are known, and the factors F_i can somehow be estimated, Equations (4) and (5) can be used to compute the pseudocontact shifts. Comparison with the experimental pseudocontact shifts will then in principle reveal, whether the single crystal g-tensor is also applicable on the time scale of the NMR experiment for the compound in solution at ambient temperature. On the other hand if the resonance positions of a sufficient number of protons and the principal values of the g-tensor are known, a complete description of the g-tensor can in principle be obtained through a process of fitting the direction of the principal axes to the NMR data. In this way the orientation of the principal g-tensor axes in ferricytochrome b_5 has been determined, as described in the following section.

At this point, with $\Delta\nu_d$ and $\Delta\nu_{pc}$ known, one can obtain $\Delta\nu_c$ from Equation (3). $\Delta\nu_c$ (in ppm) is related to the contact interaction constant A by [10]:*

$$\Delta\nu_{ci} = -A_i \frac{\gamma_e}{\gamma_I} \frac{S(S+1)}{3kT}, \tag{6}$$

where γ_e and γ_I are the gyromagnetic ratios for the electron and the nuclear spin, respectively. From the contact interaction constants A_i one can derive values for the electron spin densities on the atoms of the conjugated porphyrin ring system [11–14] and thus describe the electronic wave functions which contain fractions of the unpaired spin.

The reliability of the data depends of course on the number of nuclei for which the resonance positions can be measured. In the proton NMR spectra of hemoproteins it is mostly the limited spectral resolution which determines the amount of experimental data. We have therefore chosen cytochrome b_5 for these studies, which contains a larger number of resolved lines than most of the other proteins. In the iron porphyrin complexes the investigations were extended to include the carbon-13 resonances.

3. Proton NMR Studies of the g-Tensor in Ferricytochrome b_5

The cytochrome b_5 used for these experiments has a molecular weight of ca. 12000, and consists of one polypeptide chain with 93 amino acid residues and one protoheme IX group. The amino acid sequences of cytochromes b_5 from several species [15–17],

* In (6) it is assumed for simplicity that only the electronic ground state is populated, and the effect of the g-tensor anisotropy on $\Delta\nu_c$ [6] is not accounted for. This should not affect the general conclusions drawn in the discussion section [2].

and the molecular conformation in single crystals of the ferric [18] and the ferrous [19] protein from calf liver are known. In both oxidation states the two axial positions of the heme iron are occupied by an imidazole nitrogen atom of the histidyl residues in positions 39 and 63 of the polypeptide chain.

In the proton NMR spectrum of ferricytochrome b_5 (Figure 1) approximately 25 well resolved resonances are in the spectral regions from -30 to -10 ppm, and 0 to 8 ppm. Many of these resonances correspond to protons of protoheme IX and the

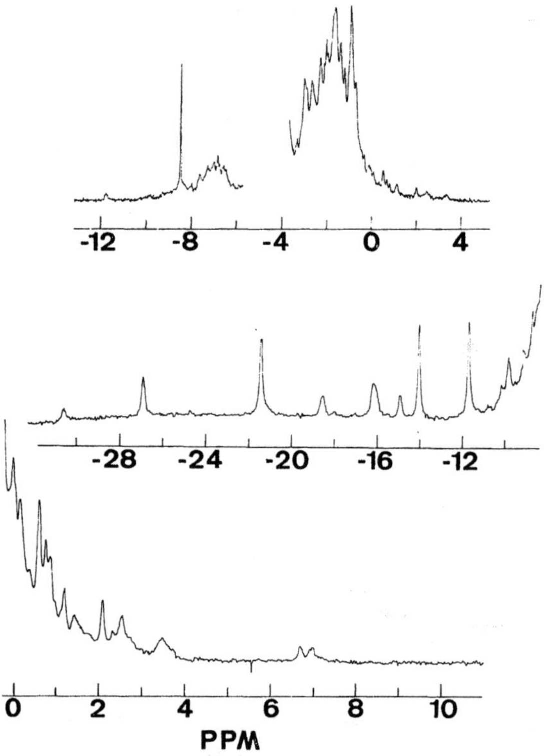

Fig. 1. Proton NMR spectrum at 220 MHz of a ca. 0.008-M solution of ferricytochrome b_5 in deuterated 0.2-M phosphate buffer, pD = 6.3 at 29°C. The spectral region from -4 to -6 ppm, which contains the resonance of HDO and its spinning side bands, has been omitted. Resonance positions are in ppm from internal DSS. The three spectral regions are represented with different horizontal and vertical scales. The sharp line at -8.4 ppm comes from a low molecular weight component which could later be removed by exhaustive dialysis.

two axial histidyl ligands, which can be shifted from their positions in the diamagnetic compounds by contact and pseudocontact interactions with the unpaired electron [2]. However, comparison of the resonance intensities in Figure 1 with the number of protons in the molecules which are bound to the iron shows that in addition to some or all of the protons of the iron ligands quite a number of amino acid resonances are also outside of the spectral region from 0 to -10 ppm. In particular there are six methyl resonances between 0.7 and 2.5 ppm at 29°C (Figures 1 and 2) which can be assigned to aliphatic amino acid side-chains. Additional methyl resonances of the polypeptide chain are in positions near 0 ppm. The positions of all these methyl lines are determined by $\Delta\nu_d + \Delta\nu_{pc}$, and they have been used to study the orientation of the g-tensor in this molecule.

Approximate values for $\Delta\nu_d$ were computed as described earlier [4, 9] and from this the high field methyl resonances in the diamagnetic ferrocytochrome b_5 (Figure 3) could be identified as shown on the left-hand side of Figure 4 [9]. Since there appear to be no major differences between the molecular conformations in the oxidized and reduced protein [19] the same values of $\Delta\nu_d$ should be valid in both oxidation states for the resonances used in this analysis, which are all located in the interior of the molecule [18].

Because only the principal values of the g-tensor have so far been determined by EPR [20], a straightforward application of Equations (4) and (5) for the computation of pseudocontact shifts was not possible. Instead, keeping the principal values fixed, the directions of the principal axes of the g-tensor were varied in search for agreement

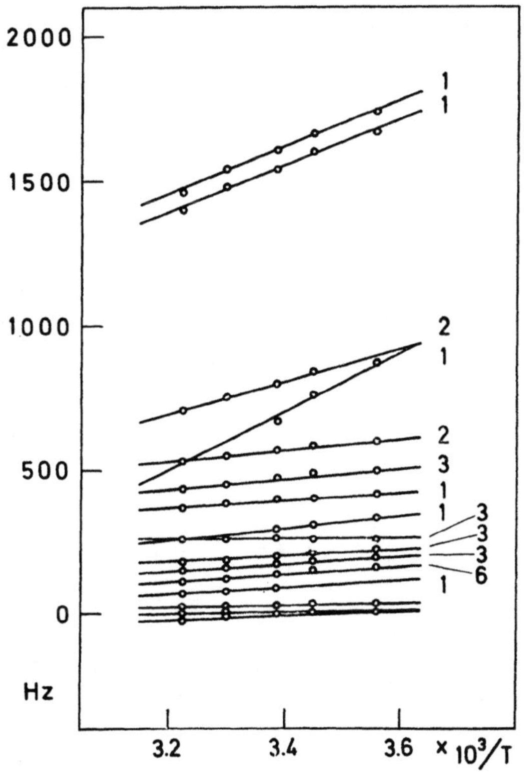

Fig. 2. Dependence on the reciprocal of temperature of the proton resonances between −0.5 and +8 ppm in ferricytochrome b_5. The number of protons corresponding to the resonance intensities are given on the right-hand side.

between calculated and observed pseudocontact shifts. In this process the principal values $g_z = 3.03$, $g_y = 2.23$, and $g_x = 1.43$ were used [20], and the dependence of the pseudocontact shifts on the orientation of the g-tensor was computed for all the methyl groups within a sphere of 14 Å around the heme iron. The directions of the principal axes were modified by variation of the Eulerian angles θ, ψ and ϕ in steps of 4°, 4°/sin θ, and 6°, respectively. Since the positions relative to the heme group must be quite similar for all the peptide methyl groups with resonances shifted to high field positions by pseudocontact coupling, the factor F_i in Equation (3) was assumed to be identical for all these resonances. For each set of θ, ψ and ϕ we then tried to fit the computed and the experimental resonance positions and their temperature dependences by further varying the reduction factor F in the range 0.6 to 1.0. It was found that only one relatively small region in the three-dimensional space spanned by θ, ψ, and ϕ gives a reasonable fit with the experiment, and on the basis of some additional details in the NMR spectral features, which were discussed elsewhere [9], the g-tensor orientation could be determined within relatively narrow limits. The z-axis is tilted

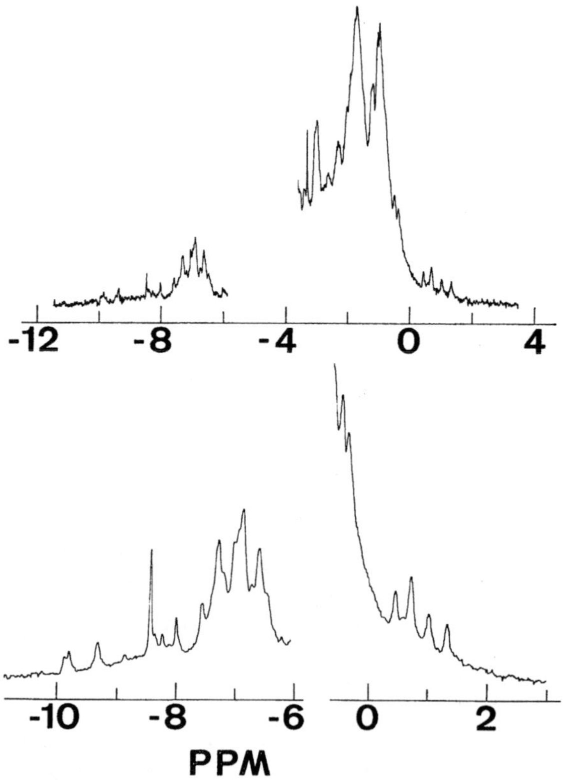

Fig. 3. 220 MHz proton NMR spectrum of 0.008-M ferrocytochrome b₅ in 0.2-M deuterated phosphate buffer, pD 7.0, at 29°C. The spectral regions from −1 to 3 ppm, and from −6 to −10 ppm are also shown on an expanded scale. The sharp line at −8.4 ppm corresponds to a low molecular weight component which could later be removed by exhaustive dialysis.

away from the heme normal by $12 \pm 3°$ and the y-axis lies in the heme plane at an angle of $13 \pm 4°$ to the line connecting the nitrogen atoms of the pyrrole rings VR and PL [21]. The x-axis lies therefore near the line connecting the nitrogen atoms of the pyrrole rings VL and PR, but is tilted out of the heme plane by 12°. This situation is illustrated in Figure 5, which shows further that g_x is roughly perpendicular to the planes of the two axial imidazole rings of His [39] and His [63].

From the above fitting procedure the high field methyl resonances could be assigned as shown in Figure 4. The three leucyl residues in positions 23, 25 and 32 belong to the β-sheet which forms the bottom of the heme crevice. Alanine (67) and valine (45) are located in the α-helices which form the walls of the heme crevice [38]. With the resonance assignments of Figure 4 the differences between calculated and observed resonance positions in ferricytochrome b_5 do not exceed 0.2 ppm if F is taken to be 0.77. In view of the approximate nature of the computations this close fit must be coincidental. On the other hand, considering the number of well resolved resonances

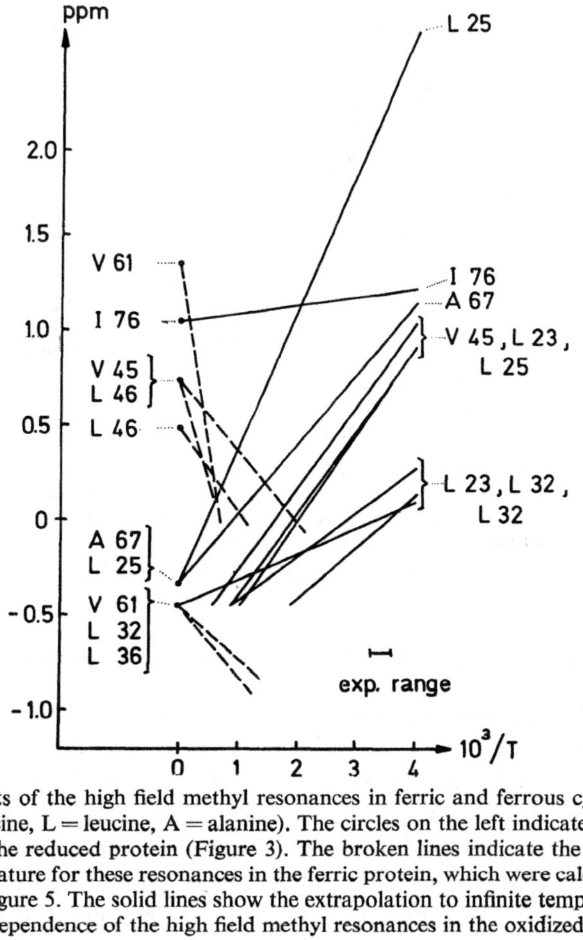

Fig. 4. Assignments of the high field methyl resonances in ferric and ferrous cytochrome b_5 (V = = valine, I = isoleucine, L = leucine, A = alanine). The circles on the left indicate the observed resonance positions in the reduced protein (Figure 3). The broken lines indicate the dependence on the reciprocal of temperature for these resonances in the ferric protein, which were calculated on the basis of the g-tensor in Figure 5. The solid lines show the extrapolation to infinite temperature of the measured temperature dependence of the high field methyl resonances in the oxidized protein (Figure 2).

available for the fitting process this result would hardly be comprehensible unless it is assumed that the conformation of the heme crevice in single crystals [18] and the principal values of the g-tensor in a frozen solution [20] are essentially maintained in the protein in solution at room temperature. Figure 4 shows that with the exception of isoleucine (76), which is located at a distance of 17.5 Å from the heme iron near the indole ring of tryptophan [22], all the high field methyl resonances in ferrocytochrome b_5 (Figure 3) are shifted to positions at low field from DSS by the pseudocontact interactions in the ferric protein.

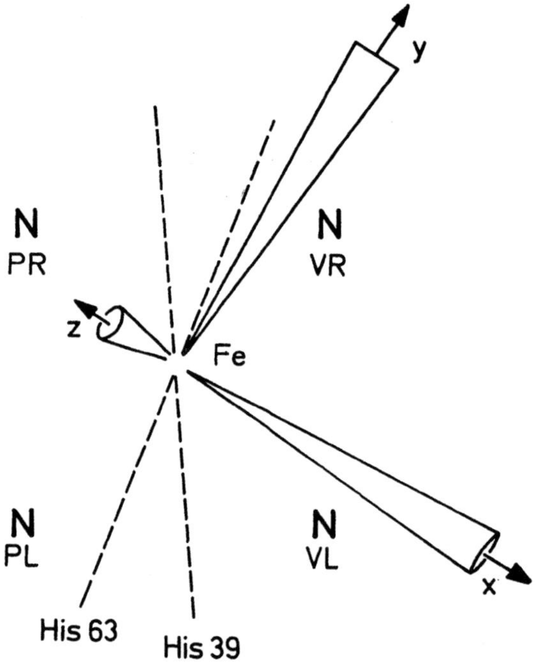

Fig. 5. Orientation of the principal axes of the g-tensor in cytochrome b_5 as derived from the NMR measurements. The figure represents the view on the heme plane from the side of the axial histidine in position 63. For each axis the range of directions for which good fits with the experimental data were obtained is indicated by a cone. The broken lines indicate the orientations of the planes of the imidazole rings in the axial histidyl residues.

4. Carbon-13 Resonances in Low Spin Iron (III) Porphyrin Complexes

Carbon-13 spectra were obtained with the Fourier transform technique. The individual resonances were identified in proton-carbon-13 heteronuclear double resonance experiments. This is illustrated in Figures 6 and 7 for the iron (III) tetraphenyl-porphin (TPP) complex. In the proton noise decoupled spectrum (Figure 6A) seven resonances are observed which correspond to the seven different types of carbon atoms in TPP (Figure 7). Selective irradiation of the proton resonance a (Figure 7) yields spectrum 6B, and irradiation of the proton line b″ gives spectrum 6C. The resonance at −91

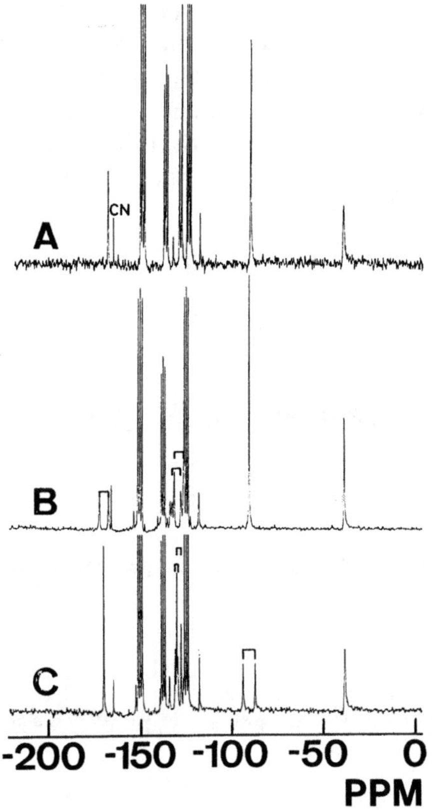

Fig. 6. Carbon-13 NMR spectra of Fe^{3+}(tetraphenylporphin)$(CN)_2$ in a mixed solvent of d_5-pyridine and D_2O 4: 1 at at 34°C. The seven resonances of TPP (see Figure 7) are at −40, −91, −118, −128, −129, −133, and −168 ppm. The three triplets at −124, −137, and −150 ppm are the resonances of d_5-pyridine, and the line at −165 ppm comes from CN^-. A. ^{1}H-noise decoupled. B. Selective double resonance irradiation at proton resonance a″ (Figure 7). C. Selective double resonance irradiation at proton resonance b″ (Figure 7).

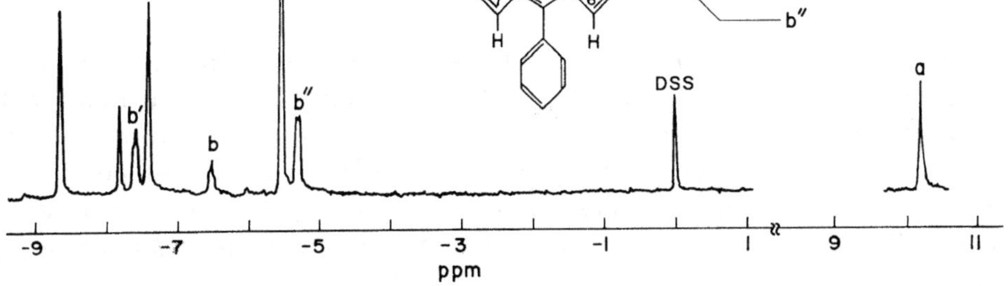

Fig. 7. Proton NMR spectrum at 220 MHz of Fe^{3+}(tetraphenylporphin)$(CN)_2$ in a mixed solvent of d_5-pyridine and D_2O 4: 1 at 26°C.

ppm corresponds thus to the β-carbon atoms (a in Figure 7), the resonances at -128, -129, and -168, which are split into doublets by the proton-carbon-13 spin-spin couplings in Figure 6B, to the carbon atoms b′, b, and b″ of the phenyl rings (Figure 7), respectively, and the lines at -40, -118, and -133 ppm to the three quarternary carbon atoms in the molecule. From comparison with the iron (III) porphin complex [22] the line at -40 ppm corresponds to the quarternary carbons of the pyrrole rings.

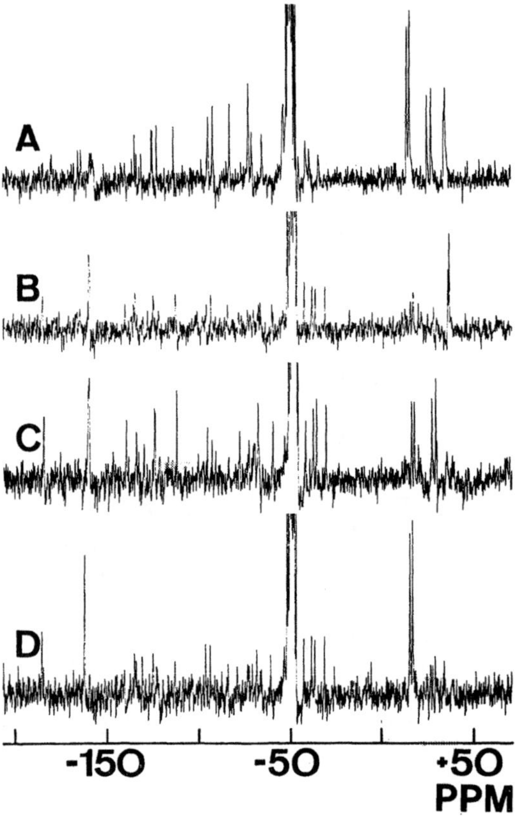

Fig. 8. Carbon-13 NMR spectra of Fe^{3+}(protoporphyrin IX)$(CN)_2$ in CD_3OD at 34° C. The strong line at -50 ppm comes from the solvent, and the resonance at -163 ppm from CN^- A. ^1H-noise decoupled. B. Selective double resonance irradiation at the two proton resonances a which are at -18 to -19 ppm in Figure 9. C. Selective irradiation of the four resonances a and f which are at -12 to -15 ppm in Figure 9. D. Selective irradiation of the two resonances c (Figure 9).

The remaining two quarternary carbons have not been individually assigned so far. The resonance at -165 ppm is also observed in a solution of KCN in a pyridine–water mixed solvent. In similar ways the carbon-13 resonance have been identified in the spectra of the iron (III) porphin complex and the Zn^{2+}-complexes with porphin and TPP [22]. Hence the hyperfine shifts are now known for all the proton resonances in the iron (III) complexes with porphin and TPP, for all three carbon-13 resonances in

TABLE I

^{1}H and ^{13}C resonances in Zn^{2+}(Porphin)[a] and Fe^{3+}(Porphin)$(CN)_2$[a]

	Zn(Porphin)[b]	Fe^{3+}(Porphin)$(CN)_2$[b]	$\Delta\nu_{hf}$
β-H	− 9.8	14.8	24.6
meso-H	− 10.6	− 1.3	9.3
β-C	−133.2	−92.6	40.6
meso-C	−105.7	−56.0	49.7
quart. C	(−150)[c]	−49.4	100

[a] In a mixed solvent of d_5-pyridine and D_2O 4:1 at 34°C.
[b] Chemical shifts in ppm from internal TMS. Negative numbers indicate shifts to lower field.
[c] From comparison with Zn^{2+}(TPP) in $CDCl_3$ [22].

the porphin complex (Table I), and for 5 of the 7 carbon-13 resonances in the TPP complex.

Carbon-13 spectra were also recorded for the cyanide complex of iron (III) protoporphyrin IX. As can be seen from Figures 8 and 9 we have so far assigned the resonances in the spectral region from 10 to 40 ppm. Four of these lines correspond to the four ring methyl-carbon atoms, and two to the methylene carbon atoms c (Figure 9) of the propionic acid side chains. From Figure 8 it would appear as if the two carbon resonances f (Figure 9) might accidentally coincide with the resonances c. Work on this and on the resonance assignments in the lower field regions of the spectrum is currently in progress.

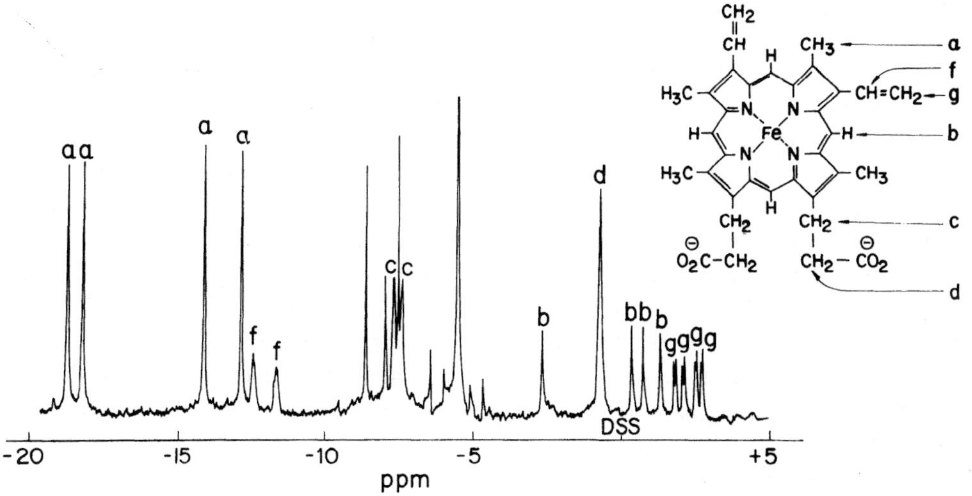

Fig. 9. Proton NMR spectrum at 220 MHz of Fe^{3+}(protoporphyrin IX)$(CN)_2$ in a mixed solven of d_5-pyridine and D_2O 4:1 at 8°C.

5. Discussion

At the present stage the data on cytochrome b_5 which are summarized in Figure 4 provide a handle for the investigation of the molecular conformations which may occur in the reduced and oxidized states of this molecule under various conditions. On a more general basis it is worth noting that the g-tensor in Figure 5 is quite closely related to those observed by EPR in single crystals of ferricytochrome c [8] and cyano-ferrimyoglobin [7]. Overall the outcome of this investigation thus appears to support that the data on the electronic g-tensor of low spin ferric hemoproteins measured in frozen solution or in single crystals are also valid under the conditions of the NMR experiments. Thus one should on the basis of the presently available data be in a position to investigate conformational changes which occur during oxidation-reduction reactions of the type of Equation (2) for a few proteins, e.g., it should be possible to decide whether the conformational differences between oxidized and reduced cytochrome c in single crystals [23] are also present in solutions of this molecule.

With the g-tensor known, the contributions $\Delta\nu_{pc}$ to the hyperfine shifts of the resonances of protoheme IX and the two axial histidines can be computed with Equations (4) and (5). From Equation (3) the contact shifts are then obtained for those resonances which can be assigned to specific protons of the iron ligands, and hence the spin density distribution in the heme group can be investigated. In the final outcome the analysis of the data on the iron ligands should also provide a check on the validity of the g-tensor in Figure 5 [24].

In contrast to the hemoproteins, the iron porphyrin complexes will in general contain no nuclear resonances which are sizeably shifted by pseudocontact coupling and yet not affected by contact interactions with the electronic spin. Furthermore no single crystal EPR data are as yet available for these complexes, and it has previously been observed [2, 24] that the g-tensor which is effective in producing the NMR pseudocontact shifts is quite different from that observed in frozen solutions of these compounds. With the data now available from the proton and carbon-13 NMR spectra it should in principle be possible to determine the contributions $\Delta\nu_d$, $\Delta\nu_{pc}$ and $\Delta\nu_c$ (Equation (3)) from the NMR results alone. On the basis of the proton and carbon-13 contact shifts a rather detailed description of the electronic wave functions should then be possible. Hopefully these findings will also contribute to a better understanding of the electronic structures in the heme groups of hemoproteins.

A preliminary interpretation of the proton and carbon-13 NMR data for the iron porphin complex (Table I), in which the relations proposed by McConnell [11] and Karplus and Frankel [13, 14] were used to derive experimental spin densities from the observed nuclear resonance shifts, has been presented elsewhere [22]. The following qualitative features appear to emerge. First, the pseudocontact shifts arising from the electron spin localized on the iron (Equations (4) and (5)) are smaller than what one might have expected from the g-values observed in the frozen solution of the complex [2]. This seems to be in qualitative agreement with earlier NMR data on the iron (III) complex with deuteroporphyrin IX dimethyl ester [25]. Second, the pseudocontact shifts due to the electron spin density localized in the π-orbitals of the carbon atoms make a substantial contribution to the hyperfine shifts of the carbon-13 resonances. Third, there is a positive spin density on all the carbon atoms of the porphin ring. Work on a more detailed interpretation of the experimental data given in this paper is in progress.

Acknowledgements

We would like to thank Dr. P. Strittmatter for a generous gift of ferricytochrome b_5, and Dr. F. Scott Mathews for the atomic coordinates of the protein. This work was supported by the Swiss National Science Foundation (Project 3.423.70).

References

[1] McDonald, C. C. and Phillips, W. D.: 1970, in *Fine Structure of Proteins and Nucleic Acids* (ed. by G. D. Fasman and S. D. Timasheff), M. Dekker, New York, p. 1.
[2] Wüthrich, K.: 1970, *Structure and Bonding* **8**, 53.
[3] Wüthrich, K. and Keller, R. M.: in *Proc. 4th Intern. Biophys. Congr.*, Moscow 1972 (in press).
[4] Shulman, R. G., Wüthrich, K., Yamane, T., Patel, D. J., and Blumberg, W. E.: 1970, *J. Mol. Biol.* **53**, 143.
[5] McConnell, H. M. and Robertson, R. E.: 1958, *J. Chem. Phys.* **29**, 1361.
[6] Kurland, R. J. and McGarvey, B. R.: 1970, *J. Magn. Res.* **2**, 286.
[7] Peisach, J., Blumberg, W. E., and Wyluda, B. J.: 1971, in *Proc. First Eur. Biophys. Congr.*, p. 109.
[8] Mailer, C. and Taylor, C. P. S.: 1972, *Biophys. Soc. Abstracts, Biophys. J.* **12**, 238a.
[9] Keller, R. M. and Wüthrich, K.: 1972, *Biochim. Biophys. Acta* **285**, 326.
[10] Bloembergen, N.: 1957, *J. Chem. Phys.* **27**, 595.
[11] McConnell, H. M.: 1956, *J. Chem. Phys.* **24**, 764.
[12] Bolton, J. R., Carrington, A., and McLachlan, A. D.: 1962, *Mol. Phys.* **5**, 31.
[13] Karplus, M. and Fraenkel, G. K.: 1961, *J. Chem. Phys.* **35**, 1312.
[14] Fraenkel, G. K.: 1962, *Pure Appl. Chem.* **4**, 143.
[15] Ozols, J. and Strittmatter, P.: 1969, *J. Biol. Chem.* **244**, 6617.
[16] Tsugita, A., Kobayashi, M., Tani, S., Kyo, S., Rashid, M. A., Yoshida, Y., Kajihara, T., and Hagihara, B.: 1970, *Proc. Nat. Acad. Sci. US* **67**, 442.
[17] Nobrega, F. G. and Ozols, J.: 1971, *J. Biol. Chem.* **246**, 1706.
[18] Scott Mathews, F., Levine, M., and Argos P.: 1972, *J. Mol. Biol.* **64**, 449.
[19] Scott Mathews, F.: private communication.
[20] Bois-Poltoratsky, R. and Ehrenberg, A.: 1967, *Europ. J. Biochem.* **2**, 361.
[21] Schoenborn, B. P., Watson, H. C., and Kendrew, J. C.: 1965, *Nature* **207**, 28.
[22] Wüthrich, K. and Baumann, R.: *Helv. Chim. Acta* (in press).
[23] Takano, T., Swanson, R., Kallai, O. B., and Dickerson, R. E.: 1971, in *Cold Spring Harbor Symposium on Quantitative Biology.*
[24] Shulman, R. G., Glarum, S. H., and Karplus, M.: 1971, *J. Mol. Biol.* **57**, 93.
[25] Wüthrich, K., Shulman, R. G., Wyluda, B. J., and Caughey, W. S.: 1969, *Proc. Nat. Acad. Sci. US* **62**, 636.

A HIGH RESOLUTION NMR STUDY OF HYDROGEN BONDED PROTONS IN tRNA

R. G. SHULMAN

Bell Laboratories, Murray Hill, New Jersey 07974, U.S.A.

and

D. R. KEARNS, B. R. REID, and Y. P. WONG

University of California, Riverside, California 92502, U.S.A.

Summary. Recent high resolution NMR studies of the hydrogen bonded protons in purified tRNA molecules are reviewed. The number of hydrogen bonds correspond very well with those predicted by the cloverleaf model and the resonances can be separated into those arising from AU and GC base pairs.

Résumé. Les pics fournis en RMN à 220 et 300 MHz par des solutions aqueuses de tRNA ont été attribués à des protons chelatés de NH cycliques et à des protons amino. Les protons du NH cyclique apparaissent à bas champ, entre −11 et −15 ppm, l'étalon interne étant le DSS. On trouve 20 ± 2 protons de ce type dans le $tRNA^{Phe}$ de levure, en excellent accord avec les prévisions du modèle en feuille de trèfle. L'étude d'un système modèle (en collaboration avec D. M. Crothers) comportant dAACAA + dTTGTT montre que les résonances U ou T N_3H apparaissent vers −13.5 ppm et la GN_1H vers −12,5 ppm. A partir de ces étalonnages pris pour première approximation, les spectres de RMN de sept tRNA purifiés ont été décomposés en un certain nombre de paires AU et GC en accord avec les modèles en feuille de trèfle construits d'après leur séquence connue. Les déplacements dus au courant de cycle de ces résonances NH ont été calculés et utilisés pour obtenir un accord convenable avec certaines caractéristiques de spectre. Ces attributions sont étayées par les spectres de la moitié TΨ et de l'extrémité 3′ d'un fragment de $tRNA^{Phe}$ de levure renfermant les 3/4 de la molécule, qui montrent le nombre prévu de résonances dans les domaines spectraux attendus.

Dans la région des amino les protons amino à liaisons hydrogène, entre −7,4 et −9,0 ppm, peuvent être distingués des protons de groupements amino libres observés entre −6,0 et 7,4 ppm.

Les intensités indiquent que les deux types de protons aminés subissent un déplacement vers les faibles champs lorsque le groupement amino présente une liaison hydrogène, ceci en accord avec des expériences d'échange du tritium.

Recently it has been possible to observe high resolution NMR peaks of the hydrogen bonded protons in tRNA. It is the purpose of this report to summarize the results of these experiments stressing the resolution obtained, the number of hydrogen-bonded protons counted, their assignments to particular base pairs and finally the inference which can be drawn about the structure of tRNA.

The low field region of the 220 MHz NMR spectrum of $tRNA^{Phe}_{Yeast}$ is shown in Figure 1. The dashed line shows that when the $tRNA^{Phe}_{Yeast}$ is dissolved in D_2O the only resonances observed in the region from 6 to 15 parts per million (ppm) below the internal standard DSS (2,2-dimethyl-2-silapentane-5-sulfonate) are the aromatic protons of the bases near −8 ppm. When the tRNA was dissolved in H_2O the spectra shown by the solid line were observed, where additional protons are observed in the four distinct regions −6.3 to −7.4, −7.4 to −9.0, −9.0 to −11.0, and −11.0 to −15.0 ppm. The protons in these regions can be integrated and assigned to particular groups in the tRNA on the basis of model system results, in particular the early

C. Sadron (ed.), Dynamic Aspects of Conformation Changes in Biological Macromolecules. 165–169.

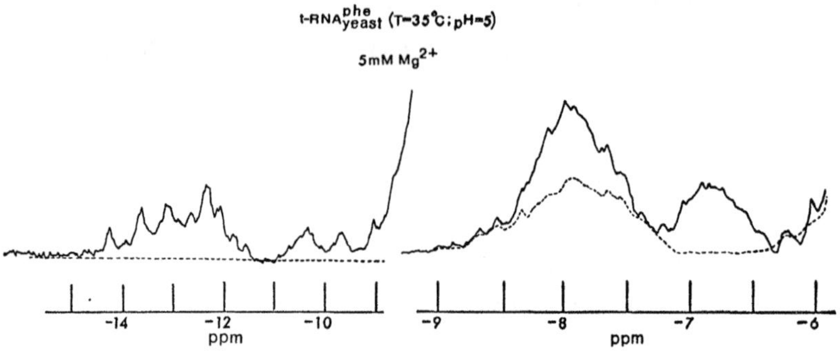

Fig. 1. Proton NMR spectrum at 220 MHz of a solution of tRNA$^{Phe}_{Yeast}$. Solid lines are resonances obtained with H_2O solvent, dotted lines with D_2O. (Temperature 35° C, concentration 2 mM, NaCl 0.1 M, Mg^{+2} 5 mM, pH 5.0.)

studies by Katz and Penman (1966) and Shoup *et al.* (1966). These numbers and assignments are presented in Table I.

The hydrogen-bonded ring NH protons are found in the −11 to −15 ppm region. Since this includes the N_1H of guanine and the N_3H of uridine there is one proton resonance per base pair in this region. Figure 2 shows the high resolution proton NMR spectrum in this region of tRNA$^{Phe}_{Yeast}$ taken at 300 MHz. The resolution is very good and it is possible to determine the intensities of the resolved lines which are given in Table II. The total intensity in this region corresponded to 18 protons when determined by these internal comparisons where the lowest field resonance, A, is assigned an intensity of one proton. In a separate experiment the region was assigned

TABLE I

A comparison of the number of exchangeable protons resonances observed in NMR with the number of resonances expected on the basis of the cloverleaf model. In calculating the number expected from a cloverleaf structure two different assumptions were used. In case A we assumed that both amino protons of a hydrogen-bonded amino group are shifted downfield. In case B we assumed that only one of the protons of a hydrogen-bonded amino group is shifted downfield from its normal position in the monomer

Type of proton	Exchangeable protons observed by NMR				Number of protons calculated from cloverleaf model	
	Total observed	H-bonded	Non-H-bonded	Spectral range (ppm)	Case A	Case B
					Both amino protons shifted	One amino proton shifted
$-NH_2$	112	67		−7.4 to −9.0	64	32
			45	−6.3 to −7.4	52	84
ring NH	19 ± 2	19		−11.0 to −15	20 ± 1	

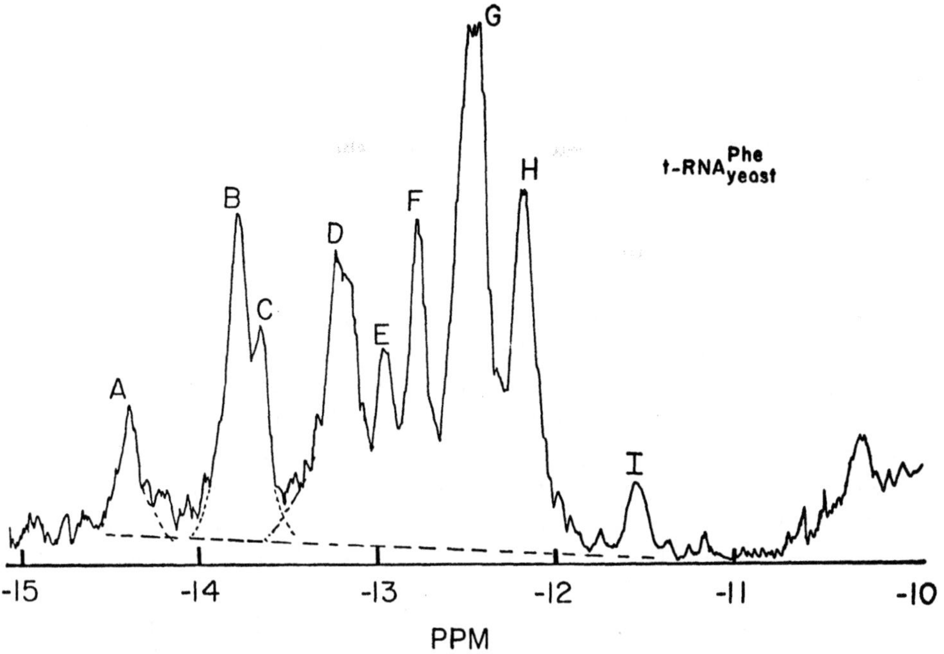

Fig. 2. Proton NMR spectrum at 300 MHz of $tRNA^{Phe}_{Yeast}$. Temperature 35°C and pH 7.5. (NaCl 0.1 M, Mg^{+2} 5 mM, tRNA concentration 2 mM.)

a total intensity of 21 ± 2 protons by comparing the integrated intensity with that of an external standard, the lowest field methyl resonance of cyanometmyoglobin. An average of these two measurements indicated that resonances from 19 ± 2 protons were observed in the -11 to -15 ppm region which agrees well with the 20 ± 1 expected from the cloverleaf model depending upon whether or not base paired resonances are expected from one G–U and one 2-methyl G–C pairs. Furthermore the number of base pairs measured and the number expected from the cloverleaf model are in excellent agreement in four other purified tRNA molecules which we have stud-

TABLE II

Line	Intensity
A	1
B	2
C	1
D	3
E	1
F	2
G	5
H	2
I	~1

ied. Hence the simplest interpretation is that the resonances in the −11 to −15 ppm region come from the cloverleaf structure with no discernible contributions from tertiary structure.

In order to assign these resonances to specific base pairs several additional steps have been taken. First model system studies have shown that the hydrogen-bonded U3 proton of the AU base pair is observed at lower fields than the G1 proton of the GC pairs. For example in the base paired form of

$$\begin{array}{c} \mathrm{AACAA} \\ |\,|\,|\,|\,| \\ \mathrm{TTGTT} \end{array}$$

a resonance of intensity four protons was found at −13.6 ppm and another narrower resonance corresponding to one proton was observed at −12.4 ppm (Crothers and Shulman, to be published). The helical geometry of the RNA and DNA helices requires that these ring NH protons would be shifted upfield by ring currents from their neighbors, particularly the neighboring purines. These ring current shifts can be calculated from approximate molecular orbitals and the helical structure. In this way it is possible to remove these contributions from the shifts in the above mentioned pentanucleotide and to obtain the positions of the ring NH protons of the AU and GC base pairs respectively in the absence of ring current shifts. From this information it is possible to calculate the resonance positions expected for the helical sections of the cloverleaf model for eight different tRNA molecules and the predicted patterns have been compared with the observed. To a first approximation the calculated spectra agree with the observed. In particular the number of proton resonances in the lower field region (i.e. below −13.3 ppm) agrees very well with the numbers of AU base pairs in the cloverleaf model, while the intensities above −13.3 ppm agree well with the numbers of GC pairs. At present, however, the agreement is generally no better than this rough division into an AU and a GC region, since the exact shapes of these regions are not predicted by the ring current calculations alone. Additional experiments on model systems as well as partially digested fragments of $tRNA^{Phe}_{Yeast}$ are underway in an attempt to assign specific resonances to particular base pairs.

We turn now to the −7.4 to −9.0 ppm region, which contains hydrogen bonded amino protons. As can be seen from Table I there are 67 exchangeable protons in this region. This agrees quite well with the number of amino protons calculated in case A where it is assumed that both of the amino protons of a hydrogen bonded amino group contribute to resonances in this region, but it does not agree very well with the number calculated in case B, where it is assumed that only one proton per amino group will resonate in this region. It can be seen that the 45 protons contributing to the −6.3 to −7.4 ppm region and which are assigned to free amino protons also agree, within experimental errors, with the number expected from case A but disagree with case B.

Finally we note that the three or four resonances observed in the −9.0 to −11.0 ppm region have not been identified with particular residues. Studies of the denaturation and aggregation of tRNA suggest that they are associated with the tertiary structure – a point which requires further study.

References

Preliminary reports of the work described here have been published in:
Kearns, D. R., Patel, D. J., and Shulman, R. G.: 1971, *Nature* **229**, 338.
Kearns, D. R., Patel, D. J., Shulman, R. G., and Yamane, T.: 1971, *J. Mol. Biol.* **61**, 265, and a longer account is in press.
Wong, Y. P., Kearns, D. R., Reid, B. R., and Shulman, R. G.: 1972, *J. Mol. Biol.* **72**, 725.

The early work on monomers is by:
Katz, L. and Penman, S.: 1966, *J. Mol. Biol.* **15**, 220.
Shoup, R. R., Miles, H. T., and Becker, E. D.: 1966, *Biochem. Biophys. Res. Commun.* **23**, 194.

DISCUSSION

Bradbury: Is there an effect of base stacking on the chemical shift of the H bonds in base pairs and what may be expected on comparing the stacks in the arms of the tRNA structure with single base pairs (which may form in tertiary structure)?

Shulman: As far as we can tell the intrinsic position of the ring NH proton resonance is independent of the base stacking and is affected by its neighbors only through their ring currents.

Wüthrich: What method did you use to record the spectra when you had only a minimal amount of a 3×10^{-4} M solution for the experiment?

Shulman: The spectra shown were accumulated for two to twelve hours.

Wüthrich: I would like to make a comment concerning the experiment in which partial sequences of the tRNA were used to identify the resonances of certain exchangeable protons. It appears to me that in the absence of some other convincing explanation the line broadening of the resonances in the partial sequences should be taken as an indication of intermolecular association. One can also expect that such intermolecular interactions would be manifested most clearly in the resonances of the exchangeable protons. Therefore much care should be exercised if these data are to be used to interpret phenomena which arise apparently from intramolecular interactions in the intact tRNA.

Shulman: One must be careful to distinguish between generalized intermolecular interactions which seem to be responsible for the broadening sometimes observed/in the fragments and the intramolecular hydrogen bonding responsible for the well resolved lines observed. The arguments in favor of assigning these resonances to the base pairs expected are first their numbers agreed, second the break-up into A–U and G–C pairs agreed and third in some cases, i.e. the anticodon stem, three different fragments were prepared and the same resonances were observed.

Karplus: (1) There appeared to be significant shifts between the fragment N–H resonances and their values in the complete tRNA molecule. Can these shifts be correlated with specific differences in secondary or tertiary structure between the fragment and the complete molecule? (2) In any of the models of tertiary structure for tRNA, do bases come sufficiently close to the observed N–H resonances to contribute to their shifts. If so, would this be useful in testing models?

Shulman: (1) Not at the present. However, when we are confident of the ring current contributions it then should be possible to see whether other physical mechanisms need to be considered and then to determine structures. (2) The most obvious contributions are from the bases just beyond the helical regions and we cannot evaluate them because we don't know the structures. Once again we expect that when the ring currents are more accurately known and any other significant contributions evaluated, it should be possible to use the shift to determine the structure. At present we are trying to use the structure to understand the shifts.

CINÉTIQUE DES CHANGEMENTS DE CONFORMATION TERTIAIRES DE L'HÉMOGLOBINE ÉTUDIÉE PAR PHOTOLYSE-LASER

B. ALPERT et R. BANERJEE

Institut de Biologie Physico-chimique, 13, rue Pierre et Marie Curie, Paris 5, France

et

L. LINDQVIST

Laboratoire de Photophysique moléculaire, Université de Paris-Sud, 91 Orsay, France

Résumé. A partir de la carboxyhémoglobine l'hémoglobine libre a pu être produite dans sa phase pré-réagissante avec le ligand par un flash laser à 529 nm.

L'évolution du spectre d'absorption pendant une période de 400 nsec montre qu'il apparaît une espèce transitoire qui se transforme en hémoglobine normale.

Ces espèces transitoires sont décelées aussi sur les deux sous-unités isolées α et β de l'hémoglobine, mais non sur le dérivé oxycarboné du ferrohème ni sur la myoglobine de cheval.

Les cinétiques de transformation de l'hémoglobine sont sensibles à la température et le pH.

Ces modifications doivent refléter surtout un changement de structure tertiaire puisqu'on les observe aussi sur les chaînes α.

Lors de la photolyse, le changement de l'état du spin du fer doit altérer le couplage du spin électronique du fer de l'hème avec les phonons de la protéine. Cette perturbation se répercuterait sur les liaisons de la chaîne polypeptidique et un changement de conformation tertiaire se produirait jusqu'à ce qu'un nouvel équilibre de l'hème non ligandé avec la globine s'établisse dans le couplage spin–phonon.

Summary. As is well known, carbonmonoxy and deoxy hemoglobin are characterized by structural differences at various levels, quaternary and tertiary. Perhaps the most important is the difference in spin state of the iron going over from $S=0$ in carboxy to $S=2$ in the deoxy form, since it is this difference that is believed to act as a trigger for the overall conformation change. Recent X-ray cristallographic results have related the two spin states of the iron to different sets of structural coordinates describing the position of the iron atom relative to the porphyrin plane as well as the distance of the Fe–N(F8) bond.

The structural transition from the carboxy to deoxyhemoglobin was obtained by photolysis with the help of a Q-switched Nd glass laser, using the second harmonic (529 nm). The total duration of the flash was 50 nsec. The compounds examined in this work were the carbonmonoxy derivatives of human hemoglobin and those of the isolated α- and β-chains, at different pH values between 5.8 and 9.5, at 3°C and at 20°C.

Kinetic spectrophotometry performed on the above compounds revealed a transmittance change during the laser pulse, then a return to a constant intermediate level over a period of 50 to 100 nsec *after* the complete extinction of the pulse. The absorption spectrum at the end of the observation period (300 nsec) as calculated from experiments at different observation wavelengths correspond to those of the corresponding deoxy derivatives; these results show that a transient deoxy species is formed immediately after the flash and changes into the stable deoxy species. The transformation follows first-order kinetics at all wavelengths studied. The kinetic constants obtained at different pH are tabulated. While those for the isolated chains do not vary with pH, the transformation kinetics for hemoglobin is pH-dependent, with a maximum at pH about 6.5.

The spectral change shown by the deoxy derivatives shortly after their formation is interpreted as corresponding to a structural change of the protein, primarily at the tertiary level. Furthermore the different effect of pH on the kinetics for the α- and β-chains on the one hand and that for Hb on the other hand leaves open the possibility that the tertiary structural change in the latter may be followed by a quaternary structural change unresolved by spectral monitoring.

The results are discussed in the context of changes of spin–phonon coupling.

C. Sadron (ed.), Dynamic Aspects of Conformation Changes in Biological Macromolecules. 171–179.

Abréviations. Hb: hémoglobine libre; HbCO: carboxyhémoglobine; Hb BME: hémoglobine ayant réagi avec le bis(maléidométhyl)éther; Hb CPA: hémoglobine dont deux résidus C terminaux de la chaîne β sont enlevés par l'action de la carboxypeptidase A.

1. Introduction

On sait que la fixation des ligands gazeux (O_2 ou CO) par l'hémoglobine produit concurremment un changement de structure de la protéine et une modification des propriétés magnétiques du fer. Les travaux de Perutz et de ses collaborateurs [1, 2], basés sur la diffraction des rayons X, donnent une description topographique de l'environnement de l'hème, aux niveaux tertiaire et quaternaire, relative à la 'structure oxy' et à la 'structure déoxy'. Nous avons cherché à étudier les aspects dynamiques de ce changement de structure, en utilisant le fait que la carboxy-hémoglobine est photodissociable [3, 4, 5]. Nos expériences ont consisté à déceler par spectroscopie d'absorption les états métastables par lesquels passe la macromolécule aprés sa photodissociation.

En soumettant de la carboxy-hémoglobine à l'éclair lumineux d'un laser à 529 nm, nous avons pu produire de l'hémoglobine libre et l'observer en des temps très brefs après irradiation, grâce à un dispositif photo-électrique. La haute résolution temporelle de ce dispositif a permis d'enregistrer l'évolution du spectre de cette hémoglobine libre dans sa phase pré-réagissante avec le ligand. Ainsi, une espèce transitoire précédant la formation de l'hémoglobine libre stable a pu être détectée. Les dérivés oxycarbonés ou oxygénés des chaînes α et β évoluent de la même manière vers les formes déligandées normales [6]. Dans le présent travail nous avons étudié ce phénomène transitoire plus en détail; en particulier, à différents pH et à deux températures différentes (3 °C et 20 °C).

2. Techniques expérimentales

Le laser, à verre dopé au néodyme (C.G.E., VD 231), fonctionnant en mode déclenché, délivre des impulsions d'une énergie maximum de 60 J à 1058 nm. L'émission est entièrement éteinte après une durée de 50 nsec. Un cristal KDP permet d'obtenir la conversion d'environ 10% de l'émission laser à la fréquence doublée (529 nm). Les études ont été faites à 529 nm, à diverses énergies limitées supérieurement à 0,7 J. On produit à l'aide de lentilles un faisceau laser de section elliptique d'une hauteur de 5 mm et d'une largeur de 8 mm.

Le système à étudier, contenu dans une cuve pour spectrophotométrie, à 4 faces polies (côtés 1 × 10 mm ou 5 × 10 mm) est soumis au rayonnement laser comme l'indique la Figure 1.

Un dispositif d'analyse spectrophotométrique, comprenant une lampe flash au xénon, un monochromateur, un photomultiplicateur et un oscilloscope, est également représenté dans la Figure 1. On mesure la transmission lumineuse à angle droit par rapport au faisceau actinique laser, sur une longueur de 2 mm et une hauteur de 4 mm. La lumière d'analyse est obtenue en faisant décharger une capacité dans la lampe au xénon (durée d'éclair d'environ 2 msec). Une ligne à retard permet de synchroniser

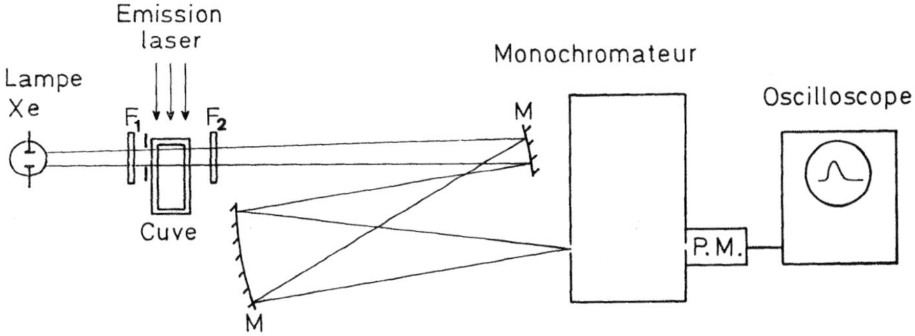

Fig. 1. Dispositif expérimental. La lumière spectrale provient d'une lampe xénon (Xe). Le filtre F_1 protège la cuve contenant la solution protéique du rayonnement ultraviolet de cette lampe. Le faisceau polychromatique d'analyse, par l'intermédiaire des deux miroirs M, converge sur la fente d'entrée du monochromateur qui sélectionne entre 200 nm et 600 nm, la longueur d'onde d'observation. La lumière verte de 529 nm, diffusée par la solution, est éliminée de ce faisceau de mesure par un autre filtre F_2. Le photomultiplicateur placé à la sortie du monochromateur détecte, à longueur d'onde fixe, les variations dans le temps de l'intensité de la lumière transmise par la cuve. L'oscilloscope reproduit ces changements rapides de transmission.

l'amorçage du flash et l'impulsion laser de façon que l'impulsion laser ait lieu pendant la période d'émission maximum de la lampe flash d'analyse. La résolution du dispositif de mesure est de 10 nsec.

L'hémoglobine humaine A et ses sous unités α, β ont été préparées par les méthodes usuelles. La régénération des groupes sulfhydriques libres des chaînes a été effectuée à l'aide de β-mercaptoéthanol [7]. Toutes les solutions hémoprotéiques utilisées ont été équilibrées aux divers pH par dialyse contre des tampons phosphate ou borate de force ionique 0,3.

Dans toutes les expériences la concentration de l'hémoprotéine était de 10^{-4} M en hème.

3. Résultats

La Figure 2 montre la variation de transmission à deux longueurs d'onde dans une expérience type où la carboxyhémoglobine a été soumise à l'action d'une impulsion laser à 529 nm.

Dans une première étude [6] nous avions montré que la transmittance à la fin de la période d'observation correspondait à celle de l'hémoglobine désoxygénée normale, et qu'une espèce transitoire, détectable juste après la photolyse, évoluait vers cette forme de déoxyhémoglobine stable. Cet état transitoire, qui apparaît aussi bien dans les concentrations importantes d'hémoprotéine libre que dans les concentrations faibles, dure moins de 200 nsec. La réaction avec le ligand, très lente comparée à ces cinétiques, n'affecte pas leur étude. En effet la forme oxycarbonée produite par association avec le ligand CO, était de 5% seulement après une période de 20 μsec.

Les chaînes α et β présentent un comportement analogue. Par contre cet état transitoire n'a jamais été décelé sur le ferrohème ou la myoglobine de cheval.

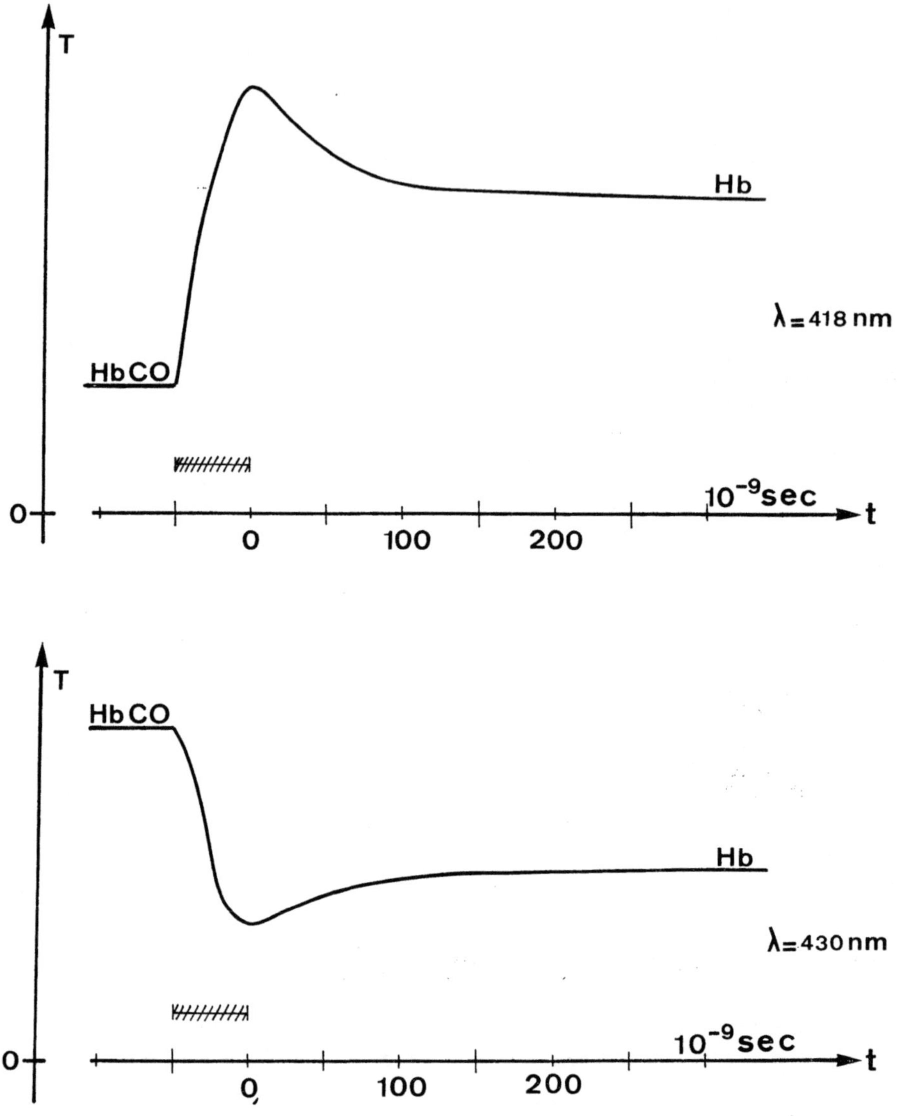

Fig. 2. Courbes observées sur l'écran de l'oscilloscope. Ces traces reproduisent les intensités lumineuses avant, pendant et après la photo-excitation du produit, aux deux longueurs d'onde d'observation: 418 nm et 430 nm. |////////| représente la durée totale de l'éclair laser pendant laquelle la photolyse se produit. Expérience réalisée sur la carboxy hémoglobine (HbCO) à 20°C et pH 6,6.

Nous avons interprété ce changement spectral comme étant dû à la restructuration moléculaire (oxy → deoxy).

Les courbes expérimentales ont été soumises à l'examen analytique afin d'obtenir les constantes cinétiques de la transconformation moléculaire. Pour l'hémoglobine A, la chaîne α et la chaîne β, ces cinétiques sont du 1er ordre. Les constantes ont été

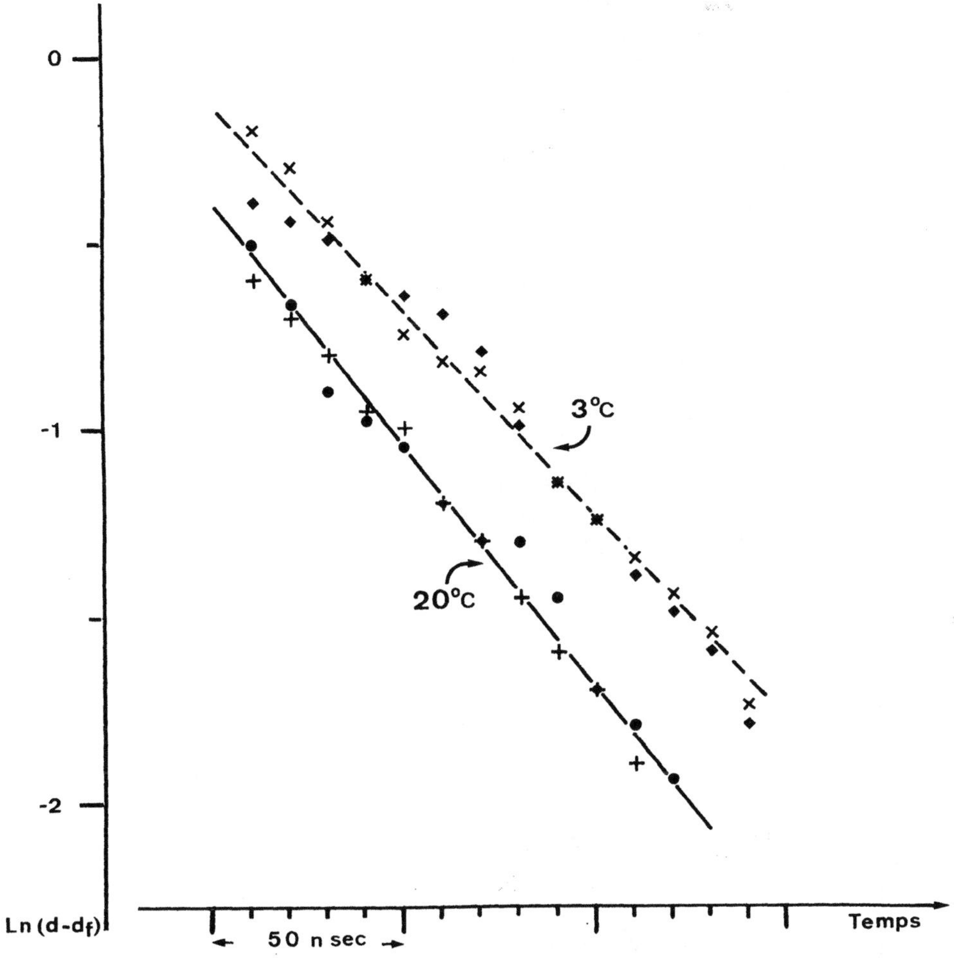

Fig. 3. Loi de la cinétique. En prenant pour référence la densité optique de la déoxyhémoglobine (d_f), le logarithme népérien des écarts à cette densité optique ($d - d_f$) est porté en fonction du temps. L'alignement des points expérimentaux montre que la cinétique observée suit une loi du 1er ordre. Expérience réalisée avec de l'hémoglobine humaine A à 20°C et 3°C; pH 7,5 et force ionique 0,3.

mesurées aux deux longueurs d'onde d'observation 418 nm et 430 nm. Les déterminations à ces deux longueurs d'onde ont donné les mêmes constantes (seules les mesures sur la chaîne β présentaient une certaine dispersion; les valeurs retenues sont les moyennes de ces mesures). Les valeurs de ces constantes, en milieux de pH différents (à $T = 20\,°C$) sont représentées dans le Tableau I. La Figure 3 donne une représentation graphique des cinétiques pour l'hémoglobine, à pH 7,5, respectivement à 3 °C et 20 °C.

Les cinétiques de transformation de l'hémoglobine sont sensibles à la température (Figure 3). La constante cinétique décroît avec la température. L'amplitude du signal, après l'impulsion laser, semble croître quand la température s'abaisse; cependant

TABLEAU I

Valeurs des constantes cinétiques d'ordre un observées sur l'hémoglobine et ses sous-unités α et β à une température de 20° C

pH	$k \times 10^{-7}$ sec^{-1}		
	Hb	α	β
5,7	1,2	↑	↑
6	1,35		
6,6	1,8	1,15	0,75
7,2	1,57[a]		
7,5	1,3[a]		
9,5	1,05	↓	↓

[a] A ces deux pH, les cinétiques pour l'Hb stripped ou avec DPG étaient indifférentiables.

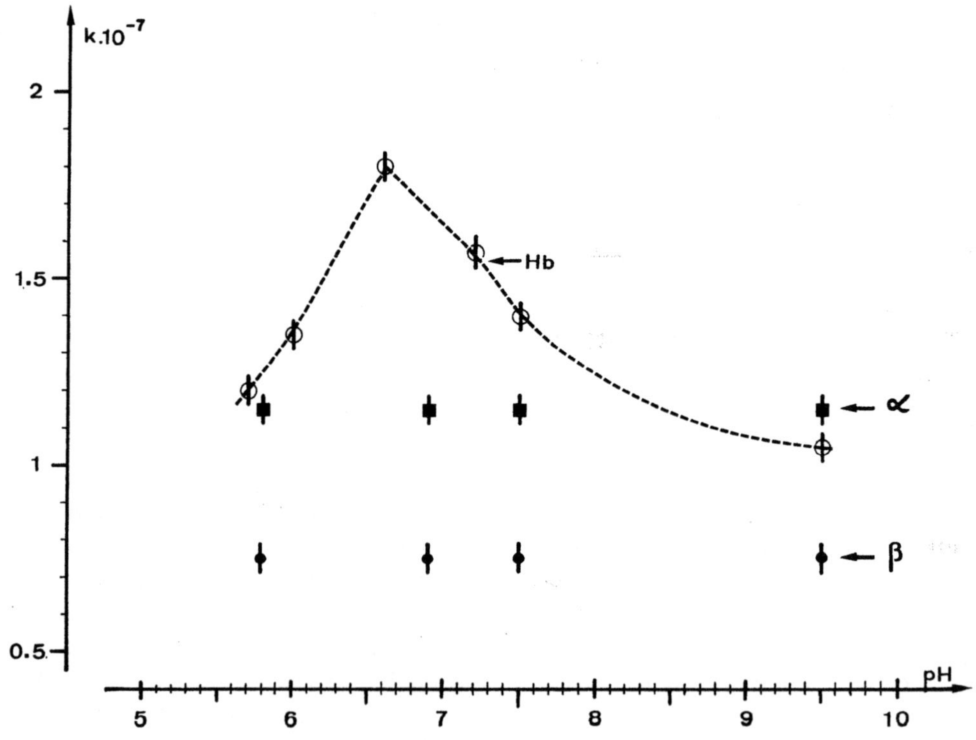

Fig. 4. Dépendances des constantes cinétiques de la transformation observée en fonction au pH du milieu. Etude faite à 20° C et force ionique 0,3.

la transmission finale de la forme dissociée n'est pas affectée par la température.

La cinétique de transformation varie avec le pH pour l'hémoglobine mais pas pour les chaînes isolées α et β (Figure 4).

4. Discussion

Nous pensons que l'évolution spectrale observée après la photolyse traduit une modification structurale de la protéine, ou tout au moins une modification de l'environnement de l'hème. La photolyse de la carboxy-hémoglobine s'accompagne du passage de l'état de spin du fer de $S=0$ à $S=2$ dans la protéine libre. Cette modification magnétique au niveau du site primaire de l'oxygène se répercute à travers la molécule et change sa conformation.

Puisqu'on observe l'effet sur les dérivés de l'hémoglobine indépendamment de leur état d'agrégation (β tétramère ou α monomère), l'état transitoire observé traduit principalement une modification tertiaire de la protéine. Le changement de l'état de spin de l'ion ferreux Fe^{++} perturberait profondément le couplage du spin électronique du fer de l'hème avec les phonons de la protéine et provoquerait dans toute la molécule des vibrations forcées. Dans la chaîne polypeptidique, des liaisons se rompraient et de nouvelles s'établiraient [8]: une restructuration globale se produirait jusqu'à ce qu'un nouvel équilibre de l'hème non ligandé avec sa globine se fasse dans le couplage spin phonon. Nous pensons que nos mesures se rapportent à la cinétique de l'établissement du nouvel équilibre spin–phonon du produit stable final: l'hémoglobine libre normale. Cette hypothése nous semble d'autant plus valable que pour l'hème seul ($Fe^{++}CO$) il ne se produit aucun état transitoire après la photolyse. Le départ du ligand produit des effets de nature très différentes pour l'hème seul et l'hème associé à sa globine.

La présence ou l'absence de 2,3-diphosphoglycérate dans la solution d'hémoglobine n'altère en rien l'évolution du phénomène. Si nos cinétiques sont effectivement corrélées aux remaniements moléculaires, le changement de structure serait déclenché sans intervention notable de cet effecteur. Il semble alors logique de penser que la transition spectrale est provoquée par des modifications à l'intérieur des sous-unités elles-mêmes. Nos cinétiques refléteraient surtout ce réarrangement tertiaire. Il faut souligner que la cinétique de la transformation observée varie avec le pH dans le cas de l'hémoglobine et non dans le cas des chaînes isolées. Ce comportement rappelle celui de ces composés lors de l'oxygénation: la constante d'équilibre d'oxygénation des chaînes ne dépend pas du pH, alors que celle de l'hémoglobine en dépend et passe par un minimum à pH 6,3 (effet Bohr). La cinétique de la transformation étudiée dans ce travail passe précisément pour l'hémoglobine par un maximum dans cette région du pH. Il nous semble encore prématuré de dire s'il existe une relation entre ces deux phénomènes. D'après Perutz, l'effet Bohr, come la coopérativité, découle d'un changement de conformation quaternaire. La chaîne β, qui existe sous forme de tétramère, possède certes une structure quaternaire, mais on admet généralement qu'à l'encontre de celle de l'hémoglobine, cette structure n'est pas modifiée lorsque la protéine réagit avec l'oxygène. Les résultats rapportés plus haut ne sont donc pas, à priori, en contradiction avec l'idée que le changement de structure tertiaire serait suivi de très près par celui de la structure quaternaire quand il y a lieu. Toutefois, une discussion approfondie sur l'effet de pH sera plus aisée lorsqu'on connaîtra le comportement d'autres composés d'hémoglobine, tels que ceux obtenus par modification chimique (Hb BME) ou enzymatique (Hb CPA) qui ne montrent ni effet Bohr d'oxygénation ni changement de structure quaternaire.

D'autre part l'absence d'état transitoire après photolyse du carboxyhème ou de la carboxymyoglobine peut s'interpréter aussi dans le contexte d'un changement de conformation protéique. Dans le cas de la myoglobine de cheval, des contraintes stériques locales empêchent sans doute la perturbation produite au niveau du fer de se propager à travers la protéine, entravant la restructuration tertiaire. Les résultats récents de Makinen [9] semblent confirmer notre interprétation. Le changement d'état de spin du fer dans la myoglobine de cheval n'est pas accompagné d'un mouvement du plan de l'hème par rapport à son voisinage protéique. En ce qui concerne l'hème isolé, la photolyse doit seulement provoquer un mouvement relatif du fer par rapport au noyau porphyrique pendant ou après le changement de l'état de spin de l'ion ferreux. On ignore quel pourrait être la vitesse de ce mouvement; il est vraisemblable qu'elle est trop grande pour être détectée avec notre appareillage.

Dans l'ensemble, les résultats rapportés dans ce travail semblent montrer que la restructuration, tertiaire et peut-être quaternaire, de l'hémoglobine est un processus relativement rapide comparé à celle présumée pour certaines autres protéines enzymatiques. Le nombre élevé des réajustements structuraux impliqués expliquerait néanmoins qu'elle soit beaucoup plus lente que le phénomène électronique primaire qui est le changement de l'état de spin.

Références

[1] Perutz, M. F., Muirhead, M., Cox, J. M., et Goaman, L. G. G.: 1968, *Nature* **219**, 131.
[2] Bolton, W. et Perutz, M. F.: 1970, *Nature* **228**, 551.
[3] Haldane, J. S. et Lorrain Smith, J.: 1896, *J. Physiol.* **20**, 497.
[4] Bücher, T. et Nagelein, E.: 1942, *Z. Biochim.* **311**, 163.
[5] Noble, R. W., Brunori, M., Wyman, J., et Antonini, E.: 1967, *Biochemistry* **6**, 1216.
[6] Alpert, B., Banerjee, R., et Lindqvist, L.: 1972, *Biochem. Biophys. Res. Commun.* **46**, 2.
[7] Geraci, G., Parkhurst, L. J., et Gibson, Q. H.: 1969, *J. Biol. Chem.* **244**, 4664.
[8] Perutz, M. F.: 1970, *Nature* **228**, 726.
[9] Makinen, M.: Communication personnelle.

DISCUSSION

Shulman: Have you varied the intensity of the laser flash? I would expect that a very weak flash should differ from a very strong one in the deoxy band, if the quaternary switch occurs during your experiment. On the other hand, if the quaternary change has not yet occurred it should be interesting to increase your observation time, so as to observe that, in addition to the faster tertiary changes which you do see.

Alpert: Dans la mesure du possible, l'effet a été observé à différentes intensités du pulse laser. Dans tous les cas, quand la photolyse était complète, l'amplitude du signal ne variait pas avec l'intensité de la raie vert. Mais l'étude systématique en fonction de l'intensité seule est chose délicate.

Karplus: The very interesting experiments of Alpert, Lindqvist and Banerjee raise certain questions concerning the mechanism of the observed effect. It seems unlikely that the photodissociation leads directly to the high spin iron, followed by motion of the iron out of the heme plane. This is true because the electronic transition involved is unlikely to be to the high spin state (a single electron *d–d* transition to the CO antibonding dz^2 orbital seems most probable) and because the high spin state is unstable (because of occupation of the dx^2-y^2 orbital) with the iron in the plane. Thus, what may be happening is that a path is followed in which the spin state (and orbital occupation) changes as the iron moves out of the plane. Whether the iron motion, relative to the protein, is to be regarded as providing a 'sudden' impulse, followed by conformational relaxation of the protein or whether the entire process happens in a more concerted fashion, is uncertain.

Alpert: Il me semble peu probable que le changement de spin du fer de $S=0$ à $S=2$ ne soit pas réalisé après 50 nsec, puisqu'il s'agit d'une nouvelle répartition électronique au niveau de Fe^{++}. Si l'état du spin se modifiait plus lentement et était la transition observée, je suppose qu'elle devrait apparaître sur l'hème Fe^{++} isolé comme sur la myoglobine de cheval. Il semble vraisemblable de croire que le changement électronique de Fe^{++} est de beaucoup plus rapide qu'une restructuration de la macromolécule.

TRANSITIONS RÉVERSIBLES D'ACTIVITÉ ET DE STRUCTURE QUATERNAIRE DE LA L(+) LACTATE: CYTOCHROME *c* OXYDORÉDUCTASE (CYTOCHROME b_2) DE LA LEVURE *HANSENULA ANOMALA*

ALAIN BAUDRAS*

Laboratoire de Chimie Biologique, Institut de Physiologie, 81 Grande Rue Saint-Michel, 31 Toulouse, France

Résumé. La L(+)lactate:cytochrome *c* oxydoréductase (cytochrome b_2) de la levure *Hansenula anomala* est une flavo-hémoprotéine tétramérique qui peut former un complexe moléculaire stable avec le cytochrome *c*. Sous forme de ce complexe, l'enzyme est fortement stabilisée dans son état de structure tétramèrique alors qu'à l'état libre elle subit, à faible force ionique, une dissociation spontanée réversible en monomères inactifs qui conservent leur flavine prosthétique associée. Nous décrivons l'action de facteurs physiques du milieu (force ionique et température) ainsi que celle d'une série d'effecteurs spécifiques (substrat et analogues, cytochrome *c*) sur les transitions d'inactivation/réactivation et les changements de structure quaternaire qui leur sont associés.

Summary. Cytochrome b_2, the L(+)lactate:cytochrome *c* oxidoreductase (EC 1.1.2.3), which we have recently obtained pure from the aerobic yeast *Hansenula anomala* contains both flavin and heme prosthetic groups. The active tetramer ($M = 236000$ g) is strongly stabilized on binding cytochrome *c* in conditions of low ionic strength which would otherwise promote its reversible dissociation into inactive monomers. Our efforts have been developped in view of correlating states of quaternary structure and activity during inactivation/reactivation transitions induced by changes of the ionic strength (phosphate buffer pH 7.2) and in the presence of several effectors (substrate and analogs, cytochrome *c*). The main features which will be described here are the following:

(1) efficient thermal control can be exerted on the rate of the inactivation reaction induced by low ionic strength (phosphate pH 7.2; $\Gamma/2 = 0.03$ M) due to its large Arrhenius activation energy (60 kcal/mole between 5 and 20°C);

(2) the logarithm of the apparent rate constant of the inactivation reaction decreases linearly as a function of $\sqrt{(\Gamma/2)}$ (phosphate buffer pH 7.2; 20°C; $0.02\ \mathrm{M} < \Gamma/2 < 0.15$ M); this thermal inactivation is practically abolished when the enzyme is incubated (at 20°C) either in phosphate buffer or in the presence of various neutral salts (pH 7.2) of sufficiently high ionic strength ($\Gamma/2 \geqslant 0.2$ M);

(3) inactivation of the enzyme at low ionic strength is prevented by L(+)lactate (substrate) or oxalate (inhibitor) and by ferricyanide or cytochrome *c* (acceptors of the electrons transferred in the catalysed reaction);

(4) this latter protection effect is dependent on the binding of only one cytochrome *c* molecule per tetramer of the enzyme (dissociation constant of the complex 0.2 μM);

(5) following thermal inactivation at low ionic strength (20°C; phosphate buffer pH 7.2; $\Gamma/2 = 0.03$ M) the activity of the enzyme is rapidly recovered to high levels ($> 80\%$) on adding salt and substrate or substrate analogs D(−)lactate or oxalate (inhibitors of the catalytic reaction); on the other hand, cytochrome *c* is ineffective as a re-activator when added to the inactivated enzyme;

(6) fluorescence measurements show that during the progressive thermal inactivation induced by low ionic strength only a small amount of the prosthetic flavin is irreversibly dissociated; the inactive and reactivable hemoprotein still binds rather firmly this prosthetic group (K_d 0.2 μM);

(7) a series of ultracentrifugal experimentation on sucrose gradients is reported which show that the inactive and reactivable enzyme is unambiguously in the monomeric state; its apparent sedimentation coefficient ($S_{20,w\,\mathrm{app}} = 4.3$ svedberg) is consistent with a compact globular shape ($f/f_0 = 1.16$);

* Nouvelle adresse: Centre de Recherche de Biochemie et de Génétique Cellulaires CNRS, 118 route de Narbonne, 31077 – Toulouse Cedex, France.

C. Sadron (ed.), Dynamic Aspects of Conformation Changes in Biological Macromolecules. 181–205.

the formation of complexes between the active tetramer or the inactive enzyme and cytochrome *c* is substantiated by the same technique;

(8) preliminary experiments based on the simultaneous measurement of activity and apparent molecular weight (light scattering) during reactivation transitions seem to indicate that the recovery of enzymatic activity is not necessarily concerted with a monomer→tetramer transition.

Abréviations utiliseés. FMN: flavine-mononucléotide; EDTA: éthylène-diamine-tétra acétate; DTT: dithiothréitol (réactif de Cleland).

1. Introduction

Les levures aérobies possèdent une lacticodéshydrogénase mitochondriale de nature flavohémoprotéique: la L(+) lactate: cytochrome *c* oxydoréductase (EC 1.1.2.3) ou cytochrome b_2. Seule l'enzyme de *Saccharomyces cerevisiae* (levure de boulangerie) était connue et étudiée jusqu'à ces toutes dernières années, essentiellement sous la forme de préparations cristallisées obtenues selon la méthode décrite par Appleby et Morton [1, 2]. L'état actif est tétramérique et chaque unité moléculaire porte 4 FMN et 4 hémes [3, 4, 5]. Le tétramère de l'enzyme cristallisée peut être dissocié, par action de l'urée ou du chlorure de guanidinium, en huit unités inactives qui sont de deux types: les unes portent l'héme, les autres, réceptrices de la flavine, ont perdu ce dernier groupe prosthétique dans ces milieux; la réassociation ainsi que la réactivation, par élimination de l'agent dissociant, ne sont que partielles [6, 7, 8, 9, 10]. En réalité, l'enzyme ne comporte originellement que quatre chaînes polypeptidiques puisque des travaux récents ont montré que ces dernières subissent chacune une coupure protéolytique au cours de la purification qui aboutit à sa cristallisation [11, 12]. La forme non protéolysée, purifiée récemment, présente des propriétés fonctionnelles qui sont virtuellement identiques à celles de l'enzyme dite 'physiologique' décrite dans les extraits bruts de levure et qui différent notablement de celles de l'enzyme cristallisée [12, 13].

L'enzyme homologue extraite de la levure *Hansenula anomala*, que nous avons obtenue à l'état pur non cristallisée [14, 15], présente le même état de structure quaternaire (quatre chaînes portant chacune un héme *et* une flavine) et, pratiquement, la même masse molaire (236000 g) [15, 11]. Un phénomène d'inactivation thermique a été observé lorsque cette enzyme est exposée à des milieux de faible force ionique [16]; ce phénomène accompagne une dissociation du tétramère en monomères qui bien qu'inactifs conservent leur flavine fixée dans un état non fluorescent [17]. L'excellente réversibilité de ces phénomènes ainsi que la simplicité de mise en oeuvre de l'agent causal (force ionique), nous ont conduit à développer un projet d'étude cinétique détaillée des transitions d'inactivation/réactivation et de dissociation/réassociation. Par la variété de ses marqueurs intrinsèques, ce matériel enzymatique est particulièrement riche et bien adapté à l'étude dynamique de corrélations entre perturbations structurales et fonctionnelles. En particulier, on peut espérer atteindre des informations nouvelles sur les caractèristiques structurales requises pour la reconnaissance du substrat (ou de ses analogues) et son oxydation ainsi que pour la reconnaissance du cytochrome *c*, l'accepteur d'électron présumé dans la réaction catalysée, avec lequel l'enzyme est susceptible de former des complexes moléculaires stables [18. 19].

Dans cette présentation, nous montrerons d'abord les caractèristiques essentielles des transitions réversibles d'inactivation/réactivation induites par simple changement de force ionique et en présence de quelques effecteurs (substrat et analogues, cytochrome *c*), puis nous chercherons à caractèriser les modifications de structure quaternaire qui accompagnent ces phénomènes.

2. Matériel et méthodes

2.1. Enzyme: préparation et dosages

La L(+) lactate: cytochrome *c* oxydoréductase de la levure *Hansenula anomala* a été préparée à l'état pur selon le protocole décrit antérieurement [15]. La concentration des solutions d'enzyme est calculée, par référence à l'héme, à partir de l'absorbance de la bande de Soret du spectre de l'enzyme réduite par le lactate; pour ce calcul, l'absorbance suivante a été adoptée: ε (423 nm) = 183 mM^{-1} cm^{-1}.

L'activité de l'enzyme est détectée par la réduction du ferricyanure de potassium, suivie au spectrophotomètre à 420 nm (Zeiss PMQ2 couplé à un enregistreur Sargent SRLG) dans le milieu standard tamponné suivant: phosphate (monopotassique/disodique) 0,1 M, EDTA 0,1 mM, ferricyanure de potassium 1,5 mM et L(+) lactate 10 mM; pH 7,20. Toutes les mesures sont faites à 25° C. Dans tous les cas, les vitesses initiales sont relevées et servent au calcul de l'activité moléculaire (exprimée par référence à l'héme: électron-équivalents transférés par mole d'héme et par seconde); l'activité moléculaire de l'enzyme pure active est 1150 s^{-1} dans ces conditions.

La teneur en flavine (libre et totale) des échantillons d'enzyme est mesurée fluorimètriquement (Fica 55): λ excitation 436 nm, λ émission 535 nm. La flavine fixée à l'enzyme n'étant pas fluorescente, son dosage est obtenu par mesure dans la guanidine 3 M qui provoque sa dissociation complète. Pour chaque expérience, une courbe d'étalonnage est établie par addition de quantités connues de FMN aux solutions d'enzyme elles-mêmes.

2.2. Ultracentrifugation analytique sur gradients de saccharose

Des gradients linéaires 5–20% en saccharose, en milieu tamponné de pH 7,20, sont réalisés grâce à un système de mélange. Les échantillons (100 ou 150 μl) sont déposés à la surface des tubes (5 ml) préalablement refroidis dans la glace et soumis à centrifugation à 4–6° C et 45000 rév./min. dans une centrifugeuse Spinco modèle L 50. Le contenu des tubes est généralement récupéré, par fractions de trois gouttes, à partir du sommet, en injectant par le fond des tubes un liquide dense à l'aide d'une pompe. La récupération est faite en chambre froide (2–4° C) et les fractions recueillies sont conservées dans la glace jusqu'à leur utilisation.

2.3. Diffusion de la lumière

Les mesures d'intensité de lumière diffusée sont effectuées à l'aide du photogoniodiffusiomètre Fica 50, à la longueur d'onde 578 nm. L'étalon protéique utilisé est la trypsine pour laquelle une masse molaire de 23700 g a été adoptée. La trypsine (Novo) est mise en solution dans HCl 0,01 N + $CaCl_2$ 0,025 M, additionnée de NaCl à la concentration finale 1 M. Après 12 heures de conservation de la solution à 4° C, le matériel insoluble est éliminé de la solution par centrifugation et le surnageant mis à dialyser pendant une nuit à 4° C contre la solution HCl 0,01 N + $CaCl_2$ 0,025 M. Cette technique de préparation permet d'éliminer la protéine dénaturée présente dans les échantillons commerciaux de trypsine [20, 21]. Tous les solvants utilisés dans ces expériences, ainsi que les solutions étalon de trypsine elles-mêmes, sont dépoussièrées par filtration sur Millipore 0,22 μm.

2.4. Produits et réactifs

Toutes les solutions sont faites dans l'eau désionisée puis distillée en appareil de quartz. Les sels courants sont de qualité RP (Prolabo) ou équivalents. Les solutions de L(+) lactate (Sigma, grade L1) sont dosées enzymatiquement par le ferricyanure de potassium en présence de cytochrome b_2 de *Saccharomyces cerevisiae* (enzyme cristallisée). Le cytochrome *c* (Sigma, Type III ou Type VI) est totalement oxydé par addition de ferricyanure de potassium (10% en équivalents) puis chromatographié, dans le tampon voulu, sur colonne de Sephadex G25. Le DTT (dithiothréitol ou réactif de Cleland) provient de Calbiochem. Le saccharose utilisé dans les expériences d'ultracentrifugation est de provenance B.D.H. (qualité Analar).

3. Résultats et discussion

3.1. *Étude des transitions réversibles d'inactivation/réactivation*

A. Inactivation

1. *Influence de la température sur l'inactivation en tampon phosphate 13,5 mM ($\Gamma/2 = 0{,}031$ M) à pH 7,20*

Lorsque l'enzyme, à concentration voisine de 3 μM (en héme), est incubée à diverses températures dans un tampon phosphate 13,5 mM contenant 20 μM EDTA, pH 7,20 ($\Gamma/2 = 0.031$ M), on note une décroissance progressive de son activité résiduelle. La vitesse initiale d'inactivation augmente très rapidement avec la température, entre 5 et 20° C (Figure 1a). L'analyse graphique de l'évolution de l'inverse de l'activité résiduelle en fonction du temps d'incubation montre que la réaction d'inactivation paraît être du second ordre (ordre dans le temps) (Figure 1b). La pente des droites ainsi obtenues permet de calculer, pour chaque température, une valeur de la constante apparente de vitesse. Dans le domaine de température étudié, la loi d'Arrhénius est respectée et on peut estimer à 60 kcal/mole la valeur de l'énergie d'activation

→

Fig. 1. Inactivation thermique à faible force ionique. (a) L'enzyme (activité moléculaire initiale 1250 sec^{-1} à 30° C) est incubée, à la concentration 2,3 μM (en hème), dans un tampon phosphate 13,5 mM, EDTA 20 μM, pH 7,20, aux températures indiquées sur les courbes. Par prélèvement d'aliquotes au cours du temps, l'activité résiduelle est mesurée dans les conditions standard; elle est exprimée en pourcentage de l'activité initiale de l'enzyme non traitée. (b) Analyse de l'ordre apparent (dans le temps) de la réaction d'inactivation: ([activité résiduelle]$^{-1}$: $\Delta[A]_{420\,nm}{}^{-1} \times$ min).

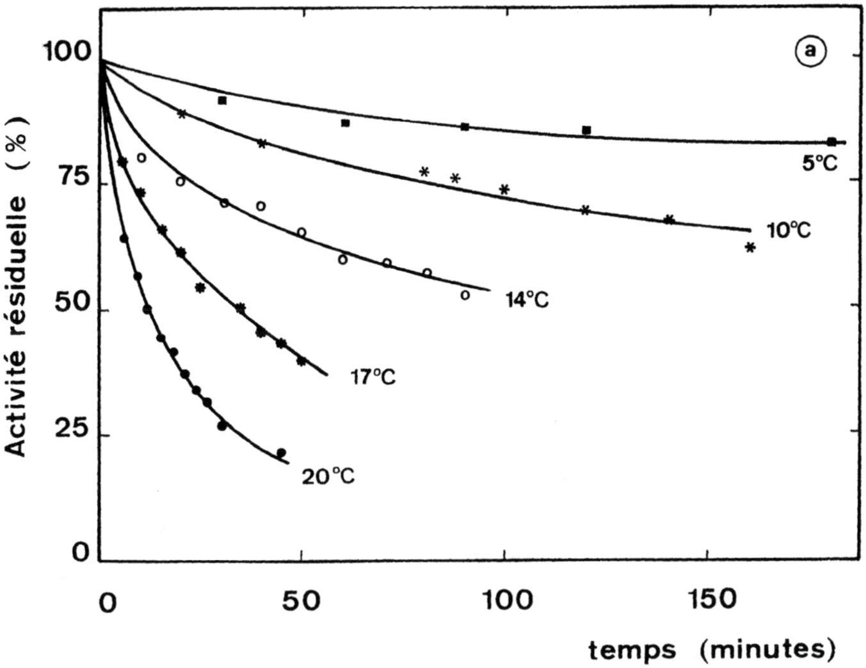
a
100
75
50
25
0
Activité résiduelle (%)
5°C
10°C
14°C
17°C
20°C
0
50
100
150
temps (minutes)

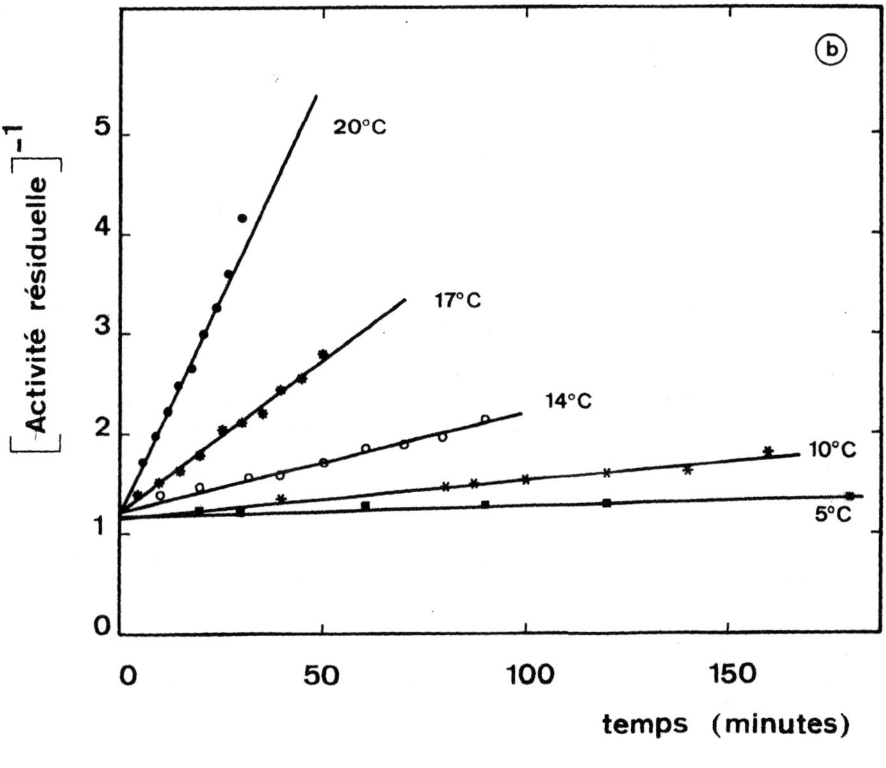
b
[Activité résiduelle]-1
5
4
3
2
1
0
20°C
17°C
14°C
10°C
5°C
0
50
100
150
temps (minutes)

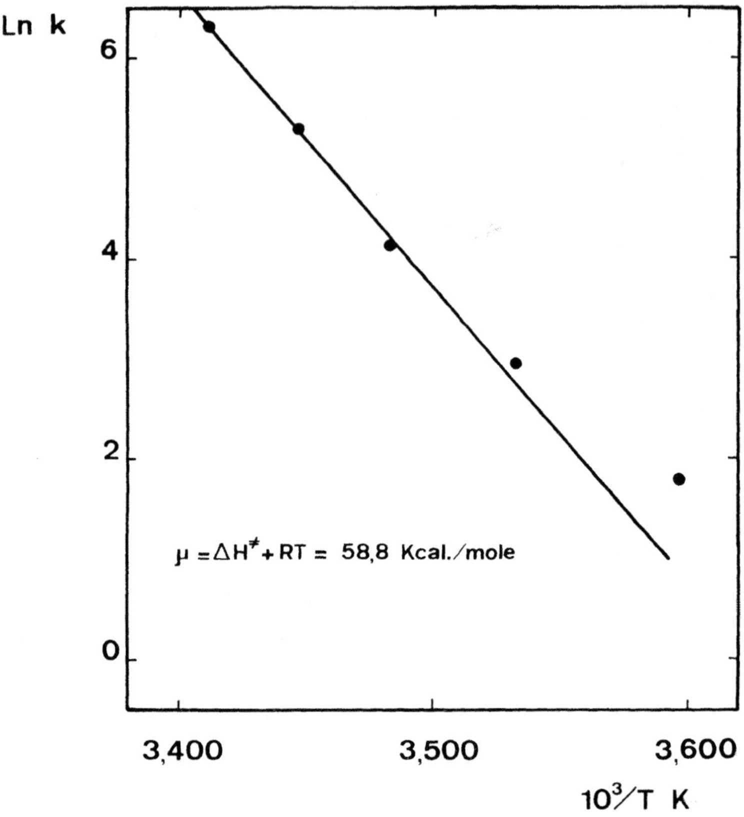

Fig. 2. Energie d'activation expérimentale de la réaction d'inactivation. La constante de vitesse apparente de la réaction, k ($M^{-1} s^{-1}$), est calculée à chaque température à partir de la pente des représentations linéarisées de la Figure 1b. En raison de l'incertitude importante sur la détermination de k à 5° C, le point correspondant a été volontairement écarté.

expérimentale du processus, dans les conditions de milieu utilisées (Figure 2). Cette observation explique qu'un contrôle thermique efficace puisse être exercé sur le processus d'inactivation, à température proche de 0° C. En effet, l'enzyme est stable sous forme active à faible force ionique et à basse température pour plusieurs heures, au moins à faible concentration (entre 5 et 20° C, la vitesse initiale d'inactivation est multipliée par 100).

→

Fig. 3. Effet de la force ionique (phosphate, pH 7,20) sur l'inactivation de l'enzyme à 20° C. (a) L'enzyme (2,6 μM en hème) est incubée à 20,0 ± 0,1° C dans une série de tampons phosphate de pH 7,20 et de concentration variable contenant 20 μM EDTA. La force ionique totale $\Gamma/2$ (M) calculée pour ces différents milieux est: 1: 0,018; 2: 0,031; 3: 0,053; 4: 0,093; 5: 0,124; 6: 0,240. On mesure et exprime l'activité résiduelle comme il a été décrit Figure 1a. (b) Par une représentation graphique analogue à celle de la Figure 1b, les constantes de vitesse apparentes de la réaction d'inactivation sont calculées; le logarithme de ces grandeurs est représenté en fonction de la racine carrée de la force ionique.

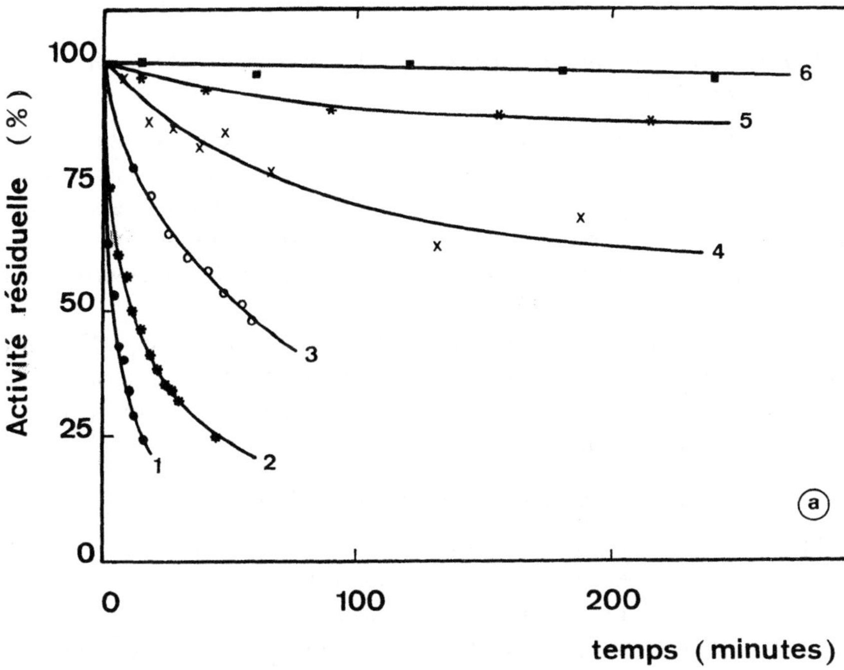

Activité résiduelle (%)
100
75
50
25
0
0
100
200
temps (minutes)
1
2
3
4
5
6
a

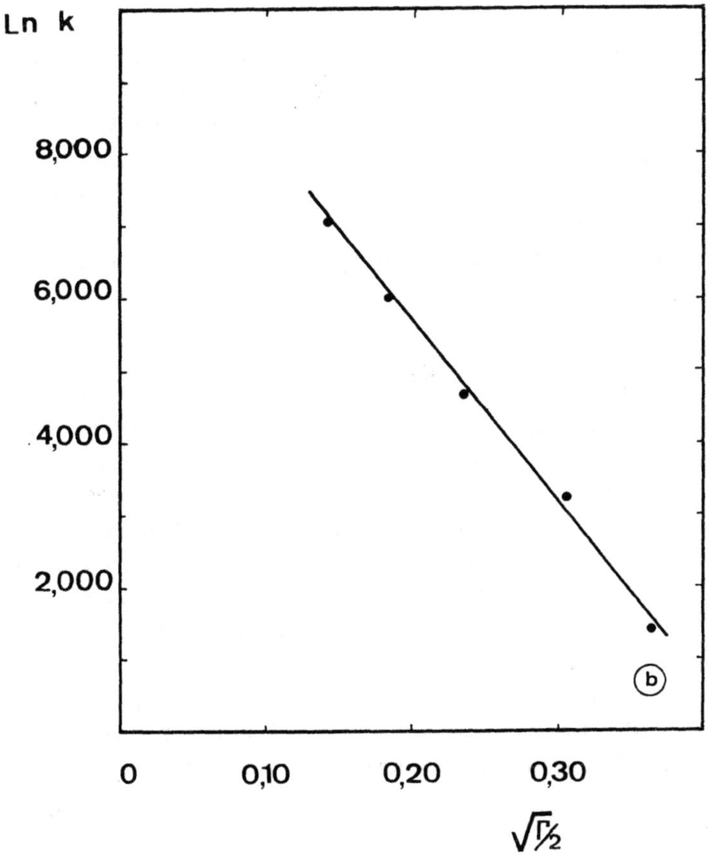

Ln k
8,000
6,000
4,000
2,000
0
0,10
0,20
0,30
√Γ/2
b

2. *Influence de la force ionique (phosphate) sur la vitesse d'inactivation à* 20°C *et pH* 7,20

L'inactivation, analysée à température constante (20°C) sur des solutions d'enzyme ($\sim$3 μM en héme) incubées en milieu tampon phosphate pH 7,20, est considérablement stimulée lorsque la concentration totale en phosphate descend en dessous de 50 mM ($\Gamma/2=0{,}12$ M). Lorsque la force ionique totale est de l'ordre de 0,02 M, le temps de demi-inactivation est voisin de deux minutes; il est d'une heure environ pour $\Gamma/2=$ $=0{,}05$ M (Figure 3a).

Le phénomène d'inactivation apparaît ainsi gouverné en grande partie par une perturbation d'interactions électrostatiques ou polaires; ce fait est matérialisé par l'existence d'une relation linéaire de variation du logarithme de la constante de vitesse apparente avec $\sqrt{(\Gamma/2)}$ (Figure 3b).

3. *Action de protecteurs ioniques et d'effecteurs spécifiques*

L'effet protecteur exercé par le phosphate à force ionique moyenne ne semble pas être une propriété particulière à caractère de spécificité. En effet, comme le montre le Figure 4a, les sels neutres courants présentent un effet de protection contre l'inactivation à 20°C dans le phosphate 13,5 mM, pH 7,20 ($\Gamma/2=0{,}031$ M) qui est pratiquement identique à celui qu'on obtiendrait avec du tampon phosphate de force ionique totale équivalente.

Les effecteurs impliqués dans l'acte catalytique stabilisent fortement la forme active: c'est le cas de L(+) lactate (substrat), de l'oxalate (inhibiteur) et du ferricyanure de potassium (accepteur d'électrons dans la réaction catalysée). Ce dernier effet ne peut être rapporté à l'incrément de force ionique ($\Gamma/2=0{,}006$ M) qui est très inférieur à celui qui serait exigé d'un sel neutre pour provoquer le même effet ($\Gamma/2=0{,}080$ M). Le cytochrome *c* est également un protecteur remarquable; son action sera décrite plus en détail au prochain paragraphe. L'ensemble de ces observations est matérialisé dans la Figure 4b.

4. *Protection par formation de complexe avec le cytochrome c*

La possibilité de former, à faible force ionique, des complexes moléculaires stables entre la L(+) lactate: cytochrome *c* oxydoréductase de levure et le cytochrome *c* a été démontrée il y a quelques années pour l'enzyme cristallisée de *Saccharomyces cerevisiae*; la stoechiométrie de la réaction est inattendue: une seule molécule de cytochrome *c* est fixée par tétramère de l'enzyme [18, 19]. La formation de ce complexe stabilise l'enzyme vis-à-vis de la dénaturation par l'urée et le phénomène a pu être mis à profit pour étudier quantitativement l'équilibre d'association; les quatre sites enzymatiques sont stabilisés en bloc par la fixation d'un seul cytochrome *c* sur le tétramère [18, 24].

Une étude analogue a été faite en utilisant le cytochrome *c* comme protecteur de l'enzyme de *Hansenula anomala* contre l'inactivation thermique réversible obtenue par incubation à faible force ionique (phosphate 10 mM, pH 7,20, 20°C). On constate (Figure 5a) que la vitesse initiale apparente d'inactivation décroit progressivement

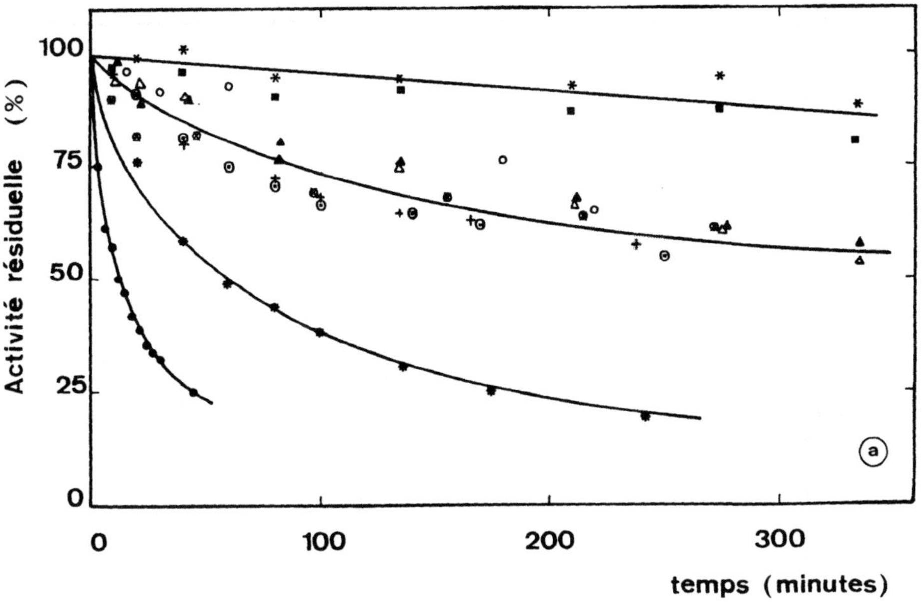

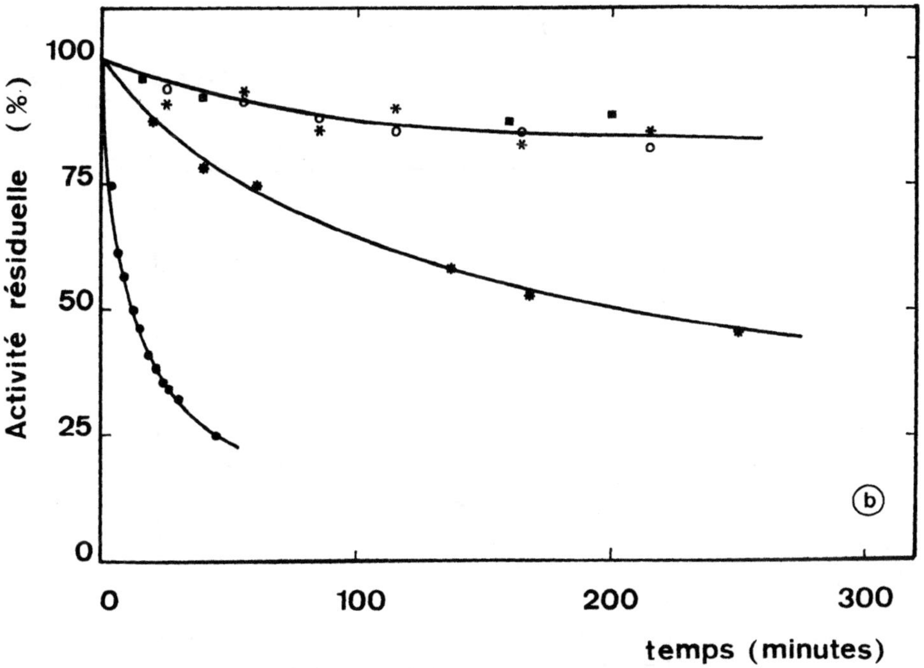

ig. 4. Influence de quelques sels et d'effecteurs spécifiques sur la vitesse d'inactivation. L'enzyme ,6 μM en hème) est incubée à 20,0 ± 0,1°C dans une série de milieux tamponnés par du phosnate 13,5 mM de pH 7,20 (contenant 20 μM EDTA) et additionnés de: (a) sels neutres à la conntration 0,1 M: ■ $SO_4(NH_4)_2$; * citrate de sodium; △ NO_3Na; ▲ ClO_4K; ○ SO_4Na_2; ⊕ ClK; CH_3COONa; ⊙ NaCl; ✱ tris-HCl; ● témoin sans addition de sel. (b) effecteurs intervenant ıns la réaction catalysée: * ferricytochrome *c* (Sigma Type III) 50 μM; ○ L(+)lactate 50 mM; ■ oxalate de sodium, 0,1 M; ✱ $Fe[(CN)_6]K_3$, 1 mM; ● témoin.

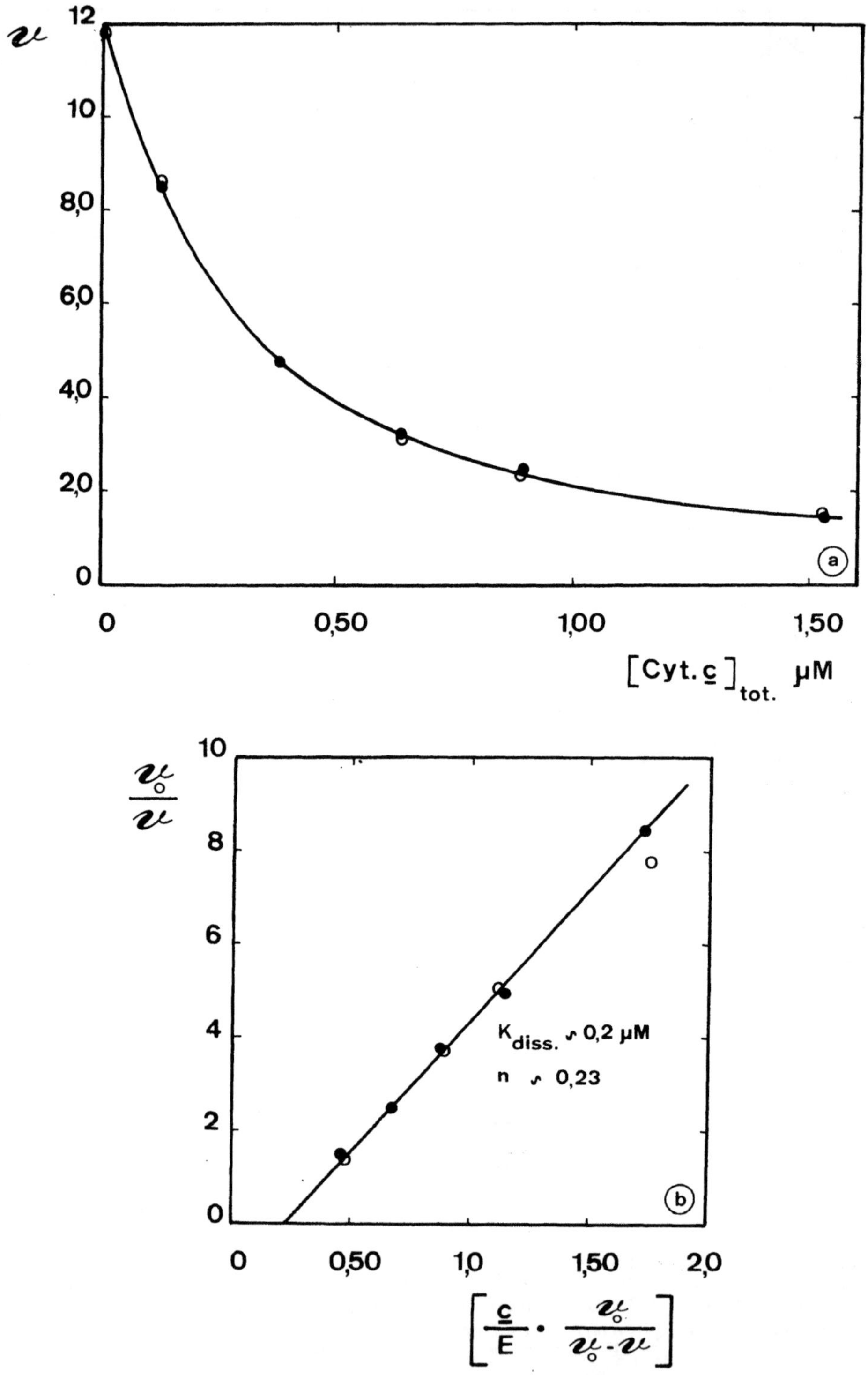
v
12
10
8,0
6,0
4,0
2,0
0
0
0,50
1,00
1,50
[Cyt. c]tot. µM
a
v₀/v
10
8
6
4
2
0
0
0,50
1,0
1,50
2,0
Kdiss. ∼ 0,2 µM
n ∼ 0,23
b
[c/E · v₀/(v₀-v)]

lorsque la concentration de cytochrome *c* augmente; elle devient quasi-nulle à saturation par cet effecteur (protection totale). Tenant compte de ce dernier fait et faisant l'hypothèse que la vitesse initiale apparente d'inactivation est proportionnelle à la concentration d'enzyme non complexée, nous avons tenté d'analyser les résultats selon une représentation de type Scatchard qui doit donner une droite si les sites récepteurs du cytochrome *c* sont identiques et indépendants. La Figure 5b montre qu'il semble bien en être ainsi. Comme pour l'enzyme cristallisée de *Saccharomyces*, le phénomène de protection s'exerce globalement sur l'unité tétramèrique dès que celle-ci a fixé un seul cytochrome *c*. La constante de dissociation du complexe est voisine de 0,2 μM. Nous savons actuellement que, dans le tétramère, chaque protomère peut fixer une molécule de cytochrome *c* ([17] et ci-dessous 3.2, A. 3); par conséquent, la perturbation des interactions produite par l'occupation d'un seul des quatre sites est suffisante pour renforcer la cohésion de l'édifice tétramèrique.

B. RÉACTIVATION

1. *Mise en évidence du phénomène de réactivation*

La possibilité de régénérer la forme active de l'enzyme nous est apparue à l'occasion de l'observation préliminaire suivante: des échantillons d'enzyme (concentration en héme voisine de 10 μM) inactivés à plus de 90% par traitement thermique à faible force ionique (phosphate 20 mM, pH 7,20, 20°C) étant additionnés de DL-lactate (0,2 M final) puis de sulfate d'ammonium solide jusqu'à précipitation de la protéine, donnaient un culot de centrifugation totalement soluble dans le milieu dit LPE (DL-lactate 0,3 M, pyrophosphate de sodium 50 mM, EDTA 0,1 mM, pH 6,8) et présentant une activité moléculaire pratiquement identique à celle de l'enzyme active initiale.

La présence de lactate est un facteur critique qui conditionne une réactivation rapide de l'enzyme en solution. La Figure 6 montre les résultats d'une expérience d'inactivation/réactivation. Cette dernière est opérée ici dans des conditions que nous utilisons comme test standard de la capacité de réactivation de l'enzyme préalablement évoluée thermiquement à faible force ionique. On constate que la récupération d'activité, dans ces conditions, est très rapide et presque complète: elle permet d'atteindre régulièrement au moins 80% de l'activité moléculaire de l'enzyme non traitée.

←

Fig. 5. Analyse de la protection par le cytochrome *c*. Une série de solutions de ferricytochrome *c* (Sigma Type III) est préparée dans le tampon phosphate 10 mM, pH 7,20 (contenant 1 mM EDTA). L'inactivation thermique est déclanchée soit par addition de l'enzyme (équilibrée dans le même tampon par passage rapide sur colonne de Séphadex G25 en chambre froide; concentration finale dans les milieux d'incubation 1,0 μM) dans ces solutions de cytochrome *c* préalablement équilibrées à 20,1 $\pm$ 0,1°C, soit en portant rapidement à cette même température les mélanges enzyme–cytochrome *c* préparés à 0°C. (a) Evolution de la vitesse initiale apparente d'inactivation (exprimée en unités arbitraires), calculée comme il est décrit Figure 1b, en fonction de la concentration totale de cytochrome *c*. (b) Analyse graphique des résultats de la Figure 5a selon une représentation dérivée de Scatchard: v_0, vitesse initiale d'inactivation à concentration nulle en cytochrome *c*; *c*, concentration totale en cytochrome *c*; *E*, concentration totale en enzyme (exprimée en hème); *n*, nombre de sites impliqués dans le phénomène de protection.

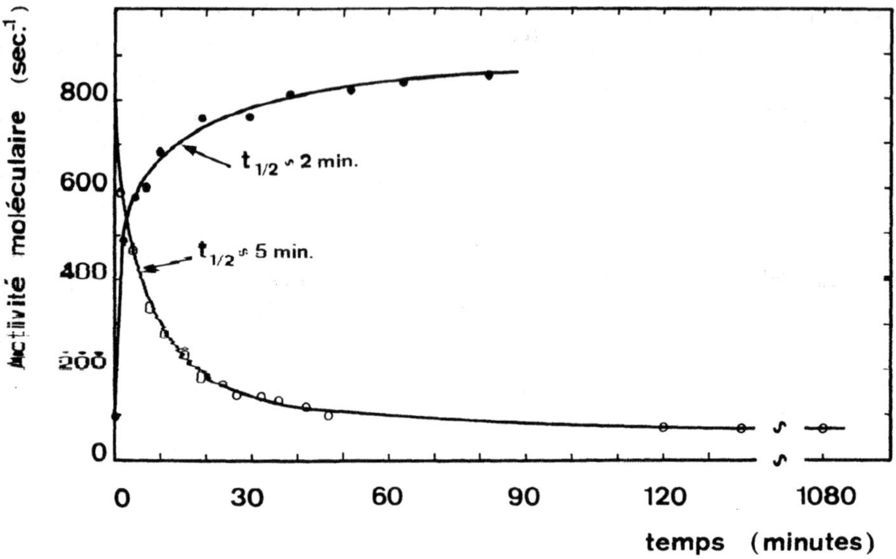

Fig. 6. Cinétiques d'inactivation et de réactivation de l'enzyme à 20° C. L'enzyme est équilibrée dans du tampon phosphate 10 mM, pH 7,20, contenant 1 mM EDTA et 0,1 mM DTT, par passage sur colonne de Séphadex G25 en chambre froide. Elle est inactivée à 20,2 ± 0,1° C, à la concentration 15,2 μM (en hème), dans ce même milieu. L'activité résiduelle est mesurée au cours du temps. Après une heure d'inactivation, une partie de la solution est diluée avec un volume égal du milieu dit LPE; la réapparition de l'activité est suivie en fonction du temps d'incubation du mélange à 20,2 ± 0,1° C.

Nous avons observé d'autre part que la vitesse de réactivation augmente beaucoup avec la température: une détermination provisoire de la valeur de l'énergie d'activation expérimentale a donné $\Delta H \sim 30$ kcal/mole (entre 0 et 20° C).

2. *Influence de la force ionique (phosphate) sur la vitesse de réactivation à 20°C et pH* 7,20

La Figure 7 montre la récupération progressive de l'activité obtenue par incubation à 20° C dans des tampon phosphate pH 7,20 de concentrations variées. On peut constater qu'en absence de lactate ou de tout autre effecteur, la réactivation est lente et partielle. Ce dernier caractère est probablement dû à évolution parallèle du monomère réactivable vers une (ou des) forme(s) dénaturée(s). La part de cette voie devient négligeable lorsque des réactivations rapides peuvent être obtenues en présence d'effecteurs.

3. *Effecteurs de la réactivation*

Nous avons montré au premier paragraphe de ce chapitre une expérience de réactivation rapide de l'enzyme dans le milieu standard dit LPE 1/2 (DL-lactate 0,15 M, pyrophosphate de sodium 0,025 M, EDTA 0,05 mM, pH 6,80). Une analyse des facteurs en jeu est donnée dans la Figure 8a. Si, dans ce milieu, le seul lactate est remplacé par du chlorure de sodium à la mêne concentration, on note une réactivation lente et partielle. Le pyrophosphate ne présente pas de propriétés particulières de ce

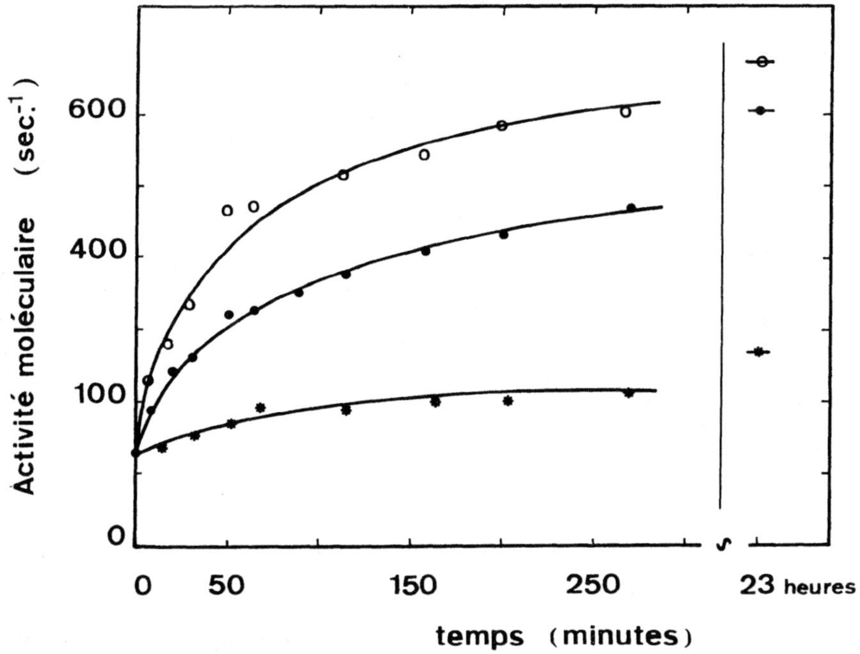

Fig. 7. Influence de la concentration en phosphate (pH 7,20) sur le phénomène de réactivation. Même protocole expérimental que celui décrit dans la Figure 6. L'enzyme est inactivée à la concentration 67,2 μM (en hème). Après une heure d'inactivation à 20,1 $\pm$ 0,1° C, des échantillons d'enzyme traitée sont prélevés et dilués dans une série de tampons phosphate de pH 7,20 (contenant 1 mM EDTA et 0,1 mM DTT) dont les concentrations finales sont: * 0,05 M; ● 0,12 M; ○ 0,20 M. Concentration d'enzyme pendant la réactivation: 3,2 μM (en hème). Température 20,1 $\pm$ 0,1° C.

point de vue par rapport au tampon Na/K-phosphate de mêmes force ionique et pH. L'action spécifique du lactate s'éxerce totalement à la concentration 10 mM en isomère L(+). Dans ce cas, on observe une inactivation secondaire lorsque l'incubation est prolongée à l'air libre. Ce fait est dû à la consommation progressive du lactate par fonctionnement de l'enzyme avec l'oxygène atmosphérique comme accepteur. Le phénomène qui est également observé avec l'enzyme native, devient très marqué lorsque sont peu à peu établies des conditions où, en présence d'oxygène, la concentration en lactate devient très faible. Cette évolution apparaît beaucoup plus tardivement dans les milieux riches en lactate (LPE ou LPE 1/2) et peut, dans tous les cas, être évitée en travaillant sous atmosphère d'azote.

L'action stimulatrice du lactate n'apparait pas stéréospécifique; en effet, le D(−) lactate, ainsi d'ailleurs que l'oxalate, tous deux inhibiteurs de l'enzyme [15], facilitent la réactivation. Cet effet est surtout marqué lorsqu'une force ionique suffisante est assurée par le tampon utilisé (Figure 8b). Par contre, dans des conditions de milieu (phosphate 10 mM, pH 7,20, à 20° C) et de concentration (25 μM) où il exercerait une protection totale contre le phénomène d'inactivation thermique, le cytochrome *c* ne provoque aucune réactivation de l'enzyme préalablement évoluée thermiquement. Un

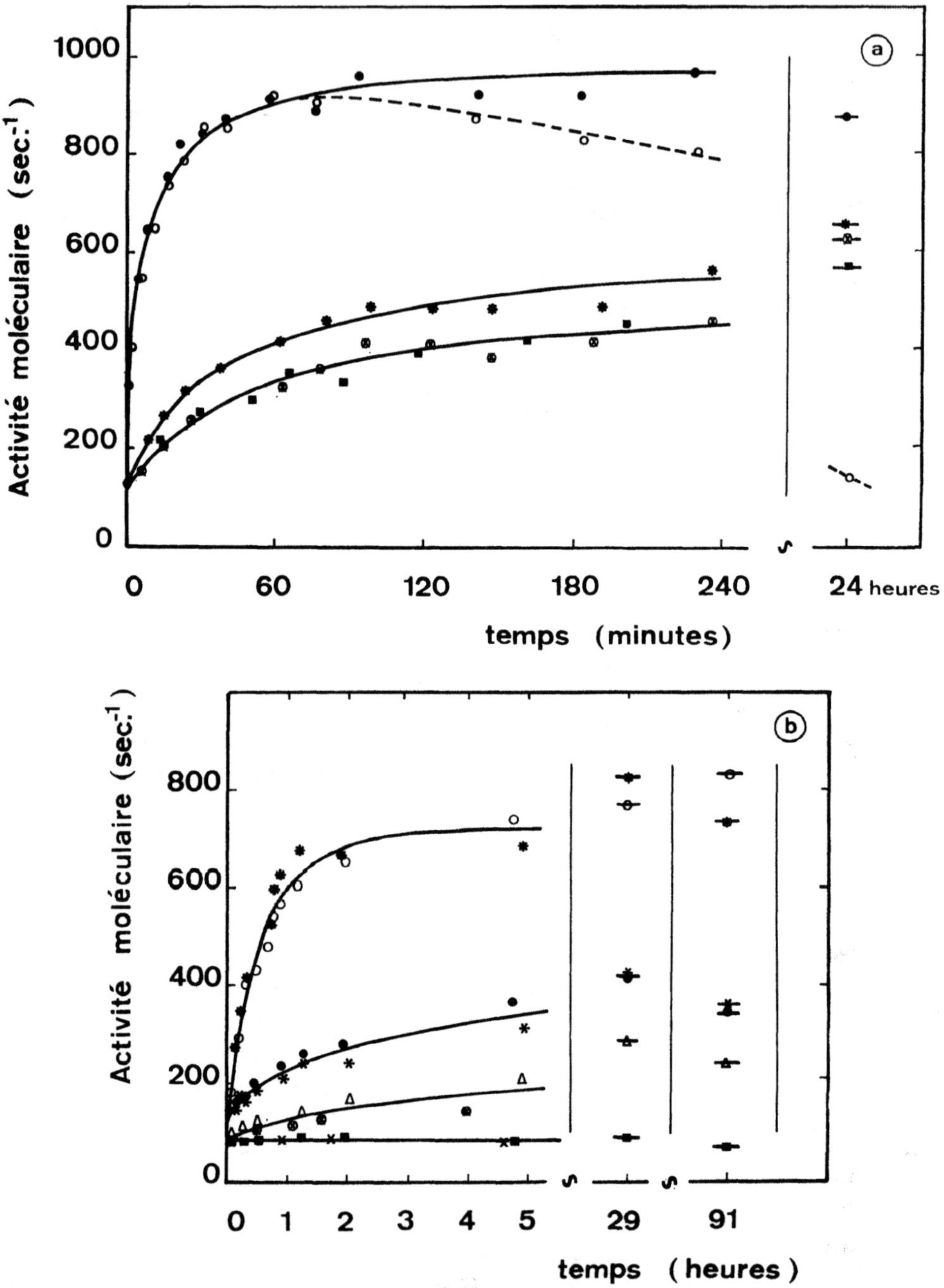

Fig. 8. Effecteurs de la réactivation. Expériences d'inactivation/réactivation à 20,1 ± 0,1°C faites selon le protocole général décrit dans les Figures 6 et 7. (a) Inactivation de l'enzyme pendant une heure, à la concentration 67,2 μM (en hème). Réactivation, à la concentration 3,2 μM (en hème), dans les milieux (contenant 1 mM EDTA et 0,1 mM DTT) de compositions finales: ● DL-lactate de sodium 0,15 M + pyrophosphate de sodium 25 mM, pH 6,8 (LPE 1/2); * phosphate 0,15 M, pH 6,8

effet positif ne peut être envisagé à des forces ioniques plus importantes puisque les complexes enzyme/cytochrome *c* qui reposent sur des interactions électrostatiques ne sont pas stables dans ces milieux [19].

C. ÉTAT D'ASSOCIATION DE LA FLAVINE PROSTHÉTIQUE AU COURS DES TRANSITIONS D'INACTIVATION/RÉACTIVATION

1. *Corrélation entre état d'association de la flavine et activité de l'enzyme*

La flavine prosthétique ne présente pas de fluorescence lorsqu'elle est fixée à l'enzyme sous sa forme active. En conséquence, toute dissociation de ce groupe, réversible ou irréversible, peut être suivie quantitativement par une apparition de fluorescence caractèristique.

L'analyse simultanée, sur une solution d'enzyme exposée au traitement thermique à 20°C en tampon phosphate 10 mM pH 7,20, de l'évolution de l'activité enzymatique et de l'apparition de fluorescence conduit à l'observation remarquable suivante: la forme inactive de l'enzyme conserve, pour l'essentiel, son groupe flavinique fixé dans un état non fluorescent (Figure 9a). En outre, deux parts peuvent aisément être faites dans la faible proportion de flavine libre apparue. D'une part, on note une libération irréversible qui est très lente et mise en évidence par prolongation du traitement thermique au-delà de la zone d'inactivation proprement dite. Il s'agit ici clairement d'une évolution secondaire irréversible de l'enzyme vers un état non réactivable et d'affinité négligeable pour la flavine. La Figure 9b montre en effet une diminution du niveau de réactivation, lorsque celle-ci est entreprise plus tardivement, qui est en corrélation directe avec une augmentation de la proportion de flavine libre finalement présente dans l'enzyme réactivée. D'autre part, lors de la réactivation en présence de lactate, on observe une phase initiale de réassociation d'une partie de la flavine qui était présente, à l'état libre, dans l'enzyme inactivée. Ce phénomène peut, à priori, être le résultat de l'un ou l'autre des processus suivants: soit une reconnaissance directe du substrat par la forme inactive avec augmentation de son affinité pour la flavine, soit un renforcement de l'affinité de l'enzyme consécutif à son retour à la forme active, en présence de lactate. Il n'est pas possible de trancher actuellement avec certitude entre ces deux alternatives; nous verrons néanmoins ci-dessous que, dans des conditions où l'enzyme reste à l'état inactif, un analogue du substrat, l'oxalate (inhibiteur), ne modifie pas la valeur de la constante de dissociation dans l'équilibre:

$$\text{hémoprotéine (inactive)} + \text{flavine} \leftrightarrows \text{flavohémoprotéine (inactive)}.$$

($\Gamma/2$ identique a ●); ⊕ chlorure de sodium 0,15 M + pyrophosphate de sodium 25 mM, pH 6,80; ○ phosphate 0,12 M + L-lactate 10 mM, pH 7,20; ■ phosphate 0,12 M + chlorure de sodium 10 mM, pH 7,20. (b) Inactivation de l'enzyme pendant une heure à la concentration 4,42 μM (en hème). Réactivation, à la concentration 2,21 μM (en hème), dans les milieux (contenant 1 mM EDTA et 0,1 mM DTT): — tampon phosphate 0,105 M, pH 7,20, contenant: ○ D-lactate de sodium 21 mM; ✱oxalate de sodium 10 mM; △ ferricytochrome *c* (Sigma Type VI) 25 μM; — tampon phosphate 10 mM, pH 7,20, contenant: ● D-lactate de sodium 21 mM; * oxalate de sodium 10 mM; ■ ferricytochrome *c* (Sigma Type VI) 25 μM. Témoins: ⊕ phosphate 0,105 M seul; × phosphate 10 mM seul.

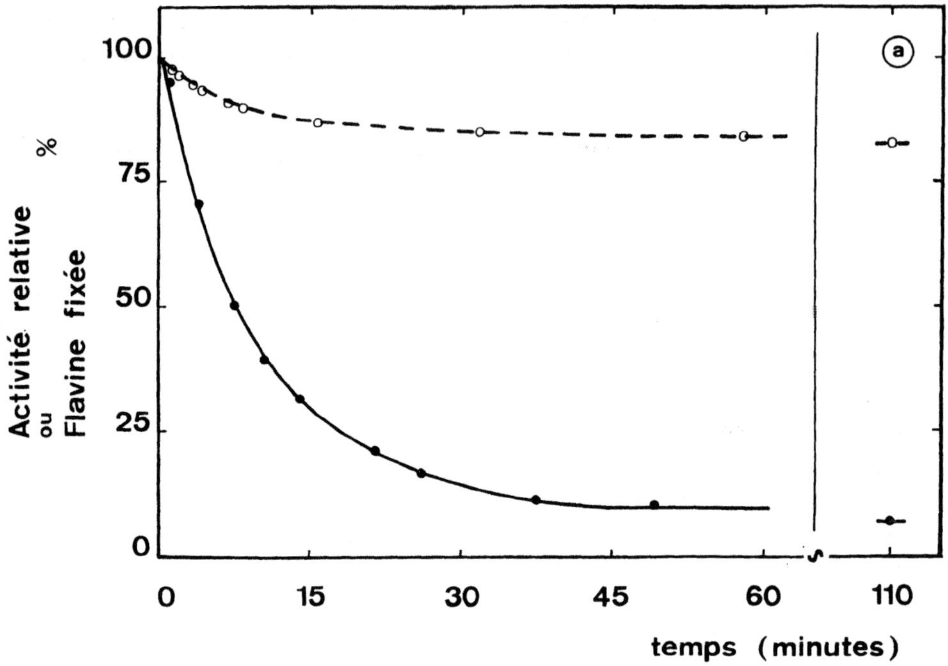

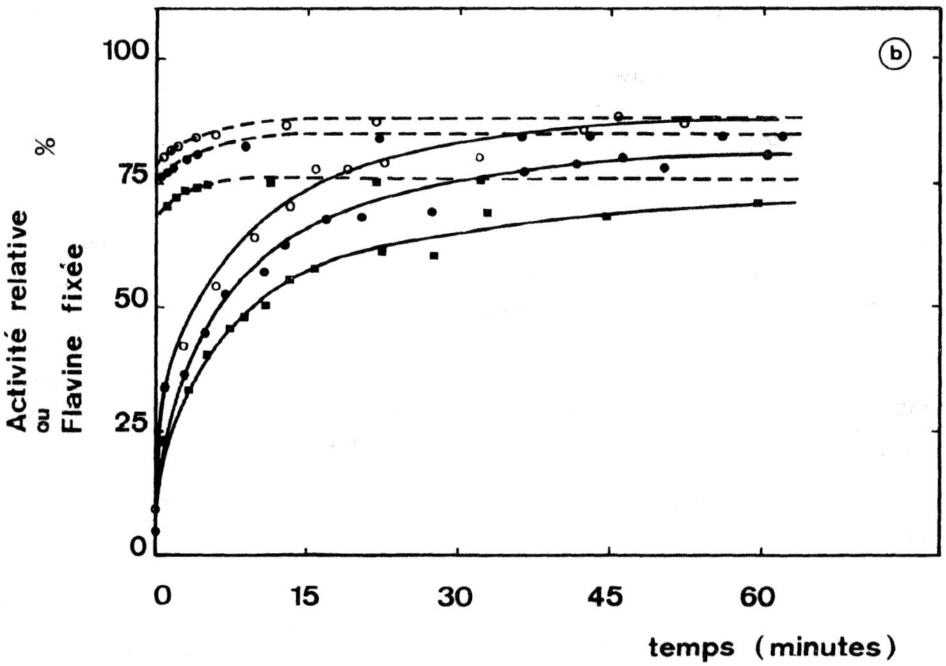

Fig. 9. État d'association de la flavine au cours de transitions d'inactivation/réactivation. Expérience d'inactivation/réactivation à 20,1 ± 0,1° C faites selon le protocole général décrit dans la Figure 6. Une cuve fluorimétrique contenant l'enzyme à la concentration 4,5 μM (en hème) est placée dans le

2. *Estimation de l'affinité de l'enzyme inactivée pour la flavine*

L'observation rapportée ci-dessus montre que l'enzyme inactivée thermiquement constitute un système en équilibre réversible d'association/dissociation relativement à la flavine prosthétique. En opérant des dilutions progressives sur une solution dans cet état, dans le domaine des concentrations micromolaires (concentration totale en héme), on doit provoquer une dissociation définie de flavine par déplacement de cet équilibre et pouvoir accéder à la valeur de la constante de dissociation.

Des mesures précises n'ont pu être faites jusqu'à présent dans le tampon phosphate 10 mM, pH 7,20, en raison d'une interférence entre la rééquilibration de cette réaction et le phénomène d'évolution irréversible décrit au paragraphe précédent. Par contre,

TABLEAU I

Mesure de la constante de dissociation du complexe flavine–enzyme inactive

	Concentration totale d'enzyme (μM)	Concentration de flavine libre (μM)	$K_{diss.}$ (μM)
Expérience 1	2,30	$0{,}57_5$	$0{,}19_4$
	$1{,}07_5$	$0{,}45_5$	0,33
	0,46	0,25	0,30
Expérience 2	0,82	0,33	0,22
	0,52	0,20	$0{,}12_5$
	$0{,}34_1$	$0{,}15_5$	$0{,}12_9$
Expérience 3	$0{,}51_1$	$0{,}25_2$	$0{,}24_5$
	$0{,}27_3$	$0{,}16_8$	$0{,}26_8$
+ 10 mM Oxalate	$0{,}18_2$	$0{,}12_6$	$0{,}28_3$

La constante de dissociation est calculée à partir de la relation $K_{diss.} = [E]\,[F]/[EF]$. La concentration de flavine libre [F] est mesurée fluorimétriquement et est prise égale à la concentration d'enzyme inactive libre [E]; cette dernière, ainsi que la concentration du complexe enzyme inactive – flavine ($[EF] = [E]_{tot.} - [F]$), est exprimée par rapport à l'hème. Conditions de milieu: phosphate 0,105 M, EDTA 1 mM, DTT 0,1 mM, pH 7,20, 24°C. Moyenne des valeurs observées: $K_{diss.} = 0{,}23$ μM. Activité résiduelle des échantillons: Exp. 1 et 2: 75 sec^{-1}, Exp. 3: 60 sec^{-1}. Enzyme inactivée thermiquement pendant une heure à 20°C dans phosphate 10 mM, EDTA 1mM, DTT 0,1 mM, pH 7,20.

compartiment correspondant, thermostaté à 20,1 ± 0,1°C, du fluorimètre. (a) L'intensité de fluorescence (○) et l'activité (●) de cette solution d'enzyme sont suivies en fonction du temps d'incubation. (b) Après des durées différentes d'inactivation, une partie de cette solution est diluée avec un volume égal du milieu dit LPE; la réapparition de l'activité (—), ainsi que la diminution de l'intensité de fluorescence (---) sont suivies en fonction du temps d'incubation du mélange à 20,1 ± 0,1°C. ○ 1 heure d'inactivation; ● 6 heures d'inactivation; ■ 24 heures d'inactivation.

dans le tampon phosphate 0,105 M, pH 7,20, l'intensité de la fluorescence flavinique atteint effectivement, à chaque dilution, un plateau stable pour plusieurs minutes après la transition de dissociation. Nous avons contrôlé que le système n'évoluait pas du point de vue de son état d'activité pendant la durée de ces mesures. Les résultats obtenus (Tableau I) donnent pour la constante de dissociation, dans les conditions décrites, une valeur voisine de 0,2 μM ($\Delta G'_{\mathrm{diss}} \sim +9$ kcal/mole).

Il faut remarquer que l'affinité' apparente de l'enzyme inactive pour la flavine n'est pas modifiée par la présence d'oxalate à la concentration 10 mM qui serait quasi-saturante pour l'enzyme active ($K_i \sim 0{,}7$ mM [15]). L'observation d'un effet d'analogue du substrat sur la constante de dissociation du complexe hémoprotéine (inactive) – flavine aurait été un indice sérieux de reconnaissance de cet effecteur. Dans cette optique, il convient de rappeler que pour le cytochrome b_2 cristallisé de *Saccharomyces cerevisiae*, le L(+)lactate et l'oxalate stabilisent considérablement le complexe actif hémoprotéine (apoenzyme tétramérique) – FMN [22, 23].

3.2. *Corrélations entre activité et états de structure quaternaire*

A. ÉTUDES HYDRODYNAMIQUES

1. *Mise en évidence de l'état monomére inactif par ultracentrifugation sur gradients de saccharose*

C'est à l'utilisation de cette technique que nous devons d'avoir obtenu les résultats initiaux qui encouragèrent le développement du présent travail. En effet, bien qu'elle ne se prête pas à l'analyse de phénomènes en évolution rapide, l'ultracentrifugation sur gradients permet une caractèrisation des propriétés variées des constituants puisque ceux-ci sont récupérés d'une manière fractionnée après leur séparation. Il nous est apparu très tôt que l'enzyme inactivée thermiquement à faible force ionique, dans les conditions précédemment décrites, présentait un coefficient de sédimentation apparent compatible avec un état monomère de la protéine. Ce fait fût rapidement établi d'une manière définitive par une étude d'ultracentrifugation selon la méthode d'approche de l'équilibre de sédimentation [11].

Dans le premier diagramme (Figure 10a) est analysée une solution d'enzyme partiellement inactivée par exposition à 20°C dans le phosphate 10 mM, pH 7,20. Par ordre des valeurs croissantes du coefficient de sédimentation, on note successivement la présence d'un constituant inactif non réactivable, celle du constituant principal également inactif *mais* réactivable, puis celle de l'enzyme tétramérique active. Cette dernière étant utilisée comme référence pour le calcul des coefficients de sédimentation, on obtient respectivement les valeurs suivantes pour les deux constituants inactifs: $S_{20,w} = 2{,}3$ et 4,3 svedberg. Par intégration, les aires sous les pics de concentration donnent les proportions respectives suivantes: 14%, 73% et 13% pour les trois constituants.

Le constituant non réactivable est probablement la forme dénaturée de l'enzyme monomérique, ayant perdu sa flavine prosthétique, dont l'existence a été invoquée pour expliquer les déficits de réactivation (3.1 C). La valeur de son coefficient de sédi-

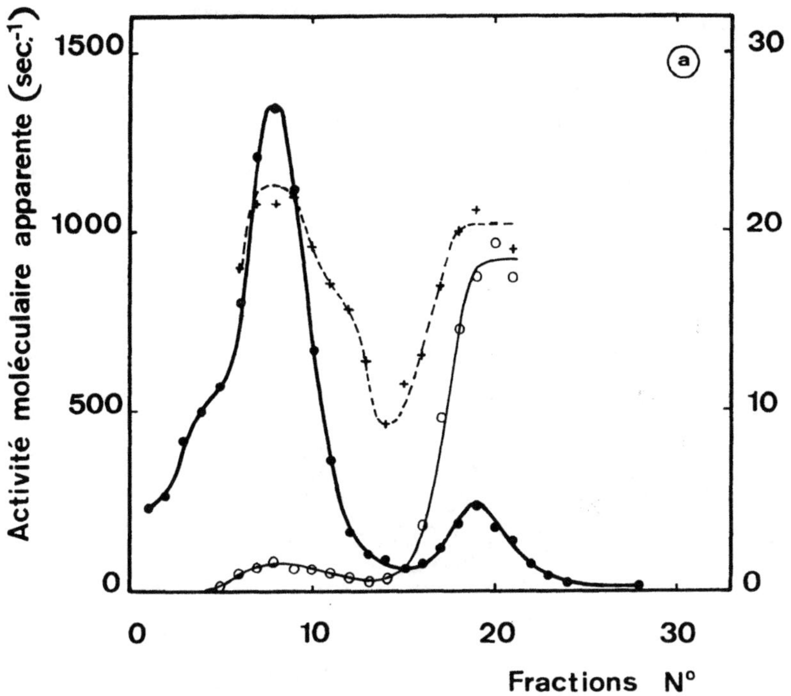

Activité moléculaire apparente (sec.$^{-1}$)
1500
1000
500
0
a
30
20
10
0
Concentrations µM
0
10
20
30
Fractions N°

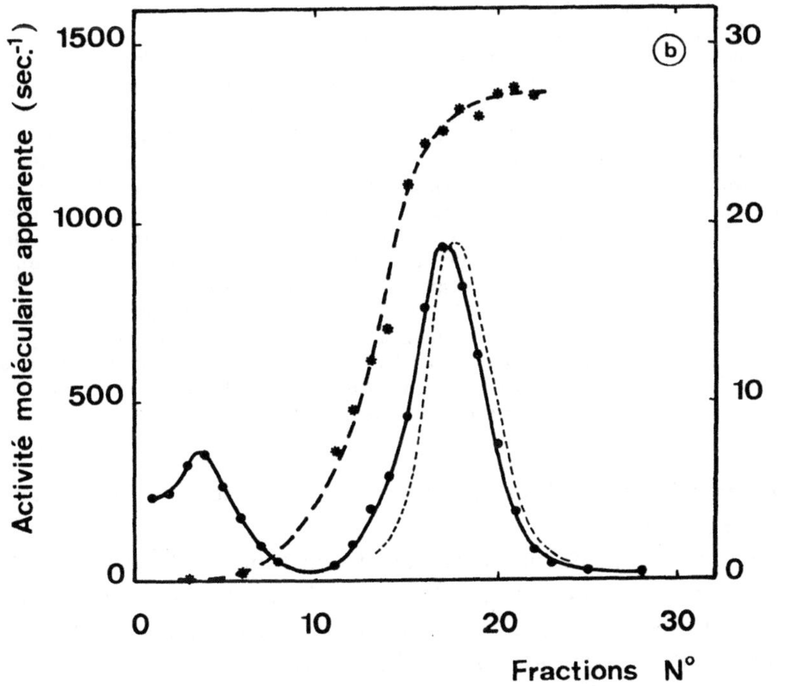

Activité moléculaire apparente (sec.$^{-1}$)
1500
1000
500
0
b
30
20
10
0
Concentrations µM
0
10
20
30
Fractions N°

Figs. 10a–b.

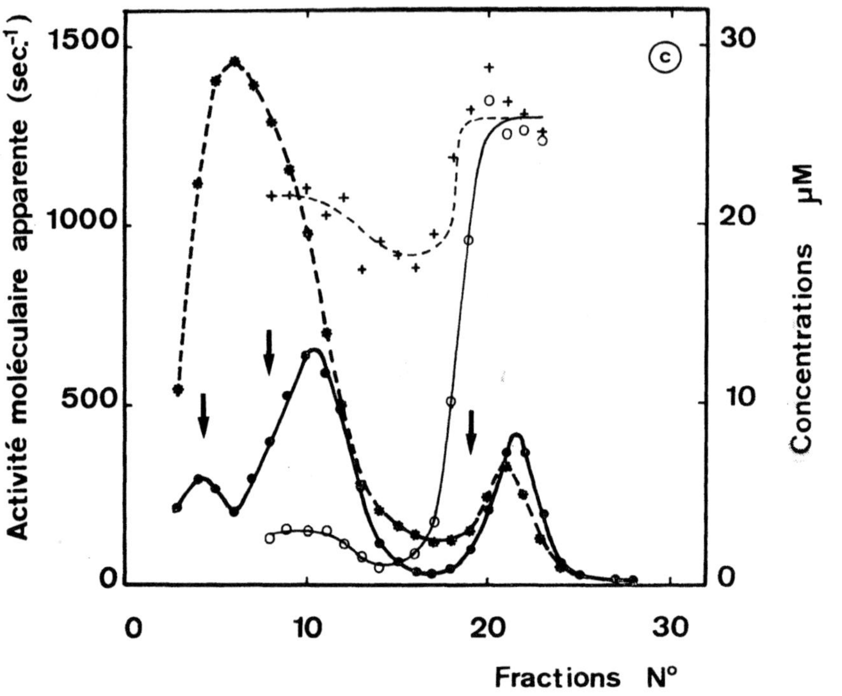

Figs. 10a–c. Étude par ultracentrifugation sur gradient de saccharose d'états de structure et d'activité impliqués dans le phénomène d'inactivation thermique. Même protocole expérimental que celui décrit dans la Figure 6. L'enzyme est inactivée à la concentration 208,6 μM (en hème), à 20,2 $\pm$ 0,1° C; au bout de 21 minutes d'incubation son activité moléculaire est de 159 sec.$^{-1}$ (a) Profils de concentration (●) et d'activité de cette solution d'enzyme déposée sur un gradient de saccharose, lequel est dissous dans du tampon phosphate 10 mM, EDTA 1 mM, DTT 0,1 mM, pH 7,20. L'activité des différentes fractions est mesurée immédiatement après leur récupération (○) et au bout de 24 heures d'incubation, à 4° C, dans du mélange dit LPE(+). (b) Profils de concentration (●) et d'activité (*) de cette solution d'enzyme réactivée par du L-lactate (100 mM), déposée sur un gradient de saccharose, lequel est dissous dans du milieu dit LPE. Profil de concentration (----) d'une solution d'enzyme non traitée thermiquement, d'activité moléculaire 955 sec^{-1}, déposée sur un gradient identique au précédent. (c) Un mélange de la même solution d'enzyme traitée et de cytochrome *c* (dissous dans du tampon phosphate 10 mM) de concentrations finales respectives: 158 μM (en hème) et 310 μM, est déposé sur un gradient de saccharose, lequel est dissous dans du tampon phosphate 10 mM, pH 7,20 (contenant EDTA 1 mM et DTT 0,1 mM). Les concentrations en cytochrome *c* (*) et en hème de cytochrome b_2 (●) sont calculées d'après [19]. Les mesures d'activité des différentes fractions sont faites dans les mêmes conditions que celles décrites dans la Figure 10a. Dans les trois cas, le sommet des tubes correspond à la fraction 0.

mentation est en effet voisine de ce que l'on peut attendre pour une pelote statistique de même masse (60000 dalton).

La réactivation du constituant majeur, par incubation dans le mélange LPE, est obtenue sans aucune addition de flavine. Cette dernière est donc bien présente associée à l'enzyme inactive en quantité quasi-stoechiométrique. Ce fait est en plein accord avec les observations décrites ci-dessus (3.1 C). La décroissance, de part et d'autre du sommet du pic, de l'activité moléculaire apparente (elle est rapportée à l'hème), après réactivation, pourrait traduire un déficit en flavine d'autant plus marqué que la concentration de protéine est plus faible (dissociation intervenant durant la migration sur le gradient). Ce point reste à préciser ultérieurement par une étude approfondie des bilans de flavine à travers tout le contenu des tubes centrifugés.

Du point de vue de la structure tertiaire, on peut affirmer que la forme monomère réactivable de l'enzyme est une entité essentiellement globulaire. Le calcul du coefficient de sédimentation théorique pour une sphère anhydre et rigide de même masse (4,94 svedberg) conduit en effet à un rapport de friction $f/f_0 = 1{,}16$. Cette observation permet de comprendre que la flavine prosthétique, bien que liée de manière non covalente, puisse rester associée à l'hémoprotéine dans cet état inactif.

2. *État tétramérique de l'enzyme réactivée en présence de lactate*

Il était important de vérifier que l'enzyme inactivée thermiquement à faible force ionique puis réactivée par incubation en présence de lactate retrouvait effectivement sa structure tétramèrique initiale. C'est bien ce que l'on observe (Figure 10b): il est pratiquement impossible de distinguer, sur la base de leur coefficient de sédimentation, les formes actives d'origine et régénérée. On notera dans le diagramme la présence du constituant irréversiblement dénaturé ($S_{20,w} = 2{,}2$ svedberg).

3. *Formation de complexes avec le cytochrome c*

Nous avons vu précédemment (3.1) que le cytochrome *c* forme un complexe avec l'enzyme tétramérique qu'il stabilise ainsi contre l'inactivation thermique à faible force ionique mais qu'il ne provoque aucune réactivation du monomère, à concentration d'enzyme voisine de 2 μM (en hème), dans le phosphate 10 mM (pH 7,20). La Figure 10c montre néanmoins que, dans ce milieu, le cytochrome *c* se combine non seulement avec le tétramère actif mais également avec l'enzyme dissociée inactive. L'analyse de tels profils de concentration peut fournir trois types d'informations: la valeur du coefficient de sédimentation apparent des constituants peut être déterminée mais on peut également apprécier la stoechiométrie et la stabilité des complexes éventuellement formés entre les deux protéines. De ce point de vue, on peut caractèriser le constituant le plus lourd ($S_{20 \cdot w} = 11{,}7$ svedberg; $f/f_0 = 1{,}22$) comme étant un complexe tétramère–(cytochrome $c)_4$. Le déficit, par rapport à la stoechiométrie 1:1 (en hèmes), notée au sommet du pic lourd ainsi que la traînée du cytochrome *c* indiquent une stabilité moyenne ($10^{-7} < K_{diss.} < 10^{-6}$ M). En ce qui concerne le constituant inactif réactivable, on peut affirmer qu'il fixe également le cytochrome *c* par le fait que le pic de concentration de ce dernier montre à la fois un entraînement net vers le fond ($S_{app} = 3{,}2$ svedberg) et une dissymétrie très marquée du front de migration, au niveau du pic de l'enzyme. La stoechiométrie de ce complexe est plus difficile à préciser en raison de la proximité du pic de cytochrome *c* libre en excès; elle est pratiquement 1:1 (en hèmes) dans la région du front du constituant inactif de l'enzyme. Le coefficient de sédimentation apparent calculé pour ce dernier, soit $S_{20,w} = 5{,}7$ svedberg, est incompatible avec un état monomère de l'enzyme. En effet, par rapport au monomére dans le même milieu (cf. Figure 10a), l'accroissement relatif observé de 4,3 à 5,7 svedberg dépasse de très loin la limite théorique admissible pour l'accroissement de masse dû à la fixation du cytochrome *c* (modèle sphérique: $S'_0/S_0 = (m'_0/m_0)^{2/3} = 1{,}142$ contre $5{,}7/4{,}3 = 1{,}326$ pour la protéine réelle). Il semble donc raisonnable de penser que, en présence de cytochrome *c*, et dans les conditions de ces expériences, l'enzyme inactive réactivable existe soit à l'état de dimères soit, plus vraissemblablement, à l'état d'un mélange monomère–dimère qui serait en équilibre rapide. Des expériences toutes récentes utilisant la technique de filtration sur tamis moléculaire conduisent à une observation similaire.

La comparaison des aires sous les pics de concentration de l'enzyme semble montrer une légère augmentation de la proportion du tétramère par rapport à ce qu'elle était pour le témoin sans cytochrome *c* (Figure 10a). Des expériences complémentaires seront nécessaires pour savoir si cette différence est réellement significative.

B. ÉTUDES PAR DIFFUSION DE LA LUMIÈRE

1. *Caractérisation des états limites monomère et tétramère*

Une étude par les méthodes de l'équilibre de sédimentation a permis de calculer des masses molaires respectives de 220000 g et 236000 g (±10000 g) pour l'enzyme dans les milieux tampon phosphate 0,1 M et 0,2 M additionnés de 50 mM DL-lactate

TABLEAU II

Étude de transitions de réactivation dans divers milieux: corrélations entre activité et poids moléculaire apparent mesuré par diffusion de la lumière

Milieu	M_w app.	% Tétramère	Activité moléculaire (sec^{-1})	Concentration μM (en hème)
	185000	70	477	1,06
LPE, pH 6,8	220000	89	690	1,06
	220000	89	795	1,06
Phosphate 0,2 M,	77000	~10	50	2,10
pH 7,20	110000	28	452	2,10
Idem +	134000	41	546	2,10
L-lactate 28 mM	162000	57	740	2,10
Phosphate 10 mM,	64500		51	4,68
pH 7,20	69000	~ 5	55	3,18
	79000	~11	51	1,13
Idem +	150000	50	339	3,70
L-lactate 28 mM	173000	64	518	3,70
	209000	83	760	3,70

L'activité enzymatique et l'intensité de lumière diffusée sont mesurées au cours du temps après addition de l'enzyme inactive (100 μl) dans les divers milieux (1750 μl) équilibrés à 20,0 ± 0,2° C. Le pourcentage de tétramère (en poids) est calculé en faisant l'hypothèse que la solution est un mélange de monomères et de tétramères.

(pH 7,2), tandis qu'après inactivation et incubation prolongée dans le phosphate 10 mM (pH 7,2), la masse molaire observée est de l'ordre de 60000 g [11].

Les mesures que nous avons éffectuées par diffusion de la lumière sont en bon accord avec ces valeurs: 210000 ± 12000 g pour le tétramère dans le tampon phosphate 0,2 M, pH 7,20 (sans lactate) et 220000 à 240000 g ± 15000 g pour l'enzyme réactivée dans le milieu LPE après traitement thermique ([25] et Tableau II). La forme inactive et réactivable, obtenue par traitement thermique court (1 heure) à 20° C dans le tampon phosphate 10 mM, pH 7,20, montre des valeurs de masse molaire apparente comprises entre 63000 et 78000 g. Il s'agit clairement de l'état monomère, les valeurs un peu élevées pouvant être dues à la présence d'une certaine proportion (5 à 10%) de tétramère.

2. *Corrélations entre états d'activité et de structure quaternaire au cours des transitions de réactivation*

Nous avons tenté de suivre, par diffusion de la lumière, quelques cinétiques de reformation du tétramère tout en mesurant parallèlement, sur la même solution, l'évolution de l'activité enzymatique. Ces expériences sont rendues délicates par la nécessité de prélever des échantillons ou d'ajouter des réactifs (lactate) dans la solution analysée.

Les observations les plus significatives dont nous disposons actuellement sont regroupées dans le Tableau II. Lorsque l'enzyme préalablement inactivée est diluée dans le milieu LPE équilibré à 20° C, on note toujours une montée très rapide de l'intensité de la lumière diffusée qu'il est pratiquement impossible d'enregistrer avec l'appareil utilisé; le niveau atteint est très proche de celui qui est attendu pour le tétramère. Dans le même temps, l'activité n'est récupérée que partiellement (400 à 600 sec^{-1}) et continue de croître alors même que l'intensité de lumière diffusée reste quasi-constante. Il semble donc que, dans ces conditions, la réassociation des monomères en tétramères précède la réapparition de l'activité. Une situation pratiquement opposée est observée à la suite de la dilution de l'enzyme inactive dans le tampon phosphate 0,2 M, pH 7,20 (sans lactate) à 20° C. On sait que dans ce milieu la réactivation est relativement lente (3.1); il en est de même pour l'évolution de la lumière diffusée. L'augmentation apparente du taux de tétramère semble ici insuffisante pour rendre compte de l'activité récupérée pendant le même temps. La reformation du tétramère n'est que partielle ($M_{w\,app} = 110000$ g) et le reste même après addition de L-lactate 28 mM ($M_{w\,app}$. final = 162000 g). La participation de dimères, qui pourraient être actifs dans ce milieu, est envisagée. Enfin, lorsqu'on rajoute du L-lactate à concentration modérée (28 mM) au monomère dilué dans le phosphate 10 mM (pH 7,20) à 20° C, on observe cette fois que récupération d'activité et formation du tétramère progressent de conserve.

Ces observations préliminaires tendent à montrer que, selon les conditions de milieu, les deux phénomènes de tétramèrisation et de réactivation ne sont pas nécessairement concertés.

4. Conclusion

En présentant cet ensemble de résultats, nous avons voulu montrer les qualités de la

L(+)lactate: cytochrome *c* oxydoréductase de la levure *Hansenula anomala* comme matériel d'étude des phénomènes ultimes d'organisation conduisant à un état fonctionnel tétramérique (constitution de la structure quaternaire). Il est en effet remarquable de pouvoir disposer d'un système structuré doté de ses groupes prosthétiques mais inactif catalytiquement dans un état monomérique relativement stable. L'amplitude des réponses observées au cours des transitions d'inactivation/réactivation par le jeu de facteurs du milieu comme la force ionique et la température ou d'effecteurs variés laisse espérer que des conditions seront trouvées où, par l'isolement de certaines étapes, le mécanisme de ces transitions pourra être peu à peu approfondi. Un objectif important sera de rechercher si les monomères inactifs possèdent des sites de reconnaissance pour les effecteurs réactivateurs ou si ces derniers exercent, en se fixant sur le tétramère actif, une action plus banale de déplacement d'équilibres. On sait en effet qu'une modulation des vitesses de processus de structuration par des effecteurs mérite d'être envisagée comme mécanisme possible de régulation des systèmes enzymatiques [26, 27, 28].

Ce travail a été initié au Centre de Génétique Moléculaire du CNRS à Gif-sur-Yvette, France, avec la collaboration de M. A. Spyridakis; il a été poursuivi à Toulouse avec la participation de M. M. Bardet, Melle B. Blazy, Mme M. Castaing et M. M. Prats.

Références

[1] Appleby, C. A. et Morton, R. K.: 1954, *Nature* **173**, 749.
[2] Appleby, C. A. et Morton, R. K.: 1959, *Biochem. J.* **71**, 492.
[3] Jacq, C. et Lederer, F.: 1970, *Eur. J. Biochem.* **12**, 154.
[4] Pajot, P. et Groudinsky, O.: 1970, *Eur. J. Biochem.* **12**, 158
[5] Monteilhet, C. et Risler, J. L.: 1970, *Eur. J. Biochem.* **12**, 165.
[6] Baudras, A. et Labeyrie, F.: 1968, dans *Structure and Function of Cytochromes* (K. Okunuki, M. D. Kamen, et I. Sekuzu, éds.), University of Park Press, Baltimore, p. 601.
[7] Forestier, J. P. et Baudras, A.: 1971, dans *Flavins and Flavoproteins* (H. Kamin, éd.), University of Park Press, Baltimore, p. 599.
[8] Lederer, F. et Simon, A. M.: 1971, *Eur. J. Biochem.* **20**, 469.
[9] Mevel-Ninio, M., Pajot, P., et Labeyrie, F.: 1971, *Biochimie* **53**, 35.
[10] Mevel-Ninio, M.: 1972, *Eur. J. Biochem.* **25**, 254.
[11] Labeyrie, F. et Baudras, A.: 1972, *Eur. J. Biochem.* **25**, 33.
[12] Jacq, C. et Lederer, F.: 1972, *Eur. J. Biochem.* **25**, 41.
[13] Somlo, M. et Slonimski, P.: 1966, *Bull. Soc. Chim. Biol.* **48**, 1221.
[14] Baudras, A.: 1971, *Biochimie* **53**, 929.
[15] Baudras, A. et Spyridakis, A.: 1971, *Biochimie* **53**, 943.
[16] Spyridakis, A. et Baudras, A.: article en préparation.
[17] Baudras, A., Prats, M., et Bardet, M.: article en préparation.
[18] Baudras, A., Capeillère-Blandin, C., Iwatsubo, M., et Labeyrie, F.: 1972, dans *Wenner-Gren Symposium on Structure and Function of Oxidation Reduction Enzymes, Stockholm 1970* (Å. Akeson et A. Ehrenberg, éds.), Pergamon Press, Oxford, p. 273.
[19] Baudras, A., Krupa, M., et Labeyrie, F.: 1971, *Eur. J. Biochem.* **20**, 58.
[20] Kunitz, M. et Northrop, J. H.: 1934, *J. Gen. Physiol.* **17**, 591.
[21] Yon, J.: 1959, *Biochim. Biophys. Acta* **31**, 75.
[22] Iwatsubo, M. et Di Franco, A.: 1965, *Bull. Soc. Chim. Biol.* **47**, 891.
[23] Baudras, A.: 1965, *Bull. Soc. Chim. Biol.* **47**, 1177.
[24] Capeillère-Blandin, C.: résultats non publiés.

[25] Castaing, M.: 1972, Rapport Stage DEA Biologie Physicochimique, Orsay.
[26] Deal Jr., W. C.: 1969, *Biochemistry* **8**, 2795.
[27] Frieden, C.: 1970, *J. Biol. Chem.* **245**, 5788.
[28] Teipel, J. W. et Koshland Jr., D. E.: 1971, *Biochemistry* **10**, 798.

DISCUSSION

Pantaloni: Comment peut-on expliquer que la cinétique d'inactivation de l'enzyme qui est observée soit du deuxième ordre, alors que d'après le modèle proposé on s'attendrait à trouver plutôt une cinétique d'inactivation du premier ordre ?

Baudras: La cinétique d'ordre deux est effectivement inattendue. Il est toujours hasardeux d'interpréter sans support expérimental un ordre apparent dans le temps. Des expériences montrant l'influence de la concentration en enzyme sur la vitesse initiale d'inactivation permettront de définir l'ordre vrai par rapport à la protéine. La réaction présente bien cet ordre deux pour toute la gamme des températures et forces ioniques utilisées (Figures 1 et 3).

Mme Buc: A-t-on pu mettre en évidence des mutants de l'enzyme qui seraient modifiés dans leurs propriétés d'association entre les sous-unités, ou avec le cytochrome *c* ?

Baudras: Nous ne disposons pas actuellement de mutants de la levure utilisée comme source d'enzyme (*Hansenula anomala*). Le seul point de comparaison, pour certaines propriétés, est le cytochrome b_2 cristallisé de *Saccharomyces cerevisiae*. Dans le cas présent elle est de peu d'utilité puisque les phénomènes décrits ne sont pas observés avec cette enzyme.

Pantaloni: Quelles sont les méthodes utilisées qui ont permis de mettre en évidence le très intéressant complexe entre le tétramère de cytochmore b_2 et quatre molécules de cytochrome *c* ?

Baudras: Actuellement nous ne disposons que des données d'expériences d'ultracentrifugation sur gradients de saccharose des mélanges enzyme/cytochrome *c*. Notre impression est qu'en matière de constante(s) d'équilibre, c'est l'occupation du site qui provoque en bloc la protection du tétramère actif qui impliquerait la plus grande affinité (cf. discussion de la Figure 10c).

ON ALLOSTERIC MODELS*

JEFFRIES WYMAN

CNR Centre for Molecular Biology at the Institute of Biochemistry, University of Rome, Italy, and Regina Elena Institute for Cancer Research, Rome, Italy

Summary. The basic idea of the allosteric interpretation of macromolecular behaviour is that the various ligand binding sites of the macromolecule interact not directly but indirectly as a result of ligand linked conformational equilibria, but within this general framework many different models are possible. It can be shown that the two models most commonly employed, namely the MWC model and the induced-fit model of Koshland and Nemethy, are in reality both special cases of a more general parent model. The MWC model is always cooperative. In contrast the induced-fit model may be either cooperative or anticooperative, depending on the choice of constants. An example of allosteric behaviour which illustrates these ideas is provided by the components of trout hemoglobin.

Résumé. L'idée de base de l'interprétation allostérique du comportement macromoléculaire est que l'interaction des sites macromoléculaires de fixation des ligands n'est pas directe mais indirecte, découlant d'équilibres conformationnels qui sont fonction des ligands. Dans ce cadre général de nombreux modèles différents sont cependant possibles. On peut montrer que les deux modèles dont l'usage est le plus courant, le modèle MWC et le modèle d'ajustement induit de Koshland et Nemethy, sont en réalité tous deux des cas particuliers d'un modèle parent plus général. Le modèle MWC est toujours coopératif. En revanche le modèle d'ajustement induit peut, selon le choix des constantes, être coopératif ou anti-coopératif. Un exemple de comportement allostérique qui illustre ces idées est fourni par les constituants de l'hémoglobine de la truite.

The underlying idea of the allosteric interpretation of macromolecular behavior is that the various ligand binding sites of a macromolecule interact, not directly, but only indirectly as a result of ligand-linked conformational equilibria. Thus any cooperativity of ligand binding is the result of cooperativity present in the conformational equilibria of the macromolecule in the absence of ligand. This is the parent concept, but within its broad shadow lie many different types of physical model, and in any attempt to interpret equilibrium, steady state, and kinetic experiments on allosteric proteins, it is important to have a clear grasp of the nature of such models. A large part of what follows is devoted to a description and comparison of the two most current allosteric models, the MWC model [1] and the induced-fit model [2, 3].

1. Some General Relations

As a helpful lead into the subject, we begin by considering the simple case of a macromolecule which exists in only one conformation and contains r binding sites for a ligand X. We postulate nothing as regards the possible composition of the macromolecule in terms of subunits, and make no assumption about the sameness or independence of the r sites. The binding behavior of the macromolecule is described by the

* Reprinted by permission from *Current Topics in Cellular Regulation*, Volume 6 (edited by B. L. Horecker and E. R. Stadtman), Academic Press, 1972.

C. Sadron (ed.), Dynamic Aspects of Conformation Changes in Biological Macromolecules. 207–223.

binding potential [4]

$$Л = RT \ln (1 + K_1 x + \cdots + K_r x^r) \equiv RT \ln N, \tag{1}$$

and the total amount of X bound by it is given by

$$\bar{X} = \frac{dЛ}{d\mu_x} = \frac{dЛ}{RT\, d\ln x} = \frac{d \ln N}{d \ln x}. \tag{2}$$

Here the polynomial in parentheses, denoted by N, is the binding polynomial of the macromolecule, and μ_x and x denote, respectively, the chemical potential and activity of X; the K's which occur in the polynomial are the overall equilibrium constants for the reaction of the macromolecule with 1, 2,... r molecules of X. It is evident that the *relative* amounts of the various macroscopic forms M, MX,... MX_r are given by the successive terms of the binding polynomial N.

We now pass on to the case where the same macromolecule, with its r binding sites, exists, as an at least partially allosteric system, in $s+1$ conformations, which, in the absence of ligand X, occur in the mole fractions ν_0, ν_1,... ν_s corresponding to the mole ratios 1, f_1,... f_s. In this case, by a simple extension of the concept of the last paragraph, we see that in the presence of ligand at activity x the relative amounts of all the macroscopically different forms are given by the elements of the $(s+1) \times (r+1)$ matrix

$$\begin{matrix} 1 & f_1 & \cdots & f_s \\ K_{01}x & f_1 K_{11}x & \cdots & f_s K_{s1}x \\ \vdots & \vdots & & \\ K_{0r}x^r & f_1 K_{1r}x^r & \cdots & f_s K_{sr}x^r, \end{matrix}$$

of which the general element of index ij is $f_i K_{ij} x^j$. Here again nothing is assumed about the presence of subunits and there is no restriction on the values of the K's beyond the fact that every $K_{i0} = 1$. This matrix, which is characteristic of the macromolecule, may be considered in relation to the expression for its binding potential. On the basis of a different and more general analysis, this may be written as [5]

$$Л = RT \ln \sum_{i=0}^{s} \nu_i \exp (Л_i/RT), \tag{3}$$

where $Л_i$ is the binding potential of the macromolecule in conformation i, or, alternatively, since dividing the sum $\sum$ by the constant ν_0 makes no difference to $\bar{X}$, as

$$Л = RT \ln \sum_{i=0}^{s} f_i \exp (Л_i/RT). \tag{3.1}$$

When each $Л_i$ in Equation (3.1) is expressed in terms of the corresponding binding polynomial N_i as in Equation (1), the result is

$$Л = RT \ln \sum_{i=0}^{s} f_i [1 + K_{i1}x + \cdots + K_{ir}x^r]. \tag{4}$$

Comparison shows that the $(s+1)(r+1)$ terms of the summation, which is nothing but the full binding polynomial of the macromolecule, are identical with the elements of the characteristic $(s+1)\times(r+1)$ matrix. We see therefore that, quite generally, the relative amounts of the various forms of an allosteric molecule, as indeed of any macromolecule, are given by the various terms of its full binding polynomial. It may of course happen that each Π_i in (3.1) is itself of the same form as (3.1), as will be the case when the allosteric macromolecule is composed of allosteric subunits (see Section 4 below), but this does not invalidate the general principle, and, since the exponential and logarithm are inverse functions, only results in an enlargement of the polynomial.

2. A Parent Model

In reality both the MWC model and the induced-fit model, at least as commonly interpreted, are only special cases of a more general parent model, which is itself a rather special case. The underlying assumptions of this model are the following. The macromolecule is composed of a set of identical subunits. Each of these subunits has a single site for a given ligand and each one can exist in only one or other of two different states (or conformations) A and B characterized by the binding constants K_A and K_B. The macromolecule itself exists in a variety of polymeric conformations, which are uniquely determined by the state of the subunits – one for A_s, one for BA_{s-1}, another for B_2A_{s-2}, and so on. All subunits are independent of one another in their reaction with ligands, and of course the whole system is in thermodynamic equilibrium. This means that the model is completely allosteric, all site-site interactions being indirect and mediated by the ligand-linked conformational equilibria.

In this special case, since each subunit contains a single ligand binding site, and since the number of polymeric conformations is the same as the number of subunits, s becomes equal to r and the characteristic matrix becomes square. At the same time, owing to the fact that the sites in any polymeric conformation bind independently of one another, the polynomial N_i corresponding to the conformation i in which i subunits are in state B and $r-i$ are in state A, becomes

$$(1 + K_A x)^{r-i} (1 + K_B x)^i .$$

Thus the full binding polynomial of the macromolecule is given by

$$N = \sum_{i=0}^{r} f_i (1 + K_A x)^{r-i} (1 + K_B x)^i . \tag{5}$$

From this it follows that element ij of the characteristic $(r+1)\times(r+1)$ matrix, whose elements give the relative amounts of all the different forms realized in the presence of X, is*

$$f_i x^j \sum_{m=0}^{j} \frac{i!K_B^m \qquad (r-i)!\, K_A^{j-m}}{m!\,(i-m)!\,(j-m)!\,(r-i-j+m)!}, \tag{6}$$

* In employing expression (6), it should be recalled that $0! = 1$, and that the factorial of a negative integer is to be taken as infinity.

where j denotes the number of molecules of ligand bound per macromolecule and m is a parameter. It is to be noted that by summing the elements in appropriate ways we can use the matrix to obtain not merely the relative amounts of the individual forms, but also of any desired combination of them as a function of x, e.g., the fraction of the molecules combined with 0, 1,... r molecules of X regardless of conformation, or the fraction of molecules in each conformation regardless of the distribution of X. The use of the characteristic matrix helps to clarify a number of points such as the concept of the 'switchover point' introduced by Hopfield *et al.* [6] in their analysis of the ligand binding process in hemoglobin.*

Equation (5) may be developed to read as an ordinary polynomial (a simple power series) in x – each term would then correspond to the sum of the elements in a row of the matrix. If in such a development the ratio of the coefficient of x^{j+1} to that of x^j $(1 \leqslant j \leqslant r-1)$ is statistical or less than statistical, then the polynomial will be completely factorable and the macromolecule will therefore behave as if it contained a set of r independent sites. The binding curve will be the same as, or less steep than, a simple titration curve, and the system will show no cooperativity – the value of n in a Hill plot will be everywhere equal to or less than unity. (Any allosteric system is of course cooperative in the sense that the total work of liganding is diminished as a result of the ligand-linked conformational changes. This is a distinction which should be kept in mind.) On the other hand, if the ratios of successive coefficients are *not* all statistical (or less than statistical), then the polynomial will *not* be completely factorable and the ligand binding of the macromolecule will not be describable in terms of a set of r independent sites – it will be necessary to invoke positive (cooperative) interactions between some at least of the sites. The value of n at any given point of a Hill plot may or may not be greater than 1. In the extreme case where the ratio of *every* coefficient to the preceding one is greater than statistical, the value of n will be *everywhere* greater than 1**, and we may describe the system as being completely cooperative.

All this is brought out by the following considerations. Suppose that the probability of a subunit, in the absence of ligand, being in state A is p and that of its being in state B is $q=1-p$, independently of the state of the $r-1$ other subunits. Then v_0, which is the same as the probability of all r subunits in the polymer being in state A, is p^r; on the same basis

$$v_i = \frac{r!}{i!\,(r-i)!}\,p^{r-i}q^i = \frac{r!}{i!\,(r-i)!}\,p^r\left(\frac{q}{p}\right)^i. \tag{7}$$

It follows that the f_i in Equation (5) have the statistical values

$$1,\ r\frac{q}{p},\ \cdots\ \frac{r!}{i!\,(r-i)!}\left(\frac{q}{p}\right)^i,\ \cdots, \tag{8}$$

* As an alternative and more geometrical way of representing such multiple equilibria we might introduce an r dimensional cube, of which the number of i dimensional subelements (corners, edges, 'faces') is given by the ith term of the expansion of $(2+1)^r$.

** Except at the two limits, $x \to 0$ and $x \to \infty$, where $n \to 1$.

and N becomes simply

$$N = \left[(1 + K_A x) + \frac{q}{p}(1 + K_B x)\right]^r = [1 + (pK_A + qK_B)\,x]^r p^{-r}. \tag{9}$$

The constant factor p^{-r} makes no difference to $\bar{Y}$ and may be omitted. In this case, therefore, the polymer binds like a simple monomer with binding constant $pK_A + qK_B$, which is an average value of the two K's the weighting factors being p and q.

If the ratios of successive f's are all less than statistical, then the polynomial in (5), again factorable, can be written in the form

$$\begin{aligned} N &= [(1 + K_A x) + b_1(1 + K_B x)]\,[(1 + K_A x) + b_2(1 + K_B x)] \cdots \\ &\qquad \cdots [(1 + K_A x) + b_r(1 + K_B x)] \\ &= (1 + b_1)(1 + b_2)\cdots(1 + b_r)\left[1 + \frac{(K_A + b_1 K_B)\,x}{1 + b_1}\right]\cdots \\ &\qquad \cdots\left[1 + \frac{(K_A + b_r K_B)\,x}{1 + b_r}\right], \end{aligned} \tag{10}$$

where the b's are real positive numbers. The polymer therefore behaves as if it contained r *different* and independent sites, each characterized by a binding constant $K_i = (K_A + b_i K_B)/(1 + b_i)$. It will be seen that $1/(1 + b_i)$ and $b_i/(1 + b_i)$ correspond respectively to the probabilities p and q introduced above.*

3. The MWC and Induced-Fit Models as Subcases

In the MWC model it is assumed not only that the state (conformation) of a given subunit is subject to influence by that of its neighbors, but that the interaction is so strong as to require all subunits to be always in the same state, so that only two polymeric conformations are permitted. This steric interaction has been interpreted

* If we were to admit complex factors, it is clear that they would have to occur in pairs of complex conjugates, one for $b = b' + ib''$ and the other for $b = b' - ib''$, in order to give a product with real coefficients. It can be shown that such a pair of factors would correspond to a pair of positively interacting sites, the probability of one site being occupied being increased by occupation of the other. By extending this principle we could, formally at least, factor any binding polynomial.

Another remark about the binding polynomial and the associated matrix concerns their relation to symmetry. It can be shown that symmetry of the binding curve, $\bar{Y}$ versus $\ln x$ (or the corresponding Hill plot), requires a corresponding symmetry of the binding polynomial represented by the following relation between its coefficients

$$K_r = x_{\frac{1}{2}}^{-r}, \qquad K_i x_{\frac{1}{2}}^i = K_{r-i} x^{r-i},$$

where $x_{1/2}$ is the activity of the ligand required to half-saturate the macromolecule. It follows from this that the curve showing the amount of each form MX_i versus $\ln x$ is the mirror image of the corresponding curve for its opposite, MX_{r-i}, reflected about the line $x = x_{1/2}$. Since each K_i is the sum of the coefficients in any row of the matrix, the above relations therefore impose a set of restrictions on these coefficients. In a cooperative system the extreme forms will be predominant; in an anti-cooperative system, the middle forms.

as representing a requirement of symmetry conservation [1]. Mathematically it means that only the first and last columns of the characteristic matrix remain, and that the polynomial resulting from the development of Equation (4) becomes simply*

$$N = (1 + K_A x)^r + f_r(1 + K_B x)^r. \tag{11}$$

In the case of the induced-fit model, no such drastic steric constraint is invoked. On the other hand, it is assumed that the values of the constants are such that the macromolecules in any given stage of ligand binding, e.g., the macromolecules MX_j, exist essentially all in only one polymeric conformation, namely that in which the subunit combined with ligand are in the high affinity state, while the others are in the low affinity state. This means that as liganding proceeds the system passes down the principal diagonal of reaction scheme I, where A represents a subunit in the low affinity state (A) and B a subunit in the high affinity state (B).

	$i = 0$	$i = 1$	$i = 2$	...
$j = 0$	A_r	BA_{r-1}	B_2A_{r-2}	
$j = 1$	A_rX	$BA_{r-1}X$	$B_2A_{r-2}X$	...
$j = 2$	A_rX_2	$BA_{r-1}X_2$	$B_2A_{r-2}X_2$	
⋮	⋮	⋮	⋮	

Scheme I

The condition for this is that for any degree of ligand binding j, i.e., in any row j of the reaction scheme, all terms except that corresponding to $i = j$ be negligible. Now it follows directly from (6) that as the ratio K_B/K_A increases indefinitely, the frequencies of the various forms in the reaction scheme approach those given in Table I.

Thus all terms *below* the principal diagonal become negligible regardless of the f_i. On the other hand, those *above* the diagonal will be negligible *if and only if*

$$\frac{i!}{(i-j)!\,j!} f_i \ll f_j \tag{12}$$

for all values of j and $i > j$. But this condition can always be satisfied if we make the ratio of each f_i to the following one sufficiently large. Let α_i and α_j be factors which relate f_i and f_j to their statistical values given by Equation (8). Then the inequality (12) becomes simply

$$\frac{(r-j)!}{(i-j)!\,(r-i)!} \ll \frac{\alpha_j}{\alpha_i}\left(\frac{p}{q}\right)^{i-j}. \tag{13}$$

* This becomes the same as the expression $(1 + x)^r + L(1 + cx)^r$ originally given in the paper by Monod, Wyman, and Changeux [1], when we choose the standard state of the ligand so that $K_A = 1$ and identify f_r with L and K_B with C.

TABLE I

Frequencies of various forms

$j \backslash i$	0	1	2	3	i
0	1	f_1	f_2	f_3	f_i
1	0	f_1K_Bx	$2f_2K_Bx$	$3f_3K_Bx$	if_iK_Bx
2	0	0	$f_2K_B^2x^2$	$3f_3K_B^2x^2$	$\frac{i(i-1)}{2}f_iK_B^2x^2$
3	0	0	0	$f_3K_B^3x^3$	$\frac{i(i-1)(i-2)}{2\cdot 3}f_iK_B^3x^3$
j	0	0	0	0	$\frac{i!}{(i-j)!\,j!}f_iK_B^jx^j$

By making p/q sufficiently large, this relation may always be satisfied for any value of the ratio α_j/α_i not equal to zero. Subject to this constraint, α_i and α_j can be given any values we please.

Provided condition (12) is satisfied, then all terms *above* as well as *below* the principal diagonal may be forgotten and the binding polynomial becomes simply

$$N = 1 + f_1K_Bx + f_2K_B^2x^2 + \cdots + f_rK_B^rx^r. \tag{14}$$

If the α's are such that the f's all have their statistical values, or that the ratios of successive f's are less than statistical, then the polynomial will be completely factorable and the system will show no cooperativity. In the opposite case where the ratios of successive f's are all greater than statistical, the binding curve will be everywhere steeper than a simple titration curve ($n > 1$) and the system will exhibit a degree of cooperativity determined by the values of the α's.

In both models the cooperativity, or better perhaps the *apparent* cooperativity, as judged from a Hill plot, is dependent on the choice of the f's, all the interactions between sites being indirect and mediated by ligand-linked conformational changes. This of course is the essence of the allosteric concept. The MWC model represents an extreme choice where all the mixed polymeric conformations are excluded, all f's except f_r and $f_0 = 1$ being set equal to zero. This model shows complete cooperativity, in the sense that the ratio of *every* term to the preceding one in the developed polynomial is greater than its statistical value so that n is everywhere > 1. This is apparent when we write the polynomial in the form

$$1 + r\bar{K}x + \cdots \frac{r!\bar{K}^i}{i!(r-i)!} + \cdots + \bar{K}^rx^r,$$

where the bars denote averages (e.g., $\overline{K^i} = \nu_0 K_A^i + \nu_r K_B^i$) and take account of the relation $\overline{K^i} > \overline{K^{i-1}}\bar{K}$.*

In contrast to the MWC model, the induced-fit model, though cooperative in the sense that the work of liganding is always less than it would be in the absence of the conformational changes, nevertheless in terms of a Hill plot may be either cooperative or anticooperative, i.e., the value of n may be either less than or greater than unity, depending on the choice of the f's.

It is easy to show that in the case of the MWC model the total *apparent* free energy of interaction realized per site in completely saturating the molecule with ligand X, which is given by the spacing of the asymptotes, is

$$\Delta F_{\mathrm{I\,total}} = RT \ln \frac{(\nu_0 K_A + \nu_r K_B)(\nu_0 K_A^{r-1} + \nu^r K_B^{r-1})}{\nu_0 K_A^r + \nu_r K_B^r}. \tag{15}$$

This may also be written as

$$\Delta F_{\mathrm{I\,total}} = -RT \ln \frac{\bar{K} \cdot \overline{K^{r-i}}}{\overline{K^r}},$$

and on the basis of the statistical relation just introduced is always positive. On the other hand, for the induced-fit model

$$\Delta F_{\mathrm{I\,total}} = -RT \ln \frac{f_1 f_{r-1}}{r^2 f_r}, \tag{16}$$

and may be either positive or negative. It is significant that in this case $\Delta F_{\mathrm{I\,total}}$ depends

* A proof of this is the following. Let $x_1, x_2, \ldots x_n$ be any set of positive numbers.

$$\begin{aligned}\overline{x^{t-1}} \cdot \bar{x} &\equiv \sum_i \frac{x_i^{t-1}}{n} \cdot \sum \frac{x_i}{n} \\ &= \sum_i \frac{x_i^t}{n^2} + \sum_i \sum_{j \neq i} \frac{x_i x_j^{t-1}}{n^2} \\ &= \frac{n}{n^2} \sum_i x_i^t - \frac{n-1}{n^2} \sum_i x_i^t + \sum_i \sum_{j \neq i} x_i x_j^{t-1}/n^2 \\ &= \overline{x^t} - \sum_i \sum_{j \neq i} \frac{(x_i - x_j)(x_i^{t-1} - x_j^{t-1})}{2n^2}.\end{aligned}$$

Provided $t \geqslant 2$, all pairs of parentheses in the double sum have the same sign. When $t = 1$, the second parenthesis vanishes. When $t \leqslant 0$, the two parentheses are of opposite sign. It follows that for our case where $t \geqslant 2$

$$\overline{x^t} > \overline{x^{t-1}}\bar{x} > \overline{x^{t-2}}\bar{x}^2 > \cdots (\bar{x})^t.$$

only on the f's and not at all on the values of the K's, K_A being assumed to be negligible in relation to K_B. This constitutes a significant difference between the two models and is closely related to the question of the shape invariance of the binding curves, which has been the subject of frequent discussion. It will be seen that in the case of the MWC model shape is determined by the two parameters K_B/K_A and f_r; in the case of the induced-fit model it is determined by the r parameters, f, but it is unaffected by the K's. This leads us to consider the effect of a second ligand.

4. The Effect of a Second Ligand

Suppose that in the general case considered in Section I the macromolecule, in addition to the r sites for ligand X, also contains t sites for a second ligand Y. Then the characteristic $(s+1)\times(r+1)$ matrix is replaced by a three-dimensional $(s+1)\times(r+1)\times(t+1)$ array of elements corresponding to the binding polynomial

$$1 + \sum_{i=0}^{s} \sum_{j=0}^{r} \sum_{k=0}^{t} f_i K_{ijk} x^j y^k . \tag{17}$$

Let us see how this carries over into the case of the two special allosteric models we have been considering, the MWC model and the induced-fit model.

Assume that in both cases each subunit carries, in addition to the site for X, a second site for Y and that the binding constant, L, of this site is, like K, different in the two states (conformations) A and B of the subunit. Then the binding polynomial for the MWC model becomes

$$N = (1 + K_A x)^r (1 + L_A y)^r + f_r (1 + K_B x)^r (1 + L_B y)^r , \tag{18}$$

and that for the induced-fit model

$$N = \sum_{i=0}^{r} f_i K_B^i (1 + L_B y)^i x^i . \tag{19}$$

It will be seen that in either case the effect of changing the activity of the second ligand is the same as that of changing the f's and hence is bound to change the shape of the binding curves. Now in many allosteric systems the binding curves appear to be invariant in shape for changes in the activity of the control ligand, the effect of the control ligand being simply to shift them back and forth along the $\ln x$ axis. This *seems* to be the situation in the case of the Bohr effect in human hemoglobin [7]. How can it be explained?

The most obvious and plausible explanation would be that the control effect arises locally within the subunits. Then of course Equations (18) and (19) would no longer be applicable, and the problem would disappear. If such a local effect were direct, not allosteric, there would be no difficulty in accounting for it, but if the effect is allosteric, as there are strong grounds for believing in the case of the Bohr effect of hemoglobin, then the situation is not so simple. We submit the following proposal. Suppose that in the low affinity form (conformation) A the subunit in reality exists in two *sub*conformations A1 and A2 characterized by binding constants K_{A1} and K_{A2} and L_{A1} and L_{A2} for ligands X and Y, respectively, and that in the absence of both ligands the mole fractions of the two subconformations are ν_{A1} and ν_{A2}. Then the binding polynomial of the subunit in state A is

$$N = \nu_{A1}(1 + K_{A1}x)(1 + L_{A1}y) + \nu_{A2}(1 + K_{A2}x)(1 + L_{A2}y). \tag{20}$$

When this is expanded and 'normalized' by dividing by the sum of all terms independent of X, we obtain the expression

$$1 + \left(\frac{1 + L''_{A}y}{1 + L'_{A}y}\right) \bar{K}_{A}x, \tag{21}$$

where

$$\begin{aligned} \bar{K}_A &= \nu_{A1}K_{A1} + \nu_{A2}K_{A2} \\ L'_A &= \nu_{A1}L_{A1} + \nu_{A2}L_{A2} \\ L''_A &= \frac{\nu_{A1}K_{A1}L_{A1} + \nu_{A2}K_{A2}L_{A2}}{\nu_{A1}K_{A1} + \nu_{A2}K_{A2}}. \end{aligned} \tag{22}$$

This shows that the subunit in the overall conformation A has a binding constant K_A for X which varies with the activity of the control ligand as a result of local, i.e., intrasubunit, conformational changes in accordance with the simple expression (21). As seen from the definitions (22), the constants $\bar{K}'_A$, L''_A, and L'' in (21) are simply averages of the various subconstants $K_{A1} \ldots L_{A2}$, since ν_{A1} and $\nu_{A2} = 1 - \nu_{A1}$ are frequencies or probabilities. An expression of exactly the same form as (21) applies to the subunits in the overall conformation B. Now if we assume that the difference between states A and B involves only the X binding properties and that the two K's in state B are simply multiples of those in state A, then it will be seen that for either of our models (and indeed for any purely allosteric model) the K_A and K_B which describe the system are simple functions of the activity of ligand Y given by

$$K = K_0 \left(\frac{1 + L''y}{1 + L'y}\right). \tag{23}$$

This would provide nicely for the shape invariance of the X binding curves. It is the same as the expression used in the past at a more phenomenological level to describe the alkaline Bohr effect of hemoglobin.

One can push this idea a little farther. We know from linkage relations (the effect of a ligand on an equilibrium constant) that in the absence of both ligands the ratio of molecules in subconformation A2 to those in A1 is given by ν_{A2}/ν_{A1}; in the presence of saturating amounts of Y, but in the absence of X, it is $\nu_{A2}L_{A2}/\nu_{A1}L_{A1}$; in the presence of saturating amounts of X but in the absence of Y it is $\nu_{A2}K_{A2}/\nu_{A1}K_{A1}$; in the presence of saturating amounts of both ligands it is $\nu_{A2}K_{A2}L_{A2}/\nu_{A1}K_{A1}L_{A1}$. Let us assume that in the absence of X the molecules are essentially all in subconformation A1, regardless of the presence or absence of Y, and that when saturated with X they are all essentially in subconformation A2, again regardless of the presence of Y. This means that

$$\nu_{A1} \to 1\,, \qquad \nu_{A1}L_{A1} \gg \nu_{A2}L_{A2}\,, \qquad \nu_{A2}K_{A2} \gg \nu_{A1}K_{A1}\,, \\ \nu_{A2}K_{A2}L_{A2} \gg \nu_{A1}K_{A1}L_{A1}\,. \tag{24}$$

In this extreme case, which represents an essentially complete subformational change induced by ligand X (say oxygen in the case of hemoglobin), the constants in (22) become simply

$$\bar{K}_A = \nu_{A1}K_{A1}\,, \qquad L'_A = L_{A1}\,, \qquad L''_A = L_{A2}\,. \tag{25}$$

If, under these extreme conditions, instead of one there were two or more independent binding sites for Y in each subunit, the single factor $(1+L''y)/(1+L'y)$ in Equation (23) would be replaced by a product of two or more such factors, one for each Y binding site. This is exactly what has been invoked more phenomenologically to explain the total Bohr effect of hemoglobin, including both the alkaline and acid part.

5. The Effect of Subunit Heterogeneity

In both the MWC model and the induced-fit model, it is assumed that the subunits are all alike. The question arises as to what sort of modification is required when they are *not* all alike, but have different ligand binding constants. This is not an idle question, for there are a number of cases of enzymes and respiratory proteins where there is reason to believe that the subunits fall into classes having different properties. Thus in human hemoglobin, recent studies indicate that the oxygen binding properties of the α and β chains, as they exist in the tetramer, are significantly different; in respect to their redox properties the chains are almost certainly different, and it appears that as a result of different local oxidation Bohr effects they become progressively more different as the pH is lowered. As we shall see in Section 6, recent studies on trout hemoglobin show that one of its components consists of two kinds of chain whose oxygen binding properties, like the redox properties of the α and β chains of human hemoglobin, are different and become progressively more different at acid pH, an effect which serves an important physiological function.

The general case is complicated, requiring as it does the introduction of different values of the binding constants K_A and K_B for each of the different classes of subunit. Without, however, attempting a general formulation it is clear that in the case of the MWC model the effect of this will be to change the binding polynomials of each of the two permitted polymeric conformations from the simple forms $(1+K_Ax)^r$ and $(1+K_Bx)^r$ to the noncooperative forms $(1+K'_Ax)\,(1+K''_Ax)\ldots$ and $(1+K'_Bx)\,(1+ +K''_Bx)\ldots$. In general this may be expected to result in a reduction in the cooperativity of the whole macromolecule. A similar result might also be expected in the more complex case of the induced-fit model. Each case of this kind is probably better treated individually. For example, if we adopt the two-state model for hemoglobin, the elements of the characteristic matrix will be the same as the terms of the expanded form of the suitably modified binding polynomial given by Equation (11), namely

$$(1 + K_{A\alpha}x)^2 (1 + K_{A\beta}x)^2 + f_4 (1 + K_{B\alpha}x)^2 (1 + K_{B\beta}x)^2 .$$

6. An Example from Trout Hemoglobin

What appears to be an almost perfect example of the principles of the two preceding sections is to be found in trout hemoglobin, which has been extensively studied in recent months by my colleagues in both Rome and Camerino. So far, results are limited to two more or less preliminary papers [8, 9], but the work is going on and a more complete account will be given soon.

Trout hemoglobin contains four different components, which can be readily separated by the modern techniques of protein fractionation and are designated as I, II,

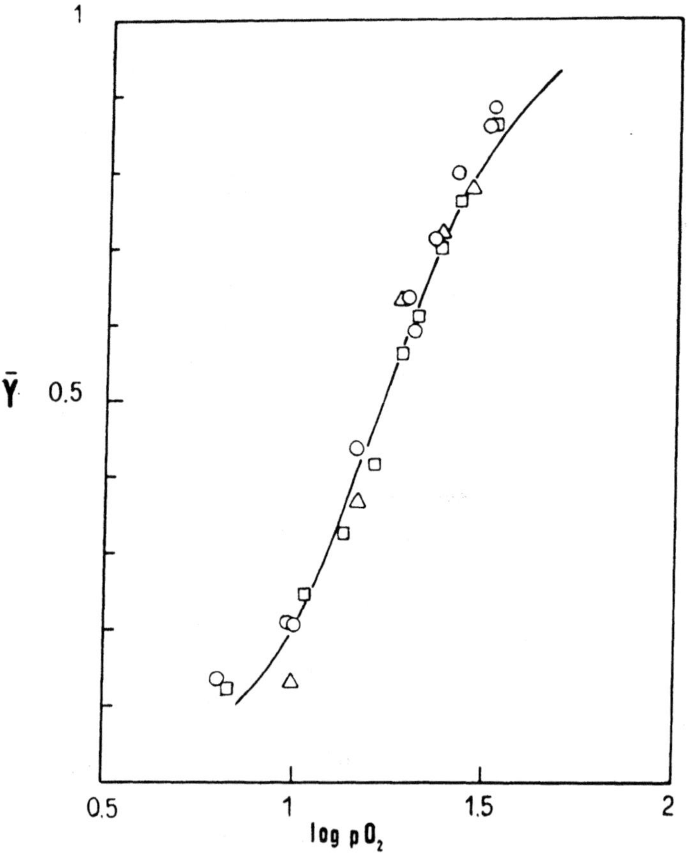

Fig. 1. Oxygen dissociation curve of component I of trout hemoglobin at 20° in 0.2 M phosphate buffer. Protein concentration 3–5 mg/ml. ○ pH 6.8; △ pH 7.2; □ pH 7.6. The smooth curve is a theoretical one corresponding to a value of $n = 2.6$. From Binotti *et al.* [8].

III, and IV in accordance with their relative speeds of migration in electrophoresis. The relative amounts of I, II, and IV are, respectively, approximately 20, 10, and 65%, component III being present as only a very small minority. Up to now, work has been concentrated on the two major components, I and IV, both of which exist under ordinary conditions as tetramers.

Component I contains two types of chain, as shown by the two bands observed on electrophoresis of the denatured material. So far, however, it has proved impossible to separate the chains in the native state, and the tetrameric molecule gives no evidence of dissociation into dimers under any of the experimental conditions that have been tried. Associated with this is the fact that the molecule fails to hybridize. It contains no C terminal histidine or tyrosine. As for functional properties, it is highly cooperative in its oxygen binding ($n \cong 3$ independently of pH), but shows no Bohr effect (see Figure 1). The oxygen affinity, as well as the cooperativity, is only very slightly if at all dependent on temperature and the maximum value of the enthalpy of oxygenation cannot exceed 3000 cal per mole of oxygen, as compared with about 12000 for human hemoglobin; thus the corresponding free energy change, including the contribution due to cooperativity, must result primarily from entropy effects.

Component IV likewise consists of two types of chain; however, unlike component I,

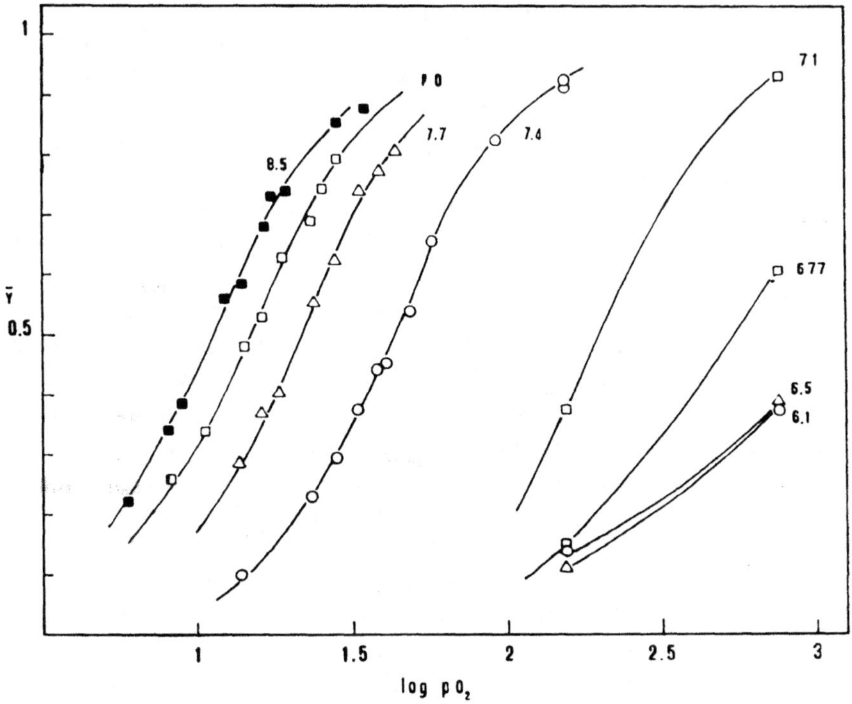

Fig. 2. Oxygen dissociation curves of component IV at different pH values and at 20° in phosphate buffers. Protein concentration 3–5 mg/ml. From Binotti *et al.* [8].

it can be readily digested by the carboxypeptidases, and the results indicate that the C terminal residue of one type of chain is probably arginine and that the other type of chain ends with tyrosine and histidine. Like component I, this component shows no tendency to dissociate and fails to hybridize. But its most interesting properties involve its liganding behavior with oxygen and carbon monoxide. At alkaline pH (8–9) it shows high cooperativity, like component I ($n=3$), but unlike component I it has an oxygen affinity which is dependent on temperature in accordance with a heat of oxygenation of about 12000 cal. As the pH is lowered to 5 or 6 there is a profound change in the liganding behavior. As shown in Figure 2, the value of n drops to 1 or less, and the oxygen affinity becomes so low that it is impossible to complete the binding curve, even with the use of pure oxygen at atmospheric pressure; at pH 6 the $p_{1/2}$ may be estimated as 1000 times or more greater than at pH 8.5.

How is this behavior to be interpreted? In my view the most plausible picture is the following. The Bohr effect is local to the chains, as in the case of human hemoglobin, but, in contrast to the case of hemoglobin in its reaction with oxygen, it is grossly different for the two types of chain. At alkaline pH the chains are essentially equivalent; however, owing to the large difference in the two Bohr effects, they become widely different at acid pH, and this, in conformity with the allosteric principles described above, leads to a large decrease in n and the development of a biphasic type of binding curve with a very high value of $p_{1/2}$.

We have looked at this behavior of component IV in terms of mechanism; it is also worthwhile to look at it teleologically. The trout, like many other fish, has the problem of maintaining his up and down position in the water. A mechanism for this is provided by the swim bladder, a gas chamber whose volume determines the buoyancy of the fish. But clearly, in the absence of some further provision, this by itself would place the fish in a position of unstable equilibrium; if he were to descend, the increase of hydrostatic pressure would diminish the volume of the chamber and hence reduce his buoyancy – just the opposite of what is required for him to keep his balance. To overcome this he must be in a position to secrete gas into the bladder, as need arises, against what may be a considerable pressure if he is at any significant depth, and in any case it will be necessary to replace the gas which is constantly going into solution. Clearly component IV provides a mechanism for doing this. As the hemoglobin passes through the gills it takes up oxygen under conditions where its oxygen affinity is high; when it passes through the rete mirabile, the vascular organ adjacent to the swim bladder, where the pH is low, its oxygen affinity is greatly reduced and the oxygen dissociation pressure becomes very high, potentially far above 1 atmosphere. Thus the hemoglobin, exploiting to the maximum the principle of the Bohr effect, acts as a pump to force oxygen into the bladder against the opposing hydrostatic pressure. The question might be asked: What is the source of the energy expended in this process? The answer is clear: it is the movement of proton, carried by the blood, from a region of high activity in the rete mirabile to one of low activity in the gills and other tissues. The metabolism of the fish maintains the pH gradient, and we may suppose that the blood flow through the secreting organ, or perhaps the local pH in the rete, is subject to nervous or hormonal control in accordance with the needs of the moment.

And what of the function of component I? This would seem to be the more prosaic one of assuring the normal transport of oxygen to the tissues. But why, in contrast to component IV and to the mammalian hemoglobins, should this molecule have been so designed as to have no heat of oxygenation? It is tempting to answer this question in the following way. It is of course essential for the actively swimming trout that oxygen transport should not be cut off or seriously reduced in cold water. In mammals body temperature is maintained constant within narrow limits, in spite of changes of temperature of the surroundings; their hemoglobin therefore operates always at constant temperature. In contrast, in the poikilothermous fish there is no such regulation of body temperature, and to make up for its lack there has been developed in the course of evolution a highly specialized form of the hemoglobin molecule which is insensitive to temperature.

There is every reason to believe that the trout does not stand alone. Tuna fish hemoglobin behaves in much the same way [10], and one may anticipate that other fast swimming fish will be found to show the same remarkable molecular adaptation. This remains to be explored in further experiments.

7. Concluding Remarks

The fish hemoglobins represent only a very special case, special even among the hemoglobins. They were introduced simply as a possible example of the concepts of Sections 4 and 5. But although other respiratory proteins and enzymes behave quite differently, a very large number of them undoubtedly fall together within the class of allosteric macromolecules. In this discussion we have attempted to clarify the common features of all such systems as represented in the various models which have become current, starting with the 'parent model' of Section 2, of which both the MWC model and induced-fit model are special cases. But it should not be forgotten that other models are possible within the framework of the allosteric concept. Of the relations we have developed, only those of Section 1 are of general applicability. They bring out the usefulness of the concept of the binding polynomial and the associated characteristic matrix in describing any allosteric macromolecule.

In conclusion, perhaps an apology is due to the reader for the somewhat abstract character of this discussion, for certainly the interest of the molecular biologist lies, and rightly so, in physical mechanism rather than in formalism. Apart from the fact that a precise mathematical formulation provides a powerful touchstone for the evaluation of any physical interpretation, a defense might be offered on the basis of a more positive and deeper philosophy. If it be true, as so persuasively maintained by Hardy [11] in his charming little essay 'A Mathematician's Apology', that mathematical objects are the most real elements of our experience, enjoying as they do a timeless and spaceless existence of their own which by far transcends that of the localized and transitory physical world around us, then surely it is in order to see how far we can go in displaying our physical models as special incarnations of these universal objects.

Acknowledgement

This work was supported by a grant from the U.S. National Science Foundation.

References

[1] Monod, J., Wyman, J., and Changeux, J. P.: 1965, *J. Mol. Biol.* **12**, 88.
[2] Koshland, D. E., Nemethy, G., and Imer, D. F.: 1966, *Biochemistry* **5**, 365.
[3] Haber, J. E. and Koshland, D. E.: 1967, *Proc. Nat. Acad. Sci. US* **58**, 2087.
[4] Wyman, J.: 1965, *J. Mol. Biol.* **11**, 631.
[5] Wyman, J.: 1967, *J. Am. Chem. Soc.* **89**, 2202.
[6] Hopfield, J. J., Shulman, R. G., and Ogawa, S.: 1971, *J. Mol. Biol.* **61**, 425.
[7] Antonini, E. and Brunori, M.: 1971, in *Hemoglobin and Myoglobin in Their Reactions with Ligands*, Res. Monogr.: Frontiers Biol. (ed. by A. Neuberger and E. L. Tatum), North-Holland Publ. Co., Amsterdam.
[8] Binotti, I., Giovenco, S., Giardina, B., Antonini, E., Brunori, M., and Wyman, J.: 1971, *Arch. Biochem. Biophys.* **142**, 274.
[9] Giovenco, S., Binotti, I., Brunori, M., and Antonini, E.: 1970, *Int. J. Biochem.* **1**, 57.
[10] Rossi Fanelli, A. and Antonini, E.: 1960, *Nature (London)* **186**, 895.
[11] Hardy, G. H.: 1967, *A Mathematician's Apology*, Cambridge Univ. Press, London and New York.

DISCUSSION

Karplus: (1) The question of the nature of the Bohr effect in hemoglobin has come up several times. It has been suggested that the linear release of protons as a function of O_2 binding implies that only tertiary structural changes are involved. In model calculations done by Dr. A. Szabo and myself, it has become clear that the observed linear relationship can hold if the groups releasing the Bohr protons are affected by quaternary as well as by tertiary structural changes. Significant deviations from linearity appear only near the end points (near 0% or 100% saturation) for which good measurements are not yet available. It would be most useful if such measurements could be made. (2) Although the formal scheme provided by Wyman includes the KNF and the MWC models for hemoglobin cooperativity as limiting cases, it appears to me that it is important to emphasize the difference in physical mechanism involved in the two models. In the KNF scheme it is tertiary structural changes which directly lead to cooperativity. In the MWC model tertiary structural changes clearly must occur, but it is the quaternary structural changes that yield the cooperative effect; that is, tertiary structural changes in a molecule whose quaternary structure is fixed (say by some chemical modification) would not yield cooperative binding.

Wyman: (1) There is of course a question of the extent to which the liberation of proton is strictly linear in oxygen uptake. It would be preferable, if it were experimentally practical, to see how far the spacing of the asymptotes (the total interaction energy) is independent of pH. Unfortunately it is not practical to do this with sufficient accuracy. (2) There is of course a real and profound difference between the physical picture of the two models. But, with the limitations of the parent model, the cooperativity in both cases depends on the values of the f's and hence on the interaction of the quaternary structures in the unliganded molecules.

Sadron: Quite recently Weber has given us a talk about the thermodynamics of the reactions which occur on different sites of a single macromolecule (i.e. a protein) when these sites are correlated. In Weber's derivation a special attention is given to the definition of a 'free energy of coupling' between the sites. Could you tell us how this way of approach can be integrated in the general treatment that you have considered at the beginning of your paper?

Shulman: You have shown in this talk how the different models of cooperativeness can be treated as special cases of a general model. In this way you have underemphasized the different physical mechanisms which have been the basis of the two-state MWC model and the alternative Koshland model. In the former the primary cause of the changing affinity of hemoglobin for oxygen was the switch from one quaternary structure to another. Within one quaternary structure the affinity should be independent of the number of oxygen molecules already bound. In the Koshland 'induced-fit' model the affinity depended upon the number of oxygen molecules already bound and it was tacitly assumed that the quaternary structure was irrelevant. Since one can experimentally change the quaternary structure of some hemoglobins without changing the number of ligands, and vice-versa, it has been possible to decide between these two models in favor of the MWC. Therefore I feel that to stress the formal equivalence of these different models at this time tends to be confusing, because it implies that the physical differences are unimportant, while in fact they have finally been distinguished.

Wyman: I would not want to underestimate the difference in the physical pictures corresponding to the two models. In the MWC model there are only two quaternary conformations; in the Koshland model there are many, through which the molecule passes on its way to complete saturation with ligand, and these involve forms each characterized by a different affinity. What I wish to emphasize is that both models result from the fundamental phenomenon of ligand-linked conformational equilibria. The common feature of both models, as I have presented them, is that the cooperativity of ligand binding is already present in the cooperativity of the conformational equilibria of the macromolecules in the absence of ligands. I am, naturally, pleased, as one of the original proponents of the MWC model, with the evidence you present for its applicability in the case of hemoglobin.

CONFORMATIONAL CHANGES IN GLYCOGEN PHOSPHORYLASE b FROM RABBIT SKELETAL MUSCLE. EQUILIBRIUM AND KINETIC STUDIES

H. BUC, M. H. BUC, M. MORANGE, and L. C. OUDIN
Département de Biologie Moléculaire, Institut Pasteur, Paris, France

and

H. WINKLER
Max-Planck Institut für biophysikalische Chemie, Göttingen, Germany

Summary. Phosphorylase b from rabbit skeletal muscle is activated by 5′-AMP and inhibited by glucose-6-phosphate. Binding studies of radioactive 5′-AMP and Glc-6-*P* and competition experiments allow us to derive an experimental partition function which is in good agreement up to 20° C with the theoretical partition function derived from the Monod-Wyman-Changeux model. Relaxation experiments performed at low temperature support this interpretation.

6-Mercaptopurine riboside 5′-phosphate is a good analog of 5′-AMP if one considers the homotropic effects in the binding and the heterotropic effects in the interaction with glucose-6-*P*. However, the effect upon the substrate during catalysis differs from that of 5′-AMP, suggesting a specific role of the position 6 of the purin ring in the correct positioning of the substrate, probably via an induced-fit mechanism.

Fast kinetic experiments performed with or without any added effector indicate the presence of several relaxation times. Their dependence on pH and temperature indicate that more than two conformations are significantly populated in phosphorylase b, even in the absence of added effector.

Résumé. La phosphorylase b du muscle de lapin est activée par le 5′-AMP et inhibée par le glucose-6-phosphate. Les études de fixation du 5′-AMP et du Glc-6-*P*, et les expériences de compétition, permettent de calculer une fonction de partition expérimentale qui est en bon accord jusqu'à 20° C avec la fonction de partition théorique déduite du modèle de Monod, Wyman et Changeux. Les expériences de relaxation soutiennent à basse température cette interprétation.

Le 6-mercaptopurine riboside 5′-phosphate est un bon analogue du 5′-AMP en ce qui concerne les effets homotropes de fixation, ainsi que les effets hétérotropes vis-à-vis du glucose-6-*P*. Cependant, l'effet exercé sur le site du substrat lors de la catalyse diffère de celui du 5′-AMP, suggérant un rôle spécifique de la position 6 du noyau purique dans le positionnement correct du substrat, sans doute via un mécanisme d'ajustement induit.

Les expériences de cinétique rapide, effectuées en présence ou en absence d'effecteur, permettent de mettre en évidence plusieurs temps de relaxation. Leur dépendance vis-à-vis du pH et de la température indiquent que plus de deux conformations majeures existent en équilibre dans la phosphorylase b, même en absence d'effecteur.

1. Introduction

Phosphorylase b catalyses the first step of glycogen breakdown in muscle according to the reaction:

Glycogen (*n* residues) + inorganic phosphate
Phosphorylase b ↓↑
Glycogen (*n* − 1 residues) + glucose-1-phosphate.

It is therefore a key enzyme whose activity must be modulated according to physi-

C. Sadron (ed.), Dynamic Aspects of Conformation Changes in Biological Macromolecules. 225–245.

ological demand. Products of the glycogenolysis pathway have been known for a long time to act as effectors of the enzyme which is practically inactive in the absence of AMP. In particular, kinetic studies have shown that interactions occur between the substrate and the effectors, and between effector molecules. For a review, see Fischer *et al.* [1a], and Madsen [1b]. The sigmoidal increase of the catalytic activity in the presence of various concentrations of 5′-AMP, the activator, was in fact the first case of an allosteric effect which was reported [2], although the term allosteric was not used at that time.

We have chosen to focus our attention on the particular interactions which occur between 5′-AMP (or analogs of AMP) and glucose-6-phosphate (glucose-6-*P*), and to relate these interactions with the conformational transitions of the enzyme observed by kinetic methods, in the absence and presence of these ligands.

(1) In Section 2 we show that, in parallel to the antagonistic effect of 5′-AMP and glucose-6-*P* on the enzymatic activity, each ligand also has an antagonistic effect on the binding of the other which occurs via a conformational transition in the phosphorylase b dimer; the results are expressed in terms of an experimental partition function which is identical to the one derived from a concerted two-state transition between a state binding almost exclusively 5′-AMP and a state binding exclusively glucose-6-*P*.

(2) A comparative study of the binding of an analog of 5′-AMP, 6-mercaptopurine riboside 5′-phosphate (SHMP), which activates phosphorylase b, has been done. It shows that positive interaction between activator and subtrate depends on the substituant at position 6 in the base. The study of an analog of 5′-AMP (a Co III-AMP complex which totally inactivates the subunit onto which it is covalently bound), shows that indeed there are interactions between the activator and the substrate in the subunit which is still active.

(3) We report relaxation experiments performed in the absence and in the presence of the activator (or analog of the activator). They show pre-existing equilibria in phosphorylase b in the absence of any added effectors. The value and the amplitude of these relaxation times, τ_1, τ_2, τ_4, are affected by temperature, pH, depend on substitutions on the coenzyme of phosphorylase, pyridoxal-phosphate, and are affected by AMP.

2. Allosteric Interactions Between AMP and Glucose-6-Phosphate

(a) Two sites of strong affinity for AMP (the activator), glucose-6-*P* and ATP (inhibitors of the catalytic activity) have been found per phosphorylase b dimer, using equilibrium dialysis [3, 4]. No interaction can be detected between ATP sites, while homotropic cooperative effects exist between AMP sites or between glucose-6-*P* sites (Figure 1a, b, c).

Phenomenological dissociation constants characteristic of the complexes of the first (K_1^0 and $K_1'^0$) or the second (K_2^0 and $K_2'^0$) molecule of effector with the enzyme, in the absence of the other effector, have been obtained from experimental saturation functions.

Their dependence upon temperature is remarkably different (Figure 2): whereas

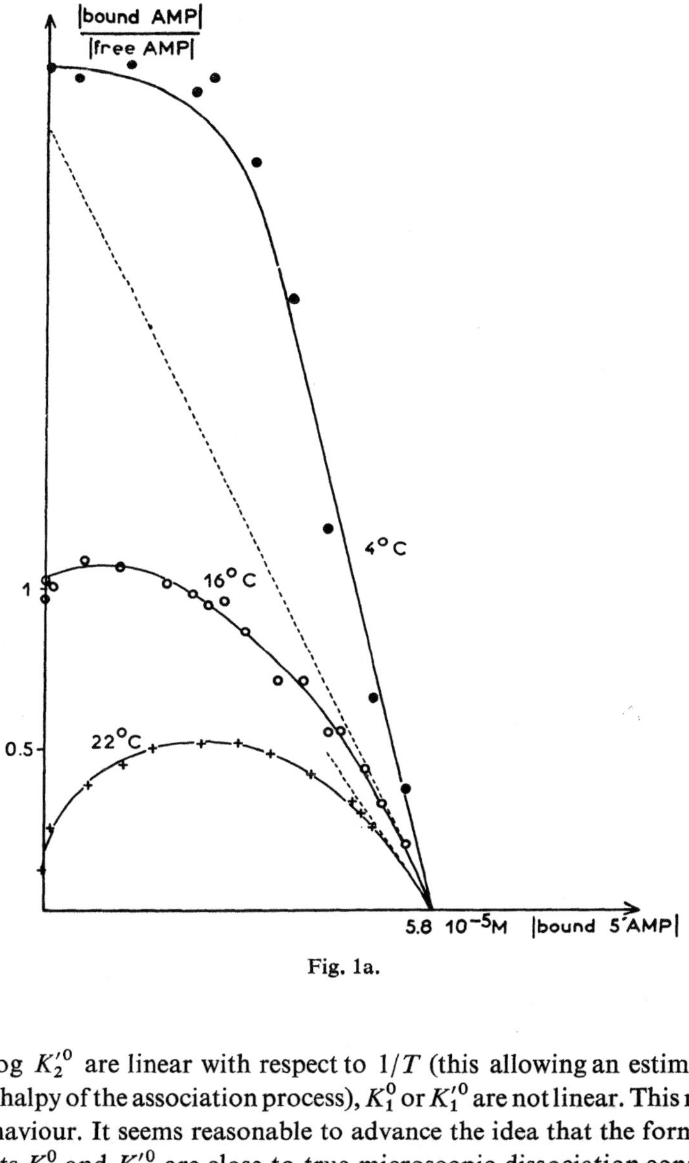

Fig. 1a.

log K_2^0 or log $K_2'^0$ are linear with respect to $1/T$ (this allowing an estimation of the standard enthalpy of the association process), K_1^0 or $K_1'^0$ are not linear. This reflects more complex behaviour. It seems reasonable to advance the idea that the formal dissociation constants K_2^0 and $K_2'^0$ are close to true microscopic dissociation constants, while K_2^0 and $K_2'^0$ reflect a combination of at least two events, transconformation and binding, at the microscopic level.

The cooperativity of binding of glucose-6-*P* and AMP depends markedly upon temperature in an opposite way; the change of cooperativity can be represented by the ratios K_2^0/K_1^0 or $K_2'^0/K_1'^0$. It can be seen in Figure 3 that the sum $K_2^0/K_1^0+K_2'^0/K_1'^0$ is close to 1 at a given temperature. These data will be used later on in the discussion, after introduction of the experimental and theoretical partition functions.

(b) The interaction between the binding of AMP and glucose-6-*P* was measured by

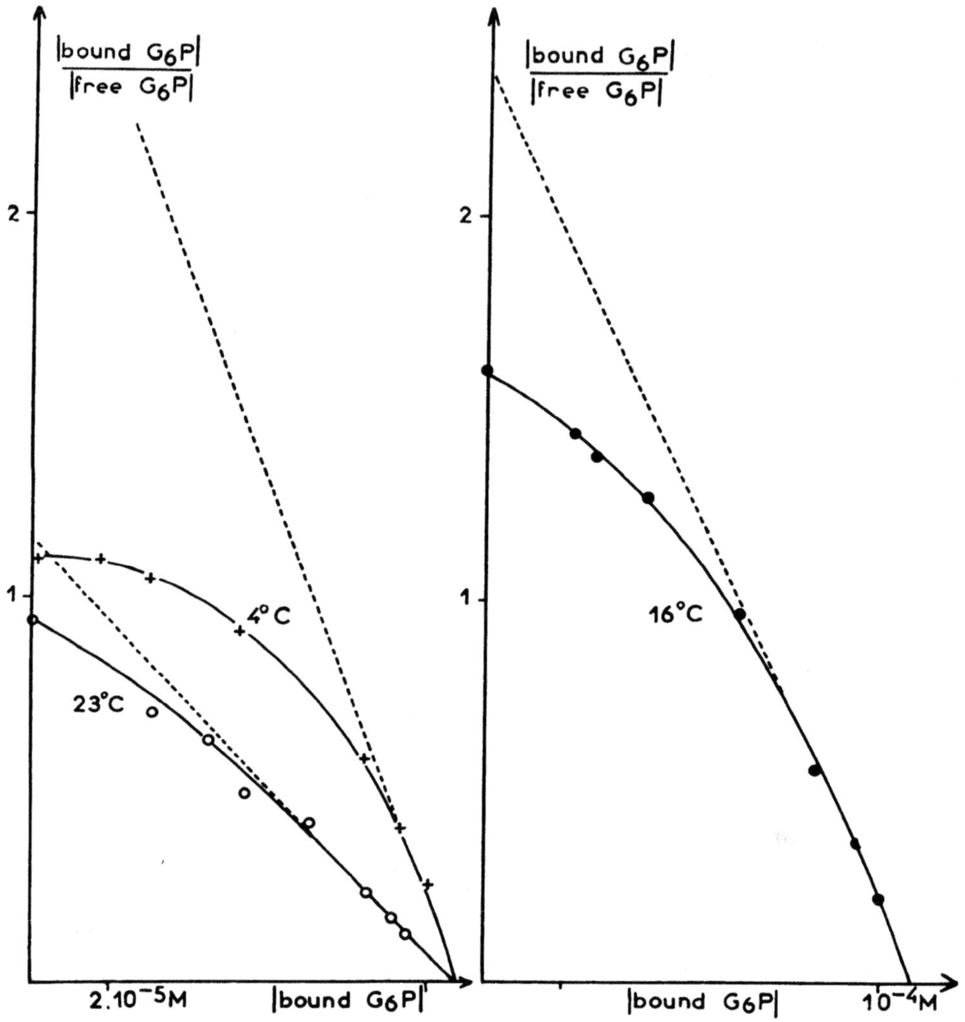

Fig. 1b.

equilibrium dialysis. It can be expressed in a similar way by phenomenological constants K_1, K_2, K_1' and K_2' which refer respectively to the binding of glucose-6-*P* in the presence of AMP, or to the binding of AMP in the presence of glucose-6-*P*.

Two functions can be studied in one experiment: the release of a given amount of one of the effectors together with the binding of the antagonistic effector. When trace amounts of one effector are used, one can get directly the variation of $1/K_1$ (or $1/K_1'$), the dissociation constant of the first enzyme-effector complex. The set of binding and release functions allows the determination of $1/K_1$ and $1/K_2$ as a function of free AMP concentration, and of $1/K_1'$ and $1/K_2'$ as a function of free glucose-6-*P* concentration.

The following results are obtained:

(1) AMP strictly precludes the formation of the first glucose-6-*P* enzyme complex:

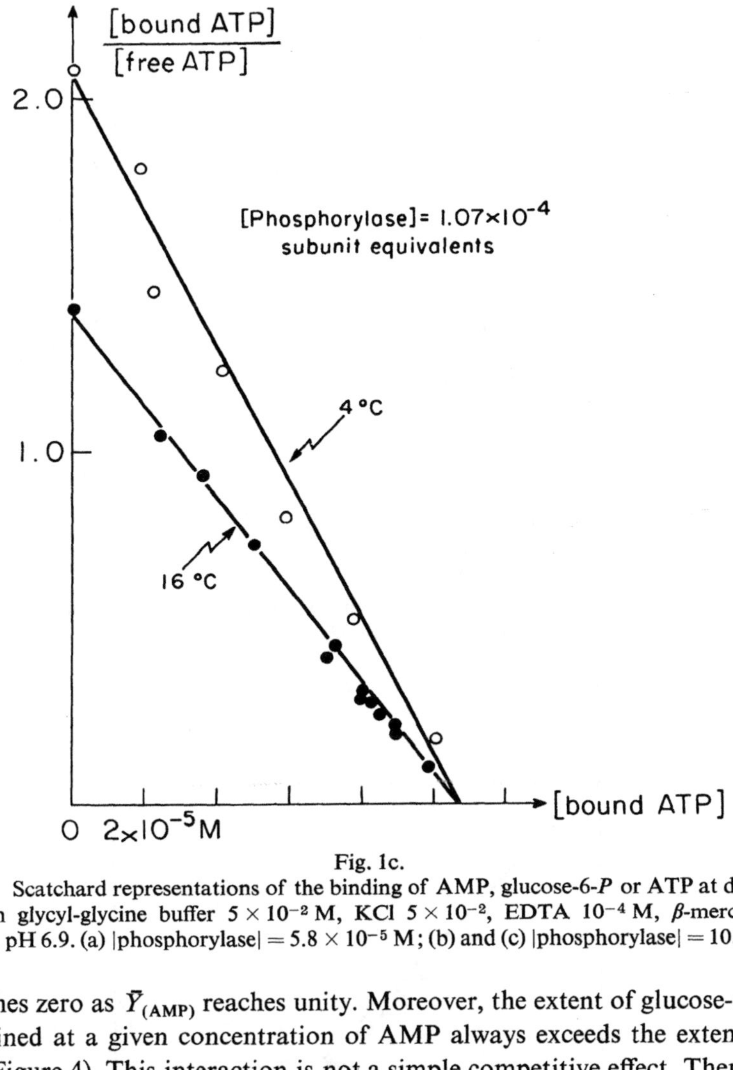

Fig. 1c.

Figs. 1a–c. Scatchard representations of the binding of AMP, glucose-6-*P* or ATP at different temperatures in glycyl-glycine buffer 5×10^{-2} M, KCl 5×10^{-2}, EDTA 10^{-4} M, β-mercaptoethanol 5×10^{-2} M, pH 6.9. (a) |phosphorylase| $= 5.8 \times 10^{-5}$ M; (b) and (c) |phosphorylase| $= 10.7 \times 10^{-5}$ M.

$1/K_1$ reaches zero as $\bar{Y}_{(\mathrm{AMP})}$ reaches unity. Moreover, the extent of glucose-6-*P* exclusion obtained at a given concentration of AMP always exceeds the extent of AMP binding (Figure 4). This interaction is not a simple competitive effect. Then, once the first glucose-6-*P* complex is formed, AMP cannot affect the association of a second molecule of glucose-6-*P* with the dimeric unit: K_2 does not depend on AMP concentration.

(2) Symmetrical observations are made for the dependence of the AMP enzyme dissociation constants K_1' and K_2' on glucose-6-*P* concentration (Figure 5).

The exclusion of the first molecule of AMP by glucose-6-*P* precedes the saturation function $\bar{Y}_{(G\text{-}6\text{-}P)}$. In this case, however, the exclusion is not absolutely complete, the final value of $1/K_1'$ at saturation of glucose-6-*P*, $1/K_\infty'$, being 20 times smaller than $1/K'$; this 'residual binding' at high glucose-6-*P* concentration seems to be non-cooperative.

The second association constant $1/K_2'$ is not absolutely independent of glucose-6-*P*, but its decrease lags behind the change of $1/K_1'$.

(c) All these results can be summarized by making use of the binding polynomial

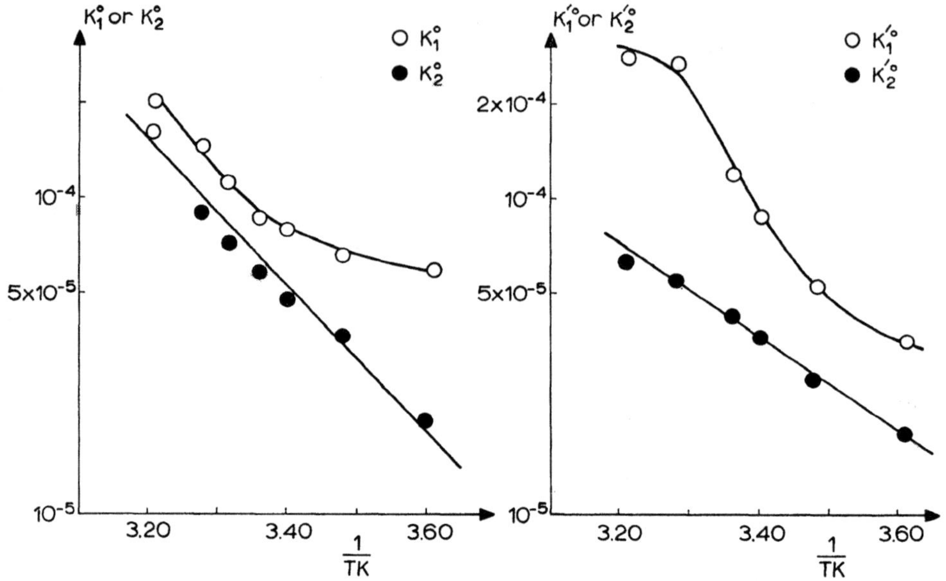

Fig. 2. Effect of temperature on the dissociation constants, $K_1{}^0$, $K_2{}^0$, $K_1{}'^0$ and $K_2{}'^0$ (log scale). $(\Delta H_{\text{ass.}})_{\text{AMP}} = -6.5$ kcal/mole; $(\Delta H_{\text{ass.}})_{G\text{-}6\text{-}P} = -11$ kcal/mole.

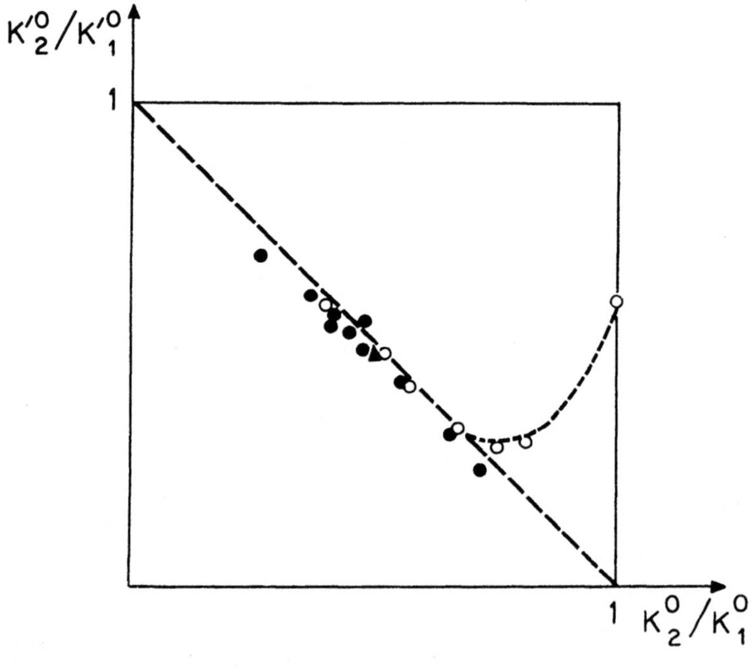

Fig. 3. Relation between $K_2{}^0/K_1{}^0$ and $K_2{}'^0/K_1{}'^0$. Open circles, constant pH. Black circles, constant buffer (pH 6.9 at 22° C). Triangle: point obtained for phosphorylase b reduced by borohydride at 16° C. Each point represents the value of $K_2{}'^0/K_1{}'^0$ and $K_2{}^0/K_1{}^0$ at a given temperature. Temperature increases from left to right along abcissa.

N which represents the sum of the concentrations of the different species present at given concentrations of glucose-6-P (y) and AMP (x) [4, 5].

$$N(x, y) = 1 + \frac{2x}{K_1'^0} + \frac{x^2}{K_1'^0 K_2'^0} + \frac{2y}{K_1^0} + \frac{y^2}{K_1^0 K_2^0} + \frac{4xy}{K_1^0 K_{12}'} + \frac{2xy^2}{K_{12}'(K_{22})^2} + \frac{2yx^2}{(K_{22}')^2 K_{12}} + \frac{x^2 y^2}{(K_{22}')^2 (K_{22})^2}.$$

At any given concentration of x and y the corresponding saturation functions are obtained by:

$$\bar{Y}_{(\mathrm{AMP})} = \frac{1}{2}\left(\frac{\mathrm{dln}\,N}{\mathrm{dln}\,x}\right)_y \qquad \bar{Y}_{(\mathrm{G\text{-}6\text{-}}P)} = \frac{1}{2}\left(\frac{\mathrm{dln}\,N}{\mathrm{dln}\,y}\right)_x,$$

$N(x, y)$ is identical with the simple formula corresponding to the concerted model proposed by Monod, Wyman and Changeux [6]:

$$N(x, y) = \left[\left(1 + \frac{x}{K_2'^0}\right)^2 + L_0\left(1 + \frac{y}{K_2^0}\right)^2 \left(1 + \frac{x}{K_\infty'}\right)^2\right] \frac{1}{L_0 + 1},$$

and all the results exposed above are completely in agreement with this formulation.

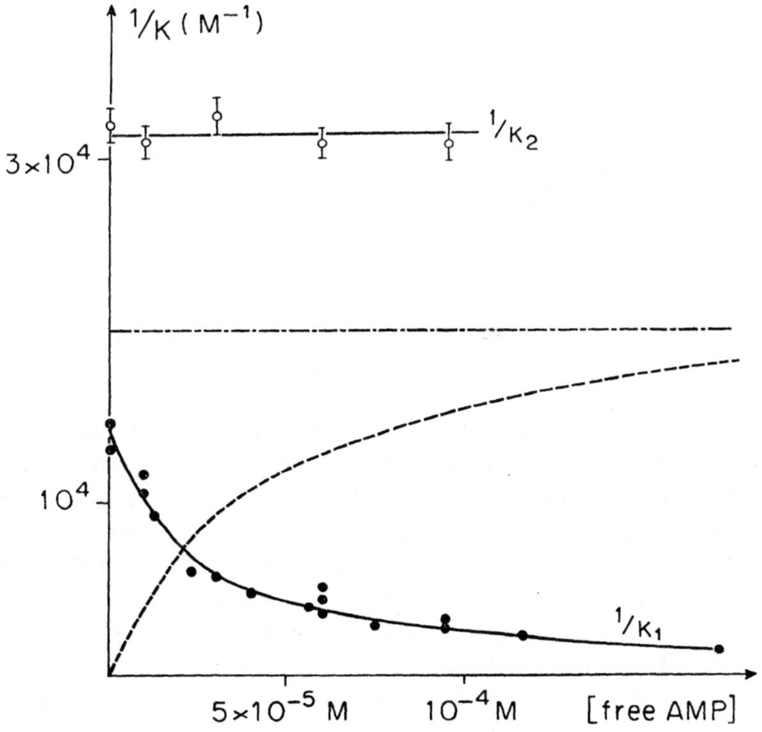

Fig. 4. Change in $1/K_1$ and in $1/K_2$, association constants of the glucose-6-P enzyme complexes, when the free AMP concentration varies (16°C, glycyl glycine, KCl, EDTA, β-Me buffer, pH 7.1). Dotted lines: saturation function in AMP and asymptotic limit of this function.

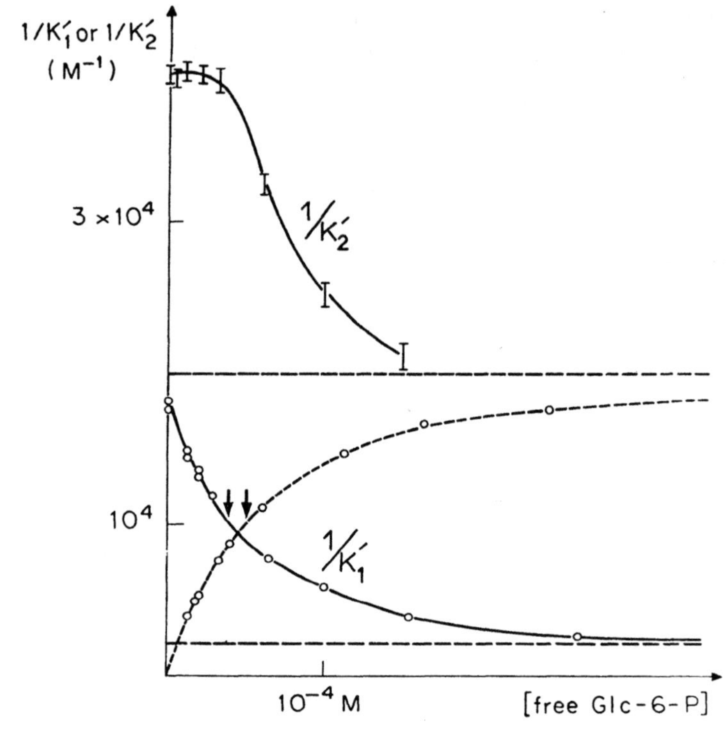

Fig. 5. Variation of $1/K_1'$ and $1/K_2'$, association constants of the AMP enzyme complexes, with the free glucose-6-*P* concentration. Dotted lines: the saturation function $\bar{Y}|_{G\text{-}6\text{-}P}|$ determined under the same conditions and asymptotic limit of this function. The first and second arrows indicate the concentrations of glucose-6-*P* corresponding respectively to half release of AMP, or half saturation in glucose-6-*P*.

3. Interactions Between Effectors or Between Effector and Substrates when AMP is Replaced by Analogs

We have seen in Section 2 that a two-state model with a concerted transition between the inactive and active states, B_2 and A_2, modulated by the antagonistic effectors, is a good first approximation. However, even on the basis of the competition experiments, the model applies only to a limited range of temperature and pH (Figure 3). Other indications of a more complex interaction pattern are numerous in the literature; in particular, the break in the Arrhenius plot of the logarithm of the maximal velocity of the catalytic reaction with respect to $1/T$ [3], the change in reactivity of one cystein with respect to DTNB when AMP and glucose-6-*P* are present together at a particular concentration [7], the changes in interaction between substrate and activator when AMP is replaced by inosine-5′-monophosphate [8].

So as to understand better the activating effect of AMP in the catalytic process we have made a systematic study of the effects of analogs of AMP.

These are base, nucleoside and nucleotide derived from adenine, adenosine and AMP by substitution of the NH_2 at the 6 position with an –SH residue. The binding properties of each of the effectors have been tested by equilibrium dialysis in the presence of AMP or glucose-6-*P*, together with their effect in catalysis [9]. We shall report here the results obtained with 6-mercaptopurine-riboside-monophosphate (SHMP) which activates phosphorylase b up to 61% of the maximal activation obtained at saturation of 5′-AMP.

A large differential spectrum in the visible, with a maximum at 335 nm, is generated when SHMP is added to phosphorylase b (Figure 6). The variation of absorbance at at 335 nm as a function of SHMP measures the binding of SHMP onto phosphorylase. One mole of SHMP is bound per enzyme subunit. Figures 7, 8 and 9, show the similarity between the homotropic and heterotropic effects of SHMP and AMP. It can be seen from Figure 7 that the binding of SHMP is cooperative, and that the saturation functions for SHMP and AMP superimpose on each other if an affinity factor of 2.9 is introduced all along the abcissa; this affinity factor is equal to the ratio of the dissociation constant for the complexes enzyme–SHMP and enzyme–AMP.

Figures 8 and 9 show that AMP or glucose-6-*P* are excluded by SHMP. The exclusion of trace amounts of radioactive AMP by cold AMP or cold SHMP gives identical curves if the affinity factor of 2.9 is introduced in the concentration of effectors. The

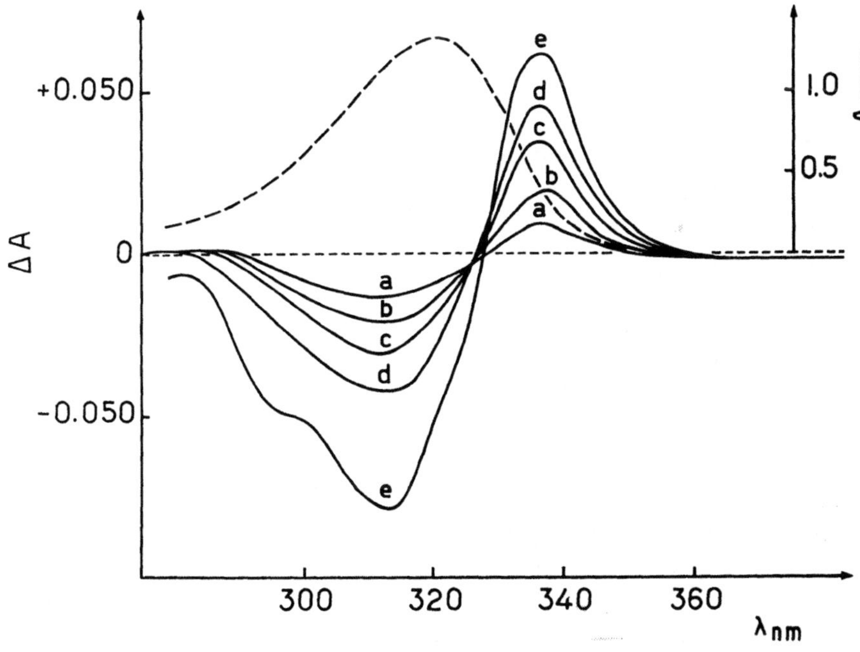

Fig. 6. Differential spectrum of phosphorylase b in the presence of various concentrations of SHMP versus free enzyme, at 4°C. |phosphorylase| = 3.7×10^{-5} M; in glycyl glycine 5×10^{-2} M, KCl 5×10^{-2} M, EDTA 10^{-4} M, pH 7.5; |SHMP| a: 1.04×10^{-5} M; b: 2.1×10^{-5} M; c: 3.5×10^{-5} M; d: 5.5×10^{-5} M; e: 1.1×10^{-4} M; optical pathway of the cell: 0.5 cm.

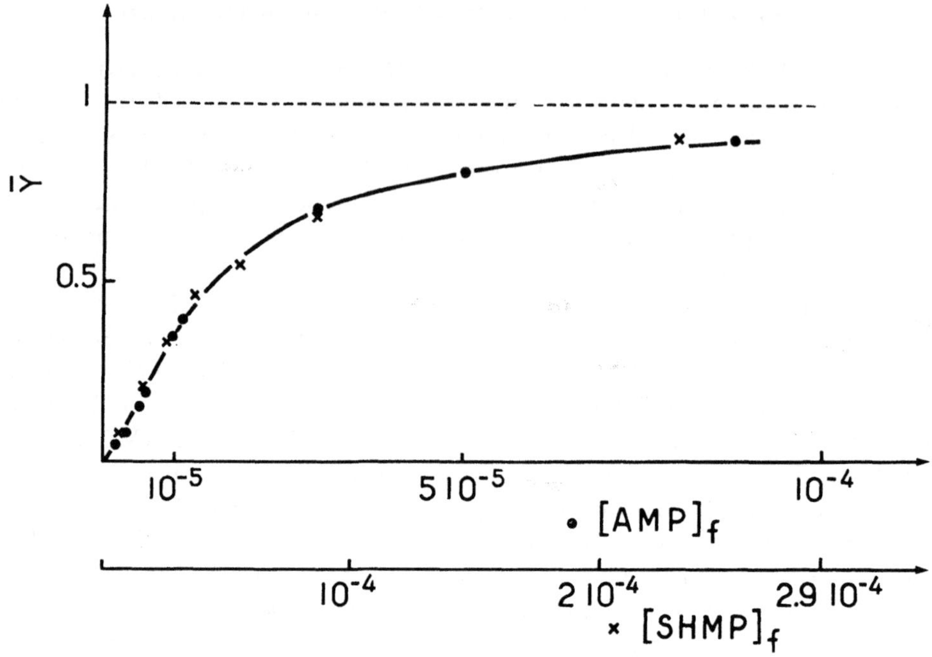

Fig. 7. Comparison between the saturation functions of SHMP, measured by spectrophotometry, and of AMP, measured by equilibrium dialysis, as functions of free ligand. Same buffer as in Figure 6 (4°C).

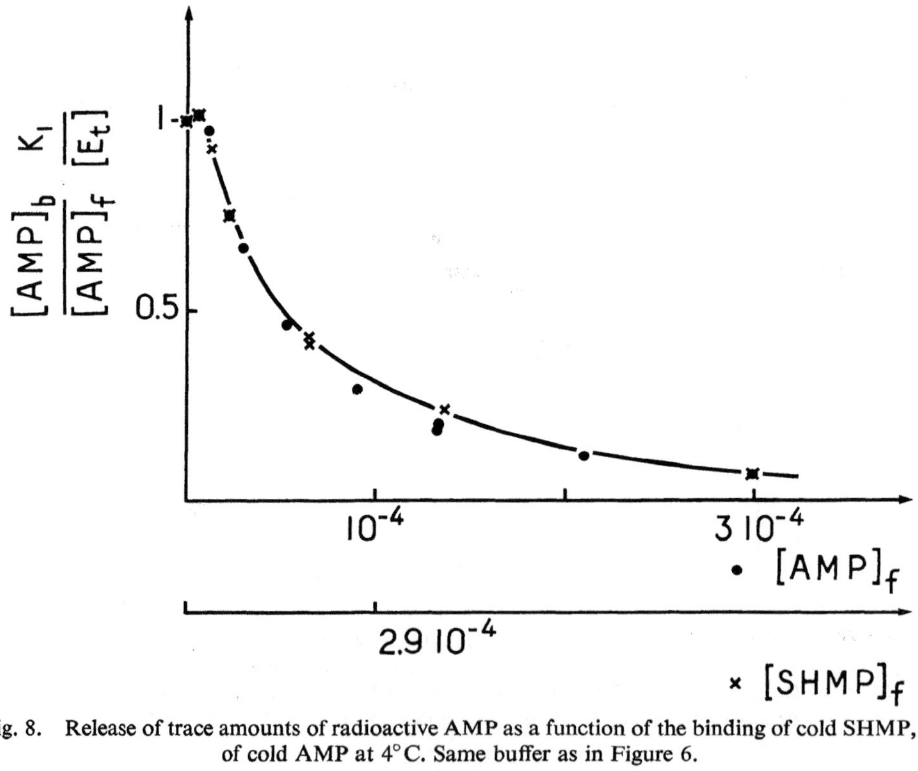

Fig. 8. Release of trace amounts of radioactive AMP as a function of the binding of cold SHMP, or of cold AMP at 4°C. Same buffer as in Figure 6.

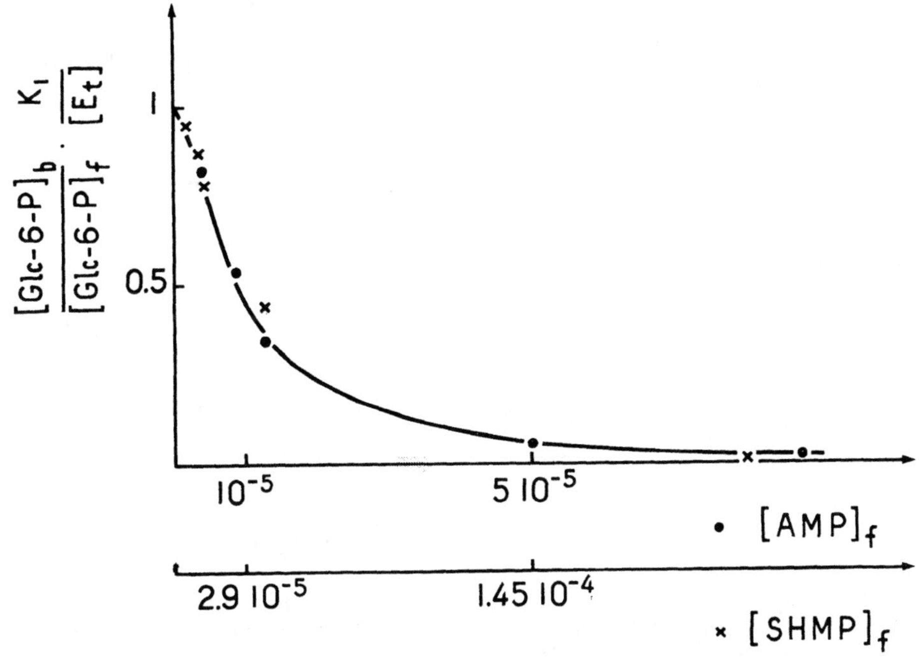

Fig. 9. Release of trace amounts of radioactive glucose-6-*P* as a function of the concentration of free SHMP or AMP at 4°C. Same buffer as in Figure 6.

exclusion of trace amounts of glucose-6-*P* by SHMP is cooperative. Here again, one can chase the same amount of bound glucose-6-*P* by a given concentration of AMP or by the same concentration of SHMP multiplied by the affinity factor 2.9. Accordingly, as in the case of the interaction between AMP and glucose-6-*P* (Section 2), the interaction between SHMP and glucose-6-*P* cannot be a simple competitive effect, but occurs via a conformational transition. It is therefore very likely that SHMP binds to the same stereospecific site as AMP, since it totally excludes AMP and generates homotropic effects and heterotropic effects towards glucose-6-*P* which are similar to those generated by AMP (if one takes into account the relative affinities of the two effectors).

However, the fact that SHMP cannot totally activate phosphorylase b implies that the heterotropic effects towards the substrate are not identical to those of AMP. Figure 10 shows that the concentration of SHMP required to obtain half the maximal activation differs markedly from that required with AMP. The ratio between these concentrations is approximately 7.

These results indicate that one must clearly distinguish two functions of an allosteric ligand analogous to AMP: it is a trigger of a concerted transition to another 'state' which has a very poor affinity for glucose-6-*P* and it can position with more or less efficiency the substrate at its catalytic site. The concerted transition is required

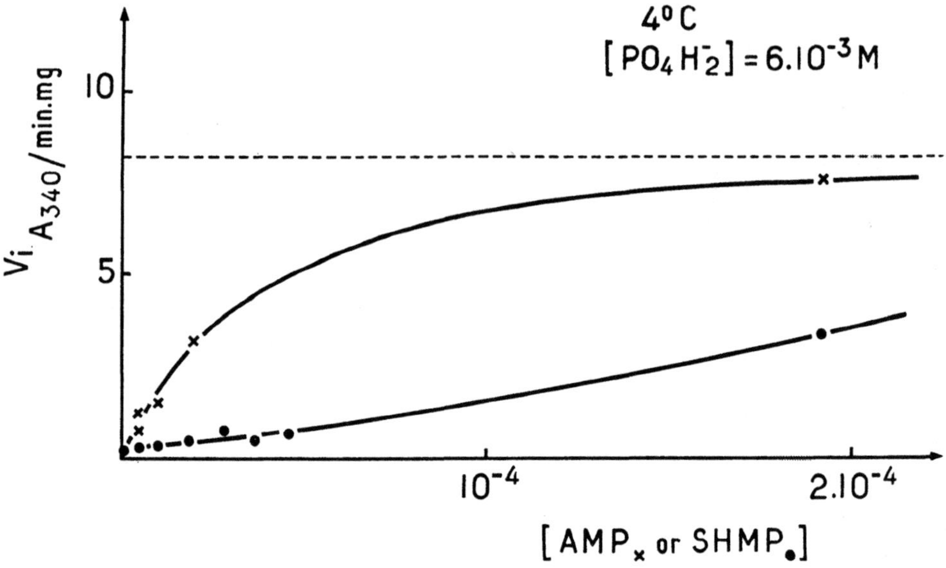

Fig. 10. Initial velocity of the enzymatic reaction measured according to Helmreich [19] in a coupled assay system. |phosphorylase| 4 to 40 γ; glycyl glycine 5×10^{-2} M, KCl 5×10^{-2} M, EDTA 10^{-4} M, pH 7.5, 4°C. The values obtained for SHMP have been multiplied by the affinity factor 100/61 corresponding to the ratio between the maximal activities with AMP or SHMP.

but is not sufficient in itself to lead to efficient catalysis. Another example of the uncoupling of these two functions is given by 2′-aminopurine riboside 5′-monophosphate, another AMP analog which binds cooperatively on phosphorylase b and non-cooperatively to phosphorylase a (as AMP or SHMP), but is a poor activator. The 6-amino position of the adenine base is therefore required to induce an efficient positioning of the substrate. (Table I gives a preliminary survey of the binding properties of some analogs of 5′-AMP and illustrates the respective role of the parts of the molecule on the cooperative binding process leading to activation.) The studies made in collaboration with A. Danchin on a stable complex between phosphorylase b and an AMP-Co(III) chelate which knocks down catalytic activity and binding of both AMP and glucose-6-*P* on the subunit where it binds, lead to the same conclusion [10].

4. Relaxation Experiments

The results given in Section 3 illustrate the complexity of the transitions leading to an active molecule. In fact, the monitoring of several probes attached at specific sites of phosphorylase b have led in particular Birkett *et al.* [11] to postulate the existence of 8 different conformations in equilibrium when the effector and the substrate bind to the enzyme. (These probes are a fluorescent label attached to a single SH group per subunit, and manganese ions: the interaction of Mn^{++} with phosphorylase

TABLE I

A preliminary survey of the binding properties of some analogs of 5′-AMP

Effectors tested	K_1 (M)	Binding index of cooperativity (K_2/K_1)	Exclusion ($K_{1/2}$) of 'traces' of		Effects on catalytic activity	
			AMP (M)	Glc-6-*P* (M)	Activation (% of V_m)	Inhibition[d] (M)
5′-AMP	2.2×10^{-5}	>1		[c]	100	
Adenine	2×10^{-4}	1	[b]	2.6×10^{-3}		2.4×10^{-4}
Adenosine	–	–	2.8×10^{-2}	$>10^{-2}$		8.6×10^{-4}
6-Mercaptopurine	10^{-4}	1	–	–		8×10^{-5}
6-Mercaptopurine riboside	–	–	7.4×10^{-4}	10^{-3}		2.2×10^{-4}
SHMP	6.3×10^{-5}	>1[a]	1.8×10^{-4}	7.8×10^{-5}	>60	
2–Amino purine	–	–	2×10^{-3}			3×10^{-4}
2-Amino purine riboside	–	–	1.5×10^{-2}	1.35×10^{-2}		3.8×10^{-4}
2-Amino purine riboside 5-phosphate	–	>1	2×10^{-4}	1.2×10^{-4}	~ 15	

All results were obtained at 4°C in glycylglycine 5×10^{-2} M, EDTA 10^{-4} M buffer (pH 7.5).

[a] The cooperativity of SHMP binding is identical to that of 5′ AMP.

[b] AMP is excluded but $K_{1/2}$ has not been accurately measured in this case.

[c] 1 AMP or 1 SHMP molecule chases more than 1 molecule of inhibitor.

[d] These values refer to the concentration of analog required to decrease the residual activity (without 5′-AMP) to 50%.

is studied by proton relaxation enhancement and electron spin resonance.) Their model takes into account discontinuities in the temperature dependences of the proton relaxation rates and of the –SH reactivity. This anomalous behaviour with respect to temperature has been noted by other groups for the enzymatic activity, for the binding of AMP and glucose-6-*P* [3, 4], and for the titration of another (buried) –SH residue [12]. Nuclear magnetic resonance studies of the non-exchangeable protons of AMP in D_2O indicate also a temperature-dependent conformational change with a transition around 14°C [13]. See also [14].

Moreover, since the distances between the stereospecific sites for glucose-1-*P* and AMP appear to be relatively short, Radda *et al.* think that the corresponding conformational changes might be localized to small parts of the subunit [15].

Kinetic studies were started in order to better understand the transition pattern under simplified conditions.

In the absence of any added effector, three relaxation times, τ_1, τ_2, τ_4, are detected by relaxation techniques between 4°C and 30°C and between pH 6 and pH 8. They do not depend on protein concentration and they must therefore correspond to transitions in the dimer molecule.

(a) The simplest situation exists at *rather alkaline pH and at low temperature*: here the slowest time τ_4 has the only significant amplitude. The binding step corresponding to association with AMP or SHMP is in the 100 μsec range, and therefore much faster than τ_4, which is slower than 1 sec. Two experiments have established that τ_4 restores under these conditions homotropic interactions between AMP molecules via a concerted transition:

(1) The initial situation being the equilibrium between unliganded phosphorylase b molecules, one makes a sudden perturbation in the AMP concentration, and follows by Colowick and Womack method the uptake of radioactive AMP in the 10 sec to 20 min range [16]. Two processes are found: a quick binding of AMP which is non-cooperative, followed by an increased uptake depending on the AMP final concentration along the relaxation time τ_4.

(2) One can also observe, at a given concentration of protein and ligand, the change in optical density at 295 nm in a slow *T* jump, the final temperature being 4°C. One gets the same dependence of $1/\tau_4$ on the free AMP concentration (Figure 11).

The corresponding scheme can be written:

$$\begin{array}{lccc|l}
\text{no AMP bound} & A_2^0 & \underset{k_{BA}^0}{\overset{k_{AB}^0}{\rightleftharpoons}} & B_2^0 & \\
 & \downarrow\uparrow & & \downarrow\uparrow & \\
\text{one molecule bound per dimer} & A_2^1 & \rightleftharpoons & B_2^1 & \text{fast binding processes} \\
 & \downarrow\uparrow & & \downarrow\uparrow & \\
\text{fully saturated dimer} & A_2^2 & \rightleftharpoons & B_2^2 & \downarrow \\
 & \longleftarrow & \text{slow process} & \longrightarrow &
\end{array}$$

Then, $1/\tau_4$ is given by:

$$1/\tau_4 = k_{AB} + k_{BA},$$

and if one assumes that all the backward steps k_{BA} are the same, one gets the same equation as in the case of yeast glyceraldehyde-3-*P* dehydrogenase [17], namely:

$$1/\tau_4 = k^0_{BA}\left[1 + L_0 \frac{(1 + |x|/K'_B)^2}{(1 + |x|/K'_A)^2}\right].$$

K'_A and K'_B are estimated from K'_2 and K'_∞ by the same identification in the partition function which has been given in Section 2 (K'_B being large with respect to K'_A) and leads to:

$$k^0_{AB} = 5.2 \times 10^{-3}\ \text{sec}^{-1} \qquad L_0 = 1.5.$$
$$k^0_{BA} = 3.5 \times 10^{-3}\ \text{sec}^{-1}$$

(3) A transition state E′ between the inactive and active states has been character-

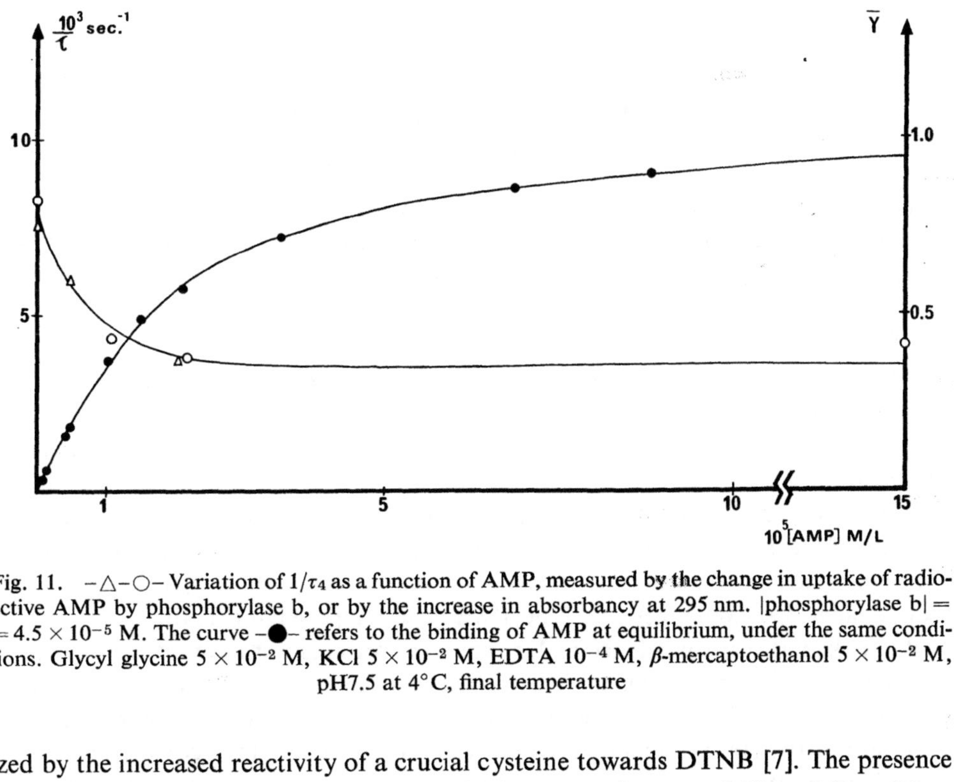

Fig. 11. –△–○– Variation of $1/\tau_4$ as a function of AMP, measured by the change in uptake of radioactive AMP by phosphorylase b, or by the increase in absorbancy at 295 nm. |phosphorylase b| = $= 4.5 \times 10^{-5}$ M. The curve –●– refers to the binding of AMP at equilibrium, under the same conditions. Glycyl glycine 5×10^{-2} M, KCl 5×10^{-2} M, EDTA 10^{-4} M, β-mercaptoethanol 5×10^{-2} M, pH7.5 at 4°C, final temperature

ized by the increased reactivity of a crucial cysteine towards DTNB [7]. The presence of a proper mixture of AMP and glucose-6-*P* increases the accessibility of the residue, but the state E′ is never significantly populated at 4° and pH 7.5. This is not the case at higher temperature or at lower pH, where the crucial SH is titrated linearly with the concentration of DTNB.

The rate of access to E′ at 4° has been measured under conditions where the titration of the cystein is zero order with respect to DTNB; it is equivalent to the sum of the isomerization constants k_{AB} and k_{BA} affected by the weight of the conformations

A_2 and B_2

$$k_{AB}\bar{A} + k_{BA}\bar{B}.$$

The value obtained by this method (around 1 minute) is smaller but close to the value of $1/\tau_4$.

(4) Change in binding of glucose-6-*P* leads also to a slow transition close to $1/\tau_4$. In this case, however, the precise dependance of $1/\tau_4$ versus glucose-6-*P* concentration has not been obtained.

(5) The displacement of SHMP (x') on the enzyme by increased concentrations of glucose-6-*P* has been studied by stopped flow. The rate of displacement is indicated in Figure 14.

The interpretation of these heterotropic effects is not easy. The release seems to follow the pathway:

$$D + 2x' + y \underset{(1)}{\rightarrow} Dx'_2y \underset{(2)}{\rightarrow} D'x'_2y \underset{(3)}{\rightarrow} D'y + 2x'.$$

The rate of displacement of x' by glucose-6-*P*, step 3, is limited by the isomerization taking place along step 2. It is worth noticing that this rate is close to the expected value at this temperature of the second relaxation time τ_2, which is found in the absence of both ligands.

Therefore, while the rate of change in homotropic interaction is limited by the slowest relaxation time τ_4, transconformation leading to the mutual exclusion of activators and inhibitors seems to depend on τ_2.

(b) *Higher Temperatures*

As the temperature is increased, and/or the pH lowered, the amplitudes of the two fast pre-existing relaxation times τ_1 and τ_2 increase in the absence of the ligand. In particular, the amplitude of τ_2 passes through a maximum around 16°C, the critical temperature where a change in the geometry of the stereospecific site for AMP is observed [13], and where a break appears in the Arrhenius plot of the maximum velocity of the enzymatic reaction versus $1/T$. As τ_1 and τ_2 start to become dominant, the amplitude of τ_4 decreases markedly and vanishes above 20°C, τ_4 being now in the second range. Concomitantly, the crucial SH is fully exposed to DTNB.

We looked at the dependence of $1/\tau_1$ and $1/\tau_2$ with respect to temperature (Figure 12) and pH. $1/\tau_1$ corresponds to the titration of a single pK on the molecule (Figures 13a, b) equal to 7.3. $1/\tau_2$, on the other hand, is also strikingly dependent on pH, the midpoint of the strongly cooperative curve of $1/\tau_2$ versus pH occurring also at pH 7.3 (Figure 13c).

The simplest scheme which can account for these data is a transition of the type:

$$\begin{array}{ll} & E_1H_2 \rightleftharpoons E_2H_2 \quad (pK > 8.5) \\ & \downarrow\uparrow \\ pK\ 7.3 & E_1H^- + H^+ \\ & \downarrow\uparrow \\ & E_1^{--} + 2H^+. \end{array}$$

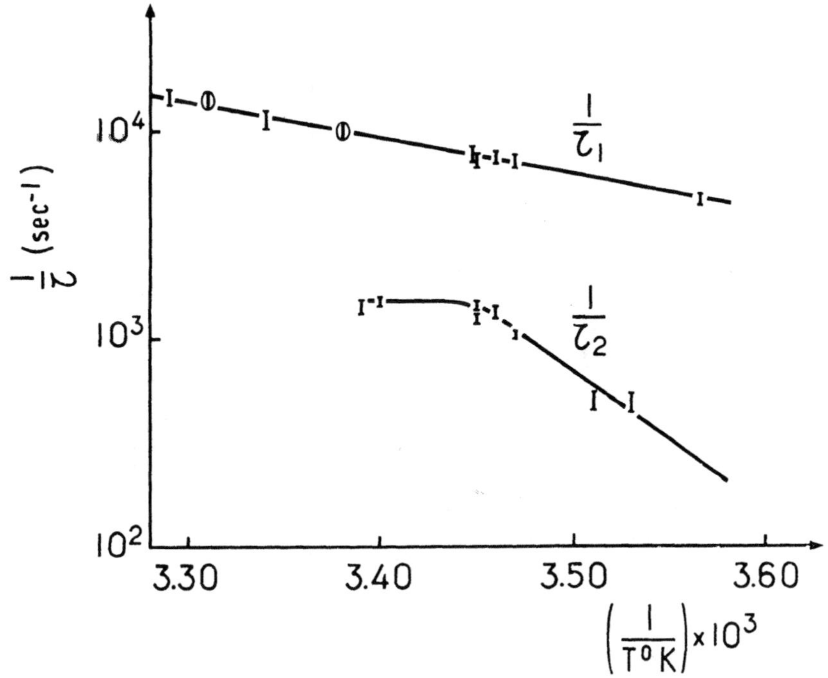

Fig. 12. Measure of the dependence of the relaxation times τ_1 and τ_2 with respect to temperature, expressed as an Arrhenius plot. $\Delta H\tau_1 = 8.6$ kcal/mole. A break in the Arrhenius plot of $1/\tau_2$ is observed around 16°C. The amplitudes of τ_1 and τ_2 are maximal around 16°–20°C. Glycerophosphate 5×10^{-2} M, KCl 5×10^{-2} M, EDTA 10^{-4}, pH 6.9. 0 symbols refer to results obtained with phosphorylase a.

$1/\tau_1$ corresponds to the relaxation rate associated with the proton uptake on E_1^{--}, and $1/\tau_2$ represents the isomerization between E_1^{--} and E_2 H_2 coupled with the proton uptake.

Increased concentrations of AMP have the same effect on $1/\tau_2$ as an increase of pH. We looked also at the relaxation spectrum of enzymes reconstituted with different analogs of the coenzyme, pyridoxal 5′-phosphate (PLP), in order to know what kind of modifications leads to the disappearance or to the enhancement of these fast relaxation times. τ_1 is present in phosphorylase a, which does not require AMP for activation, and τ_2 is absent. Any modification in the heterocyclic ring of pyridoxal phosphate which leads to an inactive enzyme of different quaternary structure [18], abolishes also τ_2. The replacement of PLP by pyridoxal phosphate 5′-methyl ester which. leads to an enzyme having a very similar structure but no activity, abolishes both τ_1 and τ_2. The main conclusion of these studies is that τ_2 is present when AMP activates these modified glycogen phosphorylases.

5. Conclusions

Glycogen phosphorylase b appears to be a rather complex enzyme. One should con-

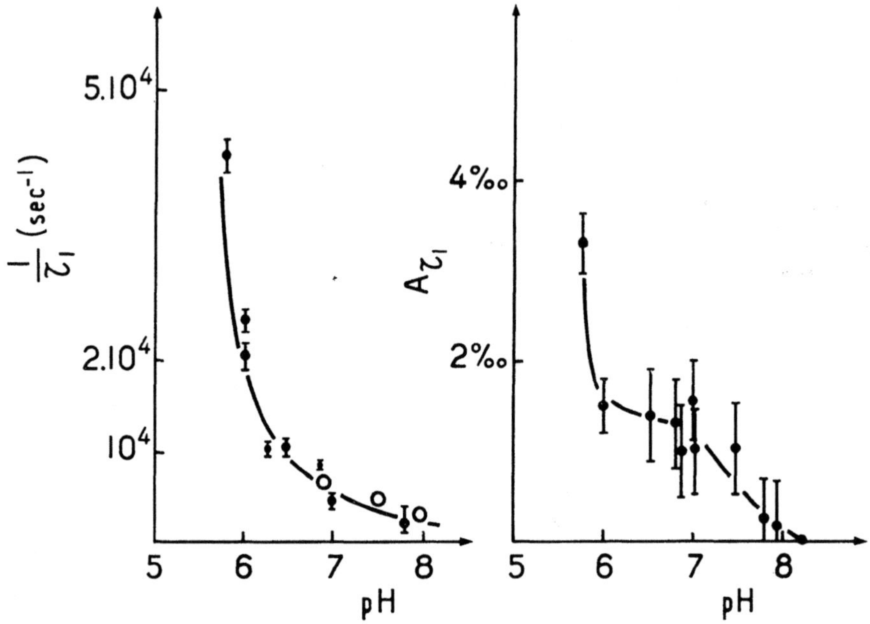

Fig. 13a.

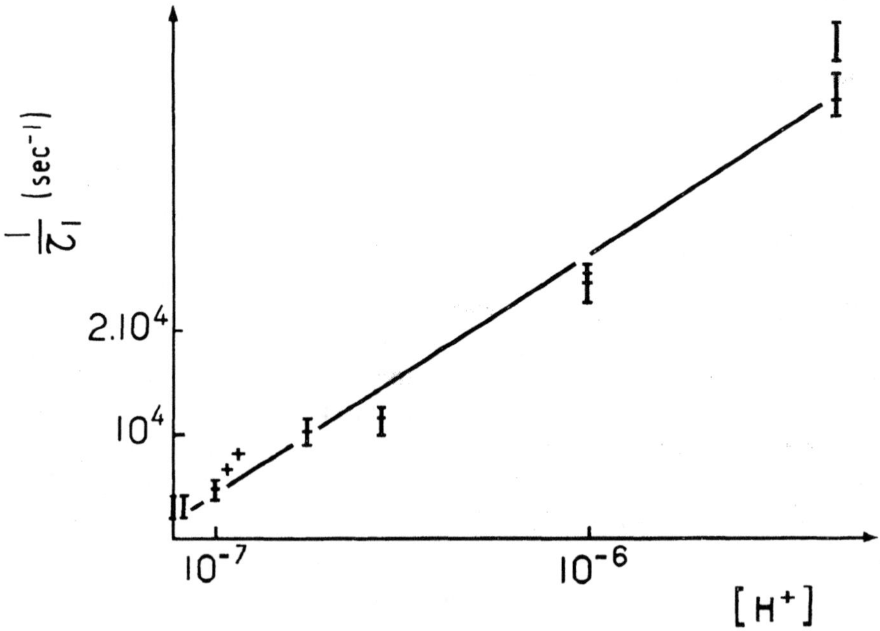

Fig. 13b.

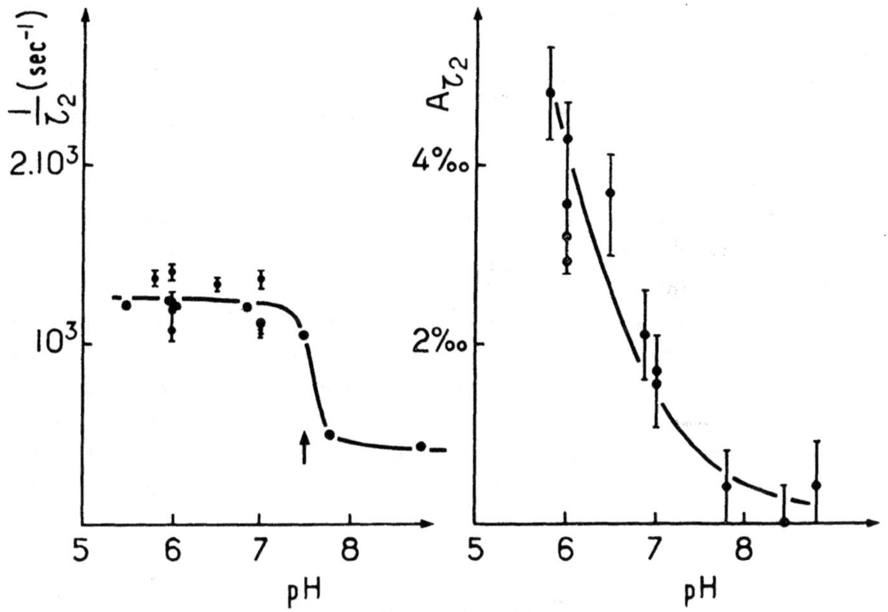

Fig. 13c.

Figs. 13a–c. Variation of $1/\tau_1$ and $1/\tau_2$ with respect to pH and corresponding amplitudes. Jump between 13°C and 20°C. |phosphorylase b| = 5×10^{-5} M adjusted at different pH's in glycerophosphate, KCl, EDTA buffer. In the presence of 10^{-3} M AMP at pH 6.9, $1/\tau_2 = 3 \times 10^2$ sec^{-1}.

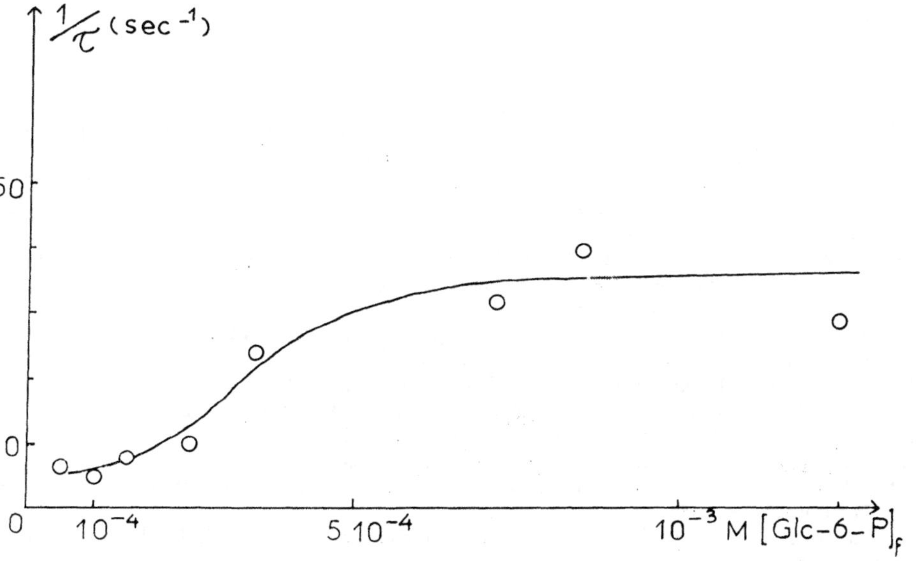

Fig. 14. Results of a stopped-flow experiment performed at 5°C. Phosphorylase in one syringe is mixed with SHMP and various amounts of glucose-6-*P* in the other syringe. Final concentrations: |enzyme| = 2.1×10^{-5} M; |SHMP| = 1.44×10^{-4} M in glycyl glycine, KCl = 5×10^{-2} M, EDTA 10^{-4} M, pH 7.5. Observation at 340 nm.

sider that AMP has two different functions, namely it triggers a proper quaternary change and it positions correctly the substrate at its catalytic site. A systematic method of comparing the role of the substitution in the AMP molecule has been proposed, and effectively applied to the comparison of 6-mercaptopurine 5′-monophosphate and AMP.

The kinetic scheme involved in the quaternary transition appears to be simple at low temperatures, and is compatible with the Monod–Wyman–Changeux model. The activation energy associated with $1/\tau_4$ (ca. 24 kcal/dimer) is very large and the transition very slow. An important structural rearrangement must take place; a predissociated conformation having a crucial SH residul freely exposed appears to be associated with the transition. At higher temperatures faster conformational changes become preponderant. They are pH-dependent, they are associated with the activation process and a structural change of the geometry of the activator site is taking place in the critical temperature range. It has not been possible up to now to present an overall plausible kinetic scheme which accounts for all the transitions triggered by pH, temperature or activator concentration.

Acknowledgements

Our research was supported by grants from the Centre National de la Recherche Scientifique, Délégation Générale à la Recherche Scientifique et Technique, Commissariat à l'Énergie Atomique and National Institutes of Health. We thank EMBO for the attribution of the block grant between Professor Monod's and Eigen's laboratories (1967–71).

References

[1a] Fischer, E. H., Pocker, A., and Saari, T. C.: 1970, in *Essays in Biochemistry* (ed. by P. N. Campbell and F. Dickens), Vol. 6, Academic Press, London, pp. 23–68.
[1b] Madsen, N. B.: 1972, in *PAABS Symposium on Molecular Basis of Biological Activity*, (ed. by K. Gaede, B. L. Horecker, and W. J. Whelan), Vol. I, Academic Press, New York, pp. 13–51.
[2] Cori, C. F. and Cori, G. T.: 1936, *Proc. Soc. Exp. Biol. Med.* **34**, 702–05.
[3] Kastenschmidt, L. L., Kastenschmidt, J., and Helmreich, E.: 1968, *Biochemistry* **7**, 4543–56.
[4] Buc, H.: 1973, submitted to *Eur. J. Biochem.*
[5] Wyman, J.: 1967, *J. Am. Chem. Soc.* **89**, 2202.
[6] Monod, J., Wyman, J., and Changeux, J. P.: 1965, *J. Mol. Biol.* **12**, 88–118.
[7] Buc, M. H. and Buc, H.: 1973, submitted to *Eur. J. Biochem.*
[8a] Black, W. J. and Wang, J. H.: 1968, *J. Biol. Chem.* **243**, 5892–98.
[8b] Black, W. J. and Wang, J. H.: 1970, *Biochim. Biophys. Acta* **212**, 257–68.
[9] Morange, M.: 1972, Diplôme d'Études Approfondies, Paris.
[10] Danchin, A. and Buc, H.: 1973, *J. Biol. Chem.* **248**, 3241.
[11] Birkett, D. J., Dwek, R. A., Radda, G. K., Richards, R. E., and Salmon, A. G.: 1971, *Eur. J. Biochem.* **20**, 494–508.
[12] Buc, M. H.: 1970, 'Étude de quelques équilibres conformationnels de la phosphorylase b du muscle de lapin', Thesis Paris.
[13] Danchin, A. and Buc, H.: 1972, *FEBS Letters* **22**, 289–93.
[14] Damjanovich, S. and Kleppe, K.: 1966, *Biochim. Biophys. Acta* **122**, 145–47.
[15] Bennick, A., Campbell, I. D., Dwek, R. A., Price, N. C., Radda, G. K., and Salmon, A. G.: 1971, *Nature New Biol.* **234**, 140–143.

[16] Remy, P. and Buc, H.: 1970, *FEBS Letters* **9**, 152–55.
[17] Kirschner, K., Eigen, M., Bittman, R., and Voigt, B.: 1966, *Proc. Nat. Acad. Sci. US* **56**, 1661.
[18] Pfeuffer, T., Ehrlich, J., and Helmreich, E.: 1972, *Biochemistry* **11**, 2136–45.
[19] Helmreich, E. and Cori, C. F.: 1964, *Proc. Nat. Acad. Sci. US* **51**, 131–38.

DISCUSSION

Vasilescu: Avez-vous une idée précise sur les conformations possibles du 5′-AMP lors des réactions d'activation que vous invoquez ?

Buc: Dans l'enzyme nous ne connaissons pas la structure stéréochimique du 5′-AMP (cf. toutefois les mesures de distances de Birkett *et al.*).

Vasilescu: Il a été montré par des calculs EHT menés en notre laboratoire (Thèse de spécialité de mcéanique quantique – J. Lespinasse, Université de Nice) que le 5′-AMP pouvait présenter deux conformations probables dont celle déterminée par cristallographie.

Pantaloni: Quelles sont les données qui permettent dans cet enzyme d'exclure de façon sûre l'existence d'interactions directes entre le substrat et les effecteurs ?

Buc: Il y a certainement une liaison dans la sous-unité entre site Glc-1-P et AMP [15], [10].

QUELQUES PROBLÈMES DE CINÉTIQUE CONFORMATIONNELLE DE SYSTÈMES COOPÉRATIFS

R. CERF

Laboratoire d'Acoustique Moléculaire, E.R.A. du C.N.R.S., Institut de Physique, Université Louis Pasteur, 4, rue Blaise Pascal, Strasbourg, France

Résumé. Ayant souligné l'importance de la coopérativité en biologie, on restreint le sujet à la cinétique des transconformations coopératives de macromolécules. Le problème théorique est alors celui de la cinétique du modèle d'Ising à une dimension.

Deux solutions sont résumées (cas des chaînes longues; cas de réduction à un modèle séquentiel).

La cinétique de l'isomérisation cis-trans de la poly-L-proline illustre certains aspects de la théorie, tandis que des mesures nouvelles de l'absorption des ultrasons dans des solutions d'acides poly-L-glutamique montrent l'importance des fluctuations du nombre de séquences hélice (ou pelote) ininterrompues.

Des modèles simples, illustrés par leurs applications, sont finalement passés en revue (modèle par tout ou rien utilisé par Eigen et Pörschke dans l'étude cinétique de l'appariement des bases des oligonucléotides; un modèle non-arrhénien à deux états; un modèle à trois états conduisant à interpréter certaines expériences de saut de température lent pour des solutions de protéines pancréatiques par l'existence d'une conformation intermédiaire).

Summary. While the utmost importance of cooperativity in biological processes is underlined, the subject is restricted to linear chain molecules and the kinetics of their transconformations, such as the helix–coil transition of polypeptides and of oligonucleotides, and protein unfolding and refolding.

The theoretical problem is connected with the kinetics of a linear Ising lattice. An exact solution for long chains, based on the triplet closure approximation, is reviewed, as well as an early calculation giving the initial change with time of the helicity, after a sudden perturbation, in terms of a mean relaxation time.

As an illustration of the latter theoretical results, experiments on the poly-L-proline cis-trans isomerization are briefly described. On the other hand, new ultrasonic studies of the kinetics of the helix–coil transition of poly-L-glutamic acid are reported, showing that fluctuations of the number of uninterrupted helix (or coil) sequences contribute to the ultrasonic absorption.

An important case of reduction of the general (Ising) problem is that of sequential processes. An example is described, for which the eigenvalues of the transition rate matrix can be obtained easily, thus leading to an exact solution.

Simple models can be quite successful in treating kinetic problems of cooperative systems. An example is provided by the all-or-none model used by Eigen and Pörschke in their investigation of the oligoribouridylic–oligoriboadenylic acid system. The nucleation process was shown to involve three A.U base pairs, as in the codon–anticodon interaction, and the activation enthalpy was found to be negative.

A non-Arrhenius two-state model, adequate for studying the kinetics of cooperative processes of the type considered, is described; it leads also to a negative activation enthalpy, in a somewhat different way.

Analysis of the kinetics of proteins unfolding and refolding, using three-state models, is finally reviewed. The values of the relaxation amplitudes show that a three-state system can exhibit one relaxation time only. Thus, with the aid of certain assumptions, one can interpret in terms of an intermediate state slow T-jump measurements on pancreatic proteins as a function of temperature. The intermediate state must then have a nonnegligible population at certain temperatures. A simple model, with two identical steps, leads to satisfactory agreement for chymotrypsin-α, thus suggesting the reality of the postulated intermediate state.

C. Sadron (ed.), Dynamic Aspects of Conformation Changes in Biological Macromolecules. 247–269.

1. Introduction

La coopérativité est une propriété qui se manifeste dans les domaines les plus divers. Elle résulte des actions mutuelles que les molécules, et plus généralement les éléments ou les individus d'une collectivité exercent les uns sur les autres, de proche en proche. Certaines propriétés de la matière découlent de tels effets coopératifs; les transformations ordre-désordre en sont un exemple bien connu en physique.

La coopérativité se manifeste en biologie moléculaire, par exemple dans la transition hélice-pelote statistique des polypeptides, et par suite dans les changements de conformation des protéines. Elle se manifeste comme un facteur d'ordre au niveau moléculaire, lors de la biosynthèse des protéines et des acides nucléiques, mais également au niveau des cellules, des tissus et des organismes. La division cellulaire, la différenciation cellulaire, l'activité nerveuse sont des processus biologiques hautement coopératifs. Il semble que l'étude de certains d'entre eux puisse être abordée en termes de structures dissipatives [1].

Les aspects cinétiques des processus précédents ne sont pas les moins importants. C'est ainsi que la fonction biologique d'une protéine exige, en même temps qu'une structure unique d'une grande stabilité, une grande aptitude à présenter des changements de conformation. C'est ainsi encore qu'au niveau supracellulaire l'activité du système nerveux central s'accompagne de phénomènes rythmiques qui lui sont propres.

Il sera nécessaire dans ce qui suit de restreindre le sujet. Les systèmes qu'on envisagera seront constitués d'atomes ou de groupements d'atomes ordonnés selon une file. Il sera donc question, essentiellement, de cinétique conformationnelle de macromolécules en chaîne.

Un système de cette nature, lorsque les interactions sont à courte portée, ne présente pas de transition de phase, pas d'états métastables, et les manifestations de nonlinéarité sont atténuées. Du point de vue théorique, les caractéristiques du système sont celles du modèle d'Ising, avec interactions entre seuls proches voisins, et la suite de l'exposé sera consacrée en partie à la cinétique de ce dernier modèle.

La Figure 1 est une représentation schématique du modèle original d'Ising, qui consiste en une chaîne linéaire de N spins, dont chacun ne peut être orienté que vers le haut ou vers le bas de la figure. Les spins voisins interagissent, et leur énergie mutuelle est plus grande lorsqu'ils sont en position antiparallèle que s'ils sont en position parallèle. Du point de vue énergétique, il serait donc avantageux que les N

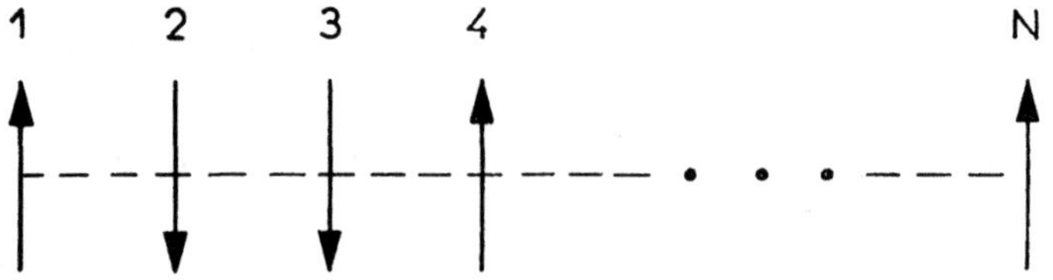

Fig. 1. Schéma d'une chaîne d'Ising de N spins. Les spins ont tendance à s'orienter parallèlement les uns aux autres.

spins fussent tous parallèles les uns aux autres. Mais cette disposition, qui entraînerait une aimantation spontanée, est exclue dans le cas d'une chaîne linéaire très longue, pour des raisons entropiques. C'est ce qui interdit au modèle d'Ising à une dimension et à interactions à courte portée, d'avoir une transition de phase. Par contre, il peut se constituer des séquences plus ou moins longues de spins de même orientation.

Le modèle précédent, adapté par Zimm et Bragg [2], permet d'étudier la transition hélice-pelote statistique des polypeptides. Chaque élément –CO–CHR–NH– de la chaîne peut se trouver dans l'un ou l'autre de deux états (comme les spins du modèle d'Ising); ce sont la 'conformation hélice' (celle qu'il affecte dans l'hélice-α), et la 'conformation pelote statistique' (qui désigne en fait une variété de conformations). Le paramètre $\sigma \ll 1$ de Zimm et Bragg permet d'assigner un poids statistique faible à l'interruption d'une zône pelote, c'est-à-dire à la création d'un noyau hélice. L'inverse σ^{-1} de ce paramètre est une mesure de la coopérativité; plus σ est petit, plus longues seront les séquences ininterrompues d'éléments de même conformation; par exemple, pour une chaîne très longue, les séquences hélice ininterrompues comportent en moyenne $\sigma^{-1/2}$ éléments (distance de coopérativité). Comme pour le modèle d'Ising, le nombre d'états possibles est de l'ordre de 2^N.

A partir de chaque état, le système peut accéder à un certain nombre d'autres états, avec des probabilités de transition qui dépendent des seuls états initial et final. Il s'agit d'un processus stochastique markovien homogène dans le temps, qu'on sait étudier en principe, pourvu que l'on puisse écrire la matrice $\mathbf{Q}$ des vitesses de transition et la diagonaliser. On rappelle que les éléments non-diagonaux de $\mathbf{Q}$ sont les vitesses de transition; ils sont indépendants du temps et non-négatifs; les éléments diagonaux résultent de ce que la somme par ligne des éléments de $\mathbf{Q}$ est nulle.

Si l'on range les probabilités des états de la chaîne en un vecteur-ligne $\mathbf{P}(t)$, celui-ci satisfait à l'équation différentielle linéaire suivante:

$$\frac{\mathrm{d}}{\mathrm{d}t}\,\mathbf{p}(t) = \mathbf{p}(t)\,\mathbf{Q}(t). \tag{1}$$

Des éléments de la solution pour un système de spins ont été donnés par Glauber [3], moyennant des hypothèses particulières concernant la matrice $\mathbf{Q}$.

Il existe toutefois une autre méthode. Au lieu de chercher à déterminer les probabilités des états de la chaîne, on peut chercher celles de certaines séquences particulières. Soit X_j une séquence de j éléments dans une chaîne infinie, par exemple pour $j=5$, ABBAA, où A et B désignent les deux états possibles de chaque élément. La probabilité $p(X_j)$ de la séquence X_j obéit à l'équation différentielle linéaire suivante [4]:

$$\frac{\mathrm{d}}{\mathrm{d}t}\,p(X_j) = \sum_{i=1}^{j+1} C_i p(Y_i), \tag{2}$$

où les C_i sont des combinaisons linéaires des éléments de $\mathbf{Q}$. L'équation différentielle pour $p(X_j)$ contient par conséquent des probabilités de séquences de $j+1$ éléments, ce qui conduirait pour une chaîne infiniment longue à une suite infinie d'équations. Le problème est encore plus complexe pour une chaîne finie, du fait que $p(X_j)$, par exemple, dépend de la position de la séquence X_j le long de la chaîne.

Les équations différentielles (2) ont pu être écrites pour une chaîne infinie jusqu'à $j=6$, ainsi que pour une chaîne courte de 7 éléments [5]. Des méthodes approchées développées par Gô [6], Silberberg et Simha [4, 5, 7] ainsi que par Schwarz [8], permettent de réduire le nombre des équations à $4N-5$ pour une chaîne finie, et à 4 pour une chaîne infinie. L'une de ces méthodes est brièvement décrite au paragraphe 2 ci-dessous. Certaines grandeurs avaient déjà été obtenues moyennant des calculs simples. C'est le cas de la variation initiale avec le temps t (après une perturbation brusque du système) de la fraction d'éléments se trouvant par exemple sous la forme A [9]. Les expressions obtenues ont été confrontées avec des mesures cinétiques pour la poly-L-proline [10]; ces résultats sont également discutés au paragraphe 2.

Des résultats de mesures de l'absorption des ultrasons dans des solutions d'acide poly-L-glutamique sont discutés au paragraphe 3.

Un cas important de réduction se présente lorsque le processus est de type séquentiel. Un tel processus comporte une série bien définie d'étapes consécutives, les états intermédiaires n'ayant pas tous des populations négligeables.

L'appariement des bases des oligonucléotides fournit un exemple de processus séquentiel. Les études d'Eigen et Pörschke [11] ont montré que la formation d'une double hélice (A.U) à pH neutre à partir d'acides oligoadénylique et oligouridylique s'effectue selon le schéma suivant:

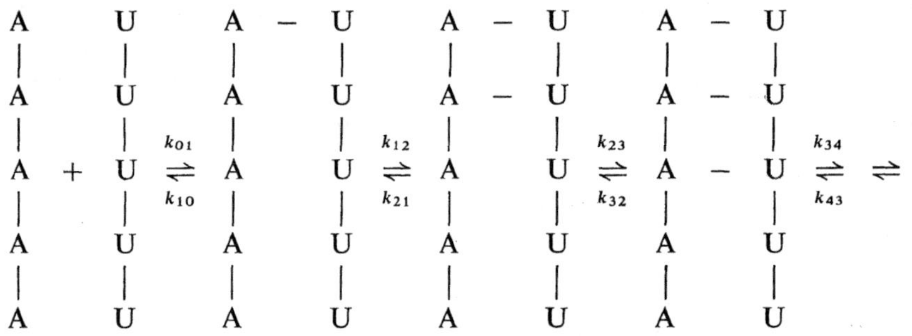

Il se forme d'abord un noyau, par appariement de trois paires de bases. Ce processus bimoléculaire, relativement lent, est suivi de l'appariement successif des 4ème, 5ème, etc. paires de bases. Dans les conditions expérimentales [11], ce dernier processus (monomoléculaire) de croissance est environ 10^4 fois plus rapide que le précédent. En outre, les doubles chaînes qui ne seraient pas appariées exactement par les bouts, ou qui n'auraient formé qu'une ou deux paires de bases appariées, ont une durée de vie très courte, et n'interviennent pas dans la cinétique. Ainsi, dans l'association comme dans la dissociation des deux chaînes, on a une série bien définie d'étapes consécutives.

D'autre part, les caractéristiques cinétiques qu'on vient de rappeler, notamment la rapidité du processus de croissance, font que pour une double hélice comportant jusqu'à 18 paires de bases, il se présente une simplification. On peut en effet, en première approximation, considérer les deux seuls états tels que, respectivement: (a) les deux chaînes soient séparées, (b) la double hélice soit entièrement formée. On est dans ce cas en présence d'un processus par 'tout ou rien' [11]. C'est ce qui a permis à Eigen et Pörschke d'aborder de façon simple la cinétique du transfert de l'information génétique (cf. paragraphe 5.1 ci-dessous).

Eigen et Pörschke ont toutefois souligné que l'approximation par tout ou rien manque de précision à partir de 18 paires de bases, et qu'il est alors nécessaire d'envisager les états intermédiaires comportant une double hélice partiellement formée. C'est la situation typique qui nécessite l'emploi d'un modèle séquentiel. Dans l'exemple actuel, pour deux chaînes de N éléments chacune, il y a $N-1$ états; d'une façon générale, le nombre d'états du modèle séquentiel est de l'ordre du nombre d'éléments de la chaîne, ou de chacune des chaînes.

Mais pour des polynucléotides dont la longueur serait supérieure à la distance de coopérativité, et qui par suite comporteraient des sections hélice et des sections désordonnées alternées, on retrouverait la situation plus complexe évoquée précédemment, et le modèle séquentiel ne serait plus valable.

On peut utiliser le modèle séquentiel dans une variété de questions. Il permet notamment d'étudier la cinétique de la transition hélice–pelote statistique des oligopeptides ([12] et le paragraphe 4 ci-dessous), et il a été proposé de l'appliquer à la cinétique de dénaturation des protéines [13]. Des modèles séquentiels ont été traités [13–15] par la première méthode, consistant à considérer les différents états de la chaîne, soit de manière approchée, moyennant l'hypothèse que les états intermédiaires ont des occupations quasi-stationnaires, soit par des calculs sur machine. J'ai constaté, pour un modèle séquentiel particulier, que le problème peut être résolu de manière complète, du fait qu'il est possible d'écrire explicitement l'équation aux valeurs propres [12]. La solution, qui se prête à diverses généralisations, est exposée brièvement au paragraphe 4.

Le paragraphe 5 est consacré à des modèles simples qui ont permis d'interpréter des cinétiques coopératives.

2. Cinétique du modèle d'Ising

En vue de réduire leur système infini d'Équations (2) à des équations en nombre fini, Silberberg et Simha ont utilisé des conditions de 'fermeture'. On voit immédiatement que dans l'étude statistique du modèle d'Ising avec interactions entre seuls voisins il y a 'fermeture aux doublets' (cf. [8]), du fait qu'il est toujours possible formellement, d'attribuer l'énergie d'interaction d'un couple d'éléments à l'un des éléments, par exemple celui de droite.

Dès lors, à l'équilibre, indiqué par l'indice supérieur zéro, la probabilité d'une séquence quelconque, par exemple ZYXV... (où chacune des lettres représente A ou B), s'exprime uniquement en fonction de probabilités de singulets et de doublets (fermeture aux doublets):

$$p^0(\mathrm{ZYXV}\ldots) = \frac{p^0(\mathrm{ZY})\, p^0(\mathrm{YX})\, p^0(\mathrm{XV})}{p^0(\mathrm{Y})\, p^0(\mathrm{X})}\cdots \tag{3}$$

On notera que $p^0(\mathrm{YX})/p^0(\mathrm{Y})$ est la probabilité conditionnelle que, le deuxième élément étant Y, le troisième soit X, etc.

Dans l'étude cinétique au contraire, la probabilité qu'un élément A devienne B, ou réciproquement, dépend de ses deux voisins. Par suite, la probabilité qu'un élément terminal d'une séquence de j éléments change d'état dépend d'un élément extérieur à cette séquence. C'est pourquoi on est conduit, pour une chaîne infinie, au système infini d'Équations (2).

Ayant écrit des équations cinétiques exactes pour les séquences comprenant jusqu'à 6 éléments, Silberberg et Simha ont pu comparer les effects de la fermeture aux triplets et de la fermeture aux quadruplets. La fermeture aux triplets, par exemple, d'après laquelle hors d'équilibre la probabilité d'une séquence quelconque s'exprime uniquement en fonction de probabilités de doublets et de triplets, s'écrit:

$$p(\mathrm{ZYXVU}\ldots) = \frac{p(\mathrm{ZYX})\, p(\mathrm{YXV})\, p(\mathrm{XVU})}{p(\mathrm{YX})\, p(\mathrm{XV})\, p(\mathrm{VU})} \cdots. \tag{4}$$

Pour comparer les effets des deux fermetures, on suppose qu'une perturbation brusque est imposée au système; les fréquences des singulets, doublets, triplets, ainsi que les longueurs moyennes de séquences sont alors calculées en fonction du temps, après la perturbation, dans les deux cas (fermeture aux triplets et fermeture aux quadruplets). Les résultats des deux fermetures sont presqu'identiques [5] dans des conditions très diverses, ce qui semble indiquer que la fermeture aux triplets, sans être une condition cinétique rigoureuse [7], constitue une bonne approximation [5].

Partant de cette approximation, Schwarz [8] a explicité les $4N-5$ équations pour une chaîne finie, ainsi que les 4 équations pour une chaîne infinie. Ces dernières donnent les dérivées par rapport au temps des 4 probabilités qu'on a choisies comme variables, par exemple $p(\mathrm{B})$, $p(\mathrm{AB})$, $p(\mathrm{ABA})$, $p(\mathrm{BAB})$. Les deux premières équations sont linéaires, les deux dernières non-linéaires, alors que les équations de départ (2) sont toutes linéaires.

La non-linéarité résulte ici de la réduction du nombre des variables consécutive à l'introduction d'une condition de fermeture. On peut noter à ce sujet, qu'on est parti d'un modèle markovien à nombre fini d'états (cf. le paragraphe 1). En fait, les formes désordonnées des systèmes considérés comportent un nombre infini d'états. Il faut alors se souvenir que, si les processus envisagés sont fondamentalement markoviens, la réduction d'un nombre infini à un nombre fini de variables peut aussi conduire à un comportement non-markovien du modèle. Le caractère markovien ne doit donc pas être considéré comme ne pouvant pas être remis en question.

Les 4 équations donnant $p(\mathrm{B})$, $p(\mathrm{AB})$, $p(\mathrm{ABA})$, $p(\mathrm{BAB})$ pour une chaîne infinie [8] sont identiques à celles que Gô [6] avait obtenues par une méthode différente. Gô avait en outre explicité les équations linéarisées valables pour une perturbation petite à partir de l'équilibre, et décrit le comportement des 4 temps de relaxation correspondants [6]. Récemment Schwarz [16] a donné des expressions algébriques de ces 4 temps de relaxation, valables lorsque la coopérativité est élevée.

On notera par ailleurs que si B désigne la forme hélice de l'élément d'un polypeptide, $p(\mathrm{B})$ n'est autre que le degré d'hélicité θ (fraction des éléments sous forme hélice). L'équation différentielle pour $p(\mathrm{B})$ avait été obtenue par Schwarz [9] antérieurement au développement de la théorie générale. Cette équation s'écrit:

$$\begin{aligned} \frac{\mathrm{d}p(\mathrm{B})}{\mathrm{d}t} &= 2k_{\mathrm{F}}p(\mathrm{AAB}) - 2k_{\mathrm{D}}p(\mathrm{ABB}) + \\ &+ f_{\mathrm{A}}[\sigma k_{\mathrm{F}}p(\mathrm{AAA}) - k_{\mathrm{D}}p(\mathrm{ABA})] + f_{\mathrm{B}}[k_{\mathrm{F}}p(\mathrm{BAB}) - \sigma k_{\mathrm{D}}p(\mathrm{BBB})] \end{aligned} \tag{5}$$

avec:

$$\begin{aligned} p(\mathrm{AAB}) &= p(\mathrm{AB}) - p(\mathrm{BAB}) \\ p(\mathrm{ABB}) &= p(\mathrm{AB}) - p(\mathrm{ABA}) \\ p(\mathrm{AAA}) &= 1 - p(\mathrm{B}) - 2p(\mathrm{AB}) + p(\mathrm{BAB}) \\ p(\mathrm{BBB}) &= p(\mathrm{B}) - 2p(\mathrm{AB}) + p(\mathrm{ABA}). \end{aligned} \tag{6}$$

Les coefficients cinétiques k_{F} et k_{D} caractérisent les croissances de séquences ininterrompues A et B:

$$\mathrm{AAB} \underset{k_{\mathrm{D}}}{\overset{k_{\mathrm{F}}}{\rightleftharpoons}} \mathrm{ABB}.$$

Ils sont liés par la condition $k_{\mathrm{F}}/k_{\mathrm{D}} = s$; s et σ sont les coefficients statistiques de Zimm et Bragg (cf. le paragraphe 1). Il reste donc dans l'expression de $\mathrm{d}p(\mathrm{B})/\mathrm{d}t$ (comme dans celles des trois autres dérivées) trois coefficients cinétiques indépendants: k_{F}, f_{A} et f_{B}.

Ces deux derniers coefficients caractérisent respectivement la formation (et la disparition) d'un noyau hélice et d'un noyau pelote:

$$\mathrm{AAA} \underset{f_{\mathrm{A}}k_{\mathrm{D}}}{\overset{\sigma f_{\mathrm{A}}k_{\mathrm{F}}}{\rightleftharpoons}} \mathrm{ABA}$$

$$\mathrm{BBB} \underset{f_{\mathrm{B}}k_{\mathrm{F}}}{\overset{\sigma f_{\mathrm{B}}k_{\mathrm{D}}}{\rightleftharpoons}} \mathrm{BAB}.$$

Après une perturbation brusque petite au temps $t=0$, on a [9], d'après les Équations (5) et (6):

$$\frac{1}{\Delta\theta}\left(\frac{\mathrm{d}\theta}{\mathrm{d}t}\right)_{t=0} = \frac{1}{\tau^*} = k_{\mathrm{F}}[(s-1)^2 + 4\sigma], \tag{7}$$

où le temps de relaxation moyen τ^* est défini par:

$$1/\tau^* = \left(\sum_l \beta_l/\tau_l\right)/\sum_l \beta_l \tag{8}$$

les τ_l étant les temps de relaxation du système, et les β_l des coefficients tels que:

$$\Delta\theta = \theta_{t=\infty} - \theta_{t=0} = -\sum_l \beta_l. \tag{9}$$

L'Équation (7) montre que τ^* passe par un maximum à mi-transition ($s=1$), d'autant plus accentué que σ est plus petit devant l'unité, donc que la transition est plus coopé-

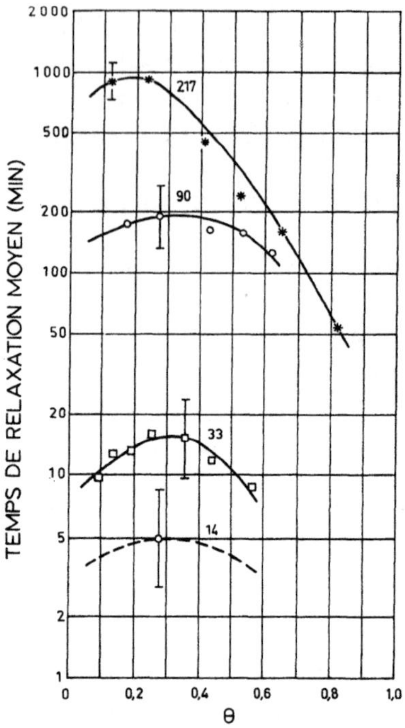

Fig. 2. Temps de relaxation moyen τ^* de la poly-L-proline en fonction du degré de conversion θ pour différentes longueurs de chaîne.

rative. Il s'agit d'un aspect particulier du *ralentissement des processus par coopérativité,* qu'illustrent les résultats expérimentaux de Winklmair, Engel et Ganser [10] représentés sur la Figure 2.

Ces mesures concernent la transition I$\rightleftharpoons$II entre les deux hélices de la poly-L-proline, dont la première est une hélice droite, avec toutes les liaisons peptidiques en configuration *cis,* et la seconde une hélice gauche avec toutes les liaisons peptidiques en configuration *trans.* Il s'agit par conséquent d'une transition ordre $\rightleftharpoons$ ordre; celle-ci peut néanmoins être étudiée à partir du modèle de Zimm et Bragg moyennant des modifications mineures. En effet, pour que ce modèle soit valable, il est seulement requis que chaque élément puisse affecter deux formes. Ce peuvent être, respectivement la 'conformation hélice' et la 'conformation pelote statistique' d'un élément de chaîne, mais également, respectivement, la forme *cis* et la forme *trans* d'une liaison peptidique.

La transition I$\rightleftharpoons$II peut être induite par une modification de la composition du solvant. Aussi dans les expériences de Winklmair, Engel et Ganser, la perturbation du système est-elle produite par une brusque modification de cette composition. Les temps de relaxation initiaux sont mesurés à des degrés de conversion θ compris entre 0,1 et 0,85, et pour des chaînes comportant entre 14 et 217 éléments. Ces temps sont compris entre plusieurs minutes et plus d'une quinzaine d'heures.

Les résultats des mesures cinétiques ont été comparés à l'expression du temps de relaxation moyen pour une chaîne de longueur finie, obtenue par Schwarz [17]. Cette expression fait intervenir deux coefficients statistiques de nucléation, et le seul coefficient cinétique $k_{\rm I}$ caractérisant la croissance d'une séquence I ininterrompue:

$$\mathrm{I} \quad \mathrm{II} \quad \mathrm{II} \underset{k_{\rm II}}{\overset{k_{\rm I}}{\rightleftharpoons}} \mathrm{I} \quad \mathrm{I} \quad \mathrm{II}.$$

Compte tenu des données déduites des courbes de transition, il a été trouvé qu'à 70° C $k_{\rm I}$ est compris entre 0,05 $\sec^{-1}$ et 10 $\sec^{-1}$, selon la composition du solvant. La nucléation d'un élément I dans une séquence ininterrompue II est environ 10^5 fois plus lente. L'introduction d'un seul coefficient cinétique [17] permet de représenter l'ensemble des résultats [10].

3. Absorption des ultrasons dans les solutions de polypeptides

La perturbation d'un équilibre conformationnel par une onde ultrasonore fournit des données sur la cinétique des transconformations moléculaires. La mesure de l'absorption des ultrasons dans les solutions de polypeptides fournit ainsi des données expérimentales susceptibles d'être comparées aux résultats de la théorie cinétique du modèle d'Ising, résumée au paragraphe 2 ci-dessus. La variation du temps de relaxation ultrasonore accompagnant celle du degré d'hélicité θ (cette dernière étant généralement obtenue en modifiant le pH), a été trouvée conforme, pour divers polypeptides, au comportement du temps de relaxation moyen τ^* considéré par Schwarz (cf. le paragraphe 2). Il faut toutefois noter:

(a) qu'on a parfois comparé à la théorie des données qui ne traduisent pas une cinétique conformationnelle (comme la souligné Zana [18]);

(b) que la mesure du temps de relaxation ultrasonore en solution aqueuse présente des difficultés particulières conduisant souvent à des incertitudes expérimentales notables.

Dans ce qui suit on considère exclusivement la variation de l'absorption ultrasonore avec θ, pour laquelle une observation nouvelle a été faite récemment [19, 20].

La Figure 3 concerne deux acides poly-L-glutamiques (1 et 2) dont les masses moléculaires sont respectivement 90000 et 12000, en solution dans H_2O, 0,2 M NaCl. Le transfert de proton sur les groupes carboxyliques ainsi que la fixation sur site de contre-ions contribuent à l'absorption ultrasonore [21, 22]. Il est donc nécessaire de séparer ces effets de l'absorption due à la perturbation de l'équilibre conformationnel. La fixation sur site du contre-ion a été éliminée [20] par une méthode que Tondre et Zana ont déjà appliquée à d'autres substances (contre-ion volumineux [22]). Michels a pu vérifier en outre la présence d'un terme conformationnel en effectuant des mesures comparatives pour un acide poly-DL-glutamique [20]. La Figure 3 représente pour les deux APLG envisagés ici la contribution conformationnelle, après élimination des autres effets (l'indice c signifie 'conformationnel'; $\Delta\alpha$ désigne l'excès d'absorption dû au soluté). Sur la Figure 3 on a également représenté, pour les deux APLG, les variations avec le pH du degré d'hélicité θ obtenu par spectrométrie ultraviolette et par dichrographie [20].

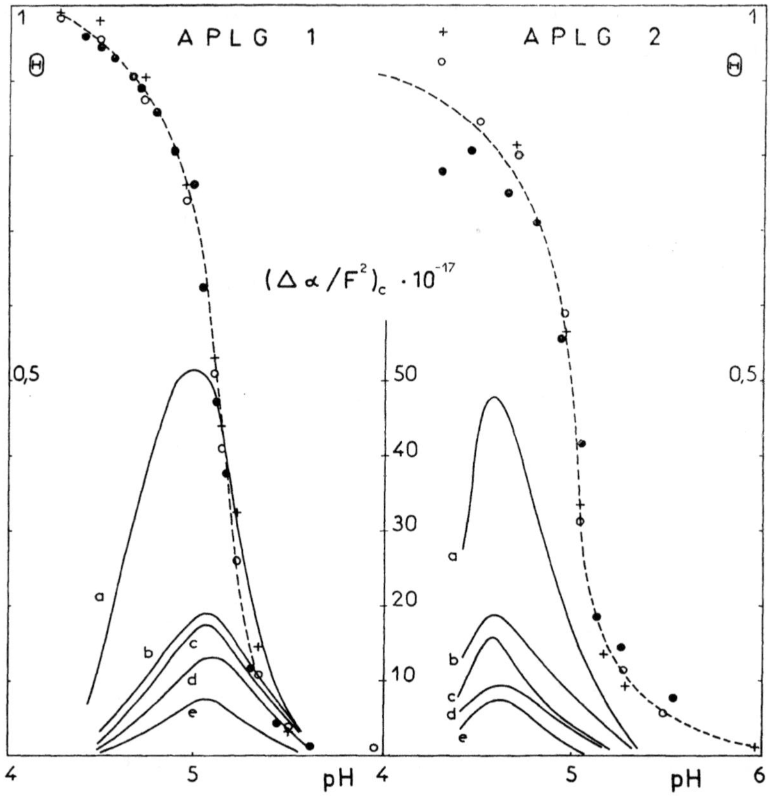

Fig. 3. Variation de $(\Delta\alpha/F^2)_c$ en fonction du pH à différentes fréquences, et courbe de transition pour deux acides poly-L-glutamique de masses moléculaires différentes dans H_2O, 0,2 M NaCl. La courbe de transition a été obtenue par spectrométrie ultraviolette (●) et par dichrographie à 222 nm (○) et 208 nm (+). Fréquences en MHz: (a) 1,04; (b) 2,82; (c) 3,55; (d) 5,04; (e) 7,96.

Compte tenu des diverses causes d'erreur dans la détermination de l'absorption ultrasonore α et du degré d'hélicité θ, le maximum de $(\Delta\alpha/F^2)_c$, où F désigne la fréquence, se présente pour l'APLG 2 à θ compris entre 0,7 et 0,85. A ces valeurs de θ, les séquences désordonnées ininterrompues sont en moyenne 2 à 3 fois plus courtes que les séquences hélice ininterrompues. On est ainsi conduit à envisager *une contribution à l'absorption ultrasonore des fluctuations du nombre de séquences ininterrompues* [19, 23].

Dans le cas de l'APLG 1, le maximum de $(\Delta\alpha/F^2)_c$ se présente pour une valeur de θ comprise entre 0,5 et 0,7, compatible à la limite avec la valeur 0,5 trouvée par d'autres auteurs [24, 25].

On se bornera ici aux indications qui précèdent, la possibilité d'une contribution à l'absorption ultrasonore des fluctuations du nombre de séquences hélice (ou pelote) ininterrompues étant discutée ailleurs plus en détail [23].

4. Un modèle séquentiel

On a indiqué précédemment que le modèle d'Ising présente un cas de réduction important lorsque le processus est séquentiel. Le modèle séquentiel le plus simple comportant une étape de nucléation est représenté par le schéma réactionnel suivant:

$$0 \underset{q}{\overset{p_0}{\rightleftharpoons}} 1 \underset{q}{\overset{p}{\rightleftharpoons}} 2 \underset{q}{\overset{p}{\rightleftharpoons}} \cdots \underset{q}{\overset{p}{\rightleftharpoons}} j \underset{q}{\overset{p}{\rightleftharpoons}} \cdots \underset{q}{\overset{p}{\rightleftharpoons}} N\,,$$

où les $0 \leqslant j \leqslant N$ désignent les états successifs de la molécule en chaîne, L'étape de nucléation est la transition $0 \to 1$, et les coefficients cinétiques p_0 et p sont de ce fait, généralement liés par l'inégalité $p_0/p = \sigma \ll 1$. Ce modèle est représenté de manière schématique sur la Figure 4b dans un diagramme énergétique. Les états zéro et N représentant, par exemple, respectivement, la forme pelote statistique et la forme hélice-α d'un oligopeptide, auquel le présent modèle s'applique [12], du fait que les oligopeptides comportent une seule séquence hélice; à l'état j, $j+2$ éléments de la chaîne ont la conformation hélice.

La matrice stochastique **Q** s'écrit ici:

$$\mathbf{Q} = \begin{vmatrix} -p_0 & p_0 & 0 & . & . & 0 \\ q & -(p+q) & p & & & . \\ . & . & . & . & . & . \\ . & & & q & -(p+q) & p \\ 0 & . & . & 0 & q & -q \end{vmatrix} \tag{10}$$

Le fait que seules soient permises les transitions entre états voisins (caractère séquentiel du modèle) entraîne que la matrice **Q** n'a d'éléments non-nuls que sur la diagonale principale et sur les deux parallèles adjacentes. Cette circonstance facilite l'obtention des valeurs propres λ, dont les inverses changés de signe sont les temps de relaxation, ainsi que celle des vecteurs propres. En outre, le fait que toute matrice **Q** possède une valeur propre nulle de multiplicité 1, permet d'écrire l'équation caractéristique sous la forme d'un produit de facteurs. En posant:

$$-\lambda_i = p + q - 2(pq)^{1/2} \cos \phi_i\,, \tag{11}$$

l'équation caractéristique s'écrit [12]:

$$[2c^{1/2} \cos \phi - (1 + c)]\,[(p - p_0) \sin N\phi - \\ - pc^{1/2} \sin (N + 1)\, \phi]\, (\sin \phi)^{-1} = 0\,, \tag{12}$$

où la constante d'équilibre p/q a été posée égale à c^{-1}.

L'annulation du premier crochet de l'Équation (12) fournit la valeur propre $\lambda_0 = 0$. L'annulation du second crochet fournit une équation en ϕ, qui possède soit N racines réelles, soit $N-1$ racines réelles et une racine imaginaire. On a toujours N valeurs de λ réelles et strictement négatives, qui fournissent les N temps de relaxation.

Lorsque $c^{1/2} < (1-\sigma)(1+1/N)$, l'Équation (12) possède une racine imaginaire $\phi_1 = i\psi_1$, à laquelle correspond le temps de relaxation le plus long τ_1. Si en outre c n'est pas trop voisin de l'unité et si $\sigma \ll 1$, τ_1 est donné par l'expression suivante [12]:

$$\tau_1^{-1} \# p(1-c)\,[\sigma + (1-c)\,c^N], \tag{13}$$

qui coïncide avec le résultat fourni par l'approximation quasistationnaire [14].

Lorsque c n'est pas voisin de 1, τ_1 est sensiblement égal au temps de relaxation τ associé au changement complet de conformation de la molécule: $0 \rightleftharpoons N$. En effet, le temps moyen de premier passage τ_{0N}, au bout duquel la molécule à l'état N parvient à l'état 0 est très voisin de $[p(1-c)^2 c^N]^{-1}$, tandis que le temps moyen τ_{N0} requis pour la transformation inverse est voisin de p_0^{-1}. On peut donc écrire [12]:

$$\tau^{-1} = \tau_{0N}^{-1} + \tau_{N0}^{-1} \sim p(1-c)\,c^N + p_0 \# \tau_1^{-1}/(1-c) \sim \tau_1^{-1}. \tag{14}$$

Le temps de relaxation le plus court, τ_N, est de l'ordre de $p^{-1}(1+c^{1/2})^{-2}$, et par suite de l'ordre du temps moyen de formation de la liaison-hydrogène. On notera qu'à l'approximation de l'Équation (13) on a la relation suivante entre le temps le plus long et le temps le plus court:

$$\tau_1^{-1} \# \tau_N^{-1}[\sigma + (1-c)\,c^N]\,(1-c^{1/2})/(1+c^{1/2}), \tag{13'}$$

qui pour $c \ll 1$ se réduit à:

$$\tau_1 \# \sigma^{-1}\tau_N. \tag{13''}$$

Dans ce dernier cas ($c \ll 1$) tous les temps $\tau_2 \ldots \tau_N$ ont sensiblement la même valeur. Dans le cas général, il satisfont à des inégalités qu'on peut expliciter facilement [12].

La signification du coefficient de nucléation σ demande à être précisée en fonction des conditions expérimentales. Si la nucléation consiste en l'appariement de deux segments appartenant à des molécules différentes, comme dans les mesures d'Eigen et Pörschke (cf. les paragraphes 1 et 5.1), il apparaît des concentrations dans l'expression de σ.

D'autre part, le fait de poser $k_{j,j+1} = p$, indépendamment de j pour $j \geqslant 1$, comme on l'a fait au début du présent paragraphe, peut constituer une approximation. Pour un oligopeptide par exemple, si la nucléation se produit à l'intérieur de la chaîne, le processus de croissance de l'hélice s'effectue des deux côtés du noyau, et l'on a, au début, $k_{j,j+1} = 2u$, où u représente le coefficient de propagation d'une hélice par un seul bout. Lorsque l'hélice est entièrement formée d'un côté du noyau, ou si la nucléation s'est produite en bout de chaîne, on a: $k_{j,j+1} = u$. On a donc: $k_{j,j+1} = \beta u$, $1 < \beta < 2$, ce qui montre la nature de l'approximation précédente. Les coefficients $k_{j,j-1}$, de leur côté, sont tous égaux à $q = 2v$, où v désigne le coefficient de propagation d'une chaîne par un seul bout.

Enfin, pour un oligopeptide, le coefficient de nucléation σ considéré ici est supérieur au coefficient de nucléation de Zimm et Bragg. En effet, ce dernier coefficient fournit le poids statistique d'un noyau hélice formé en un point donné d'une chaîne, tandis que la nucléation peut se produire ici en un point quelconque de la chaîne envisagée.

Les paragraphes précédents ont fourni des exemples différents (chaînes longues, paragraphe 2; *chaînes courtes, paragraphe* 4*) de problèmes de cinétique conformationnelle entièrement solubles.*

D'un autre côté, des modèles simples sont utilisés avec profit, et l'un d'entre eux, qui sera considéré dans un instant, a permis de résoudre un problème important (nucléation d'une double hélice (A.U)). Le paragraphe 5 fournit des exemples de tels modèles et de leurs applications.

5. Modèles à deux et trois états – Applications

5.1. Nucléation dans la formation d'une double hélice (A.U)

L'étude déjà mentionnée de la cinétique d'appariement des bases des oligonucléotides a conduit Eigen et Pörschke [11] à préciser les conditions de formation d'un noyau d'hélice double (A.U). Il est apparu que la formation d'un tel noyau requiert l'appariement de trois paires de bases. Cette conclusion a été obtenue par une analyse très simple des résultats, fondée sur l'approximation par tout ou rien, ou si l'on veut, sur un modèle à deux états.

Les mesures cinétiques, réalisées par saut de température à pH 6,9, en solution 0.05 M cacodylate de sodium, ont fourni essentiellement 1 temps de relaxation. Il est associé à l'équilibre:

$$A_N + U_N \underset{k_D}{\overset{k_R}{\rightleftharpoons}} (A + U)_N$$

et donné par l'expression suivante:

$$\tau^{-1} = k_R (C_A + C_U) + k_D, \tag{15}$$

où C_A et C_U sont les concentrations des acides oligoadénylique et oligouridylique non appariés; k_R et k_D sont respectivement les constantes de vitesse de recombinaison et de dissociation. Ces deux constantes de vitesse sont obtenues en portant τ^{-1} en fonction de la somme des concentrations d'acides non appariés.

Les coefficients cinétiques ont les propriétés suivantes. Les constantes de vitesse de dissociation k_D augmentent considérablement lorsque la température augmente, et les enthalpies d'activation correspondantes augmentent avec la longueur des chaînes. Au contraire, les constantes de vitesse de recombinaison k_R sont presqu'indépendantes de la longueur des chaînes, et diminuent lorsque la température augmente. L'enthalpie d'activation correspondante est donc négative; elle est de l'ordre de -9 kcal/mole.

Le fait que les constantes de vitesse de recombinaison dépendent peu de la longueur des chaînes résulte de ce que la cinétique de recombinaison est déterminée par l'étape de nucléation, le processus ultérieur de croissance de la double hélice étant rapide.

De l'existence d'une enthalpie d'activation négative dans le processus de recombinaison, et de sa valeur numérique, Eigen et Pörschke déduisent que la nucléation comporte plusieurs étapes, plus précisément trois étapes. Si tel est le cas, on doit avoir:

$$k_R = \frac{k_{01}}{k_{10}} \cdot \frac{k_{12}}{k_{21}} \cdot k_{23}, \tag{16}$$

où la signification des coefficients cinétiques du deuxième membre résulte du schéma réactionnel (cf. le paragraphe 1 ci-dessus). Aux constantes d'équilibre k_{01}/k_{10} et k_{12}/k_{21} caractérisant la formation des deux premières paires de bases, correspondent des variations d'enthalpie évaluées (d'après l'étude thermodynamique de Pörschke [26]) respectivement à -4 kcal/mole et $-5{,}5$ kcal/mole. En évaluant à 1 kcal/mole l'enthalpie d'activation dans la formation de la troisième paire de bases, Eigen et Pörschke obtiennent une enthalpie d'activation totale de $-8{,}5$ kcal/mole, voisine de la valeur expérimentale -9 kcal/mole.

La Figure 4a représente de manière schématique le diagramme énergétique de la réaction envisagée. Les états zéro et N représentent respectivement les deux chaînes séparées et la double hélice. On peut noter que toutes les particularités du potentiel ne

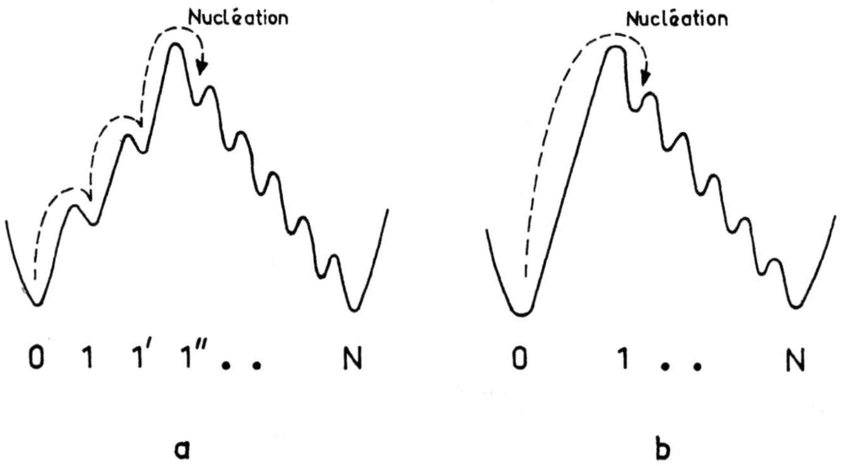

Fig. 4. (a). Diagramme des variations de l'énergie libre de Gibbs (vue schématique) dans la formation d'une double hélice (A.U); 1, 1′, 1″ sont les 3 étapes de la nucléation. (b) Même diagramme pour un processus sequentiel avec une seule étape de nucléation.

sont pas nécessaires à l'analyse. Dans détermination des constantes de vitesse de recombinaison et de dissociation k_R et k_D à partir des valeurs mesurées du temps de relaxation, seuls sont retenus les états zéro et N (emploi d'un modèle à deux états). Ensuite, la valeur de l'enthalpie d'activation de recombinaison obtenue de cette manière, avec l'aide de données thermodynamiques sur les étapes de la nucléation, permet de conclure que ces étapes sont au nombre de trois.

Il est intéressant de comparer le schéma de la Figure 4a à celui de la Figure 4b relatif au modèle séquentiel considéré au paragraphe 4. Pour ce dernier modèle, les étapes du processus de croissance sont prises en considération, tandis que la nucléation est considérée comme s'effectuant en une seule étape.

Les effets coopératifs conduisant à la consititution d'un noyau de la double hélice (A.U) fournissent ainsi une longueur de nucléation (trois paires de bases) égale aux longueurs du codon et de l'anticodon. Eigen et Pörschke [11] en déduisent que le processus de transfert d'information par triplets de bases appariées ne présente pas seulement un avantage de nature 'alphabétique'. Il s'y ajoute un autre avantage, de nature cinétique. Un nombre inférieur d'éléments appariés entraînerait en effet une stabilité insuffisante, tandis qu'un nombre d'éléments appariés supérieur produirait un ralentissement du transfert d'information.

Le système de triplets de bases appariées, de son côté, possède à la fois la stabilité et la flexibilité qui permettent au transfert d'information de s'effectuer avec un taux d'erreur acceptable ainsi qu'avec la rapidité requise.

5.2. Modèle non-arrhénien à deux états

La formation d'une double hélice (A.U) présente deux caractéristiques qui ont facilité l'étude résumée au paragraphe 5.1 ci-dessus:

(a) l'étape initiale est bimoléculaire, de sorte que les constantes de vitesse de dissociation et de recombinaison ont pu être déduites directement des variations du temps de relaxation avec la somme des concentrations d'acides libres (Équation (15));

(b) des données thermodynamiques ont pu être obtenues sur les étapes de la recombinaison.

Au contraire, lorsqu'on étudie une transconformation monomoléculaire, et sauf si l'on dispose des valeurs de la constante d'équilibre, comme pour certaines protéines considérées au paragraphe 5.3 ci-dessous, on est contraint de recourir à une analyse plus ardue [27, 28]. C'est la situation que nous avons trouvée dans l'étude des transconformations par isomérie de rotation de certaines molécules en chaînes.

Avec Candau et Zana nous avions établi, il y a une dizaine d'années, que l'absorption des ultrasons dans les solutions de polymères vinyliques comporte une contribution des transconformations locales, qu'on avait supposées coopératives et affectant des segments d'une dizaine de chaînons [cf. 28].

Malgré la difficulté mentionnée à l'instant, on peut obtenir, outre les caractéristiques thermodynamiques et cinétiques des 'systèmes relaxants' (segments de chaînes subissant la transconformation), une évaluation du nombre de ces systèmes par molécule. Il est nécessaire à cet effet de connaître le temps de relaxation, ainsi que l'amplitude correspondante; ce problème est mentionné ici parce qu'il conduit à considérer une enthalpie d'activation négative d'un point de vue montrant certaines différences avec le précédent, exposé au paragraphe 5.1.

On rappellera tout d'abord deux faits expérimentaux mis en évidence avec Lemaréchal et Fünfschilling [29].

(a) Deux relaxations ont été observées entre 1 et 150 MHz pour le polystyrène, le polyméthacrylate de méthyle et la poly-2-vinylpyridine.

(b) Certaines études avaient fourni pour le polystyrène un temps de relaxation unique variant de manière considérable avec la température [30]; mais ces données résultaient de mesures effectuées dans un domaine de fréquence étroit. Au contraire, les deux temps de relaxation que nous avons mesurés dépendent peu de la température (comportement non-arrhénien).

Si l'on admet que les deux temps de relaxation observés proviennent de deux transconformations $1 \rightleftharpoons 2$ pratiquement indépendantes l'une de l'autre, chacune d'elles est caractérisée par un temps de relaxation donné par l'expression suivante:

$$\tau^{-1} = k_{12} + k_{21}, \tag{17}$$

où k_{12} et k_{21} sont les vitesses de réaction associées à la transconformation considérée.

Pour exprimer que le temps de relaxation τ varie peu avec la température T, on peut écrire que $\mathrm{d}\tau/\mathrm{d}T = 0$, ou mieux que $\mathrm{d}(T\tau)/\mathrm{d}T = 0$ (parce que le résultat est plus simple). On obtient ainsi la condition suivante [27, 28]:

$$\exp(-\Delta G/RT) = -\Delta H_{21}/\Delta H_{12}, \tag{18}$$

où ΔH_{12} et ΔH_{21} sont les enthalpies d'activation, et $\Delta G = \Delta G_{12} - \Delta G_{21}$ l'enthalpie libre de réaction (énergie libre de Gibbs).

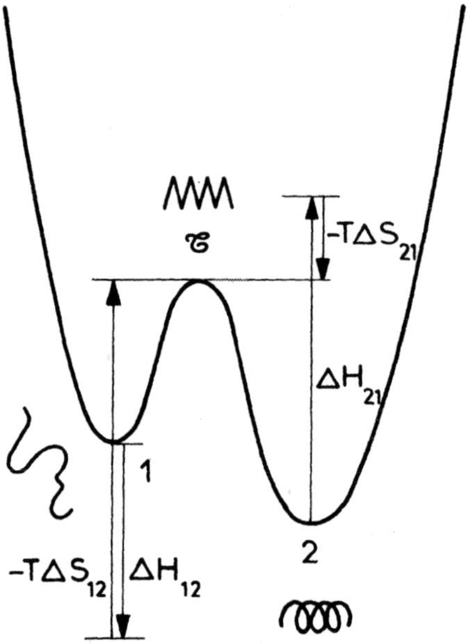

Fig. 5. Modèle non-arrhénien à deux états. Une enthalpie d'activation est négative.

Pour que la condition (18) soit remplie, il faut que les deux enthalpies d'activation soient de signes contraires, donc que l'une d'elles soit négative. La Figure 5 représente une situation où cette dernière condition est remplie. Supposons qu'un segment de la chaîne subisse un changement de conformation coopératif. La coopérativité peut nécessiter un certain degré d'ordre de la conformation 'activée' ($\mathfrak{T}$ sur la figure), donc une perte notable d'entropie dans le passage de la forme 1 à la forme 'activée' ($\Delta S_{12} < 0$). Dès lors, l'énergie libre de Gibbs d'activation ΔG_{12} peut être positive, ainsi qu'il est nécessaire, même si l'enthalpie d'activation ΔH_{12} est négative; cette dernière circonstance peut effectivement se présenter si la conformation 'activée', relativement ordonnée, est aussi relativement compacte.

Si l'état 2 est dégénéré D fois il faut multiplier $\exp(-\Delta G/RT)$ par D dans l'Équation (18), ainsi que dans les autres relations établies pour le présent modèle [27, 28]. Si l'état 1 est dégénéré D fois, $\exp(-\Delta G/RT)$ doit être multiplié par D^{-1}. Sur la Figure 5 l'état 1 est dégénéré. En fait, lorsque la dégénérescence n'intervient que pour augmenter l'entropie configurationnelle d'un état, on peut se passer d'introduire le facteur D, et considérer que l'entropie de l'état envisagé est augmentée de la quantité $R \ln D$, où R est la constante des gaz.

Les remarques suivantes permettront de préciser la signification physique du modèle envisagé.

(a) Il est connu que le modèle à deux états fournit une transition brusque entre les états 1 et 2 à la température $T_t = \Delta H/\Delta S$ (où ΔH et ΔS sont les enthalpie et entropie de réaction), lorsque $\Delta H/RT \gg 1$, $\Delta S/R \gg 1$ [31].

(b) Si l'on appelle $\mathfrak{a}$ l'amplitude de la relaxation ultrasonore du système envisagé, la quantité $T^2\mathfrak{a}/\tau$ passe par un maximum pour $T = T_t$; à cette température on a [27]:

$$\frac{\mathrm{d}^2}{\mathrm{d}T^2} \ln \frac{T^2 \mathfrak{a}}{\tau} = -\frac{1}{2}\left(\frac{\Delta H}{RT^2}\right)^2 \tag{19}$$

de sorte que, dans les conditions envisagées, $T^2\mathfrak{a}/\tau$ montre un maximum accentué. On notera que ce maximum n'est pas comparable à celui qui résulte d'une anomalie de chaleur spécifique de Schottky, comme pour le modèle classique à deux états.

En outre, à la température $T = T_s$ telle que la condition (18) soit satisfaite, on a [27]:

$$T\tau \frac{\mathrm{d}^2}{\mathrm{d}T^2} \frac{1}{T\tau} = -\frac{\Delta H_{12} \cdot \Delta H_{21}}{(RT^2)^2} \tag{20}$$

de sorte que si $\Delta H_{12} \,.\, \Delta H_{21} < 0$, et si $\Delta H_{12} \,.\, \Delta H_{21}/(RT^2)^2 \gg 1$, τ passe par un maximum accentué pour $T = T_s$.

Par conséquent, le modèle est susceptible de présenter les propriétés essentielles des systèmes coopératifs particuliers envisagés dans cet exposé.

Dans la pratique, les conditions précédentes ne sont pas nécessairement toutes remplies, puisqu'un système peut être plus ou moins coopératif et plus ou moins non-arrhénien.

(c) Comparé à celui d'Eigen et Pörschke, le présent modèle conduit à une analyse applicable à une transconformation monoléculaire, même en l'absence de données sur la constante d'équilibre.

En outre, la relation (18) montre qu'une enthalpie d'activation négative peut apparaître dans une situation différente de celle envisagée au paragraphe 5.1.

Dans cette dernière, l'apparition d'une enthalpie d'activation 'apparamment' négative, est due en effet à l'existence d'étapes, dont chacune fait intervenir une enthalpie d'activation positive. Mais on peut imaginer également que le passage $1 \rightarrow \mathfrak{F}$ sur la Figure 5 soit impossible sans la coopération de plusieurs mouvements simultanés, et fasse intervenir un état 'activé' $\mathfrak{F}$ ayant une enthalpie réellement inférieure à celle de l'état 1.

5.3. Cinétique conformationnelle de protéines

La structure que présente une protéine dans l'espace résulte d'instructions contenues dans la séquence des acides aminés qui la composent. Le mécanisme par lequel l'information génétique, sous la forme de cette structure primaire, conduit à la conformation biologiquement active n'est pas connu en détail. L'étude cinétique de la dénaturation et de la renaturation des protéines présente de l'intérêt à cet égard [32–34].

La dénaturation réversible de protéines pancréatiques comme la trypsine, la chymotrypsine et la ribonucléase A aux pH voisins de 2 (transition I) a souvent été décrite comme un changement de conformation très coopératif par 'tout ou rien' entre la forme native et la forme dénaturée.

Des expériences cinétiques par 'saut de température lent' de Pohl [32], fournissant des temps de relaxation de la seconde à la minute, semblaient confirmer ce point de vue. En effet, des mesures très complètes effectuées pour la chymotrypsine-α [32] ont montré qu'on observe un seul temps de relaxation, que celui-ci est indépendant de la longueur d'onde du rayonnement UV utilisé pour suivre l'évolution du système, et se trouve inchangé si l'on observe une variation de pouvoir rotatoire. Ces résultats venaient donc confirmer les conclusions d'études dans lesquelles l'existence de conformations intermédiaires avait été envisagée, en vue d'expliquer certaines caractéristiques (non-cinétiques) des transitions [35, 36], mais avait finalement été écartée.

Cependant, le fait que la constante de vitesse de renaturation k_R déterminée par Pohl (à partir des valeurs du temps de relaxation et de la contante d'équilibre) montre un comportement anormal avec la température ($\ln k_R$ n'est pas une fonction linéaire de T^{-1}, cf. Figure 6b) conduit à reconsidérer le problème. J'ai examiné la question de savoir si le comportement anormal de k_R observé par Pohl peut s'expliquer par un modèle à 3 états $1 \rightleftharpoons 2 \rightleftharpoons 3$ à enthalpies et entropies de réaction et d'activation indépendantes de T [37, 28].

Il faut noter qu'une étape rapide a été mise en évidence récemment dans la dénaturation de la ribonucléase-A [38]. Pour cette protéine on est donc en présence de deux temps de relaxation, et non plus d'un seul (cf. également [33, 34]).

Des étapes rapides existent également, selon toute probabilité, pour la chymotrypsine-α, qu'on envisage plus spécialement ici. L'attention sera néanmoins portée sur le processus lent caractérisé par un seul temps de relaxation, en vue d'examiner si la relaxation correspondante n'est pas complexe.

La relation (21) donne la variation de la grandeur mesurée α, consécutive à un saut de température petit ΔT:

$$(\lambda_2 - \lambda_1)(1 + k + K^{-1})\left(\frac{\alpha - \alpha_\infty}{\Delta T}\right) =$$

$$= \frac{\lambda_2 + q(1 + k)}{\lambda_1}\left[-Q\,\frac{q + \lambda_1}{QK + \lambda_1}\,\Gamma + qk\gamma\right] \times \tag{21}$$

$$\times\left[(\alpha_1-\alpha_3)+k^{-1}\frac{q+\lambda_1}{q}(\alpha_2-\alpha_3)\right]\exp(-\lambda_1 t)-$$
$$-\frac{\lambda_1+q(1+k)}{\lambda_2}\left[-Q\frac{q+\lambda_2}{QK+\lambda_2}\Gamma+qk\gamma\right]\times$$
$$\times\left[(\alpha_1-\alpha_3)+k^{-1}\frac{q+\lambda_2}{q}(\alpha_2-\alpha_3)\right]\exp(-\lambda_2 t), \qquad (21)$$

où α_∞ désigne la valeur de α atteinte au bout d'un temps long; λ_1 et λ_2 sont les inverses changés de signes des temps de relaxation τ_1 et τ_2, dont l'expression est connue [39]; q et Q désignent les vitesses des transitions $1\rightarrow 2$, $2\rightarrow 3$, k, K les rapports des populations 1 et 2, 2 et 3. En outre, $\gamma=-\Delta h/(RT^2)$, $\Gamma=-\Delta H/(RT^2)$, où Δh et ΔH désignent les différences d'enthalpie entre les états 2 et 1, 3 et 2. Enfin α_i $(i=1, 2, 3)$ est la constante optique du système à l'état i, c'est-à-dire par exemple l'absorption ultraviolette ou le pouvoir rotatoire, selon la méthode employée pour suivre l'évolution du système après le saut de température. (Dans le problème actuel, seule est utilisée [37] la valeur du rapport des deux amplitudes de relaxation figurant dans l'expression (21); concernant d'autres calculs d'amplitudes pour deux systèmes couplés, cf. Thusius [40].)

Considérons un système simple, dans lequel les deux étapes $1\rightarrow 2$ et $2\rightarrow 3$ sont identiques des points de vue thermodynamique et cinétique $(k=K,\ q=Q)$.

Il reste une hypothèse à formuler pour la variation du coefficient α_i avec la conforma-

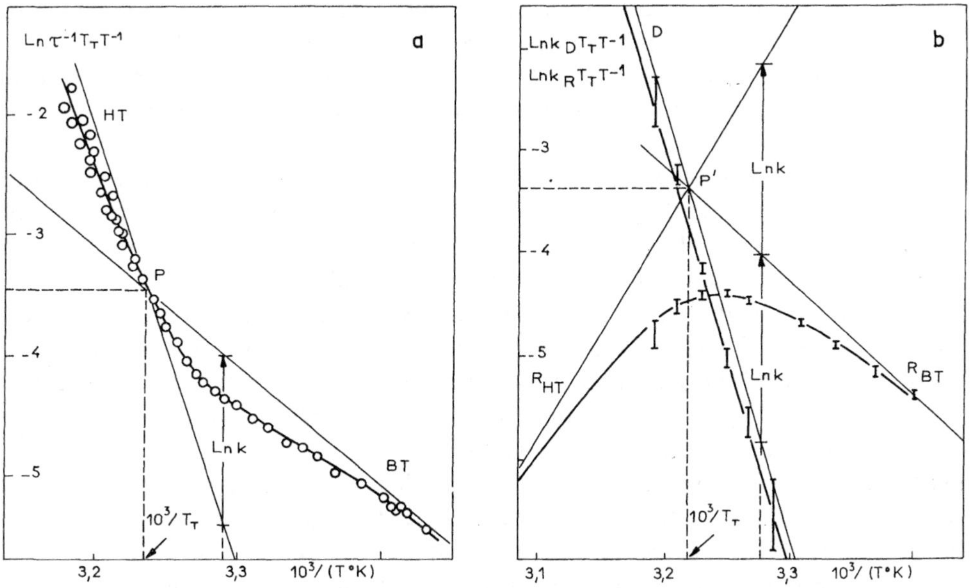

Fig. 6. (a). Chymotrypsine-α. Variations du temps de relaxation avec la température; ○ mesures de Pohl; — théorie. (b) Chymotrypsine-α. Variations des constantes de vitesse de dénaturation et de renaturation k_D et k_R avec la température; I mesures de Pohl; — théorie.

tion de la protéine. Supposons que les trois conformations soient caractérisées par deux valeurs seulement de α_i, la première α_1 pour la conformation native 1, la deuxième $\alpha_2=\alpha_3$ pour les conformations 2 et 3. On aboutit alors aux conclusions suivantes [37]:

(a) Si l'on appelle τ_2 le temps de relaxation le plus long, A_1 et A_2 les amplitudes du premier et du deuxième terme de l'expression (21), on a $|A_1| \ll |A_2|$ lorsque τ_2/τ_1 est voisin de sa valeur maximale 3; d'autre part, lorsque $|A_1|=|A_2|$ on a aussi $\tau_1=\tau_2$. Il en résulte que le système envisagé donne effectivement l'apparence d'une relaxation unique, seul étant observé $\tau_2=q^{-1}(1-k^{1/2}+k)^{-1}$.

(b) La courbe de transition, la variation du temps de relaxation avec la température, la variation des constantes de vitesse de dénaturation et de renaturation k_D et k_R avec la température, présentent des caractéristiques remarquables, qui sont toutes observées avec une précision étonnante pour la chymotrypsine-α [37]. Les Figures 6a et 6b montrent la superposition des courbes théoriques et expérimentales donnant la variation de τ, k_R et k_D avec T^{-1}. On peut noter en particulier, sur la Figure 6a deux asymptotes se croisant en P sur la courbe représentative, et sur la Figure 6b trois asymptotes concourant en un point P' de mêmes coordonnées que P sur la Figure 6a, conformément aux caractéristiques du modèle.

Les points P et P' représentent la mi-transition ($k=K=1$), pour laquelle les trois conformations postulées sont présentes en quantités égales. La comparaison des courbes théoriques et expérimentales fournit en outre les quatre paramètres thermodynamiques et cinétiques [37].

Il est aisé de traiter des systèmes à trois états moins particuliers que le précédent [28], mais leur description requiert des paramètres plus nombreux.

Je me servirai finalement des résultats précédents pour présenter quelques remarques, dont la validité n'est pas limitée au système particulier qu'on vient d'envisager.

(a) Sauf au voisinage de la température de transition T_T, le système précédent présente un seul temps de relaxation au lieu de 2, du fait que sauf pour $T \# T_T$ on a $\tau_1 \# \tau_2$.

Une telle dégénérescence est fréquente. L'exemple le plus connu est celui d'une enzyme allostérique, la D-glycéraldéhyde-3-phosphate-déhydrogénase, étudiée par Eigen et Kirschner (cf. [11]), pour laquelle la fixation du substrat ne fournit que 3 temps de relaxation au lieu des 9 qu'on pourrait attendre.

De même, le modèle d'Ising, lorsqu'il comporte un grand nombre d'éléments, présente non pas 2^N-1 mais seulement 4 temps de relaxation (cf. le paragraphe 2).

(b) Même au voisinage de la température de transition T_T, le système précédent ne montre qu'un seul temps de relaxation, par suite du comportement des amplitudes de relaxation discuté au début de ce paragraphe.

De même, il a été montré récemment [16] pour le modèle d'Ising que, si N est grand, le temps de relaxation le plus long est identique au temps de relaxation moyen τ^* (cf. le paragraphe 2). Dans ce cas, le comportement des amplitudes de relaxation entraîne qu'un seul processus de relaxation contribue à la variation initiale de l'hélicité [16] consécutive à une perturbation brusque (cf. le paragraphe 2).

C'est une règle générale que la connaissance des temps de relaxation doit être complétée par celle des amplitudes. Dans l'étude de l'absorption des ultrasons par les solutions de polypeptides (cf. le paragraphe 3), il a même été possible de parvenir à des conclusions par la seule considération de l'amplitude de la relaxation, alors que les valeurs des temps de relaxation étaient affectées d'une incertitude considérable.

(c) Il est bien connu qu'un accord entre la théorie et l'expérience, réalisé par ajustement d'un nombre élevé de paramètres, est de peu de valeur. De ce point de vue, l'accord illustré par les Figures 6a et 6b n'étant réalisé que si les constantes optiques α_2 et α_3 de la protéine sont égales, ou du moins très voisines l'une de l'autre, il serait utile d'avoir une preuve indépendante de cette égalité.

(d) Si l'on veut néanmoins admettre provisoirement, au vu des Figure 6a et 6b, la validité du modèle à trois états envisagé ci-dessus, *on peut noter que, outre leur intérêt bien connu dans l'étude de problèmes de cinétique, les méthodes dynamiques sont susceptibles de révéler l'existence de conformations intermédiaires stables de protéines.* On peut enfin se demander si de telles conformations sont susceptibles d'intervenir dans certains processus biologiques.

Références

[1] Prigogine, Y.: 1969, dans *Theoretical Physics and Biology* (ed. by Marois), North-Holland Publ. Co., Amsterdam, p. 23.
[2] Zimm, B. H. et Bragg, J. K.: 1959, *J. Chem. Phys.* **31**, 526.
[3] Glauber, R. J.: 1963, *J. Math. Phys.* **4**, 294.
[4] Silberberg, A. et Simha, R.: 1968, *Biopolymers* **6**, 479.
[5] Rabinowitz, P., Silberberg, A., Simha, R., Loftus, E.: 1969, *Adv. Chem. Phys.* **XV**, 281.
[6] Gô, N.: 1967, *J. Phys. Soc. Japan* **22**, 416.
[7] Silberberg, A. et Simha, R.: 1972, *Macromolecules* **5**, 332.
[8] Schwarz, G.: 1971, *Ber. Bunsengesell. Phys. Chemie* **75**, 40.
[9] Schwarz, G.: 1964, *Ber. Bunsengesell. Phys. Chemie* **68**, 843.
[10] Winklmair, D., Engel, J., et Ganser, V.: 1971, *Biopolymers* **10**, 721.
[11] Eigen, M.: 1967, dans *Fast Reactions and Primary Processes in Chemical Kinetics* (ed. by S. Claesson), Interscience, New York, p. 333; Pörschke, D.: 1968, Thèse, Braunschweig; Pörschke, D. et Eigen, M.: 1971, *J. Mol. Biol.* **62**, 361.
[12] Cerf, R.: 1969, *C. R. Acad. Sci Paris* **269-D**, 2140.
[13] Tsong, T. Y., Baldwin, R. L., et Mc Phie, P.: 1972, *J. Mol. Biol.* **63**, 453.
[14] Baldwin, R. L.: 1968, dans *Molecular Associations in Biology* (ed. by B. Pullman), Academic Press, New York, p. 145.
[15] Crothers, D. M., Kallenbach, N., et Davidson, N.: 1968, *J. Am. Chem. Soc.* **90**, 3560.
[16] Schwarz, G.: 1972, *J. Theoret. Biol.* **36**, 569.
[17] Schwarz, G.: 1968, *Biopolymers* **6**, 873.
[18] Zana, R.: 1972, *Suppl. J. Phys.* **33C6**, 108.
[19] Michels, B. et Cerf, R.: 1972, *C.R. Acad. Sci. Paris* **274-D**, 1096.
[20] Michels, B.: 1972, Thèse, Strasbourg.
[21] Zana, R. et Lang, J.: 1970, *J. Phys. Chem.* **74**, 2734.
[22] Tondre, C. et Zana, R.: 1971, *J. Phys. Chem.* **75**, 3367.
[23] Cerf, R.: 1972, *Suppl, J. Phys.* **33C6**, 99.
[24] Saksena, T. K., Michels, B., et Zana, R.: 1968, *J. Chim. Phys.* **65**, 597.
[25] Barksdale, A. D. et Stuehr, J. E.: 1972, *J. Amer. Chem. Soc.* **94**, 3334.
[26] Pörschke, D.: 1971, *Biopolymers* **10**, 1989.
[27] Cerf, R.: 1970, *C.R. Acad. Sci. Paris* **270-C**, 1075.
[28] Cerf, R.: 1971, *Suppl. J. Phys.* **32C5a**, 275.
[29] Fünfschilling, O., Lemaréchal, P., et Cerf, R.: 1970, *C.R. Acad. Sci. Paris* **270-C**, 569; 1971,

Chem. Phys. Letters **12**, 365; Lemaréchal, P.: 1971, Thèse, Strasbourg; Fünfschilling, O.: 1972, Thèse, Strasbourg.
[30] Bauer, H. et Hässler, H.: 1968, dans *6ème Congr. Acoust. Tokyo* J-5-11.
[31] Schellman, J.: 1955, *C.R. Lab. Carlsberg* **29**, 230.
[32] Pohl, F.: 1968, *Eur. J. Biochem.* **4**, 373; 1968, **7**, 146; 1969, *FEBS Letters* **3**, 60.
[33] Schechter, A. N., Chen, R. F., et Anfinsen, C. B.: 1970, *Science* **167**, 886.
[34] Ikai, A. et Tanford, C.: 1971, *Nature* **230**, 100.
[35] Lumry, R., Biltonen, R., et Brandts, J.: 1964, *Biopolymers* **4**, 917.
[36] Tanford, C.: 1968, *Advan. Protein Chem.* **23**, 121.
[37] Cerf, R.: 1970, *C. R. Acad. Sci. Paris* **271-D**, 2403; 1971, **272-D**, 747.
[38] Tsong, T. Y., Baldwin, R. L., et Elson, E. L.: 1971, *PNAS* **68**, 2712.
[39] Eigen, M. et De Maeyer, L.: 1963, dans *Techniques of Organic Chemistry*, (S. F. Friess, E. S. Lewis, and A. W. Weissberger, eds.) Interscience, New York, VIII-2, p. 895.
[40] Thusius, D.: 1972, *J. Am. Chem. Soc.* **94**, 356.

DISCUSSION

H. Buc: Un modèle à trois états qui dégénère en un seul temps de relaxation exige des restructures très fortes qui ne devraient pas être maintenues à tous les pH et toutes les températures. En est-il ainsi?

Cerf: Le fait qu'un seul temps de relaxation soit observable (le plus long, τ_2) est une propriété générale d'un système à trois états lorsque les deux étapes sont identiques des points de vue thermodynamique et cinétique. Moyennant cette condition d'identité des étapes, les deux temps de relaxation τ_1 et τ_2 sont égaux dans le domaine des 'hautes températures' comme dans celui des 'basses températures', et de plus (à l'exception du cas où l'on aurait $\alpha_1 = \alpha_3$), l'amplitude de relaxation A_1 associée à τ_1 est négligeable devant A_2 lorsque le rapport τ_2/τ_1 est voisin de sa valeur maximale 3. Dans ces conditions il serait très difficile d'observer plus d'un temps de relaxation.

Une condition qui me paraît beaucoup plus restrictive est que les constantes optiques α_2 et α_3 des deux formes dénaturées de la protéine soient égales ou du moins très voisines l'une de l'autre. Cette condition, que j'ai envisagée pour la chymotrypsine-α et dont j'ai dit qu'il est souhaitable d'en obtenir une confirmation, est déterminante, non pas pour la possibilité d'observer un ou deux temps de relaxation, mais pour le comportement de la constante de vitesse de renaturation à 'haute température'.

M. Lazdunski: L'interprétation des données de Pohl sur la chymotrypsine me paraît assez difficile. Les expériences sont conduites à pH 2,1. A pH 2,1 il y a équilibre entre trois formes de la protéine: deux formes globulaires I et II et une forme 'dénaturée' D; $I + H^+ \rightleftharpoons II$. La transition entre les deux formes globulaires est très probablement contrôlée par le 'démasquage' de la chaîne latérale carboxylique de Asp 194. Il y a, à pH 2,1, environ 50% de forme I et 50% de forme II. L'élévation de température transforme ce mélange de formes globulaires en forme dénaturée D. La transition $I \rightleftharpoons II$ se fait en quelques millièmes de seconde, la transition forme 'native' N(I ou II) $\rightarrow$ forme 'dénaturée' D se fait en des temps plus longs (de la seconde à la minute).

Cerf: Votre remarque, de portée générale, ne me semble pas affecter l'interprétation des résultats de Pohl, dans le cas particulier que j'ai envisagé. J'ai souligné d'une part que je considérais le seul temps long (de l'ordre de la seconde à la minute) et d'autre part que j'envisageais plus particulièrement les résultats concernant la chymotrypsine-α. D'après la valeur numérique que vous donnez pour la constante de vitesse associée à la transition $I \rightleftharpoons II$ de cette protéine, il est très probable que les deux états précédents sont 'vus' dans les expériences de Pohl comme une seule conformation.

Si mes conclusions déduites des données cinétiques de Pohl sont correctes, il y a donc, dans les conditions expérimentales envisagées, essentiellement deux conformations autres que les formes I et II. Je ne considère pas pour autant l'interprétation des résultats de Pohl comme définitivement acquise, même dans le cas de la chymotrypsine-α, puique j'ai fait moi-même des restrictions à ce sujet.

Lifson: In calculating $\partial v/(\partial \ln s)$ you used the partition function in its form appropriate for infinitely ong chains. By using the matrix method, it is possible to calculate the above expression as a function of the chain length for small and intermediate chain lengths. Since you applied the calculations for the study of chains of finite length, do you estimate the difference negligible or is it possible that it would be significant enough to justify the extra calculational work involved in the more cumbersome expressions for finite chains?

Cerf: The experimental fact is the shift in the pH scale with respect to the mid-transition of the maximum of the conformational contribution to the ultrasonic absorption in solutions of polypeptides. Michels's measurements seem to establish clearly this shift for the sample of molecular weight 12000, while the effect is at the limit of experimental errors for the sample of molecular weight 90000. I have given theoretical reasons to think that the shift is due to a contribution of the fluctuations in the number of uninterrupted helical sequences per molecule. If this explanation is correct, one would anticipate the effect to be larger at lower molecular weights, as it is the case. Thus, it would be highly valuable to have a complete kinetic theory for finite molecular weights; but this would involve more calculations than only using the expression of $\partial\nu/(\partial \ln s)$ for finite chains.

Note ajoutée sur l'épreuve au sujet de la réponse à Lifson. I made a false statement when saying that I had used the expression of the number of uninterrupted helical sequences per molecule, ν, valid at infinite chain length, and had applied it to a sample of low molecular weight. In fact, Lifson and Roig's result, derived from the matrix method, which is valid at any chain length, has been used. Thus the plot of $\partial\nu/(\partial \ln s)$ versus s that had been shown on a slide is directly comparable with the measurements. The corresponding figure is not contained in the present paper, but in reference [23, (Figure 10)]. Otherwise, my reply to Dr Lifson remains valid.

THE ANALYSIS OF CHEMICAL RELAXATION AMPLITUDES AND SOME APPLICATIONS TO REACTIONS INVOLVING MACROMOLECULES

DARWIN THUSIUS, GEORGES FOUCAULT, and FLORENT GUILLAIN

Laboratoire d'Enzymologie Physico-chimique et Moléculaire, 91405 Orsay, France

Summary. Chemical relaxation amplitudes contain information on elementary molecular and thermodynamic parameters which often cannot be obtained from static experiments. In this communication an experimental strategy is outlined for the practical exploitation of relaxation amplitude data. The method is based on a theoretical formalism which considerably reduces the computational labour involved in the derivation of amplitude expressions. Applications involving the binding of small ligands to biomacromolecules are given.

Résumé. Les amplitudes de relaxation chimique contiennent des informations sur des paramètres moléculaires et thermodynamiques qui souvent ne peuvent pas être obtenues à partir des expériences statiques. Dans cette communication une stratégie expérimentale pour l'exploitation pratique des amplitudes est presentée. La méthode est basée sur un formalisme théorique qui simplifie beaucoup le développement des expressions algébriques des amplitudes. Des applications concernant la fixation des ligands sur des biomacromolecules sont données.

1. Introduction

In recent years chemical relaxation techniques have been employed to elucidate the dynamics of a variety of reactions of biological interest [1–3]. In these techniques the reactants are mixed and the solution is allowed to reach chemical equilibrium; some external parameter such as temperature or pressure is then rapidly changed and the rate of approach of the reactant concentrations to their new equilibrium values is followed. This method permits the direct observation of reaction intermediates and the study of elementary reactions characterized by half lives as small as 10^{-9} s.

A relaxation experiment yields two different types of parameters: exponential decay times known as relaxation times, and initial signal changes, known as relaxation amplitudes. To date most applications of chemical relaxation have been concerned exclusively with the interpretation of relaxation *times*. Since a mechanism can be proposed and specific rate constants estimated from decay times alone, it is often concluded that the supplementary information contained in the amplitudes is not worth the effort of deriving and solving the numerically complex analytical expressions required for an analysis of initial signal changes.

In this communication we discuss the types of information which can be gleaned from relaxation amplitudes and we outline a method which allows one to conveniently obtain this information in practice. The experimental strategy is based on a theoretical formalism which considerably reduces the amount of mathematical labour involved in the calculation of amplitude expressions [4, 5]. The method is applied to several systems involving the binding of a small ligand to a biological macromolecule.

C. Sadron (ed.), Dynamic Aspects of Conformation Changes in Biological Macromolecules. 271–284.

2. One-Step Reactions

We first consider the amplitude analyses of simple one-step reaction systems. Table I gives the time-dependent absorbance change following the instantaneous physical perturbation of a reversible one-step chemical transformation.

TABLE I

Time-dependent absorbance change following the rapid physical perturbation of a general one-step reaction. The exponential term τ is the relaxation time and the initial signal change δOD^0 is the relaxation amplitude. Equilibrium concentrations are denoted by bars

$$\nu_1 B_1 + \nu_2 B_2 \underset{k_{21}}{\overset{k_{12}}{\rightleftharpoons}} \nu_3 B_3 + \nu_4 B_4$$

$$\delta OD_t = \delta OD^0 e^{-t/\tau}$$

$$1/\tau = k_{12} [\nu_1^2/\bar{B}_1 + \nu_2^2/\bar{B}_2] \bar{B}_1^{\nu_1} \bar{B}_2^{\nu_2} + k_{21} [\nu_3^2/\bar{B}_3 + \nu_4^2/\bar{B}_4] \bar{B}_3^{\nu_3} \bar{B}_4^{\nu_4}$$

It will be noted that the relaxation time is a function of specific rate constants and equilibrium concentrations only; it is independent of the perturbation technique and the means employed to detect the relaxation. In practice the relation between the relaxation time and equilibrium concentrations is established experimentally, and the two specific rate constants are estimated with the appropriate form of the indicated analytical expression.

TABLE II

Observed relaxation amplitude of a general one-step reaction. The optical path is assumed to be 1 cm

$$\nu_1 B_1 + \nu_2 B_2 \rightleftharpoons \nu_3 B_3 + \nu_4 B_4$$

$$\delta OD^0 = (\Delta\varepsilon \cdot \delta K/K) \Gamma$$

$$\Delta\varepsilon = \nu_3 \varepsilon_3 + \nu_4 \varepsilon_4 - \nu_2 \varepsilon_2 - \nu_1 \varepsilon_1$$

$$\delta K/K = (\Delta H/RT) \delta T/T$$

$$\Gamma = [(\nu_1^2/\bar{B}_1) + (\nu_2^2/\bar{B}_2) + (\nu_3^2/\bar{B}_3) + (\nu_4^2/\bar{B}_4)]^{-1}$$

The amplitude corresponding to the scheme of Table I is given in Table II. The relaxation amplitude is a function of molecular parameters, thermodynamic parameters and equilibrium concentrations. The expressions of Table II follow from the general formalism discussed below, or may be derived directly from mass action and mass conservation relations. Unlike the relaxation time, the relaxation amplitude is defined in terms of the experimental technique.

It is clear that the amplitude of a one-step process is the product of a concentration-independent factor, $\Delta\varepsilon \cdot \delta K/K$, and a concentration-dependent factor, Γ. The term $\Delta\varepsilon$ will be referred to as the 'specific signal change' of the chemical reaction. The specific signal change of a given model is a weighted algebraic sum of individual molar extinction coefficients, which can be written explicitly by an examination of the stoichiometry. The term $\delta K/K$ is proportional to an extensive thermodynamic variable. In a temperature jump experiment $\delta K/K$ is proportional to the enthalpy change. Finally, the factor Γ gives the concentration dependence of the observed amplitude [6]. Like the specific signal change, the Γ factor of a given model may be calculated by inspection.

It is a fortunate circumstance for the experimenter that the amplitude is a linear function of a concentration variable. Assume that the overall stoichiometry and equilibrium constant of a chemical transformation have been determined with a static technique. Assume in addition that a single relaxation process is observed. If the Γ factors are calculated with the static equilibrium constant, the amplitude data can be numerically fit to the macroscopic model with a linear least-squares method. In this approach the amplitude data yield a statistically refined value for the 'regression constant', $\Delta\varepsilon \cdot \delta K/K$. The regression constant allows one to calculate a theoretical line representing the 'best fit' of the amplitudes to the model. If the dispersion of the experimental points about the calculated line is homogeneous, it can be said that the relaxation amplitudes are consistent with the overall stoichiometry.

It is well known that one cannot 'prove' a reaction mechanism. One can only hope to show that all available experimental data are consistent with a chemically reasonable model. Even for relatively simple systems it is therefore important to demonstrate that the same mechanism can be satisfactorily fit to equilibrium parameters, relaxation amplitudes and relaxation times.

If a good fit of the amplitude data is obtained with the above procedure, one can attempt to estimate either the specific signal change or the enthalpy change from the experimental regression constant. If, for example, $\Delta\varepsilon$ has been determined by spectrophotometric titration, the amplitude data yield an estimate of ΔH. If ΔH is already known from a calorimetric study, $\Delta\varepsilon$ can be evaluated. Specific signal changes and enthalpy changes provide insight into the molecular nature of the chemical transformation being studied.

The above strategy has been used in a temperature-jump study of the interaction between proflavin and trypsin [7]. A substantial shift occurs in the visible spectrum of proflavin when it binds to the enzyme [8, 9]. A spectrophotometric titration has shown that a 1:1 complex is formed, and the overall equilibrium constant has been determined [7].

TABLE III

Relaxation amplitude expression for 1:1 complex formation

$$\mathrm{A} + \mathrm{B} \rightleftharpoons \mathrm{AB}$$
$$\delta \mathrm{OD}^0 = (\Delta\varepsilon \cdot \delta K/K)\,\Gamma$$
$$\Delta\varepsilon = \varepsilon_{\mathrm{AB}} - \varepsilon_{\mathrm{A}} - \varepsilon_{\mathrm{B}}$$
$$\delta K/K = (\Delta H/RT)\,\delta T/T$$
$$\Gamma = [(1/\bar{A}) + (1/\bar{B}) + (1/\overline{AB})]^{-1}$$

At neutral pH the relaxation spectrum consists of a single exponential. The concentration dependence of the relaxation time revealed that (1) the kinetics are second-order, (2) the association rate constant is close to a diffusion-controlled value and (3) the ratio between the association and dissociation rate constants is in good agreement with the static equilibrium constant. These results suggest that ligand binding takes place in a simple one-step process and imply that the enzyme–ligand complex exists in essentially one conformational state.

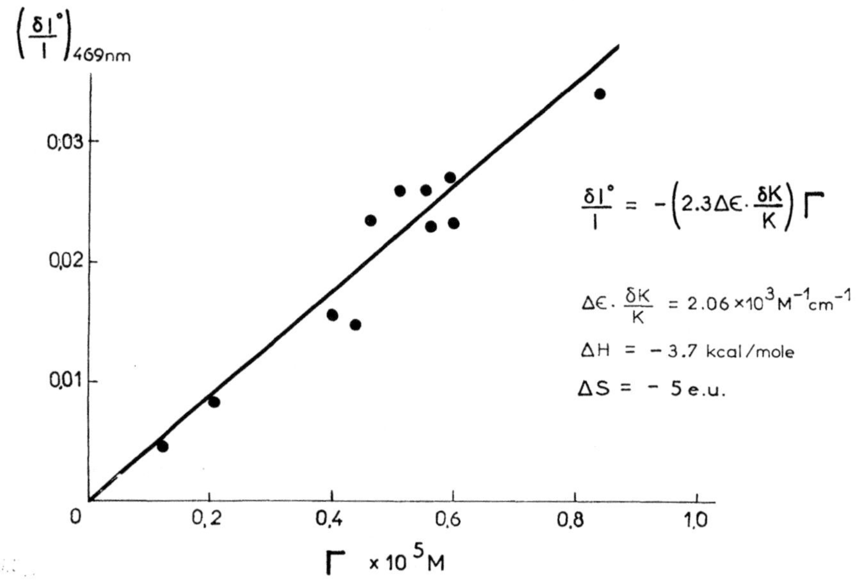

Fig. 1. Relaxation amplitude analysis of a one-step reaction system: the binding of proflavin to trypsin at neutral pH. Experimental amplitudes are expressed as initial transmission changes. The line was calculated with the least-squares estimate of $\Delta\varepsilon \cdot \delta K/K$. The specific signal change was obtained by spectrophotometric titration.

This conclusion is considerably strengthened when the relaxation amplitudes are taken into account. The pertinent Γ factors (Table III) were calculated with the static equilibrium constant and the data were treated numerically with a least-squares analysis. Figure 1 shows that the experimental amplitudes are in very good agreement with the model suggested by the kinetic data. Furthermore, the amplitude-derived enthalpy change is close to the value estimated from a classical van 't Hoff plot [10]. Finally, the values of ΔH and ΔS are entirely compatible with a simple ligand binding reaction.

3. Overall Relaxation Amplitudes: Theory

In the general case where the reaction mechanism consists of several elementary steps, the relaxation spectrum will in principle be characterized by several exponentials.

$$\delta \mathrm{OD}_t = \delta \mathrm{OD}_1^0 e^{-t/\tau_1} + \delta \mathrm{OD}_2^0 e^{-t/\tau_2} + \cdots + \delta \mathrm{OD}_R^0 e^{-t/\tau_R}, \tag{1}$$

where R is the number of thermodynamically independent elementary steps.

The experimenter who wants to quantitatively interpret the amplitudes of a multi-step reaction system is obliged to derive equations relating experimental signal changes to the specific signal changes and enthalpy changes of R elementary steps. Although numerically complex, these expressions may be obtained in a straightforward way with the schematic method developed in our laboratory.

Let us first consider the calculation of the *overall* amplitude of a mechanism involving an arbitrary number of elementary steps. The overall amplitude is the sum of all individual amplitudes, or

$$\delta \mathrm{OD}_{\mathrm{tot}}^0 = \sum_{i=1}^{R} \delta \mathrm{OD}_i^0, \tag{2}$$

and is therefore a function of the enthalpy changes and specific signal changes of all elementary reactions.

It can be shown that the overall amplitude may be written as the linear sum appearing in Table IV [5]. The constants b are related to elementary specific signal changes and thermodynamic functions; the X's are functions of equilibrium concentrations. Rules for calculating the X's are given elsewhere [5].

The formalism of Table IV suggests the following experimental procedure for fitting overall relaxation amplitudes to a multi-step model. In the relaxation experiment the reactant concentrations are varied over the widest possible concentration ranges and the corresponding overall amplitudes are recorded. If the microscopic equilibrium constants are known, the X's are calculated directly from analytical concentrations.

TABLE IV

Overall amplitude of a general reaction mechanism. The quantities $\Delta\varepsilon_\alpha$ and $\delta K_\alpha/K_\alpha$ denote the specific signal change and thermodynamic parameter of the αth elementary reaction

$$\begin{aligned} \delta \mathrm{OD}_{\mathrm{tot}}^0 = b_{11}X_{11} + b_{12}X_{12} + &\cdots + b_{1R}X_{1R} \\ + b_{22}X_{22} + &\cdots + b_{2R}X_{2R} \\ &\;\;\vdots \\ &+ b_{RR}X_{RR} \end{aligned}$$

$$b_{\alpha\alpha} = \Delta\varepsilon_\alpha \cdot \Delta K_\alpha / K_\alpha$$
$$b_{\alpha\beta} = \Delta\varepsilon_\alpha \cdot \Delta K_\beta / K_\beta + \Delta\varepsilon_\beta \cdot \Delta K_\alpha / K_\alpha$$

1 cm path length.
R = number of independent elementary reactions.

The regression constants b are then estimated numerically from the amplitudes and calculated concentration variables with a computer-assisted multilinear least-squares treatment. Finally, a theoretical set of amplitudes is calculated with the regression coefficients, and the differences between the calculated and observed data are examined.

When a satisfactory fitting is achieved, a relaxation amplitude analysis offers the possibility of estimating the elementary parameters $\Delta\varepsilon$ and $\delta K/K$ from the regression coefficients b. However, the estimation of these constants requires an independent experiment, as already seen for one-step reactions. Even if a complete analysis cannot be realized, it is of course still of interest to employ the present method with the objective of demonstrating that the concentration dependence of the amplitudes is consistent with the assumed mechanism.

If the microscopic equilibrium constants required in the above strategy are not known, they may often be estimated directly from the concentration dependence of the relaxation amplitudes with nonlinear regression [11, 12]. Another approach, discussed later in this paper, is to calculate the equilibrium constants from specific rate constants obtained in a kinetic analysis.

4. Overall Amplitudes: A Practical Application

A two-step reaction mechanism of general interest is the system of two competing second-order reactions illustrated in Table V. For example, D and I could be inhibitors which bind to the same site of enzyme E. The analytical expression of Table V follows directly from the general formalism of Table IV, where the concentration variables have been computed by examination [4, 5]. If the two equilibrium constants are known, the model can be fit to the overall amplitudes with a three-term least-squares method.

A limiting case of the Table V scheme which has been discussed recently in the literature [4, 7, 12, 13], is the system where the only absorbing species are D and ED.

TABLE V

System of two coupled second-order reactions. The Γ factors refer to individual steps: $\Gamma_1 = [(1/\bar{E}) + (1/\bar{D}) + (1/\overline{ED})]^{-1}$ and $\Gamma_2 = [(1/\bar{E}) + (1/\bar{I}) + (1/\overline{EI})]^{-1}$. The specific signal changes are $\Delta\varepsilon_1 = \varepsilon_{ED} - \varepsilon_E - \varepsilon_D$ and $\Delta\varepsilon_2 = \varepsilon_{EI} - \varepsilon_E - \varepsilon_I$

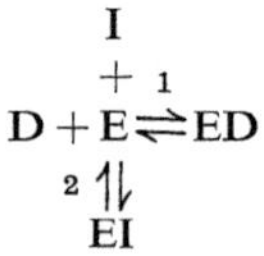

$$\delta OD_{tot}{}^0 = b_{11}\left[\frac{\Gamma_2{}^{-1}}{\Gamma_1{}^{-1}\Gamma_2{}^{-1} - \bar{E}^{-2}}\right] - b_{12}\left[\frac{\bar{E}^{-1}}{\Gamma_1{}^{-1}\Gamma_2{}^{-1} - \bar{E}^{-2}}\right] + b_{22}\left[\frac{\Gamma_1{}^{-1}}{\Gamma_1{}^{-1}\Gamma_2{}^{-1} - \bar{E}^{-2}}\right]$$

$$b_{11} = \Delta\varepsilon_1 \cdot \delta K_1/K_1$$
$$b_{12} = \Delta\varepsilon_1 \cdot \delta K_2/K_2 + \Delta\varepsilon_2 \cdot \delta K_1/K_1$$
$$b_{22} = \Delta\varepsilon_2 \cdot \delta K_2/K_2$$

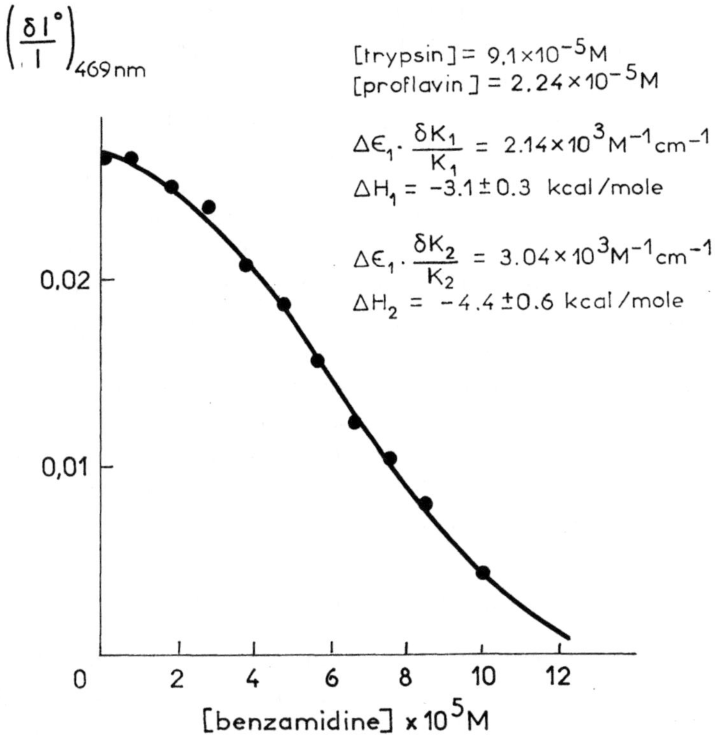

Fig. 2. Relaxation amplitude titration of the trypsin–proflavin complex by benzamidine. Conditions: 0.14 M tris-Cl^-, pH 8.1, 0.020 M $CaCl_2$, 469 nm, $T = 14°C$, $\delta T = 4.4°C$.

That is, a system where a chromophoric inhibitor D acts as an indicator for the binding of a nonchromophoric inhibitor, I. In terms of the general expression of Table V, this corresponds to the condition $\Delta\varepsilon_2 \to 0$. The overall amplitude equation then simplifies to the expression of Table VI.

An example of two competing second-order reactions of this type is where D = pro-

TABLE VI

System of two coupled second-order reactions in which D and ED are the only absorbing species

$$\begin{array}{c} \mathrm{I} \\ + \;\; {\scriptstyle 1} \\ \mathrm{D} + \mathrm{E} \rightleftharpoons \mathrm{ED} \\ {\scriptstyle 2} \upharpoonleft\downharpoonright \\ \mathrm{EI} \end{array}$$

limiting case: $\Delta\varepsilon_1 = \varepsilon_{ED} - \varepsilon_D, \qquad \Delta\varepsilon_2 = 0$

$$\delta OD_{tot}{}^{o} = \Delta\varepsilon_1 \cdot \delta K_1/K_1 \left[\frac{\Gamma_2{}^{-1}}{\Gamma_1{}^{-1}\Gamma_2{}^{-1} - \bar{E}^{-2}}\right] - \Delta\varepsilon_1 \cdot \delta K_2/K_2 \left[\frac{\bar{E}^{-1}}{\Gamma_1{}^{-1}\Gamma_2{}^{-1} - \bar{E}^{-2}}\right]$$

flavin, I = benzamidine and E = trypsin. Figure 2 gives the results of a recently reported [4] 'relaxation amplitude titration' of the proflavin–trypsin complex by benzamidine. A least-squares analysis of the experimental data according to the equation of Table VI gave the indicated values of the regression constants $\Delta\varepsilon_1 \cdot \delta K_1/K_1$ and $\Delta\varepsilon_1 \cdot \delta K_2/K_2$. These latter parameters were then employed to calculate the curve of Figure 2.

It is obvious from Figure 2 that the two-step model of Table VI is in excellent agreement with the amplitude results. In general one would not expect a good correlation if ED or EI existed in more than one conformational state or if the ternary complex EDI was present at appreciable concentration. Furthermore, the enthalpy changes of Figure 2 are in good agreement with the value of ΔH_1 estimated in the absence of benzamidine (Figure 1) and a value of ΔH_2 obtained by direct calorimetry.

Actually the concentration dependence of the relaxation *times* of the trypsin–proflavin–benzamidine reaction was the initial experimental evidence leading to the model of Table VI [7]. However, we note that the overall amplitudes provide a rigorous mechanistic test in the sense that they can be measured over a much larger concentration range than the time constants [4].

5. Individual Relaxation Amplitudes

In favorable cases the perturbation of a multi-step system will yield R individual amplitudes (cf. Equation [1]). Individual amplitudes are determined by spreading out the total signal change on the time axis; they therefore often contain mechanistic information which cannot be gleaned from the overall amplitude alone.

It has been shown that the concentration dependence of widely separated individual amplitudes ($\tau_{i-1} \ll \tau_i \ll \tau_{i+1}$) can be computed from linear relations analogous to the formalism of Table IV [5]. If the amplitudes are numbered $\delta\mathrm{OD}_1^0$, $\delta\mathrm{OD}_2^0, \ldots, \delta\mathrm{OD}_R^0$, where 1 denotes the fastest and R the slowest relaxation phenomenon, $\delta\mathrm{OD}_i^0$ will in general be a linear sum of $i(i+1)/2$ terms written as:

$$\begin{aligned}\delta\mathrm{OD}_i^0 = b_{11}x_{11} &+ b_{12}x_{12} + \cdots + b_{1i}x_{1i} \\ &+ b_{22}x_{22} + \cdots + b_{2i}x_{2i} \\ &\quad \ddots \qquad \vdots \\ &\qquad\qquad + b_{ii}x_{ii}\,,\end{aligned} \tag{3}$$

where the $b_{\alpha\beta}$ are defined in Table IV. The concentration variables $x_{\alpha\beta}$ are somewhat more complex than for overall amplitudes, but their computation is still straightforward.

We now consider the individual relaxation amplitudes of a specific example. The mechanism of Table VII is of quite considerable interest in biology since it may represent the rapid binding of a ligand followed by a conformational change. The formal kinetic analysis of this model has been discussed often in the literature, and a number of practical examples have been cited [2, 3, 6, 14]. Although several theoretical discussions on the amplitudes of this system have also appeared [3, 4, 6, 14], detailed practical applications are rare [14b].

The mechanism of Table VII is in general characterized by two relaxations. The amplitude of the fastest effect ($\tau_1 \ll \tau_2$) is

$$\delta \mathrm{OD}_1^0 = b_{11}\Gamma_1 = \Delta\varepsilon_1 \cdot \delta K_1/K_1 \frac{A^0}{(1 + B^0K_{12})(1 + (B^0K_{12})^{-1})}, \tag{4}$$

and the amplitude of the second effect is given in Table VII. These expressions are readily derived with our schematic method and assuming component B to be in excess over component A.

TABLE VII

Rapid ligand binding followed by a slow isomerism. The analytical expression refers to the amplitude of the slow relaxation phenomenon. In fluorescence measurements the constants $b_{\alpha\beta}$ are formally equivalent to those of Table V, but here $\Delta\varepsilon_\alpha$ is replaced by $\Delta\phi_\alpha$, where $\Delta\phi_1 = \gamma(q_{AB}\varepsilon_{AB} - q_A\varepsilon_A - q_B\varepsilon_B)$, $\Delta\phi_2 = \gamma(q_{AB^*}\cdot\varepsilon_{AB^*} - q_{AB}\cdot\varepsilon_{AB})$. The terms q_i and ε_i denote respectively the quantum yield and molar extinction coefficient (excitation wavelength) of species i; γ is an instrument factor. These definitions of the $\Delta\phi_\alpha$ are valid for ideal measuring conditions which are not always realized in practice [16]

$$\mathrm{A} + \mathrm{B} \underset{\text{very fast}}{\overset{1}{\rightleftharpoons}} \mathrm{AB} \underset{\text{slow}}{\overset{2}{\rightleftharpoons}} \mathrm{AB}^*$$

limiting case: $B^0 \gg A^0$

$$\delta F_2{}^0/A^0 = b_{11}X_{11} + b_{12}X_{12} + b_{22}X_{22}$$

$$X_{11} = \{(1 + K_{ap}B^0)[1 + (K_{ap}B^0)^{-1}]\}^{-1} - \{[1 + (1 + K_{23})(K_{ap}B^0)^{-1}](1 + K_{ap}B^0)\}^{-1}$$

$$X_{12} = \{(1 + K_{23}{}^{-1})(1 + K_{ap}B^0)[1 + (K_{ap}B^0)^{-1}]\}^{-1}$$

$$X_{22} = (K_{ap}B^0 + K_{23} + 1)\{(1 + K_{23})(1 + K_{23}{}^{-1})(1 + K_{ap}B^0)[1 + (K_{ap}B^0)^{-1}]\}^{-1}$$

Although numerically complex, the concentration variables of Table VII contain only three parameters, i.e., B^0, K_{ap} and K_{23}, where B^0 is the analytical concentration of species B, K_{ap} is the overall equilibrium constant and K_{23} is the isomerization equilibrium constant.

$$\begin{aligned} K_{ap} &= (\overline{AB} + \overline{AB}^*)/\bar{A}\cdot\bar{B} = K_{12}(1 + K_{23}) \\ K_{23} &= \overline{AB}^*/\overline{AB} \\ K_{12} &= \overline{AB}/\bar{A}\cdot\bar{B}. \end{aligned} \tag{5}$$

It is well known that the microscopic equilibrium constants K_{12} and K_{23} are not accessible to static experiments. Unlike the preceding examples, the fitting of amplitudes in this application requires supplementary data which are beyond the scope of a classical titration.

In those cases where a complete kinetic analysis is feasible, ratios between specific rate constants lead to estimates of all microscopic equilibrium constants. This concept is illustrated for the present model with the theoretical profiles of Figure 3. It is clear

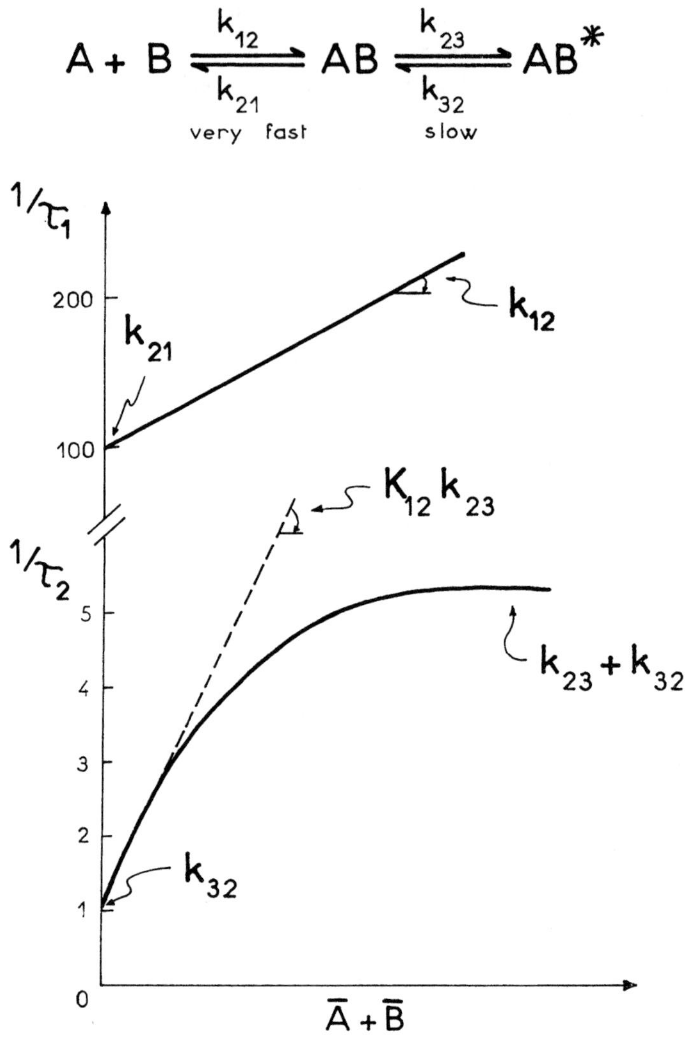

Fig. 3. Relaxation times for a rapid binding reaction followed by an isomerism.

that the constants K_{12} and K_{23} required in the fitting of the amplitudes can in principle be estimated from the concentration dependence of the relaxation times.

A reaction which very probably takes place by the mechanism of Table VII is the binding of proflavin to DNA. A large body of experimental evidence indicates that the binding of proflavin involves the formation of an 'intercalated' complex in which the planar dye molecule is bound between two adjacent base pairs.

$$P + DNA \underset{k_{21}}{\overset{k_{12}}{\rightleftharpoons}} P_{out} \underset{k_{32}}{\overset{k_{23}}{\rightleftharpoons}} P_{in}. \tag{6}$$

Li and Crothers [15] have shown that when changes in proflavin absorbance are followed, the temperature-jump relaxation spectrum consists of two resolvable exponentials. The concentration dependence of the relaxation times are consistent with the mechanism of Equation (6), and the four specific rate constants have been evaluated [15]. We are presently investigating the proflavin–DNA reaction with a fluorescence temperature-jump method. Fluorescence detection allows us to work with a large excess of DNA over proflavin, which simplifies the problem of stoichiometry arising from the uncertainty in the number of proflavin binding sites [15].

One of the main objectives of our work on this system is to estimate elementary thermodynamic and molecular parameters from the concentration dependence of the relaxation amplitudes. For example, structural considerations of the model of Equation (6) suggest that the quenching of proflavin fluorescence observed in static binding studies is due entirely to the formation of the intercalated species [17, 18], i.e., $q_P \sim q_{P_{out}} \gg q_{P_{in}}$, where q denotes quantum yield. This hypothesis can be tested directly by analyzing the amplitudes of the kinetic decay curves.

Fluorescence detection reveals at least two relaxation phenomena. However, to date we have only been able to resolve the slowest process. Figure 4 presents the con-

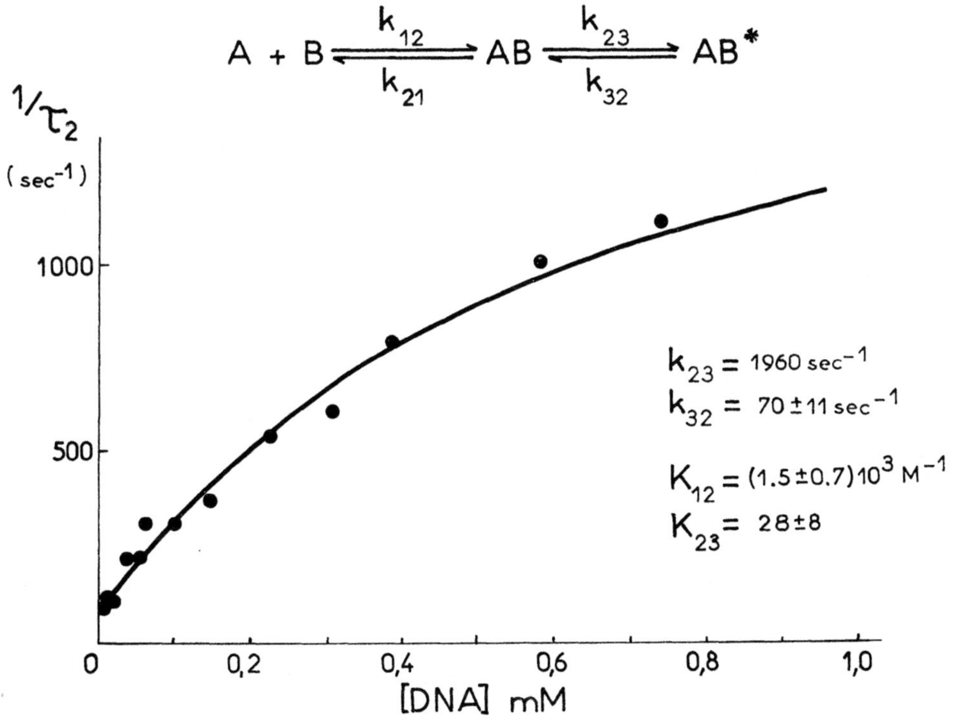

Fig. 4. Relaxation times of the slow kinetic phase observed in the binding of proflavin to DNA. The rate and equilibrium constants were evaluated with nonlinear regression. The line is theoretical. Conditions: 0.2 M $NaNO_3$, 0.01 M phosphate, pH 7.0, 1 mM EDTA, final temperature 13.5° C, excitation wavelength 436 nm.

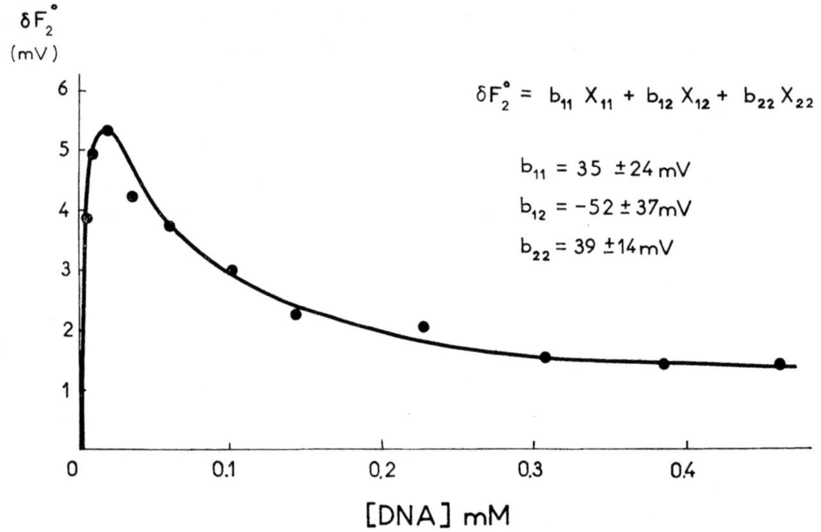

Fig. 5. Relaxation amplitudes of the slow kinetic phase observed in the binding of proflavin to DNA. The line is theoretical. Amplitude values refer to a proflavin concentration of 5×10^{-7} M.

centration dependence of the measured relaxation time. The calculated curve is in very good agreement with the assumed model, and the elementary parameters are in reasonable agreement with the values reported by Li and Crothers [15].

The amplitude data appear in Figure 5. The linear least-squares fitting was performed with the K_{23} value of Figure 4 and a K_{ap} value of $6 \times 10^4\ \mathrm{M}^{-1}$ determined by fluorometric titration. It is clear from Figure 5 that the experimental amplitudes are also entirely consistent with the intercalation model.

The large standard errors of the regression coefficients indicate that a range of b values would fit the data equally well. This observation is readily rationalized in terms of the assumed model. It will be noted from Figure 4 that the intercalated form of proflavin is much more populated than the 'outside' form, or $K_{23} \gg 1$. Substitution of this limiting condition in the general expression of Table VII reveals that in the present experiments the second amplitude is defined to a first approximation by only two linearly independent concentration variables (Table VIII). The uncertainties in the $b_{\alpha\beta}$

TABLE VIII

Relaxation amplitude expression for the limiting case in the experiments of Figures 4 and 5

$$\mathrm{A} + \mathrm{B} \underset{\text{very fast}}{\overset{1}{\rightleftharpoons}} \mathrm{AB} \underset{\text{slow}}{\overset{2}{\rightleftharpoons}} \mathrm{AB}^*$$

limiting case: $B^0 \gg A^0$

$$K_{23} = \overline{AB}^*/\overline{AB} \gg 1$$

$$\delta F_2{}^0/A^0 \sim (b_{11} + b_{12} + b_{22})\,\{(1 + K_{\mathrm{ap}}B^0)\,[1 + (K_{\mathrm{ap}}B^0)^{-1}]\}^{-1} + b_{22}\,\{[1 + (K_{\mathrm{ap}}B^0)^{-1}]^2 (K_{23})\}^{-1}$$

are therefore due to a 'degenerate' form of the amplitude equation which obtains for large K_{23} values.

The regression coefficients of Figure 5 contain information on quantum yields. extinction coefficients and elementary enthalpy changes (cf. caption to Table VII), However, in view of the sizable standard errors, the estimation of these parameters from the present amplitude data is rather tenuous.

It is known that at ionic strengths lower than employed in the present experiment, K_{23} approaches unity [15]. By studying the dynamics of proflavin binding at lower ionic strength we hope to eliminate the degeneracy illustrated in Table VIII and realize a complete amplitude analysis of this system.

Acknowledgement

This work was supported by grants from the Centre National de la Recherche Scientifique awarded to the laboratory of Professor Jeannine Yon.

References

[1] Eigen, M.: 1967, in *Fast Reactions and Primary Processes in Chemical Kinetics* (ed. by S. Claesson), Wiley, New York.
[2] Hammes, G. G.: 1968, *Advan. Protein Chem.* **23**, 1.
[3] Hammes, G. G. and Schimmel, P. R.: 1970, in *The Enzymes*, Part 2 (ed. by P. D. Boyer), Academic Press, New York, p. 67.
[4] Thusius, D.: 1972, *J. Am. Chem. Soc.* **94**, 356.
[5] Thusius, D.: submitted for publication.
[6] Eigen, M. and De Maeyer, L.: 1963, in *Techniques of Organic Chemistry* (ed. by S. F. Friess, E. S. Lewis, and A. W. Weissberger), 2nd ed., Vol. 8, Wiley, New York, Part 2.
[7] Guillain, F. and Thusius, D.: 1970, *J. Am. Chem. Soc.* **92**, 5534.
[8] Glazer, A. N.: 1965, *Proc. Nat. Acad. Sci. US* **54**, 171.
[9] Bernhard, S. A. and Gutfreund, H.: 1965, *Proc. Nat. Acad. Sci. US* **53**, 1238.
[10] Thusius, D.: unpublished.
[11] Thusius, D.: 1973, *Biochimie* **55**, 277.
[12] Eigen, M. and Winkler, R.: 1970, in *The Neuro-Sciences: Second Study Program* (ed. by F. O. Schmitt, Rockefeller University Press, New York, p. 685.
[13] Czerlinski, G. H.: 1969, in *Theoretical and Experimental Biophysics*, Vol. 2 (ed. by A. Cole), M. Dekker, New York, p. 106.
[14a] Czerlinski, G. H.: 1966, *Chemical Relaxation*, M. Dekker, New York.
[14b] Czerlinski, G. H. and Schreck, G.: 1964, *J. Biol. Chem.* **239**, 913. A practical amplitude analysis of the Table VII reaction is found in this reference.
[14c] Czerlinski, G. H.: 1964, *J. Theor. Biol.* **7**, 463.
[15] Li, H. J. and Crothers, D. M.: 1969, *J. Mol. Biol.* **39**, 461.
[16] For example, if the excitation light is partially polarized, anisotropy terms figure in the $b_{\alpha\beta}$. Tom Jovin, to be published.
[17] Heilweil, H. G. and Van Winkle, Q.: 1955, *J. Phys. Chem.* **59**, 939.
[18] Tubbs, R. K., Ditmars, Jr. W. E., and Van Winkle, Q.: 1964, *J. Mol. Biol.* **9**, 545.

DISCUSSION

Dourlent: Avez-vous fait l'analyse des amplitudes dans le cas où la relaxation chimique des complexes proflavine–ADN est détéctée par l'etude des variations d'absorption lumineuse (méthode utilisée par Li et Crothers)?

Thusius: Nous n'avons pas répété les expériences en absorbance de Li et Crothers. Les amplitudes mesurées en absorbance sont néanmoins essentielles à une analyse exhaustive des amplitudes observées en fluorescence. Plus précisément, si l'on veut obtenir les rendements quantiques des différentes espèces de proflavine à partir de ces derniers paramètres, il faut les coefficients d'extinction qui sont estimés à partir des constantes de régression caractérisant les amplitudes mesurées en absorbance.

H. Buc: Avez-vous fait des expériences calorimétriques sur vos systèmes?

Thusius: Nous avons déterminé directement par calorimétrie l'enthalpie de formation du complexe benzamidine–trypsine. La valeur calorimétrique est en bon accord avec celle estimée à partir des amplitudes de relaxation. Il est intéressant de noter qu'une analyse des amplitudes offre la possibilité de relier des données cinétiques avec les données purement thermodynamiques.

MÉCANISMES ENZYMATIQUES DE TYPE FLIP-FLOP

M. LAZDUNSKI, C. LAZDUNSKI, C. PETITCLERC, D. CHAPPELET
M. FOSSET, C. GACHE, et M. DELAAGE

Centre de Biologie Moléculaire, C.N.R.S., 31, chemin Joseph Aiguier, Marseille, France

Résumé. Les mécanismes de type flip-flop introduisent la notion de couplage dinétique entre les différentes sous-unités d'une structure enzymatique oligomère (dimère fonctionnel). Cet article présente (a) La démonstration de l'existence des mécanismes de type flip-flop avec les phosphatases alcalines intestinales et de *E.* coli. (b) Les relations qui existent entre mécanismes flip-flop et la réactivité de la moitié des sites actifs d'un enzyme polymère. (c) Les propriétés cinétiques des mécanismes flip-flop. (d) Les avantages fonctionnels et évolutionaires.

Summary. Enzymes with flip-flop mechanisms are polydimers (tetramers: double dimers, hexamers; triple dimers etc.). The subunits in the basic functional dimer do not act independently, kinetic coupling is necessary. Such a mechanism with generate in most cases Michaelis-Menten kinetics although subunits are strongly dependent in the catalytic process.

This paper presents:

(a) A demonstration of flip-flop mechanisms with intestinal and *E.* coli alkaline phosphatases.

(b) A discussion of the relationships which exist between flip-flop mechanisms and the half-of-the-sites reactivity of polymeric enzymes.

(c) A discussion of the kinetic properties of flip-flop mechanisms.

(d) An analysis of the functional and evolutionary advantages of such mechanisms.

Nos connaissances concernant la structure des protéines se sont beaucoup améliorées au cours des dernières années. Des progrès spectaculaires ont pu être observés dans le domaine de l'élucidation des séquences, la détermination des structures sous-unitaires, la cristallographie aux rayons X et la microscopie électronique. Ils nous ont appris que la grande majorité des enzymes intracellulaires se présentent sous la forme de polymères constitués de plusieurs sous unités qui la plupart du temps sont identiques. La grande caractéristique de ces enzymes pourvus de sous-unités multiples est leur haut degré de symétrie.

Du point de vue des propriétés cinétiques, ces enzymes polymères appartiennent en général à l'un ou l'autre des deux grandes classes suivantes:

(a) Les enzymes allostériques pour lesquelles la vitesse suit une loi de variation sigmoïdale avec la concentration de substrat.

(b) Les enzymes à cinétique de caractère Michaelien.

Alors que la signification fonctionnelle des structures à sous-unités multiples est tout à fait claire pour les enzymes allostériques, tel n'est pas le cas des enzymes 'michaéliens' pourtant considérablement plus nombreux. L'interprétation la plus généralement adoptée pour expliquer ces cinétiques de Michaelis classiques, analogues à celles communément observées avec les enzymes monomères, consiste à considérer que les sous-unités (ou protomères) sont absolument indépendantes les unes des autres dans leur fonctionnement à l'intérieur de l'enzyme oligomère. Il est bien évident que cette interprétation est peu satisfaisante; elle ne donne en effet aucune signification fonctionnelle aux structures quaternaires.

C. Sadron (ed.), Dynamic Aspects of Conformation Changes in Biological Macromolecules. 285–299.

Nous allons démontrer dans la suite de ce texte que cinétique de Michaelis ne signifie pas obligatoirement indépendance fonctionnelle des sous-unités. Le modèle utilisé pour décrire les couplages de fonctionnement entre sous-unités, qui constituent la base des mécanismes flip-flop, est celui des phosphatases alcalines. Les phosphatases utilisées sont celles d'*E. coli* et de la membrane intestinale.

La phosphatase alcaline d'*E. coli* est un dimère de poids moléculaire 86000 [1] formé de deux sous-unités identiques [2]. La phosphatase intestinale est aussi un dimère (glycoprotéine) mais de poids moléculaire 140000 et formé de deux sous-unités apparemment identiques. Les deux enzymes possèdent quatre atomes g de zinc par mole d'enzyme, soit deux zinc par centre actif.

Le mécanisme de base des phosphatases alcalines est décrit par l'équation qui suit:

$$E + ROPO_3^{=} \rightleftarrows ES \rightarrow \begin{array}{c} ROH \\ {}^{+}E \\ | \\ OPO_3^{=} \end{array} \rightarrow E + PO_4H^{=}.$$

Le mécanisme procède par la formation intermédiaire d'un phosphoryl-enzyme [3]. Bien qu'on représente habituellement l'étape de phosphorylation comme irreversible, il est possible de former un phosphoryl-enzyme en utilisant non pas le substrat mais le produit de la réaction, le phosphate inorganique. On a alors une situation d'équilibre représentée par l'équation suivante:

$$E + PO_4H^{=} \rightleftarrows E...PO_4H^{=} \rightleftarrows E-OPO_3^{=}$$

où $E...PO_4H^{=}$ et $E–OPO_3^{=}$ représentent respectivement le complexe non covalent et le phosphoryl-enzyme.

La phosphorylation covalente de la phosphatase par le phosphate inorganique n'intervient qu'à pH acide; seul le complexe non covalent à la possibilité de se former à pH alcalin [4].

Cette fixation non-covalente de l'orthophosphate à l'enzyme est représentée dans la Figure 1. La Figure 1 indique que malgré l'identité des chaînes de la phosphatase alcaline d'*E. coli*, les deux centres actifs ne sont pas équivalents et indépendants. La fixation du phosphate inorganique à l'enzyme présente une propriété très importante, la coopérativité négative. La fixation de la première molécule de phosphate inorganique sur l'un des sites 'désensibilise' le deuxième site qui fixe alors la seconde molécule de substrat avec énormément de difficultés et seulement à grande concentration. La formation du complexe binaire contenant deux molécules de phosphate inorganique par molécule d'enzyme n'est pas favorisée thermodynamiquement.

La comparaison des constantes de dissociation obtenues d'une part par représentation de Scatchard, d'autre part par analyse cinétique (Figure 1) indique que la seule fixation qui ait une valeur cinétique est celle qui correspond à la fixation de la première molécule d'orthophosphate sur le premier site. Cette coopérativité négative apparaît d'ailleurs aussi bien pour la phosphorylation covalente à pH acide que pour la fixation non covalente à pH alcalin [4].

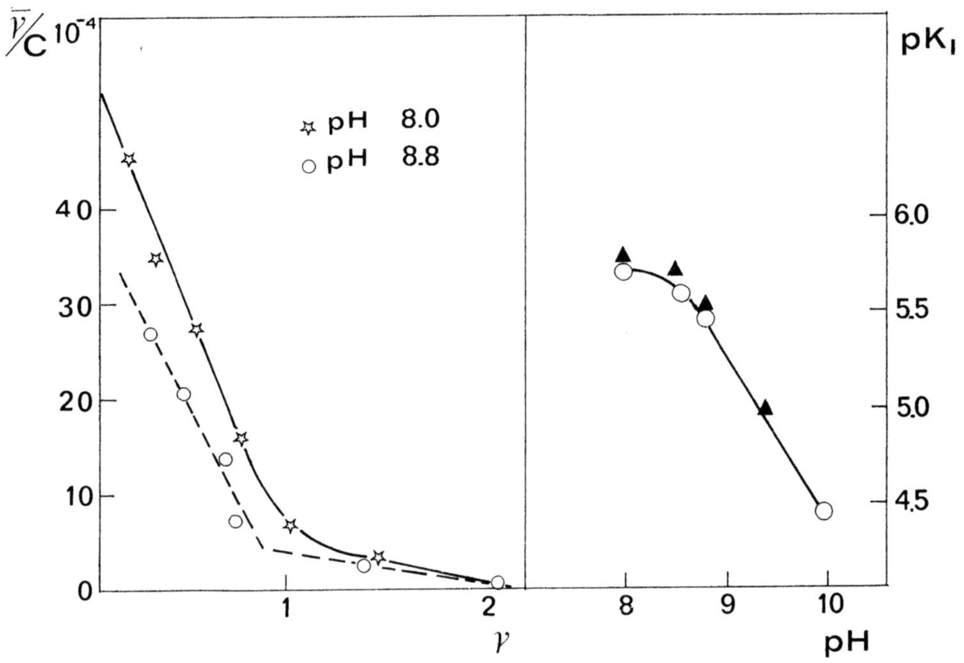

Fig. 1. *Gauche.* La courbe de Scatchard décrivant la fixation de l'orthophosphate à la phosphatase à Zn^{2+}. $\bar{\nu}$: nombre de moles de Pi fixées/dimère; *c*: concentration d'orthophosphate libre, 25°C, 0,4 M NaCl, * pH 8,0, ○ ph 8,8. *Droite.* Variation de pK_I en fonction du pH (K_I = constante de dissociation du complexe enzyme–orthophosphate). ▲ Déterminations cinétiques; ○ pK_I calculé à partir de la pente ($1/K_I$) des courbes ($\bar{\nu}/c - \bar{\nu}$) pour la saturation de premier site [4].

La Figure 2 indique la coopérativité négative est déjà extrêmement marquée à pH 4 où la saturation covalente du premier site intervient à des concentrations de l'ordre de 0.1 mM alors que la phosphorylation du second site intervient à des concentrations de l'ordre de 10 mM pour la phosphatase à Zn^{2+}. Le rapport des constantes de dissociation correspondant à la formation des complexes sur le 1er et le 2ème site est de 40. Il correspond à une énergie libre d'interaction négative de 2,16 kcal/M. La coopérativité négative est considérablement accentuée lorsque le pH est plus élevé. A pH 5 (Figure 2), le rapport entre les deux constantes K_1 et K_2 est de 700, ce qui correspond à une énergie apparente d'interaction de 3,84 kcal/M. La coopérativité négative peut également être mise en évidence par voice cinétique [5]. L'échange des deux phosphates non marquées du composé diphosphorylé $E\genfrac{}{}{0pt}{}{\diagup P}{\diagdown P}$ par des phosphates radioactifs se fait en deux étapes. L'un des phosphates s'échange extrêmement rapidement alors que l'autre s'échange considérablement plus lentement.

La Figure 3 donne un instantané de ce qui se passe à l'état stationnaire à différents pH et avec différents substrats choisis à dessein pour la différence considérable qui existe entre leurs structures chimiques et par conséquent entre leurs réactivités. Les

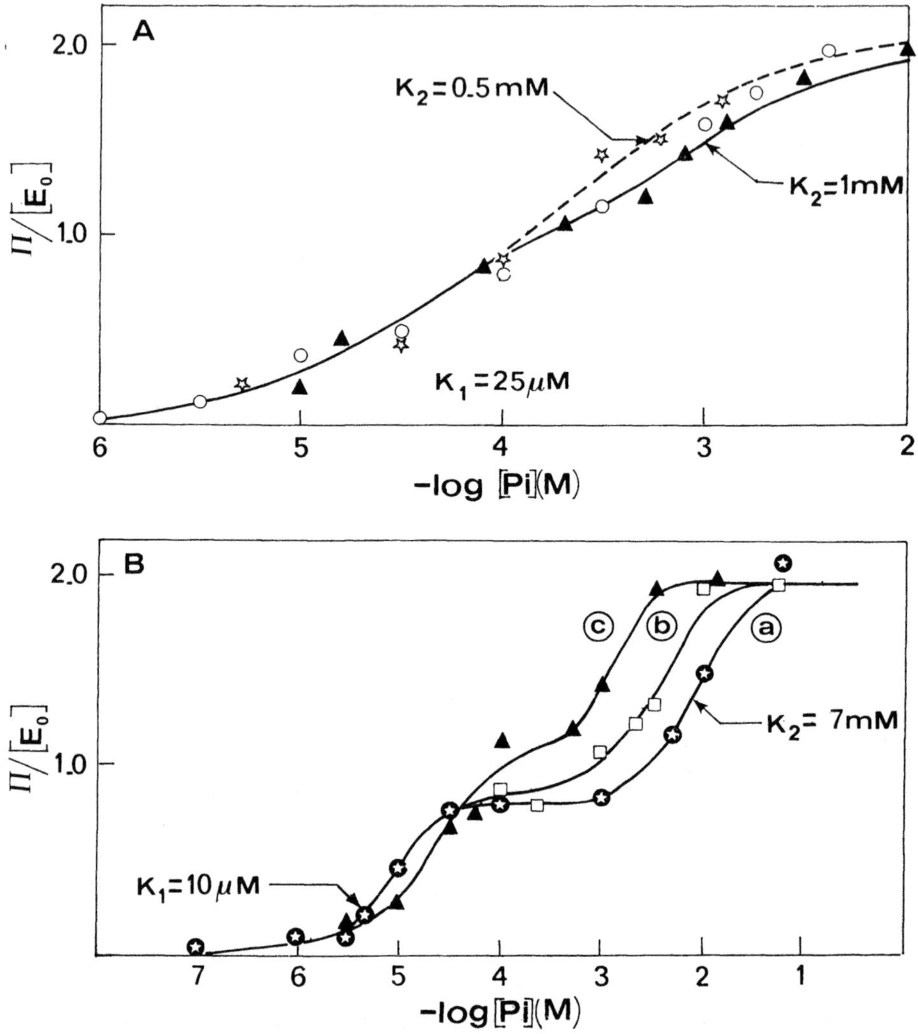

Fig. 2. Influence de la concentration de [^{32}P]orthophosphate sur la phosphorylation covalente des phosphatases à Zn^{2+} et à Co^{2+} à 20°C [5]. (A) pH 4,2, force ionique 0,05 △, ⊙ phosphatase à Zn^{2+}, * phosphatase à Co^{2+}. Les courbes en traits pleins ou en pointillés sont calculées en supposant deux centres actifs avec des affinités différentes pour l'orthophosphate. (B) Comportement de la phosphatase à Zn^{2+} à pH 5,0, 20°C, force ionique 0,05 (a), 0,15 (b), 0,5 (c). π/E_0 = moles de [^{32}P] phosphate (π) liées de façon covalente par mole d'enzyme (E_0).

deux sites de l'enzyme de coli peuvent être phosphorylés à grande concentration de substrat à pH acide. Toutefois cette phosphorylation est encore une fois anticoopérative. Le premier site est phosphorylé à faibles concentrations de substrat, le second à fortes concentrations de substrat seulement [5]. A pH alcalin, que ce soit avec l'enzyme à Zn^{2+} ou avec l'enzyme à Co^{2+}, elle aussi catalytiquement active [6], il n'est jamais possible d'incorporer plus d'un phosphate covalent par mole d'enzyme.

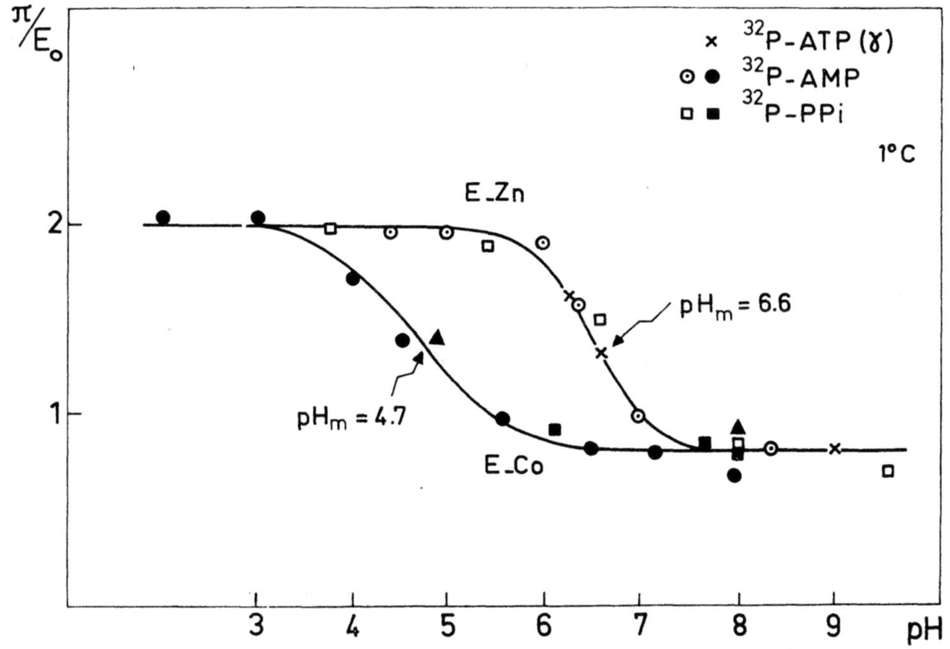

Fig. 3. Influence du pH sur la phosphorylation des centres actifs des phosphatases alcalines à Zn^{2+} et à Co^{2+} avec les substrats et à l'état stationnaire. Phosphorylation maximum de la phosphatase à Zn^{2+} avec [^{32}P]AMP ○, [^{32}P]ATP ×, [^{32}P]pyrophosphate □. Phosphorylation maximum de la phosphatase à Co^{2+} avec [^{32}P]AMP ●. [^{32}P]pyrophosphate ■ et *p*-nitrophénylphosphate ▲ [5].

C'est la première indication que dans les conditions de pH où l'enzyme opère normalement, un seul des deux centres actifs est phosphorylé à l'état stationnaire. L'autre site se trouve donc dans un état différent, impliqué au même moment, comme nous le verrons plus loin, dans une opération catalytique qui n'est pas la déphosphorylation. Il faut remarquer enfin que tous les substrats se comportent de manière identique en dépit de leur différence de réactivité chimique.

Tous les résultats précédemment discutés indiquent clairement que les deux centres actifs de la phosphatase alcaline d'*E. coli* ne sont pas indépendants. Cette constatation semble bien être en contradiction avec l'interprétation classique d'indépendance des sites qu'on pourrait faire à partir du comportement cinétique parfaitement Michaelien observé pour l'enzyme avec des concentrations de substrat allant de 10^{-7} à 10^{-12} M. Il est bien évident, qu'au moins dans ce cas particulier l'interpretation classique est inadequate.

Le mécanisme proposé pour cette enzyme apparait dans la Figure 4 [5]. L'un des deux centres actifs fixe une première molécule de substrat. Cette fixation a un effet intra-sous-unité dû à un changement de conformation de la sous-unité fixatrice, changement de conformation qui 'prépare' ou 'active' le centre actif pour l'opération de phosphorylation. Elle a également un effet inter-sous-unité puisque l'autre centre actif est alors incapable de fixer la seconde molécule de substrat.

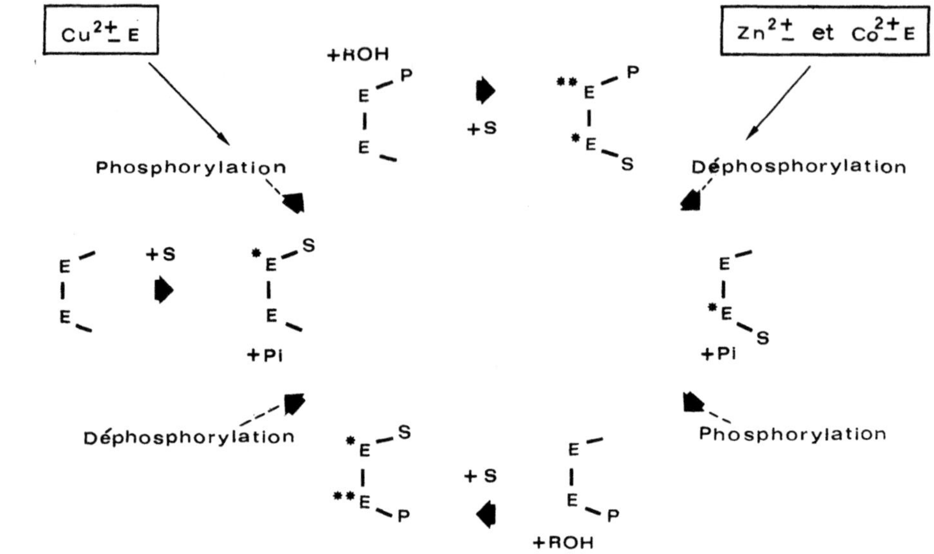

Fig. 4. Représentation simple du mécanisme flip-flop de la phosphatase alcaline. E* et E** indiquent les activations intra et inter sous unités respectivement. E*–S représente la fixation non-covalente du substrat tandis que E–P ou E**–P représentent la phosphorylation covalente. L'activation de la sous-unité est nécessaire à la fois à la phosphorylation (E*) et à la déphosphorylation (E**). Position des étapes déterminantes à pH alcalins pour les différentes métallophosphatases actives, $k_{cat} = 40 \sec^{-1}$ pour l'enzyme à Zn^{2+}, 4 $\sec^{-1}$ pour l'enzyme à Co^{2+}, 0,44 $\sec^{-1}$ pour l'enzyme à Cu^{2+}. Les valeurs sont pratiquement indépendantes de la nature chimique du substrat [5].

Le dérivé monophosphorylé E(–P)|E— est stable et permet d'autre part la fixation d'une 2ème molécule de substrat sur le second site. Cette fixation de substrat sur le second site permet la déphosphorylation du premier site. Le mécanisme fait intervenir deux types de couplages: couplage intra-sous-unité entre la fixation du substrat sur le premier site et la première étape chimique: la phosphorylation du même site; couplage inter-sous-unité entre la fixation du substrat sur le second site et la déphosphorylation du premier. Ces deux opérations de couplage sont réalisées par le biais de changements de conformation qui sont déterminants.

Tous les complexes qui apparaissent dans cette figure ont été isolés, sauf le complexe E(–P)|E(–S). On sait toutefois que ce complexe contient une phosphate covalent par mole d'enzyme (Figure 3). D'autre part en utilisant l'arséniate radioactif et l'enzyme à Co^{2+} nous avons pu isoler une complexe contenant deux moles d'arséniate par mole d'enzyme, l'une d'entre elles est fixée de façon non covalente et peut être facilement éliminée par dilution ou échangée contre de l'arséniate froid, l'autre est fixée de façon stable à l'enzyme à Co^{2+} par l'intermédiaire d'une liaison covalente [7].

La Figure 5 démontre le couplage entre la fixation du substrat sur le second site et la déphosphorylation du premier. Le dérivé monophosphorylé ne se déphosphoryle que lentement à pH 5.0. Si l'on sature le second site par un analogue de substrat qui n'a pas la possibilité de phosphoryler l'enzyme, la déphosphorylation du premier site est accélérée par un facteur d'environ 10.

La Figure 6 démontre que la fixation du substrat à l'enzyme induit un changement de conformation du complexe métallique qu'est le centre actif. Ce changement de conformation visualisé ici par la différence de spectre EPR entre l'enzyme libre et le complexe contenant une mole d'orthophosphate par mole de phosphatase à Cu^{2+} [8] peut aussi être mis en évidence en suivant les changements du spectre visible ou ORD de la phosphatase à Co^{2+} [9], les changements du spectre visible de la phosphatase à Cu^{2+} ou les changements du spectre EPR de la phosphatase à Mn^{2+} [10].

La Figure 4 indique la position des étapes déterminantes pour les différents métalloenzymes à pH alcalins. L'étape déterminante est la déphosphorylation c'est à dire l'étape correspondant à la coopérativité cinétique inter-sous-unité pour les phosphatases à Zn^{2+} et à Co^{2+}. L'étape déterminante est la phosphorylation c'est à dire l'étape correspondant à la coopérativité intra-sous-unité pour la phosphatase à Cu^{2+};

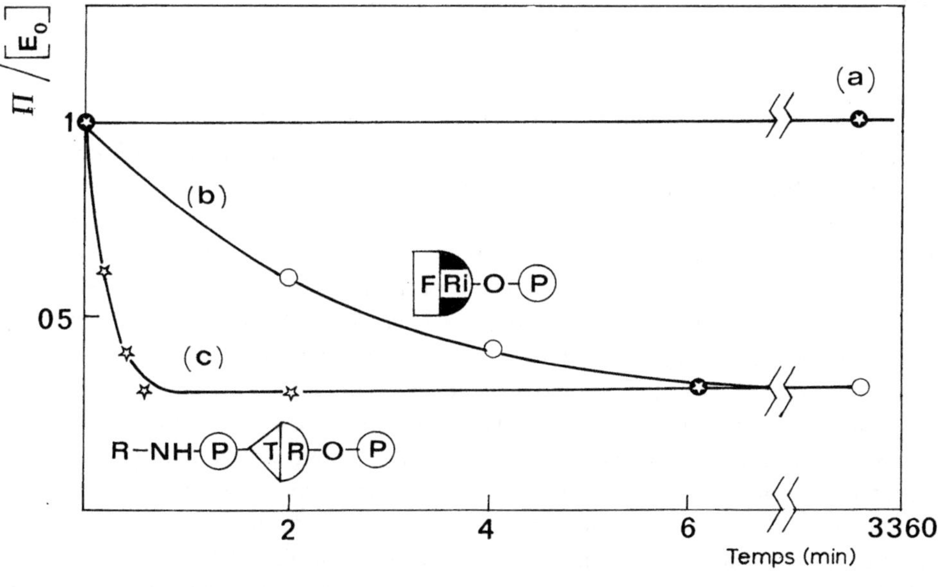

Fig. 5. La stabilité du dérivé monophosphorylé de la phosphatase à Zn^{2+}, pH 5,0, 0° C, NaCl 0,4 M. Courbes: (a) stabilité de la phosphatase monophosphorylée à une concentration de 0,1 mM (constante de dissociation du complexe 0,01 mM); (b) déphosphorylation du dérivé monophosphorylé à une concentration de 0,5 μM; (c) déphosphorylation du dérivé monophosphorylé à une concentration de 0,1 mM en présence de 1,7 mM de p-chloro-anilido-phosphate.

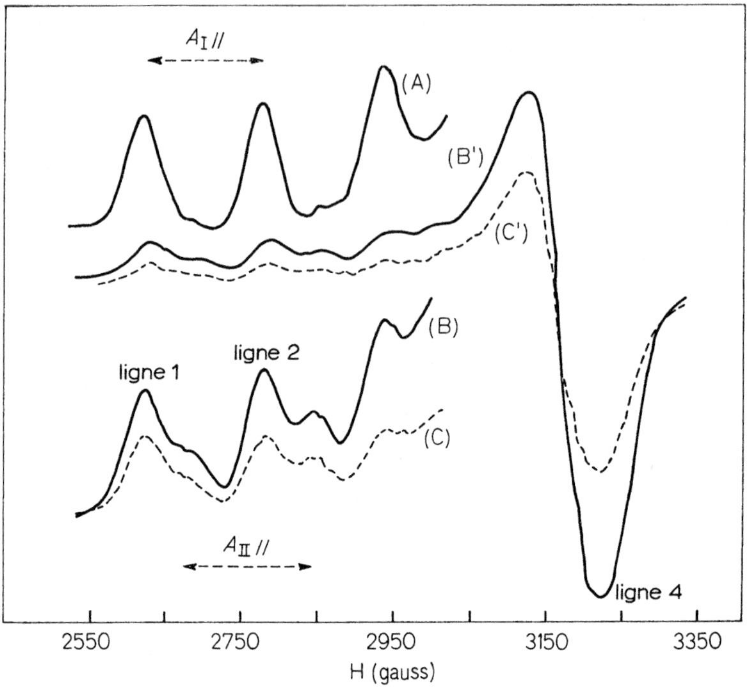

Fig. 6. Spectre EPR de la phosphatase à Cu^{2+}. (A) apophosphatase + 2 atomes g de Cu par mole de protéine; (B) et (B′) apophosphatase + 4 atomes g de Cu par mole de protéine; (C) et (C′) spectre du complexe 1: 1 formé entre la phosphataseà Cu^{2+} ($Cu^{2+}/E_0 = 4$) et le phosphate inorganique.

chacune de ces étapes a une vitesse qui est indépendante de la nature chimique du substrat utilisé [5] ce qui indique bien que le contrôle de la vitesse est dû au changement de conformation de l'enzyme dans les deux cas.

On doit naturellement s'attendre à ce qu'un changement de métal au centre actif qui favoriserait l'existence de complexes symétriques (c'est-à-dire contenant deux moles de substrat fixées de façon covalente ou non par mole d'enzyme) fasse 'sortir' l'enzyme du cycle stationnaire représenté dans la Figure 4. Dans ce cas la vitesse de fonctionnement de l'enzyme devrait être considérablement diminuée. La phosphatase à Cd^{2+} possède aussi la propriété d'anticoopérativité pour la phosphorylation covalente du centre actif par l'ATP et l'AMP. Mais dans ce cas, à la différence de ce qui se passait avec les phosphatases à Zn^{2+} et à Co^{2+} (Figure 3), le complexe formé en présence de grandes concentrations de substrat est le composé diphosphorylé [11]. La phosphatase à Cd^{2+} (4 atomes g de Cd^{2+} par mole de protéine) est dotée d'une activité 1000 fois moins élevée au moins que celle de la phosphatase à Zn^{2+}.

Les propriétés de la phosphatase intestinale sont, pour l'essentiel, similaires à celles observées pour la phosphatase d'*E. coli.* L'enzyme possède une coopérativité négative accentuée pour la fixation non-covalente de l'orthophosphate (Figure 7). L'enzyme possède une réactivité apparente d'un site sur deux pour l'opération de phosphory-

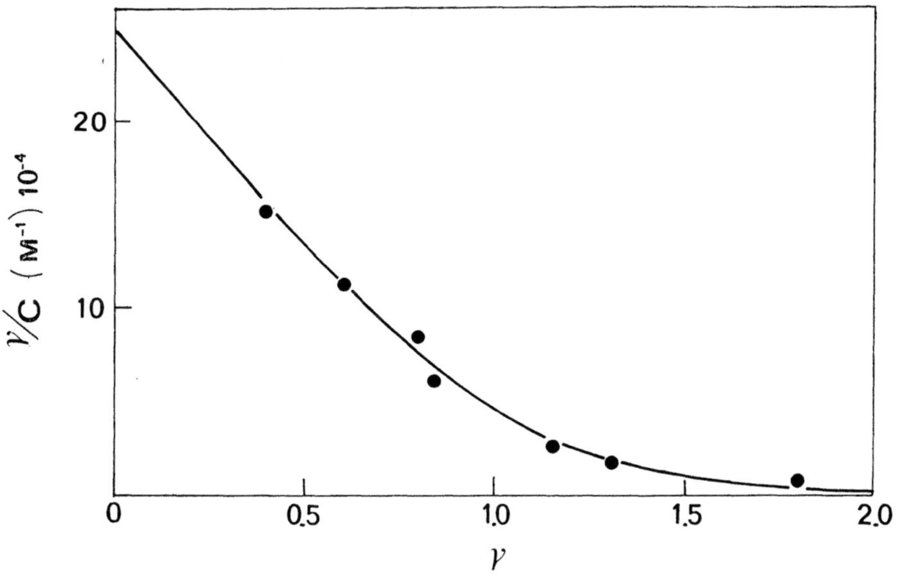

Fig. 7. Courbe de Scatchard pour la fixation non covalente de l'orthophosphate à l'enzyme intestinale (voir Figure 1) pH 8, 15°. Le tracé a été calculé pour deux sites qui forment avec le phosphate des complexes ayant des constantes de dissociation: $K_1 = 4$ μM, $K_2 = 0{,}13$ mM.

lation, à pH alcalins. Le mécanisme (Figure 4) proposé pour la phosphatase d'*E. coli* enzyme périplasmique, vaut donc aussi pour la phosphatase intestinale, enzyme membranaire. La phosphorylation (coopérativité intra sous unité) est déterminante aux pH supérieurs à 7,0. Là encore l'analogie des vitesses maxima mesurées à pH 8,0 et à pH 10,0 avec des substrats de réactivités très différentes indique que les étapes correspondant aux couplages intra-sous-unités et inter-sous-unités ont leurs vitesses contrôlées par un changement de conformation de l'enzyme. Le mécanisme d'action des phosphatases leur permet de catalyser l'hydrolyse de substrats de réactivités variées qui formeront à la même vitesse le produit intéressant, le phosphate inorganique; cette propriété correspond pour la cellule à une nécessité physiologique essentielle [5, 11].

Il est bien évident toutefois que ce mécanisme est utilisable pour n'importe quel type d'enzyme qu'elle soit très spécifique ou aspécifique comme la phosphatase, qu'elle fasse ou non intervenir un complexe covalent. Les données présentées dans le tableau suggèrent que le mécanisme puisse être tout à fait général. Un grand nombre d'enzymes présentant une cinétique à caractère michaélien sont des enzymes qui présentent une coopérativité négative absolue au niveau de la fixation du substrat ou une réactivité apparente de la moitié des sites dans la réaction de catalyse (Tableau I). Cette dernière propriété indique de toute évidence que les sous-unités ne sont pas indépendantes dans l'opération de catalyse. Le lecteur pourra se reporter à une revue récente pour la discussion des propriétés des enzymes qui figurent dans le Tableau I

TABLEAU I

Enzymes 'demi-sites' avec cinetiques michaéliennes
(*Current Topics in Cellular Regulation*, Vol. VI, pp. 267–309 et références citées)

Enzyme	Nombre de sous-unités	Nombre de 'sites fonctionnels'	Substrat ou analogues testés
m-Malate déhydrogénase	2	1	NAD^+
Alcool déhydrogénase	2	1	NADH, aldéhydes, *o*-nitroso, N-N-diméthyl aniline
Transaldolase	2	1	Dihydroxyacétone
UDP-Gal-épimérase	2	1	NAD^+
Phosphatase alcaline d'*E. coli*	2	1	Pi arséniate, phosphates organiques, PPi
Phosphatase alcaline intestinale	2	1	Pi, PPi, phosphates organiques
Glyceraldéhyde-3*P* déhydrogénase du muscle	4	2	3-Phosphoglyceroylphosphate, *β*-(2-furyl) acryloylphosphate
β-Galactosidase	4	2	3-Nitro-4-amino-phényl-éthylthio-*β*-D-galactoside
Méthionine t-RNA synthétase *E. coli*	4	2	L-Méthionine, L-méthioninyl adénylate
Acétyl cholinesterase	4	2	*o*-Nitrophényl diméthyl carbamate
Glutamate déhydrogénase	6	3	NADP
Glutamine synthétase	8	4	Méthionine sulfoximine
Acétoacétate décarboxylase	12	6	Anhydride acétique, 2,4-dinitrophényl acétate

[5]. Tous ces enzymes sont des dimères ou des polydimères (tétramère = double dimère, hexamère = triple dimère, etc).

Il est extrêmement facile d'écrire une très grande variété de mécanismes impliquant le couplage de fonctionnement entre les deux sous-unités d'un dimère; tout ces mécanismes conduisant naturellement à une loi de variation de v en fonction de S qui est du type michaélien. On peut par exemple écrire 32 mécanismes différents pour un enzyme fait de dimères fonctionnels qui utilisent 2 substrats et forment 2 produits sans former d'intermédiaire covalent selon que l'on couple la fixation du substrat sur une sous-unité et la réaction chimique sur l'autre; la fixation du substrat et l'exclusion du produit, etc.

L'analyse de ces mécanismes amène à discuter certains d'entre eux qui permettent de démystifier l'idée qu'il existe des voies indiscutables et bien établies qui permettent de démontrer l'indépendance des sous-unités dans une enzyme donné.

Le mécanisme présenté dans la Figure 8 indique qu'un mécanisme d'action catalytique peut fort bien impliquer un fonctionnement couplé des sous-unités et néanmoins

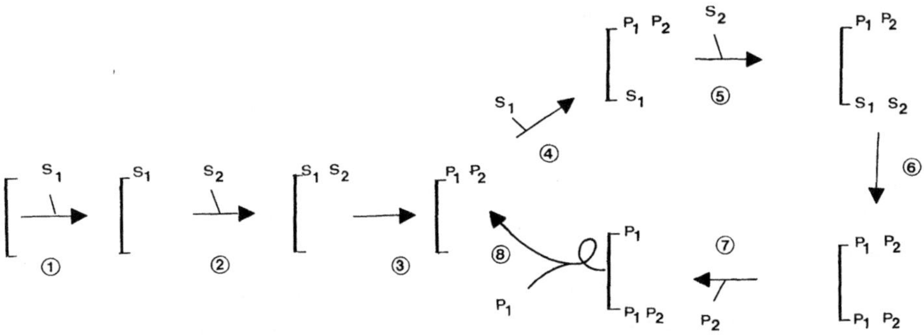

Fig. 8. Exemple de mécanisme flip-flop pour un dimère fonctionnel avec deux substrats et deux produits. La fixation de S_1 est anticoopérative, la fixation de S_2 est couplée (couplage intra-sous-unité) à la fixation de S_1 sur la même sous-unité. Lorsque S_1 et S_2 se sont fixés sur la sous-unité, ils peuvent réagir et former P_1 et P_2 (étape (3)). C'est là le point d'entrée dans le cycle stationnaire. Les étapes (1), (2) et (3) sont donc des étapes qu'on trouve uniquement dans l'état prestationnaire. On accumule dans un tel mécanisme un complexe $\left[\begin{smallmatrix} P_1P_2 \\ P_1P_2 \end{smallmatrix}\right.$. Dans l'étape (7) on indique que la réaction chimique sur la 2ème sous-unité est couplée à l'exclusion de P_2 à partir de la première. Le produit P_1 est exclu à l'étape (8). La flèche avec une boucle indique qu'à ce moment on retombe sur un complexe avec une seule molécule de P_1 et une seule molécule de P_2. Il faudra un autre tour de cycle pour exclure les produits à partir du second centre actif. Chaque boucle stationnaire nécessite l'introduction d'une seule molécule de S_1 et d'une seule molécule de S_2; le mécanisme donne une cinétique michaélienne par rapport à S_1 et S_2. Si l'étape (7) est l'étape déterminante, le complexe qui s'accumule à l'état stationnaire est $\left[\begin{smallmatrix} P_1P_2 \\ P_1P_2 \end{smallmatrix}\right.$. Il y a dans ces conditions 2 moles de P_1 et de P_2 produites par mole de dimère à la fin de la période prestationnaire ('burst'). Si l'étape la plus lente de l'état prestationnaire est l'étape (3), la cinétique prestationnaire est évidemment du premier ordre. Avec le même mécanisme et un autre substrat S_1 ou S_2 on pourrait très bien avoir l'étape (5) par exemple comme étape déterminante. Dans ce cas le 'burst' prestationnaire de P_1 et de P_2 serait seulement de 1 mole par mole d'enzyme.

présenter une cinétique stationnaire de type michaélien, une cinétique prestationnaire du 1er ordre et une formation de produits dans l'état prestationnaire qui correspond à deux moles de P_1 ou de P_2 libérés par mole de dimère. On obtiendrait évidemment un comportement cinétique absolument identique, si les deux sous-unités étaient indépendantes. Un tel mécanisme permettrait par exemple d'expliquer sans difficulté la situation apparemment paradoxale qui existe pour l'alcool déshydrogénase: les chercheurs utilisant l'acétaldéhyde comme substrat trouvent un 'burst' de deux moles de NAD par mole d'enzyme dans la période prestationnaire [12] alors que les chercheurs utilisant la benzaldéhyde (ou substrats apparentés) [13, 14] trouvent seulement un burst de 1 mole de NAD formé par mole d'enzyme dimère dans l'état préstationnaire. Le cycle stationnaire est le même pour les deux types de substrat, mais l'étape (4), (5) ou (6) est limitante pour les substrats type benzaldéhyde et l'étape (7) ou (8) pour les substrats type acetaldéhyde.

La démonstration qu'un des substrats de la réaction se fixe de façon indépendante sur les deux sites du dimère fonctionnel n'est pas non plus la preuve que les sous-unités agissent de façon indépendante dans la catalyse.

Les mécanismes de type flip-flop ont un certain nombre d'avantages d'ordre fonctionnel:

(a) Avantages thermodymamiques liés au couplage [5].

(b) Exclusion automatique du produit de la réaction [11].

Cette propriété est particulièrement bien illustrée dans le cas de la phosphatase: le substrat arrive sur *une* sous-unité et exclue le produit de l'*autre* sous-unité (Figure 5).

(c) Intérêt dans le fonctionnement des complexes multi-enzymatiques tels que le complexe de la pyruvate déshydrogénase et l'acetoacétyl carboxylase [5].

(d) Faculté d'autorégulation de la réaction catalytique liée à la possibilité de créer facilement des mécanismes faisant apparaître des inhibitions ou des activations par excès de substrat. La littérature scientifique est riche d'exemples de telles inhibitions ou activations qui jusqu'ici, et dans la majorité des cas, ont été interprétées en introduisant des 'sites secondaires' pour la fixation du substrat. Une telle interprétation est bien entendu peu satisfaisante. La Figure 9 présente un exemple d'explication simple des effects cinétiques de l'excès de substrat.

Le principal avantage nous apparaît cependant pouvoir être d'ordre évolutionnaire. Les enzymes intracellulaires font généralement partie des chemins métaboliques. Leurs valeurs de K_m et V_m ne peuvent être des paramètres 'au hasard'. Le K_m et le V_m d'un enzyme donné doivent être ajustés aux K_m et au V_m de l'enzyme qui précède et de l'enzyme qui suit. Les deux paramètres fondamentaux, K_m et V_m, doivent être discutés indépendamment; ils sont liés à différents événements du processus catalytique. Les

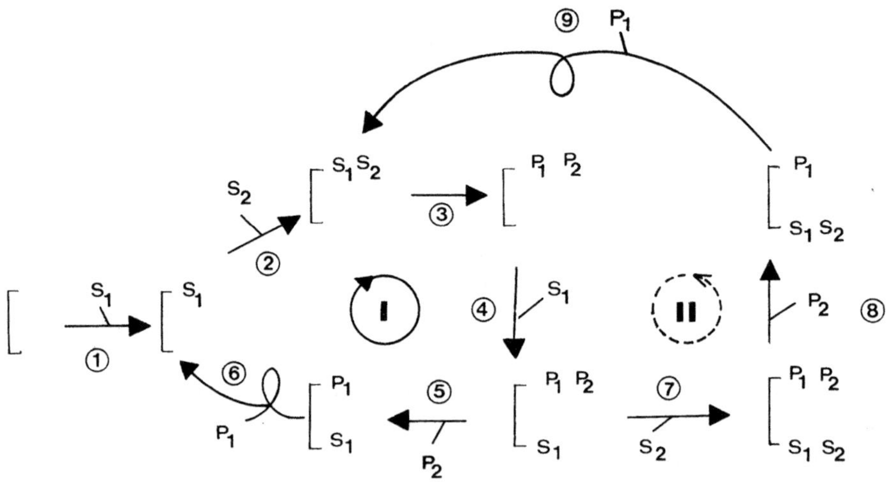

Fig. 9. Exemple de mécanisme flip-flop expliquant facilement l'activation ou l'inhibition par excès de substrat. Le cycle I est un cycle flip-flop classique. Nous supposerons que l'étape déterminante de ce cycle est l'étape (5). Ceci revient à dire que la fixation du substrat S_1 sur la 2ème sous-unité produit un changement de conformation inter-sous-unité qui exclue P_2 de la première sous-unité. Les caractéristiques cinétiques du seul cycle I sont de type michaélien. Toutefois ce cycle peut être branché sur un autre cycle indiqué II. Ce cycle ne peut être amorcé qu'à grandes concentrations de substrat S_2 lorsque l'étape (7) est possible. Dans ces conditions, si l'étape (8) est plus lente que l'étape (5), il y a inhibition par excès de substrat; si elle est plus rapide, il y a activation.

valeurs de V_m caractérisent l'efficacité catalytique. Dans une suite métabolique, le produit d'un enzyme se trouve être nécessairement le substrat de l'enzyme suivant. Ceci implique des analogies dans la géométrie des centres actifs des enzymes successifs. C'est pourquoi l'ajustement des valeurs de K_m par mutations ne semble pas devoir présenter de grands problèmes. (au moins lorsque les K_m représentent les constantes de dissociation vraies des complexes enzyme–substrat). La situation est tout à fait différente pour les valeurs de V_m. Les transformations successives qui apparaissent dans un chemin métabolique impliquent en effet des réactions chimiques de caractère très différent. On peut prendre pour exemple fort illustratif, quatre étapes successives de la glycolyse: une transphosphorylation catalysée par la phosphofructokinase, une aldolisation par l'aldolase, une isomérisation par la triose phosphate isomérase, et une oxydation par la glyceraldéhyde-3-phosphate déshydrogénase.

Si l'étape la plus lente de chacune de ces réactions correspond à la vitesse de la transformation chimique du substrat, l'ajustement rapide des valeurs de V_m pour les enzymes consécutifs aura peu de chances de se réaliser; il n'y a en effet aucune analogie mécanistique entre une aldolisation et une réaction d'oxydation. Au contraire, le problème de l'ajustement mutuel des valeurs de V_m par mutations serait facilement résolu si les étapes les plus lentes étaient toujours directement associées aux changements structuraux affectant les complexes enzyme–substrat. La catalyse de type flip-flop offre une possibilité de ce genre. Les changements de conformation inter-sous-unités sont essentiels dans ce type de mécanisme. Celà signifie par exemple que la fixation d'une molécule de substrat sur une sous-unité d'un dimère fonctionnel induit la formation d'une géométrie correcte du centre actif dans l'autre sous-unité, de telle manière que la catalyse puisse se produire.

La catalyse dépend de ce réarrangement inter-sous-unité. Il est bien possible que, dans la plupart des cas, la réorganisation structurale soit beaucoup plus lente que le processus catalytique qui suit. Un changement de conformation limitant permettrait une modulation facile des valeurs de V_m. Toute mutation d'un résidu donné dans la chaîne polypeptidique monomère apparaît deux fois dans un dimère symétrique. L'effet catalytique d'une telle mutation sera considérablement amplifié par la propriété de symétrie si le résidu est impliqué dans le changement de conformation inter-sous-unité du dimère fonctionnel à mécanisme flip-flop.

Pour que cet exposé soit à la fois suffisamment clair et succinct, nous avons insisté tout particulièrement sur les enzymes présentant des cinétiques à caractère michaélien. Il faut mentionner qu'il n'y a aucune incomptabilité entre mécanisme flip-flop et la cinétique sigmoïdale qui caractérise les enzymes allostériques. Il est même probable qu'un certain nombre d'enzymes allostériques puissent avoir un comportement fonctionnel de base du type flip-flop, tel semble bien être le cas de la CTP synthetase. On pourra trouver dans une revue récente une discussion détaillée sur ce sujet [5].

Références

[1] Schlesinger, M. J.: 1965, *Brookhaven Symp. Biol.* **17**, 66.
[2] Rothman, F. et Byrne, R.: 1963, *J. Mol. Biol.* **6**, 330.

[3] Engström, L.: 1962, *Biochim. Biophys. Acta* **56**, 606.
[4] Lazdunski, C., Petitclerc, C., Chappelet, D., et Lazdunski, M.: 1969, *Biochem. Biophys. Res. Commun.* **37**, 744.
[5] Lazdunski, M.: 1972, dans *Current Topics in Cellular Regulation*, Vol. VI, pp. 267–309.
[6] Lazdunski, C. et Lazdunski, M.: 1969, *Eur. J. Biochem.* **7**, 294.
[7] Petitclerc, C., Lazdunski, C., Chappelet, D., Moulin, A., et Lazdunski, M.: 1970, *Eur. J. Biochem.* **14**, 301.
[8] Lazdunski, C., Chappelet, D., Petitclerc, C., Leterrier, F., Douzou, P., et Lazdunski, M.: 1970, *Eur. J. Biochem.* **17**, 239.
[9] Simpson, R. T. et Vallee, B. L.: 1968, *Biochemistry* **7**, 4343.
[10] Lazdunski, C., Petitclerc, C., Chappelet, D., Leterrier, F., Douzou, P., et Lazdunski, M.: 1970, *Biochem. Biophys. Res. Commun.* **40**, 589.
[11] Lazdunski, M., Petitclerc, C., Chappelet, D., et Lazdunski, C.: 1971, *Eur. J. Biochem.* **20**, 124.
[12] Shore, J. D. et Gutfreund, H.: 1970, *Biochemistry* **9**, 4655.
[13] Bernhard, S. A., Dunn, M. F., Luisi, P. L., et Schack, P.: 1970, *Biochemistry* **9**, 185.
[14] Luisi, P. L. et Favilla, R.: 1972, *Biochemistry* **11**, 2303.

DISCUSSION

Ptak: Je n'ai pas très bien compris le rôle des métaux tant du point de vue conformationnel que du point de vue de la cinétique de réaction. Par exemple, les spectres RPE à basse température du Cu^{2+} signifient-ils que le cuivre est redistribué ou qu'il se forme un complexe ternaire? Les différences entre Zn^{2+}, Cd^{2+} et Mn^{2+} sont-elles dues uniquement à des différences de constante de stabilité, les sites de fixation étant supposés les mêmes?

M. Lazdunski: Il y a quatre atomes-grammes de métal par mole d'enzyme. Les mêmes sites de fixation attachent soit Zn^{2+}, Co^{2+}, Mn^{2+}, Cd^{2+}. Il y a deux classes de sites de fixation sur chaque sous-unité: un site fort qui attache le métal fortement, un site faible qui attache le métal plus faiblement. Le spectre RPE de l'enzyme à Cu^{2+} permet de séparer la contribution des sites de type I (qui fixent Cu^{2+} fortement) et des types II (qui fixent Cu^{2+} avec moins d'affinité). Lorsqu'on ajoute le phosphate inorganique, les deux types de signaux sont perturbés. Il se forme dans ces conditions un complexe non-covalent 1:1 entre l'enzyme à Cu^{2+} et le phosphate inorganique. L'un des deux centres actifs reste donc 'vide'. La fixation du phosphate inorganique sur l'un des centres actifs change la structure des chélates formés entre la protéine et les cuivres de type I et II. Les différences d'activité entre les enzymes à Zn^{2+}, Cd^{2+}, Mn^{2+}, Cu^{2+} ne sont apparemment pas directement liées aux différences d'affinité du métal pour l'apoprotéine. Le cadmium et le manganése sont, comme l'atome de zinc, fortement liés à l'apoprotéine mais l'activité des phosphatases à cadmium et à manganèse est négligeable. Le cobalt n'est pas très fortement attaché et donne un enzyme à Co^{2+} très active.

Pantaloni: J'ai beaucoup apprécié cet exposé clair et convaincant sur le mécanisme flip-flop de la phosphatase alcaline. La généralisation de ce mécanisme à d'autres, voire à tous les enzymes oligomériques paraît très séduisante. En effet, si on élargit le mécanisme flip-flop du couplage catalytique indispensable entre deux sous-unités à la notion d'interactions catalytiques positives entre deux sous-unités d'une enzyme oligomère, la généralisation paraît très raisonnable et il semble étonnant qu'elle n'ait pas été proposée plus tôt. Cependant, dans une première étape, il y a lieu d'être prudent en citant une liste d'enzymes à mécanisme flip-flop qui serait basée uniquement sur le fait que le nombre de sites actifs est apparemment égal à la moitié du nombre de sous-unités. Car si cela est un critère valable, cela ne constitue évidemment pas une preuve suffisante. D'autre part, certaines déterminations du nombre de sous-unités enzymatiques ont été remises en question récemment pour quelques enzymes de la liste présentée. Dans le cas de la glutamate déshydrogénase, le mécanisme flip-flop ne doit être considéré que comme une des trois ou quatre hypothèses qui pourraient rendre compte des résultats expérimentaux et que nous sommes actuellement en train de tester. Ces quelques remarques ne devraient diminuer en rien la portée générale de ce mécanisme flip-flop et le nouveau concept introduit s'avère déjà stimulant et fructueux pour l'enzymologie.

M. Lazdunski: Les mécanismes de type flip-flop permettent d'expliquer assez simplement et de la même manière des 'anomalies' cinétiques observées avec des enzymes catalysant des réactions fort différentes.

Mme Labeyrie: Il s'agit d'une remarque sur cet ajustement nécessaire des vitesses des différents processus enzymatiques constituant une chaîne réactionnelle; il faut que cette cohérence des vitesses se maintienne, chez certains organismes, dans une échelle de température considérable, par exemple de 0 à 50°.

M. Lazdunski: Les changements de conformation inter-sous-unités permettent de corréler facilement (grâce aux propriétés de symétrie) les vitesses k_{cat} mais aussi les énergies d'activation.

Karplus: It is certainly possible that the flip-flop mechanism could serve the function of adjusting the rates of enzyme-catalyzed reactions so that a process involving several consecutive steps, each catalyzed by a separate enzyme, could occur in a balanced way (i.e. without accumulation of intermediates due to one reaction being much faster than the others in the chain). However, it is not valid to argue that such a mechanism is required because it is difficult to construct the enzymes in the chain so that they themselves catalyze each of the different reactions at the balanced rate. If such a chain has significant survival value for the organism, there is little question that evolutionary pressures could adjust the rates as necessary by the same mutation mechanisms as created the enzymes in the first place. If one needs to adjust the rate of an enzyme reaction during the life of a single individual or if a single enzyme participates in many different chains, a flip-flop, allosteric, or other control mechanisms seems more necessary.

M. Lazdunski: Such a mechanism is probably not required. However, rapid adjustment of maximum rates between successive enzymes appears easier if rates are controlled by inter-subunit conformational changes. The modulation of the rate of such a conformational change will necessitate, in most cases, a number of mutations which is much lower than that required for a complete reconstruction of a new active site. Maybe, the main advantage of giving an evolutionary role to flip-flop-type mechanisms is that they make use of the very important property of symmetry.

Lifson: At least it is easy to see how an enzyme could be 'speeded down' if its speed does not constitute a selective advantage.

STRUCTURE TRIDIMENSIONNELLE DES ACIDES RIBONUCLÉIQUES DE TRANSFERT

J. P. EBEL

Laboratoire de Chimie Biologique, Université Louis Pasteur, Rue Descartes, 67000 Strasbourg, France

Résumé. Dans cette revue seront décrites les diverses approches chimiques, enzymatiques et physiques employées pour étudier le problème de la structure tridimensionnelle des tRNA. Les divers modèles proposés, ainsi que des modifications éventuelles de la conformation des tRNA au cours de ses interactions, seront discutés à la lumière des résultats expérimentaux précis disponibles.

Summary. The aim of this review is to describe the present state of the problem of the three-dimensional structure of the tRNA's.

The comparison of the known primary structures of tRNA's leads to a model of secondary structure, which is common to all tRNA's: the clover-leaf structure. However, this structure cannot explain all the properties of the tRNA's. Multiple and reversible conformational states, which are recognized or not by several enzymes and which are dependent on the ionic strength, have been described. Furthermore, studies on the accessibility of tRNA's to several enzymes or chemical agents, on the complementary binding of oligonucleotides and tritium exchange experiments have shown that several regions, which should be exposed in the clover-leaf structure, seem to be masked.

The cocrystallisation of several tRNA's suggests similarities in the three-dimensional structure of tRNA's. This has been confirmed by small-angle X-ray scattering, fluorescence spectra in the presence of dyes or high resolution electronmicroscopic studies, which are in agreement with a compact asymmetric structure. Up to now only limited results have been obtained from the X-ray diffraction studies on several cristallized tRNA's.

Several models of tertiary structure of tRNA's have been proposed, which are all derived from the clover-leaf structure. They will be discussed taking into account some features which seem to be common to all tRNA's, for instance: (i) the extended structure with a giration radius of 24 to 25 Å; (ii) the fact that 4-thiouridine in position 8 is close to the cytidine residue in position 13, as shown by the formation of a covalent bond between the 2 nucleosides in several *E. coli* tRNA's following a photochemical reaction; (iii) the fact that the purine in position 15 (from the 5′-end) is complementary to the pyrimidine in position 29 (from the 3′-end).

Finally, the problem of possible conformational changes of tRNA's, when they interact with various enzymes, will be discussed.

1. Introduction

Les acides ribonucléiques de transfert (tRNA) jouent un rôle fondamental dans la traduction du message génétique. Le rôle de ces molécules est de fixer spécifiquement les aminoacides et de les transférer dans la chaîne peptidique dans une position déterminée par la séquence des codons le long du RNA messager. Ce processus est en réalité extraordinairement complexe et nécessite toute une série d'interactions. On peut distinguer, des interactions RNA–RNA comme celles qui ont lieu entre le codon du RNA messager et l'anticodon du tRNA, et des interactions RNA–protéines, comme celles qui ont lieu avec les enzymes de modification des tRNA, les aminoacyl-tRNA synthétases, les facteurs d'initiation ou d'élongation dans la synthèse protéique, sans oublier celles qui ont lieu à la surface du ribosome.

C. Sadron (ed.), Dynamic Aspects of Conformation Changes in Biological Macromolecules. 301–326.

L'un des problèmes fondamentaux est de comprendre les relations qui existent entre les nombreuses activités dont cette molécule multifonctionnelle est le siège et sa structure. Les tRNA représentent la classe de RNA dont la taille est la plus faible: leur poids moléculaire varie de 25000 à 30000, ce qui correspond à une longueur d'environ 75 à 85 nucléotides. Des progrès considérables ont été réalisés dans l'élucidation de la séquence des nucléotides dans ces molécules. Mais la connaissance de la structure primaire de ces composés ne peut suffire à comprendre les interactions multiples et hautement spécifiques qui ont lieu au niveau des tRNA. Ces interactions nécessitent l'existence dans ces molécules d'une structure tridimensionnelle bien définie dont la connaissance est nécessaire à la compréhension de la fonction des tRNA.

Notre propos est, dans cette revue, de faire le point des connaissances actuelles sur ce problème de la configuration spatiale des tRNA. Nous nous limiterons à la description et à la discussion des méthodes qui ont permis de préciser cette structure et des modèles de structure tridimensionnelle qui ont été proposés. Nous discuterons également le problème de savoir s'il existe plusieurs conformations correspondant à des fonctions différentes, c'est-à-dire le problème du polymorphisme des tRNA. Par contre nous ne rentrerons pas dans l'étude des sites de reconnaissance des tRNA au cours des interactions diverses auxquelles ils donnent lieu: sites de reconnaissance par les aminoacyl-tRNA synthétases, par d'autres enzymes ou facteurs, par les ribosomes, etc. Ce dernier problème est si vaste et si complexe qu'il nécessiterait une étude particulière.

2. Structure primaire

La connaissance de la structure tridimensionnelle des tRNA nécessite d'abord celle de leur structure primaire. Depuis l'établissement de la séquence nucléotidique du $tRNA^{ala}$ de levure par Holley et ses collaborateurs [1, 2], la structure primaire de près de 35 tRNA provenant soit de procaryotes (*E. coli*, *Salmonella typhimurium*, *Staphylococcus aureus*, *Staphylococcus epidermidis*), soit d'eucaryotes (*Saccharomyces cerevisiae*, *Torulopsis utilis*, germe de blé foi de rat) a été déterminée (pour une revue de la question, voir [3–6]). A titre d'exemple, nous montrons dans les Figures 1 et 2 les structures de deux tRNA dont les séquences nucléotidiques ont été déterminées dans notre laboratoire [7–9] et d'autres structures sont représentées dans les Figures 4 à 7.

3. Caractéristiques communes, structure secondaire: le modèle en feuille de trèfle

L'examen de la structure de l'ensemble des tRNA de structure connue laisse apparaître un certain nombre de caractéristiques communes. Quoique des différences de séquence remarquables existent dans le même organisme entre les tRNA spécifiques de divers aminoacides et dans des organismes différents entre tRNA spécifiques du même anoacide, on observe des homologies structurales remarquables entre les divers tRNA. On peut distinguer dans tous les tRNA plusieurs régions présentant des caractéristiques communes, en particulier des séquences identiques et des régularités dans la localisation de certaines bases et dans les distances entre certaines séquences.

Une autre caractéristique remarquable apparaît dans la localisation des régions susceptibles d'être appariées. L'existence dans les tRNA d'une structure secondaire, c'est-à-dire de régions où les bases sont appariées avec formation d'une structure en double hélice, avait été postulée avant même la connaissance de leur structure primaire à la suite d'expériences de dénaturation thermique, qui avaient mis en évidence une forte hypochromicité. L'élucidation de la séquence des tRNA a confirmé l'existence de régions possédant un haut degré de complémentarité et il s'est avéré qu'un modèle de structure secondaire commun à tous les tRNA pouvait être déduit de la structure primaire: c'est le modèle en feuille de trèfle proposé pour la première fois par Holley *et al.* [1] pour le tRNA^{ala} de levure.

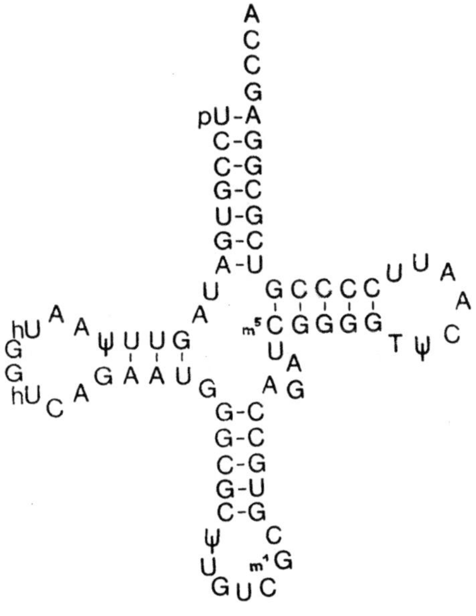

Fig. 1. Structure du tRNA^{asp} de levure [7, 8].

Les tRNA montrés dans les Figures 1, 2, 4, 5, 6 et 7 sont représentés avec leur structure en feuille de trèfle. La Figure 3 en donne une représentation schématique avec les principales caractéristiques communes à l'ensemble des tRNA de structure connue. Le modèle met en évidence l'existence de plusieurs régions appelées 'bras', comportant des séquences appariées et des séquences non appariées représentant des 'boucles'.

Le bras I ou bras de l'aminoacide comprend les extrémités 3′ et 5′ terminales, en particulier la séquence 3′ terminale commune CCA sur laquelle vient se fixer l'aminoacide. Elle comporte dans tous les tRNA 7 paires de bases appariées, à l'exception du

$tRNA^{met}_f$ de *E.coli* qui n'en renferme que 6 [10] et du $tRNA^{his}$ de *Salmonella typhimurium* qui en renferme 8 [11].

Le bras II comprend d'abord, à partir de l'extrémité 5′, deux bases non appariées dont la première en position 8 est un U ou un 4-thiouracile. La région appariée de ce bras comporte dans tous les tRNA 3 ou 4 paires de bases et la boucle renferme dans presque tous les tRNA un ou plusieurs dihydro uracile, d'où le nom de bras du dihydrouracile donné à cette région. Le nombre de nucléotides contenus dans cette boucle varie de 7 à 12. En position 14–15 à partir de l'extrémite 5′, on rencontre toujours une séquence AG ou AA. S'il y a un G en position 15, il lui correspond un C ou m^5C en position 29 à partir de l'extrémité 3′, dans la boucle variable, à l'emplacement précédant le bras du TΨC. S'il y a un A en position 15, il lui correspond un U à cet emplacement. Cette correspondance, postulée pour la première fois par Levitt [12] joue un rôle important dans l'établissement de la structure tertiaire des tRNA.

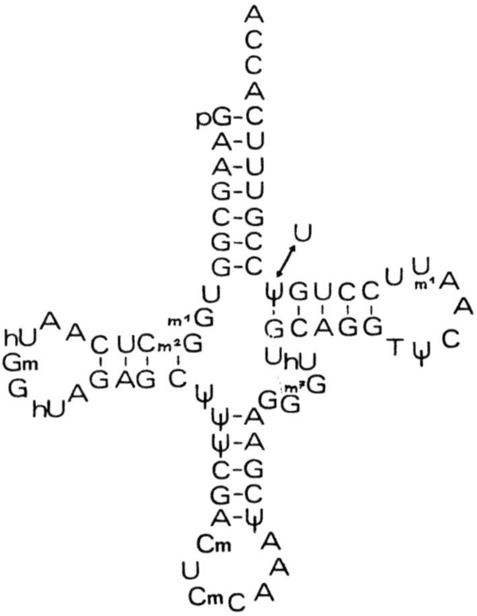

Fig. 2. Structure du $tRNA^{trp}$ de levure [9].

Dans la boucle du dihydrouracile, on trouve une autre caractéristique générale des tRNA: la présence de la séquence G–G ou Gm–G, à l'exception des $tRNA^{gly}$ des *Staphylococcus epidermidis* [13] et *aureus* [14]. Il est intéressant de signaler que ces tRNA ne fonctionnent pas dans la synthèse des protéines, mais participent à celle du peptidoglycane de la paroi bactérienne [13].

Le bras III est celui de l'anticodon. Il comprend toujours 17 nucléotides répartis en général entre 5 paires de bases appariées et 7 nucléotides dans la boucle. La boucle de l'anticodon comprend du côté 5′ une séquence pyrimidine-U. L'anticodon est toujours suivi du côté 3′ d'une purine le plus souvent modifiée, sauf dans le cas des $tRNA^{gly}$ de *Staphylococcus* qui, rappelons-le, n'interviennent pas dans la synthèse des protéines.

La région suivante IV présente des variations de longueur considérables, c'est pourquoi nous l'appelons région variable. Sa longueur varie en effet de 3 à 16 nucléotides. Lorsque sa longueur est élevée, par exemple dans le cas des $tRNA^{ser}$ de levure [15] (Figure 6), une structure pouvant comporter jusqu'à 5 paires de bases complémentaires peut exister. Le dernier nucléotide de la région variable est toujours une pyrimidine C, m^5C ou U. Rappelons qu'il lui correspond en position 15 dans la boucle du dihydrouracile une purine complémentaire.

Le bras V, appelé bras du TΨC, comporte, comme celui de l'anticodon, une région appariée de taille constante de 5 paires de bases et une boucle de 7 nucléotides. Il renferme la séquence GTΨC commune à tous les tRNA de séquence connue, sauf dans les $tRNA^{gly}$ de *Staphylococcus* où elle est remplacée par la séquence GUGC [13]. La séquence GTΨC est toujours suivie d'une purine, puis d'un A ou d'un m^1A. On trouve toujours une pyrimidine comme dernière base de la boucle et elle est suivie d'un C apparié au G du GTΨC.

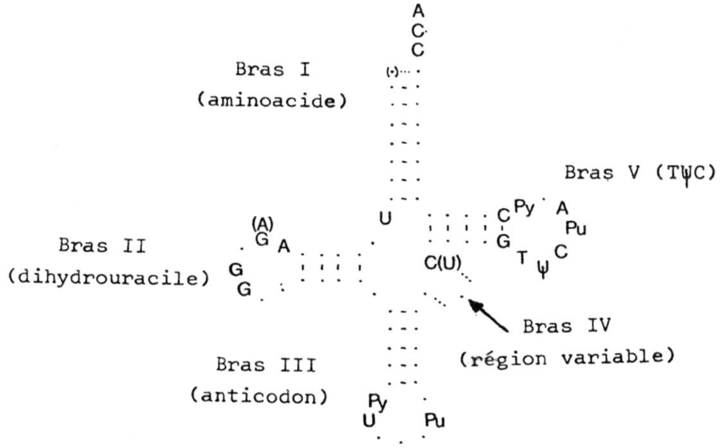

Fig. 3. Nucléotides communs à tous les tRNA de structure connue intervenant dans la synthèse des protéines. (Les modifications des nucléotides n'ont pas été prises en considération dans cette figure.)

Si l'on étudie la localisation dans le modèle en feuille de trefle des bases communes à l'ensemble des tRNA, on constate que ces dernières se trouvent essentiellement dans les régions non appariées. C'est là également que se trouvent le plus de nucléotides mineurs. La région qui présente le plus de ressemblances d'un tRNA à l'autre est la boucle du TΨC. Par contre les régions appariées presèntent le plus de différences d'un tRNA à l'autre.

Le fait que tous les tRNA peuvent être représentés par une structure en feuille de trèfle peut déjà être considéré comme un argument en faveur de cette structure. Cependant diverses expériences ont été réalisées pour vérifier la réalité de ce modèle. L'expérience qui l'appuie le plus est celle de Mirzabekov *et al.* [16], qui, après avoir découpé le $tRNA_1^{val}$ de levure [17, 18] en 4 fragments par action ménagée de diverses endonucléases (Figure 4), ont montré qu'il n'est possible de réassocier que les fragments Q_1 et Q_4, Q_2 et Q_3, Q_3 et Q_4 et enfin le mélange Q_1+Q_2+ $+Q_3+Q_4$, à l'exclusion de toute autre combinaison de fragments. Le fait que Q_1 et Q_2 ne peuvent se réassocier peut s'expliquer par le fait que la région appariée dans le bras

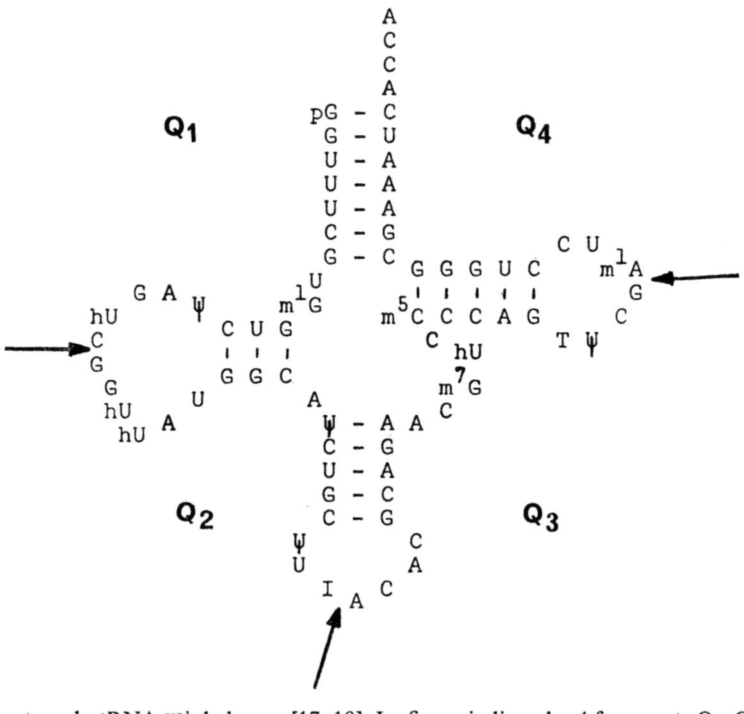

Fig. 4. Structure du $tRNA_1^{val}$ de levure [17, 18]. La figure indique les 4 fragments Q_1, Q_2, Q_3 et Q_4 de l'expérience de Mirzabekov *et al.* [16].

du dihydrouracile ne renferme dans ce tRNA que 3 paires de bases. Un autre argument en faveur du modèle en feuille de trèfle est la plus grande réactivité des régions non appariées prévues par cette structure à l'action d'enzymes ou de réactifs chimiques ou leur appariement plus facile à des oligonucléotides complémentaires. Cette approche va être décrite et discutée plus loin.

Cependant toute une série de résultats d'expériences biologiques, chimiques et physiques montrent que la structure en feuille de trèfle ouverte ne peut à elle seule expliquer toutes les propriétés des tRNA et amènent à envisager une configuration tridimensionnelle plus complexe.

4. Mise en évidence d'une structure tertiaire des tRNA

4.1. Mise en évidence d'états conformationnels multiples et réversibles, reconnus ou non par des enzymes

4.1.1. *Mise en évidence de formes dénaturées des tRNA, biologiquement inactives*

Le concept d'une configuration tridimensionnelle des tRNA a été fortement appuyé par la mise en évidence chez divers tRNA de deux états conformationnels réversibles, l'un natif et biologiquement actif, l'autre dénaturé et inactif. Cette observation, effectuée d'abord avec le tRNA$_3^{leu}$ de levure [19] puis avec le tRNA^{trp} de *E. coli* [20], a ensuite été élargie à de nombreux autres tRNA [21]: en chauffant des tRNA natifs en présence d'EDTA, ils deviennent incapables de fixer les aminoacides en présence de l'aminoacyl-tRNA synthétase correspondante ou de refixer le triplet CCA en présence de la tRNA-nucléotidyltransférase. Les tRNA ainsi dénaturés peuvent ensuite être complètement renaturés en les chauffant en présence de Mg^{++}.

Dans le cas du tRNA^{phe} de levure, le traitement précédent ne conduit pas à une renaturation complète du tRNA. Mais nous avons pu montrer [22] que la préincubation du tRNA incomplètement renaturé en présence de la phénylalanyl-tRNA synthétase permet d'obtenir une aminoacylation totale. La phénylalanyl-tRNA synthétase est donc capable d'induire une transconformation du tRNA d'une forme active.

Tous ces résultats montrent que l'ativité biologique et plus particulièrement la capacité de fixer spécifiquement les aminoacides en présence des aminoacyl-tRNA

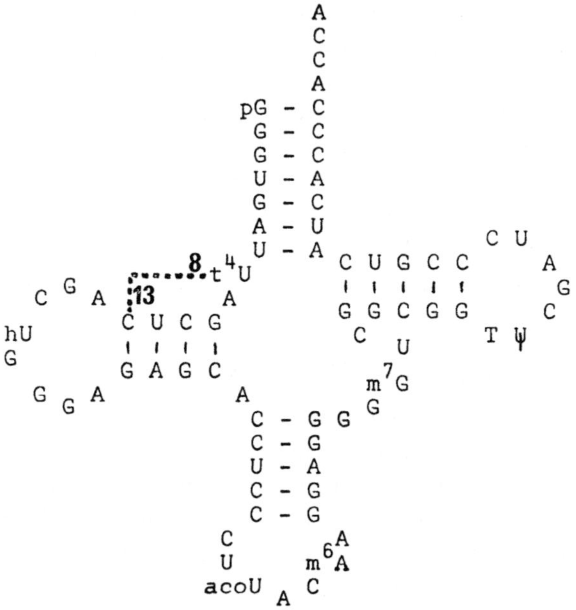

Fig. 5. Structure du tRNA$_1^{val}$ de *E. coli* [45]. La figure indique le pontage entre 4 thiouridine en 8 et cytidine en 13 obtenu par rèaction photochimique [46].

synthétases correspondantes, est étroitement dépendante de leur structure tertiaire. D'autres experiénces le confirment: lorsque l'on augmente la température, on observe une diminution de l'aminoacylation qui peut être mise en relation directe avec la courbe de fusion [23]. Par ailleurs, il a été montré, en particulier dans notre laboratoire [24], qu'il est possible de provoquer des erreurs d'aminoacylation en utilisant des conditions spéciales (augmentation de la concentration des ions Mg^{++}, présence de solvants organiques, etc.) qui sont susceptibles de modifier la structure tertiaire des tRNA.

Nous verrons plus loin que les deux formes inactive et active présentent des caractéristiques physiques différentes, ainsi qu'un comportement différent vis-à-vis d'agents enzymatiques et chimiques. De faibles modifications dans la structure primaire peuvent influencer considérablement la facilité du passage de la forme native à la forme dénaturée: Un $tRNA^{trp}$ extrait d'un mutant de *E. coli* et différent du $tRNA^{trp}$ normal par le remplacement d'un G dans le bras du dihydrouracile par un A s'est révélé être beaucoup plus stable que le $tRNA^{trp}$ normal (25).

4.1.2. *Reconnaissance par la polynucléotide phosphorylase*

L'action de la polynucléotide phosphorylase a permis de mettre en évidence l'existence d'états conformationnels multiples des tRNA en solution. Cet enzyme dégrade les chaînes polynucléotidiques en présence d'orthophosphate avec formation de nucléosides diphosphates (réaction de phosphorolyse) par un mécanisme de dégradation récurrente à partir de l'extrémité 3'-OH et sans dissociation du complexe entre le polynucléotide et l'enzyme, avant que la dégradation de la chaîne ne soit totale.

Il a été observé [26, 27] qu'une fraction seulement des molécules de tRNA peut être phosphorolysée à basse température: à une température donnée, certaines molécules sont complètement dégradées, tandis que les autres restent entièrement résistantes. Il a pu être montré que cette différence de comportement des deux classes de tRNA était dûe à des différences de conformation, dépendantes de la température et de l'environnement ionique. Ces conformations sont cependant différentes des conformations native et dénaturée définies en fonction de leur aptitude à fixer l'aminoacide en présence des aminoacyl-tRNA synthétases. La position de l'équilibre entre ces deux conformations dépend de la température et varie également avec la nature du tRNA.

4.1.3. *Reconnaissance par la polynucléotide kinase*

L'existence de conformations multiples des tRNA est également suggérée par des expériences de phosphorylation par la polynucléotide kinase en position 5'-OH de divers tRNA de levure préalablement déphosphorylés par action de la phosphomonoestérase. On a observé [28] une rephosphorylation incompléte de ces tRNA dans des conditions où des oligonucléotides sont complètement phosphorylés: 20% sont rapidement phosphorylés, tandis que les 80% restants fixent les phosphates à une vitesse beaucoup plus lente, et des concentrations d'enzyme 100 fois supérieures sont nécessaires pour obtenir la phosphorylation totale. Il a été montré que cette différence de comportement était dûe à des conformations différentes des tRNA se traduisant dans les molécules résistantes par une protection de l'extrémité 5'-OH dans la structure tridimensionnelle du tRNA. Le phénomène est dépendant de la température: en augmentant la température, le rapport des deux conformations est modifié en faveur de la forme phosphorylable.

4.2. Étude de l'accessibilité des diverses regions du tRNA

Les résultats précédents suggèrent l'existence d'états conformationnels multiples des tRNA dépendant de la température et de l'environnement ionique. Les changements de structure tertiaire correspondants doivent se traduire par une accessibilité différente de certaines régions des tRNA. Cette accessibilité a été étudiée par l'emploi de réactifs enzymatiques ou chimiques, par des essais de liaison d'oligonucléotides complémentaires au niveau de certaines régions du tRNA et par des expériences d'échange d'hydrogène.

4.2.1. *Action de diverses nucléases*

Des expériences de digestion enzymatique partielle par la ribonucléase pancréatique ou par la ribonucléase T_1 ont été effectuées dans le but de localiser les zônes accessibles à ces enzymes (pour une revue de la question, voir [3, 29, 30]). D'une manière

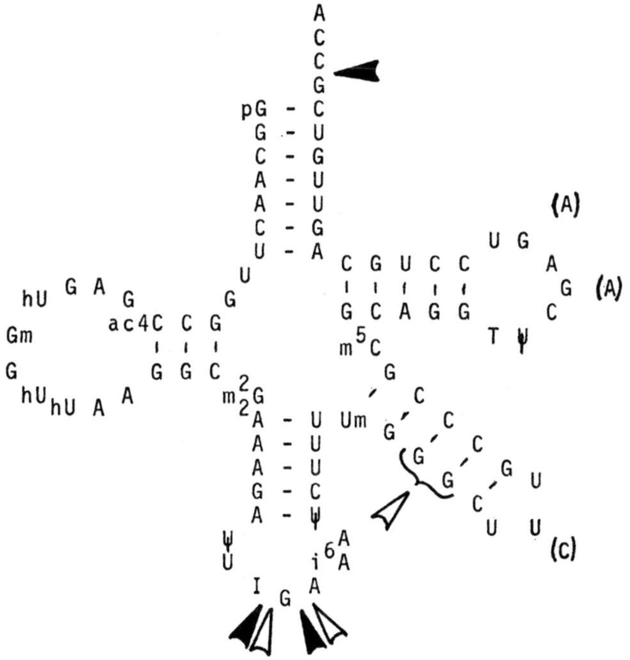

Fig. 6. Structure du tRNA$_{I}$ et $_{II}$^{ser} de levure [15]. La figure indique les coupures obtenues par hydrolyse ménagée par la ribonucléase T_1 sur la forme native (flèches pleines) et la forme dénaturée (flèches blanches) [21].

générale, ces expériences ont mis en évidence une réactivité plus grande des régions situées dans les boucles par rapport aux régions appariées, mais on observe des différences de réactivité marquées entre les boucles prévues par le modèle en feuille de trèfle. La boucle de l'anticodon est la plus sensible à l'action des nucléases, suivie de celle du dihydrouracile et de la région 3′ terminale renfermant le triplet CCA. La boucle de TΨC est par contre nettement moins reactivé. Les Figures 6 et 7 montrent les liaisons coupées par hydrolyse ménagée par la ribonucléase T_1 dans le $tRNA_{I}^{ser}$ et le $tRNA^{phe}$ de levure [21].

Il convient cependant d'être prudent dans l'interprétation des expériences d'attaque par les nucléases quant à l'accessibilité des boucles. D'autres paramétres que l'exposition de la zône étudiée peuvent intervenir dans l'interaction entre protéines et tRNA, telles que des considérations stériques. C'est ainsi que le $tRNA^{lys}$ de levure n'est pas coupé par la ribonucléase pancréatique au niveau de la boucle de l'anticodon, alors que cette boucle est riche en pyrimidines [32]. Inversement la boucle du TΨC du $tRNA^{phe}$ de levure, qui est résistante à l'action des ribonucléases pancréatique et T_1, est attaquée par une autre ribonucléase, la ribonucléase N_1 [33].

Des différences de réactivité envers diverses nucléases ont été observées entre les

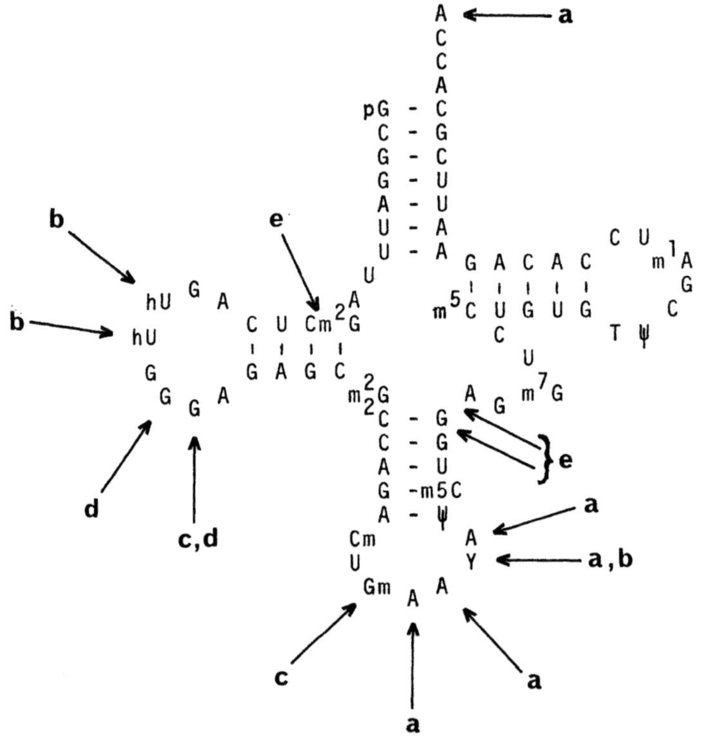

Fig. 7. Structure du $tRNA^{phe}$ de levure [31]. La figure indique les modifications obtenues par action de divers réactifs chimiques (a) action de l'acide monoperphtalique [57]; (b) action du borohydrure de sodium (58); (c) action du kéthoxal [59] et les coupures obtenues par hydrolyse ménagée par la ribonucléase T_1 sur la forme native (d) et sur la forme dénaturée (e) [21].

formes native et dénaturée des tRNA. C'est ainsi que la forme dénaturée est plus sensible à l'attaque de la ribonucléase pancréatique [34] et de la polynucléotide phosphorylase [35]. De plus, les zônes appariées de la région variable (IV) du $tRNA^{ser}$ de levure et de la région du dihydrouracile du $tRNA^{phe}$ de levure sont exposées à l'action de diverses nucléases dans les formes dénaturées, alors qu'elles sont protégées dans les formes natives [21] (Figures 6 et 7). Ceci confirme l'existence de conformations différentes dans les formes native et dénaturée des tRNA.

Les régions appariées sont peu sensibles à l'action des nucléases. Cependant, un résultat surprenant a été obtenu par Philippsen et Zachau [36], qui ont constaté que la nucléase de rein de veau, enzyme qui attaque spécifiquement les régions non appariées dans les polynucléotides [37], coupe le $tRNA^{phe}$ de levure natif au niveau des nucléotides 69 et 70, c'est-à-dire dans le brin situé du côté 3′ terminal, au milieu du bras de l'aminoacide, alors que cette région est appariée dans le modèle en feuille de trêfle. Ce résultat pourrait signifier que la structure en double hélice de ce bras est moins stable qu'on ne le suppose généralement.

D'autres expériences montrent cependant une difficulté d'attaque par certains enzymes des nucléotides de la région terminale du bras de l'aminoacide. Par action de la phosphodiestérase de venin de serpent sur le $tRNA^{phe}$ de levure, l'AMP terminal est rapidement enlevé, le CMP adjacent est déjà coupé avec une vitesse plus lente, tandis que le CMP suivant n'est enlevé que par un traitement prolongé en présence d'une concentration élevée d'enzyme [38]. De même l'enlèvement par la phosphatase alcaline du phosphate 5′ terminal du $tRNA^{phe}$ de *E. coli* n'est pas possible à 37°; il n'a lieu qu'à température voisine de la température de fusion [39]

4.2.2. *Action de réactifs chimiques*

De nombreux réactifs chimiques ont été utilisés pour déterminer les régions accessibles des tRNA (pour une bibliographie de la question, voir [3, 29, 40, 41]). En fait, cette approche intéressante présente quelques difficultés, en raison de l'altération fréquente de la structure secondaire ou tertiaire entraînée par beaucoup de ces réactifs. Pour donner des résultats interprétables quant à la conformation tridimensionnelle des tRNA, les réactifs chimiques devront maintenir l'intégrité de cette structure, ce qui suppose qu'ils ne provoquent pas de coupures et qu'aucune ouverture des régions hélicoïdales n'ait lieu. Ce résultat sera atteint d'autant plus facilement que le réactif attaquera un nombre limité de nucléosides, si possible un seul. En fait, seul un petit nombre de réactifs sont spécifiques d'une seule base, ceux touchant spécifiquement les nucléosides mineurs présentant un intérêt particulier.

Une analyse détaillée des résultats obtenus pour chaque tRNA avec ces divers réactifs n'est pas possible ici. D'une manière générale, les résultats sont compatibles avec le modèle en feuille de trêfle, en ce sens que les régions situées dans les boucles sont plus réactives que les régions appariées. Mais on trouve des différences de réactivité considérables entre certaines régions qui, selon le modèle en feuille de trêfle, devraient être exposées de la même façon. Comme dans l'attaque enzymatique, la région la plus réactive est la boucle de l'anticodon, suivie de la boucle du dihydrouracile et de la région 3′ terminale contenant le triplet CCA. La quasi-totalité des résultats montrent une réactivité faible des bases de la boucle du TΨC que le modèle en feuille de trèfle ne permet pas de prévoir. La Figure 7 montre les nucléosides attaqués par plusieurs réactifs dans le $tRNA^{phe}$ de levure [31], qui a été l'un des plus étudiés.

La technique des modifications chimiques a aussi permis de retrouver la différence de réactivité de certaines régions des tRNA lorsquil's sont sous la forme native ou sous la forme dénaturée. Par exemple, il a été montré [42] que les résidus 4-thiouracile présents dans de nombreux tRNA de E. coli sont inaccessibles à l'oxydation périodique lorsque ces tRNA sont sous la forme native, alors qu'ils deviennent accessibles à cette modification lorsqu'ils sont sous la forme dénaturée.

En fait, il convient dans les expériences de modification chimique de vérifier si l'inaccessibilité aux réactifs n'est pas une propriété intrinsèque de la séquence étudiée plutôt qu'une conséquence de la protection au sein d'une structure secondaire ou tertiaire. Il a été montré [43] que la réactivité des résidus cytosine vis-à-vis de la méthoxyamine peut être fortement modifiée pour le nucléoside voisin dans des nucléoside diphosphates de structure CpN. Ce résultat pourrait expliquer que le C de l'anticodon dans le $tRNA_f^{met}$ de *E. coli* n'est pas modifié par action de l'hydroxylamine [44].

Une réaction particulièrement intéressante est la transformation photochimique de la 4-thiouridine par irradiation à 335 mm. Lorsque cette réaction est appliquée aux tRNA renfermant ce nucléoside mineur en position 8, par exemple le $tRNA_{I_1}^{val}$ de *E. coli* [45] (Figure 5), il se forme une liaison covalente entre cette base et la cytidine en position 13 située dans la région hélicoïdale du bras du dihydrouracile [46]. La nature de cette liaison a été déterminée sur le produit formé par irradiation du mélange des 2 nucléosides correspondants [47]. Il a été montré que le tRNA irradié contenant ce pontage fonctionnait normalement dans la synthèse protéiqie, ce qui suggère que la structure du tRNA modifié n'a pas subi de grands changements par rapport à la forme native et que, par voie de conséquence, la formation de cette liaison interne reflète la proximité de ces deux nucléosides dans le tRNA.

Une méthode prometteuse permettant d'obtenir des pontages entre oligonucléotides voisins dans la structure tridimensionnelle des tRNA a été développée par Danchin [48], en utilisant le marquage par les ions Co^{+++}. La chimie structurale de l'ion cobaltique est voisine de celle de l'ion Mg^{++}, mais à la différence de ce dernier, les ions Co^{+++} forment des liaisons pratiquement covalentes de sorte qu'ils peuvent être utilisés comme marqueurs spécifiques des tRNA.

4.2.3. *Essais de liaison d'oligonucléotides complémentaires de certaines séquences du tRNA*

Une autre manière d'obtenir des informations sur les régions exposées des tRNA est de mesurer les constantes d'association d'oligonucléotides avec des séquences complémentaires dans la molécule des tRNA. Si une région du tRNA n'est ni appariée, ni enfouie dans la structure tertiaire, elle devrait être capable de lier l'oligonucléotide.

A l'inverse, des régions en double hélice ne devraient pas en être capables.

Cette approche a été spécialement explorée par Uhlenbeck [49, 50], qui a étudié plus particulièrement les $tRNA^{tyr}$ et M_f^{met} de *E. coli* et le $tRNA_3^{leu}$ de levure, en déterminant les constantes d'association avec divers oligonucléotides. Les 3 tRNA présentent des caractéristiques d'association semblables. Dand tous les cas une partie de la boucle de l'anticodon, une partie de la boucle du dihydrouracile, la séquence ACCA 3′ terminale et la boucle variable se sont avérées capables de lier des oligonucléotides (la boucle variable avec des constantes d'association faibles), tandis que la boucle du TΨC n'a donné lieu à aucune interaction. Les régions en double hélice du modèle en feuille de trèfle se sont également montrées inaccessibles, sauf la région hélicoïdale du bras du dihydrouracile dans la forme native du $tRNA_3^{leu}$ qui a lié des tétranucléotides complémentaires, ce qui suggère une stabilité de cette hélice exceptionnellement faible. Ce résultat doit être mis en relation avec la stabilité particulière de la forme dénaturée de ce tRNA: en effet, alors que la plupart des tRNA dénaturés se renaturent spontanément en présence d'ions Mg^{++} à température ordinaire, le $tRNA_3^{leu}$ de levure nécessite un chauffage en présence de Mg^{++} à 60°.

Dans l'interprétation des résultats de cette approche intéressante, il convient cependant de tenir compte de divers faits qui peuvent la compliquer, par exemple la possibilité d'appariements parasites en triple hélice ou encore le fait qu'une structure en simple brin très relachée devrait en principe s'apparier difficilement à des trinucléotides. De plus, il peut y avoir compétition des oligonucléotides avec des appariements existant normalement, mais instables. Enfin l'absence de réactivité peut, au moins partiellement, être liée à la conformation même de la région étudiée. Des expériences préliminaires dans lesquelles l'association d'oligonucléotides avec la boucle TΨC a été étudiée comparativement dans des molécules de tRNA entières et dans la moitié 3′ du $tRNA^{asp}$ de levure [51] ont permis de trouver des réactivités assez semblables, ce qui tendrait à montrer que l'absence de réactivité de cette boucle résulte de propriétés intrinsèques de cette séquence de nucléotides et non de sa protection dans la structure tertiaire.

4.2.4. *Expériences d'échange d'hydrogène*

L'échange de tritium a été utilisé comme une approche pour étudier la structure tertiaire des tRNA. On peut prévoir que cet échange est plus lent lorsque les H sont impliqués dans un appariement de bases. Diverses expériences [52, 53] ont permis de mettre en évidence les faits suivants: (1) Le nombre de liaisons H trouvé dans tous les tRNA étudiés est plus grand que celui prévu par le modèle en feuille de trèfle, par exemple, 115 H sont échangés dans le $tRNA_f^{met}$ de *E. coli* contre environ 95 prévus dans le modèle en feuille de trèfle. (2) A faible force ionique, les liaisons H sont moins nombreuses et parfois différentes de celles qui existent dans les conditions normales. (3) Les formes dénaturées présentent moins d'appariements de bases que les formes natives.

4.2.5. *Conclusion*

On peut essayer, à la suite des résultats obtenus en utilisant les diverses approches que nous venons d'examiner, de tirer quelques conclusions sur la plus ou moins grande accessibilité des régions définies par le modèle en feuille de trèfle. Dans l'interprétation de ces résultats, il conviendra d'avoir toujours présente à l'esprit l'objection qu'une absence de réactivité d'une region dans le tRNA peut être une propriété intrinsèque consécutive à sa séquence nucléotidique et par conséquent ne pas signifier obligatoirement que c'est une région protégée.

Tout concourt à admettre que les appariements prévus dans le modèle en feuille de trèfle existent réellement dans la structure des tRNA. C'est en particulier le cas du bras de l'aminoacide. Cependant l'accessibilité de cette région à l'endonucléase de rein de veau [36] pourrait signifier qu'il existe un équilibre entre une forme ouverte et une forme appariée.

Des résultats quelque peu contradictoires ont été obtenus concernant l'accessibilité de l'extrémité CCA 3′ terminale. Des expériences faisant appel à une hydrolyse ménagée par les endonucléases ou à l'emploi de modifications chimiques montrent une grande accessibilité de ce triplet. Mais d'autres résultats suggèrent au contraire une protection au moins partielle. Nous avons déjà mentionné l'enlèvement difficile du deuxième C sous l'influence de la phosphodiestérase de venin de serpent. La variation du spectre de fluorescence d'un tRNA dont le A terminal a été remplacé par la formycine [54] suggère que ce nucléoside terminal est impliqué dans une structure tertiaire. Il a par ailleurs été montré dans notre laboratoire [55] qu'une modification de conformation lors de l'attachement du A terminal dans la réaction de la tRNA nucléotidyl transférase est également suggérée par le fait que le tRNA-CC dépourvu du A terminal est un inhibiteur compétitif dans la réaction d'attachement du CMP au tRNA sans CCA, alors que le tRNA-CCA se comporte comme un inhibiteur non compétitif dans la même réaction. Cependant la démonstration que les trinucléotides UGG et GGU se fixent très bien sur la séquence terminale ACCA [49] est difficilement conciliable avec un appariement de ce triplet. Elle est en faveur de l'accessibilité du triplet, tout en suggérant que cette extrémité ne flotte pas librement, mais qu'elle possède une conformation relativement rigide.

En ce qui concerne la boucle du dihydrouracile, il est probable qu'elle est exposée dans la structure tertiaire, compte tenu de sa vulnérabilité aux réactifs enzymatiques et chimiques. Cependant la fixation d'oligonucléotides complémentaires ne se fait facilement qu'avec une partie de cette boucle [49]. Rappelons que la complémentarité constante de la purine en 15 avec la pyrimidine de la boucle variable suggère que la boucle du dihydrouracile doit être au voisinage de la boucle variable.

L'exposition de la boucle de l'anticodon dans la structure tertiaire peut être considérée comme établie. Le fait que seules les 5 bases de la boucle du côte 5′ sont susceptibles d'appariement [49] suggère qu'il existe une discontinuité entre ce groupe de bases et les deux autres de l'anticodon situées du côté 3′, ce qui n'est pas compatible avec le modèle de Fuller et Hodgson proposé pour la conformation de la boucle de l'anticodon [56].

La conformation de la région variable dépend évidemment de sa taille. Là où le nombre de nucléotides est suffisamment élevé et que le modèle en feuille de trèfle prévoit une région hélicoïdale supplémentaire, il y a toutes raisons de penser que ces appariements sont effectivement réalisés [60], mais il y a des indications tendant à montrer que la dernière paire de cette boucle est relativement labile, puisque la cytosine impliquée dans cette paire est susceptible d'être modifiée [61].

La valeur relativement faible des associations obtenues avec des oligonucléotides complémentaires suggère que cette région, ou au moins des parties de cette région, sont impliquées dans la structure tertiaire du tRNA [49]. Nous avons mentionné l'appariement probable de la dernière pyrimidine de cette région avec la purine en 15 de la boucle du dihydrouracile. De plus, la région variable contient souvent une 7-méthylguanosine. Dans le $tRNA^{phe}$ de levure, cette base est inaccessible à l'action du borohydrure de Na [62]. Il faut remarquer qu'à pH 7 cette 7-méthylguanosine porte une charge positive qui pourrait interagir avec un phosphate chargé négativement.

La boucle du TΨC pose un problème. Un nombre considérable de données semble montrer son inaccessibilité à de nombreux réactifs enzymatiques et chimiques. Ceci est particulièrement net avec les nucléosides mineurs de cette boucle, comme le Ψ qui n'est pas modifié par l'acrylonitrile [63] et comme la 1-méthyladénosine qui ne réagit pas avec le borohydrure de Na [62]. De plus, on n'observe aucune liaison d'oligonucléotides complémentaires au niveau de cette boucle [49]. Mais le fait que le même phénomène s'observe avec la moitié 3′ de la molécule [51], par conséquent dans des conditions où la protection par la structure tertiaire ne peut intervenir, suggère que cette inaccessibilité est en réalité liée à une conformation particulière de cette boucle.

4.3. Études physiques

L'existence d'une structure tertiaire plus complexe que le modèle en feuille de trèfle ouvert est largement confirmée par de nombreuses études physiques. Les forces qui peuvent intervenir dans la stabilisation d'une telle structure sont multiples. Il faut tenir compte: (1) de l'appariement classique par liaison hydrogène de certaines bases situées en-dehors des régions hélicoïdales du modèle en feuille de trèfle; (2) de l'empilement ('stacking') des bases; (3) d'attractions ou de répulsions provoquées par les anions de la chaîne polynucléotidique et les cations qui lui sont associés; (4) des restrictions de rotation imposées au squelette sucre-phosphate et à la liaison de la base au sucre.

Il faut cependant remarquer que l'interaction avec les voisins immédiats dans la chaîne polynucléotidique peut contribuer d'une manière importante à la conformation de la région. Des études cristallographiques sur des dinucléotides [64, 65] apportent des renseignements sur la flexibilité de la liaison phosphodiester et de la liaison base–sucre. Une telle approche permettra peut-être de prévoir la structure des boucles. Mais dans l'état actuel de nos connaissances, la complexité des interactions est telle qu'il n'est pas possible de prédire à partir de la structure primaire quelle va être la structure tertiaire d'un tRNA. Un grand nombre de méthodes physiques ont été appliquées à ce matériel en vue d'abord de confirmer l'existence d'une telle structure, puis d'en déterminer les caractéristiques. Il ne nous est pas possible de passer en revue toutes les techniques utilisées. Nous nous contenterons d'indiquer celles qui ont donné les résultats les plus significatifs et vont nous permettre de discuter la validité de quelques modèles de structure proposés.

Nous rappelons d'abord les résultats d'expériences de variation de la viscosité et de la densité optique en fonction de la température [66]. La variation de la viscosité se traduit par une courbe diphasique. Entre 20° et 40°, on observe un changement important de cette propriété, alors que la variation de la densité optique à 260 mm est faible. Au-delà de 45°, on observe une nouvelle variation de la viscosité, parallèle cette fois-ci à celle de la densité optique. La première variation de la viscosité a été attribuée à la perte de la structure tertiaire, la seconde à celle de la structure secondaire.

Des études plus détaillées de dénaturation thermique [67, 68] effectuées sur des tRNA spécifiques en suivant la variation de l'absorption d'un nucléoside mineur, par exemple la 4-thiouridine, ont permis de mettre en évidence toute une série de conformations intermédiaires entre la forme native présentant une structure tertiaire et la forme dénaturée. Ces conformations sont aussi dépendantes de la concentration saline, les fortes concentrations salines favorisant l'établissement d'une structure voisine de l'état natif.

Des expériences de saut de température ('*T*-jump') ont été réalisées [69, 70]. Elles permettent de mettre en évidence d'éventuels couplages entre bras différents de la molécule, la principale difficulté étant d'attribuer chaque signal à un bras particulier.

L'influence des ions Mg^{++} sur la structure tertiaire a donné lieu à de nombreux travaux. Il ressort de ces études qu'ils causent une stabilisation importante de la structure tertiaire, mais sans conduire à une conformation différente de celle observée à forte concentration d'ions Na^{+}. Les expériences d'équilibre de dialyse [71, 72], de résonance magnétique nucléaire du Mn^{++} [73, 74], de dichroïsme circulaire [75] et de titration de fluorescence [76], montrent l'existence de deux catégories de sites de fixation: des sites forts, au nombre de 4 à 6, et des sites faibles, entre 20 et 30.

La caractérisation de la nature de cette structure tertiaire des tRNA a donné lieu à de nombreux travaux. Il faut d'abord remarquer qu'elle doit présenter de grandes analogies pour l'ensemble des tRNA. Ceci est en particulier suggéré par la cocristallisation de plusieurs tRNA différents dans une structure cristalline donnant des diagrammes de diffraction des rayons X correspondant à un cristal unique [77].

Ceci est d'ailleurs compatible avec le fait que, à côté de la capacité de fixer les aminoacides qui est spécifique de chaque tRNA et qui implique donc des différences structurales, les tRNA présentent des propriétés biologiques communes, impliquant une certaine homologie de structure: interactions avec les ribosomes, avec la tRNA nucléotidyl transférase, qui attache le groupement CCA à tous les tRNA, avec les enzymes de modification des tRNA, avec le facteur d'élongation *Tu*.

L'analogie dans la structure tertiaire des tRNA est confirmée par les expériences de diffusion centrale des rayons X, qui ont permis d'avoir des informations sur les dimensions et la forme des molécules en solution. La méthode a été appliquée à toute une série de tRNA de levure ou de *E. coli* (pour une revue de la question voir [40]). Pour tous ces tRNA, des rayons de giration entre 23.5 Å et 25 Å ont été trouvés. On peut donc en conclure que la dimension des tRNA en solution est très semblable.

L'étude des courbes de diffusion exclut définitivement tout arrangement en trèfle ouvert [96]. Le calcul des dimensions des tRNA par divers auteurs a conduit aux valeurs suivantes: $22 \times 36 \times 92$ Å [78], $25 \times 35 \times 85$ Å [80] et $40 \times 40 \times 100$ Å [79]. Les écarts trouvés entre ces valeurs résultent non pas de divergences dans l'interprétation des données qui sont voisines, mais dans la manière dont chaque auteur définit la dimension de la molécule. Toutes ces valeurs sont en accord avec une molécule allongée. Si on l'assimile à un ellipsoïde, on trouve un rapport grand axe/petit axe voison de 3.

La forme asymétrique des tRNA découle également de travaux de polarisation de fluorescence [81–84], ainsi que de l'étude des spectres de fluorescence obtenus soit en fixant des colorants comme la proflavine [85] ou le bromure d'éthidium [86], soit en utilisant la présence dans la molécule de tRNA de nucléosides naturellement fluorescents. C'est ainsi que le $tRNA^{phe}$ de levure contient, à côté de son anticodon, un nucleoside mineur Y dérivé de la guanosine fortement fluorescent. Quand on introduit dans le $tRNA^{phe}$ une deuxième molécule fluorescente, on peut mesurer la distance entre les deux molécules fluorescentes par transfert d'énergies de fluorescence [87]. Plusieurs dérivés acridiniques ont été fixés à l'extrémité 3′-OH du $tRNA^{phe}$ après oxydation périodique. Le transfert d'énergie observé par 'quenching' de la fluorescence du nucléoside Y et activation des molécules acceptrices est faible, ce qui suggère que l'anticodon est à plus de 40 Å de l'extrémité CCA.

Signalons enfin que des expériences de microscopie électronique en haute résolution suggèrent une dimension de $85 \times 42 \times 36$ Å [88].

Toutes ces méthodes ne fournissent cependant que des informations très globales sur la structure tridimensionnelle des tRNA. Les progrès énormes que l'étude par diffraction aux rayons X des protéines cristallisées a fait faire à la connaissance de la structure tridimensionnelle de ces macromolécules a amené à essayer d'appliquer la même technique aux tRNA. Un grand progrès dans cette voie a été réalisé par la réussite de la cristallisation de plusieurs tRNA. Actuellement une douzaine de tRNA ont été obtenus cristallisés (pour une revue de la question, voir [40, 89]) et il est actuellement possible de dégager quelques règles et caractéristiques générales pour la préparation et les propriétés de ces cristaux.

La cristallisation peut être réalisée, en présence d'ions monovalents ou divalents, par addition soit d'un agent relargant comme le sulfate d'ammonium, soit de solvants organiques comme l'éthanol, le *t*-butanol, le dioxane, l'isopropanol. Jusqu'a présent, des cristaux appartenant à des systèmes cristallographiques variés ont été obtenus. Très souvent les dimensions de la maille cristalline et le groupe d'espèce ont pu être interprétés comme résultant de l'empilement de dimères de la molécule de tRNA, les deux molécules de tRNA dans le dimère étant en contact sur des longueurs variables et leurs axes étant parallèles. Une grande variabilité du volume de la maille et du nombre de molécules présentes dans cette maille a été observée. Pour les mailles qui ont été déterminées jusqu'à présent, le volume disponible par molécule de tRNA hydraté semble avoir une valeur minimale d'environ 67000 $Å^3$, qui correspondrait donc à l'empilement le plus compact possible de ces molécules. Cette valuer est très voisine de la valeur la plus fréquente observée dans les différents cristaux qui est de 74000 $Å^3$, ce qui semblerait indiquer qu'une limite à leur empilement est atteinte et que cette limite est la même pour tous les tRNA examinés. C'est là un indice supplémentaire d'une similitude de structure pour les divers tRNA.

Dans l'état actuel de la question, aucun renseignement précis n'a encore pu être tiré de l'étude des tRNA par diffraction des rayons X sinon la conclusion [90] de la présence dans la structure des tRNA de courts segments en double hélice parallèles. Des études sont actuellement effectuées pour rechercher des méthodes de substitution isomorphe avec des substituants renfermant des métaux lourds (voir [40]).

5. Modèles

En l'absence de données plus précises de l'analyse cristallographique, on en est réduit à ne pouvoir faire que des spéculations sur la structure tridimensionnelle des tRNA. Ces spéculations se sont traduites par l'élaboration de plusieurs modèles de structure.

L'élaboration d'un modèle de structure valable dans ses grandes lignes pour l'ensemble des tRNA est justifiée, en raison des arguments que nous avons déjà évoqués précédemment: la cocristallisation de nombreux tRNA dans un cristal unique, la valeur identique des paramètres trouvés pour des tRNA différents dans les expériences de diffusion centrale des rayons X, l'accessibilité des mêmes régions à divers agents enzymatiques ou chimiques dans des tRNA différents, le voisinage identique de certaines bases dans ces tRNA, la reconnaissance de l'ensemble des tRNA par des enzymes ou des facteurs uniques, la liaison enfin de l'ensemble des tRNA d'une manière identique au ribosome. Tous ces arguments militent en faveur d'une structure tridimensionnelle très voisine pour l'ensemble des tRNA.

5.1. Caractéristiques structurales auxquelles doivent se conformer les modèles

Si la construction d'un modèle peut donc être considérée comme justifiée, elle doit cependant tenir compte d'un certain nombre de données précises sur la structure des tRNA qui se dégagent des études que nous avons mentionnées précédemment:

(1) Le modèle en feuille de trèfle est la base du modèle de structure tridimensionnelle. Nous avons vu que ce modèle de structure secondaire peut être déduit de toutes les structures primaires des tRNA jusqu'à présent déterminées et que, sauf quelques rates exceptions, les résultats d'experiénces physiques, chimiques ou enzymologiques sont compatibles avec ce modèle.

(2) Le tRNA est une molécule compacte et asymétrique, présentant une structure allongée, avec un rayon de giration, déduit des expériences de diffusion centrale des rayons X, de 23,5 à 25 Å. Dans l'utilisation de cette donnée, il faudra cependant tenir compte du fait que à masse égale le rayon de giration ne dépend pas tant de la longueur de la molécule que de l'asymétrie dans la distribution de la matière. A longueur égale, une molécule enflée au centre aura un rayon de giration plus petit qu'une molécule enflée aux extrémités.

(3) La thiouridine qui existe en position 8 dans plusieurs tRNA de *E. coli* doit se trouver à proximité immédiate de la cytidine en position 13 dans la région hélicoïdale du bras du dihydrouracile, puisque dans la réaction photochimique décrite plus haut, il se forme une liaison covalente entre ces deux bases. Le fait que les tRNA contenant ce pont continuent à fonctionner normalement dans la synthèse protiqéue montre que cette liaison n'a pu induire que des modifications de structure faibles et qu'e'lle na pas été formée dans un état transitoire de la molécule, mais dans un état proche de sa forme native. La structure du photoproduit [47] indique la manière dont les résidus 8 et 9 doivent se replier dans le grand sillon du bras du dihydrouracile [91]. Compte tenu de la similitude de structure tertiaire que nous avons admise pour tous les tRNA, on peut postuler que la relation de voisinage du nucléoside en 8 (qui est toujours un U ou un dérivé de U) et de la région hélicoïdale du bras du dihyroudracile est un caractère général des tRNA.

(4) Dans tous les tRNA on trouve la complémentarité entre la purine située en 15 à partir de l'extrémité 5′ dans la boucle du dihydrouracile et la pyrimidine située en 29 à partir de l'extrémité 3′, dans la boucle variable. Levitt [12] a le premier postulé que l'appariement entre ces deux bases pouvait être une contrainte générale à tous les tRNA. Un argument en faveur de cette conception est apporté par le fait que dans le $tRNA^{tyr}_{III\,A15}$ extrait d'un mutant de *E. coli* [92], le changement de G_{15} en A inactive la molécule, cependant que C_{57}, inaccessible à la méthoxyamine dans les $tRNA^{tyr}$ normaux, le devient dans ce mutant [93]. L'existence de cet appariement suggère qu'il s'agit d'une paire antiparallèle insérée dans des chaînes sucre-phosphate parallèles [94].

Les quatre données précédentes sont les seules données précises dont on dispose actuellement pour juger de la valeur d'un modèle de structure tridimensionnelle. Comme nous venons de le voir, on dispose de beaucoup d'autres données, mais elles apportent des renseignements moins précis. Par exemple, les études de transfert d'énergie de fluorescence entre deux marqueurs fluorescents confirmant simplement la forme allongée de la molécule de tRNA. Les études de RMN ou d'échange de tritium confirment que les appariements prévus par le modèle en feuille de trèfle ne peuvent être les seules liaisons H présentes dans la molécule, mais ne nous renseignent pas sur la nature des appariements supplémentaires [95]. Enfin les expériences d'accessibilité des régions à divers réactifs conduisent à des interprétations souvent discutables, comme nous l'avons souligné plus haut.

5.2. Les différents modèles proposés

Une dizaine de modèles (12, 57, 80, 96 à 104) ont été proposés tenant plus ou moins compte, souvent en raison de la date où ils ont été établis, des critères communs que nous venons de définir. Ces modèles sont schématisés dans la Figure 8. Nous ne décrirons pas les caractéristiques détaillées de chacun d'entre eux. Nous nous contenterons de discuter leur compatibilité avec les critères généraux précédents.

La compatibilité de certains modèles avec les données de la diffusion centrale des rayons X a été discutée par certains auteurs. C'est ainsi que, selon Ninio *et al.* [105],

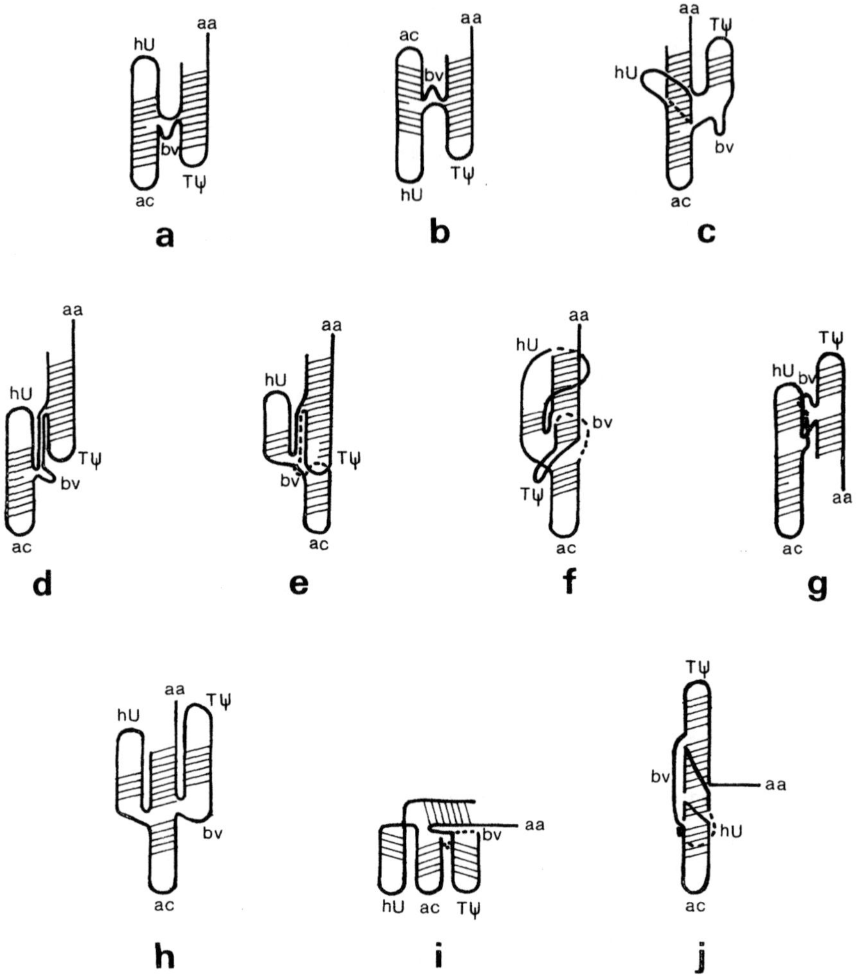

Fig. 8. Schémas des modèles proposés pour la structure tridimensionnelle des tRNA, (a) Lake et Beaman [96]; (b) Doctor *et al.* [97]; (c) Ninio *et al.* [98, 99]; (d) Levitt [12]; (e) Fuller *et al.* [100]; (f) Connors *et al.* [80], Danchin [101]; (g) Yang [102]; (h) Cramer *et al.* [57]; (i) Melcher [103]; (j) Abraham [104].

le rayon de giration du modèle de Cramer *et al.* [57] n'atteindrait que 18 Å au lieu des 24–25 Å déterminés expérimentalement et celui des modèles du type Levitt [12], Fuller *et al.* [100], Connors *et al.* [80], Danchin [101] et Yang [102], qui présentent tous un renflement central, et à fortiori celui de Melcher [103], qui a une forme très compacte, n'atteindraient au maximum que 23 Å. Cependant, pour accorder une confiance absolue aux données de la diffusion centrale des rayons X, il faudrait qu'il y ait davantage de résultats où l'on dispose simultanément de données fournies par cette méthode et par la diffraction des rayons X.

La proximité des résidues uridine ou 4-thiouridine en 8 et cytidine en 13, nécessaire pour expliquer la formation de la liaison covalente par photoréaction, n'est pas respectée dans une certain nombre de modèles, tels ceux de Cramer *et al.* [57] ou d'Abraham [104], ni dans les modèles du type H (Lake et Beeman [96], Doctor *et al.* [97]).

La contrainte résultant de la complémentarité entre la purine en 15 dans la boucle du dihyroudracile et la pyrimidine dans la boucle variable n'est pas respectée dans tous les modèles qui ont précédé le modèle de Levitt [12], par exemple celui de Cramer *et al.* [57] ou celui de Ninio *et al.* [98]. Ce dernier modèle a par la suite été raffiné [99], de manière à le rendre compatible avec cette exigence.

Plusieurs appariements supplémentaires ont été postulés dans un certain nombre de modèles. L'appariement A_{14}–G_{15} avec Ψ_{55}–C_{56} proposé par Cramer *et al.* [57] ne peut être généralisé étant donné que dans plusieurs tRNA G_{15} est remplacé par A_{15}. L'appariement de GG de la boucle du dihydrouracile avec CC du triplet CCA, proposé dans le modèle de Cramer *et al.* [57] et dans la forme 'peptide' du modèle de Danchin [101], ne pourrait être, s'il existait, qu'un appariement labile, étant donné l'accessibilité facile aux réactifs chimiques et la forte interaction avec les oligonucléotides complémentaires du triplet CCA.

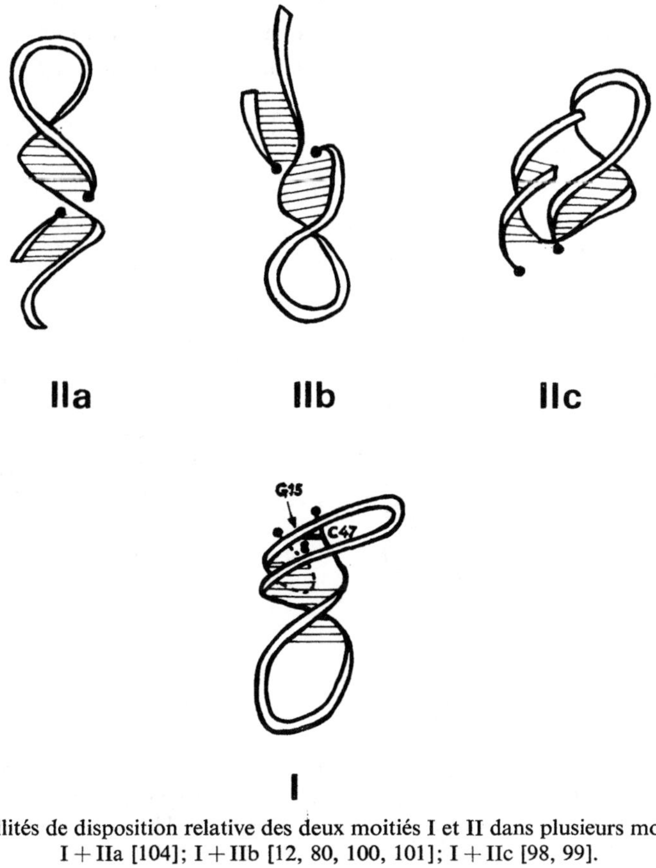

Fig. 9. Possibilités de disposition relative des deux moitiés I et II dans plusieurs modèles de tRNA; I + IIa [104]; I + IIb [12, 80, 100, 101]; I + IIc [98, 99].

Si l'on fait la synthèse des données concernant la compatibilité des modèles avec les deux critères de la proximité U_8–C_{13} et purine 15-pyrimidine de la boucle variable, ainsi qu'avec la nécessité d'une molécule compacte et allongée, sans cependant s'arrêter à une valeur trop précise du rayon de giration, on constate un accord plus grand avec ces critères des modèles de Levitt [12], de Fuller *et al.* [100], de Connors *et al.* [80], de Danchin [101], de Yang [102] et de la version remaniée du modèle de Ninio *et al.* [99] que les autres modèles proposés.

En fait, si l'on compare entre eux les modèles que nous venons de mentionner, on constate [99] qu'en gros, on peut considérer le tRNA comme formé de 2 moitiés (Figure 9). La première moitié comprend les bras du dihydrouracile (y compris le petit segment 8–9 contenant U ou 4 thio U) et de l'anticodon et la région variable; la deuxième moitié comprend le bras de l'aminoacide et le bras du TΨC. Compte tenu de la liaison thio U–C et de la contrainte de Levitt, les deux points de raccordement de ces deux moitiés doivent être voisins dans l'espace. Les différences entre les modèles se placent au niveau de la position de la deuxième moitié par rapport à la première. Les différentes positions proposées sont schématisées dans la Figure 9.

Nous montrons dans la Figure 10 la représentation plus détaillée de 4 modèles. Il ressort de l'analyse précédente que, parmi tous les modèles proposés, certains ont plus de chances de correspondre à la réalité que d'autres. Mais, dans l'état actuel de nos connaissances, ils ne peuvent être considérés que comme des hypothèses de

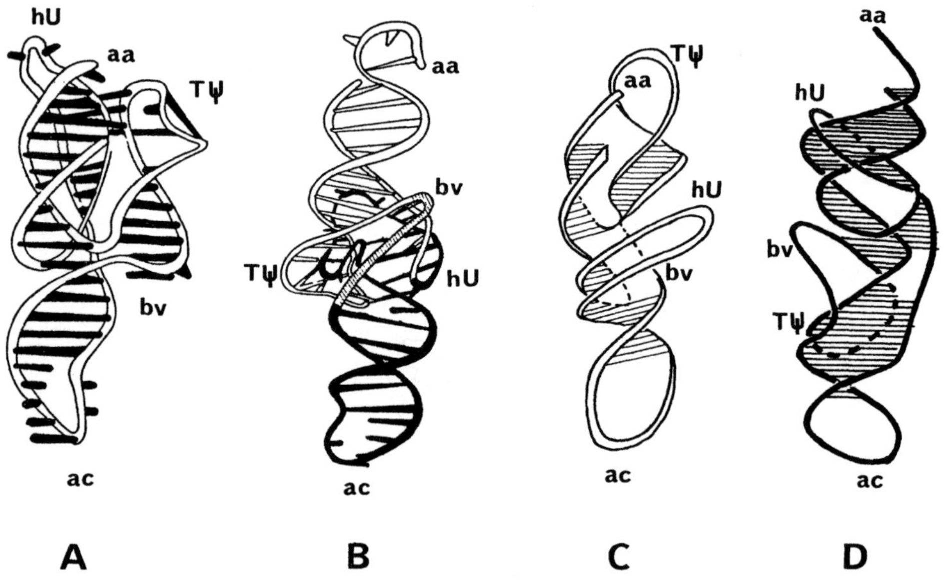

Fig. 10. Représentation détaillée de plusieurs modèles de structure tertiaire des tRNA. A: Cramer *et al.* [57]; B: Levitt [12]; C: Ninio *et al.* [99] (représentation modifiée du modèle décrit en [98]); D: modèle pour la structure A de Danchin [101]; ce modèle reprend les caractéristiques topologiques du modèle de Connors *et al.* [80] en y ajoutant certains appariements spécifiques.

travail. Seuls les progrès dans l'étude des cristaux de tRNA pourront apporter une réponse définitive au problème de la structure tridimensionnelle des tRNA.

6. Polymorphisme des tRNA

On sait que les tRNA sont des molécules multifonctionnelles et on peut se demander si, dans les diverses activités qu'ils exercent (fixation spécifique des aminoacides, reconnaissance des codons du mRNA, fixation au ribosome, reconnaissance des facteurs d'initiation ou d'élongation dans la synthèse protéique), ils présentent toujours la même conformation.

Nous avons déjà évoqué précédemment la possibilité d'une modification de conformation des tRNA lors de l'attachement du triplet CCA terminal [55]. Les recherches les plus nombreuses ont été effectuées pour déceler une éventuelle modification de conformation au cours de la réaction d'aminoacylation. Il est connu que plusieurs composants du système de synthèse protéique sont capables de distinguer les amino-acyl-tRNA non chargés [106, 107]. Mais des expériences effectuées avec le $tRNA_I^{val}$ de *E. coli*, dans lequel le pontage décrit précédemment entre 4-thiouracile en 8 et cytosine en 13 a été réalisé par irradiation à 335 nm [45, 46], ont montré que cette contrainte n'empêchait aucunement ce tRNA d'être actif dans toutes les étapes de la synthèse protéique [108, 109]. Ce résultat montre que, si le tRNA change de conformation au cours des réactions auxquelles il participe, cette déformation ne peut être considérable puisqu'elle est compatible avec la contrainte précédente.

Cependant des différences dans les propriétés physico-chimiques des tRNA aminoacylés et non aminoacylés ont été observées: par exemple dans les propriétés de sédimentation [110], dans la liaison avec l'oligo C [111], peut-être dans les courbes de fusion et de dichroïsme circulaire [112, 113]. Par contre, ni les études d'accessibilité aux nucléases [114] et à divers agents de modification chimique [115], ni les expériences d'échange de tritium [52] n'ont permis de déceler des différences entre les deux formes. Par ailleurs la cinétique d'échange de tritium n'est pas modifiée lors de l'interaction du $tRNA^{ile}$ de *E. coli* avec l'isoleucyl-tRNA synthétase [116]. Une étude par diffusion centrale des rayons X a été effectués [105] sur le $tRNA_1^{val}$ de *E. coli* dans trois états: la molécule non acylée, la molécule non acylée et comportant le pontage thio U–C, et la molécule acylée avec le groupement NH_2 bloqué par le groupement phénoxyacétyle. Les paramètres morphologiques (poids moléculaire, rayons de giration) ont été trouvés pratiquement identiques dans les 3 états, ce qui suggère que la forme générale de la molécule n'a subi que peu de variations lors des différentes étapes de son activité. Par contre, dans une étude de diffusion aux grands angles, une différence a été observée entre les formes non acylée et acylée, qui est interprétée par les auteurs comme pouvant résulter d'un faible changement de conformation, lors de la fixation de l'aminoacide, provoquant une altération des interactions soluté-solvant.

Remerciements

Nous remercions tout particulièrement MM. Jacques Ninio et Jean Witz des conseils qu'ils nous ont prodigués pour la rédaction de ce mémoire. Nos remerciements vont également à MM. Guy Dirheimer, Jacques Bonnet et Gérard Keith pour les discussions fructueuses que nous avons eues sur divers aspects de ce travail et l'aide qu'ils nous ont apportée dans son élaboration.

Références

[1] Holley, R. W., Apgar, J., Everett, G. A., Madison, J. P., Marquisee, M., Merrill, S. H., Penswick, J. R., et Zamir, A.: 1965, *Science* **147**, 1462.
[2] Merril, C. R.: 1968, *Biopolymers* **6**, 1727.
[3] Zachau, H. G.: 1969, *Angew. Chem.* **81**, 645.
[4] Staehelin, M.: 1971, *Experientia* **27**, 1.
[5] Dirheimer, G., Ebel, J. P., Bonnet, J., Gangloff, J., Keith, G., Krebs, B., Kuntzel, B., Roy, A., Weissenbach, J., et Werner, C.: 1972, *Biochimie* **54**, 127.
[6] Ebel, J. P.: 1972, 'Structure des acides ribonucléiques', dans *Problèmes actuels de Biochimie Générale*, Masson, Paris, p. 89.
[7] Keith, G., Gangloff, J., Ebel, J. P., et Dirheimer, G.: 1970, *C.R. Acad. Sci. Paris* **271**, 613.
[8] Gangloff, J., Keith, G., Ebel, J. P., et Dirheimer, G.: 1971, *Nature New Biol.* **230**, 125.
[9] Keith, G., Roy, A., Ebel, J. P., et Dirheimer, G.: 1971, *FEBS Letters* **17**, 306.
[10] Dube, S. K., Marcker, K. A., Clark, B. F. C., et Cory, S.: 1968, *Nature* **218**, 232.
[11] Singer, C. E., et Smith, G. R.: 1972, *J. Biol. Chem.* **247**, 2989.
[12] Levitt, M.: 1969, *Nature* **224**, 759.
[13] Roberts, R. J.: 1972, *Nature New Biol.* **237**, 44.
[14] Roberts, R. J.: 1972, in *Symposium tRNA*, Princeton.
[15] Zachau, H. G., Dutting, D., et Feldmann, H.: 1966, *Hoppe-Seylers Z. physiol. Chem.* **347**, 212.
[16] Mirzabekov, A. D., Lastity, D., Levina, E. S., et Bayev, A. A.: 1970, *FEBS Letters* **7**, 95.
[17] Bayev, A. A., Venkstern, T. V., Mirzabekov, A. D., Krutilina, A. I., Li, L., et Akselrod, V. D.: 1967, *Mol. Biologyia* **1**, 754.
[18] Bonnet, J., Ebel, J. P., et Dirheimer, G.: 1971, *FEBS Letters* **15**, 286.
[19] Lindahl, T., Adams, A., et Fresco, J. R.: 1966, *Proc. Nat. Acad. Sci. US* **55**, 941.
[20] Gartland, N. J. et Sueoka, N.: 1966, *Proc. Nat. Acad. Sci. US* **55**, 948.
[21] Streek, R. E. et Zachau, H. G.: 1971, *FEBS Letters* **13**, 329.
[22] Renaud, M., Bollack, C., et Ebel, J. P.: 1971, in *Symposium de la Société de Chimie Biologique sur les tRNA*, Strasbourg, 1971.
[23] Johnson, L. et Söll, D.: 1971, *Biopolymers* **10**, 2209.
[24] Giege, R., Kern, D., Ebel, J. P., et Taglang, M.: 1971, *FEBS Letters* **15**, 281.
[25] Hirsh, D.: 1971, *J. Mol. Biol.* **58**, 439.
[26] Thang, M. N., Beltchev, B., et Grunberg-Manago, M.: 1971, *Eur. J. Biochem.* **19**, 184.
[27] Beltchev, B., Thang, M. N., et Portier, C.: 1971, *Eur. J. Biochem.* **19**, 194.
[28] Hänggi, U. J., Streek, R. E., Voigt, H. P., et Zachau, H. G.: 1970, *Biochim. Biophys. Acta* **217**, 278.
[29] Ebel, J. P.: 1968, *Bull. Soc. Chim. Biol.* **50**, 2255.
[30] Reid, B. R., Einarson, B., et Schmidt, J.: 1972, *Biochimie* **54**, 325.
[31] Rajbhandary, U. L., Chang, S. H., Stuart, A., Faulkner, R. D., Hoskinson, R. M., et Khorana, H. G.: 1967, *Proc. Nat. Acad. Sci. US* **57**, 751.
[32] Madison, J. P., Bogulawski, S. J., et Taetor, G. H.: 1972, *Sciences* **176**, 687.
[33] Harbers, K., Thiebe, R., et Zachau, H. G.: 1972, *Eur. J. Biochem.* **26**, 132.
[34] Adams, A., Lindahl, T., et Fresco, J. R.: 1967, *Proc. Nat. Acad. Sci. US* **57**, 1684.
[35] Thang, M. N., Guschlbauer, W., Zachau, H. G., et Grunberg-Manago, M.: 1967, *J. Mol. Biol.* **26**, 403.
[36] Philippsen, P., et Zachau, H. G.: 1972, in *Symposium tRNA*, Princeton.
[37] Kasai, K. et Grunberg-Manago, M.: 1967, *Eur. J. Biochem.* **1**, 152.
[38] Zubay, G. L.: 1968, *Methods Enzymol.* **12B**, 227.
[39] Stern, M., Sutra, I. E., Littauer, U. Z.: 1969, *Biochemistry* **8**, 313.
[40] Cramer, F.: 1971, *Progr. Nucleic Acid Res. Mol. Biol.* **11**, 391.

[41] Miura, K.: 1967, *Progr. Nucleic Acid Res. Mol. Biol.* **6**, 39.
[42] Blake, R. D., Ziff, E., Kao, S., et Fresco, J. R.: 1972, dans *Symposium on tRNA*, Princeton, 1972.
[43] Steinschneider, A. et Lesham, B.: 1972, *J. Mol. Biol.* **67**, 333.
[44] Kawamura, Y. et Mizuno, Y.: 1972, *Biochim. Biophys. Acta* **277**, 323.
[45] Yaniv, M. et Barrell, B. G.: 1969, *Nature* **222**, 278.
[46] Favre, A., Michelson, A. M., et Yaniv, M.: 1971, *J. Mol. Biol.* **58**, 367.
[47] Leonard, N. J., Bergstrom, D. E., et Tolman, G. L.: 1971, *Biochem. Biophys. Res. Commun.* **44**, 1524.
[48] Danchin, A.: 1972, *Biochimie* **54**, 333.
[49] Uhlenbeck, O.: 1972, *J. Mol. Biol.* **65**, 25.
[50] Uhlenbeck, O., Chirikjian, J. G., et Fresco, J. R.: 1972, *Fed. Proc.* **31**, 422.
[51] Ninio, J., Unger, F., et Gangloff, J.: communication personnelle.
[52] Englander, J. J., Kallenbach, N. R., et Englander, S. W.: 1972, *J. Mol. Biol.* **63**, 153.
[53] Webb, P. K. et Fresco, J. R.: 1972, dans *Symposium on tRNA*, Princeton, 1972.
[54] Ward, D. C., Reich, E., et Stryer, L.: 1969, *J. Biol. Chem.* **244**, 1228.
[55] Rether, B. et Ebel, J. P.: 1971, dans *Symposium de la Société de Chimie Biologique sur les tRNA*, Strasbourg, 1971.
[56] Fuller, W. et Hodgson, A.: 1967, *Nature* **215**, 817.
[57] Cramer, F., Doepner, H., Von der Haar, F., Schlimme, E., et Seidel, H.: 1968, *Proc. Nat. Acad. Sci. US* **61**, 1384.
[58] Igo-Kemenes, T., et Zachau, H. G.: 1969, *Europ. J. Biochem.* **10**, 549.
[59] Litt, M. et Greenspan, C. M.: 1972, *Biochemistry* **11**, 1437.
[60] Coutts, S. M.: 1971, *Biochim. Biophys. Acta* **232**, 94.
[61] Cashmore, A. R., Brown, D. M., et Smith, J. D.: 1971, *J. Mol. Biol.* **59**, 359.
[62] Igo-Kemenes, T. et Zachau, H. G.: 1971, *Eur. J. Biochem.* **18**, 292.
[63] Yoshida, M., Furuichi, Y., Kaziro, Y., et Ukita, T.: 1968, *Biochim. Biophys. Acta* **166**, 636.
[64] Rubin, J., Brennan, T., et Sundraralingham, M.: 1972, *Biochemistry* **11**, 3112.
[65] Gray, D. M., Tinocco, I., et Chamberlin, M. J.: 1972, *Biopolymers* **11**, 1235.
[66] Henley, D. D., Lindahl, T., et Fresco, J. R.: 1966, *Proc. Nat. Acad. Sci. US* **55**, 191.
[67] Cole, P. E., Yang, S. K., et Crothers, D. M.: 1972, in *Symposium on tRNA*, Princeton, 1972.
[68] Römer, R., Riesner, D., et Maas, G.: 1970, *FEBS Letters* **10**, 352.
[69] Riesner, D., Römer, R., et Maas, G.: 1970, *Eur. J. Biochem.* **15**, 85.
[70] Dourlent, M., Yaniv, M., et Hélène, C.: 1971, *Eur. J. Biochem.* **19**, 108.
[71] Danchin, A.: 1972, *Biopolymers* **11**, 1317.
[72] Yarus, M. et Rashbaum, S.: 1972, *Biochemistry* **11**, 2043.
[73] Cohn, M., Danchin, A., et Grunberg-Manago, M.: 1969, *J. Mol. Biol.* **39**, 199.
[74] Danchin, A. et Gueron, M.: 1970, *Eur. J. Biochem.* **16**, 532.
[75] Willick, G. E. et Kay, C. M.: 1971, *Biochemistry* **10**, 2216.
[76] Robison, B. et Zimmerman, T. P.: 1971, *J. Biol. Chem.* **246**, 110.
[77] Blake, R. D., Fresco, J. R., et Langridge, R.: 1968, *Nature* **220**, 5174; 1970, **225**, 32.
[78] Pilz, I., Kratky, O., Cramer, F., Von der Haar, F., et Schlimme, E.: 1970, *Eur. J. Biochem.* **15**, 401.
[79] Ninio, J., Favre, A., et Yaniv, M.: 1969, *Nature* **223**, 1333.
[80] Connors, P. G., Labanauskas, M., et Beeman, W. W.: 1969, *Science* **166**, 1528.
[81] Kisselev, N. A. et Kisselev, L. L.: 1961, *Dokl. Akad. Nauk. SSSR* **141**, 980.
[82] Millar, D. B. et Steiner, R. F.: 1966, *Biochemistry* **5**, 2289.
[83] Borisova, O. F., Kisselev, L. L., Surovaja, A., Tumerman, L. L., et Frolova, L. Y.: 1964 *Dokl. Akad. Nauk. SSSR* **159**, 1154.
[84] Dembo, A. T., Susfenov, N. I., et Feigin, L. A.: 1967, *Soviet Phys. Crystallographia* **11**, 508.
[85] Grosjean, H., Werenne, J., et Chantrenne, H.: 1968, *Biochim. Biophys. Acta* **166**, 616.
[86] Tao, T., Nelson, J. H., et Cantor, C. R.: 1970, *Biochemistry US* **9**, 351.
[87] Beardsley, K. et Cantor, C. R.: 1970, *Proc. Nat. Acad. Sci. US* **65**, 39.
[88] Abermann, R. J. et Yoshikami, D.: 1972, *Proc. Nat. Acad. Sci. US* **69**, 1587.
[89] Arnott, S.: 1971, *Progr. Biophys. Mol. Biol.* **22**, 181.
[90] Sakurai, T., Rao, S. T., Rubin, J., et Sundaralingam, M.: 1971, *Science* **172**, 1134.
[91] Bergstrom, D. E. et Leonard, N. J.: 1972, *Biochemistry* **11**, 1.

[92] Abelson, J. N., Gefter, M. L., Barnett, L., Landy, A., Russell, R. L., et Smith, J. D.: 1970, *J. Mol. Biol.* **47**, 15.
[93] Cashmore, T.: 1971, *Nature New Biol.* **230**, 236.
[94] Ninio, J.: communication personnelle.
[95] Gueron, M.: 1971, *FEBS Letters* **19**, 264.
[96] Lake, J. A. et Beeman, W. W.: 1968, *J. Mol. Biol.* **31**, 115.
[97] Doctor, B. P., Fuller, W., et Webb, N. L. W.: 1969, *Nature* **221**, 58.
[98] Ninio, J., Favre, A., et Yaniv, M.: 1969, *Nature* **223**, 1333.
[99] Ninio, J.: 1971, Thèse Doct. Sciences, Université Paris 7.
[100] Fuller, W., Arnott, S., et Creek, J.: 1969, *Biochem. J.* **114**, 269.
[101] Danchin, A.: 1971, *FEBS Letters* **13**, 152.
[102] Jang, C. G.: 1972, in *Symposium on tRNA*, Princeton, 1972.
[103] Melcher, G.: 1969, *FEBS Letters* **3**, 185.
[104] Abraham, D. J.: 1971, *J. Theor. Biol.* **30**, 83.
[105] Ninio, J., Luzzati, V., et Yaniv, M.: 1972, *J. Mol. Biol.* 71, 217.
[106] Lipmann, F.: 1969, *Science* **164**, 1024.
[107] Lengyel, P. et Söll, D.: 1969, *Bact. Rev.* **33**, 264.
[108] Yaniv, M., Chestier, A., Gros, F., et Favre, A.: 1971, *J. Mol. Biol.* **58**, 381.
[109] Chaffin, L. J., Omilianowski, D. R., et Bock, R. M.: 1971, *Science* **172**, 854.
[110] Chatterjee, S. K. et Kaji, H.: 1970, *Biochim. Biophys. Acta* **224**, 854.
[111] Danchin, A. et Grunberg-Manago, M.: 1970, *FEBS Letters* **9**, 327.
[112] Watanabe, K. et Imahori, K.: 1971, *Biochem. Biophys. Res. Commun.* **45**, 488.
[113] Melcher, G., Paulin, D., et Guschlbauer, W., *Biochimie* **1**, 43.
[114] Hänggi, U. J. et Zachau, H. G.: 1971, *Eur. J. Biochem.* **18**, 496.
[115] Litt, M.: 1971, *Biochemistry* **10**, 2223.
[116] Yarus, M.: 1972, *J. Biol. Chem.* **247**, 2738.

DISCUSSION

De Coen: Parmi les 4 modèles que vous avez dit être très semblables un de ceux-ci me paraît néanmoins assez différent par le fait qu'il implique une proximité entre la boucle de l'anticodon et le bout CCA alors que les 3 autres placent ces deux groupes aux deux extrémités les plus éloignées. Est-ce que les expériences de 'quenching' de la fluorescence de la base Y de la boucle de l'anticodon par un chromophore accroché au bout CCA ne permettent pas d'écarter le modèle impliquant une proximité entre l'anticodon et le bout CCA?

Ebel: Si les interprétations de ces expériences de 'quenching' de fluorescence sont correctes, le modèle de Jang ne répond effectivement pas à ces exigences.

M. Lazdunski: Sait-on s'il y a des différences de structure spatiale entre tRNA et précurseurs de tRNA?

Ebel: Aucun travail expérimental n'a été fait à ma connaissance sur ce point.

THE RATE OF CARBON MONOXIDE BINDING TO HEMOGLOBIN KANSAS

J. J. HOPFIELD

Department of Physics, Princeton University, Princeton, N.J. 08540, U.S.A.

and

S. OGAWA and R. G. SHULMAN

Bell Laboratories, Murray Hill, N.J. 07974, U.S.A.

Summary. Hemoglobin Kansas is a mutant G4(102)Asn→Thr which has a lower affinity and Hill constant ($n = 1.3$) than Hemoglobin A. Since the Monod–Wyman–Changeux (MWC) model is a good description of hemoglobin A the oxygen affinity of Hb Kansas was accommodated by adjusting just one MWC parameter, L. Consequently the switch from the deoxy (T) form to the oxy (R) form should occur much later in Hb Kansas than in HbA. This later switch can be used to predict that Hb Kansas should not show any increase in the CO combination rate constant during the course of ligation, in contrast to HbA, where it increases by a factor of ~ 2. Stopped-flow measurements of the rate of CO combination with Hb Kansas confirm these predictions.

Résumé. L'hémoglobine Kansas est un mutant G4(102)Asn→Thr dont l'affinité et la constante de Hill ($n = 1{,}3$) sont inférieures à celles de l'hémoglobine A. Le modèle Monod–Wyman–Changeux (MWC) décrivant bien l'hémoglobine A, l'affinité pour l'oxygène de Hg Kansas a été adaptée en ajustant le seul paramètre L du modèle MWC. Il s'ensuit que le passage de la forme deoxy (T) à la forme oxy (R) devrait se produire beaucoup plus tard dans Hb Kansas que dans HbA. Le changement plus tardif permet de prédire que Hb Kansas ne doit montrer aucune augmentation de la vitesse de combinaison au CO pendant son association à des ligandes, au contraire de HbA où cette vitesse croît avec un facteur d'environ 2. Ces prévisions sont confirmées par des mesures d'écoulement interrompu.

1. Introduction

It is becoming clear from a variety of experiments that the two-state MWC model (Monod, Wyman, and Changeux [1]) provides a good first order description of oxygen and carbon monoxide binding to hemoglobin A. The NMR experiments on half ligated mixed state hemoglobin showed that changing solution conditions could switch these hemoglobins from one quaternary structure to another [2] and these findings have been supported by other assays of quaternary structure [3, 4]. Furthermore studies of the CO binding kinetics show that the half ligated material in the oxy quaternary structure, as determined from NMR, has a fast CO combination rate and in the deoxy quaternary structure a slow rate [5]. This suggested strongly that the quaternary structure was essentially all important in determining the kinetics, and by inference, the affinities. Furthermore it has been shown that the two state MWC model could satisfactorily describe the extensive kinetic data [6] of Hemoglobin A.

Hemoglobin Kansas is an interesting mutant hemoglobin which has been found in mother and son of one family [7]. It has a lower oxygen affinity and degree of co-operativity than HbA ($P_{1/2} = 21$ mm at 20°C, pH 7.4, $n = 1.3$), and normal Bohr Effect ($P_{1/2} = 3.2$ mm at 20°C and pH 9.0 [8]). The altered amino acid G4(102)β

C. Sadron (ed.), Dynamic Aspects of Conformation Changes in Biological Macromolecules. 327–331.

Asn→Thr has been shown to be at the $\alpha_1\beta_2$ interface in fully ligated hemoglobin where it makes a hydrogen bond with Asp G1 (94)α forming the only hydrogen bond across the $\alpha_1\beta_2$ interface [9]. The Fourier difference synthesis of deoxyhemoglobin A and Kansas show no evidence for the existence of a polar intersubunit bond in either [10]. Hence these authors have pointed out that the effects of this mutation upon the stability of the oxygenated quaternary structure should be considerably stronger than that of the deoxygenated quaternary structure which should explain the observed tendency of Hb Kansas to split into dimers. Greer has reported widespread differences between the tertiary structures of the subunits HbA and Hb Kansas in the unligated forms, particularly in the altered β subunits. The long range object of the present experiments is to understand whether Hb Kansas, like HbA, can be described by the MWC model, and if so, which of its parameters are different from those of HbA. Edelstein [11] has shown the shape of the equilibrium oxygenation curve of Hb Kansas can be explained by assuming one MWC parameter alone (L) is altered, but there is insufficient information in the equilibrium data to demonstrate the uniqueness of this supposition. Predictions of parameter alterations based on the X-ray structure are ambiguous, for it is not *a priori* obvious whether the widespread tertiary structural differences between HbA and Hb Kansas are of energetic significance in the binding of ligands.

2. Experimental Methods and Results

Hemoglobin Kansas was obtained from the freshly drawn blood of the same man originally discussed by Reissman, Ruth and Nomura, by the procedure described by Bonaventura and Riggs. Hemoglobin A was obtained from the freshly drawn blood of another individual. Both were dissolved in 0.1 M phosphate buffer (pH 7.0) to heme concentrations of $\sim 30 \times 10^{-6}$ Molar as HbO_2. Just before measuring these samples they were diluted 20% by the addition of a freshly prepared buffered dithionite solution to a final dithionite concentration of 0.01 Molar. The CO solutions were made from buffer solutions, 0.01 M in dithionite, freshly saturated at 20°C and diluted 1:10 or 1:5 in different experiments before mixing. At the saturation temperature the CO concentration in solution was 1.01×10^{-3} Molar.

The kinetic measurements were carried out using a model D-110 Durrum–Gibson [12] stopped-flow spectrophotometer. The absorbance change at a wavelength of 432 mμ was observed in a 2 mm cell, and the output of the spectrophotometer was temporarily stored in a Biomation model 802 transient recorder. The traces used were directly transferred from the transient recorder to the memory of a digital computer and were processed to yield the apparent bimolecular rate constant. Corrections were made for the finite HbCO concentration during the reacion. The derived rate constants of Hb Kansas and HbA are compared in Figure 1. Error bars indicate the sum of statistical errors, which dominate at large values of Y, and possible consistent sources of error such as concentrations and temperature which dominate at low values of Y. Several features are noteworthy. First the results shown for HbA are in complete accord with previous published results, the extrapolated value at $Y=0.0$ of agreeing well with the previous value of $1.0\times10^5\ M^{-1}\ sec^{-1}$ while the increase of the rate during the reaction, about 70% in the present case, is also in agreement with the reported value of $1.0\times10^5\ M^{-1}\ sec^{-1}$ [13] for essentially identical solution conditions. Second the extrapolated rates at $Y=0$ are very similar for HbA and Hb Kansas. Although an accurate extrapolation is not possible without knowing the form of the curves, they seem to be equal to within $\pm15\%$. Third, in contrast to HbA, there is no discernible change of the combination rate of Hb Kansas over the range $0\leqslant Y\leqslant0.8$. The absence of any acceleration of the rate as a function of Y is in qualitative agreement with the alteration of the MWC parameter L proposed by Edelstein.

As discussed previously, the switch over point i_s where the energies in the R and T states are equal is given by the expression $i_s=-\log L/\log c$, where $L=T_0/R_0$, the relative populations of the two states with no ligands bond, and $c=K_R/K_T$ where K_R, K_T are the ligand binding constants in the R and T states respectively. For

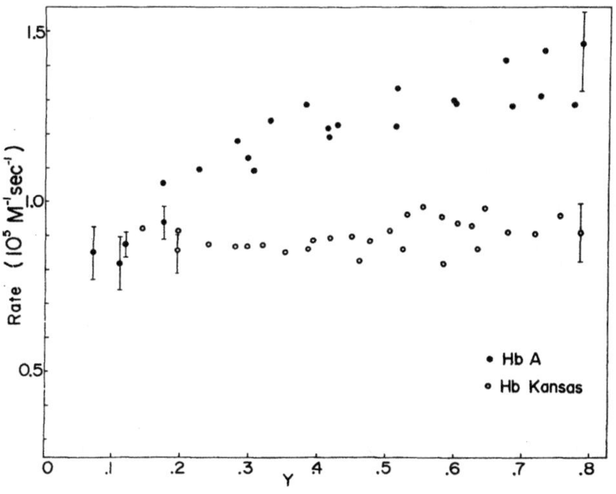

Fig. 1.

Hba $i_s\approx2.5$, as shown in Figure 2, while the equilibrium value of $P_{1/2}$ for Hb Kansas can be fitted by changing L so that $i_s\approx3.5$. The energy level diagram comparing HbA and Hb Kansas on these assumptions is shown in Figure 2 and provides the basis for understanding the data of Figure 1. In HbA the switch to the rapidly reacting R form occurs after 2 or 3 ligands are bound. Under the assumption of fast T→R conversion it has been shown [6] that this can account for the increase of the forward rates. However in Hb Kansas the T→R switch only occurs after approximately all four ligands are bound so that the quickly reacting R form does not contribute.

The initial rate in both HbA and Hb Kansas measures k'_T, which are equal in the two molecules. Since k'_T is not diffusion limited, it is thus likely that K_T is also the same in the two molecules.

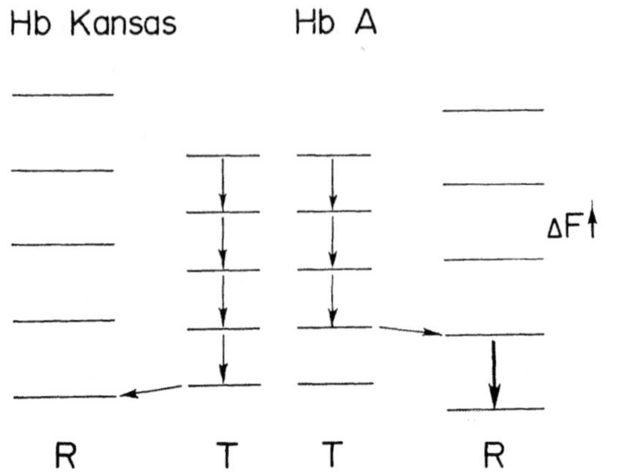

Fig. 2.

Since the ligation steps all occur in T quaternary structure for Hb Kansas, the absence of heterogeneity implies that there is no major difference in 'on' rates for α and β subunits in this molecule.

In summary, the present experimental information is consistent with Hb Kansas following an MWC model with its value of L different from HbA, its value of k'_T (and probably K_T) the same as that of HbA, and the value of K_R not yet certain. We are continuing kinetic experiments to determine other MWC parameters of Hb Kansas, for it seems important to understand how ligand binding parameters are influenced by the structural changes observed in X-ray and NMR experiments.

References

[1] Monod, J., Wyman, J., and Changeux, J. P.: 1965, *J. Mol. Biol.* **12**, 88.
[2] Ogawa, S. and Shulman, R. G.: 1971, *Biochem. Biophys. Res. Commun.* **42**, 9.
[3] Ogata, R. and McConnell, H. M.: 1971 *Cold Spr. Harb. Symp. Quant. Biol.* **36**, 337.
[4] Nagel: 1972, to be published.
[5] Cassoly, R., Gibson, Q. H., Ogawa, S., and Shulman, R. G.: 1971, *Biochem. Biophys. Res. Commun.* **44**, 1015.
[6] Hopfield, J. J., Shulman, R. G., and Ogawa, S.: 1971, *J. Mol. Biol.* **61**, 425.
[7] Reissman, K. R., Ruth, W. E., and Nomura, T.: 1961, *J. Clin. Invest.* **40**, 1826.
[8] Bonaventura, J. and Riggs, A.: 1968, *J. Biol. Chem.* **243**, 980.
[9] Perutz, M. F. and Lehmann, H.: 1968, *Nature* **214**, 902.
[10] Greer, J.: 1971, *J. Mol. Biol.* **59**, 99.
[11] Edelstein, S.: 1971, *Nature* **230**, 224.
[12] Gibson, Q. H.: 1959, *Progr. Biophys. Chem.* **9**, 1.
[13] Antonini, E., Chiancone, E., and Brunori, M.: 1967, *J. Biol. Chem.* **242**, 4360.

DISCUSSION

Goldberg: Is there not a contradiction between the allosteric model (spontaneous equilibrium) and the 'switch' you report between the R to T conformations as revealed by NMR studies which reflect a conformational change? According to what you said, this switch would occur at only one saturation level.

Shulman: There is in the MWC model of allosteric proteins always a degree of ligation for which the energies of the T and R manifold are equal. We have been able to change that cross-overpoint by lowering the energy of the entire T manifold by binding organic phosphates to the hemoglobin. In this way the equilibrium between the T and R forms has been changed.

Cassoly: Ne croyez-vous pas qu'on simplifie beaucoup trop le problème en ne considérant, comme vous l'avez fait, que deux états (R et T) pour la structure quaternaire de l'hémoglobine? Que pensez-vous de structures intermédiaires? Cela me semble prouvé expérimentalement au moins dans un cas, celui des hybrides de valence, si l'on considère leur très faible dissociation en dimères sous leur forme désoxygénée. Cette propriété semble en effet caractéristique de la forme R, alors que dans les expériences de RMN et en cinétique, les hybrides sont dans l'état T.

Shulman: The dissociation into dimers will also be a function of the strains and changes at the interfaces, which are known to depend upon ligation. Hence, a different dissociation behavior does not necessarily indicate a different quaternary structure.

INTERACTIONS DE LA PROFLAVINE AVEC DIFFÉRENTS ADN ET POLYNUCLÉOTIDES DE SYNTHÈSE

J. RAMSTEIN, J. F. HOGREL, M. DOURLENT, M. LENG, et C. HÉLÈNE

Centre de Biophysique Moléculaire, 45045, Orléans Cédex, France

Résumé. Une étude comparative de la fixation de la proflavine sur différents acides nucléiques (ADN de composition variable, Poly A, Poly U) a été entreprise à l'aide de plusieurs techniques; spectrophotométrie d'absorption, relaxation chimique, viscosité. Dans le cas de l'ADN on observe une influence de la nature des sites sur la géométrie des complexes. Dans le cas du Poly A en simple hélice, les études à l'équilibre démontrent l'existence de deux complexes non indépendants. Pour tous les polymères, on met en évidence une phase de relaxation qui peut être associée à un processus apparemment bimoléculaire, malgré l'existence de différences assez importantes dans la forme de signaux, attribuées à la structure du polymère. La constante de vitesse apparente de recombinaison du processus bimoléculaire dépend assez peu de la nature du polymère. Un mécanisme unique peut servir de base à l'interprétation des résultats: une forme partiellement intercalée du colorant joue le rôle d'intermédiaire entre la molécule liée électrostatiquement à l'extérieur du polymère et la molécule complètement intercalée entre les paires de bases de la double hélice.

Summary. The visible absorption band of a mixture containing proflavine and a nucleic acid can be quantitatively resolved by the consideration of three different spectroscopic species whose absorption spectra do not depend significantly on the nature of the polymer [1, 2]: ADN, Poly A, Poly U, Poly (A.U.), tRNA. These spectra are assigned to monomeric free dye (F), externally bound and stacked dye (complex I) and to dye interacting with bases (complex II). When type II spectrum is predominant, the apparent binding constant is approximately independent on the G.C content of the DNA double helix but the location of the bound dye with respect to the bases depends on the nature of the binding sites: the binding is more external in G.C-rich regions than in A.T-rich regions as revealed by viscosity and relaxation data [6]. In the case of single-stranded Poly A the binding isotherms are in better agreement with a model in which the two binding processes are not independent than with a model assuming two independent classes of binding sites [1, 2]. This favours the idea that in Poly A each nucleotide can bind only one dye molecule which belongs either to complex I or to complex II. The optical properties of complex II are consistent with the intercalation hypothesis. Temperature-jump experiments on single-stranded Poly A proflavine systems show three relaxation phases. The fastest is not measurable and attributed either to the helix–coil transition of the polymer or to a fast equilibrium between the monomeric electrostatically bound form and the intercalated form of the complexed dye. The slowest phase has a rather broad relaxation spectrum which can be assigned to the cooperative aggregation process in complex I [8]. The intermediate phase can be associated to an apparently bimolecular process

$$\text{proflavin} + \text{polymer} \underset{k_D}{\overset{k_R}{\rightleftharpoons}} \text{complex},$$

where k_R/k_D is equal to the apparent equilibrium constant measured in diluted systems where no external stacking occurs. With the double helical form of Poly U, at low temperatures, the fast initial part of the relaxation process can also be accounted for by this mechanism. The slow phase likely reflects a conformational change of the polymer [12]. For all polymers (including DNA) the values of the bimolecular recombination rate constants k_R are within one order of magnitude while k_D depends more strongly on the nature and the structure of the macromolecule (Table II).

The results are interpreted on the basis of a common general mechanism, in which the intercalation in A.T-rich double-stranded regions is preceded by two intermediate steps. The first one is a weak electrostatic external binding. The second one consists in the intercalation of only a part of the aromatic ring between two consecutive bases. This partly intercalated form corresponds to the '$D_{(out)}$' complex in the mechanism proposed by Li and Crothers [7].

C. Sadron (ed.), Dynamic Aspects of Conformation Changes in Biological Macromolecules. 333–347.

1. Analyse quantitative des spectres d'absorption

Dans un grand nombre de cas, le spectre d'absorption visible d'un mélange proflavine–acide nucléique peut être interprété quantitativement [1, 2] par l'existence de trois espèces spectroscopiquement distinctes (Figure 1).

Le spectre F correspond à celui de la proflavine libre.

Le spectre I correspond à la fixation coopérative de molécules de colorant empilées le long de la chaîne phosphodiester. Les interactions impliquées dans ce mode d'association (complexe I) sont essentiellement des interactions électrostatiques entre le polymère et le colorant, et des interactions de Van der Waals–London entre les

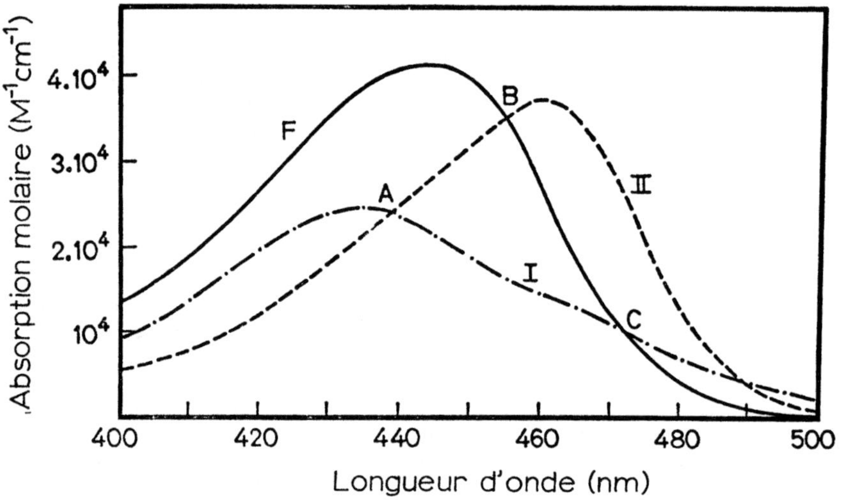

Fig. 1. Spectres d'absorption des différentes espèces spectroscopiques présentes dans un mélange proflavine – acide nucléique. F: proflavine libre; I: association coopérative sur la chaîne poly anionique; II: interaction colorant–bases.

molécules de colorant voisines. Ce type de spectre se rencontre dans toutes les associations coopératives proflavine–polyanions (ex.: polyphosphates, polysaccharides, acide poly-L-glutamique).

Le spectre II correspond à un mode d'association dans lequel le colorant interagit avec les bases de l'acide nucléique. Dans le cas de l'ADN natif ce mode de fixation s'accompagne d'un accroissement de longueur de la double hélice et a été interprété par une intercalation du colorant entre deux paires de bases consécutives [3].

Pour un mélange proflavine–acide nucléique présentant un spectre donné, il est possible [2] de calculer indépendamment les proportions relatives γ_F, γ_I, γ_{II} des trois espèces en se plaçant successivement au trois longueurs d'ondes isobestiques (A, B, C, Figure 1). La somme S de ces trois quantités doit être égale à 1 si les hypothèses de départ sont correctes. On obtient une bonne vérification ($S = 1 \pm 0{,}1$) avec les polymères suivants:

(a) ADN natif ou dénaturé thermiquement [1];
(b) Poly A Poly U en double hélice [1];
(c) ARN de transfert [2];
(d) Poly A en simple brin [2] (voir Figure 4);
(e) Poly U en simple brin [2].

Avec le Poly U en double hélice la vérification est moins bonne ($1 \leqslant S \leqslant 1{,}30$).

Pour une concentration donnée D en proflavine, la fraction γ_{I}, lorsqu'elle n'est pas négligeable, présente un maximum pour des rapports P/D compris entre 1 et 2 [1, 2]. A l'exception du Poly A en simple brin, la fraction γ_{I} peut être négligée aux grandes valeurs de P/D. Il a été montré que pour l'ADN natif et le Poly A Poly U en double hélice la thermodynamique du complexe II pouvait être représenté par un modèle anticoopératif dit du site exclu [4, 5].

Ces analogies spectrales nous ont incités à poursuivre l'étude comparative de ces systèmes en concentrant notre attention sur la nature des interactions impliquées dans le domaine où l'espèce spectroscopique de type II est prépondérante.

2. Influence de la composition en bases de l'ADN

Dans le cas des ADN natifs, si le spectre de type II correspond à un complexe d'intercalation, la longueur L de la double hélice est donnée par:

$$L/L_0 = 1 + 2r,$$

r étant le nombre de molécules de proflavine fixées par nucléotide; L_0 étant la longueur en absence de colorant fixé.

En utilisant des ADN de faible poids moléculaire l'étude de la viscosité permet de calculer L/L_0 [6]. Les résultats obtenus avec des ADN de compositions en bases différentes sont indiqués sur la Figure 2. Dans tous les cas on obtient une droite d'équation

$$L/L_0 = 1 + \alpha r,$$

mais la pente α augmente avec le pourcentage en paires A.T.

La cinétique de relaxation des complexes a été étudiée par la méthode des sauts de température [6]. Pour l'ADN de thymus de veau les résultats obtenus sont analogues à ceux de Li et Crothers [7]. On observe deux temps de relaxation (Figure 3):

(a) l'inverse du temps rapide τ_1 varie linéairement avec la somme des concentrations en colorant et sites libres et doit être associé à un processus bimoléculaire;

(b) l'inverse du temps lent τ_2 tend vers un plateau pour les concentrations élevées.

Avec l'ADN de *Micrococcus lysodeikticus*, on observe essentiellement un temps de relaxation τ_1 dont les variations (Figure 3) indiquent qu'il doit être associé à une étape bimoléculaire. L'existence d'une phase plus lente est visible mais l'amplitude associée est faible.

Pour interpréter les résultats obtenus avec l'ADN de thymus de veau, Li et Crothers

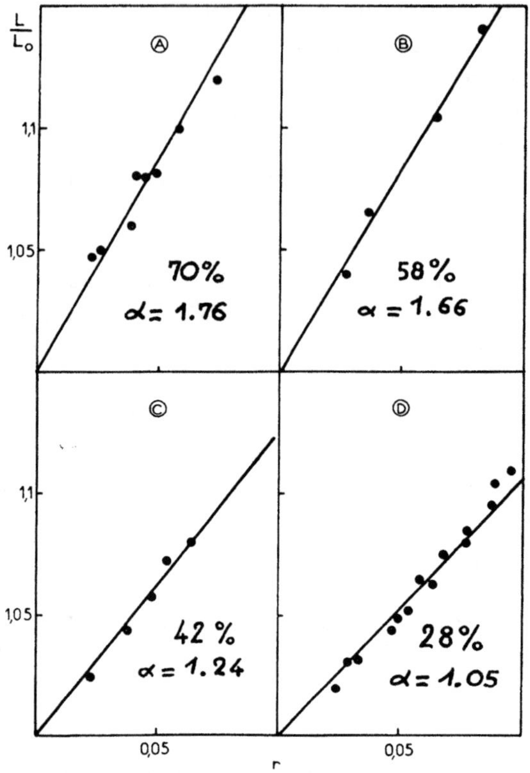

Fig. 2. Allongements de la double hélice en fonction du nombre moyen de molécules de colorant fixé par nucléotide, pour différents ADN. [Na^+] 0,1 M; pH 5; $T = 20°$C. A. *Clostridium perfringens* 70 % A.T; B. *Thymus de veau* 58 % A.T; C. *Serratia marcescens* 42 % A.T; D. *Micrococcus lysodeikticus* 28 % A.T.

[7] ont proposé le mécanisme suivant:

$$\text{Pro} + \text{DNA} \underset{k_{21}}{\overset{k_{12}}{\rightleftharpoons}} (\text{D})_{\text{ext}} \underset{k_{32}}{\overset{k_{23}}{\rightleftharpoons}} (\text{D})_{\text{inter}}, \tag{I}$$

dans lequel la 1ère étape est plus rapide que la seconde. Les temps de relaxation sont donnés par

$$\frac{1}{\tau_1} = k_{12}(\overline{\text{Pro}} + \overline{\text{DNA}}) + k_{21} \tag{1}$$

$$\frac{1}{\tau_2} = k_{32} + \frac{k_{23}(\overline{\text{Pro}} + \overline{\text{DNA}})}{(\overline{\text{Pro}} + \overline{\text{DNA}}) + 1/K_{12}}. \tag{2}$$

Pro représente la proflavine libre, $(\text{D})_{\text{inter}}$ représente le complexe d'intercalation. La forme intermédiaire $(\text{D})_{\text{ext}}$ est considérée comme étant un complexe externe qui précède la réaction d'intercalation. Les propriétés spectrales de cet intermédiaire, sont très

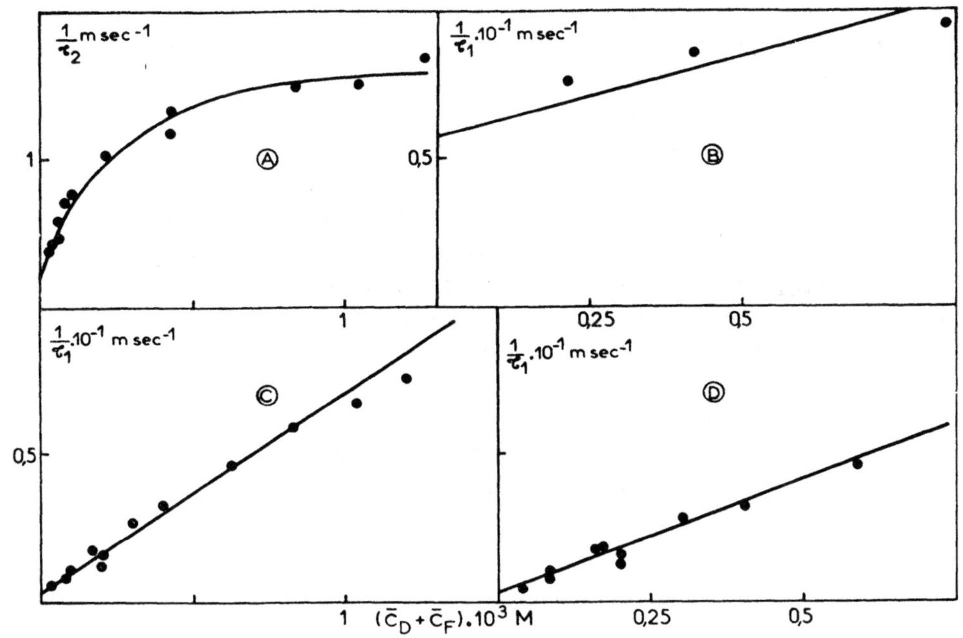

Fig. 3. Variations des temps de relaxation en fonction des concentrations à l'équilibre. $[Na^+]$ 0,1 M; pH 5; $T = 16$°C; $r = 0{,}037$; A et B: ADN de thymus de veau; C et D: ADN de *Micrococcus lysodeikticus*.

voisines de celles de $(D)_{inter}$ [7] et la valeur élevée de K_{12} (Tableau I) indiquent que ce complexe n'est pas dû simplement aux interactions électrostatiques PO_4^--colorant mais implique un certain couplage avec les bases [7].

Dans le cas de l'ADN de thymus de veau la constante K_{23} est égale à 9,6 dans une force ionique 0,1 M (Na^+). A partir de cette valeur on peut calculer que 90% du colorant fixé se trouve sous la forme intercalée. Si l'on applique le même modèle à l'ADN de *Micrococcus lysodeikticus* la constante K_{23} peut être estimée à 0,25 dans les

TABLEAU I

Application du modèle de Li et Crothers à deux ADN de compositions différentes. $[Na^+] = 0{,}1$ M; 10^{-2} M acétate de sodium, pH 5; 10^{-4} M EDTA; $T_f = 16$°C. Les concentrations des polymères sont exprimées en équivalents nucléotide par litre

	k_{12} $(M^{-1}\,s^{-1})$	k_{21} (s^{-1})	k_{23} (s^{-1})	k_{32} (s^{-1})	K_{12} (M^{-1})	K_{23}	K_{app}[a] (M^{-1})
Micrococcus lysodeikticus	$0{,}72 \times 10^7$	$0{,}3 \times 10^3$			$2{,}4 \times 10^4$	0,25	3×10^4
Thymus de veau	$0{,}56 \times 10^7$	$0{,}57 \times 10^4$	$1{,}54 \times 10^3$	$1{,}6 \times 10^2$	$4{,}4 \times 10^3$	9,6	$4{,}4 \times 10^4$ 4×10^4

[a] Constantes d'équilibre apparentes $K_{app.} = K_{12}(1 + K_{23})$ déterminées selon la méthode de Li et Crothers [7].

mêmes conditions de force ionique. Environ 20% à 50% du colorant fixé serait réellement intercalé entre les paires de bases, ce qui est en accord avec le faible allongement de la double hélice observé par viscosité.

Les conclusions de cette étude sont les suivantes [6]:

Pour les deux ADN les énergies libres d'interaction sont très voisines (puisque les constantes d'équilibre globales sont du même ordre de grandeur, voir Tableau I), les déplacements spectraux sont analogues et le modèle du site exclu s'applique. Cependant les résultats des études cinétiques et de viscosité montrent que la position du colorant fixé dépend de la nature des bases impliquées dans le site. Pour les sites riches en A+T la fixation est plus interne que pour les sites riches en G+C.

En même temps qu'ils mettent en évidence une sélectivité des bases dans la géométrie du complexe, ces résultats posent le problème de la structure du complexe $(D)_{ext}$ qui doit répondre aux critères suivants:

(a) spectre d'absorption du type II (interaction avec les bases);
(b) la fixation sur un site exclut les deux sites voisins;
(c) ne modifie que très peu la longueur de la double hélice.

3. Interactions avec Poly A et Poly U

Avec le Poly A en simple brin, les résultats d'études à l'équilibre ont été analysés par comparaison avec deux modèles théoriques résultant de l'extension du modèle d'Ising à la compétition entre deux processus de fixation [1, 8].

Le premier modèle considère que le polymère possède deux classes de sites indépendantes. Il correspond au mécanisme suivant

$$\begin{aligned} \text{Pro} + (\text{Poly A})_{\text{I}} &\overset{K_{\text{I}},\,K'_{\text{I}}}{\rightleftharpoons} \text{complexe I} \\ \text{Pro} + (\text{Poly A})_{\text{II}} &\overset{K_{\text{II}}}{\rightleftharpoons} \text{complexe II} \end{aligned} \qquad \text{(II)}$$

dans lequel Pro désigne la proflavine libre, $(\text{Poly A})_{\text{I}}$ et $(\text{Poly A})_{\text{II}}$ dont les sites de types I et II (qui peuvent être en nombres différents); K_{I} et $K'_{\text{I}} < K_{\text{I}}$ sont deux constantes d'équilibre qui dans le modèle d'Ising sont associées respectivement aux étapes de nucléation et de propagation du processus I qui est coopératif [8]. Pour le second équilibre les résultats peuvent être décrits aussi bien par un modèle non coopératif (une seule constante d'équilibre k_{II}) que par un modèle d'Ising anticoopératif (K_{II} et $K'_{\text{II}} < K_{\text{II}}$).

Le second modèle considère une seule classe de sites à deux états d'occupation: chaque nucléotide ne peut fixer qu'une molécule de proflavine, soit sous la forme de complexe I, soit sous la forme de complexe II. Le mécanisme le plus général correspondant à ce modèle est schématiquement

$$\text{Pro} + \text{Poly A} \begin{cases} \overset{K_{\text{I}},\,K'_{\text{I}}}{\rightleftharpoons} \text{complexe I} \\ \qquad\quad \upharpoonleft\downharpoonright \\ \underset{K_{\text{II}}}{\rightleftharpoons} \text{complexe II} \end{cases} \qquad \text{(III)}$$

Compte tenu du fait que les forces électrostatiques opèrent à plus grande distance que les forces de Van der Waals–London on peut raisonnablement supposer que ce mécanisme général se réduit à

$$P + \text{Poly A} \underset{}{\overset{K_I,\, K'_I}{\rightleftharpoons}} \text{complexe I} \rightleftharpoons \text{complexe II}. \qquad \text{(IV)}$$

Pour les valeurs du rapport (P/D) des concentrations en Poly A et colorant supérieures à 2 les deux modèles sont en accord satisfaisant avec les résultats expérimentaux (Figure 4). Par contre, aux faibles valeurs de P/D, comprises entre 0 et 2, on constate que le modèle du site à deux états d'occupation (second modèle) est en meilleur accord avec les résultats expérimentaux, en particulier pour l'isotherme γ_{II} (Figure 4).

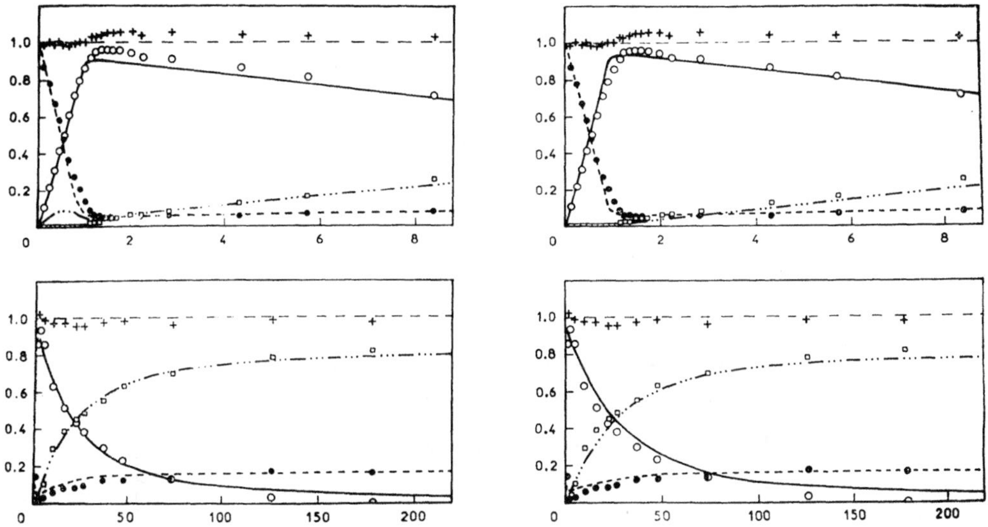

Fig. 4. Comparaison des isothermes de fixation de la proflavine sur le poly A deux modèles théoriques décrits dans le texte. $[Na^+]$ 2mM; pH 6,9; $T = 0°C$; concentration totale en colorant 20 μM. En abscisse: rapport [nucléotides]/[Proflavine]; En ordonnées: valeurs des fractions des différentes espèces spectroscopiques. Symboles: colorant engagé dans des aggrégats de type I: ○ valeur expérimentale, ___ valeur théorique; colorant engagé dans le complexe II: □ valeur expérimentale, - ··· - valeur théorique; colorant libre + colorant engagé sous forme monomère dans le complexe I (cf. [2]): ● valeur expérimentale, ---- valeur théorique; sommes des différentes fractions: + valeur expérimentale, — — — valeur théorique. *A gauche*: modèle à deux classes de sites indépendants; $K_I = 7 \times 10^3$ M^{-1}, $K'_I = 10^6$ M^{-1}, $K_{II} = 5K_I$. *A droite*: modèle à une classe de site possédant deux états d'occupation; $K_I = 6 \times 10^3$ M^{-1}, $K'_I = 10^6$ M^{-1}, $K_{II} = 5K_I$.

Les propriétés optiques du complexe II (absorption, fluorescence, dichroïsme circulaire) sont en accord avec l'hypothèse d'une intercalation entre deux adénines consécutives [1, 9]. Cette interprétation a été également proposés pour le complexe poly U – bromure d'éthidium à partir d'études de résonance magnétique nucléaire; [10]. Dans le domaine des valeurs élevées de P/D, le complexe II est prépondérante. Le signal de relaxation observé par la méthode des sauts de température n'est pas

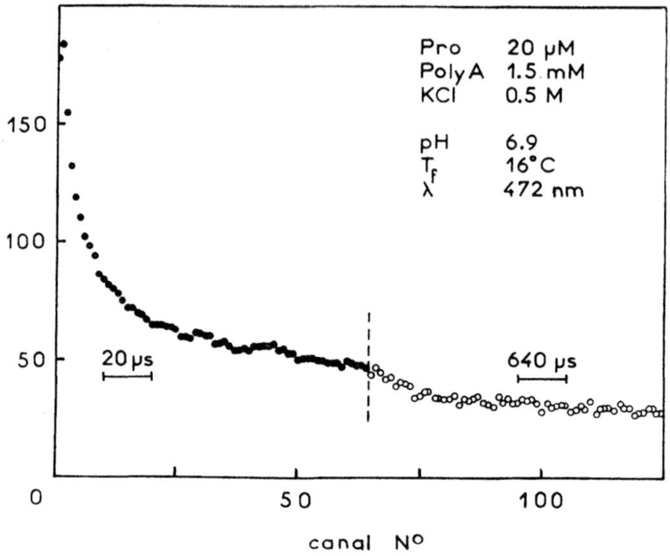

Fig. 5. Signal de relaxation typique obtenu en présence de Proflavine et Poly A (moyenne sur 8 sauts de température mémorisés à l'aide d'un analyseur multicanaux).

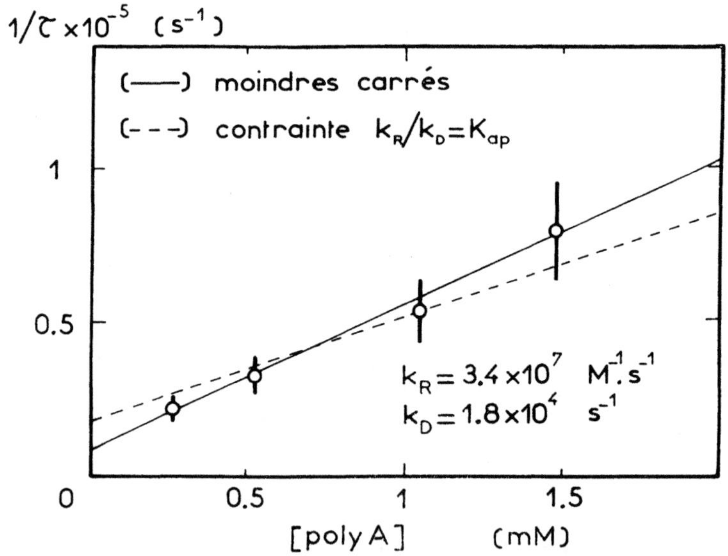

Fig 6. Variation du temps de relaxation de la phase intermédiaire dans le système proflavine–Poly A, en fonction de la concentration totale en polymère. La concentration totale en Poly A étant toujours beaucoup plus grande que la concentration totale en proflavine (20 μM), on a [Poly A] $\simeq$ [$\overline{\text{Poly}}$ A + $\overline{\text{Pro}}$]. Les valeurs de k_R et k_D indiquées dans la figure sont celles qui correspondent à la contrainte $k_R/k_D = K_{app.}$.

simple et le rapport signal/bruit est assez mauvais. Les informations que l'on peut recueillir des expériences sont donc essentiellement qualitatives. On peut distinguer trois phases dans le processus de relaxation (Figure 5). La phase initiale est très rapide et le (ou les) temps de relaxation correspondant(s) est (sont) plus court(s) que la constante de temps de chauffage de la cellule ($\sim 2\ \mu s$). La phase finale, la plus lente, a une amplitude assez faible et n'est pas monoexponentielle. La phase intermédiaire peut être décrite par un seul temps de relaxation dont l'inverse dépend linéairement des concentrations dans le domaine étudié (Figure 6). Si l'on compare les résultats cinétiques au modèle du site à deux états retenu précédemment, on peut interpréter les différentes phases de la relaxation de la manière suivante:

(a) Il a été montré qu'une fixation coopérative conduit à un spectre de relaxation assez large [11]. La phase lente peut donc être attribuée à l'existence, en quantité non négligeable de molécules empilées dans le complexe I.

(b) La phase rapide peut être associée soit à la transition hélice–pelote du polymère, soit à l'équilibre

$$\text{complexe I}_{,\text{monomère}} \underset{k_{21}}{\overset{k_{12}}{\rightleftharpoons}} \text{complexe II}\,.$$

En effet, dans le cas d'une macromolécule en simple hélice cette étape peut être très rapide car la structure du polymère est beaucoup moins rigide que celle d'une double hélice, ce qui peut abaisser considérablement la barrière d'énergie d'activation.

(c) Pour le Poly A en simple brin la probabilité d intercalation à partir d'une molécule appartenant à un aggregat de type I est très faible et l'on pourra négliger ce processus en première approximation. Si, dans le mécanisme cyclique III on désigne par k_{01} et k_{10} les constantes de vitesse associées au processus I, et par k_{02} et k_{20} les constantes de vitesse associées au processus II, le temps de relaxation τ de la phase intermédiaire est donné par une expression de la forme:

$$1/\tau = k_{\mathrm{R}}(\overline{\mathrm{Pro}} + \overline{\mathrm{Poly\,A}}) + k_{\mathrm{D}} \tag{3}$$

avec

$$\begin{aligned} k_{\mathrm{R}} &= k_{01} + k_{02} \\ k_{\mathrm{D}} &= (k_{10} + K_{12}k_{20})/(1 + K_{12})\,. \end{aligned} \tag{4}$$

Dans le cas du mécanisme linéaire (IV) l'Équation (3) est encore valable avec:

$$\begin{aligned} k_{\mathrm{R}} &= k_{01} \\ k_{\mathrm{D}} &= k_{10}/(1 + K_{12})\,. \end{aligned} \tag{5}$$

Les deux mécanismes rendent compte de la variation de $1/\tau_2$ obtenue expérimentalement (Figure 6) et aucun d'eux ne peut être éliminé sur la base des résultats cinétiques puisque dans chaque cas on obtient:

$$k_{\mathrm{R}}/k_{\mathrm{D}} = K_{01}(1 + K_{12}) = K_{\mathrm{I}} + K_{\mathrm{II}} = K_{\mathrm{apparent}}\,.$$

On peut déterminer K_{app} à l'équilibre en opérant avec une solution diluée de colorant (2×10^{-6}M) afin déliminer les empilements dans le complexe I [9]. On obtient une valeur plus faible que celle déduite du rapport k_R/k_D obtenu à partir de la méthode des des moindres carrés. Cependant, compte tenu de la faible précision avec laquelle le temps de relaxation est mesuré on peut considérer que cette différence n'est pas significative (cf. Figure 6).

Lorsqu'on utilise le Poly U sous la forme de pelote statistique, les constantes K_I et K_{II} sont plus faibles que dans le cas du Poly A en simple hélice [2]. D'après le modèle précédent de temps de relaxation de la phase intermédiaire est:

$$1/\tau = k_R(\overline{\text{Pro}} + \overline{\text{Polymère}} + 1/K_{app}). \tag{6}$$

Si l'on admet que la constante k_R dépend peu de la nature du polymère (voir Tableau II) la diminution de K_{app} conduit à une augmentation de la vitesse de relaxation. C'est effectivement ce que l'on observe: le signal obtenu avec les mélanges proflavine–poly U sous forme pelote statistique est trop rapide pour pouvoir être analysé.

TABLEAU II

Constantes de vitesse apparentes pour la phase de relaxation bimoléculaire observée avec tous les polymères. Les concentrations des polymères sont exprimées en équivalents nucléotides par litre. La constante apparente d'équilibre $K_{app.}$ a été déterminée selon la méthode de Li et Crothers [7]

Polymère	$k_R \times 10^{-7}$ ($M^{-1}\,s^{-1}$)	$k_D \times 10^{-3}$ (s^{-1})	$k_R/k_D \times 10^{-3}$ (M^{-1})	$K_{app.} \times 10^{-3}$ (M^{-1})	T (°C)	pH	(M)
DNA (T.V.)	0,56	5,7	4,4	40	15	5	0,1 Na^+
DNA (M.L.)	0,72	0,3	24	30	15	5	0,1 Na^+
Poly A	3,4[a]	18[a]		1,7	16	6,9	0,5 K^+
Poly U	2,1	37	0,57	0,41	3	6,9	0,5 Cs^+

[a] k_R et k_D sont déterminés avec la contrainte $k_R/k_D = K_{app.}$. (cf. Figure 6).

Avec le Poly U en double hélice [12] le signal de relaxation se compose d'une partie rapide descriptible par un seul temps de relaxation suivie d'un effet plus lent, d'amplitude opposée, et ne pouvant être décrit par un seul temps de relaxation (Figure 7). La vitesse moyenne de cette phase lente semble indépendante des concentrations et est l'ordre de grandeur de celle obtenue au début de la zone de fusion du Poly U [12]. La température à laquelle les expériences de relaxation ont été effectuées se situe dans le domaine de prétransition, où la conformation du polymère est déjà sensible à la température comme le montre l'étude de sa viscosité [12]. On peut donc vraisemblablement attribuer l'effet lent à une redistribution des molécules de colorant consécutive au changement de conformation du polymère.

Le temps de relaxation associé à l'effet rapide est connu avec une précision meilleure que dans le cas du Poly A et suit une loi de la forme

$$1/\tau = k_R(\overline{\text{Pro}} + \overline{\text{Poly U}}) + k_D.$$

La valeur du rapport k_R/k_D est très voisine de la constante apparente d'équilibre mesurée par spectrophotométrie d'absorption (Figure 8). Il semble donc qu'un mécanisme du type IV puisse encore rendre compte des phénomènes observés si la deuxième étape est plus rapide que la première, ou si la concentration en complexe I est négligeable.

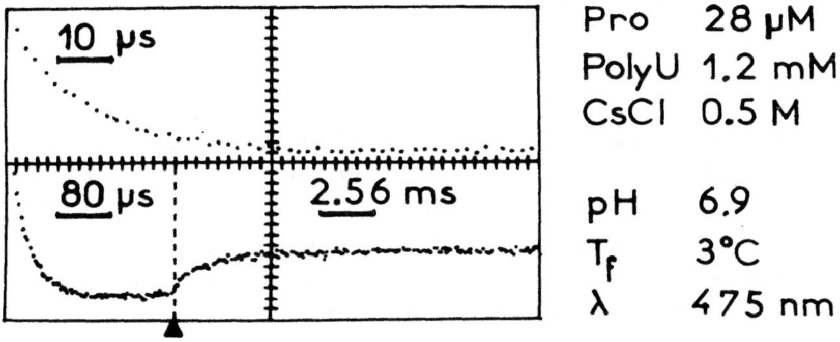

Fig. 7. Signal de relaxation de la proflavine en présence de Poly U sous forme ordonnée. Concentrations totales en Proflavine 28 μM, en Poly U 1,2 mM (nucléotides). Force ionique 0.5 M Cs^+, pH = 6,9. Température 3°C. Longueur d'onde 475 nm. Ce signal correspond à une moyenne sur 8 sauts de température.

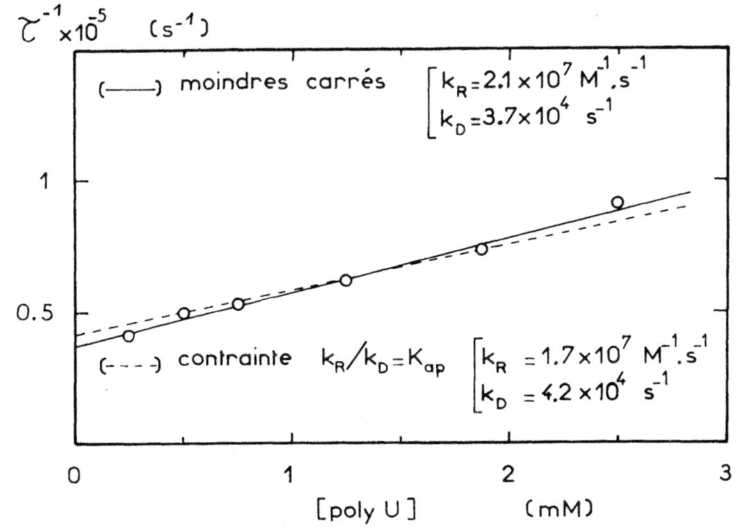

Fig. 8. Variation du temps de relaxation de la phase rapide dans le système proflavine–Poly U sous forme ordonnée, en fonction de la concentration totale Poly U. Comme dans la Figure 6, on a [Poly U] $\#$ [$\overline{\text{Pro}}$ + $\overline{\text{Poly}}$ U].

4. Discussion et conclusions

Les études cinétiques effectuées avec différents polymères montrent que dans le domaine où le spectre de type II est prépondérant, la forme des signaux de relaxation est assez variable d'un polymère à l'autre. Cependant, il est toujours possible de mettre en évidence une phase pouvant être décrite par un temps de relaxation dépendant des concentrations selon une loi du type Equation (6).

Du point de vue formel ceci peut être associé à un processus bimoléculaire

$$\text{Pro} + \text{Polymère} \underset{k_D}{\overset{k_R}{\rightleftharpoons}} \text{complexe},$$

dans lequel les constantes k_R et k_D sont des grandeurs apparentes qui peuvent englober plusieurs étapes élémentaires. Si l'on compare les valeurs de k_R et k_D obtenues avec les différents systèmes, on constate que k_R varie assez peu avec la nature du polymère alors que k_D en dépend beaucoup (Tableau II). Ceci suggère que la constante k_R est très probablement associée dans tous les cas à un même phénomène qui pourrait être l'étape initiale de la réaction: formation d'un complexe purement électrostatique entre le colorant et la chaîne phosphate. Ce processus est contrôlé par la diffusion, mais du fait de la compétition avec les contre-ions de la solution la constante apparente de recombinaison (k_R) peut être nettement inférieure à la constante de diffusion (k_i) [8]:

$$k_R \sim k_i/(1 + K_{Na}[Na^+]).$$

Si l'on fait abstraction des phénomènes associés aux changements conformationnels intrinsèques du polymère (c'est-à-dire non liés à la présence du colorant), tous les résultats cinétiques peuvent s'interpréter par des variantes d'un même mécanisme de base

$$\begin{array}{l} \text{Pro} + \text{Polymère} \underset{k_{10}}{\overset{k_{01}}{\rightleftharpoons}} \text{complexe I}_{,\text{mono}} \underset{k_{21}}{\overset{k_{12}}{\rightleftharpoons}} \text{complexe II, 1} \underset{k_{32}}{\overset{k_{23}}{\rightleftharpoons}} \text{complexe II, 2} \\ \qquad\qquad\qquad + \text{Pro} \ \upharpoonleft\downharpoonright \\ \qquad\qquad\qquad\quad \text{complexe I}_{,\text{di}} \\ \qquad\qquad\qquad + \text{Pro} \ \upharpoonleft\downharpoonright \\ \qquad\qquad\qquad\quad \text{complexe I}_{,\text{tri}} \\ \qquad\qquad\qquad + \text{Pro} \ \upharpoonleft\downharpoonright . \\ \qquad\qquad\qquad\qquad \vdots \end{array}$$

La lère étape consiste en la fixation externe et essentiellement électrostatique d'une molécule de colorant le long de la chaîne phosphate (complexe I, monomère). Ce complexe peut donner naissance soit au processus coopératif d'empilement (chaîne d'équilibres verticale) par fixation de nouvelles molécules de colorant sur les sites voisins (cette réaction ne se manifeste qu'en présence de quantités relativement abondante de colorant libre et si la constante K_{12} n'est pas trop grande) soit au complexe

II,1 dont le spectre serait du type II. Cans le cas des ADN ce complexe correspondrait à la forme $(D)_{ext}$ du modèle de Li et Crothers [7]. Dans le cas des polynucléotides en simple brin le complexe II,1 correspondrait à la forme intercalée, forcément partiellement pour des raisons géométriques, et constituerait l'étape ultime de la réaction ($k_{23}=0$ ou très petit).

La troisième étape correspond au processus d'intercalation dans les hélices doubles et plus particulièrement les paires A.T. Cette étape est la plus lente de la chaîne d'équilibres horizontale. Dans le domaine où la forme spectroscopique II est prépondérante, le processus d'empilement est soit négligeable (ADN), soit plus lent que les autres étapes. Si l'on suppose que les complexes I, mono et II, 1 sont en équilibre très rapide, les temps de relaxation mesurables pour la chaîne horizontale sont donnés par

$$\frac{1}{\tau_1} = k_{01}(\overline{\text{Pro}} + \overline{\text{Polymère}}) + \frac{k_{10}}{1 + K_{12}} \tag{7}$$

$$\frac{1}{\tau_2} = k_{32} + \frac{k_{23}K_{01}K_{12}(\overline{\text{Pro}} + \overline{\text{Polymère}})}{1 + K_{01}(\overline{\text{Pro}} + \overline{\text{Polymère}})(1 + K_{12})}, \tag{8}$$

en accord avec les résultats expérimentaux.

Si l'on fait l'hypothèse que le complexe I, mono est en quantité suffisamment petite pour pouvoir être négligée on obtient

$$\frac{1}{\tau_1} = \frac{k_{01}}{1 + S}(\overline{\text{Pro}} + \overline{\text{Polymère}}) + \frac{k_{10}}{K_{12}(1 + S)} \tag{9}$$

avec

$$S = 1 + k_{10}/k_{12}$$

et

$$\frac{1}{\tau_2} = k_{32} + \frac{k_{23}K_{01}K_{12}(\overline{\text{Pro}} + \overline{\text{Polymère}})}{1 + K_{01}K_{12}(\overline{\text{Pro}} + \overline{\text{Polymère}})} \tag{10}$$

également en accord avec les résultats expérimentaux.

La question se pose de savoir si les complexes II,1 ont ou non des structures analogues dans les simples et doubles hélices. Dans l'état actuel de nos connaissances il n'existe pas de preuves expérimentales permettant de répondre nettement à cette question. Cependant, certains indices peuvent suggérer que des analogies structurales sont effectivement possibles:

(a) dans le cas des ADN, le complexe II,1 (forme $(D)_{ext}$ du mécanisme de Li et Crothers [7]) possède un spectre d'absorption très voisin de celui de la forme intercalée et l'on sait que dans l'ADN de *Micrococcus lysodeikticus*, où cette forme est prédominante, le modèle du site exclu continue de s'appliquer. Ces résultats semblent compatibles avec un mécanisme dans lequel la forme $(D)_{ext}$ dite 'externe' serait en fait un intermédiaire entre une forme purement externe (complexe I) et la forme intercalée (complexe II,2), c'est-à-dire correspondrait à une intercalation d'une partie de la molécule de colorant entre deux bases adjacentes et dont la charge positive resterait en interaction électrostatique avec le groupement phosphate situé entre ces deux bases. Très récemment, Galley et Purkey [13] sont arrivés à une conclusion analogue à partir de l'étude des propriétés de phosphorescence des complexes proflavine–poly (dA.BrdU).

(b) Dans le cas des chaînes en simple brin l'hypothèse d'une intercalation partielle du colorant dans la forme II,1 est également en excellent accord avec les résultats expérimentaux obtenus sur le Poly A.

Il apparaît ainsi que les deux modèles d'intercalation proposés respectivement par Lerman [3] et Pritchard *et al.* [14] pour l'ADN ne seraient pas contradictoires, mais, en quelque sorte, complémentaires. L'intercalation partielle (complexe II,1) se rencontrerait essentiellement dans les structures en simple brin (dont l'ADN dénaturé), dans les régions riches en G+C de l'ADN natif et comme précurseur de l'intercalation entre les paires A.T. Le modèle de Lerman s'appliquerait au complexe II,2, c'est-à-dire à la forme intercalée dans les régions riches en A+T. Ces idées rejoignent et précisent celles d'Ichimura *et al.* [15].

Remerciements

Nous remercions le Professeur Ch. Sadron de l'intérêt qu'il porte àce travail.

Nous tenons à remercier également l'éditeur de *Biochem. Biophys. Res. Commun.* (**47**(1972) 874–82) pour la permission de reproduire les Figures 2 et 3, et l'éditeur de *Eur. J. Biochem.* pour la permission de reproduire les Figures 1 et 4.

Références

[1] Dourlent, M.: 1973, Thèse de Doctorat, Orléans.
[2] Dourlent, M. et Hélène, C.: 1971, *Eur. J. Biochem.* **23**, 86.
[3] Lerman, L. S.: 1961, *J. Mol. Biol.* **3**, 18; 1963, *Proc. Mat. Acad. Sci. US* **49**, 94.
[4] Armstrong, R. W., Kurucsev, T., et Strauss, U. P.: 1970, *J. Am. Chem. Soc.* **92**, 3174.
[5] Bauer, W. et Vinograd, J.: 1970, *J. Mol. Biol.* **54**, 281.
[6] Ramstein, J., Dourlent, M., et Leng, M.: 1972, *Biochem. Biophys. Res. Commun.* **47**, 874.
[7] Li, H. J. et Crothers, D. M.: 1969, *J. Mol. Biol.* **39**, 461.
[8] Schwarz, G.: 1970, *Eur. J. Biochem.* **12**, 442.
[9] Hogrel, J. F.: 1973, Thèse de 3è Cycle, Orléans.
[10] Kreishmann, G. P., Chan, S. I., et Bauer, W.: 1971, *J. Mol. Biol.* **61**, 45.
[11] Schwarz, G. et Balthasar, W.: 1970, *Eur. J. Biochem.* **12**, 461.
[12] Thrierr, J. C., Dourlent, M., et Leng, M.: 1971, *J. Mol. Biol.* **58**, 815.
[13] Galley, W. C. et Purkey, R. M.: 1972, *Proc. Nat. Acad. Sci. US* **69**, 2198.
[14] Pritchard, N. J., Blake, A., et Peacocke, A. R.: 1966, *Nature* **212**, 1260.
[15] Ichimura, S., Zama, M., Fujita, H., et Ito, T.: 1969, *Biochim. Biophys. Acta* **190**, 116.

DISCUSSION

Guschlbauer: La liaison de la proflavine sur les DNA dépend du contenu G.C. Est-ce que vous avez testé des DNA du même contenu G.C qui sont différents, par exemple T_7 et *E. coli*, en vue de définir l'effet des plus proches voisins?

Dourlent: Nous n'avons pas effectué les expériences que vous évoquez. Par ailleurs, je ne pense pas que la méthode expérimentale utilisée soit assez fine pour pouvoir mettre en évidence les effets de séquence, si ceux-ci existent. A l'équilibre, par absorption optique, nous voyons un effet global qui montre que la constante apparente de fixation forte du colorant dépend peu de contenu moyen en G + C (voir Table I). Dans les expériences de saut de température et dans les études de viscosité nous voyons également un phénomène global qui, lui, dépend beaucoup du contenu moyen en G.C.

Frédéricq: Ne pensez-vous pas qu'il soit dangereux de déterminer les proportions des différents types de complexes sur la base unique de mesures de spectre d'absorption? Cela postule en effet qu'il n'existe dans le milieu qu'un petit nombre d'espèces spectrales discrètes. En fait, je pense qu'il est nécessaire de combiner les résultats de différentes méthodes telles que spectres de fluorescence, dichroïsme cellulaire, etc. pour obtenir des résultats plus précis. Des interactions variées entre molécules libres et molécules liées peuvent en effet provoquer des variations progressives des propriétés spectrales des complexes. Dans le cas d'interactions DNA–acridine orange en particulier, nous avons pu mettre en évidence ce genre de modification par l'utilisation combinée de différentes méthodes spectrales.

Dourlent: Puisque la méthode d'analyse des spectres d'absorption donne des résultats cohérents, cela nous incite à penser que les hypothèses formulées constituent un nombre minimal admissible pour pouvoir rendre compte des phénomènes observés, du moins en première approximation. Les études cinétiques nous ont ensuite permis de montrer que le spectre de type II pouvait correspondre à deux complexes différents ($C_{II,1}$ et $C_{II,2}$). Ce que l'on voit à l'équilibre est donc la somme $C_{II,1} + + C_{II,2}$.

Je suis en accord avec votre remarque concernant l'évolution progressive des propriétés spectrales des complexes. Par exemple, dans le complexe I, le spectre d'absorption dépend probablement du nombre n de molécules contenues dans une séquence ininterrompue. Cependant les études de Li et Crothers (*Biopolymers* **8** (1969) 217) ont montré que le changement spectral le plus important dans la formation des agrégats est observé lors du passage monomère→dimère. Cela justifie dans une certaine mesure l'approximation qui consiste à admettre que le spectre des molécules empilées le long de la chaîne phosphate ne dépend pas de $n \geq 2$. (Voir aussi ref. 9.)

Pour l'interaction proflavine – Poly A, nous avons des données de fluorescence et de dichroïsme [1] qui sont compatibles avec les résultats obtenus en absorption mais qui restent au même niveau d'approximation. Nous n'avons pas abordé l'étude des interactions ADN–acridine orange, car ce colorant, même en l'absence de polymère, a une très forte tendance à l'auto-association, ce qui complique énormément l'interprétation des données expérimentales. Enfin, je voudrais faire remarquer que l'interprétation et le mécanisme que nous proposons résultent d'une étude faite sur un seul type de colorant en présence de différents polymères. Ce mécanisme n'est peut-être pas généralisable aux autres colorants. Löber et Achtert (*Biopolymers* **8** (1969) 595) ont étudié l'interaction d'un ADN avec différents colorants et montré que la structure des complexes est probablement très dépendante de la nature du ligande.

De Coen: Dans vos études de l'interaction entre la proflavine et le Poly A, je n'ai pas très bien compris les arguments qui sont à la base de l'hypothèse d'existence de deux sites. Lorsque le Poly A est complètement saturé, il semble que la moitié du nombre total de sites seulement soient occupés. Quel est le pourcentage d'occupation de chacun des sites dans ces conditions de saturation?

Dourlent: L'hypothèse des deux sites indépendants constitue le modèle le plus simple (voire le plus simpliste) qui permette de tenter l'interprétation de nos résultats. Formellement, ce modèle correspond au partage des molécules de ligande entre deux polymères P_I et P_{II} mélangés en quantités égales, avec pour seule différence le fait que P_I et P_{II} sont ici portés par la même chaîne (ce qui ne modifie en rien les résultats). Ce modèle ne permet pas de rendre compte des résultats expérimentaux de façon totalement satisfaisante, car il prévoit l'existence de complexe II dans un domaine où celui-ci n'est pas observé expérimentalement. Pour expliquer ce désaccord il faut envisager un modèle dans lequel chaque nucléotide ne peut fixer qu'une molécule de colorant soit dans l'état I, soit dans l'état II.

Lorsque le Poly A est saturé, tous les sites sont occupés. L'état d'occupation est presque exclusivement de type I. Cela résulte de la coopérativité positive du processus I et du fait que la constante K_{II} est comprise entre K_I et K'_I. Dans le cas de l'ADN à faible force ionique nous avons trouvé que K_{II} est toujours plus grand que K'_I. Il en résulte que, même aux faibles valeurs de P/D, le complexe II l'emporte sur le complexe I.

INTERACTIONS DES ACIDES AMINÉS AROMATIQUES AVEC LES ACIDES NUCLÉIQUES

Structure des complexes et changements de conformation

C. HÉLÈNE, H. BORAZAN*, J. L. DIMICOLI, J. C. MAURIZOT, M. DURAND, et J. J. TOULME

Centre de Biophysique Moléculaire, 45045, Orléans Cédex, France

Résumé. Les interactions des cycles indoliques et phénoliques avec les acides nucléiques ont été étudiées par différentes techniques physiques: fluorescence, absorption, résonance magnétique nucléaire, dichroïsme circulaire, centrifugation. Les composés étudiés sont, d'une part, des amines aromatiques (tryptamine, sérotonine, tyramine), d'autre part, des oligopeptides contenant des résidus aromatiques (Lys Tyr Lys, Lys Trp Lys...).

La fluorescence des noyaux indole et phénol est éteinte dans les complexes formés avec le poly A, l'ADN ou les ARNs de transfert. La formation de complexes d'empilement du noyau indole avec les bases des acides nucléiques en simple et en double hélice est démontrée par des études de RMN. Le noyau phénol ne semble former ce type de complexes qu'avec les polynucléotides en simple brin. Un autre type de complexe (probablement par liaison hydrogène) est observé avec les dérivés de la tyrosine et la sérotonine. La formation des complexes s'accompagne de changements de conformation du polynucléotide ou de l'acide nucléique qui sont détectés par RMN et dichroïsme circulaire.

Summary. Complex formation between aromatic amines (tryptamine, serotonin, tyramine...) or oligopeptides containing aromatic amino acids (Lys Tyr Lys, Lys Trp Lys...) and nucleic acids (poly A, DNA, tRNA's) has been investigated by physical methods: fluorescence, absorption, proton magnetic resonance (PMR), circular dichroism (CD) and centrifugation.

The fluorescence of indole and phenol rings is quenched in the complexes. Indole derivatives are able to form stacked complexes with bases both in single-stranded and double-stranded nucleic acids, as shown by PMR studies. Phenol derivatives as well as serotonin (5-hydroxytryptamine) can form at least two types of complexes with the bases of nucleic acids involving either stacking or hydrogen-bond interactions. Competition between these two modes of binding depends on the secondary structure of the nucleic acid (single-stranded polymers favour stacking interactions) and on the environment of the phenol ring in oligopeptides. In all cases, electrostatic contribution to the interaction is demonstrated by the induced dissociation of the complexes upon increasing the ionic strength.

Interaction of aromatic amino acids with single-stranded poly A leads to a change in the polymer conformation. The CD amplitude decreases and the PMR spectrum of poly A indicates a decreased stacking of adenine residues. Interactions with poly C are rather weak as compared to poly A.

The CD spectrum of double-stranded DNA is affected by the binding of aromatic amines and oligopeptides. These modifications are ascribed to a change in the conformation of DNA induced by the indole and phenol rings. In tRNA's, oligopeptides bind both to single-stranded and double-stranded regions. The binding also induces changes in the tRNA conformation.

These results are discussed with respect to the possible role of aromatic residues in the binding of proteins or enzymes to nucleic acids.

1. Introduction

L'étude de la nature et de la spécificité des interactions des acides aminés ou d'oligopeptides avec les acides nucléiques, les polynucléotides ou leurs constituants doit permettre

* Adresse actuelle: Department of Pharmaceutical Chemistry, Baghdad University, Irak.

C. Sadron (ed.), Dynamic Aspects of Conformation Changes in Biological Macromolecules. 349–366.

de déterminer le rôle joué par différents acides aminés dans la reconnaissance sélective des acides nucléiques par les protéines ou les enzymes. L'étude des changements de conformation des acides nucléiques induits par la fixation de protéines peut, d'une part, contribuer à la détermination de la structure des complexes formés et, d'autre part, permettre de comprendre le comportement des complexes au cours de leurs réactions ultérieures.

Les acides aminés possédant un cycle aromatique sont susceptibles de jouer un rôle particulier dans les interactions entre les protéines et les acides nucléiques à cause des possibilités d'interaction avec les cycles aromatiques des bases des acides nucléiques [1]. De telles interactions ont été mises en évidence au cours de l'étude de différents systèmes. En solution aqueuse congelée, des études de luminescence ont démontré l'existence de complexes entre le tryptophane et les nucléosides, complexes caractérisés par un nouveau spectre de fluorescence et une extinction pratiquement totale de la fluorescence des deux molécules constituant le complexe [2]. En milieu fluide, des études de résonance magnétique nucléaire (RMN) [3–5] et d'absorption [3] ont permis de conclure à la formation de complexes de stoechiométrie 1:1 dans lesquels les deux cycles aromatiques sont parallèles l'un à l'autre. Une interaction des acides aminés aromatiques avec des nucléotides fixés sur un support rigide a également été mise en évidence [6]. La fixation (non covalente) de la tryptamine [7, 8], de la sérotonine [7, 8] ou de la tyramine [9] sur les polynucléotides ou les acides nucléiques a permis, dans certaines conditions, de démontrer l'existence d'interactions directes entre le cycle aromatique de l'amine et les bases puriques ou pyrimidiques. Des études de fluorescence et de RMN ont montré que le tryptophane ou la tyrosine, inclus dans des oligopeptides, pouvaient interagir directement avec les bases de l'acide nucléique [10]. Des dipeptides contenant le tryptophane stabilisent également l'ADN en double hélice vis-à-vis de sa dénaturation thermique [11].

Au cours de l'étude effectuée dans notre laboratoire, nous avons été conduits à déterminer l'influence que les acides aminés aromatiques pouvaient avoir sur la structure des polynucléotides ou des acides nucléiques au cours de la fixation soit des amines aromatiques soit d'oligopeptides contenant les acides aminés aromatiques. Après avoir brièvement rappelé les expériences permettant de mettre en évidence la formation des complexes, nous présenterons les résultats relatifs à la structure des complexes formés et aux changements de conformation induits dans les acides nucléiques ou les polynucléotides. Les techniques expérimentales, décrites dans des publications récentes [1, 3, 5, 8, 9], ne seront pas exposées ici.

2. Résultats

2.1. Mise en évidence de la formation des complexes

Les résultats obtenus en solution aqueuse congelée ont démontré que la formation de complexes entre les acides aminés aromatiques et les bases des acides nucléiques s'accompagne d'une extinction de la fluorescence de l'acide aminé [2, 9]. L'addition d'une amine aromatique (tryptamine, sérotonine, tyramine) au poly A en simple hélice ou au DNA en double hélice à pH 7 s'accompagne d'une diminution de l'intensité de fluorescence du cycle indole ou phénol [7, 8]. L'étude du déclin de fluorescence sous excitation par flash et l'analyse de la diminution de l'intensité de fluorescence pour différentes concentrations en acide nucléique a permis de montrer que le rendement de fluorescence des amines aromatiques fixées à l'acide nucléique est réduit à zéro [8]. Ce résultat reste valable lorsque des oligopeptides comme Lys Trp Lys ou Lys Tyr Lys interagissent avec le ploy A en simple brin: le rendement de fluorescence des résidus Trp ou Tyr est nul dans le complexe. Il n'en est pas de même lorsque ces oligopeptides interagissent avec l'ADN en double hélice. Des expériences de centrifugation démontrent que la quantité de complexe formé est nettement supérieure à celle déterminée d'après la diminution de l'intensité de fluorescence en supposant l'existence d'un seul complexe dont le rendement de fluorescence serait nul. Dans les mêmes conditions initiales, la concentration en Lys Trp Lys fixée déterminée par centrifugation est plus grande que celle de Lys Tyr Lys. Au contraire, l'extinction de fluorescence est plus grande pour ce dernier tripeptide.

L'interaction des amines aromatiques et des oligopeptides avec le poly C en simple brin est beaucoup plus faible qu'avec le poly A.

Dans les cas où il ne se forme qu'un seul complexe non fluorescent, les mesures de fluorescence peuvent être utilisées pour déterminer la constante d'association K et le nombre n de sites de fixation de l'amine aromatique ou de l'oligopeptide. En supposant que ces sites sont indépendants et équivalents, l'équation de Scatchard doit être vérifiée:

$$r/C_f = K(n - r), \tag{1}$$

où r représente le nombre de molécules d'amine ou d'oligopeptide fixées par nucléotide

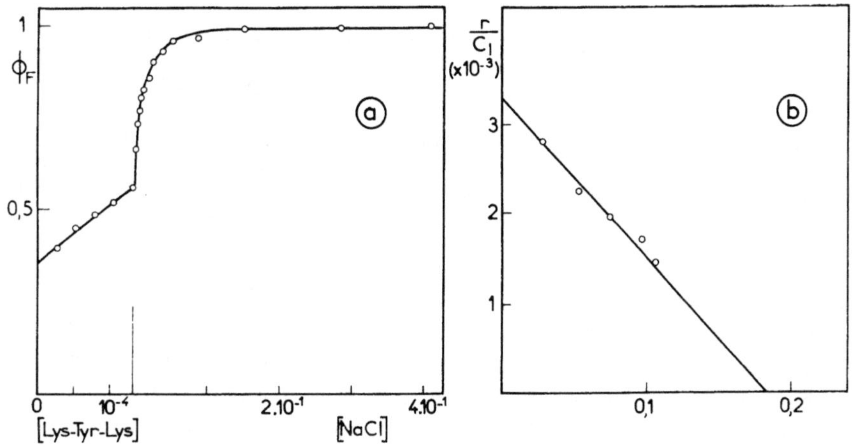

Fig. 1. (a) Variation du rendement de fluorescence de Lys Tyr Lys en présence de poly A dans un tampon cacodylate de sodium 1 mM à pH 7 et effet de l'addition de NaCl. Le rendement de fluorescence est mesuré relativement à celui du peptide libre pris pour unité. (b) Représentation de Scatchard (Équation (1)) pour la fixation de Lys Tyr Lys sur le poly A à pH 7 en utilisant les résultats de la Figure 1a.

$(r = C_b/P)$; C_b et C_f sont les concentrations d'amine ou d'oligopeptide liées et libres, respectivement. La Figure 1 montre que la fixation des oligopeptides Lys Trp Lys et Lys Tyr Lys sur le poly A en simple brin obéit bien à l'Équation (1).

La compétition entre un oligopeptide et une amine aromatique comme la 5-méthoxytryptamine peut être facilement étudiée par fluorescence. En effet, la 5-méthoxytryptamine peut être excitée à 310 nm où ni l'acide nucléique, ni l'oligopeptide n'absorbent. La fixation de l'oligopeptide s'accompagnera d'une dissociation du complexe de la 5-méthoxytryptamine. Puisque l'amine fixée a un rendement de fluorescence nul, la compétition pourra être suivie par l'augmentation de l'intensité de fluorescence qui en résultera. On peut ainsi étudier par fluorescence la fixation d'oligopeptides non fluorescents comme Lys Ala Lys (Figure 2). Cette méthode per-

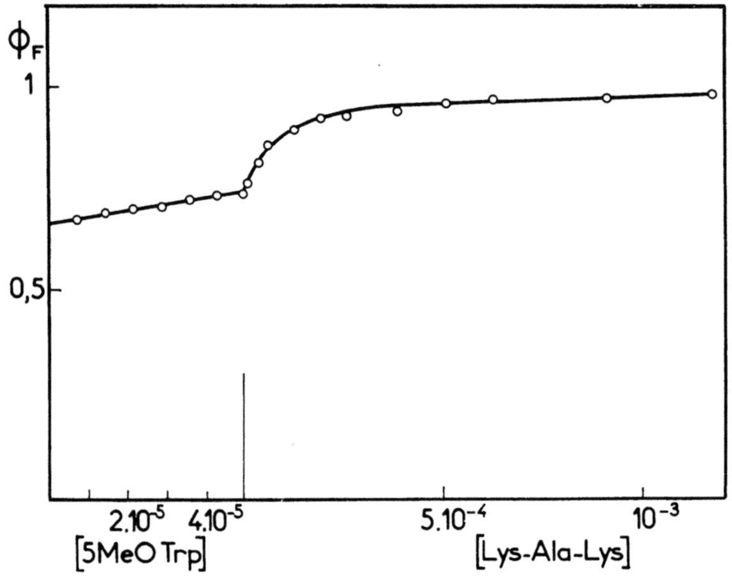

Fig. 2. Compétition entre la 5-méthoxytryptamine et Lys Ala Lys pour la fixation sur l'ADN de thymus de veau à pH 7 (cacodylate de sodium 1 mM). Le rendement relatif de fluorescence de la 5-méthoxytryptamine est mesuré à 340 nm sous excitation à 310 nm. Le rendement de fluorescence de la 5-méthoxytryptamine libre est pris pour unité.

met actuellement de comparer les affinités relatives de différents peptides. D'autres études, actuellement en cours, sont nécessaires pour rendre cette méthode quantitative.

La formation des complexes entre les oligopeptides et les acides nucléiques peut être démontrée non seulement par fluorescence et centrifugation mais également par des mesures de sédimentation. Par exemple, le coefficient de sédimentation du complexe poly A – poly U en double brin augmente de façon similaire en présence de Lys Tyr Lys et Lys Trp Lys. Le spectre d'absorption des cycles aromatiques est également perturbé par la formation des complexes.

Enfin, les résultats de RMN et de dichroïsme circulaire seront discutés dans les sections suivantes qui traitent de la structure des complexes formés et des changements de conformation qui en résultent.

2.2. STRUCTURE DES COMPLEXES FORMÉS

La force essentielle liant les amines aromatiques et les oligopeptides Lys X Lys à l'acide nucléique est d'origine électrostatique. Elle s'exerce entre les groupements $\overset{+}{N}H_3$ de l'amine ou des résidus lysine de l'oligopeptide et les groupements PO_4^- de l'acide nucléique. Ceci peut être démontré par deux types d'expériences: une élévation de la force ionique s'accompagne d'une dissociation du complexe donc d'une augmentation de l'intensité de fluorescence (Figure 1); la neutralisation de la charge positive du groupement $\overset{+}{N}H_3$ terminal des oligopeptides Lys X Lys (pK $\sim$8) s'accompagne également d'une diminution importante de l'affinité de l'oligopeptide pour l'acide nucléique [12].

La résonance magnétique du proton est une méthode de choix pour obtenir des renseignements sur la structure des complexes formés. Les interactions peuvent en effet modifier la position, la largeur des raies et les couplages des différents protons soit de l'oligopeptide soit de l'acide nucléique lorsque celui-ci est en simple brin (puisque les acides nucléiques en double hélice ne présentent pas de spectres RMN haute résolution).

2.2.1. *Dérivés possédant le noyau indole*

La Figure 3 montre que les raies de résonance des protons du cycle indole de la sérotonine subissent des déplacements importants vers les champs élevés en présence de poly A et d'ADN. Des élargissements importants sont observés avec l'ADN. Des résultats analogues sont obtenus avec les autres dérivés à noyau indole étudiés, qu'il s'agisse des amines aromatiques ou des oligopeptides contenant le tryptophane. Seules les amplitudes des déplacements varient d'un composé à l'autre. A concentration fixe en poly A ou en ADN, les déplacements augmentent quand la concentration du dérivé indolique diminue. L'extrapolation à concentration nulle permet d'obtenir le déplacement chimique des protons du cycle indolique dans le complexe ou la moyenne pondérée des déplacements chimiques de différents complexes s'il en existe plusieurs. L'hypothèse d'un échange rapide entre les molécules libres et les molécules fixées sur la macromolécule a toujours été vérifiée dans nos expériences. Les observations suivantes peuvent être faites (Table I). Dans les complexes avec le poly A en simple hélice, les déplacements chimiques des protons du cycle indole sont toujours très voisins, que l'on étudie les amines aromatiques ou les oligopeptides. Ces déplacements sont du même ordre de grandeur que ceux observés dans les complexes 1:1 obtenus avec les monomères [3–5]. Dans les complexes avec l'ADN, les déplacements chimiques des protons de la tryptamine et de la 5-méthoxytryptamine sont comparables à ceux observés dans le cas du poly A alors que ceux de la sérotonine et de Lys Trp Lys sont beaucoup plus faibles.

Les déplacements importants (de l'ordre de 0,5 ppm), vers les champs élevés par

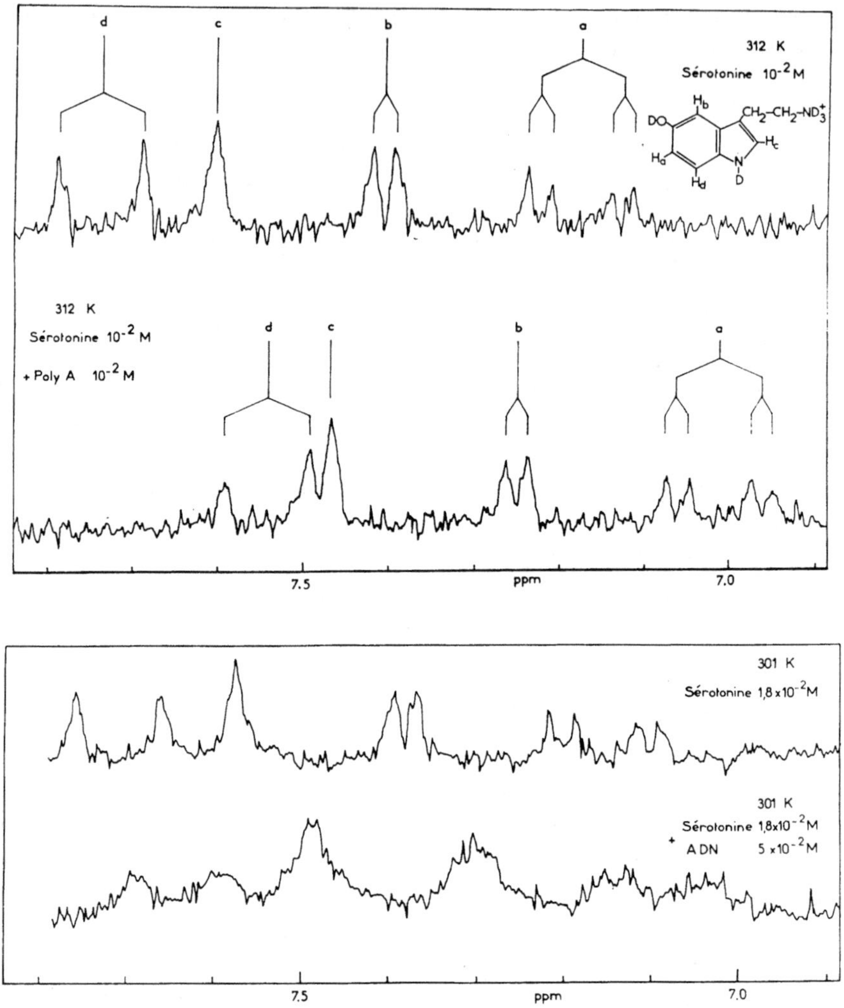

Fig. 3. Spectre RMN des protons du cycle de la sérotonine (5-hydroxytryptamine) en présence (a) de poly A et (b) d'ADN de thymus de veau.

rapport aux molécules libres sont attribués à l'effet d'anisotropie magnétique des bases nucléiques, essentiellement des bases puriques, et traduisent la formation d'un empilement des bases et du noyau indolique. Dans la plupart des complexes étudiés, à l'exception des complexes sérotonine – ADN et Lys Trp Lys – ADN, les variations de déplacement chimique sont analogues à celles obtenues pour les complexes 1:1 entre monomères, démontrant ainsi que tous les cycles indoliques forment le même type de complexe d'empilement. Les variations nettement plus faibles obtenues pour les complexes formés par la sérotonine et Lys Trp Lys avec l'ADN peuvent s'interpréter par la formation de deux types de complexes: l'un donnant lieu à un déplace-

TABLE Ia

Différences entre les déplacements chimiques des différents protons de la sérotonine et de la méthoxytryptamine dans différents complexes (obtenus par extrapolation à concentration nulle) et les déplacements chimiques des protons des molécules non complexées. Les mesures ont été effectuées à 315 K pour l'ADN et à 299 K pour le Poly A. (Voir Figure 3 pour la nomenclature des différents protons.)

Complexe	H_d	H_c	H_b	H_a	CH_3
Sérotonine–purine	63	49	82	82	
Sérotonine–AMP	56	39	43	36	
Sérotonine–Poly A	61	37	42	52	
Sérotonine–ADN	21	28	22	22	
5-méthoxytryptamine–Poly A	70	37	50	70	22
5-méthoxytryptamine–ADN	56	68	54	56	40

TABLE Ib

Différence entre les déplacements chimiques du proton H_c de la tryptamine ou du tryptophane dans différents complexes et les déplacements chimiques du proton H_c des molécules non complexées. Les conditions sont identiques à celles de la Table Ia

Tryptamine AMP	Tryptamine Poly A	Tryptamine ADN
39	32	56

Lys Trp Lys Poly A	Lys Trp Lys ADN	Trp Lys Poly A	Trp Lys ADN
35	20	28	17

ment vers les champs élevés et correspondant au complexe d'empilement; l'autre ne donnant lieu à aucune (ou une très faible) variation de déplacement chimique par rapport à la molécule libre. Nous avons déjà donné plus haut des preuves de l'existence de plusieurs types de complexes lors de la comparaison des résultats de fluorescence et de centrifugation.

La 5-méthoxytryptamine et la 5-hydroxytryptamine (sérotonine) ont des comportements très voisins vis-à-vis du poly A ou de l'ADN lorsque la fluorescence est utilisée comme méthode d'investigation. La différence de comportement observée en RMN vis-à-vis de l'ADN permet de suggérer qu'une partie du complexe formé par la sérotonine fait intervenir une liaison hydrogène du groupement hydroxyle soit avec les bases soit avec les groupements phosphate de l'ADN. Ce type de complexe dans lequel il n'y aurait pas d'empilement du noyau indole avec les bases aurait les mêmes déplacements chimiques que la sérotonine libre, sa fluorescence pouvant être éteinte par un mécanisme analogue à celui démontré pour les composés à noyau phénol [26]. La méthylation du groupement hydroxyle aurait pour effet de supprimer ce type de complexe au profit du complexe d'empilement. Cette hypothèse est en accord avec les résultats obtenus par dichroïsme circulaire (section suivante) et également avec ceux déduits de la comparaison entre la tyramine et son dérivé méthylé (voir ci-dessous). Bien que la quantité totale de complexe formé diminue lorsque l'ADN se dénature thermiquement [8], la variation de déplacement chimique des protons de la sérotonine augmente (contrairement à ce qui est observé avec la 5-méthoxytryptamine ou la tryptamine). Ce résultat met en évidence la compétition entre deux types de complexes dont le résultat dépend de la structure secondaire de l'acide nucléique. La structure en simple brin (poly A ou ADN dénaturé) favorise le complexe d'empilement alors que la structure en double brin (ADN natif) en défavorisant l'empilement permet la formation de complexes par liaison hydrogène quand cela est possible (sérotonine). Si l'on admet que les déplacements chimiques des protons de la sérotonine engagée par liaison hydrogène dans un complexe sont identiques à ceux de la sérotonine libre, la comparaison des résultats obtenus avec la 5-méthoxytryptamine et la sérotonine (Table I) permet de conclure que 40% des molécules de sérotonine fixées à l'ADN forment un complexe d'empilement et 60% un complexe par liaison hydrogène.

2.2.2. *Dérivés possédant le noyau phénol*

Les dérivés à noyau phénol donnent lieu à la formation de complexes analogues à ceux déjà décrits ci-dessus dans le cas de la sérotonine. L'analogie chimique entre la tyramine et la sérotonine est évidente puisque ces deux composés possèdent un cycle aromatique substitué par un groupement hydroxyle. Ainsi, la tyramine forme seulement un complexe d'empilement avec le poly A en simple hélice, complexe caractérisé par des déplacements chimiques voisins de ceux déterminés pour les complexes 1:1 entre monomères. Ce résultat reste valable si l'on remplace la tyramine par son dérivé méthylé ou par des oligopeptides comme Lys Tyr amide ou Lys Tyr Lys.

Dans le cas de l'ADN en double hélice, on observe qualitativement la même différence entre la tyramine et la *p*-méthoxyphényléthylamine (dérivé *o*-methylé de la tyramine) qu'entre la sérotonine et la 5-méthoxytryptamine: les déplacements chimiques extrapolés sont beaucoup plus faibles pour les dérivés hydroxylés et peuvent être interprétés par la compétition entre deux types de complexes: empilement et liaison hydrogène, la méthylation du groupement OH ayant pour effet de supprimer le deuxième type de complexe et de favoriser le premier. Dans le cas des oligopeptides contenant la tyrosine, les déplacements des raies de résonance des protons du cycle phénolique vers les champs élevés en présence d'ADN sont toujours faibles par rapport à ceux observés avec le poly A. En outre, leur amplitude dépend de l'environnement immédiat du résidu tyrosine: des déplacements appréciables vers les champs élevés sont obtenus avec Lys Tyr amide; seuls des élargissements de raies (pas de déplacements) sont observés avec Lys Tyr Lys (Figure 4). Le complexe d'empilement étant favorisé dans les acides nucléiques en simple brin comme le poly A, le déplace-

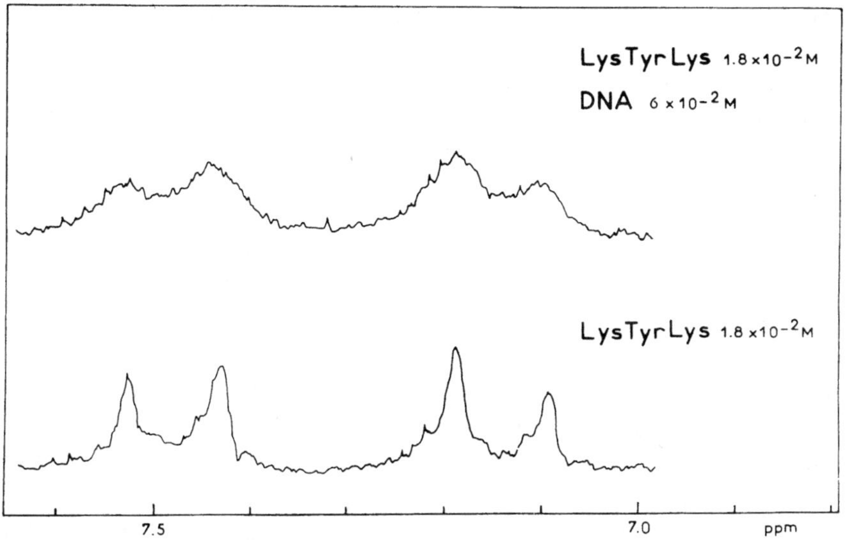

Fig. 4. Spectre RMN des protons de la tyrosine dans l'oligopeptide Lys Tyr Lys ($1{,}8 \times 10^{-2}$M) en l'absence et en présence d'ADN de thymus de veau (6×10^{-2}M) à pH 7.

ment des raies de résonance des protons du cycle phénolique vers les champs élevés augmente au cours de la dénaturation thermique de l'ADN. Ce résultat est analogue à celui déjà décrit pour la sérotonine (voir ci-dessus). Néanmoins on doit remarquer que les déplacements des raies de résonance du cycle de la *p*-méthoxyphényléthylamine vers les champs élevés augmentent également au cours de la dénaturation thermique de l'ADN (contrairement à ceux de la 5-méthoxytryptamine). Cette différence de comportement reflète probablement la différence du bilan énergétique de formation des complexes qui fait intervenir à la fois une modification de l'empilement des bases (dans la simple et la double hélice) et l'interaction du cycle aromatique de l'amine avec les bases (l'énergie de cette dernière interaction est certainement plus élevée dans le cas du noyau indolique que dans celui du noyau benzénique). Une étude plus approfondie est nécessaire pour analyser tous les paramètres qui entrent en jeu.

2.3. Changements de conformation induits dans les acides nucléiques

Plusieurs techniques (viscosité, RMN, dichroïsme circulaire) ont été utilisées pour obtenir des informations sur les changements de conformation d'un acide nucléique ou d'un polynucléotide lors de la fixation d'amines aromatiques ou d'oligopeptides.

Dans certaines conditions, les mesures de viscosité ou de coefficient de sédimentation peuvent fournir des renseignements sur la forme globale du complexe. L'augmentation de viscosité due à la fixation de colorants sur les acides nucléiques [13] a ainsi été interprétée par un allongement de la double hélice dû à l'intercalation des molécules de colorants entre les plateaux de bases. La viscosité de l'ADN augmente en présence de 5-méthoxytryptamine. Cependant, les effets sont faibles (20%) et difficiles à interpréter, car aux faibles forces ioniques utilisées (10^{-3}), la viscosité de l'ADN est très sensible à l'addition de cations monovalents qui provoquent une diminution de viscosité (Leng, résultats non publiés). Dans le cas des oligopeptides Lys X Lys, une diminution de la viscosité est observée quelle que soit la nature de X (Trp, Tyr ou Lys). Cependant, la diminution est moins grande si X=Trp bien que la quantité de complexe formé soit sensiblement la même.

2.3.1. *RMN*

La RMN permet de suivre l'évolution des polynucléotides en simple brin, comme le poly A, qui présentent un spectre haute résolution (ce qui n'est pas le cas avec les acides nucléiques en double hélice) [14–16]. Les raies de résonance des protons des bases du poly A en simple hélice à pH 7 sont déplacées vers les champs élevés et élargies par rapport au monomère correspondant (AMP). Une diminution de l'interaction entre les bases – obtenue, par exemple, par une élévation de température – conduit à un déplacement de ces raies vers les champs plus faibles [14, 15]. La formation d'un complexe d'empilement entre les bases du poly A et un acide aminé aromatique va produire deux effets de sens opposés: l'éloignement des cycles puriques provoquera un déplacement des raies de l'adénine vers les champs faibles; les protons de l'adénine vont simultanément subir l'effet d'anisotropie magnétique du cycle aromatique de

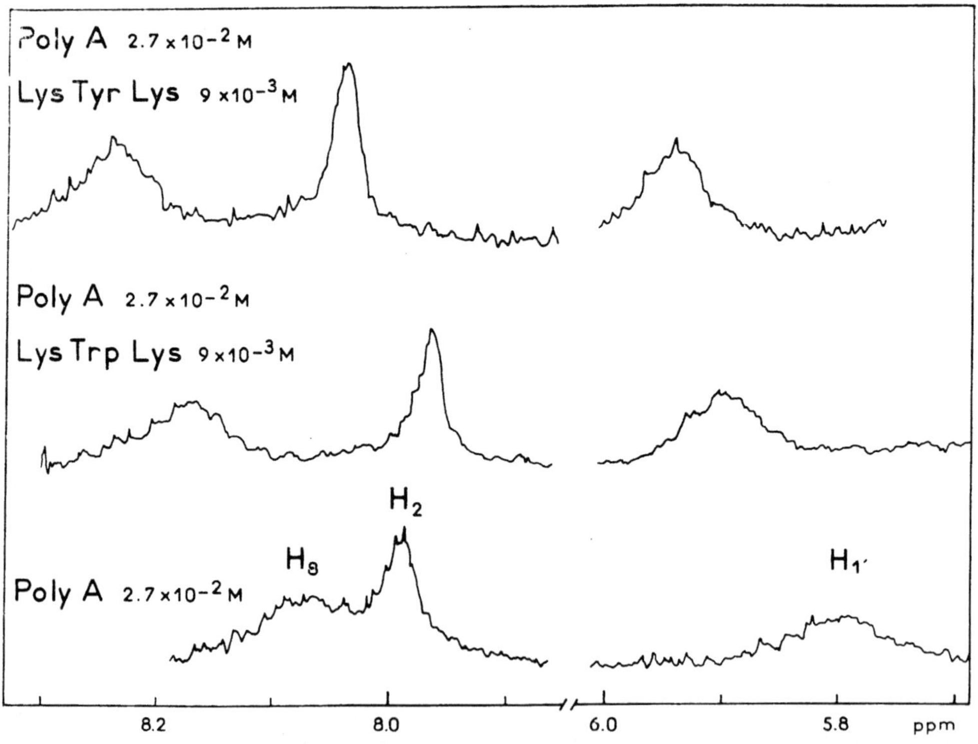

Fig. 5. Spectre RMN des protons H_2 et H_8 de l'adénine dans le poly A à pH 7 en l'absence et en présence de Lys Tyr Lys et Lys Trp Lys.

l'acide aminé, effet qui provoquera un déplacement vers les champs plus élevés. Le déplacement finalement observé sera la résultante de ces deux effets.

Le cycle phénolique possédant une anisotropie magnétique plus faible que celle de l'adénine, on peut prévoir que les raies de résonance du poly A interagissant avec la tyramine ou un des oligopeptides contenant la tyrosine subiront des déplacements vers les champs faibles. C'est ce que l'on observe expérimentalement (Figure 5).

Dans le cas des dérivés à noyau indolique, le résultat est plus difficilement prévisible car l'anisotropie magnétique de l'indole est comparable à celle de l'adénine. Le déplacement observé dépendra cependant de la température puisque l'interaction adénine–adénine dans le poly A diminue quand la température s'élève alors que l'effet d'anisotropie du cycle indolique sera indépendant de la température si la structure stéréochimique du complexe n'est pas changée. Les résultats obtenus peuvent être résumés de la façon suivante (Figure 5): la raie de résonance du proton H_2 de l'adénine se déplace vers les champs élevés quelle que soit la température lorsque le poly A interagit avec la tryptamine, la sérotonine et les oligopeptides Lys Trp Lys ou Trp–Lys. La résonance du proton H_8 du poly A se déplace vers les champs faibles à basse température alors qu'elle se déplace vers les champs élevés à température plus élevée.

Enfin, il a déjà été montré par RMN que le poly A subissait une transition de structure entre 0 et 20°C se traduisant par l'apparition progressive des raies de résonance des protons de l'adénine, les raies étant élargies de telle façon à 0°C que le spectre haute résolution du poly A ne peut plus être enregistré [15]. La fixation des amines aromatiques et des oligopeptides entraîne une diminution de la température à laquelle se fait cette transition, traduisant une plus grande désorganisation de la structure hélicoïdale en présence des molécules fixées [17, 18].

2.3.2. *Dichroïsme circulaire (D.C.)*

L'asymétrie intrinsèque des acides nucléiques donne naissance à des spectres de dichroïsme circulaire caractéristiques de la conformation de ces macromolécules. Une interprétation théorique satisfaisante du dichroïsme des polynucléotides en simple brin comme le poly A a été proposée antérieurement. On peut donc déduire la nature des changements de conformation à partir des modifications du spectre de dichroïsme circulaire provoquées par la fixation des molécules étudiées ici. Dans le cas de l'ADN en double hélice, la faible intensité du dichroïsme circulaire ne permet pas d'établir facilement une corrélation entre les modifications de spectres et les changements de conformation. Dans tous les cas, la contribution éventuelle d'un dichroïsme induit dans les bandes d'absorption de l'acide aminé aromatique est à prendre en considération.

L'interaction entre les amines aromatiques et le poly A peut être observée par D.C. Dans le poly A l'interaction entre les bases donne un spectre dichroïque composé d'une bande positive et d'une bande négative d'égales intensités [19]. Par interaction avec les amines aromatiques on observe une diminution de l'intensité de ces deux bandes. Cet effet observé avec la tyramine, la tryptamine, la sérotonine et l'histamine n'existe pas avec l'isopropylamine. L'addition de NaCl redonne aux deux bandes leur in-

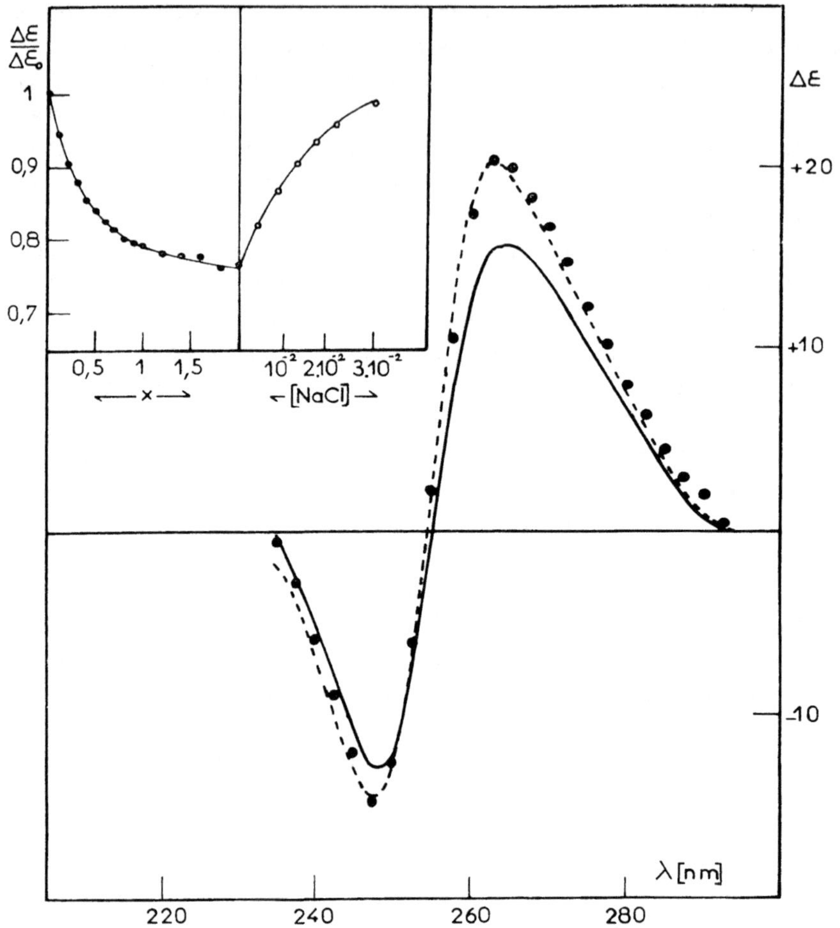

Fig. 6. Changement du spectre de dichroïsme circulaire du poly A en simple brin à pH 7; $5{,}6 \times 10^{-4}$ M dans 1 mM cacodylate de sodium et 1 mM NaCl à 0° C: --- poly A seul; —— poly A + $1{,}12 \times 10^{-3}$ M Lys Phe Lys; ●●● poly A + $1{,}12 \times 10^{-3}$ M Lys Phe Lys + 3×10^{-2} NaCl. Encart: Variation de l'intensité relative de la bande positive à 262 nm fonction du rapport × = [peptide]/[poly A]; puis de la concentration en NaCl.

tensité initiale démontrant la contribution de l'interaction électrostatique dans la formation des complexes.

Des effets similaires ont été observés avec les différents peptides possédant un acide aminé aromatique: Lys Tyr Lys, Lys Phe Lys, Lys Trp Lys. Sur la Figure 6, nous avons illustré le cas de Lys Phe Lys. La diminution maximum d'intensité de la bande positive est de 25% alors que celle de la bande négative n'est que de 15%. Ces deux bandes retrouvent leur intensité initiale lorsque la concentration en NaCl est augmentée. Par contre avec le peptide Lys Ala Lys ne comportant pas d'acide aminé aromatique, une modification extrêmement faible est observée. La différence de comportement entre ces deux peptides illustre le rôle important que joue le noyau aromatique dans l'interaction avec le poly A et dans les changements de spectres de dichroïsme circulaire.

La modification observée dans le spectre de dichroïsme circulaire est compatible avec une diminution de l'interaction entre les bases du Poly A en raison de l'intercalation des résidus aromatiques.

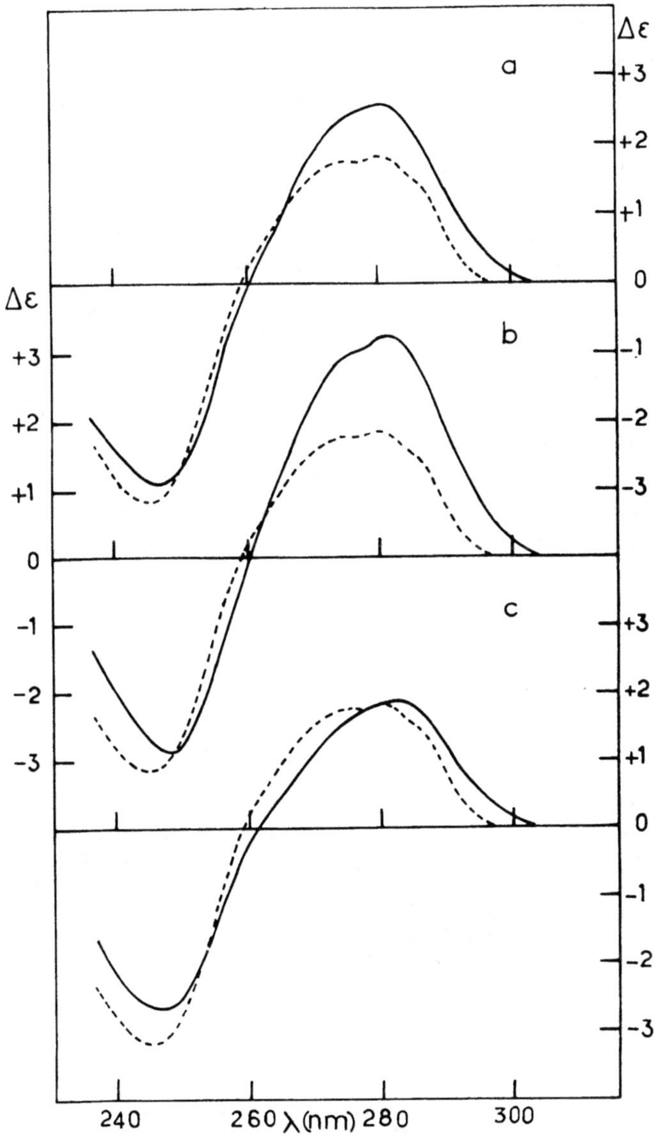

Fig. 7. Spectre de dichroïsme circulaire de l'ADN ($C = 6{,}8 \times 10^{-4}$ M) en l'absence (----) et en présence (——) d'amines aromatiques dans 1 mM cacodylate de sodium et 1 mM NaCl à 0°C: (a) ADN + $6{,}1 \times 10^{-4}$ M tryptamine; (b) ADN + $8{,}0 \times 10^{-4}$ 5-méthoxytryptamine; (c) ADN + $6{,}2 \times 10^{-4}$ M sérotonine.

Aucune modification du spectre de dichroïsme circulaire du Poly C en simple hélice à pH neutre n'a été détectée en présence des amines aromatiques et des peptides Lys Tyr Lys et Lys Trp Lys, confirmant les résultats de fluorescence indiquant la très faible interaction de ces composés avec le Poly C.

2.3.3. *ADN en double hélice*

En présence d'amine aromatique le spectre dichroïque de l'ADN subit des modifications importantes. Sur la Figure 7 nous avons illustré le spectre de l'ADN en présence et en l'absence de tryptamine, de 5-méthoxytryptamine et de sérotonine. Sur la Figure 8 nous avons représenté la différence entre les spectres dichroïques du complexe et de l'ADN seul. Dans tous les cas le spectre différence est important. Cependant, selon la nature du composé, des différences existent. Pour la tryptamine et la 5-méthoxytryptamine il y a une forte augmentation de l'intensité dichroïque de la bande positive, 70% pour la méthoxytryptamine et 40% pour la tryptamine. Cette augmentation s'accompagne d'un déplacement vers les plus fortes longueurs d'onde. La bande négative subit également un tel déplacement alors que son intensité est peu affectée. Avec la sérotonine, on note une augmentation de l'intensité du spectre dans la région 280–300 nm et contrairement aux deux autres composés une diminution vers 270 nm. La bande négative subit une diminution d'intensité accompagnée d'un déplacement vers les plus fortes longueurs d'onde.

Les amines aromatiques ne possèdent pas de spectre dichroïque. Néanmoins après fixation sur l'ADN un dichroïsme induit peut apparaître dans la région où elles absorbent. Dans les conditions expérimentales utilisées, les modifications que nous ob-

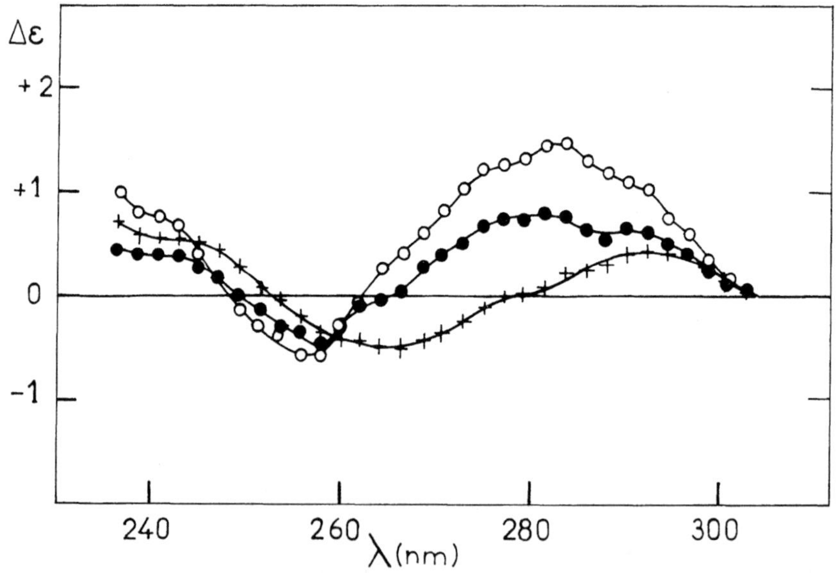

Fig. 8. Spectre de dichroïsme différence entre le complexe et l'ADN seul. Les conditions sont celles de la Figure 7. ● ● ● tryptamine, ○ ○ ○ méthoxytryptamine, + + + sérotonine.

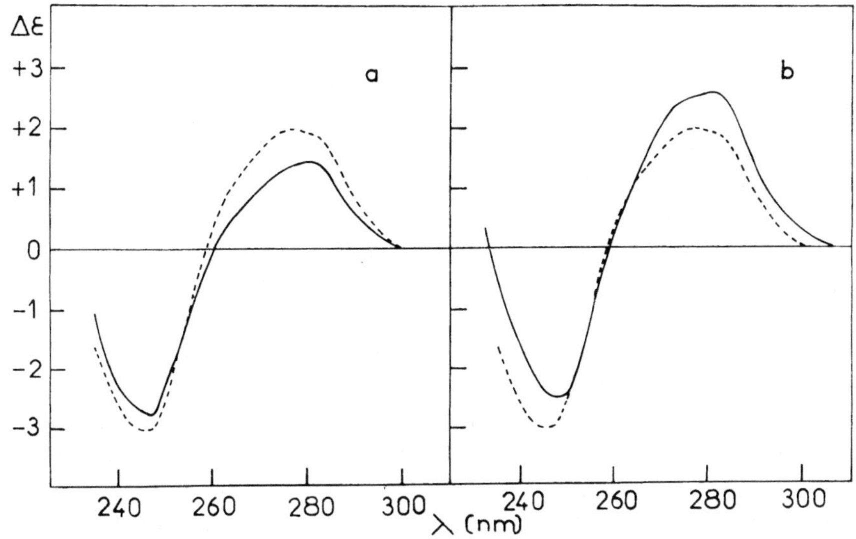

Fig. 9. Spectre de dichroïsme circulaire de l'ADN seul (---) et en présence d'amines aromatiques (——). Les conditions expérimentales sont les mêmes que pour la Figure 7; (a) ADN + $2,36 \times 10^{-3}$ M tyramine; (b) ADN + $3,4 \times 10^{-3}$ M *p*-méthoxyphényléthylamine.

servons ne semblent pas dues à ce phénomène; les spectres différence obtenus avec la tryptamine et avec la méthoxytryptamine présentent la même forme et sont situés dans la même région spectrale, bien que le spectre d'absorption de la méthoxytryptamine soit situé à plus grandes longueurs d'onde que celui de la tryptamine. Il semble donc que l'on puisse attribuer les modifications de spectre à un changement dans la conformation de l'ADN. Comme seule l'intensité du spectre différence varie entre la tryptamine et la 5-methoxytryptamine on peut penser que l'ADN adopte le même type de conformation lorsqu'il y a fixation. Il faut d'ailleurs noter que par fluorescence le produit $K \times n$ obtenu pour la fixation de la méthoxytryptamine est d'environ 1000 alors qu'il n'est que de l'ordre de 260 pour la tryptamine. Pour la sérotonine c'est un autre type de courbe différence que l'on obtient ce qui doit refléter un mode d'interaction différent. Ce résultat est à rapprocher de ceux obtenus par RMN et discutés dans la section précédente.

Sur la Figure 9 nous avons représenté les spectres de dichroïsme circulaire de l'ADN avec et sans *p*-méthoxyphényléthylamine et tyramine. La tyramine induit une diminution de l'intensité des deux bandes positive et négative, sans modification de leur position. Au contraire avec la *p*-méthoxyphényléthylamine, on observe une augmentation d'intensité de la bande positive accompagnée d'un déplacement vers les grandes longueurs d'onde. L'intensité de la bande négative diminue et subit un déplacement dans le même sens.

Il est frappant de constater que comme pour la sérotonine et la méthoxytryptamine, des changements différents de conformation apparaissent avec la tyramine et son homologue *O*-méthylé. On peut penser que pour la sérotonine et la tyramine la présence du groupement hydroxyle pouvant donner une liaison hydrogène peut modifier le mode d'interaction, et par conséquent la conformation adoptée par l'ADN.

Il est difficile à partir des seuls spectres de dichroïsme circulaire de déterminer quelles sont les conformations adoptées par l'ADN en présence des amines aromatiques. Il est condiséré en général que l'ADN en solution adopte une conformation correspondant à la forme B. Néanmoins, des études par dichroïsme circulaire ont montré que des changements peuvent exister avant la température de dénaturation [20, 21]. D'autres perturbations telles que l'addition de solvant organique [22, 23], la présence de super-hélice [24], l'interaction avec des histones ou des fractions d'histones [25] modifient la forme du spectre de dichroïsme circulaire de l'ADN. Dans le cas de l'intercalation de colorants, il n'est pas possible de déterminer quelle est la modification du spectre dichroique de l'ADN en raison de l'existence d'un fort dichroïsme induit dans la région d'absorption des bases. Les changements observés sont très proches pour la tryptamine, la méthoxytryptamine et la *p*-méthoxyphényléthylamine et peuvent indiquer le même type d'interaction; le spectre dichroïque correspondrait dans ce cas au spectre d'un ADN dont les bases forment un complexe d'empilement avec le cycle aromatique de l'amine.

La fixation du peptide Lys Trp Lys sur l'ADN entraîne une augmentation de l'intensité de la bande positive du spectre dichroïque du DNA, de la même manière que la tryptamine. Par contre l'interaction avec Lys Tyr Lys conduit à une diminution comparable à celle obtenue avec la tyramine.

2.3.4. *ARNs de transfert*

La fixation des oligopeptides Lys Tyr Lys et Lys Trp Lys sur les ARNs de transfert spécifiques de la phenylalanine (levure), de la formylméthionine (*E. coli*) et de la valine (*E. coli*) a été étudiée à la fois par fluorescence et par dichroïsme circulaire. Ces ARNs ont été au préalable dialysé contre de l'EDTA pour éliminer les cations bivalents. Les résultats de fluorescence permettent de distinguer plusieurs types de sites de fixation: des sites forts (2 ou 3 par molécule d'ARN) et des sites plus faibles (entre 15 et 20). La fixation des oligopeptides s'accompagne d'une modification importante du spectre de dichroisme circulaire de l'ARN caractérisés en particulier par l'apparition d'une bande négative de faible intensité vers 295 nm [12]. Une telle bande apparaît également lorsque des cations bivalents comme Mg^{2+} se fixent sur l'ARN. Ces résultats peuvent donc être interprétés par un changement de la structure secondaire et/ou tertiaire de l'ARN de transfert consécutive à la fixation de l'oligopeptide [12].

3. Conclusion

De l'ensemble des résultats présentés ci-dessus on peut tirer les conclusions suivantes:

(a) Dans les acides nucléiques en simple brin, comme le poly A ou l'ADN dénaturé, les cycles aromatiques des amines (tryptamine sérotonine, tyramine) et des oligopeptides contenant la lysine et un acide aminé aromatique, forment des complexes d'empilement avec les bases. Ceci se traduit par une perte de structure secondaire (hélicoïdale) de l'acide nucléique, nécessaire pour permettre l'intercalation du cycle indolique ou phénolique.

(b) Dans les acides nucléiques en double hélice, comme l'ADN natif, plusieurs types de complexes peuvent se former. La tryptamine, la 5-méthoxytryptamine, la *p*-méthoxyphenyléthylamine ne forment apparemment qu'un complexe d'empilement avec les bases. La sérotonine, la tyramine et les oligopeptides contenant la tyrosine donnent naissance à au moins deux types de complexes en compétition: un complexe d'empilement et un complexe par liaison hydrogène. L'importance relative de ces deux complexes dépend non seulement de la structure secondaire de l'acide nucléique (le simple brin favorisant l'empilement), mais également de l'environnement du résidu aromatique dans un oligopeptide.

(c) Enfin il faut noter l'existence de complexes probablement purement électrostatiques dans lesquels le cycle aromatique n'interagit pas avec l'ADN en double brin.

L'ensemble de ces résultats démontre que les acides aminés aromatiques des protéines peuvent participer à la formation des complexes protéines–acides nucléiques par une interaction directe avec les bases (empilement, liaison H). Cette possibilité d'interactions dépend à la fois de la structure locale de l'acide nucléique et de l'environnement du résidu aromatique dans la protéine. Ces interactions entraînent des changements de structure de l'acide nucléique qui peuvent être importants pour l'activité ultérieure du complexe. L'effort doit être porté maintenant vers l'étude de la sélectivité éventuelle des interactions des acides aminés aromatiques avec les bases dans les acides nucléiques, sélectivité qui pourrait intervenir au niveau de la reconnaissance spécifique des acides nucléiques par les protéines ou les enzymes.

Remerciements

Nous tenons à remercier M. le Professeur Ch. Sadron pour l'intérêt qu'il porte à ce travail, M. Joël Rossi, pour la synthèse de certains des oligopeptides utilisés au cours de cette étude et Melles F. Brun et A. Millet pour leur aide technique efficace.

Références

[1] Hélène, C.: 1971, *Nature* **234**, 120.
[2] Montenay-Carestier, T. et Hélène, C.: 1971, *Biochemistry* **10**, 300.
[3] Dimicoli, J. L. et Hélène, C.: 1971, *Biochimie* **53**, 331.
[4] Lawaczeck, R. et Wagner, K. G.: 1972, *J. Magn. Res.* **8**, 164.
[5] Dimicoli, J. L. et Hélène, C.: 1972, *J. Am. Chem. Soc.* **95**, 1036.
[6] Saxinger, C., Ponnamperuma, C., et Woese, C.: 1971, *Nature* **234**, 172.
[7] Smythies, J. R. et Antun, F.: 1969, *Nature* **223**, 1061.
[8] Hélène, C., Dimicoli, J. L., et Brun, F.: 1971, *Biochemistry* **10**, 3802.
[9] Hélène, C., Montenay-Garestier, T., et Dimicoli, J. L.: 1971, *Biochim. Biophys. Acta* **254**, 349.
[10] Hélène, C. et Dimicoli, J. L.: 1972, *FEBS Letters* **26**, 6.
[11] Brown, P. E.: 1970, *Biochim. Biophys. Acta* **213**, 282.
[12] Hélène, C., Dimicoli, J. L., Borazan, H. N., Durand, M., Maurizot, J. C., et Toulmé, J. J.: 1973, dans *Symposium on Conformation of Biological Molecules and Polymers*, Jerusalem, p. 361.
[13] Cohen, G. et Eisenberg, H.: 1969, *Biopolymers* **8**, 45.
[14] McDonald, C. C., Philips, W. D., et Penman, S.: 1964, *Science* **144**, 1234.

[15] McTague, J. P., Ross, V., et Gibbs, J. H.: *Biopolymers* **2**, 163.
[16] McDonald, C. C., Philips, W. D., et Lazar, J.: 1967, *J. Am. Chem. Soc.* **89**, 4166.
[17] Razka, M. et Mandel, M.: 1971, *Proc. Nat. Acad. Sci. US* **68**, 1190.
[18] Dimicoli, J. L.: 1972, Thèse de Doctorat, Orléans, France.
[19] Van Holde, K. E., Brahms, J., et Michelson, A. M.: 1965, *J. Mol. Biol.* **12**, 726.
[20] Brahms, J. et Mommaerts, W. F. H. M.: 1964, *J. Mol. Biol.* **10**, 73.
[21] Gennis, R. B. et Cantor, C. R.: 1972, *J. Mol. Biol.* **65**, 381.
[22] Brahms, J., Mommaerts, W. F. H. M., et Sadron, C.: 1964, *C. R. Acad. Sci. Paris.* **258**, 2203.
[23] Green, G. et Mahler, H. R.: 1970, *Biochemistry* **9**, 368.
[24] Maestre, M. F. et Wang, J. C.: 1970, *Biopolymers* **10**, 1021.
[25] Shih, T. Y. et Fasman, G. D.: 1972, *Biochemistry* **11**, 398, et références citées.
[26] Feitelson, J.: 1964, *J. Phys. Chem.* **68**, 391.

DISCUSSION

De Coen: D'après les études de Kerckx il est peut-être possible de donner une précision concernant le mouvement des bases lors de l'interaction du résidu aromatique avec l'acide nucléique. On peut étirer légèrement la chaîne phosphate-sucre et faire tourner les bases ensemble autour des liaisons C1′–N en les gardant parallèles l'une à l'autre. On obtient de cette manière une cage hydrophobe ayant exactement les dimensions du noyau aromatique à intercaler. Ce mécanisme me paraît compatible avec les variations spectroscopiques observées.

Hélène: Les résultats de Kerckx concernent le poly A en simple hélice et l'explication que vous donnez est fort plausible. Dans le cas de l'ADN en double brin, si l'on maintient l'appariement des bases, les possibilités de distortion sont certainement plus réduites. Il est néanmoins possible d'intercaler des molécules aromatiques comme les colorants de type acridine entre les plateaux des bases.

Wüthrich: Pouvez-vous nous indiquer l'ordre de grandeur des temps de vie des complexes formés dans vos systèmes?

Hélène: Nous n'avons pas, à l'heure actuelle, de résultats permettant de déterminer le temps de vie des complexes formés avec le poly A ou l'ADN. Les expériences de relaxation après un saut de température n'ont pas donné de résultats concluants. Pour un équilibre de type $A+B \rightleftharpoons AB$, si l'on suppose que la formation des complexes est contrôlée par la diffusion ($k \sim 10^8$ M^{-1} sec^{-1}), une constante d'équilibre de 10^4 M^{-1} correspond à un temps de vie de 10^{-4} sec. Nous serions donc, pour les mesures RMN, dans la zone d'échange rapide (ce qui est en accord avec l'observation d'une seule raie de résonance pour chaque proton). L'élargissement des raies de résonance des protons des cycles aromatiques liés à l'ADN est dû, dans ces conditions, à une mobilité réduite de ces cycles dans le complexe.

Grosjean: Qu'est-ce qui vous porte à éviter explicitement dans votre exposé le terme 'intercalation' et à utiliser celui d'empilement' des résidus tryptophane dans l'acide nucléique?

Hélène: Le terme 'intercalation' doit être réservé aux molécules qui sont insérées entre les plateaux de base d'un ADN en double hélice (en conservant l'appariement des bases). Les résultats de RMN que j'ai présentés permettent de conclure que les cycles de bases puriques et des résidus tryptophane sont supperposés (effets de courants annulaires). Ils ne nous permettent pas de conclure que le tryptophane est inséré dans la double hélice. La détermination de la mobilité du cycle indole dans le complexe est nécessaire pour répondre à cette question (voir réponse à Wüthrich).

De Coen: Le mécanisme d'intercalation que je viens de mentionner n'exige nullement un allongement considérable de la molécule. Un très léger allongement de quelques % suffit mais une rotation autour des liaisons C′1–N assure pratiquement seule un écart double des paires de bases.

ENERGY TRANSFER STUDY IN THE DNA–ETHIDIUM BROMIDE COMPLEX BY MEANS OF ANISOTROPY DECAY

D. GENEST and PH. WAHL

Centre de Biophysique Moléculaire, 45045, Orléans Cédex, France

Summary. The fluorescence anisotropy decay of the DNA–ethidium bromide complex has been studied for different values of the ratio *P/D*, of the molar concentrations of DNA and EB. An increase of the kinetics of the fluorescence depolarization has been observed when *P/D* decreases. This phenomenon is attributed to energy transfers between molecules of dye.

A quantitative interpretation of this phenomenon using the Förster's theory of energy transfers, permits to determine the geometrical parameters defining the chromophores positions, included in the DNA matrix.

Résumé. Le déclin d'anisotropie de fluorescence du complexe bromure d'éthidium (BET)–DNA a été étudié en fonction du rapport *P/D* des concentrations molaires de DNA et de BET. Il a été observé expérimentalement une accélération de la cinétique de dépolarisation de fluorescence lorsque *P/D* diminue. Ce phénomène est attribué au transfert d'énergie entre molécules de colorant.

Une interprétation quantitative du phénomène au moyen de la théorie des transferts d'énergie de Förster, permet de préciser les paramètres géométriques définissant la position des molécules de colorants inclus dans la matrice de DNA.

1. Introduction

The dye ethidium bromide (EB) which has a weak fluorescence in water solution becomes strongly fluorescent when it interacts with the DNA molecule. There is some evidence [1, 2, 3] in favour of an intercalation of this dye between two base pairs of the DNA double helix.

The fluorescence anisotropy decay of the EB–DNA complex has been analysed in previous publications of our laboratory [4, 5]. At high *P/D* (the ratio of the number of nucleotides per bound dye), this decay corresponds to a torsional brownian motion. For smaller *P/D*, the slope of the anisotropy decay increases, and this is ascribed to energy migration between the dyes fixed on the same DNA molecule.

In this communication, the experimental results showing the influence of *P/D* on the anisotropy is given. These results are interpreted on the basis of Förster's theory of transfers by dipolar resonance.

In order to take into account the geometrical arrangement of the dye molecules in the macromolecular matrix, two approaches are used. The first consists in an analytical calculus valuable for high values of *P/D*. The second used a Monte Carlo method and is well suited for low values of *P/D*. In this last calculation, one can take into account a possible angular deformation of the DNA helix induced by the EB molecules. This problem has two opposite answers. On one hand, Waring and Fuller [6], Bauer and Vinograd [7] are in favour of an unwinding of the DNA helix induced by EB molecules. On the other hand, Paoletti and Le Pecq [8] think rather that the EB molecules

C. Sadron (ed.), Dynamic Aspects of Conformation Changes in Biological Macromolecules. 367–379.

provoke a winding of the DNA helix. This last work is based on results obtained by steady-state fluorescence polarization measurements, and an analysis by a Monte Carlo method.

2. Material and Methods

The determination of the decays are made by the single photoelectron counting technique [9, 10]. A simplified scheme of the apparatus is given in Figure 1. The polarizer P_2 can rotate around the Ox axis and permits to measure alternatively the

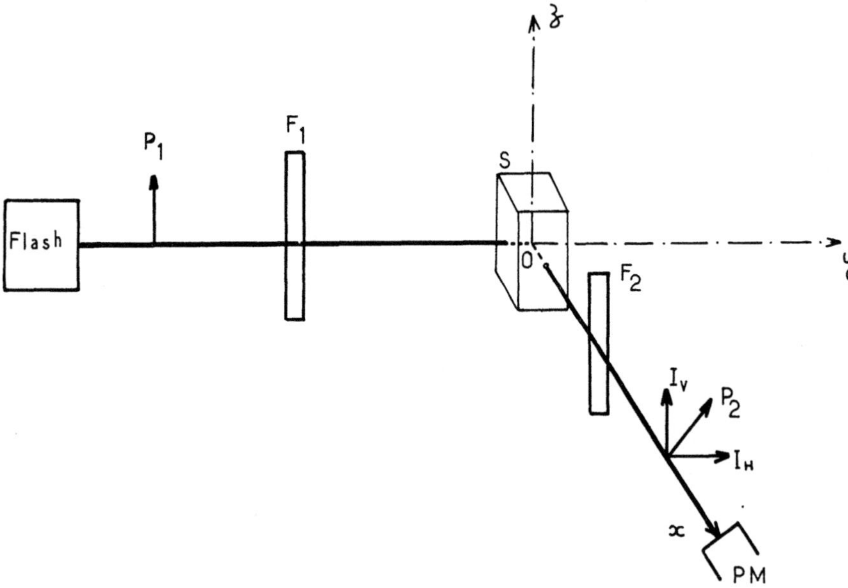

Fig. 1. The light source gives short light pulses. The incident beam passes through a polarizer P_1 which polarizes vertically the electric field, then through a filter F_1 which permits to choose the excitation wavelength. The sample is placed in the cuvette S and the fluorescent beam is observed in the direction Ox perpendicular to the exciting beam. The fluorescence light passes through a filter F_2 permitting to choose the emission wavelength then through a rotating polarizer P_2, and is received by the photomultiplier PM.

two principal components of the polarization $I_V(t)$ and $I_H(t)$. From these experimental curves one derives the following ones:

$$s(t) = I_V(t) + 2I_H(t) \tag{1}$$

$$d(t) = I_V(t) - I_H(t). \tag{2}$$

$s(t)$ is the decay of the whole fluorescence and $d(t)$ describes the state of polarization of the fluorescence light.

These experimental curves are related to the true decays $S(t)$ and $D(t)$ by the following products of convolution:

$$s(t) = \int_0^t g(T)\, S(t - T)\, \mathrm{d}T \tag{3}$$

$$d(t) = \int_0^t g(T)\, D(t - T)\, \mathrm{d}T, \tag{4}$$

where $g(t)$ is the temporal distribution of light intensity of the flash which can be determined if the fluorescence cuvette is replaced by a reflector.

The anisotropy decay $r(t)$ is defined by:

$$r(t) = D(t)/S(t). \tag{5}$$

3. Experimental Data

Different solutions with varying P/D have been studied. Some of the $s(t)$ and $d(t)$ curves are given in Figure 2. It shows that the slope of $d(t)$ increases with decreasing values of P/D, while the shape of $s(t)$ remains the same. Therefore this variation must

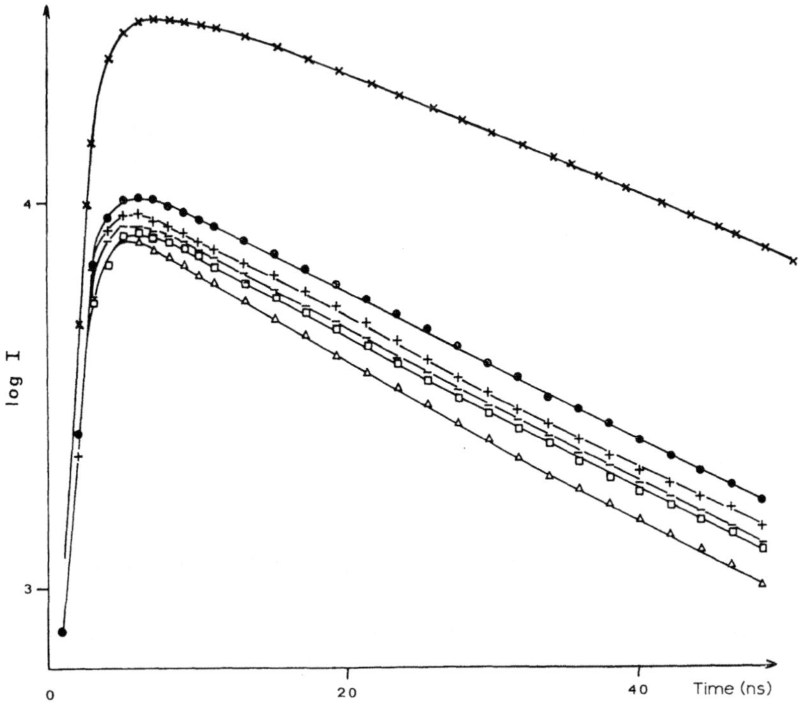

Fig. 2. Experimental curves $s(t)$ and $d(t)$ following different values of P/D. × × × represents the $s(t)$ curve which is identical for all ratios P/D. The different $d(t)$ are: ●●● $P/D = 200$; + + + $P/D = 42$; --- $P/D = 30$; □ □ □ $P/D = 20$; △ △ △ $P/D = 16$ (taken from [5]).

come from an increase in the slope of $r\ (t)$. It seems reasonable to relate this phenomenon to the energy transfer between like molecules, the so-called energy migration.

At high values of P/D ($P/D=200$), $r\ (t)$ depends only on the rotational brownian motions. It is found that the experimental curve fits the following expression:

$$r_B(t) = \tfrac{1}{2}r_0\left(e^{-t/\theta} + 1\right), \tag{6}$$

with $\theta = 25$ n sec and $r_0 = 0.32$, in good agreement with a previous determination [4]. This behaviour is attributed to a torsional deformation of the DNA molecule.

4. Theoretical Interpretation

4.1. *General Principle of the Analysis*

If we admit that the brownian disorientation is independent on the disorientation by transfers one may write:

$$r(t) = r_B(t)\cdot r_T(t), \tag{7}$$

where $r_B(t)$ is the brownian anisotropy decay, and $r_T(t)$ is the decay of anisotropy due to energy migrations. $r_B(t)$ is given by expression (6) and $r_T(t)$ is computed by mean of the relation [11]:

$$r_T(t) = \frac{3\ \overline{\cos^2 \Phi\,(t)} - 1}{2}, \tag{8}$$

where Φ is the angle between the dipolar transition momentum of the primary excited chromophore at time $t=0$ and the chromophore excited at time t. The bar indicates the average of all possible values of Φ at time t.

The following expressions are calculated:

$$D'(t) = r(t)\cdot S(t) \tag{9}$$

$$d'(t) = \int_0^t g(T)\, D'(t-T)\, \mathrm{d}T\,. \tag{10}$$

At last $d'(t)$ is compared to $d(t)$, the experimental curve.

4.2. *The Theory of Transfers*

In order to interprete this phenomenon quantitatively, Förster's theory has been applied.

In this theory one assumes that the energy may migrate from a chromosphore A to a chromophore B. The rate of energy transfer following Förster is given by the expression [12, 13]:

$$V_T = \frac{C}{n^4}\cdot\frac{\chi^2}{R^6}\cdot\frac{J}{\tau_0}. \tag{11}$$

$C = 9000 \log 10/128\pi^5 N$ is an universal constant, N being Avogadro's number; n is the refractive index of the medium, which is here the DNA molecule ($n = 1.4$).

J/τ_0 only depends on the spectral characteristics of the chromophore. τ_0 is the natural lifetime of fluorescence, and J is the overlap integral defined as

$$J = \int \frac{F(\bar{\nu})\,\varepsilon(\bar{\nu})}{\bar{\nu}^4}\,\mathrm{d}\bar{\nu},$$

where F is the quantum fluorescence spectrum and ε is the molar extinction coefficient at wavenumber $\bar{\nu}$.

χ^2/R^6 only depends on the geometry of the two chromophores A and B. R is the distance between A and B. χ is an angular factor defined by

$$\chi = \cos\theta - 3\cos\theta_A \cos\theta_B,$$

where θ is the angle between the dipolar transition momenta $\vec{\mu}_A$ and $\vec{\mu}_B$ of A and B, and θ_A and θ_B are respectively the angle between $\vec{\mu}_A$ and $\vec{\mu}_B$ with the axis joining A and B.

J has been computed from the corrected emission spectra and the absorption spectra. It is found to be equal to 1.7×10^{-15} cm^6/mole. The value of τ_0 (73 nsec) is obtained from the equation [14]:

$$\frac{1}{\tau_0} = 2.88 \times 10^{-9} n^2 \frac{1}{\int (F(\bar{\nu})/\bar{\nu}^3)\,\mathrm{d}\bar{\nu}} \int \frac{\varepsilon(\bar{\nu})}{\bar{\nu}}\,\mathrm{d}\bar{\nu}.$$

4.3. *Calculation of the Anisotropy of Fluorescence*

This process of transfer between like molecules causes an energy migration which in turn produces a fluorescence depolarization. In order to calculate the kinetics of the energy migration one has to introduce the spatial distribution of the chromophores. The rate constant V_T of an elementary transfer is indeed function of the geometrical factor χ^2/R^6.

To solve this problem, we have used two kinds of calculation.

(a) An analytical computation based on an approximation introduced by Förster for the determination of the concentration depolarization in homogeneous solutions.

(b) A Monte Carlo calculation.

A. ANALYTICAL COMPUTATION

An expression of $r_T(t)$ has been established in a previous work, which takes into account the linear repartition of chromophores on DNA [5]. Two main approximations of Förster's calculation [12, 13] have been kept for the establishment of the expression of $r_T(t)$:

(a) the energy migrates only between the primary excited molecule and its nearest neighbour;

(b) an average value of χ^2 is taken for the angular factor.

The results of the comparison between the experimental curves $d(t)$ and the calculated ones are given in Figure 3. The curves 3, 4, 5 are obtained with $\chi^2 = \frac{1}{2}$ corresponding to a motion of chromophores in a plane perpendicular to the double-helix axis, and the curves 3′, 4′, 5′ with $\chi^2 = \frac{2}{3}$ correspond to a motion of dye molecules in

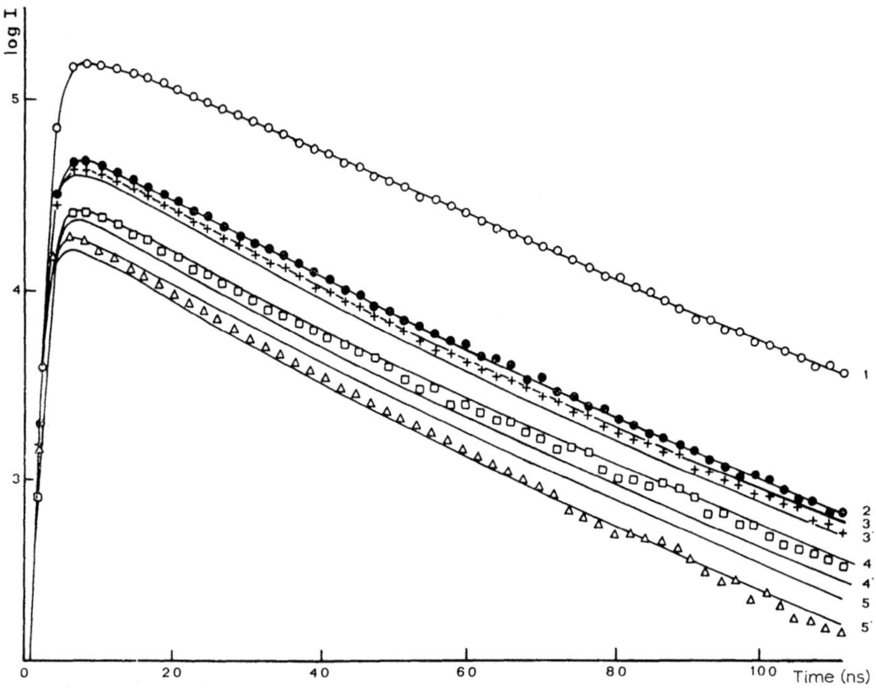

Fig. 3. Comparison between the experimental curves $d(t)$ and the curves computed with the analytical approximation. Curve 1 represents the whole fluorescence decay (○–○ experimental curve; — product of convolution). The representation of experimental curves $d(t)$ is the same as in Figure 2. To have a more clear figure, the intensities of curves corresponding to $P/D = 25$ and $P/D = 16$ have been divided respectively by 1.5 and 2. The curves drawn in full lines represent the computed curves (taken from [5]).

the whole space. For high values of P/D, the calculated curves fit the experimental ones well. For example, curves 3 and 4 are in good agreement with the experiment. But for $P/D = 16$ the theoretical curve is above the experimental one.

From this study it can be thought that ethidium bromide molecules are moving in a plane perpendicular to the DNA axis. We think that the disagreement for $P/D = 16$ is due essentially to the hypothesis that only two chromophores are involved in the energy migration process. This assumption obviously restricts the possibility of energy migration and leads to a too high anisotropy in the large time range of the fluorescence emission.

B. A METHOD OF SIMULATION

In order to go ahead with the analysis of the process of energy migration, a statistical method of simulation has been used to compute the time-dependent fluorescence anisotropy [15]. Before describing these calculations, we will specify the model of the DNA–EB complex [6, 7, 8]. It is shown in Figure 4. It corresponds to the intercalation model in which two successive sites cannot be occupied. This last statement is based on the experimental fact that the maximum ratio of dye per site is $\frac{1}{2}$ [1–3]. If no EB is intercalated, the distance between two adjacent base pairs is normally 3.4 Å, the angle α

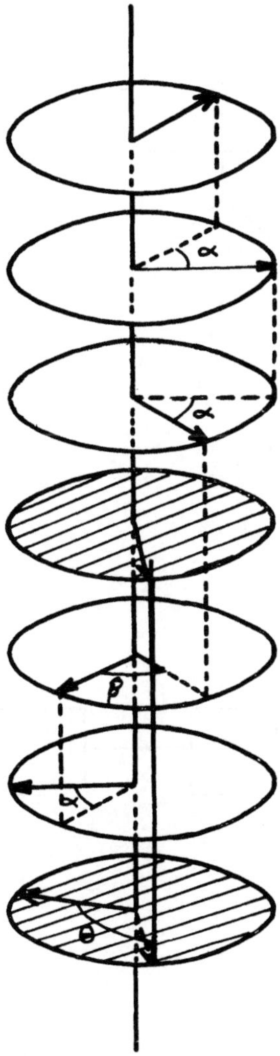

Fig. 4. The model of the DNA–dye complexes. The clear discs schematize the base pairs of DNA. The dark discs are the dye molecules. The inter-disc distance is 3.4 Å.

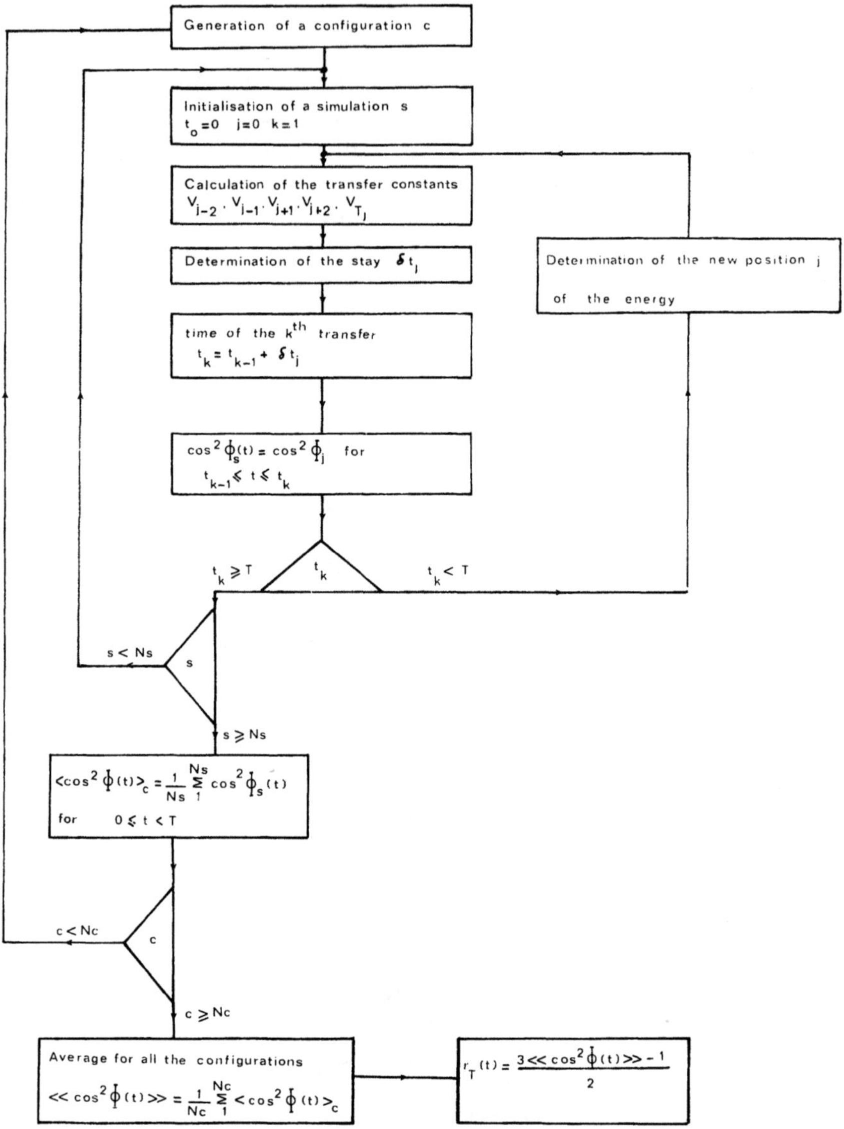

Fig. 5. Scheme of the calculation of $r_T(t)$ by the Monte Carlo method. c = number of the used configuration; N_c = maximum number of configurations; s = number of a step for a given configuration; N_s = maximum number of steps for a configuration; k = index of a transfer in a step; j = index of the position of the energy.

being 36°. When a chromophore is intercalated the distance between the adjacent base pairs becomes 6.8 Å. The angle can be modified, its new value is β. The winding angle is defined by the relation $\delta=\beta-\alpha$. The DNA helix undergoes a winding or an unwinding according to whether the value of β is greater or smaller than 36°, or the value of δ is positive or negative.

The angle between the dipolar transition momenta of two chromophores is θ. From X-ray diffraction measurements Fuller and Waring [6] determined $\beta=24°$ ($\delta=-12°$). Polarization studies under continuous excitation lead Paoletti and Le Pecq [8] to assume that $\beta=49°$ ($\delta=+13°$).

Under these conditions, the calculation proceeds along the scheme of Figure 5.

1. *Generation of a Configuration*

This is realized following the principle described by Paoletti and Le Pecq [8]. A DNA segment of 200 sites is considered. The EB molecules are distributed among these sites taking the average number of intercalated dye per nucleotide into account, under the condition that two adjacent sites cannot be occupied.

2. *Simulation of the Energy Migration for a Given Configuration*

In this section we describe the computation of the energy migration during a fixed time T.

(a) *Initiation of the migration.* It is assumed that at time zero the dye molecule occupying the most central site absorbs a photon. The moment of transition of this dye defines the origin of the angle Φ which characterizes the angular position of all other dyes.

(b) *A step in a migration.* Rate constant of migration: Let us assume that at time t_k, the excitation is on a site j. In order to simplify the computation it is assumed that this excitation may be transferred only to the two nearest preceding and the two nearest following molecules. The corresponding transfer rates V_{j-2}, V_{j-1}, V_{j+1}, V_{j+2} may be calculated by Förster's formula (11) as a function of the distance and the angle θ between the central molecule and its neighbours. The value of θ depends on β which determines the deformation angle induced on the DNA helix.

The total rate constant of transfer from the site j is then:

$$V_{Tj}=V_{j-2}+V_{j-1}+V_{j+1}+V_{j+2}\,.$$

The time of a migration step: The probability density function of a stay δt_j in the site j is

$$p(\delta t_j)=\exp(-V_{Tj}\delta t_j)\,.$$

A random number η, equidistributed on [0, 1] is picked and δt_j is obtained by setting $p(\delta t_j)=\eta$.

The time of the $(k+1)$-th transfer step of the simulation is then:

$$t_{k+1}=t_k+\delta t_j\,.$$

In particular, for the first step we have obviously

$$t_1 = \delta t_0 ,$$

since

$$t_0 = 0 .$$

Location of the energy after the (k+1)-th migration step: The relative probabilities of migration to each possible site are $P(j-i) = V_{j-i}/V_{Tj}$, where i has the different values -2, -1, $+1$, $+2$.

The five numbers 0, $P(j-2)$, $P(j-2)+P(j-1)$, $P(j-2)+P(j-1)+$ $+P(j+1)$, 1 define four intervals. Then a random equidistributed number between 0, 1 is picked, and according to whether its value is in the first, second, third or fourth interval, the energy will be ascribed to be on site $j-2$, $j-1$, $j+1$ or $j+2$.

Calculation of $\cos_s^2 \Phi(t)$: In this expression index s indicates a given simulation. During the time interval δt_j the energy remains in the site j, then one must put

$$\cos_s^2 \Phi(t) = \cos^2 \Phi_j$$

for $t_k < t < t_{k+1}$, where Φ_j is the angle occurring between the dye molecule in site j and the primary excited molecule at site 0.

In particular:

$$\cos^2 \Phi_s(t) = 1 \quad \text{for} \quad 0 < t < t_1 .$$

(c) *Completion of the migration.* The excitation site determined in the previous section is taken as the starting site for a new step in the migration. The process described in (b) is repeated. The migration is ended when t_k happens to be equal to or greater than the fixed time T.

(d) *Determination of the* $\langle\cos^2 \Phi(t)\rangle_c$ *characteristic of a configuration.* This is done by repeating several complete energy migrations from (a) and then to calculate the average for each value of t.

$$\langle\cos^2 \Phi(t)\rangle_c = \frac{1}{N_s} \sum_{s=1}^{N_s} \cos^2 \Phi_s(t) ,$$

where Ns is a fixed number of migrations, chosen to give a good statistical result.

3. *Determination of the Anisotropy Decay of Transfer*

To determine this quantity, it is necessary to repeat all the calculations of the previous section for a set of configurations generated according to subsection 1. Then, one may compute the average value

$$\langle\langle\cos^2 \Phi(t)\rangle\rangle = \sum_{c=1}^{N_c} \frac{\langle\cos^2 \Phi(t)\rangle_c}{N_c} ,$$

where N_c is the predetermined number of configurations.

The contribution of the energy migration to the decay of anisotropy is then given by:

$$r_T(t) = \frac{3\langle\langle\cos^2 \Phi(t)\rangle\rangle - 1}{2}.$$

5. Results and Discussion

For these calculations three different values of the angle β have been tried:

(1) $\beta=50°$ corresponding to Paoletti's and Le Pecq's winding angle of 14° ($\delta= = +14°$);

(2) $\beta=24°$ corresponding to Fuller's and Waring's unwinding angle of 12° ($\delta= = -12°$);

(3) $\beta=36°$ corresponding to no angular distorsion of the DNA helix ($\delta=0°$).

The different curves $r_T(t)$ computed are shown in Figure 6 for a P/D value of 8.7.

To compare the theoretical computations with the experimental data, we proceed as indicated in Section 4.1. That is to say that the theoretical curves $d'(t)$ are determined by application of Equations (6), (7), (9), (10) and by utilisation of the r_T value obtained by the simulation calculation.

In Figure 7 the computed curves $d'(t)$ may be compared to the experimental curve $d(t)$. It is seen that the theoretical curve corresponding to $\delta=-12°$ is in good agreement with the experimental curve, in a time interval from 0 to 20 nsec. For higher times, the theoretical curve is below the experimental one. On the contrary,

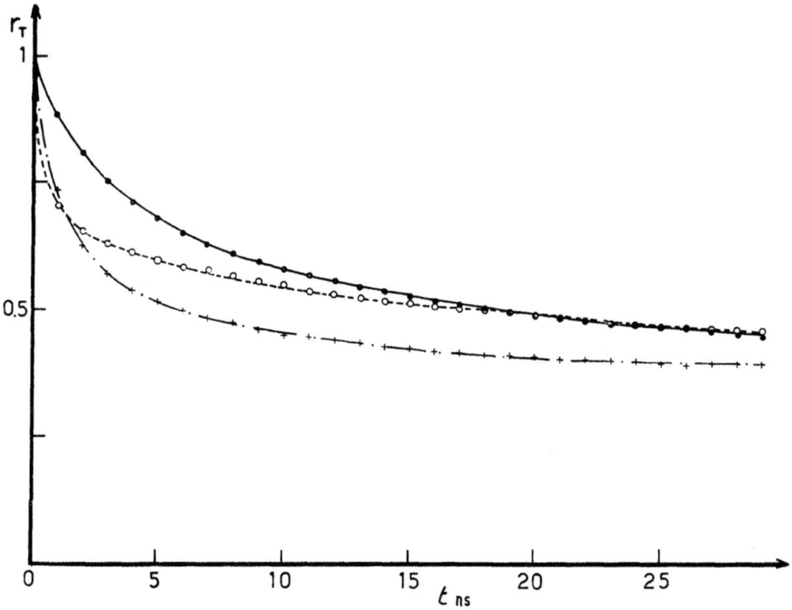

Fig. 6. Curves $r_T(t)$ for $P/D=8.7$ computed by method of simulation. ●●● corresponds to a winding angle $\delta=+14°$; ○ ○ ○ corresponds to an unwinding angle $\delta=-12°$; +++ corresponds to $\delta=0°$.

the experimental curve never fits the theoretical one corresponding to $\delta = +14°$, which is above at the beginning, and then below. Consequently, the result of our analysis is in contradition with the conclusions of Paoletti and Le Pecq, based on the determination of the time average anisotropy and a Monte Carlo calculation.

Let us notice that the theoretical time-average anisotropy $\langle r \rangle$ is obtained by integration from 0 to ∞ of $r(t)$. Under these conditions, the computed value of $\langle r \rangle$ could be close to the experimental value, since the curve $d'(t)$ corresponding to $\delta = +14°$ crosses $d(t)$. Therefore, the integration could entail a casual coincidence of the theoretical and experimental values of $\langle r \rangle$.

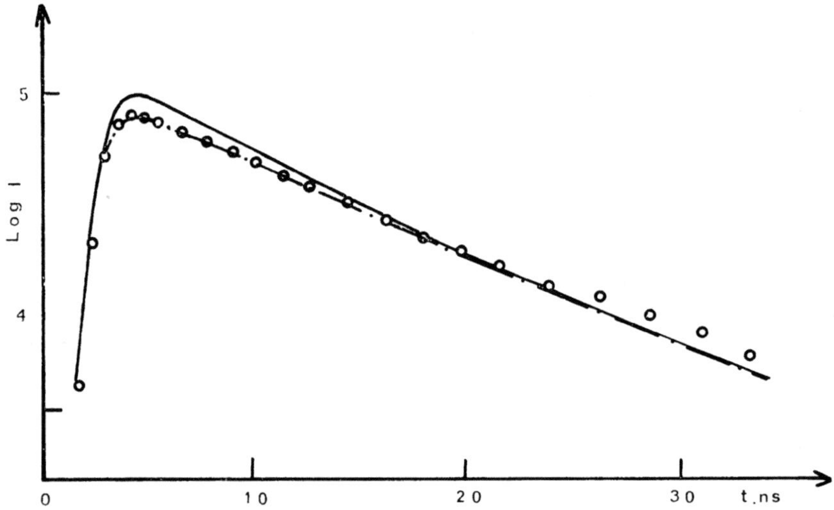

Fig. 7. Comparison between experimental curves $d(t)$ and curves $d'(t)$ computed by methods of simulation corresponding to P/D = 8.7. ○ ○ ○ experimental curve; ___ computed curve with $\delta = 14°$; –·–·– computed curve with $\delta = -12°$.

This would explain the results of Paoletti and Le Pecq. However, it has to be noticed, that the rate constant of transfer V_T (formula (11)), which has been determined in this work, is about 4 times higher than Paoletti's and Le Pecq's value. This arises from the difference between our values of J and τ_0, and the values used by these authors. This difference increases again the disagreement between the conclusions of Paoletti and Le Pecq and ours.

This discussion is a good illustration of the interest of the anisotropy decay measurements comparatively to the anisotropy determination obtained by continuous excitation.

It is evident that the principles which have been used in this work can be applied to other systems. Using this method it will be possible to obtain some information, sometimes very accurate, about the spatial repartition of a set of identical fluorescent chromophores.

References

[1] Waring, M. J.: 1965, *J. Mol. Biol.* **13**, 269.
[2] Le Pecq, J. B. and Paoletti, C.: 1967, *J. Mol. Biol.* **27**, 87.
[3] Li, H. J. and Crothers, D. M.: 1969, *J. Mol. Biol.* **39**, 461.
[4] Wahl, Ph., Paoletti, J., and Le Pecq, J. B.: 1970, *Proc. Nat. Acad. Sci. US* **65**, 417.
[5] Genest, D. and Wahl, Ph.: 1972, *Biochim. Biophys. Acta* **259**, 175.
[6] Fuller, W. and Waring, M. J.: 1964, *Phys. Chem.* **68**, 805.
[7] Bauer, W. and Vinograd, J.: 1968, *J. Mol. Biol.* **33**, 141.
[8] Paoletti, J. and Le Pecq, J. B.: 1971, *J. Mol. Biol.* **59**, 43.
[9] Koechlin, Y.: 1961, Thesis, University of Paris.
[10] Wahl, Ph.: 1969, *Biochim. Biophys. Acta* **175**, 55.
[11] Perrin, F.: 1936, *Acta Phys. Polon.* **5**, 335.
[12] Förster, Th.: 1951, *Fluoreszenz organischer Verbindungen*, Vandenhoek und Ruprecht, Göttingen, Germany.
[13] Förster, Th.: 1948, *Ann. Phys.* **2**, 55.
[14] Strickler, S. J. and Berg, R. A.: 1962, *J. Chem. Phys.* **37**, 814.
[15] Hammersley, J. M. and Handscomb, D. C.: 1967, *Les méthodes de Monte Carlo*, Dunod, Paris.

DISCUSSION

Lami: (1) Comment expliquez-vous le désaccord entre courbes théoriques et expérimentales en fin de déclin d'anisotropie? (2) Avez-vous tenu compte d'un éventuel transfert de l'énergie en retour sur le chromophore initialement excité?

Genest: (1) Le désaccord peut être dû au fait que nous supposons qu'il n'y a pas de changement de l'anisotropie brownienne. (2) Oui, l'energie peut retourner au chromophore initialement excité. Il en est tenu compte dans le calcul.

Alpert: Dans vos corrections le temps intervient uniquement du fait de l'agitation due aux mouvements browniens. Le transfert de l'énergie d'une molécule A à une molécule B se produit instantanément. Or il faut tenir compte du temps de transfert dans les interactions de type Forster. Je pense que dans ces ordres de temps la correction peut s'avérer utile.

Genest: Nous avons calculé le déclin de l'anisotropie de transfert indépendamment des mouvements browniens, en déterminant la position de l'énergie en fonction du temps. C'est le temps mis par l'énergie pour aller de A à B qui est négligé. Il est de l'ordre de 10^{-15} sec, donc non compétitif avec les phénomènes que l'on mesure qui sont de l'ordre de 10^{-9} sec.

Bourgarit: Ma question est relative au recouvrement entre les spectres d'absorption et d'émission de la molécule de BET attachée au DNA. Pouvez-vous donner des précisions?

Genest: L'intégrale de recouvrement a été mesurée et trouvée égale à $1{,}7 \times 10^{-15}$ cm^6/M.

Bourgoin: Le Pecq n'avait-il pas avancé des arguments de nature topologique faisant intervenir les figures de torsion d'ADN circulaires?

Genest: Oui (J. M. Saucier, B. Festy, and J. B. Le Pecq, *Biochimie* **53** (1971) 973); mais dans le cas du bromure d'éthydium, ces arguments sont valables aussi bien pour le modèle d'intercalation de Lerman que pour celui proposé par ces auteurs. D'autre part, nos résultats ne mettent pas en cause le modèle même proposé par Le Pecq, mais uniquement l'angle de distorsion puisque, dans son modèle, l'angle de torsion de l'hélice peut être comrpis entre $-15°$ et $+16°$, ce qui est compatible avec un déroulement de 12°.

Frédéricq: Que pensez-vous de la théorie de Weill et Calvin suivant laquelle les transferts d'énergie entre molécules de colorant liées expliqueraient non seulement la dépolarisation mais aussi l'extinction de fluorescence, qui croît avec le taux de fixation?

Genest: Ces auteurs ont utilisé la proflavine et l'acridine orange comme marqueur fluorescent. La fluorescence de ces chromophores est inhibée par les paires de bases G–C. S'il y a transfert de A–T vers G–C, il y aura donc extinction de fluorescence.

Lami: Le transfert d'énergie allonge la durée de vie de l'excitation et par là même augmente la probabilité de désactivation non radiative.

ÉTUDE DES INTERACTIONS ENTRE L'ACIDE POLYRIBO-URIDYLIQUE OU L'ACIDE POLYRIBOBROMO-URIDYLIQUE ET DES POLYCATIONS

M. LENG, V. DEUBEL, et J. C. THRIERR

Centre de Biophysique Moléculaire, 45045, Orléans Cédex, France

Résumé. Dans ce travail, il est montré que la forme ordonnée du poly U et du poly BrU, en présence de sel ou de spermine, résulte de repliement du polymère sur lui-même (modèles en épingle à cheveux). La forme ordonnée de ces polymères peut être aussi induite par des oligopeptides, mais la réaction dépend de la nature des acides aminés de l'oligopeptide.

Summary. In Section 3.1, we have compared some properties of poly U (polyribouridylic acid) and poly U – spermine complex. From spectroscopic and hydrodynamic data, it is concluded that, in the ordered form of poly U in the presence of salt or spermine, the polymer is folded on itself (hairpin model). Some properties of poly BrU (the bromuration occurred on the C_5) are reported. It is also found that the ordred form of poly BrU results from the folding of the chain on itself (in the presence of salt or spermine).

In Section 3.2 the interactions between oligopeptides X–Y–X (where X is the amino acid lysine and Y an uncharged amino acid) and poly U or poly BrU are studied. Most of these oligopeptides interact with the disordered form of the polymers. In definite conditions, they can also induce the ordered form of the polynucleotides. However, the effects of these oligopeptides are different, depending upon the nature of the amino acid Y (Y being alanine, tyrosine, tryptophan or phenylalanine).

1. Introduction

De nombreuses réactions biologiques font intervenir des interactions spécifiques entre acides nucléiques et protéines. Cette spécificité doit être due, au moins partiellement, à des interactions préférentielles entre certains acides aminés et certaines bases. Pour essayer de mettre en évidence ces interactions, il est commode d'étudier des modéles relativement simples, comme les complexes entre polypeptides et polynucléotides. Parmi les polynucléotides de synthèse, l'acide polyuridylique (poly U) nous a semblé particulièrement intéressant.

Un ensemble de résultats permet de conclure qu'à pH neutre, le poly U peut exister, suivant la température, sous une forme ordonnée ou sous une forme désordonnée [1]. Au-dessus de 25° C, le poly U a les propriétés d'un pelote statistique. Une étude récente, en solvant θ montre cependant que la chaîne de poly U est relativement rigide [2]. Cette rigidité provient probablement des répulsions entre les phosphates bien que les expériences aient été faites en force ionique élevée. En refroidissant une solution de poly U (entre 0–10° C) le poly U change de conformation suivant un processus coopératif. Le T_m dépend de la concentration et de la nature du sel. Une étude systématique du poly U et des oligo U nous a conduit à suggérer que la forme ordonnée du poly U résulte du repliement de ce polymére sur lui-même avec formation de liaisons 'hydrogène' entre les bases [3, 4]. Cette hypothèse est en accord avec les résultats de Millar et McKenzie [5].

C. Sadron (ed.), Dynamic Aspects of Conformation Changes in Biological Macromolecules. 381–391.

Il avait été aussi observé que l'addition de diamines, polyamines ou polylysine à une solution de poly U induisait une structure ordonnée du polymère [6, 7, 8].

Dans ce travail, nous montrons que dans les complexes poly U – spermine, le poly U se replie sur lui-même. De plus nous avons trouvé que des tripeptides du type (HLysPheLysOMe) ou (HLysAlaLysOMe) peuvent, en interagissant avec le poly U, induire le repliement du polymère sur lui-même. Finalement, nous comparons l'effet de ces différents oligopeptides sur le poly U et sur le poly BrU.

2. Matériel et méthodes

Le poly U provient de la Schwarz Cie. Il a été soigneusement purifié suivant une méthode précédemment décrite, et, dans certains cas, fractionné [3].

Le poly BrU a été obtenu par action du brome sur le poly U dissous dans la formamide (communication personelle de Dr. M. Mathias et Dr. A. M. Michelson). La substitution a lieu sur le C_5. Le rendement est peu différent de I. En effet, aucune trace d'uridine monophosphate (après traitement par la RNAase) n'a pu être détectée par chromatographie (solvant acétate d'ammonium 1 M, éthanol, 6/4 *v*/*v*). Le maximum du spectre d'absorption ultraviolet est à 279 nm [9]. Enfin, la comparaison des spectres de résonance paramagnétique nucléaire du poly U et du poly BrU confirme d'une part, que tous les résidus uracile sont bromés, d'autre part, que la bromuration se fait sur le C_5 (nous remercions le Dr. J. L. Dimicoli, qui a effectué ces mesures). Il est probable qu'au cours de la bromuration, il y a rupture d'un certain nombre de liaisons phosphodiesters. Deux échantillons ont été préparés et il a été trouvé, dans 0,1 M NaCl, pH 6,7, à 20°C des coefficients de sédimentation respectivement égaux à 3 et à 4,2. Dans les mêmes conditions, le coefficient de sédimentation du poly U est égal à 5,6.

La spermine est un produit Fluka et a été recristallisée [10]. Les peptides HLysTyrLysOH et HLysTrpLysOH sont des produits Mann Research Cie. Les autres peptides ont été synthétisés au laboratoire par le Dr. J. Rossi.

Les techniques physico-chimiques utilisées dans ce travail ont déjà été décrites [3].

3. Résultats

3.1. Forme ordonnée du poly U

(a) *Propriétés spectroscopiques*

La stabilité de la forme ordonnée du poly U dépend de la concentration et de la nature du sel (nous ne considérons ici que les sels monovalents). A une force ionique donnée, le T_m est plus élevé en présence de ClCs qu'en présence de ClNa. En présence de spermine, le T_m dépend du rapport phosphate/spermine. Il est de l'ordre de 30°C dans 10^{-3} M Na cacodylate (pH 6,6) pour un rapport phosphate/spermine $\simeq 4$ (en moles) alors que le T_m de la forme la plus stable dans le ClCs (0,5 M) est de l'ordre de 10°C. L'hyperchromicité et la pente de la variation de $\varepsilon_{260\ \mathrm{nm}} = f(T)$ sont plus grandes

TABLEAU I

Concentration CsCl	Poly U + CsCl			Poly U + spermine			
	T_m	H (%)	σ	P/S	T_m	H (%)	σ
0,1 M	4,3°	50	3,5°	5,5	28,4°	47	4,7
0,2 M	7,6°	49	4,8°	4,4	31,2°	42	2,2
0,5 M	11,1°	47	7				
1 M	11,2°	47	8,5°				
2 M	7,8°	41	8°				

Les complexes poly U – spermine ont été étudiés dans 10^{-3} M Na cacodylate (pH 6,6). P/S: phosphate/spermine (les concentrations sont exprimées en moles); H(%) est l'hyperchromicité au maximum d'absorption; σ représente l'intervalle de température correspondant à 75% de fusion de part et d'autre du T_m.

en présence de spermine que dans le ClCs. Les principaux résultats sont rassemblés dans le Tableau I.

Un point important est que la valeur du T_m est indépendante de la concentration en poly U ou en oligo U (expérience faite en ClCs). La réaction est donc intramoléculaire.

D'autre part, l'étude des courbes de fusion en fonction du rapport phosphate/spermine permet de conclure que la fixation de la spermine n'est pas fortement coopérative. Ceci veut dire qu'il n'y a pas deux sortes de molécules, poly U – spermine et poly U seul, mais que les molécules de spermine sont également réparties sur les molécules de poly U.

La réaction forme ordonnée⇋forme désordonnée peut être suivie par dispersion optique rotatoire ou dichroïsme circulaire [3, 5, 7, 10]. Les spectres de dichroïsme circulaire à 0°C dans 0,5 M Cls ou dans 10^{-3} M Na cacodylate en présence de spermine sont très similaires. Seules de faibles variations d'intensité peuvent être notées.

L'ensemble de ces résultats ne permet pas de mette en évidence de différences importantes entre la forme ordonnée du poly U en présence de sel ou de spermine.

(b) *Masse moléculaire*

L'étude par diffusion de la lumière de l'intensité diffusée par une solution de poly U à différentes températures montre que le degré de polymérisation du poly U est le même de part et d'autre du T_m. Il en est de même si de la spermine est ajoutée à la solution de poly U [3, 10]. Donc, la forme ordonnée du poly U ne résulte pas de l'association de plusieurs brins de poly U.

(c) *Forme*

La masse moléculaire étant constante, la forme du poly U peut être étudiée par viscosité. Les expériences de diffusion de la lumière n'apportent aucun renseignement sur la forme, les molécules étant petites devant la longueur d'onde du faisceau incident ($\lambda = 546$ nm).

Les principaux résultats sont rassemblés sur la Figure 1. Dans le ClCs (0,1 M) ou

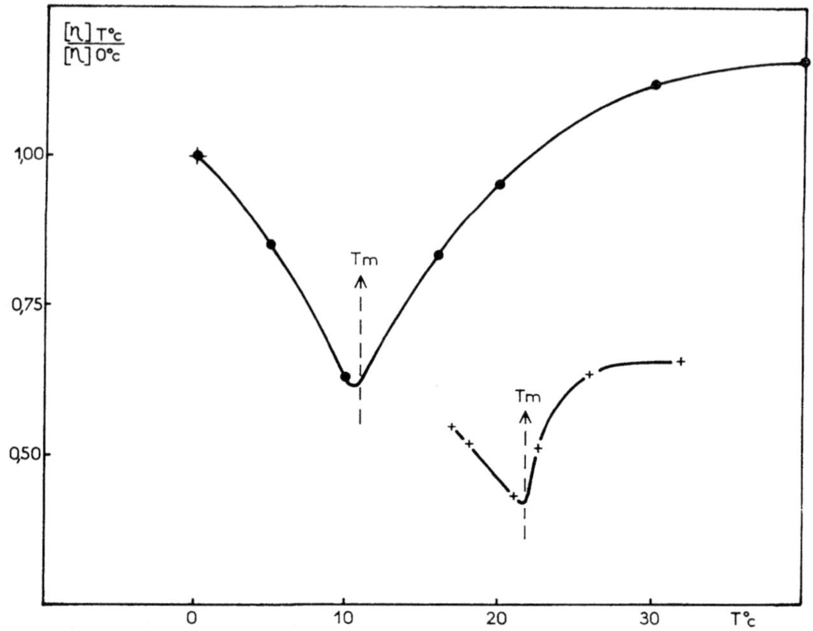

Fig. 1. Variation du rapport $\eta_T/\eta_{T=0}$ en fonction de la température. η_T et $\eta_{T=0}$ sont respectivement les viscosités intrinsèques à la température T et à 0°C. ● — ● poly U dans 0,5 M CsCl + — + complexe poly U – spermine, solvant 0,1 M NaCl, 0,05 M Na cacodylate, pH 6,6.

dans 10^{-3} M Na cacodylate en présence de spermine, l'allure de la variation de la viscosité intrinsèque η en fonction de la température est la même. Les deux courbes passent par un minimum, à une température proche du T_m. Nous avons porté $\eta_T/\eta_{T=0°}$. La viscosité de la forme désordonnée varie beaucoup avec la force ionique alors que la viscosité de la forme ordonnée en dépend beaucoup moins.

L'allure de ces courbes est carateristique de tout polymère linéaire qui peut former une double hélice avec une boucle à une extrémité (modèle en épingle à cheveux) à basse température et en chauffant, donner des structures ramifiées correspondant à plusieurs 'épingles à cheveux' reliées par des régions désordonnées. A la fin de la transition, le polymère est sous forme de pelote statistique [11]. Ces résultats sont à rapprocher de ceux obtenus sur le poly dAT.dAT [12].

La similitude des résultats obtenus sur le poly U en présence de sel ou en présence de spermine permet d'affirmer que la forme ordonnée du poly U, dans les deux cas est analogue. Cette forme ordonnée résulte du repliement du poly U sur lui-même.

D'autre part, la constante apparente d'équilibre (poly U+spermine, solvant 0,15 M Na^+, T=4°C) est égale à $1{,}2\times10^5$ M^{-1} et le nombre de sites 0,12 (ces quantités ont été obtenues à partir d'expériences d'équilibre de dialyse). Nous avons aussi construit un modèle de la forme ordonnée du poly U en appariant les bases par liaisons hydrogène (N_3H–C_4O, N_3H–C_2O) et en admettant la conformation anti. Sur ce modèle, la spermine interagit par ses deux charges extrêmes avec deux phosphates situés sur les deux brins de poly U, alors que les deux autres charges de la spermine n'interagissent pas directement avec les phosphates. Il est bien évident que la distance entre les charges jouent un rôle important. Ceci avait déjà été observé par Szer qui avait fait des mesures destabilité thermique du poly U en présence de différentes polyamines [6]. Il nous a semblé intéressant d'essayer d'obtenir d'autres renseignements en étudiant les interactions entre des oligopeptides et le poly U.

3.2. Interactions poly U – oligopeptides

Les oligopeptides peuvent interagir avec la forme désordonnée du poly U. La formation de complexe a été étudiée par des mesures du coefficient d'extinction ε (ε dépend de la force ionique) [13, 14], et par dichroïsme circulaire. Il est bien évident que si ces deux paramètres ne varient pas, cela n'implique pas que les produits n'interagissent pas. De plus, la simple observation d'une variation de ε n'apporte que peu de renseignements sur la nature de l'interaction. Enfin, les oligopeptides peuvent, comme la spermine, induire le repliement du poly U sur lui-même. Il est alors facile de suivre la formation du complexe.

(a) *Interactions poly U – HLysLysLysOH*

Aussi bien à 20°C qu'à 0°C, $\varepsilon_{260\ \mathrm{nm}}$ (poly U) est pratiquement constant jusqu'à un rapport phosphate/$(Lys)_3$ de l'ordre de 2. On peut conclure que $(Lys)_3$ interagit peu ou pas du tout avec le poly U et n'induit pas la forme ordonnée du poly U.

(b) *Interactions poly U – HLys(Lys)$_3$LysOH*

A 0°C, dans 10^{-3} M Na cacodylate, pH 6,6, en ajoutant des quantités croissantes d'oligolysine à une solution de poly U, on observe une diminution importante du $\varepsilon_{260\ \mathrm{nm}}$. Pour un rapport phosphate/$(Lys)_5 = 5{,}5$ le T_m du poly U est de 18°, l'hyperchromicité de 48% et la largeur de transition (définie dans la légende du Tableau I) de 3°.

(c) *Interactions poly U – HLysAlaLysOMe et HLysPheLysOMe*

Les deux oligopeptides interagissent avec la forme désordonnée à 30°C et, à plus basse température, induisent le repliement du poly U. Par exemple, en présence de HLysAla-LysOMe, pour un rapport phosphate/peptide $\simeq 1{,}8$, le T_m est de 9°, l'hyperchromicité de l'ordre de 50% et la larguer de la transition de 2°. Le T_m dépend du rapport phosphate/peptide (les T_m étant égaux à 0° 5 et 5° pour des rapports de 4,1 et 6,2 respectivement). L'allure des courbes est en faveur d'une distribution statistique des oligopeptides sur les molécules du poly U.

Si l'on opère à pH 8,5 (le pK du groupement α-NH_3 est de l'ordre de 8) le T_m du complexe est abaissé. Pour un rapport phosphate/peptide de 1,8, le T_m est de 4° ($T_m = 9°$ à pH 6,6), mais les valeurs de l'hyperchromicité et de σ sont les mêmes que précédemment (pH 6,6).

Si au lieu d'utiliser le dérivé estérifié, on utilise le dérivé avec la fonction carboxyle terminale, dans les mêmes conditions, il n'y a plus formation de la structure ordonnée.

Les réactions peuvent être suivies par dichroïsme circulaire. Il n'a pas été observé de différences importantes dans les spectres de la forme ordonnée du poly U, seul ou en présence d'oligopeptides.

Cependant, les courbes de titration du poly U par les oligopeptides dépendent de la nature de l'acide aminé non chargé. Sur la Figure 2, il est clair que le coefficient d'extinction ($\lambda = 260$ nm) varie différemment suivant qu'on ajoute au poly U une solution de HLysAlaLysOMe ou de HLysPheLysOMe.

(d) *Interactions poly U – HLysTyrLysOH et HLysTrypLysOH*

Même pour un rapport phosphate/oligopeptide de l'ordre de 1,4, aucune variation importante du $\varepsilon_{260\ \mathrm{nm}}$ du poly U n'a pu être observée en présence de HLysTyrLysOH. Par contre, à 30°C, HLysTrypLysOH interagit avec le poly U. En abaissant la température, $\varepsilon_{260\ \mathrm{nm}}$ diminue mais d'une façon progressive. (La transition est beaucoup plus large que dans les cas précédents.)

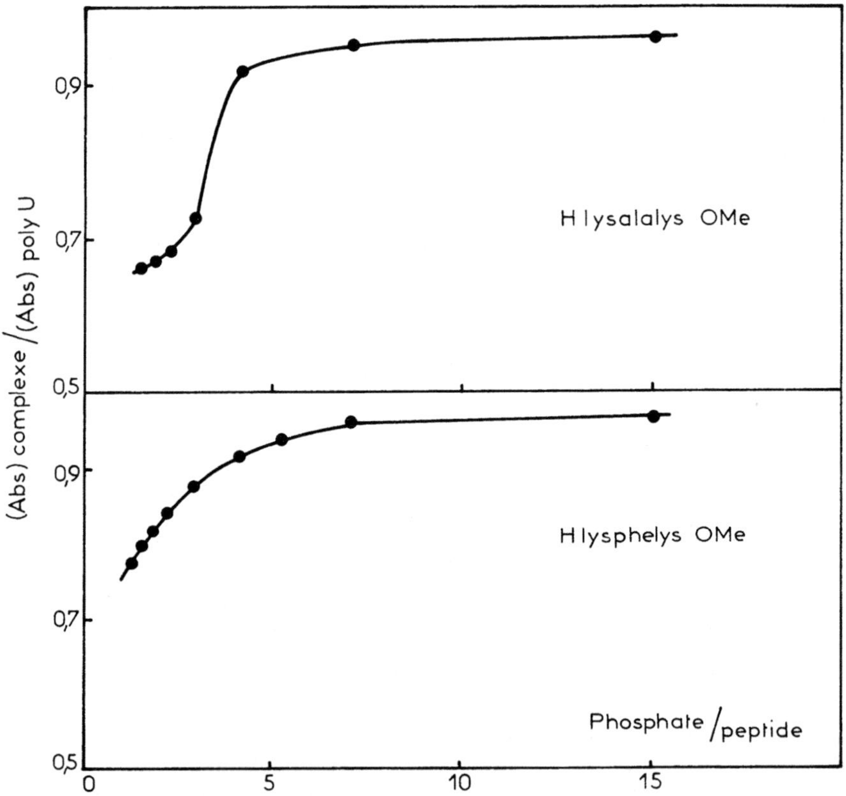

Fig. 2. Variation du rapport absorption du complexe/absorption du poly U en fonction du rapport phosphate/oligopeptide (les concentrations sont en moles); $\lambda = 260$ nm. Solvant 10^{-3} M Na cacodylate, pH 6,6. Concentration poly U $\simeq 10^{-4}$ M.

3.3. Poly BrU

(a) *Courbe de fusion*

Comme le poly U, le poly BrU peut exister sous forme ordonnée et sous forme désordonnée. La stabilité thermique de la forme ordonnée dépend de la concentration et de la nature du sel (le poly BrU étant plus stable dans le ClCs que dans le ClNa).

La pente de la courbe de fusion ($\varepsilon_{279\ \mathrm{nm}} = f(T)$) diminue lorsque la force ionique augmente. Il est à noter que le T_m dans 0,2 M ClCs est indépendant de la concentration en poly BrU (dans le domaine 10^{-4}–10^{-3} M). La spermine interagit avec le poly BrU. Les principaux résultats sont rassemblés dans le Tableau II.

TABLEAU II

Concentration CsCl	Poly BrU + CsCl			Poly BrU + spermine			
	T_m	H (%)	σ	P/S	T_m	H (%)	σ
0,1 M	20°	40	11°	17	27°	35	19,5°
0,2 M	24°	37	13,5°	10	39°	39	15,5°
0,5 M	26,8°	37	18°	7	46°	40	14°
1 M	23°	31	20°	5			
2 M	15,5°	28	20°				

Les complexes poly BrU – spermine ont ètè ètudiès dans 10^{-3} M cacodylate pH 6,6. P/S, $H(\%)$ et σ sont définis dans la légende du Tableau I.

Le changement de conformation peut être mis en évidence par dichroïsme circulaire (Figure 3). L'intensité de la bande positive à 0°C est très inférieure à celle du poly U dans les conditions où les deux polymères sont structurés. On observe, comme dans le cas du poly U, que le maximum de la bande positive se déplace vers le bleu lorsque la température diminue. Enfin, le spectre en présence de spermine est semblable à celui obtenu en présence de sel.

(b) *Masse moléculaire*

L'intensité diffusée par une solution de poly BrU (0,2 M CsCl) ne varie pas avec la température, de part et d'autre du T_m. Il n'a pas été vérifié que le dn/dc est indépendant de la température. Il avait été montré que le dn/dc du poly U ne variait pas avec la température. Nous admettrons que ceci est aussi vrai dans le cas du poly BrU. On peut alors conclure que la forme ordonnée ne résulte pas de l'association de plusieurs molécules de poly BrU. Par analogie avec le poly U, nous supposons que le poly BrU se replie sur lui-même. Des mesures préliminaires de viscosité en sont faveur de cette hypothèse, mais une étude systématique n'a pu être faite par suite d'un manque de produit.

(c) *Interactions poly BrU – oligopeptides*

Le poly BrU, comme le poly U interagit avec certains oligopeptides.

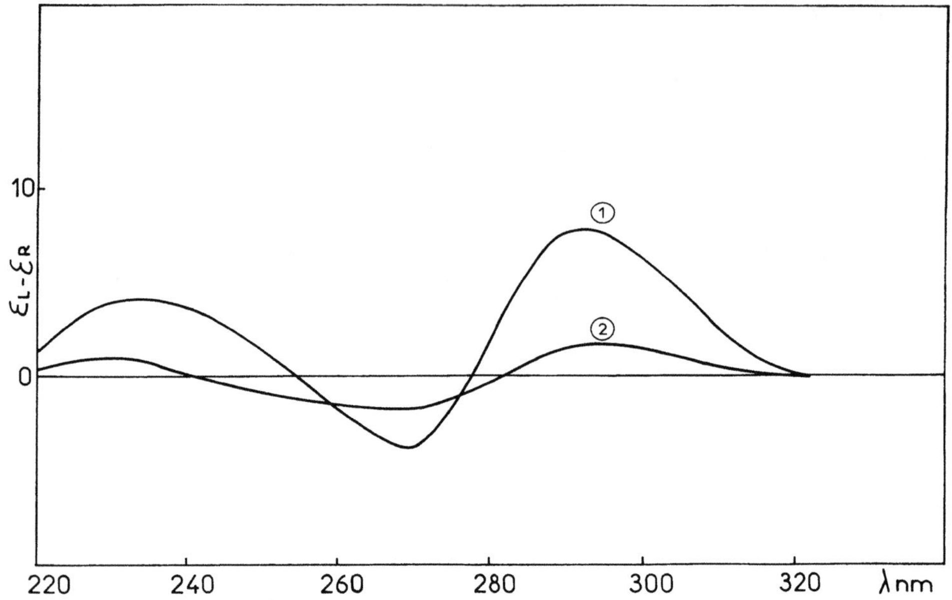

Fig. 3. Dichroïsme circulaire du poly BrU. Solvant 0,2 M CsCl, 0,01 M Na cacodylate pH 6,6. (1) $T=0°$C; (2) $T=40°$C.

Interactions poly BrU – $(Lys)_3$ et $(Lys)_5$: Aucune variation de $\varepsilon_{279\ \mathrm{nm}}$ n'a pu être mise en évidence avec $(Lys)_3$ jusqu'à un rapport phosphate/$(Lys)_3$ de l'ordre de 2,2. Par contre, $(Lys)_5$ induit la formation de la structure ordonnée. Ce complexe est thermiquement stable puisque le T_m est de 42° (phosphate/$(Lys)_5=7$). L'hyperchromicité est de 36%, mais la fusion se produit dans un large intervalle de température ($\sigma \simeq 13°$). Le spectre de dichroïsme circulaire du complexe est, en première approximation, analogue à celui du poly BrU structuré.

Interactions poly BrU – HLysAlaLysOMe, HLysPheLysOMe. Les principaux résultats sont rassembles dans le Figure 4. Tous les oligopeptides forment un complexe avec le poly BrU et dans tous les cas, il y a structuration du poly BrU. Cependant, l'allure des courbes de fusion dépend de la nature de l'oligopeptide. En particulier, en présence de HLysTrypLysOH, la courbe est beaucoup plus large qu'en présence des autres oligopeptides. La stabilité des complexes dépend aussi de la charge portée par l'oligopeptide comme le montrent les courbes avec HLysAlaLysOH et HLysAlaLysOMe.

Les spectres de dichroïsme circulaire des complexes poly BrU avec les différents polypeptides ressemblent à celui du poly BrU (forme ordonnée).

3.4. Interactions entre le poly U et le poly BrU

Il avait été rapporté que le poly U et le poly BrU interagissaient pour former des complexes à deux brins et à trois brins. En nous plaçant dans des conditions identiques à celles des auteurs [5], (0,1 M NaCl, 0,02 M $MgCl_2$, 0,05 M Na cacodylate, pH 6,7) par des expériences de courbes de mélanges, nous n'avons pas pu mettre en évidence

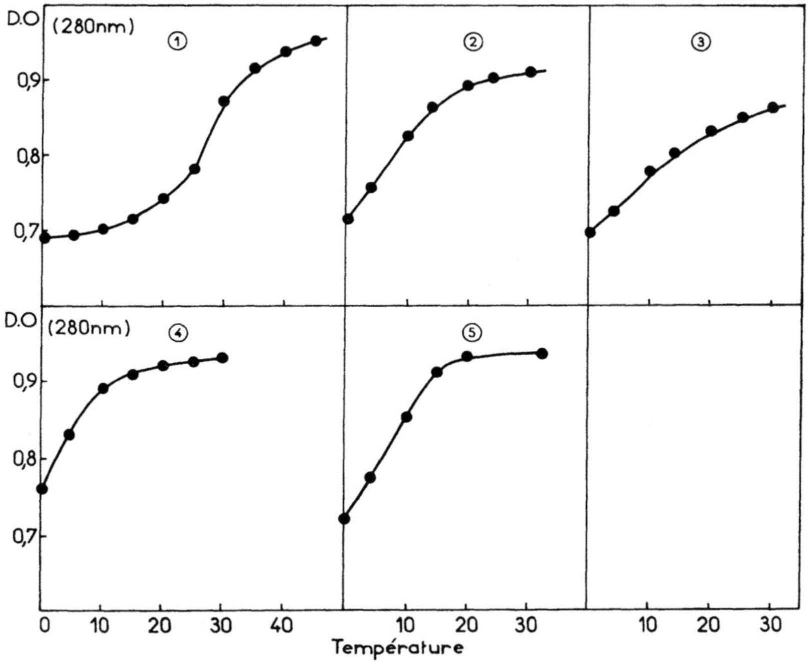

Fig. 4. Variation de la densité optique ($\lambda = 280$ nm) du poly BrU en fonction de la température, en présence de (1) HLysAlaLysOMe (phosphate=peptide = 3,6); (2) HLysTyrLysOH (phosphate/peptide = 1,4); (3) HLysTrypLysOH (phosphate/peptide = 1,6); (4) HLysPheLysOH (phosphate/peptide = 2,5); (5) HLysAlaLysOH (phosphate/peptide = 2,5). Solvant 10^{-3} Na cacodylate, pH 6,6.

la formation de complexe. L'addition de spermine ou d'oligopeptide ne favorise pas l'association des deux polymères.

4. Discussion et conclusion

En présence de sel, le poly U, comme le poly BrU peuvent former une structure ordonnée qui résulte du repliement du polymère sur lui-même avec liaisons hydrogène entre les bases. La substitution d'un hydrogène par un brome sur le C_5 conduit à une structure ordonnée thermiquement plus stable. Cependant, la fusion du poly BrU s'effectue dans un intervalle de température plus grand que celui correspondant au poly U. Dans le cas du poly U, il avait été montré que la largeur de la transition dépendait de la force ionique. Il avait été postulé que la longueur des régions en double hélice diminuait quand la concentration en sel augmentait. Si, on ultilise les mêmes arguments dans le cas du poly BrU, on doit conclure que régions en double hélice du poly BrU sont plus courtes que celles du poly U à force ionique égale. En fait, des expériences préliminaires de viscosité (résultats non publiés) sont en faveur de cette interprétation. La courbe de viscosité en forme de V a un minimum beaucoup plus arrondi avec le poly BrU qu'avec le poly U. Des expériences complémentaires devraient nous permettre de vérifier ces hypothèses.

Les différentes expériences faites sur le poly U et les complexes poly U – spermine permettent de conclure que dans les deux cas, les formes ordonnées sont identiques. Quoique notre étude sur le poly BrU et les complexes poly BrU – spermine soit moins exhaustive, il est probable que la même conclusion peut être émise pour le dérivé bromé. Un point intéressant est que la stabilité thermique du complexe poly U – spermine est beaucoup plus grande que celle du poly U seul (la différence maximale observée entre les T_m des deux produits est de l'ordre de 20°C). Il est sûr qu'il existe entre le poly U et la spermine des interactions électrostatiques. Il est possible que d'autres forces entrent en jeu dans le complexe, mais il est difficile de les préciser puisqu'on ne connaît pas encore la position relative de la spermine et des bases.

Les oligopeptides peuvent aussi induire la formation de la double hélice. En plus de la distance et de la position des charges les unes par rapport aux autres (paramètres importants comme l'avait déjà noté Szer) la nature de l'acide aminé non chargé joue un rôle. Suivant que cet acide aminé est l'alanine, la tyrosine, le tryptophane ou la phénylalanine, les variations du coefficient d'extinction avec la température, sont légèrement différentes. Ces différences sont encore plus marquées avec le poly BrU. Bien que les spectres de dichroïsme circulaire des complexes soient en première approximation identiques, une étude plus systématique par d'autres techniques physiques devrait nous permettre de préciser le rôle du résidu non chargé (effet sur le solvant, interaction avec les bases...).

Références

[1] Steiner, R. F. et Millar, D. B.: 1970, *Biological Polyelectrolytes*, Vol. 3 (A. Veis, éd.), Marcel Dekker, New York.
[2] Inners, D. et Felsenfeld, G.: 1970, *J. Mol. Biol.* **50**, 573.
[3] Thrierr, J. C., Dourlent, M., et Leng, M.: 1971, *J. Mol. Biol.* **58**, 815.
[4] Dourlent, M., Thrierr, J. C., Brun, F., et Leng, M.: 1970, *Biochem. Biophys. Res. Commun.* **41**, 1590.
[5] Millar, D. B. et McKenzie, M.: 1970, *Biochim. Biophys. Acta* **204**, 82.
[6] Szer, W.: 1970, *Ann. N.Y. Acad. Sci.* **171**, 801.
[7] Rogers, G. T., Ulbricht, T. L. V., et Szer, W.: 1967, *Biochem. Biophys. Res. Commun.* **27**, 372.
[8] Caroll, D.: 1972, *Biochemistry* **11**, 426.
[9] Massoulié, J., Michelson, A. M., et Pochon, F.: 1966, *Biochim. Biophys. Acta* **114**, 16.
[10] Thrierr, J. C., Deubel, V., et Leng, M.: 1972, *Biochimie* **54**, 1115.
[11] De Gennes, P. G.: 1968, *Biopolymers* **6**, 715.
[12] Imman, R. B. et Baldwin, R. L.: 1962, *J. Mol. Biol.* **5**, 172.
[13] Michelson, A. M. et Monny, C.: 1966, *Proc. Nat. Acad. Sci. US* **56**, 1528.
[14] Simpkins, H. et Richards, E. G.: 1967, *Biopolymers* **5**, 551.

DISCUSSION

Schulz: Je voudrais des informations sur l'aspect cinétique des changements de conformation exposés, à savoir: le système poly U plus spermine et le système poly U seul.

Leng: Aucune expérience de cinétique n'a été faite sur le complexe poly U – spermine. Par contre, nous avons étudié le poly U et les oligo U [3, 4].

Bourgarit: Les spectres de dichroïsme circulaire que vous avez présentés, manifestent l'apparence de deux points isosbestiques. En concluez-vous qu'il n'y a que deux espèces en équilibre?

Leng: Nous trouvons bien deux points isosbestiques sur les spectres du poly U et du complexe poly U – spermine. Mais ce résultat ne justifie pas une analyse des courbes en admettant deux espèces en équilibre.

Frédéricq: Je suppose que les interactions entre spermine et poly U sont essentiellement de nature électrostatique. A ce propos, existe-t-il une correspondance stérique entre les groupes positifs de la spermine et les groupes phosphate du polynucléotide? Quelle est la veleur des constantes de fixation?

Leng: Il est probable que les interactions entre le poly U et la spermine sont essentiellement du type électrostatique, mais on ne peut exclure que d'autres forces entrent en jeu. Les différences observées entre les complexes poly U – oligopeptides montrent bien que l'acide aminé non chargé joue un rôle important. La constante d'équilibre (poly U – spermine) est de $1{,}2 \times 10^5$ M^{-1} dans 0,15 M NaCl à 4°C. Nous n'avons pas déterminé les constantes d'équilibre dans les systèmes poly U – oligopeptides.

Fuchs: (1) Dans le repliement du poly U pouvez-vous trancher entre le modèle en épingle à cheveu unique et le modèle à boucles multiples? (2) Je pense qu'une façon de trancher entre ces deux modèles serait de soumettre votre acide nucléique ordonné à l'action de la formaldéhyde et d'étudier la cinétique de cette réaction. En effet, dans le cas d'une épingle à cheveu unique vous aurez une vitesse initiale de réaction beaucoup plus faible que dans le cas du modèle à épingles multiples. Par dénombrement des sites d'initiation de la réaction de la formaldéhyde on aurait une idée du nombre de bras en double hélice dans la forme ordonnée du poly U.

Leng: La viscosité intrinsèque du poly U dans deux solvants (0,2 M CaCl et 0,5 M CaCl) tend vers la même limite, ce qui laisse supposer que l'on a un modèle en épingle à cheveu unique. Il se peut que les expériences avec la formaldéhyde permettent de vérifier cette hypothèse.

ACIDE POLY-URIDYLIQUE À BASSE TEMPÉRATURE

F. TRAVERS, A. M. MICHELSON, et P. DOUZOU
Institut de Biologie Physico-Chimique, Paris, France

Résumé. Il est étudié l'influence sur le T_m du poly U d'une addition progressive de méthanol accompagnée d'une diminution de la force ionique; on obtient pour 50% de méthanol et une concentration en Na^+ inférieure à 10^{-3} M un T_m de $-32°$C. Il existe une apparente continuité entre ce phénomène et celui observé en solution aqueuse dans les conditions usuelles. Si la concentration diminue, il existe une plage restreinte pour laquelle cette transition à basse température disparaît, permettant d'envisager l'hypothèse de la formation d'une structure ordonnée à partir de brins séparés.

Summary. Aqueous-organic solutions can be used as 'antifreeze' at sub-zero temperatures and then permit to experiment biopolymers in an unusual range of temperatures.

The present work deals with the behaviour of poly U: It is established that its T_m can be shifted towards sub-zero temperatures (from $+3°$C to $-32°$C); this 'transition' disappears at a certain concentration, indicating that the structure of poly U might be made in these conditions of several strands. Such a conclusion, which is in contradiction with the present ones, obtained by hydrodynamic methods, is discussed.

1. Introduction

L'addition de solvants organiques à des solutions aqueuses de biopolymères permet d'obtenir, sous réserve d'une non-dénaturation de celui-ci, des solutions fluides à des températures bien inférieures à 0°C. Ce procédé a déjà été appliqué à des enzymes dont la réactivité est avantageusement ralentie par les basses températures, permettant la stabilisation d'éventuels intermédiaires dont l'étude se trouve facilitée [1–2]. Dans les mêmes conditions il a été possible d'induire des changements de structure et de phase sur l'acide polyglutamique: les basses températures ont fait apparaître des effets de concentration inobservés dans la gamme usuelle de température permettant de conclure au caractère intermoleculaire des transitions observées [3–4].

Le comportement de quelques polynucleotides synthétiques dans ces conditions a également été envisagé [5], le cas de l'acide polyribouridylique (poly U) a été particulièrement approfondi et constitue l'objet du présent exposé.

Lorsqu'on refroidit une solution aqueuse de poly U (pH 7, 0,5 M Na^+), en dessous de $+10°$C l'absorbance à 260 nm diminue et décrit une transition coopérative dont le T_m se situe vers $+3°$C [6]. Cette hypochromicité traduit l'élaboration d'une structure secondaire dont la nature (simple brin ou multibrins) est encore l'objet de controverses. Nous avons diminué le contenu ionique et ajouté jusqu'à 50% de méthanol de manière à amener le T_m du poly U le plus bas possible au-dessous de 0°C puis dans ces conditions nous avons fait varier la concentration en polynucleotide.

2. Matériel et méthodes

Le poly U est obtenu par polymérisation enzymatique utilisant la polynucleotide phosphorylase [7]. Il est d'abord dissout en solution concentrée (quelques mg/ml)

C. Sadron (ed.), Dynamic Aspects of Conformation Changes in Biological Macromolecules. 393–400.

dans l'eau pure ou dans le tampon cacodylate 0,15 M (pH 7) suivant les besoins. Au moment de l'utilisation il est dilué en phase aqueuse puis le méthanol est ajouté lentement au voisinage de 0°C, à un pourcentage (en volume) variant de 0 à 50%. La concentration est ajustée pour avoir une absorbance finale à 260 nm voisine de 1.

La force inonique est ajustée par le chlorure de sodium. Elle doit être diminuée au fur à mesure qu'augmente le pourcentage en méthanol pour éviter la précipitation du poly U. Celle-ci se produit inévitablement au-dessus de 50% de méthanol quelque soit la force ionique. Toutes les solutions utilisées ont une activité protonique restant voisine de la neutralité, l'étude systématique de ce facteur n'est pas envisagée dans le travail présenté ici.

Les absorbances sont mesurées sur un spectrophotomètre Cary 15 spécialement équipé par un dispositif à température variable entre +100 et −100°C avec une précision de régulation de 1/100°C [8]. Le signal proportionnel à l'absorbance ainsi que celui donné par un thermocouple plongeant dans la cellule de mesure sont envoyés sur les deux voies d'un enregistreur XY Hewlett Packard permettant ainsi un enregistrement direct de la courbe $A = f(T)$. La vitesse de variation de la température est de l'ordre de 1°C/mn.

3. Resultats

3.1. Influence du méthanol

A pH 7, 0,5 M Na^+ le poly U a un T_m de +3°C, l'hypochromicité accompagnant la structuration est de 30%, ces conditions seront prises comme référence.

TABLEAU I

Paramètres de la transition du poly U ($A_{260} = 0,5$) en fonction du % en méthanol et de [Na^+]

% méthanol	[Na^+]	T_m (°C)	$\Delta\varepsilon/\varepsilon$ (%)
0	5×10^{-1}	+3	30
0	10^{-1}	−2	–
20	8×10^{-2}	+1	27,5
20	$1,05 \times 10^{-2}$	−10	28
30	$0,95 \times 10^{-2}$	−9	29
30	6×10^{-3}	−12	30
40	$5,5 \times 10^{-3}$	−11	29
40	$3,7 \times 10^{-3}$	−16	28
50	$2,5 \times 10^{-3}$	−18	25
50	$2,5 \times 10^{-4}$	−33,5	20

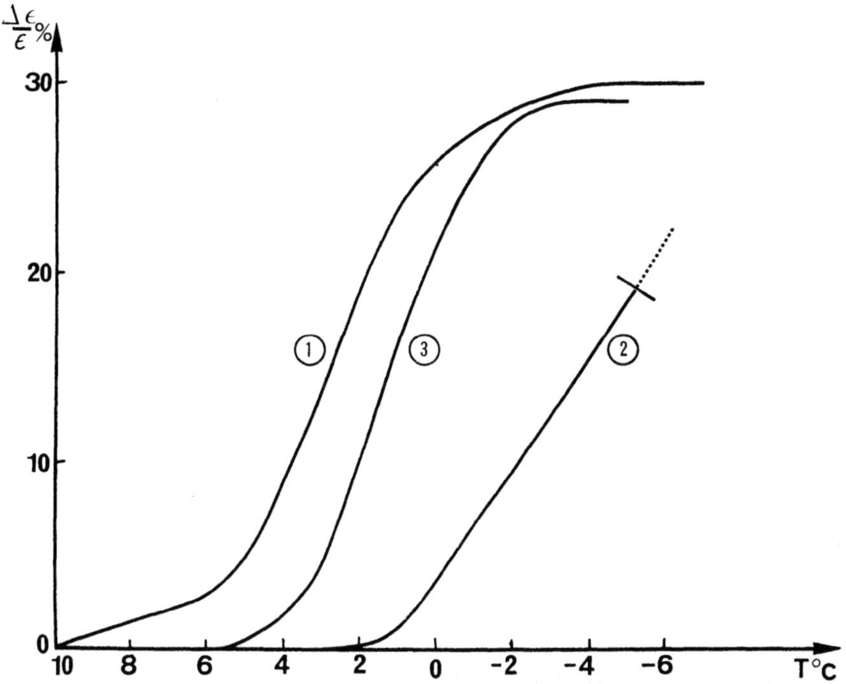

Fig. 1. Variations relatives du coefficient d'extinction à 260 nm, en fonction de la température, du poly U ($A_{260} = 0{,}8$ à $+20°C$): (1) tampon cacodylate, pH 7, 10^{-2} M, 0,5 M Na^+, (2) idem (1) 1,01 M Na^+, (3) idem (2) plus 20% de méthanol.

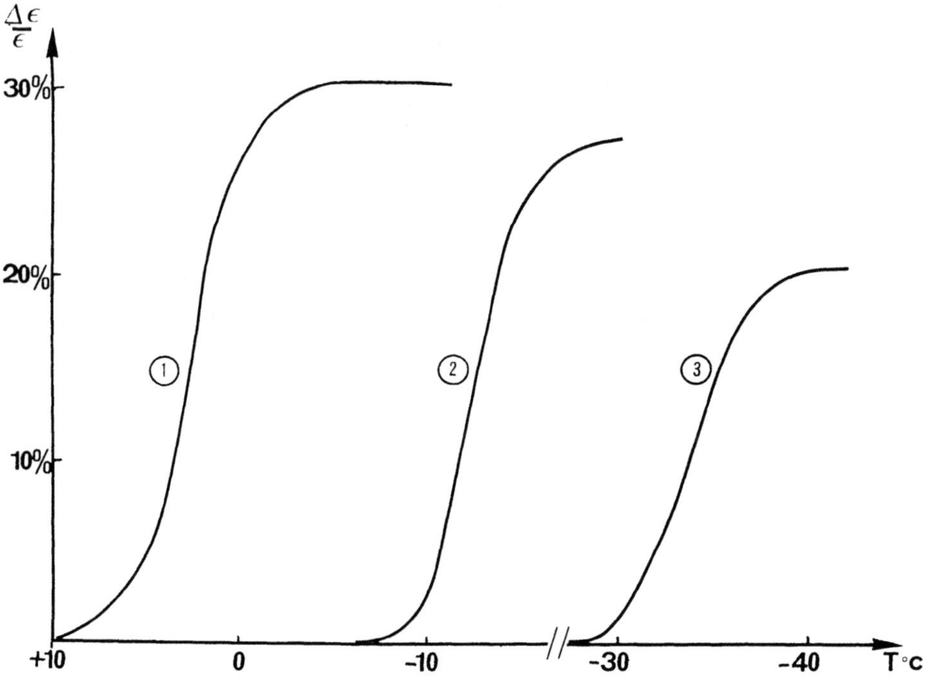

Fig. 2. Variations relatives du coefficient d'extinction à 260 nm, en fonction de la température, du poly U ($A_{260} = 0{,}8$ à $+20°C$): (1) tampon cacodylate, pH 7, 10^{-2} M, 0,5 M Na^+, (2) 40% méthanol 5×10^{-3} M Na^+, (3) 50% méthanol $2{,}5 \times 10^{-4}$ M Na^+.

Si la concentration saline (Na^+) est abaissée à 10^{-1} M la transition est moins coopérative et bien qu'elle soit interrompue par la congélation on peut évaluer le T_m à -3°C (Figure 1). Lorsqu'on ajoute 20% de méthanol à la solution précédente le T_m remonte à $+1$°C cependant que la coopérativité initiale est rétablie (Figure 1).

En diminuant la force ionique et en augmentant le pourcentage en méthanol le T_m est rejeté vers les basses températures. L'ensemble des résultats est figuré sur le Tableau I cependant que les étapes principales sont représentées sur la Figure 2. On constate que le T_m diminue régulièrement en même temps que la force ionique (nous reviendrons sur ce point au paragraphe suivant) cependant que les additions de méthanol augmentent la stabilité de la forme structurée: lorsque la concentration en Na^+ passe de 10^{-2} à $5{,}5 \times 10^{-3}$ M et simultanément le pourcentage de méthanol de 20 à 40% le T_m se maintient entre -10°C et -11°C.

Entre $[Na^+]=0{,}5$ M sans méthanol et $[Na^+]=2{,}5 \times 10^{-4}$ avec 50% de méthanol le T_m varie de $+3$°C à $-33{,}5$°C avec une coopérativité comparable cependant que le taux d'hypochromicité passe de 30% à 20%.

3.2. Influence du contenu ionique

Pour 50% de méthanol, si la concentration en Na^+ dépasse 10^{-1} M le poly U précipite. Pour ce même pourcentage en alcool et pour une concentration en poly U de 8×10^{-5} M en monomère, lorsqu'on diminue le contenus en Na^+ on observe une diminution du T_m représentée sur la Figure 3: cette représentation est linéaire en fonction de $\log[Na^+]$.

Pour les faibles valeurs de Na^+ la précision est faible en particulier à cause des ions cristallisés avec le poly U: si l'on veut conserver la linéarité de la loi précédente il faut admettre que le contenu en Na^+ apporté par le poly U est de $2{,}5 \times 10^{-4}$ M ce qui correspond environ à 3 Na^+ par résidu de nucleotide.

3.3. Influence de la concentration en poly U

Lorsqu'on diminue la concentration en poly U en présence de 50% de méthanol et à très faible force ionique la transition observée évolue (Figure 4). Jusqu'à une concentration en poly U correspondant à $A_{260}=0{,}3$ le pourcentage d'hypochromicité aussi bien que le T_m ne varient pas, cependant pour une absorbance située entre 0,3 et 0,2

TABLEAU II

$[Na^+]$	T_m à forte concentration (°C)	Absorbance de disparition
$2{,}5 \times 10^{-4}$	-32	0,2–0,3
$1{,}25 \times 10^{-3}$	-20	0,1–0,3
$5{,}25 \times 10^{-3}$	-14	0,1–0,2
10^{-2}	-10	0,06

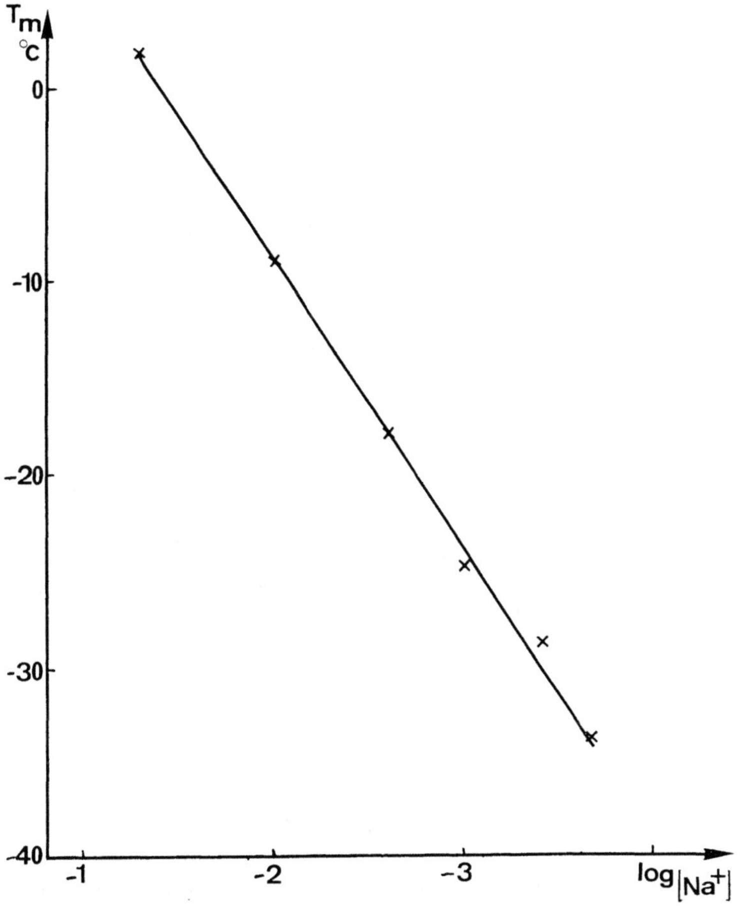

Fig. 3. T_m du poly U ($A_{260} = 0{,}8$ à $+20°$C) dans 50% de méthanol en fonction de log[Na^+].

ces deux paramètres évoluent brusquement jusqu'à une disparition complète de la transition alors que les mesures sont toujours possibles (Figure 5).

Lorsqu'on augmente la force ionique le T_m augmente et dès que celui-ci atteint $-10°$C la concentration de disparition tombe en dessous de toute valeur mesurable (Tableau II).

4. Discussion

A force ionique égale l'addition du méthanol confère à la forme structurée du poly U une plus grande stabilité thermique se traduisant par une augmentation du T_m, ce phénomène a déjà été observé [9] et il est contraire à ce qui est observé pour de nombreux polynucleotides comme par exemple le poly C.

En phase aqueuse (0,15 M Na^+), conditions usuelles d'observation, le T_m du poly U est de $+3°$C, par le jeu de l'abaissement de la force ionique et de l'augmentation du méthanol nous l'avons amené vers $-30°$C. Il semble bien exister une continuité entre

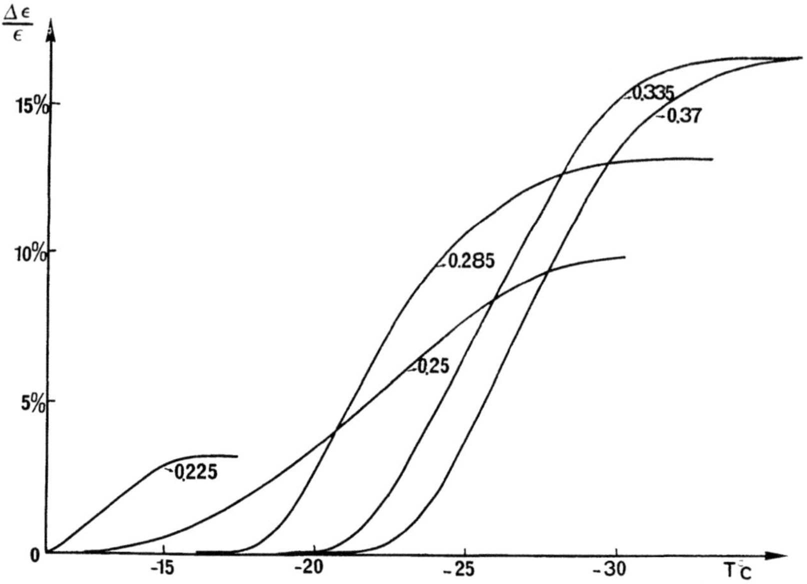

Fig. 4. Variations relatives du coefficient d'extinction à 260 nm en fonction de la température du poly U (50% de méthanol, 2×10^{-4} M Na^{+}) pour différentes concentrations: les nombres indiquent l'absorbance correspondante à 260 nm pour $+20°C$.

les deux phénomènes: la coopérativité est conservée cependant que le taux d'hypochromicité passe de 30% à 20%, effet peut-être dû au méthanol qui assurerait dans ces conditions un couplage plus faible des bases. Pour 50% de méthanol la loi de variation du T_m en fonction de Na^+ est celle classiquement observés en solution aqueuse pour plusieurs polynucleotides. De par ces observations il semble à priori raisonnable de penser que le phénomène observé dans nos conditions est de même nature que celui obtenu dans la gamme des températures usuelles.

La disparition à basse concentration de la formation de la structure ordonnée du poly U à basse température est un fait nouveau non détecté dans les solutions aqueuses. Il en résulte que la seule hypothèse envisageable est la présence d'un phénomène intermoléculaire, c'est-à-dire l'organisation d'une structure entre plusieurs brins de poly U sans qu'il soit possible pour l'instant de préciser leur nombre

Cette disparition peut être interprétée par l'existence d'une valeur critique de la distance moyenne entre les molécules au-delà de laquelle toute interaction est impossible. Au-dessus de 0°C la concentration conduisant à une distance moyenne proche de la valeur critique doit être inférieure à toute valeur mesurable par les techniques classiques. L'abaissement de la température diminuant l'agitation moléculaire supprime cette impossibilité. Ce phénomène a déjà été observé et étudié sur des interactions entre molécules plus simples du type cyanine [11].

Il faut également envisagé comme pouvant intervenir dans ce phénomène le rôle de la constante diélectrique et de la viscosité bien que dans cette gamme de température et pour ce solvant les variations restent faibles [12, 13].

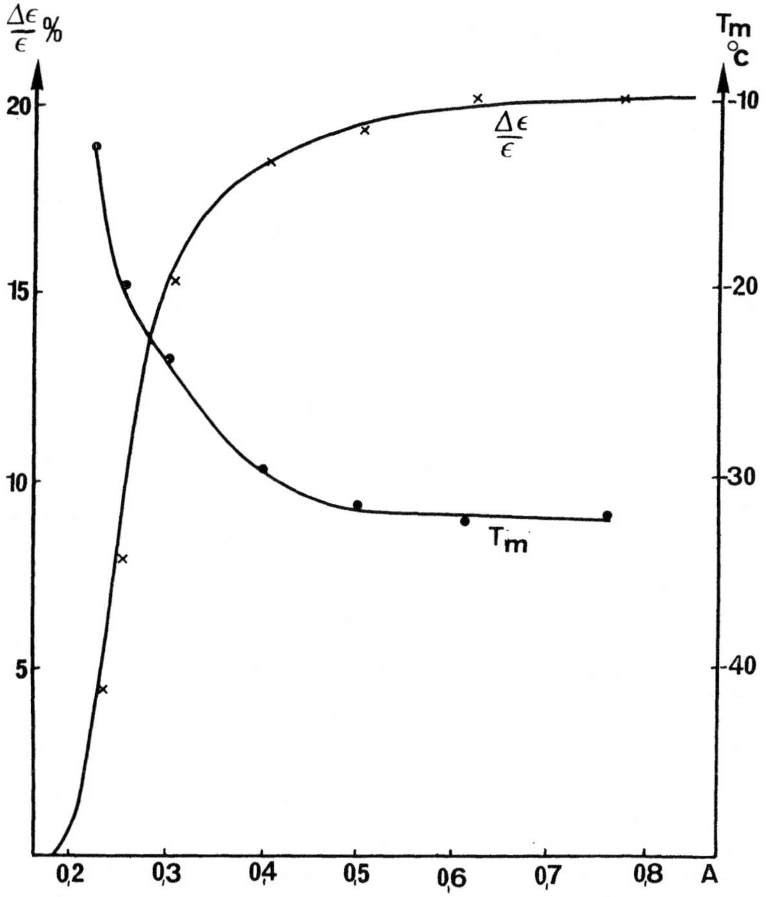

Fig. 5. Variations relatives totales du coefficient d'extinction à 260 nm accompagnant la structuration et T_m pour le poly U dans 50% de méthanol (2×10^{-4} M Na^+) en fonction de la concentration représentée par l'absorbance initiale à $+20°$C (260 nm).

Les conclusions résultant de ce travail quant à la structure ordonnée du poly U sont en contradiction avec des travaux récents [14, 15] qui concluent à la formation de structures intramoléculaires plus ou moins complexes, ces résultats ayant été obtenus par des méthodes indirectes (viscosité et ultracentrifugation). Cependant, il se peut que l'apparente continuité observée entre les deux phénomènes cache un jeu subtil de déplacement d'équilibre de transition de phase et de structure permettant de passer d'un simple brin replié sur lui-même à une structure à plusieurs brins. Avec nos seuls résultats on ne peut aller plus loin dans cette hypothèse seules des mesures de viscosité et d'ultracentrifugation dans nos conditions de température permettraient peut être d'aboutir.

Références

[1] Douzou, P.: 1971, *Biochimie* **53**, 17.

[2] Douzou, P.: 1971, *Biochimie* **53**, 1135.
[3] Travers, F. et Coupechoux, E.: 1972, *Biochimie* **54**, 457.
[4] Travers, F.: 1972, Thèse de Doctorat, Université Paris 7.
[5] Travers, F., Michelson, A. M., et Douzou, P.: 1970, *Biochim. Biophys. Acta* **199**, 29.
[6] Lipsett, M.: 1960, *Proc. Nat. Acad. Sci. US* **46**, 445.
[7] Grunberg-Manago, M.: 1963, *Advan. Nucleic. Acid. Res.* **1**, 93.
[8] Balny, C., Hui Bon Hoa, G., et Travers, F.: 1971, *J. Chim. Phys.* **68**, 366.
[9] Brown, R. A.: 1966, *Arch. Biochem. Biophys.* **115**, 102.
[10] Travers, F., Michelson, A. M., et Douzou, P.: 1970, *Biochim. Biophys. Acta* **217**, 1.
[11] Hui Bon Hoa, G. et Balny, C.: 1969, *J. Chim. Phys.* **66**, 1528.
[12] Debey, M. P.: 1970, Thèse 3e Cycle, Paris.
[13] Travers, F. et Douzou, P.: 1970, *J. Phys. Chem.* **74**, 2243.
[14] Thrierr, J. C. et Leng, M.: 1969, *Biochim. Biophys. Acta.* **182**, 575.
[15] Millar, D. B. et Mackenzie, M.: 1970, *Biochim. Biophys. Acta.* **204**, 82.

DISCUSSION

Dourlent: Comment expliquez-vous le fait que la variation d'hypochromicité d'un même poly U dépende de la concentration?

Travers: Nous avons vérifié que ce n'était pas dû à une cinétique très lente; par conséquent, cette baisse d'hypochromicité ne peut s'expliquer que par une disparition progressive de la structuration.

Leng: Peut-on éliminer la présence d'agrégats?

Travers: Cette possibilité ne peut être écartée; cependant, dans ces conditions de faible force ionique, les mesures optiques n'ont jamais laissé supposer le présence d'agrégats importants.

Daune: (1) Le méthanol diminue la stabilité du DNA en solution aqueuse dans la zone des températures ordinaires. A-t-on une explication de l'influence exactement inverse à basse température dans le cas des solutés de poly U?

(2) Est-ce que l'effet observé à basse température en fonction de la concentration en poly U n'est pas lié à un phénomène de démixtion ou à un changement de phase (repliement du poly U sur lui-même)?

Travers: (1) Le phénomène n'est pas spécial aux basses températures. Il est propre au poly U; à ma connaissance, on ne sait pas la raison de ce cas particulier par rapport à la plupart des autres polynucléotides.

(2) Rien ne permet de rejeter cette hypothèse; je ne vois cependant pas pourquoi un tel phénomène apparaîtrait à basse concentration.

Maurizot: Avez-vous étudié l'évolution du pouvoir rotatoire en fonction de la concentration en poly U dans les conditions où vous observez la disparition de l'hypochromisme? Constatez-vous un parallélisme entre le comportement du pouvoir rotatoire et celui de l'hypochromisme?

Travers: Les mesures ont été effectivement effectuées et il existe un parallélisme parfait entre les deux réponses spectroscopiques.

PHÉNOMÈNES D'HYSTÉRÈSE DANS LES POLYNUCLÉOTIDES*

WILHELM GUSCHLBAUER[a], DANIELLE THIELE[a],
MARIE-THÉRÈSE SAROCCHI[a]**, et CHRISTIAN MARCK[b‡]

[a] *Service de Biochimie, et* [b] *Groupe de Biophysique Moléculaire, Département de Biologie, Centre d'Études Nucléaires de Saclay, B. P. 2, F-91190, Gif-sur-Yvette, France*

Résumé. Les spectres DC de deux complexes polynucléotidiques ont été étudiés en fonction du pH dans 0,15 M NaCl.

La titration acide montre deux transitions pour chacun des systèmes. La titration retour montre deux étapes pour le poly (rI).poly (rC) et une seule étape sans passage par la forme intermédiaire pour le poly (dG)·poly (dC).

L'hystérèse à 25°C du poly (rI).poly (rC) est faible (Δ(pH) $= 0{,}5 \pm 0{,}1$) alors que celle du poly (dG).poly (dC) est très importante (Δ(pH) $\simeq 4{,}7$), et conduit à l'existence d'une forme acide métastable à pH neutre. Cette forme donne irréversiblement le spectre de la forme neutre après chauffage. Il est alors nécessaire de refaire un cycle de titration acide pour obtenir le spectre de la forme acide métastable.

Les aspects thermodynamiques de l'hystérèse de ces systèmes binaires sont mis en parallèle avec la théorie de l'information afin de décrire les phénomènes observés.

Summary. The CD spectra of two polynucleotide complexes poly (rI).poly (rC) and poly (dG).poly (dC) have been studied as a function of pH in 0.15 M NaCl. The acid titration shows two transitions for both systems. The alcaline back titration also shows two steps for the system poly (rI).poly (rC) and only one step for the system poly (dG).poly (dC).

While in poly (rI).poly (rC) the hysteresis at 25°C is weak for the two steps (Δ(pH) $\simeq 0.5 \pm 0.1$), that of poly (dG).poly (dC) is exceptionally large (Δ(pH) $\simeq 4.7$, at 25°) and leads to the formation of an acid metastable form at neutral pH; this metastable hysteretic form can be transformed into the neutral form by heating. This heating curve is irreversible and it is necessary to perform the whole titration loop in order to obtain the metastable protonated form again.

In view of the existence of this binary system an attempt is made to correlate thermodynamics and information theory to describe the phenomena observed.

1. Introduction

Une différence essentielle entre les protéines et les acides nucléiques réside dans le fait que pour les unes la structure et la fonction sont déterminées par un petit nombre d'interactions entre quelques groupements fonctionnels laissant une grande liberté à la molécule qui peut exister sous plusieurs formes. Pour les acides nucléiques, par contre, tous les groupements fonctionnels contribuent à la structure, le choix des possibilités de liaison est très restreint et ils ont une conformation essentiellement unique et stable. Le t-RNA et probablement le RNA ribosomal font exception et ressemblent plus aux protéines globulaires qu'au DNA.

Cependant la fonction du DNA nécessite que quelques paires de bases soient différentes du reste de la molécule. Le problème de la reconnaissance des sites qui servent de marqueurs pour les processus cellulaires spécifiques (initiation et termi-

* Partie 13 de la série *Polynucléotides protonés*; partie 12, réf. [12].

** Boursière de Roussel–UCLAF (1971–73).

‡ Boursier de thèse du CEA.

C. Sadron (ed.), Dynamic Aspects of Conformation Changes in Biological Macromolecules. 401–410.

naison de la transcription, liaison du répresseur, mécanismes de contrôle, etc.) revêt une grande importance. L'existence de telles régions a été montrée par des études génétiques, et aucune description physique de ces sites n'a été possible jusqu'à maintenant. La raison en est qu'ils ne constituent qu'une très faible fraction de la molécule totale (1% ou moins). Ainsi de toute mesure physicochimique qui n'a pas la spécificité d'un enzyme ou d'un represseur, il faut déduire 99% de bruit de fond.

Szybalski *et al.* [1] ont, les premiers, mis en évidence des séquences répétitives qui pourraient avoir une structure différente du reste de la molécule de DNA et ainsi servir de marqueur. Ces auteurs ont montré par hybridation sélective avec poly (UG), l'existence de 'clusters' de déoxycytidine d'environ 15 nucléotides de long dans le DNA de T_7. La possibilité que de telles séquences répétitives aient une structure différente a été envisagée et montrée par plusieurs auteurs en utilisant des polynucléotides synthétiques [2].

De nombreux travaux existent également au sujet des transitions structurales que des agents extérieurs peuvent induire au niveau de ces séquences [2].

De plus, Hoogsteen [3] a montré par cristallographie aux rayons X qu'il existe, pour les paires A–T, une possibilité d'appariement différent de l'appariement Watson–Crick. L'appariement Hoogsteen semble le plus probable pour les paires de nucléosides cocristallisés [4]. Seul Crick [5] a envisagé sérieusement cette alternative de paire Hoogsteen comme marqueur possible dans le DNA.

Les agents extérieurs les plus simples sont les protons et les ions hydroxyle. Alors que l'ion OH dénature les polynucléotides et les acides nucléiques, les protons induisent des transstructurations. De nombreux travaux dans notre laboratoire [6–13] ont permis de préciser les interactions possibles entre polynucléotides au cours de la titration acide. Les ions H^+ sont les agents chimiques les plus simples et sont susceptibles d'être libérés facilement.

Il ressort de ces études que seules les paires G–C (ou I–C) peuvent utiliser des protons pour former des structures nouvelles, soit par réarrangement intermoléculaire pour former un complexe à trois brins, soit par transconformation intramoléculaire. Ces complexes protonés comportent probablement des paires Hoogsteen et ont alors une structure différente de celle des complexes à pH neutre dont l'appariement est du type Watson–Crick.

Des phénomènes d'hystérèse ont été observés dans le DNA [14–15], le RNA [15–17] et les systèmes polynucléotidiques [18–20] et suggèrent que des variations induites par la capture d'un proton pourraient être stables à des pH proches de la neutralité. Mais les systèmes contenant des paires I–C ou G–C n'ont jamais été étudiés. Il apparaît que ces paires de bases montrent des hystérèses bien définies et parfois assez large, au cours de la titration acide.

Dans cette communication nous présentons une étude de deux systèmes modèles de polynucléotides montrant une telle hystérèse: Le poly (rI).poly (rC) et le poly (dG).poly (dC). Ces deux complexes se titrent l'un et l'autre en deux étapes qui présentent chacune une hystérèse; cependant la titration en retour ne se fait pas selon le même mécanisme pour les deux systèmes.

2. Matériel et méthodes

Les polynucléotides synthétiques utilisés, poly (rI), poly (rC), poly (dG), poly (dC) ont été synthétisés par Miles Inc. Elkhart, Indiana, U.S.A.

Les solutions ont été faites dans NaCl 0,15 M pour les titrations et dans Na cacodylate 0,15 M pour les expériences de chauffage.

Les concentrations ont été déterminées par spectrophotométrie UV (Cary 15) en utilisant les coefficients d'absorption suivants: poly (C): $\varepsilon_{268}=6300$; poly (I): $\varepsilon_{248}=10400$, pH 6.8 [8]; poly (dG).poly (dC): $\varepsilon_{260}=7300$ [22].

Les titrations ont été faites par dichroïsme circulaire sur un dichrographe II Roussel–Jouan et registrées sur calculateur PDP-12. Les spectres ont été corrigés et normalisés.

Les courbes de chauffage ont été enregistrées sur un autre dichrographe II Roussel–Jouan équipé d'une cuve thermostatée.

3. Résultats et discussion

3.1. Hystérèse du poly (rI) · poly (rC)

Au cours d'une étude précédente [7, 8, 11] nous avons mis en évidence la trans-

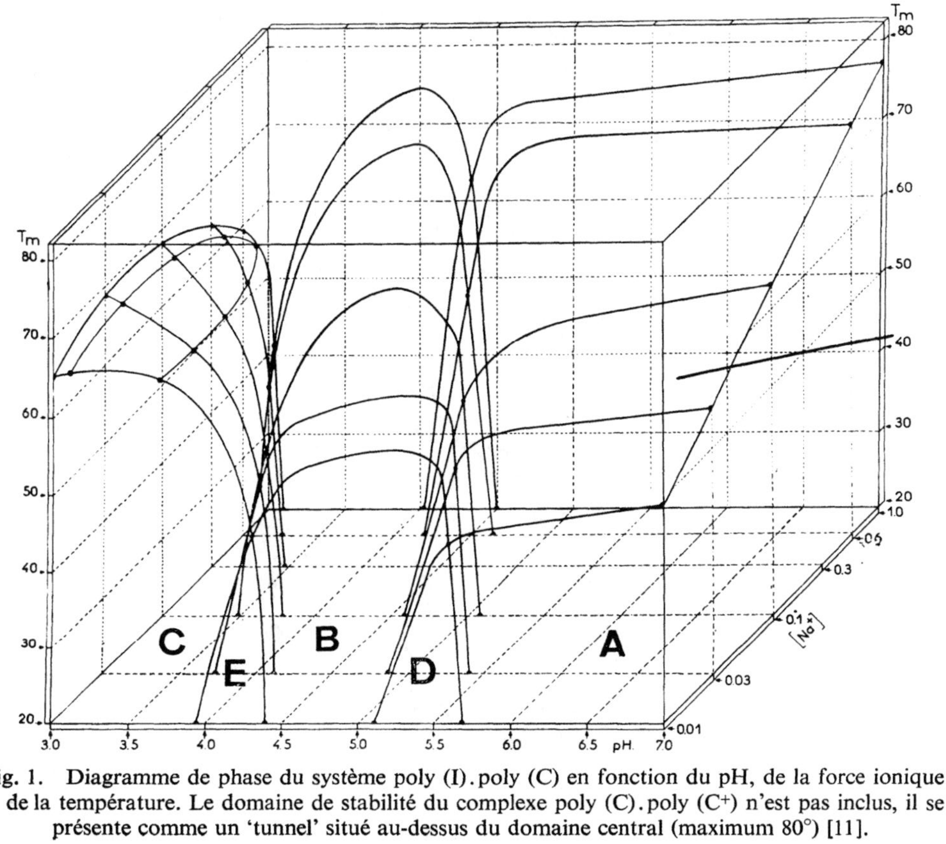

Fig. 1. Diagramme de phase du système poly (I).poly (C) en fonction du pH, de la force ionique et de la température. Le domaine de stabilité du complexe poly (C).poly (C^+) n'est pas inclus, il se présente comme un 'tunnel' situé au-dessus du domaine central (maximum 80°) [11].

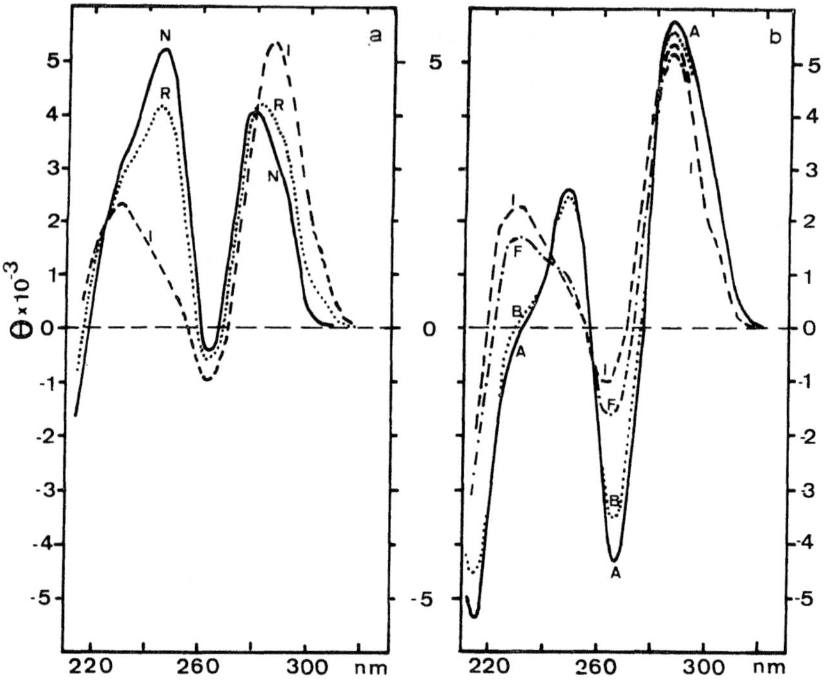

Fig. 2. Spectres DC d'un mélange 1:1 de poly (I) et poly (C) à différents pH. (a) *Première étape de titration.* Titration acide N: pH 8 à pH 5; I: pH 8 4,2; titration alcaline R: pH 5,0 (après être passé par pH 3,3) (b) *Deuxième étape de titration.* Titration acide I: pH 4,2; F: pH 3,6; A: pH 3,3; titration alcaline B: pH 3,6 (en partant de pH 3,3).

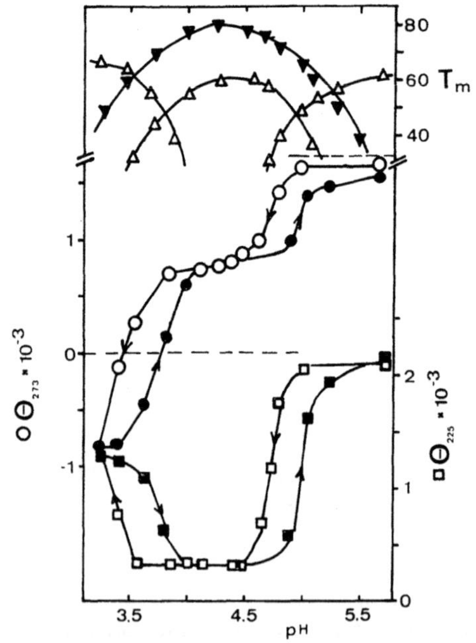

Fig. 3. Variations du DC (partie inférieure) et de la température de fusion (partie supérieure) du système poly (I).poly (C). ○: θ_{273}, □: θ_{255}. Symboles creux: titration acide; symboles pleins: titration alcaliné retour. △ T_m des complexes entre poly (I) et poly (C); ▼ T_m du poly (C).poly (C^+).

conformation suivante:

$$2 \times \text{poly}(\text{I})\cdot\text{poly}(\text{C}) \overset{H^+}{\rightarrow} \text{poly}(\text{I})\cdot\text{poly}(\text{C})\cdot\text{poly}(\text{C}^+) + \text{poly}(\text{I})$$
$$\downarrow H^+$$
$$2 \times \text{poly}(\text{I})\cdot\text{poly}(\text{C}^+)$$

Cette transition en deux étapes a été définie en fonction du pH, de la force ionique et de la température (Figure 1). On notera dans cette figure que le 'tunnel' central, domaine (B) du complexe ternaire poly (I).poly (C).poly (C^+), chevauche les régions de stabilité du poly (I).poly (C) (domaine A) aux pH élevés et celui du poly (I).poly (C^+) (domaine C) à bas pH [11]. Les zones D et E sont des domaines métastables où deux complexes peuvent exister simultanément ou non selon les conditions expérimentales et 'l'histoire' de l'échantillon. Les trois complexes ont des spectres DC très caractéristiques [7, 11] qui permettent de les distinguer (Figure 2).

La titration du poly (I).poly (C), d'abord de pH neutre à pH acide (aller), puis de pH acide à pH neutre (retour), montre un phénomène d'hystérèse, qui se traduit par des spectres différents, donc des structures différentes, pour un même pH à l'aller et au retour. Cette observation s'applique à la première transition (formation du complexe à trois brins, spectres N et R dans la Figure 2a) ainsi qu'à la deuxième transition (formation du complexe poly (I).poly (C^+), spectres F et B dans la Figure 2b). La titration retour se fait en deux étapes, comme la titration aller.

Si l'on superpose les résultats des titrations (Figure 2) et ceux des transitions thermiques [8] on constate que l'hystérèse coïncide parfaitement avec les zones métastables (Figure 3). D'autre part les transitions de titration sont très coopératives et les boucles d'hystérèse ne sont pas très larges ($\Delta(\text{pH}) \simeq 0{,}5 \pm 0{,}1$).

3.2. Hystérèse du poly (dG)·poly (dC)

La titration du poly (dG).poly (dC) présente deux étapes (Figure 4). La première conduit à une forme intermédiaire I, la seconde à une forme acide A et chacune d'elles

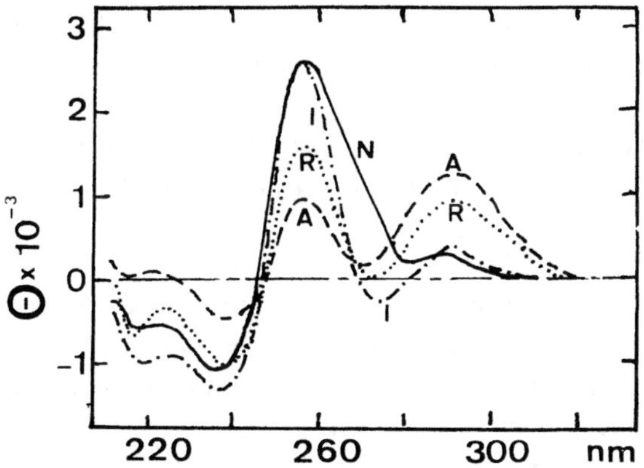

Fig. 4. Spectres DC du poly (dG).poly (dC) dans 0,15 M NaCl et à différents pH. Titration acide —— N: pH 6,8; –·–· I = pH 4,95; ---- A: pH 2,0; Titration alcaline .... R: pH 6,95.

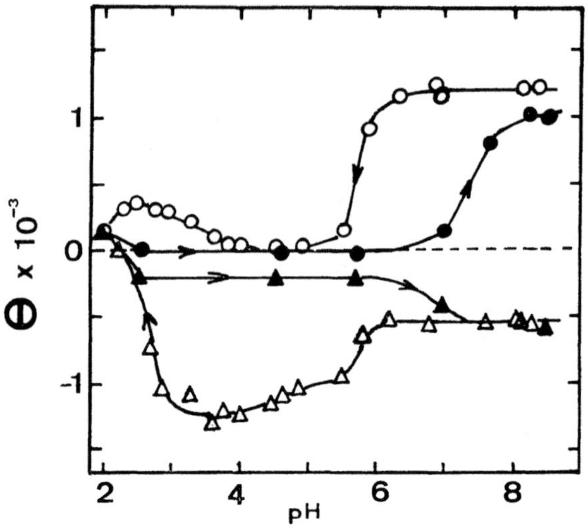

Fig. 5. Titration dichroïque du poly (dG).poly (dC) à deux longueurs d'onde différentes. △ 220 nm; ○ 269 nm. Symboles creux: titration acide; symboles pleins: titration alcaline. Les flèches indiquent le sens de la titration. Des résultats analogues ont èté obtenus à d'autres longueurs d'onde.

comporte une hystérèse. Ces deux étapes déjà observées par DOR [7] sont probablement analogues à celles du système poly (rI).poly (rC) avec formation d'un complexe intermédiaire à trois brins précédant l'apparition de la forme acide bicaténaire poly (dG)-poly (dC^{+}). Cependant, à la difference du poly (rI).poly (rC), la titration en retour de pH 2 à pH 7 ne repasse pas par la forme intermédiaire et le spectre R obtenu est très différent du spectre neutre N et analogue au spectre acide. Les spectres DC de ce polymère précédemment publiés par Wells *et al.* [22] ne concordent pas avec les nôtres et semblent être ceux d'un mélange.

Selon la longueur d'onde envisagée, on peut observer plus précisémment l'une ou l'autre étape de la titration (Figure 5), ou les deux simultanément.

D'autre part, la courbe de chauffage de la forme métastable n'est pas réversible. En effet (Figure 6), une solution 'en hystérèse' (à pH 6,8) chauffée au-dessus de 50°C passe dans une zone de stabilité et ne redonne pas le spectre initial (R) après refroidissement [25].

On peut envisager d'appliquer à ces résultats les travaux de Katchalsky et coll. [18–21], concernant la théorie des phénomènes d'hystérèse.

3.3. Considérations théoriques

Les résultats précédents montrent que certains polynucléotides peuvent assumer deux structures ou conformations différentes très stables dans les mêmes conditions de pH et de force ionique. Selon 'l'histoire' de l'échantillon on obtient l'une ou l'autre forme, ce qui suggère un phénomène de mémoire.

Un parallèlisme entre les phénomènes d'hystérèse et de mémoire a déjà été fait, en particulier par l'équipe de Katchalsky [18–21]. Dans le cas des acides nucléiques, ces

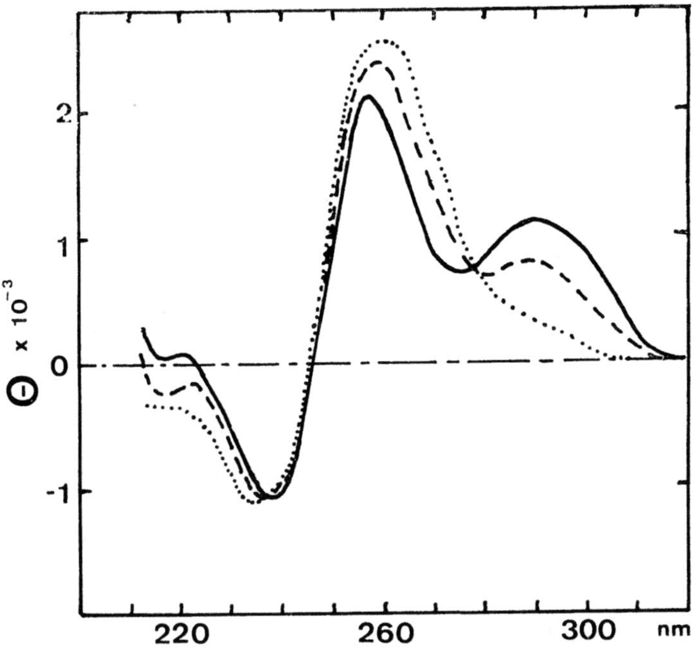

Fig. 6. Spectres CD du poly (dG).poly (dC) en hystérèse, à pH 6,8 à différentes températures. —— 20°; ---- 37°; .... 56° et retour à 20°.

auteurs ont envisagé des phénomènes de mémoire à court terme dus à des hystérèses au niveau des brins protonés. Selon Katchalsky et coll. [19–20] la perte d'énergie libre au cours du cycle d'hystérèse est donnée par l'équation:

$$\Delta G_{\text{irr.}} = - T\Delta S_{\text{irr.}} = - 2{,}3RT \int \Delta\alpha \, \mathrm{d}(\mathrm{pH}) . \tag{1}$$

Il s'ensuit pour le gain d'entropie:

$$\Delta S_{\text{irr.}} = 2{,}3R \int \Delta\alpha \, \mathrm{d}(\mathrm{pH}) , \tag{2}$$

où $\Delta\alpha$ est le degré moyen d'ionisation et $\int \Delta\alpha \, \mathrm{d}(\mathrm{pH})$ la surface de la boucle d'hystérèse. En cas de protonation totale et de coopérativité parfaite $\Delta\alpha \to 1$, ce qui donne, pour une hystérèse de une unité de pH:

$$\Delta S_{\text{irr.}}/\Delta(\mathrm{pH}) = 2{,}3R = 4{,}57 \ \mathrm{cal \cdot mole^{-1} \cdot deg.^{-1}} . \tag{3}$$

D'autre part l'idée des phénomènes de mémoire ressort aussi des propos de Szilard [23] et Brillouin [24] sur l'équivalence de l'entropie et de l'information. Selon Brillouin [24], on peut tirer du principe de négentropie la relation suivante:

$$I = - S = nK \ln(P_0/P) , \tag{4}$$

où I est le contenu d'information, S l'entropie, n le nombre de sélections, K une constante et P_0 et P le nombre de complexions (selon Planck), dans les deux populations du système.

Pour un système binaire (comme le nôtre), $P_0/P = 2$. Si l'on exprime I en unités informatiques (bits), $K = \lg_2 e = 1{,}38$ et donc:

$$I = -S = 1{,}38n \ln 2 = n \text{ bits}. \tag{5a}$$

Si I est exprimé en unités thermodynamiques, on a, $K = k = 1{,}38 \times 10^{-16}$ erg.molécules^{-1}.deg.$^{-1}$, (constante de Boltzmann), et:

$$I = -S = 1{,}38n \times 10^{-16} \cdot \ln 2 = 10^{-16} n \text{ erg.molecule}^{-1}.\text{deg.}^{-1}, \tag{5b}$$

donc par mole (N: nombre d'Avogadro):

$$\begin{aligned} I = -S &= 10^{-16} nN \,(\text{erg.mole}^{-1}.\text{deg.}^{-1}) \\ &= 6n\,(\text{mole}^{-1}.\text{deg.}^{-1}) = 1{,}4n\,(\text{cal.mole}^{-1}.\text{deg.}^{-1}). \end{aligned} \tag{5c}$$

Il apparaît donc entre les Equations (3) et (5c) un facteur $n = 3{,}3$, qui correspond au nombre d'unités de sélections impliquées dans un phénomène d'hystérèse d'une unité de pH pour tenir compte du gain d'entropie à la fin du cycle. Dans notre système, les unités de sélection sont les paires de bases.

Pour le poly (dG).poly (dC), il résulte de la Figure 5 que d(pH) = 4,7. Le pK acide extrêmement bas ne permettant pas une titration potentiométrique, nous avons repris les données de Neumann et Katchalsky [21] sur des systèmes similaires et en tenant compte de la forme du cycle d'hystérèse nous obtenons une valeur $\Delta\alpha = 0{,}6 \pm 0{,}1$. En portant cette valeur dans l'Équation (3) on obtient $\Delta S_{\text{irr}} = 13$ à 15 cal.mole^{-1}.deg.$^{-1}$, et de l'Équation (5c) on tire $n = 10$. On peut donc estimer qu'un minimum de 10 paires de bases successives en hystérèse sont nécessaires dans le poly (dG)·poly(dC) pour rendre compte des observations de la Figure 5. Il faut noter que ces résultats ne constituent qu'un ordre de grandeur pour la valeur n. Des mesures plus détaillées et une analyse plus poussée permettront de mieux comprendre ce phénomène.

3.4. Conclusions

Les résultats obtenus pour les polymères synthétiques, systèmes modèles simplifiés, suggèrent que les mêmes phénomènes d'hystérèse sont probables dans les acides nucléiques naturels. En effet, on peut imaginer que sous l'influence de l'environnement (systèmes donneurs de protons) des régions du DNA puissent changer localement la conformation. Des études spectrales récentes [12] sur les DNA naturels indiquent que de telles séquences homologues peuvent être impliquées dans des variations structurales.

Références

[1] Szybalsky, W., Kubinski, H., et Sheldrick, P.: 1966, dans *Cold Spring Harbour Symp. Quant. Biol.* **21**, 123–27.

[2] Michelson, A. M., Massoulié, J., et Guschlbauer, W.: 1967, *Progr. Nucleic Acid Res. Mol. Biol.* **6**, 83–141.
[3] Hoogsteen, K.: 1959, *Acta Crystallog.* **12**, 822–23.
[4] Haschemeyer, A. E. V. et Rich, A.: 1967, *J. Mol. Biol.* **27**, 369–80.
[5] Crick, F. H. C.: 1964, dans *6th Intern. Congr. Biochem.* **33**, 109.
[6] Courtois, Y., Fromageot, P., et Guschlbauer, W.: 1968, *Eur. J. Biochem.* **6**, 493–501.
[7] Courtois, Y., Thiele, D., et Guschlbauer, W.: 1968, *Bull. Soc. Chim. Biol.* **50**, 2373–83.
[8] Thiele, D. et Guschlbauer, W.: 1969, *Biopolymers* **8**, 361–78.
[9] Sarocchi, M. Th., Courtois, Y., et Guschlbauer, W.: 1970, *Eur. J. Biochem.* **14**, 411–21.
[10] Thiele, D. et Guschlbauer, W.: 1971, *Biopolymers* **10**, 143–57.
[11] Thiele, D., Guschlbauer, W., et Favre, A.: 1972, *Biochim. Biophys. Acta* **272**, 22–26.
[12] Sarocchi, M. Th. et Guschlbauer, W.: 1973, *Eur. J. Biochem.* **34**, 232–240.
[13] Guschlbauer, W.: 1972, dans *The Purines, Jerusalem Symp. Quant. Chem. Biochem.* **4**, 297–310.
[14] Cox, R. A. et Peacocke, A. R.: 1956, *J. Chem. Soc.* 2499–2505.
[15] Cox, R. A.: 1963, *Biochim. Biophys. Acta* **68**, 401–10.
[16] Cox, R. A.: 1966, dans *III. Jenear Symp.*, Akademie Verlag, Berlin, p. 119.
[17] Cox, R. A. et Katchalsky, A.: 1972, *Biochem. J.* **126**, 1039–54.
[18] Katchalsky, A., Oplatka, A., et Litan, A.: 1966, dans *Symp. Soc. Gen. Physiol., Molecular Architecture in Cell Physiology* (Hayashi and Szent-Györgyi, éds.), Prentice-Hall, pp. 3–25.
[19] Neumann, E. et Katchalsky, A.: 1970, *Ber. Bunsenges. Phys. Chem.* **74**, 868–79.
[20] Neumann, E. et Katchalsky, A.: 1972, *Intern. J. Neurosci.* **3**, 175–82.
[21] Neumann, E. et Katchalsky, A.: 1972, *Proc. Nat. Acad. Sci. US* **69**, 993–97.
[22] Wells, R. D., Larson, J. E., Grant, R. C., Shortle, B. E., et Cantor, C. R.: 1970, *J. Mol. Biol.* **54**, 465–97.
[23] Szilard, L.: 1929, *Z. Phys.* **53**, 840.
[24] Brillouin, L.: 1962, *Science and Information Theory*, Academic Press, New York, pp. 1–4, 119–20, 134, 152, 287, 289.
[25] Thiele, D., Sarocchi, M.-Th., Marck, C., Lezius, A., et Guschlbauer, W.: en préparation.

DISCUSSION

Ptak: Etes-vous sûr qu'en effectuant plusieurs fois de larges variations de pH, la force ionique reste rigoureusement constante?

Guschlbauer: A l'exactitude près de la mesure, oui, car nous travaillons dans 0,15 M NaCl. A plus basse force ionique, il faudrait faire des corrections, évidemment.

Wurmser: Pouvez-vous préciser dans quelle intention vous faites usage d'unités d'information plutôt que d'unités d'entropie physique?

Guschlbauer: J'utilise les unités d'information et de thermodynamique en parallèle. Un système binaire contient une unité d'information (bit) par sélection qui correspond justement à $\sim 1{,}4$ cal (mole K)$^{-1}$ en unités thermodynamiques. Nous essayons d'exprimer un système moléculaire binaire en termes thermodynamiques.

Daune: L'un des critères d'hystérèse est de pouvoir boucler des cycles plus petits à l'intérieur du cycle principal. Est-ce que ceci a été vérifié?

Guschlbauer: Oui. Une étude détaillée est en cours et sera publiée prochainement.

Bourgoin: La surface du cycle d'hystérèse dépend-elle du temps de séjour du système soit à pH très bas, soit à pH très élevé?

Guschlbauer: Non, pas à 20°. A température élevée ($\sim 60°$) on trouve des cinétiques lentes que nous n'avons pas encore étudiées en détail.

Spach: Dans le cas de l'agrégation de l'acide polyglutamique (T.M. Schuster, B. R. Jennings et G. Spach, *11ème Symposium Nobel*, Almqvist et Wicksell, Stockholm, 1968, p. 213) nous avions attribué, sur la base des théories d'Everett, l'existence du phénomène d'hystérèse à la formation de petits domaines (agrégats de quelques molécules) légèrement différents entre eux par une certaine propriété (la plus simple nous paraissant être la polydispersité). Y a-t-il, dans votre cas, une influence de l'origine des échantillons (ex. poids moléculaire) sur la forme des courbes d'hystérèse? Quelle est votre interprétation au niveau moléculaire du phénomène que vous observez?

Guschlbauer: Il est actuellement difficile de répondre à votre première question. Pour le cas du poly (I).poly (C) nous n'avons eu aucun problème avec des échantillons fournis par cinq ou six laboratoires différents. Ces polymères sont toujours assez longs (> 100000 PM) et le traitement préalable (dialyse sous choc osmotique, etc.) élimine toutes les traces de petits oligomères. Pour le poly (dG).poly (dC), nous avons, après six ans d'attente, finalement trouvé un lot convenable. Ce lot, après le traitement décrit ci-dessus, ne change pas de comportement. Il est assez difficile de déterminer le PM d'une structure bicaténaire et je vois mal comment préciser cette question. Même après séparation des brins dans CsCl alcalin les difficultés persistent car le poly (dG) s'auto-associe dans des structures à plusieurs brins.

En ce qui concerne votre deuxième question, nous avons des idées assez précises pour le poly (I). .poly (C), qui a fait l'objet de plusieurs publications [7, 8, 11]. La forme acide est probablement une paire Hoogsteen protonée. La forme intermédiaire poly(I).poly(C).poly(C^-) est un complexe à trois brins avec une paire Watson–Crick et une Hoogsteen. On peut évidemment envisager une solution analogue pour le poly (dG).poly (dC) et nous nous efforçons d'en fournir la preuve. Nous n'avons pas vu d'indication d'agrégation pour les deux systèmes.

ÉTUDE DE LA TRANSCONFORMATION THERMIQUE DU Na–DNA EN SOLUTION PAR LA SPECTROGRAPHIE DE BRUIT

DANE VASILESCU, MAURICE TEBOUL, HUBERT KRANCK et BERNARD CAMOUS

Laboratoire de Biophysique, U.E.R. Domaine Méditerranéen, Université de Nice, 06034, Nice Cédex, France

Résumé. On décrit un dispositif construit au laboratoire, à l'aide duquel il est possible de mesurer la résistance équivalente de bruit R_b d'un diélectrique liquide conducteur.

On montre que, pour des solutions de DNA dans NaCl, la mesure de R_b renseigne sur l'atmosphère ionique et que, dans un diagramme du type Arrhenius, il existe une cassure correspondant à la transition hélice → pelote.

Summary. Measuring the equivalent noise resistance R_b of Na–DNA solutions in NaCl provides information about the free ion atmosphere. The noise spectrograph, which was built in our laboratory, measures the spectral density $\langle e_\nu^2 \rangle$ of the noise voltage of liquid samples of impedance Z. This quantity, which is the mean quadratic value of the fluctuation voltage $e(t)$ for a frequency ν, is connected to the equivalent noise resistance of the sample by the Nyquist–Johnson relation:

$$\langle e_\nu^2 \rangle = 4kTR_b,$$

where k is Boltzmann's constant, T is the absolute temperature and R_b is the real part of Z.

In the absence of an electric field, the statistical fluctuations in the ion atmosphere of a solution of Na–DNA in NaCl are due, primarily, to the brownian motion of the Na^+ and Cl^- ions and of the water molecules. The fluctuations of DNA macromolecules, of very large mass, result in mean quadratic speeds far below of those of the free ions. Within the range of NaCl concentrations investigated (10^{-4} M to 5×10^{-3} M), the ions form a plasma and can be considered in the same way as the molecules of a perfect gas, providing the Coulomb electrostatic energy is very much less than their kinetic energy.

The equivalent noise resistance R_b of DNA solutions, measured to within 5%, remained constant, at fixed temperatures, when the analysis frequency varied in the range 200 Hz – 50 kHz.

The values of $R_b(T)$, obtained at fixed analysis frequency, are plotted in an Arrhenius type diagram ($\log(R_b/T)$ versus $1/T$). These graphs show breaks in the case of native DNA and are characteristic of a phase transition (thermal transconformation helix → coil).

Finally, the calculation of the number of free Na^+ ions in the solution, as a function of the temperature, reveals once more the process of ejection of compensating Na^+ ions from the macromolecules during the thermal transconformation.

1. Introduction

L'éjection d'ions compensateurs Na^+ hors des macromolécules de Na–DNA, lors de la transconformation thermique,* a été mise en évidence par des mesures diélectriques en haute fréquence [1], et confirmée par la potentiométrie à l'aide d'électrodes sélectives au cation Na^+ [2, 3]. Ce phénomène est d'ailleurs prévu par un modèle théorique de Daune [4]. Nous allons décrire une méthode s'appuyant sur la technique du bruit, permettant de percevoir cette thermodénaturation du DNA.

En l'absence de champ électrique, les fluctuations statistiques de l'atmosphère

* Nous entendons, par transconformation thermique, le changement de conformation subi par le DNA dans la transition hélice → pelote induite par chauffage.

C. Sadron (ed.), Dynamic Aspects of Conformation Changes in Biological Macromolecules. 411–420.

ionique d'une solution de Na–DNA dans NaCl sont dues, essentiellement, à l'agitation brownienne des ions Na^+, Cl^- et des molécules d'eau. Les fluctuations des macromolécules de DNA, de masse molaire très grande, donnent lieu à des vitesses quadratiques moyennes qui sont très inférieures à celles des ions libres.

Dans la gamme de concentrations en NaCl étudiée (10^{-4} M à 5×10^{-3} M), les ions forment un plasma et peuvent être considérés comme les molécules d'un gaz parfait, l'énergie électrostatique coulombienne étant très inférieure à leur énergie cinétique [5]. Aux bornes d'un condensateur contenant un tel plasma, le mouvement brownien des ions Na^+ et Cl^- crée une tension électrique fluctuant avec le temps $U(t)$, correspondant au bruit thermique de la solution et telle que:

$$\langle U_\nu^2 \rangle = 4kTR_e(Z) \quad \text{(critère de Nyquist–Johnson)} \quad [6, 7],$$

où $\langle U_\nu^2 \rangle$ est la densité spectrale de bruit ou tension quadratique moyenne, k est la constante de Boltzmann, T est la température absolue, $R_e(Z)$ est la partie réelle de l'impédance complexe Z de l'échantillon, ν est la fréquence d'analyse du bruit. Dans ce qui suit, $R_e(Z)$, résistance 'équivalente de bruit', sera représentée par R_b.

C'est la mesure de ce bruit thermique, en fonction de la température, qui conduit à la mise en évidence de la transition hélice→pelote.

2. Dispositif experimental

Le principe du spectrographe de bruit construit en notre laboratoire est le suivant:

Le bruit thermique de la solution à étudier est comparé à celui fourni par un générateur de bruit blanc. Pour cela, la tension de bruit de l'échantillon est amplifiée dans une grande gamme de fréquences, puis filtrée à une fréquence ν, avec une largeur de bande $\Delta\nu$, et enfin détectée quadratiquement et moyennée.

La chaîne complète de mesures [8–10], schématisée sur la Figure 1, comprend:

(a) Une cellule de mesure [1] contenant la solution de DNA à étudier. Le liquide constitue le diélectrique d'un condensateur cylindrique dont les armatures sont en or massif. La cellule contient 7 cm^3 de solution et possède une capacité à vide de 1 pF; l'isolement électrique est réalisé en téflon. Un liquide issu d'un thermostat assure, en circulant dans une jaquette, une régulation de température de 0,05°C, pour une gamme allant de -30°C à $+100$°C.

(b) Une enceinte soumise à un vide primaire, enfermant 28 résistances étalons, de valeurs comprises entre 100 Ω et 20 MΩ, joue le rôle de générateur de bruit blanc. Cette enceinte est thermorégulée de manière analogue à la cellule de mesure.

(c) Un amplificateur de tensions alternatives, à faible bruit propre, antimicrophonique, à grand gain (130 dB), et à large bande passante (20 Hz – 130 kHz).

(d) Un analyseur harmonique, permettant d'isoler une bande $\Delta\nu$ de 3 ou 30 Hz autour d'une fréquence ν comprise, en pratique, entre 200 Hz et 50 kHz. Un amplificateur sélectif peut remplacer l'analyseur entre 10 Hz et 20 kHz, la largeur de bande $\Delta\nu$ variant, dans ce cas, avec la fréquence centrale ν.

(e) Un détecteur quadratique, constitué essentiellement par un thermocouple

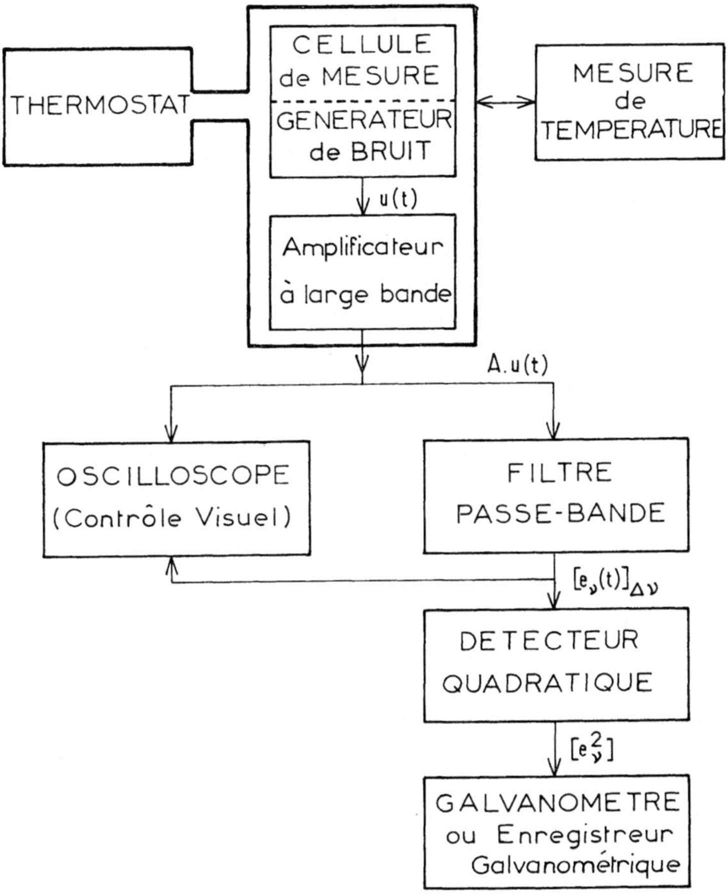

Fig. 1. Schéma de principe du spectrographe de bruit. A est le gain de l'amplificateur à large bande.

fournissant, à la sortie, un signal proportionnel au carré de l'amplitude du signal d'entrée.

(f) Un galvanomètre de grande sensibilité, $1{,}4 \times 10^{-9}$ A/mm, possédant un cadre apériodique de grande constante de temps $\tau = 3$ sec, ou bien un enregistreur galvanométrique de sensibilité 10^{-9} A/mm. La lecture finale, moyennée par l'observateur, correspond à la densité spectrale de bruit $\langle e_\nu^2 \rangle$.

3. Principe des mesures

La mesure de la résistance équivalente de bruit R_b de l'échantillon se fait par comparaison, grâce au générateur de bruit blanc étalonné. La comparaison s'effectue par interpolation linéaire, au moyen d'un ordinateur [10].

La résistance de bruit R_b ainsi déterminée est, en fait, la résistance-série R_s du condensateur constitué par la cellule de mesure; toutefois, dans les conditions de nos mesures (analyse basse fréquence, capacité à vide de la cellule de l'ordre de 1 pF), R_s peut être confondue avec la résistance parallèle R_p de l'échantillon.

4. Résultats expérimentaux

4.1. Matériaux utilisés

Nous avons opéré sur du DNA sous forme de sel de sodium extrait, soit du thymus de veau (échantillons notés V7, V320, V378, V464), soit de l'érythrocyte de poulet (P295). Les solutions ont été préparées à partir de fibres de DNA à 0°C, par dissolution progressive dans NaCl 'supra pur' pour analyses (Merck). L'eau employée a été obtenue par bi-distillation sur quartz, et présente une conductivité inférieure à $7 \times 10^{-4}\ \Omega^{-1}\ \text{m}^{-1}$.

Les concentrations de DNA utilisées se trouvent dans la gamme allant de 0,4 à 1,0 g/l, dans du NaCl de concentration de l'ordre de 10^{-3} M.

Le pH des solutions se situe entre 6,5 et 7, et varie peu au cours d'une thermodénaturation.

Le DNA, provenant de ces deux sources, présente un rapport en bases puriques et pyrimidiques sensiblement du même ordre:

$$(A+T)/(G+C) \simeq 1{,}3.$$

Ces solutions de Na–DNA ont été contrôlées par spectrophotométrie et par polarimétrie (polarimètre électronique digital Jouan). L'effet hyperchrome à 2600 Å, après dénaturation thermique à 100°C, a toujours été supérieur à 30%. Le pouvoir rotatoire spécifique, mesuré à 20°C, pour la raie D du sodium, est $+125° < (\alpha)_D^{20°C} < 150°$, ce qui est un indice de bonne conformation hélicoïdale [11].

4.2. Cinétique thermique

Les mesures peuvent être effectuées en cinétique continue, la variation de température étant de l'ordre de 0,6°C/mn, ou en cinétique semi-continue. Dans ce dernier cas, on élève la température de l'échantillon de 3 à 4°C, puis on stabilise la température pendant le temps des mesures. Les enregistrements sont obtenus à l'aide de la cinétique continue. Nous avons constaté que les transitions thermiques, que l'on observe sur le DNA, étaient plus nettes en cinétique semi-continue, l'équilibre statistique étant, sans doute mieux atteint dans ces conditions.

4.3. Comportement du solvant NaCl

Nous avons, tout d'abord, étudié la variation de la résistance-équivalente de bruit R_b de solutions aqueuses de chlorure de sodium (NaCl), en fonction de la température T, à différentes concentrations. Les courbes $R_b(T)$ présentent une décroissance d'allure hyperbolique lorsque T croît. Dans le domaine de températures étudiées (15–80°C), on constate (Figure 2) que R_b suit une loi du type Arrhenius:

$$R_b = R_0\,(T/T_0)\exp(E_G/kT),$$

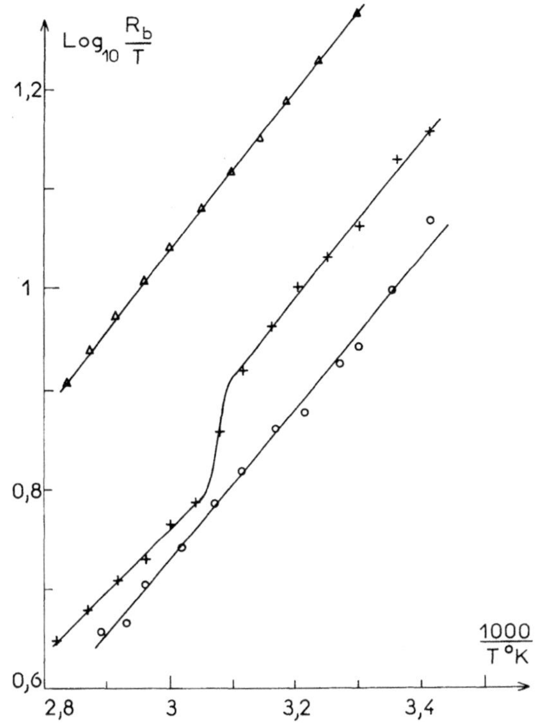

Fig. 2. Variation de la résistance équivalente de bruit, dans un diagramme du type Arrhenius. △ Solution de NaCl 10^{-3} M; + solution de DNA natif; ○ solution de DNA dénaturé thermiquement à 100° C, puis stabilisé à la température de la glace fondante. La solution de DNA (échantillon P295) possède une concentration de 0,9 g/l dans NaCl 10^{-3} M.

où E_G est l'énergie d'activation du processus et R_0 la valeur de R_b à haute température T_0.

Par analogie avec la grandeur 'conductivité électrique', que l'on mesure en appliquant un champ électrique, nous définirons une 'conductivité spécifique équivalente de bruit' σ_b selon:

$$\sigma_b = C R_b^{-1},$$

où σ_b est en $\Omega^{-1}\ \mathrm{m}^{-1}$, R_b en Ω, et C une constante géométrique. (Pour notre cellule $C = 77{,}2\ \mathrm{m}^{-1}$.) En poursuivant cette analogie, il est possible de calculer la mobilité moyenne de bruit $\langle\mu_b\rangle$ des ions de la solution; cela suppose que nous puissions étendre la notion de mobilité μ, définie quand un échantillon est soumis à un champ électrique, au cas du mouvement brownien, en l'absence de champ, d'où:

$$\sigma_b \simeq (n_{\mathrm{Na}^+} + n_{\mathrm{Cl}^-})\, e \langle\mu_b\rangle,$$

et, en tenant compte de $n_{\mathrm{Na}^+} = n_{\mathrm{Cl}^-}$, l'électrolyte étant considéré comme entièrement dissocié, nous poserons:

$$\sigma_b \simeq \langle n\rangle\, e \langle\mu_b\rangle,$$

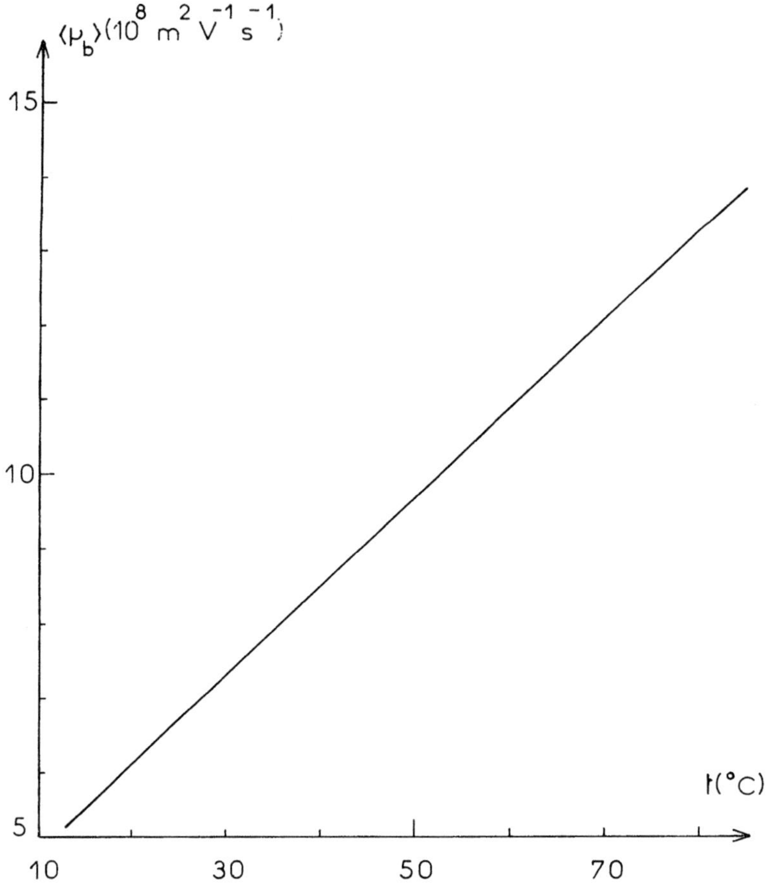

Fig. 3. Variation de la mobilité équivalente de bruit du NaCl, en fonction de la température, à 10 kHz. (La courbe donnée est la moyenne obtenue à partir de diverses concentrations en NaCl. Les valeurs sont constantes pour des fréquences comprises entre 100 Hz et 20 kHz.)

où e = charge de l'électron en coulombs, n_{Na^+} = nombre d'ions Na^+ libres dans la solution, n_{Cl^-} = nombre d'ions Cl^- libres dans la solution, $\langle\mu_b\rangle$ = mobilité moyenne de bruit, en $\mathrm{m}^2\ \mathrm{V}^{-1}\ \mathrm{sec}^{-1}$, $\langle n\rangle = n_{\mathrm{Na}^+} + n_{\mathrm{Cl}^-}$ = nombre moyen d'ions libres dans la solution de NaCl.

Dans la gamme de températures étudiées (15–80°C), la mobilité moyenne de bruit $\langle\mu_b\rangle$, analysée à une fréquence de 10 kHz, varie à peu près linéairement avec la température (Figure 3). Cette droite est une partie d'une courbe exponentielle.

A partir des valeurs de $\langle\mu_b\rangle$ ainsi déterminées, et des mesures de R_b, donc de σ_b, on déduit le nombre moyen d'ions libres $\langle n\rangle$ dans une solution de NaCl, selon

$$\langle n\rangle = n_{\mathrm{Na}^+} + n_{\mathrm{Cl}^-} = C/e\,\langle\mu_b\rangle\, R_b\,.$$

Cela suppose que la conductivité σ_b est essentiellement due au mouvement des ions Na^+ et Cl^-. On peut constater, sur la Figure 4, que, pour une solution de NaCl 10^{-3} M, $\langle n\rangle$ mesuré est bien constant, lorsque la température croît.

4.4. Etude de la transconformation thermique du Na–NDA

De précédentes mesures [8, 12] ont montré que la résistance équivalente de bruit R_b de solutions de DNA, mesurée à 5% près, était constante, à température fixe, lorsque la fréquence d'analyse ν variait dans le domaine 200 Hz – 50 kHz. Nous pouvons donc fixer la fréquence, arbitrairement, entre ces limites.

Les valeurs de R_b, obtenues par chauffage des solutions de DNA en cinétique semi-continue, sont reportées dans un diagramme de type Arrhenius (Figure 2). Les cassures observées, pour le DNA natif, sont caractéristiques d'une transition de phase [13].

Les énergies d'activation, correspondant aux droites des régions précédant la température de fusion (T_m), et suivant le T_m, sont, en première approximation, égales à la valeur moyenne de l'énergie d'activation des solutions de NaCl de conductivité correspondante.

5. Interprétation

5.1. Calcul du taux d'éjection de contre-ions Na^+ lors de la thermodénaturation du DNA

La variation de $\langle n \rangle$ en fonction de la température, pour une solution de DNA natif,

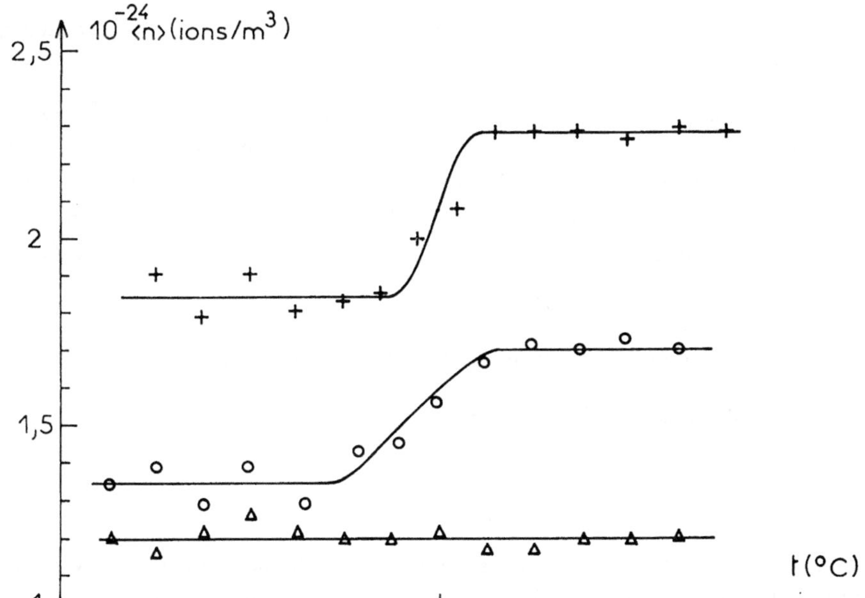

Fig. 4. Variation du nombre $\langle n \rangle$ d'ions libres dans des solutions de NaCl et de DNA, en fonction de la température. △ NaCl 10^{-3} M; ○ solution de DNA (échantillon V7) à 0,61 g/l dans NaCl 10^{-3} M; + solution de DNA (échantillon P295) à 0,9 g/l dans NaCl 10^{-3} M.

montre clairement l'existence de deux plateaux horizontaux correspondant à un nombre $\langle n \rangle$ constant, avant et après la transformation thermique (voir Figure 4). La différence de niveaux $\Delta\langle n \rangle$ entre les deux paliers corrspond à une augmentation du nombre d'ions libres dans la solution, après franchissement du T_m. Si l'on exclut le rôle des co-ions (Cl^-) dans ce processus [3], cette augmentation correspond à une éjection de contre-ions Na^+ hors des sites phosphates du DNA. Dès lors $\Delta\langle n \rangle = = \Delta\langle n \rangle_{Na^+}$.

Pour les courbes de la Figure 4, nous obtenons:

$$\Delta\langle n \rangle_{Na^+} = 0{,}363 \times 10^{24} \text{ ions/m}^3, \text{ pour le DNA V7 0,61 g/l}$$
$$\Delta\langle n \rangle_{Na^+} = 0{,}436 \times 10^{24} \text{ ions/m}^3, \text{ pour le DNA P295 0,9 g/l.}$$

Le pourcentage α d'ions Na^+ éjectés, lors de la transconformation thermique, sera donné par [1]:

$$\alpha = 100\, \Delta c_{Na^+}/c_i,$$

où c_i = concentration molaire en sites phosphates = $m/331$ (m = concentration du Na–DNA en g/l), et $\Delta c_{Na^+} = \Delta\langle n \rangle_{Na^+}/\mathcal{N}$ ($\mathcal{N}$ = nombre d'Avogadro.)

Dans l'exemple choisi sur la Figure 4, on obtient: $\alpha = 33\%$, pour le DNA V7, et $\alpha = 27{,}5\%$, pour le DNA P295.

Ces résultats sont en accord avec ceux obtenus par potentiométrie à l'aide d'électrodes spécifiques au Na^+ [2], et par des mesures diélectriques [1].

5.2. Remarque sur les énergies d'activation mesurables dans un diagramme du type Arrhenius

Egelstaff [14], dans un modèle phénoménologique, suppose que la diffusion d'une particule au sein d'un liquide s'opère par saut entre deux sites séparés par une barrière d'énergie W. L'énergie d'activation E_G du processus tient compte, simultanément, de la barrière de potentiel W entre deux sites, et de la probabilité de trouver un site libre à côté d'un site occupé. La constante de diffusion D suit une loi d'Arrhenius pure:

$$D = D_0 \exp(-E_G/kT).$$

En présence d'un champ électrique, la conductivité spécifique σ s'écrit:

$$\sigma = \sigma_0 (T_0/T) \exp(-E_G/kT),$$

et la résistance ohmique:

$$R = \sigma^{-1} = R_0 (T/T_0) \exp(E_G/kT).$$

Le comportement observé pour la résistance équivalente de bruit de nos solutions est donc analogue à celui de la résistance ohmique.

L'énergie d'activation moyenne de bruit, obtenue à partir de diverses concentrations en NaCl, est de l'ordre de 0,160 eV. Elle est légèrement inférieure à celle calculée à partir des mesures de résistivité ohmique de Gunning et Gordon [15], soit 0,180 eV.

Dans le cas de solutions de Na–DNA, les deux segments de droite observés (Figure

2) conduisent à des énergies d'activation du même ordre de grandeur. La moyenne effectuée sur toutes nos manipulations a donné:

$$\langle E_G \rangle = 0{,}155 \text{ eV}.$$

6. Conclusion

Par la spectrographie de bruit en basses fréquences, nous avons mis en évidence le phénomène de transconformation thermique du DNA; nous avons calculé le taux d'éjection des ions sodium hors des macromolécules, lors de la dénaturation thermique, et certaines énergies d'activation liées au processus.

Nos résultats sont en accord avec ceux obtenus par les méthodes diélectriques en hautes fréquences, et par les méthodes potentiométriques à l'aide d'électrodes spécifiques aux ions Na^+.

Il apparaît ainsi que le domaine des basses fréquences peut être exploré utilement par les méthodes de bruit, l'échantillon n'étant soumis à aucun champ électrique, ce qui supprime, en grande partie, la polarisation d'électrodes.

Remerciements

Ce travail a été supporté, en partie, par la R.C.P. du C.N.R.S. no. 196 (Interactions cations métalliques et molécules biologiques) et la R.C.P. no. 195 (Molécules et Macromolécules modèles).

Nous adressons nos plus vifs remerciements au Docteur G. Aubel-Sadron, Maître de Recherches au C.N.R.S., du Centre de Biophysique Moléculaire d'Orléans, et à Monsieur le Professeur M. Daune, du Centre de Recherches sur les Macromolécules de Strasbourg, qui nous ont fourni aimablement les échantillons de DNA.

Références

[1] Grassi, H. et Vasilescu, D.: 1971, *Biopolymers* **10**, 1543.
[2] Vasilescu, D. et Rix, M. A.: 1970, *Biochim. Biophys. Acta* **199**, 553.
[3] Vasilescu, D., Rix, M. A., et Grassi, H.: 1971, *C. R. Acad Sci. Paris* **D273**, 1154.
[4] Daune, M.: 1969, *Biopolymers* **7**, 659.
[5] Landau, L. et Lifchitz, E.: 1967, *Physique Statistique*, Mir. Moscou, p. 349.
[6] Johnson, J. B.: 1928, *Phys. Rev.* **32**, 97.
[7] Nyquist, H.: 1928, *Phys. Rev.* **32**, 110.
[8] Viani, R.: 1969, Thèse de spécialité, Nice.
[9] Kranck, H.: 1971, Thèse d'Université, Nice.
[10] Camous, B.: 1972, Thèse de spécialité, Marseille.
[11] Ts'o, P. O. P., Helmkamp, G., et Sander, G.: 1962, *Biochim. Biophys. Acta* **55**, 584.
[12] Vasilescu, D., Teboul, M., Viani, R., et Grassi, H.: 1968, *C. R. Acad. Sci. Paris* **B266**, 1005.
[13] Kumamoto, J., Raison, J., et Lyons, J.: 1971, *J. Theor. Biol.* **31**, 47.
[14] Egelstaff, P. A.: 1967, *An Introduction to the Liquid State*, Academic Press, New York, p. 151.
[15] Gunning, H. E. et Gordon, A. R.: 1941, *J. Chem. Phys.* **10**, 126–31.

DISCUSSION

Bourgoin: Les fluctuations du degré d'association des contre-ions n'ajoutent-elles pas un terme non négligeable dans l'expression du facteur de bruit?

Vasilescu: Il est possible qu'un tel facteur intervienne. Cependant sa contribution au bruit thermique total est négligeable eu égard aux incertitudes expérimentales ($\Delta R_b/R_b \simeq 3$ à 4%). Il existe un article de Jackson et Corriel (*J. Chem. Phys.* **40** (1964) 1 – 60) dans lequel on analyse le blocage des trajectoires du mouvement brownien des contre-ions au niveau des sites ionisables du DNA.

Bourgoin: Les fluctuations de densité des ions ne créent-elles pas localement des champs électriques importants au niveau des macromolécules?

Vasilescu: Une analyse de ces phénomènes a été faite récemment par Osawa (*Biopolymers* **9** (1970) 677).

Hanss: Vos observations ne mettent pas en évidence la relaxation diélectrique basse fréquence que nous avons observée.

Vasilescu: Un phénomène de relaxation diélectrique n'est observable que si l'accroissement de conductivité aux basses fréquences est suffisamment important pour être décelé à l'aide de notre technique. D'autre part, le diélectrique doit être dipolaire intrinsèquement, c'est-à-dire que l'on doit exclure le cas où des moments dipolaires sont induits exclusivement par un champ électrique externe.

Alpert: Quel est le temps pour faire une mesure de σ_b? Peut-on envisager une méthode de corrélation qui permettrait d'étendre ces mesures aux expériences faites avec saut de température?

Vasilescu: Actuellement la mesure de σ_b est effectuée par analyse harmonique classique. L'emploi d'un analyseur harmonique ne permet que l'étude des phénomènes stationnaires. Par contre, si l'on remplace l'analyseur harmonique par un corrélateur, il est possible de mesurer les fonctions d'autocorrélation (analogues de la densité spectrale) en temps réel. En ce cas, des cinétiques sont observables à condition d'être relativement lentes (quelques minutes ou quelques dizaines de minutes).

Zana: Pourriez-vous indiquer la fréquence maximale pour laquelle les mesures sont encore possibles ainsi que la quantité de liquide nécessaire?

Vasilescu: Cela dépend de l'amplificateur utilisé ainsi que de l'analyseur. Des mesures de bruit sont actuellement possibles dans le domaine des hautes fréquences (quelques dizaines de MHz). En ce qui concerne la quantité de liquide nécessaire, avec la cellule que nous utilisions, il faut 6 à 7 cm^3. Avec cette géométrie nous obtenons des résistances équivalentes de bruit de l'ordre de 3000 ou 4000Ω pour des solutions de NaCl 10^{-3} M. A priori on peut réduire les dimensions de la cellule afin de travailler avec de faibles volumes de liquide, à condition de trouver un compromis entre les valeurs de R_b mesurables et la géométrie de la cellule.

Frédéricq: Pensez-vous que l'on pourrait utiliser votre méthode pour déterminer la quantité de petits ions libérés lors de la combinaison du DNA avec les histones?

Vasilescu: Oui, à condition que le nombre d'ions libérés soit suffisamment grand pour produire un incrément de R_b mesurable. Dans le cas de l'interaction DNA–histones, cela semble très possible.

RECENT DEVELOPMENTS IN THE CONSISTENT FORCE FIELD CALCULATIONS

SHNEIOR LIFSON

Department of Chemical Physics, Weizmann Institute of Science, Rehovot, Israel

Summary. The general ideas underlying the consistent force field (CFF) method are described, and some new advances are briefly reviewed. The relations of the CFF to conventional empirical energy calculations and to quantum-chemical studies of large molecules are discussed.

Résumé. Nous présentons en premier lieu les idées générales constituant la base de la méthode CFF ainsi que quelques-uns de ses plus récents développements. Nous discutons ensuite des relations entre cette méthode, les procédés classiques du calcul empirique de l'énergie et les méthodes de chimie quantique en ce qui concerne leur application aux grosses molécules.

The dynamics of conformational changes is controlled by the energy as a function of the molecular geometry around and away from the molecular equilibrium conformations. Transitions from one local minimum to another occur across potential 'saddle points', where the energy is a minimum with respect to most coordinates, but a maximum with respect to some of them. The most probable path between adjacent minima passes through the saddle point of lowest energy, which constitutes the potential barrier for such transitions. When an enzyme interacts with a substrate, the dynamics of conformational change will similarly depend on the molecular energy surface, as modified by the enzyme–substrate interactions. Therefore the study of the molecular energy as a function of the molecular geometry or conformation is relevant to this conference.

The problem of evaluating the energy surface of biological macromolecules may be approached either by quantum-mechanical methods or by empirical methods, and both methods have been amply described at this conference [1, 2]. Here we shall analyze in some detail the empirical method, and show how it can be perfected gradually to a level where it may describe faithfully the energy of molecules, including biomolecules, as a function of their detailed geometry. The method by which such a systematic improvement of the choice of empirical functions is effected has been coined by our group as the 'consistent force field' (CFF) method; a few recent advances, by the CFF method, in the study of alkenes and of the NH... OC hydrogen bond, will be briefly reviewed. Finally, a rationale for our preference of the CFF method over the quantum-mechanical approach, will be presented.

The most critical steps in the empirical method are the choice of the empirical energy functions, and the determination of their parameters. (We shall use the term *parameters* for the various adaptable constants of the empirical energy functions. Molecular internal coordinates, as well as atomic coordinates, will be termed as *variables*.) One needs functions for bond stretching, bond bending, torsional rotations, repulsive and attractive nonbonded interactions between atoms, electrostatic interactions between localized partial charges, and so on. Then one has to determine the

C. Sadron (ed.), Dynamic Aspects of Conformation Changes in Biological Macromolecules. 421–429.

parameters of these functions for different types of atoms, bonds, angles, etc. Since this is a formidable task, one tends naturally to reduce the number of functions and parameters to the essential minimum.

The information used for the choice of the energy functions and the determination of their parameters is varied and heterogeneous. In vibrational spectroscopy, the theoretical analysis yields force constants and interaction- or cross-terms. The various quadratic potentials of empirical calculations use the spectroscopic force constants as their energy parameters. Other major sources are X-ray, electron and other diffraction methods which yield the geometric parameters, which are often used as the reference bond lengths, bond angles or Van der Waals radii. Measurements of virial coefficients, dipole moments, refractive indices, as well as various thermodynamics properties are used as supplementary information. Finally, theoretical derivations are involved, such as the London theory of the dispersion force and the Slater–Kirkwood approximation for the constants of this force.

There are several weak links in the elaborate procedure of assembling an empirical force field. It is important to recognize them in order to evaluate the merits and limitations of the empirical method, and in order to devise ways and means for its improvement:

(1) The compilation of empirical and semi-empirical energy functions and their parameters is eclectic by its very nature. This raises a crucial question of consistency of the various components of the empirical force field. Indeed it can easily be seen where inconsistencies do arise. The various experimental disciplines from which functions and their parameters are taken may often use different estimates for the same molecular forces or introduce different approximations in the analysis of the data. Thus when some of the force constants, taken from vibrational spectroscopy, are assembled together with nonbonded interactions taken from thermodynamic or other sources, some parts of the total energy may be counted twice while others may be neglected. Consequently, when the empirical force field (set of empirical functions) is applied to recalculate the various observable quantities from which it was assembled, one finds, as a rule, that the theoretical values generally disagree with their corresponding data. One may ignore this inconvenient fact as irrelevant, arguing that a force field assembled, say, to calculate conformations of proteins is not supposed to give correct vibrational spectra of other molecules. However, what is lost on the way is the reliability of the empirical method. Without a reliable check on the choice of energy functions and their parameters, different authors are inclined to use functions that differ widely, as was pointed out clearly by Williams, Stang and Schleyer in their review on conformational analysis [3].

(2) The various force constants and cross-terms of spectroscopic analysis are not independent of each other. For example, the Urey–Bradley force field and the valence field yield quite different values for some of the force constants [4], because they treat the cross-terms differently. Similarly, various force constants are modified when more cross terms are added to a force field. The general practice of using the force constants

while neglecting the cross-terms, when an empirical force field is assembled, is therefore a doubtful procedure. An empirical force field thus constructed would give very bad agreement with experimental spectroscopic data, although the original spectroscopic force field, on which it is partly based, might have been very good in reproducing those data. Similar inconsistencies may occur when force constants are transferred from different sources for different atoms.

(3) Conformational and thermodynamic data are generally insufficient to determine the analytic form of the nonbonded potentials, nor do they yield unequivocally the numerical values of the constants of these functions. Thus, within a wide range of choice of the strength of the attractive part of H... H or C... C interactions, one can always choose a repulsive potential and fit its parameters to give, say, the sum of Van der Waals radii as that distance for which the nonbonded interaction potential is at minimum. This ambiguity is the source of much confusion in the literature. Actually, even this latter assumption that the sum of Van der Waals radii in crystals correspond to the minimum of atom–atom interactions is rather crude [5]. An equilibrium structure of, say, a crystal of some aliphatic substance, may well be the result of short-range repulsions between the nearest hydrogen atoms of neighbour molecules, offset by the longer range attraction between all other pairs of atoms in the crystal.

The CFF method [6, 7] has been designed to provide methodical remedies to the weaknesses of the conventional empirical methods. Its basic assumption, based on a large body of experiment, is that the energy surface of a molecule in the sense of the Born-Oppenheimer approximation, i.e., the electronic ground state as a function of the atomic coordinates, *can* be expressed as a sum over elementary functions

$$V_{\text{total}} = \sum V_{nb} + \sum V(s_i) + \sum V(s_i, s_i') + \sum V(s_i, s_i', s_i'') + \cdots . \tag{1}$$

Here V_{nb} are various functions for nonbonded interactions, which may be assumed to depend on interatomic distances, unless more sophisticated models are shown to be advantageous; s_i are various internal coordinates, like bond lengths, bond angles, torsional angles, out-of-plane deformations of double bonds etc., and $V(s_i)$ are the various functions of single internal coordinates to be attributed to various atomic groupings; $V(s_i, s_i')$ and $V(s_i, s_i', s_i'')$ are functions of two and three adjacent internal coordinates, respectively. Great simplifications in some theoretical considerations and in the computational procedures are obtained when V_{total} and all elementary functions of which it is composed are transformed to cartesian coordinates. The reverse transformations from cartesian to internal coordinates can be performed by standard computer subroutines whenever deemed necessary.

V_{total} may be generalized to represent the total energy per unit cell of a crystal provided the term $\sum V_{nb}$ in Equation (1) is generalized to include the lattice sums over intermolecular interactions in the crystal.

The elementary functions are assumed to have the following properties:

(1) The various elementary functions are the same, and therefore transferable,

among molecules which contain the same molecular grouping to which the elementary functions are related.

(2) The sum in Equation (1) is quickly convergent, so that a limited number of functions is sufficient to represent the energy surfaces of large families of molecules to the desired approximation.

(3) The observable properties of molecules which can be calculated by using the elementary functions are so diverse and numerous that on the one hand it would be impossible to calculate correctly all of them with ill-chosen elementary functions, and on the other hand it is possible to use the abundance of experimental data to select the appropriate elementary functions and to determine their parameters.

All the above assumptions seem to agree in general with the accumulated knowledge and experience in the various fields of physical chemistry. However, the ultimate justification for their combined application in the CFF procedure will be judged by the degree of sucess of the CFF method to give a correct quantitative representation of molecular energy surfaces.

Among the various observables which depend in one way or another on V_{total} a particularly convenient role is played by those which are directly related in a simple mathematical manner to the different terms in the expansion of V_{total} in a Taylor series around the equilibrium conformation. Denote the $3n$-dimensional vector of the cartesian coordinates of the n atoms of a molecule by $\mathbf{r}$, and its equilibrium value by $\mathbf{r}_{\text{eq}}$. The Taylor expansion of V_{total} is

$$V_{\text{total}}(\mathbf{r}) = V_{\text{total}}(\mathbf{r}_{\text{eq}}) + \operatorname{grad} V \cdot \delta\mathbf{r} + \tfrac{1}{2}\delta\mathbf{r}^t \mathbf{F} \delta\mathbf{r} + \cdots, \tag{2}$$

where $\delta\mathbf{r}^t$ is the transpose of the increment $\delta\mathbf{r}$, and $\mathbf{F}$ is the matrix of second derivatives of V_{total} with respect to the components of $\mathbf{r}$, taken at $\mathbf{r}_{\text{eq}}$.

An inspection of Equation (2) indicates immediately that the second term must vanish at equilibrium, thus yielding the set of $3n$ equations

$$\operatorname{grad} V = 0, \tag{3}$$

the solution of which gives the calculated equilibrium conformation $\mathbf{r}_{\text{eq}}$.

The third term is related to the harmonic frequencies of the normal modes of molecular and lattice vibrations. The secular equation

$$(\mathbf{F} - \lambda\mathbf{M})\,\delta\mathbf{r} = 0, \tag{4}$$

where $\mathbf{M}$ is a $3n$-dimensional diagonal matrix with the atomic masses as the non-zero elements, has eigenvalues λ_α which give the harmonic frequencies ν_α ($\lambda_\alpha = 4\pi^2\nu_\alpha^2$) and eigenvectors $\delta\mathbf{r}_\alpha$ which give the cartesian coordinates of the normal vibrations. The $\delta\mathbf{r}_\alpha$ may be conveniently transformed finally to internal coordinates, where the symmetry properties of the normal modes are easily recognized. Equation (4) must have six zero eigenvalues corresponding to translations and rotations of the whole molecule. The eigenvalues and eigenvectors are functions of the equilibrium coordinates $\mathbf{r}_{\text{eq}}$. In fact, if $\mathbf{F}$ is calculated for an $\mathbf{r}$ which is not *exactly* $\mathbf{r}_{eq}$, Equation (4) does not hold, in which case the six zero eigenvalues fail to be derived and the λ_α do not yield the right frequencies. Note that this procedure of normal mode calculations is different from the conventional one in many important ways:

(1) It is particularly suitable for large computers. It requires one and the same program for any molecule, provided there is enough space in the memory of the computer for the $3n \times 3n$ matrices.

(2) It considers the 'force constants' of $\mathbf{F}$ as functions of $\mathbf{r}_{eq}$, and makes it possible to calculate the vibrational frequencies of molecules of different geometries, like the strained cyclic alkanes and the non-strained linear alkanes, with the same force field, i.e., with the same set of elementary functions.

(3) It *derives* the symmetry properties of the vibrational modes and of the equilibrium conformation rather than *assumes* them. The application of group-representation methods is not needed any more, and while its beauty is thus wasted, much labour is saved.

(4) The study of the effect of conformational change on vibrational spectra is straightforward, as the latter is given as a function of the first.

The first term of Equation (2), combined with the third term, is related to the energy of the equilibrium conformation. The thermodynamic functions like free energy, internal energy, enthalpy etc. are derived by adding to $V_{\text{total}}(\mathbf{r}_{eq})$ the corresponding contributions of the calculated vibrational frequencies to the various thermodynamic functions, which are obtained by standard statistical-mechanical formulae. For the crystalline state one adds the contribution of lattice vibrations, which can be derived from a modification of Equation (4). For the gaseous state one must add the standard contributions of translations, rotations and pressure. Thus it is possible to calculate from V_{total} heats of sublimation, strain energies, heats of hydrogenation, and various other thermodynamics properties.

As noted already, the simultaneous calculations of all the above mentioned observable quantities for a large family of molecules is not expected to produce a reasonable agreement with experiment unless the elementary functions included in Equation (1) describe correctly the various contributions to the total energy surface. To obtain the desired set of elementary functions, the CFF method uses the following algorithm:

(1) Choose the largest available set of experimental data, belonging to many different molecules of the same family, for which the same set of elementary functions are believed to hold, and covering as many different molecular and crystal properties as are available. Give preference to data whose interpretation is least ambiguous. For example, it is not sufficient to have precise measurements of vibrational spectra. The identification of fundamental vibrations in distinction from overtones and combinations, and the satisfactory assignment of the fundamental frequencies to the different classes of normal modes, are essential.

(2) Choose the smallest set of elementary functions which seem essential to calcu-

late all observable quantities for which the experimental data are well established. Consider this set as a preliminary trial set, to be accepted or modified according to the degree of agreement with experiment it produces.

(3) Optimize the parameters of the elementary functions by least-squares techniques to give the *best* possible simultaneous agreement with *all* available experimental data. The relative statistical weights attributed to the different experimental data in the least-squares procedure should be chosen such that the agreement with different kinds of experimental data will be balanced according to their experimental accuracy, reliability of interpretation, and intrinsic importance as information sources about the molecular energy. The resulting set of elementary functions with optimized parameters is what we call the *consistent force field* (CFF).

(4) Examine the calculated values of the observable properties derived by the CFF, asking do they agree with experimental data, and if not, why? Find out which are the *least satisfactory* ones. Find out which are the elementary functions which are mainly responsible for the disagreement, and try to replace these functions by better ones. The 'survival of the fittest' among the energy functions is one of the basic concepts of the CFF method, and upon its success depends the ability of the CFF method to obtain the electronic ground state of polyatomic molecules as a function of their atomic coordinates.

(5) Once the derived CFF is judged to be sufficiently reliable, one may use it to predict the results of experiments to be performed, to analyse particular situations of interest etc. Whenever new observations indicate shortcomings of the existing CFF, a new CFF should be constructed, where the set of experimental data should now comprise the old data plus the new observations.

We have applied this algorithm first to alkanes as the simplest family of organic molecules, with particular emphasis on cycloalkanes [6, 7], because the condition of ring closure imposes significant deviations of the internal coordinates from their common values observed in unstrained molecules. We then extended the CFF to include some amides and lactams [8, 9]. Since the results have already been published, only one interesting result will be mentioned here. It was found that parameter fitting of the Lennard–Jones potential in the study of single molecules in the gaseous phase and of crystals (including longer-range intermolecular interactions) produced different constants. Further analysis led us to the conclusion that the 9-6-1 potential is preferable over the 12-6-1 potential, since it gives somewhat better results, and in particular since it produces about the same values of the parameters for intra- and intermolecular interactions.

Recently an extensive work has been performed to include nonconjugated double bonds [10]. More than 300 data on vibrations, conformations and heats hydrogenations of linear and cyclic alkenes have been included in the optimization process. It was found that the requirement of consistency of the force field for open and closed chains of alkenes is very useful in examining and improving the assignments of the vibrational modes. Overcrowding around the double bond was found to cause, in decreasing order, opening of the bond angles adjacent to the double bond, twisting of the adjacent torsional angles, and out-of-plane deformation of the double bond: a number of predictions have been suggested that await experimental examinations [10].

An interesting 'computer experiment' is being conducted with respect to the nature of the NH... OC hydrogen bond and the dependence of its energy on its geometric variables [11]. We give the computer a trial potential for the hydrogen bond which is composed of (a) an electrostatic part in the form of partial charges, positive on the H of the NH bond and negative on the O of the C=O bond, with their opposite charges attached to the N and the C respectively, and (b) a potential of the Morse type, which gradually goes over to the Lennard–Jones potential as the NH... O line is bent. The parameters of these functions are fitted by least squares to the available experimental data on the unit cell parameters and heats of sublimation of NH... OC hydrogen-bonded crystals. The rest of elementary energy functions are transferred from the previously determined CFF. Preliminary results indicate that electrostatic and Lennard–Jones functions seem to be sufficient to account for the energy and its dependence on the geometry of the hydrogen bonds, while the parameters of the Morse potential are reduced to the level of insignificance. It may be noted that these conclusions are dictated by the least-squares procedure, unbiased by any preconceived assumption, and seem to be in agreement with theoretical conclusions long recommended [12] though not generally accepted.

Biopolymers are not a major source of precise experimental data for the construction of a CFF. However, a systematic assembly of elementary energy functions in a consistent force field including alkanes, alkenes, olefins, amides etc. may lead to a CFF which might be applicable to the study of enzymes and their substrates, polynucleotides, and various other systems of great importance in molecular biology. At present it is too early to foresee how the effect of water as a solvent might be accounted for, since this is a very complex problem without the solution of which even a perfect CFF might come short of making many significant predictions about the behaviour of biopolymers. However, just because of the complexity of the water problem, it would be very instructive to isolate it by having a clear quantitative picture of the nature of all other forces operating in biopolymers, a task which the CFF can perform in due course.

In conclusion, we wish to make some general remarks on the CFF in relation to quantum-mechanical methods of calculating energy surfaces of biopolymers and their simpler molecular analogues. Proponents of empirical methods often point out that quantum-mechanical calculations are lengthy and complex, and involve too many approximations, whose effect it is difficult to estimate. Such arguments have only limited validity, since quantum-mechanical methods are constantly improving in both accuracy and efficiency. In principle quantum-mechanical methods have the advantage over empirical methods, in that they are based on the generally accepted basic laws of nature, represented by the Schroedinger equations. However, the CFF as an empirical method has some indispensible advantages absent in the quantum-mechanical methods.

First, quantum-mechanical calculations, even if they were very accurate, could only give a numerical mapping of the energy surface of molecules, calculating point by point the electronic ground state for each molecular conformation **r**.

Second, quantum-mechanical calculations yield the values $V_{total}(\mathbf{r})$ as points on the Born–Oppenheimer surface, but are not suited to separate V_{total} to components originating from stretching, bending, twisting, nonbonded interactions of various types, etc. Thus much of what constitutes our qualitative physical picture of the forces acting in polyatomic molecules is disguised and can only partly be recovered by various indirect considerations.

Third, which is a consequence of the two former points, each molecule requires an independent quantum-mechanical calculation of its energy surface. There are no common, transferable results from one molecule to another, similar to the elementary empirical energy functions.

Thus even if we could very accurately scan the quantum-mechanical energy surfaces of many molecules, point by point, we would need some simple empirical interpolation formulae to help us see through the forest of numbers what are the regularities of energy changes as functions of the internal coordinates and interatomic distances. In other words we would have to extract empirical, analytic elementary energy functions from the quantum-mechanical calculations. The CFF, instead, tries to derive these functions directly from experiment.

References

[1] Pullman, B.: this volume, pp. 1–22.
[2] Ptak, M.: this volume, pp. 23–55.
[3] Williams, J. E., Stang, P. J., and Schleyer, P. v. R.: 1968, *Ann. Rev. Phys. Chem.* **19**, 531.
[4] Snyder, R. G. and Schachtschneider, J. H.: 1965, *Spectrochim. Acta* **21**, 169.
[5] Hirschfelder, J. O., Curtis, C. F., and Bird, R. B.: 1954, *Molecular Theory of Gases and Liquids*, Wiley, New York, p. 1036.
[6] Lifson, S. and Warshel, A.: 1968, *J. Chem. Phys.* **49**, 5116.
[7] Warshel, A. and Lifson, S.: 1970, *J. Chem. Phys.* **53**, 582.
[8] Warshel, A., Levitt, M., and Lifson, S.: 1970, *J. Mol. Spect.* **33**, 84.
[9] Karplus, S. and Lifson, S.: 1971, *Biopolymers* **10**, 1973.
[10] Ermer, O. and Lifson, S.: *J. Am. Chem. Soc.*, in press.
[11] Hagler, A., Huler, E., and Lifson, S.: to be published.
[12] Ref. [5], p. 989.

DISCUSSION

Magat: (1) Avez-vous tenu compte de la polarisation mutuelle? Ce n'est pas une grande contribution, mais elle n'est pas négligeable. (2) N'est-il pas possible d'utiliser un potentiel d'interaction dû à la liaison H comme celui développé par Ben Naim et Stillinger pour le cas de l'eau? (3) L'exposant 9 pour la répulsion H↔H a déjà été proposé vers 1931 par Wohl (*Z. Phys. Chem.*).

Lifson: It is very probable that the introduction of a polarization tensor for each bond would improve agreement with a variety of experiments. Indeed we found that atom–atom potentials are insufficient to calculate satisfactorily the unit-cell parameters of crystals of alkanes. However, this has not been done yet, mainly for computational difficulties, and the many new parameters this would require.

Ptak: In the hydrogen bond, what are the values of the electrical charges on O and N atoms in this monopole approximation?

Lifson: About a quarter of an electron charge (with positive sign) on the H bonded to N, and about 0,4 for the O. These numbers are, however, preliminary, since we are still continuing the least-square calculations.

De Coen: Pour apporter une information à la question posée par M. Magat, je voudrais signaler que nous avons effectué des calculs de l'énergie de polarisation dans le cas de l'unité dipeptique. Dans ce cas il apparaît que cette contribution apporte seulement des changements mineurs à l'allure de la surface d'énergie et comme elle double le temps de calcul à cause de la non-additivité des différents termes, nous ne l'avons pas retenue dans la suite de nos calculs. Cependant, il serait erroné de conclure que cette contribution est toujours négligeable et il est clair que, dans le cas où des groupes aromatiques sont en interaction avec des charges, elle devient certainement très importante.

Karplus: In answering the question of the correct exponent for nonbonded repulsions (i.e. in a term of the form NR^n, where R is the atom–atom distance), one has to take account of the fact that there is no single 'correct' exponent n. Because the actual potential is not well represented by such a simple functional form over its entire range, different experiments (e.g. intermolecular distances in molecular crystals, NMR relaxation times of simple molecules in the gas phase, scattering cross-section measurements) that are dominated by the interaction at different distance ranges, are interpreted by using widely different n-values. Thus, considerable care must be used in employing an A/R^n potential with a single n value to indicate the distances for which it is valid.

Magat (à Karplus): Est-ce que les différences que vous mentionnez ne proviennent pas du fait que la forme potentielle pour la répulsion ne repose sur aucune base théorique, la vraie forme étant pour le moins exponentielle. Prigogine a montré que l'approximation potentielle était une assez bonne approximation pour les propriétés thermodynamiques, mais évidemment elle ne peut pas rendre compte de collisions avec une grande énergie cinétique.

Karplus: The source of the difficulty I mentioned is that to approximate the potential of the two nonbonded atoms for all distances, a complicated expansion involving exponentials and powers of R is required, taking account of the correct $R \to \infty$ and $R \to 0$ limiting values.

Lifson: This is very true and indicates that the experimental information necessary for the choice of appropriate functions for the conformation analysis cannot be simply carried over from, say, high-energy collision experiments.

Rassat: In your last representation of the H-bond potential, it seems that it would be better to give a symmetrical structure to your potential if you want to choose between Morse and LJ representation, i.e. to write

$$V = A_{\text{electrostatic}} + B_{\text{Morse}} \exp(-\alpha\theta^2) + C_{\text{LJ}} \exp(-\alpha\theta^2)$$

and to look for the best A, B, C (and other) parameters, rather than to use your nonsymmetrical formula. Or did I misunderstand your point?

Lifson: Your proposed form could lead to an anomaly: where $\exp(-\alpha\theta^2)$ is very small, there is no repulsion and electrostatic charges alone tend to reduce the energy indefinitely by pulling opposite charges to very small distances. The form (LJ) $(1-\exp(-\alpha\theta^2))$ helps to replace the Morse potential continuously by the LJ potential.

Ten Eyck: I would like to ask a question about a future possibility of this method. Eventually you will be able to calculate conformations of proteins which are essentially *in vacuo*. Proteins are in fact studied in a solvent. It is well known that perturbations of the solvent can change protein conformations, often to the extent of complete denaturation. The sensitivity of different proteins to varying denaturing agents varies a great deal, and the actual mode of action of most denaturing agents is unknown. Would it be possible to use these methods to estimate the sensitivity of a protein to a given denaturing agent, and (probably more important) determine how a given denaturing agent probably functions?

Lifson: The consideration of the solvent effect is a rather difficult one, particularly for water. One would have to evaluate protein-water correlation function, or otherwise find a simplified but workable model for the water–protein interaction. It seems that there will be one day an answer, but I would not dare to guess what it will be.

ÉTUDE CINÉTIQUE DES CHANGEMENTS CONFORMATIONNELS DE LA GLUTAMATE DÉSHYDROGÉNASE PAR SUBSTRATS ET EFFECTEURS

M. IWATSUBO, J. M. JALLON, et A. DI FRANCO

Centre de Génétique Moléculaire, Gif-sur-Yvette, France

Résumé. L'effecteur GTP provoque, en présence du coenzyme oxydé ou réduit NADP(H), une dépolymérisation totale rapide de la GDH. La présence d'un substrat accélère encore la vitesse de ce processus. Durant la réaction catalytique, l'inhibition par le GTP apparaît lentement comparativement à la dépolymérisation, donc doit être associée à un changement conformationnel vraisemblablement de l'hexamère. Nous caractérisons ce changement structural par l'accessibilité réduite du site actif pour les substrats, manifestée par des modifications de mobilité locale (marqueurs de spin immobilisés) et des paramètres cinétiques.

Summary. A parallel kinetic study of molecular weight changes and coenzyme consumption was performed in the reaction catalyzed by GDH in the presence of the inhibitor GTP. The complete depolymerization of GDH induced by GTP in the presence of the coenzyme NADP(H) is accelerated by the presence of a substrate, either L-glutamate or 2-oxoglutarate. This fact along with other experimental evidence presented here suggest concerted interactions between the different ligand binding sites which induce a change in the association zones of the hexamers.

During the catalytic process, the inhibition after GTP addition appears comparatively more slowly than the depolymerization, thus can be associated with a subsequent conformational change. However, if this conformational change is performed before starting the reaction (by preincubation of the enzyme with both GTP and coenzyme), the catalytic process is significantly modified. We analyze this conformational change in terms of restricted accessibility of the active site for the substrates as reflected by reduced local mobility (immobilized spin labels) and decreased binding rates. We propose a mechanism to explain the GTP action suggesting that the depolymerization provokes the inhibitory structural change.

1. Introduction

La L-glutamate déshydrogénase du foie (GDH) catalyse la réaction à trois substrats suivante:

$$\text{L-glu} + \text{NAD(P)}^+ + H_2O \rightleftharpoons \text{2-oxoglu} + NH_4^+ + \text{NAD(P)H} + H^+ .$$

Son activité catalytique est modulée par deux groupes d'effecteurs: activateurs (ADP et ses analogues; leucine), inhibiteurs (GTP et ses analogues; ions bivalents: Zn^{++}, Co^{++}; certaines hormones) [1].

L'unité minimale catalytiquement active a un poids moléculaire voisin de 310000 [2, 3] et est constituée de six chaînes apparemment identiques de 505 acides aminés [4]. Cet hexamère peut former des polyhexamères de structure linéaire et indéfinie. L'équilibre héxamère–polyhexamère dépend de la concentration d'enzyme et de divers paramètres (pH, force ionique, température) [1]. Mais le phénomène le plus remarquable est l'action concertée des ligands spécifiques: coenzymes, substrats et effecteurs, sur l'état de polymérisation de la GDH. Un exemple bien connu est la dissociation totale du polyhexamère en héxamères qui est induite par le couple effecteur GTP + NAD (P)H,

C. Sadron (ed.), Dynamic Aspects of Conformation Changes in Biological Macromolecules. 431–445.

étudié de manière approfondie, surtout chez Frieden [5, 6] et à Gif-sur-Yvette [7, 8].

Les problèmes posés par les relations causales entre l'état de polymérisation et l'état fonctionnel de l'enzyme ont été discutés depuis longtemps. Nous apportons ici un certain nombre de données précises sur les changements structuraux de la protéine provoqués par les ligands spécifiques en correlation avec la régulation de la catalyse enzymatique. Nous avons en particulier étudié les vitesses de dépolymérisation de l'enzyme, les accessibilités des ligands spécifiques (substrats–coenzymes) reflétées par leurs paramètres cinétiques (constantes de vitesse d'association et de dissociation), les remaniements locaux autour de marqueurs de spin, et leurs modifications en présence de l'effecteur GTP.

Nous n'avons utilisé que le coenzyme NADP(H) car il ne possède qu'un seul site de fixation par protomère, alors que NAD(H) en possède deux [9]. Chacun des autres ligands spécifiques ne possède également qu'un site de fixation par protomère.

2. Matériel et techniques

La L-glutamate déshydrogénase cristallisée est préparée par la méthode de Iwatsubo [10] à partir du foie de bœuf. L'enzyme est recristallisé six fois en présence de Na_2SO_4. La concentration de l'enzyme est calculée sur la base de 56000 daltons par chaîne [4].

Toutes les solutions de réactifs ont été filtrées sur millipore (0,23 μ) pour éliminer les poussières. La solution d'enzyme dissout dans du tampon 0,1 M phosphate (pH 7,8) a été centrifugée 20 min à $43000 \times g$, juste avant l'expérience pour éviter la formation d'agrégats. Les spectres d'absorption ont été effectués sur un Cary 15. Les mesures statiques de fluorescence ont été réalisées sur un spectrofluorimètre construit au laboratoire [11]. Les spectres de résonance paramagnétique électronique ont été effectués sur un spectrographe E3 Varian.

Les mesures cinétiques ont été effectuées à l'aide d'un appareil 'stopped-flow' commercial (Durrum), amélioré, permettant les mesures d'absorbance, de fluorescence et de diffusion de la lumière. Le temps mort de cet appareil est de 2,5 ms.

3. Étude des cinétiques de dépolymérisation en présence de couples effecteurs variés

La glutamate déshydrogénase active se trouve à l'état hexamérique (poids moléculaire voisin de 310000), soit à très faible concentration (inférieure à 0,1 mg/ml), soit en présence du coenzyme et de l'inhibiteur GTP, soit après certaines modifications chimiques spécifiques (acétylation, pyridoxilation) [12, 13]. A forte concentration d'enzyme (supérieure à 1 mg/ml), la glutamate déshydrogénase est essentiellement sous forme polyhexamérique. Une simple dilution déplace l'équilibre de polymérisation; la dissociation est alors partielle et lente avec une constante de vitesse $k \sim 10\ s^{-1}$ à 10°C et une énergie d'activation de l'ordre de 10 kcal/mole, comme l'a montré une étude au 'stopped flow' des variations de turbidité à 450 nm [7].

Lorsque l'on provoque la dissociation, non par dilution, mais par le couple effecteur NADPH+GTP, une étude parallèle [7] a indiqué que la dissociation est beaucoup

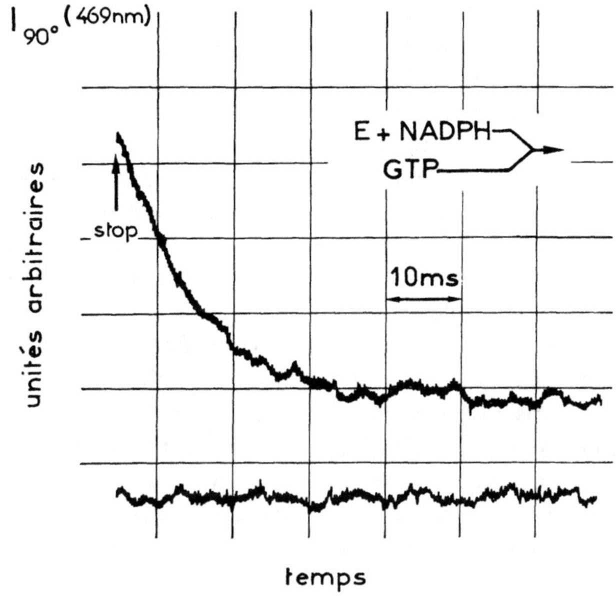

Fig. 1a.

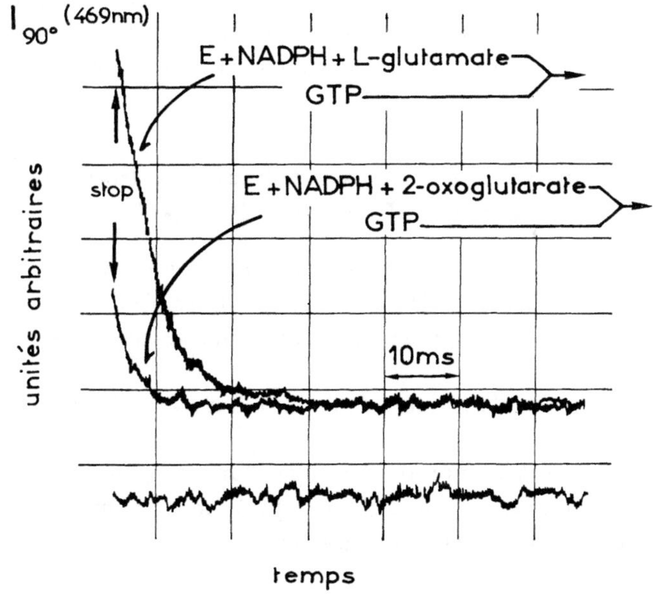

Fig. 1b.

Fig. 1. Dépolymérisation de la GDH induite par le couple effecteur GTP + NADPH. Cinétique de variation de diffusion de la lumière à 90° ($\lambda = 469$ nm), (a) en absence (b) en présence d'un ligand substrat, L-glutamate ou 2-oxoglutarate. L'arrêt de l'écoulement est indiqué sur l'enregistrement par une flèche 'stop'. Le niveau de diffusion du tampon est figuré. Conditions: E: 22 μM; NADPH: 107 μM; GTP: 505 μM; L-glutamate: 98 mM; 2-oxoglutarate: 18,2 mM; tampon tris-HCl: 0,1 M, phosphate: 10 mM, EDTA: 0,5 mM, pH 7,5; $T = 10$ °C.

plus rapide ($k \sim 70\ s^{-1}$), avec une énergie d'activation beaucoup plus faible (2 kcal/mole). Nous avons repris l'analyse cinétique de la dissociation par une méthode plus exacte, diffusion orthogonale de la lumière à 469 nm. Les résultats sont présentés sur la Figure 1a. La constante de vitesse est $100\ s^{-1}$ à 10° C valeur voisine et vraisemblablement plus précise que les déterminations antérieures.

Nous avons cherché à voir comment ce type de dissociation provoquée par le couple effecteur est sensible à la présence d'autres ligands, en particulier les substrats. Comme on le voit sur la Figure 1b, la dissociation est fortement accélérée par le L-glutamate et le 2-oxoglutarate ($k \sim 200\ s^{-1}$). Par contre les analogues D-glutamate ($k \sim 70\ s^{-1}$) et glutarate ($k \sim 15\ s^{-1}$) la ralentissent.

Le type d'action du coenzyme au sein du couple effecteur avec GTP ne semble pas dépendre spécifiquement de son état d'oxydation. On observe en effet avec NADP+ + GTP une dépolymarisation avec une vitesse assez lente ($k \sim 15\ s^{-1}$ à 10° C), accélérée, comme précédemment avec NADPH, par les substrats ($k \sim 100\ s^{-1}$).

Si l'on étudie maintenant ce qui se passe *en l'absence du coenzyme*, on constate que le GTP à lui seul, sans effet sur la dépolymérisation, devient dissociant en présence de 2-oxoglutarate, ou de L- ou D-glutamate, ces couples provoquant alors une dissociation partielle et lente.

Dans la troisième situation possible, c'est-à-dire *en l'absence de GTP*, mais en présence de NADPH, nous avons trouvé un phénomène encore non décrit: l'addition de 2-oxoglutarate induit une dépolymérisation, partielle et lente (des expériences préliminaires indiquent pour k environ $15\ s^{-1}$). Cette dépolymérisation explique la différence des niveaux initiaux dans la Figure 1b. Il semble en effet que le L-glutamate n'a pas d'effet sensible.

4. Étude, par fluorescence, des cinétiques de transitions structurales en présence du couple effecteur NADPH + GTP

La fluorescence spécifique du NADPH fixé à l'enzyme est une fonction très sensible de la présence d'autres ligands. Elle est beaucoup plus faible dans le complexe quaternaire E–NADPH–GTP–L-glu. que dans les deux complexes ternaires E–NADPH–GTP et E–NADPH–L-glu.

Les transitions qui s'établissent lors de l'addition du 4ème ligand sont relativement lentes (Figure 2).

L'interprétation n'est pas claire. Il peut s'agir d'une transition au sein du complexe quaternaire formé. En fait, on peut utiliser ce phénomène pour suivre cinétiquement la fixation lente des substrats (voir paragraphe 6).

5. Analyse de l'effet de GTP au niveau des complexes transitoires observables dans la rotation catalytique de l'enzyme

Nous avons étudié l'effet de l'inhibiteur GTP sur certaines des étapes élémentaires résolues [14–16] qui composent la réaction globale.

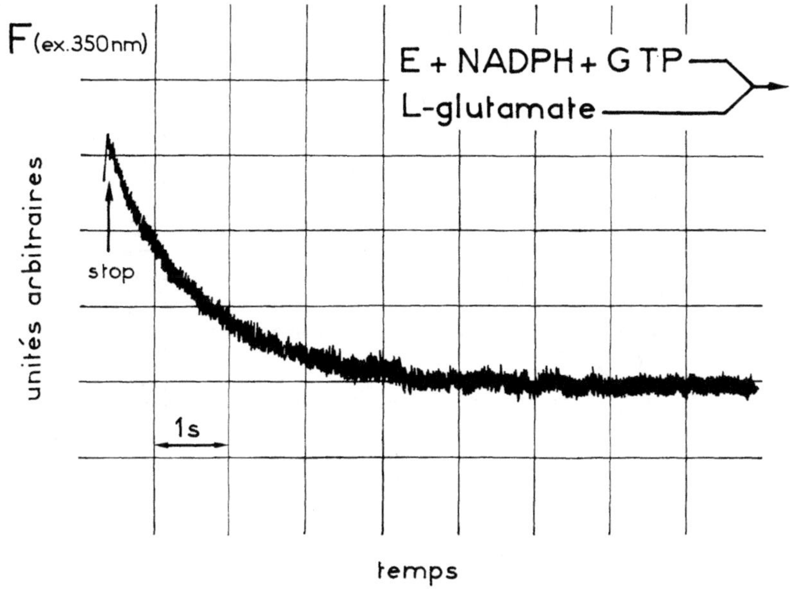

Fig. 2. Remaniement structural au site actif de la GDH. Variation au cours du temps de la fluorescence du NADPH (excitée à 350 nm) après addition de L-glutamate au complexe ternaire E–NADPH–GTP. Conditions: E: 23,8 μM; NADPH: 21,4 μM; GTP: 505 μM; L-glutamate: 0,1 M; tampon tris-HCl: 0,1 M, phosphate: 10 mM, EDTA: 0,5 mM, pH 7,5; $T = 10\,°C$.

En présence de GTP, la cinétique transitoire de la réaction est modifiée differemment suivant l'ordre de fixation des ligands spécifiques sur l'enzyme. Le fait marquant est que l'addition du couple GTP–NADP(H) à la GDH, préablement à l'addition de l'autre réactif (substrats) qui déclanche la réaction enzymatique, se traduit par une inhibition immédiate, tandis que l'addition de GTP à l'enzyme au moment du départ de la réaction se traduit par une inhibition qui s'installe progressivement.

A. PROCESSUS TRANSITOIRES DANS LE SENS DE LA RÉDUCTION DU COENZYME

Lorsque la réaction catalytique est suivie dans le sens de la réduction du coenzyme NADP par le L-glutamate, plusieurs complexes intermédiaires figurant sur le schéma suivant ont pu être mis en évidence, caractérisés et identifiés. Il s'agit des complexes II et IV. Ce schéma a été proposé comme le plus simple rendant compte à la fois des caractères des complexes observés et de l'ensemble très détaillé et complet de données cinétiques [15, 16].

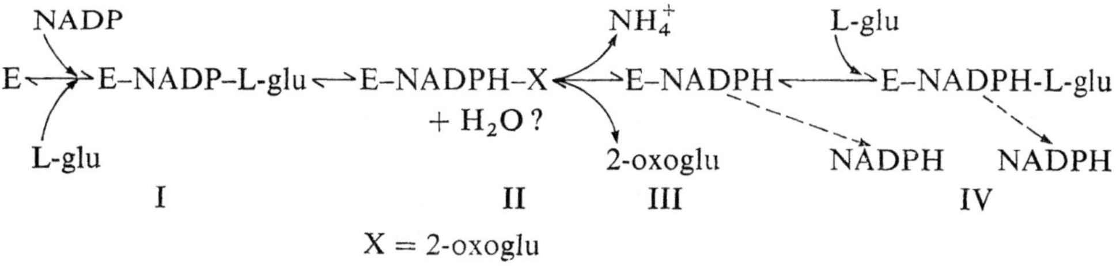

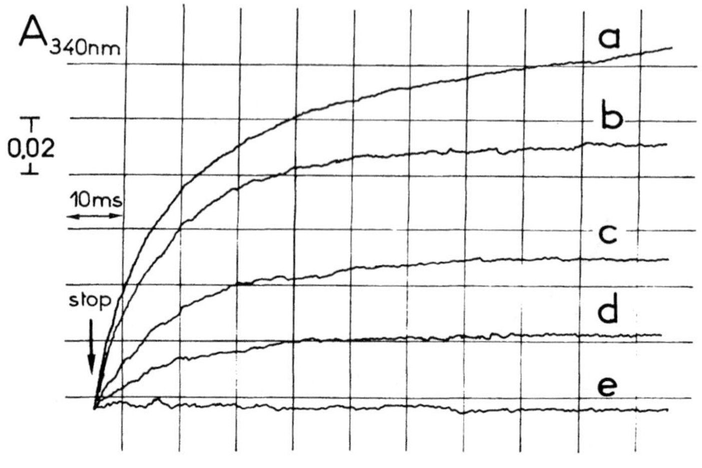

Fig. 3a.

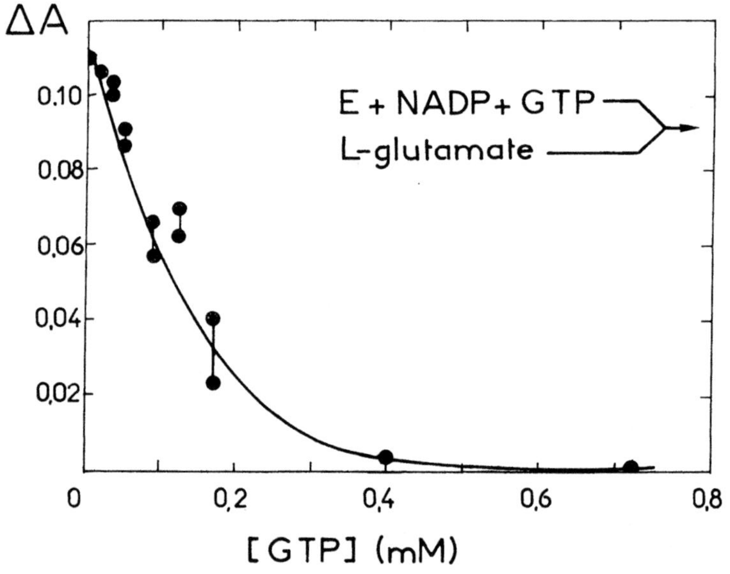

Fig. 3b.

Fig. 3. Réduction catalytique du NADP par le L-glutamate en présence de GTP lorsque le complexe E–NADP–GTP est préformé. (a) Cinétique de variation d'absorption à 340 nm. Conditions: E: 17 μM; NADP: 1,35 mM; L-glutamate: 0,1 M; GTP: concentration variable a: 0; b: 36 μM; c: 88 μM; d: 171 μM; e: 720 μM; tampon tris-HCl: 0,1 M, phosphate: 10 mM, EDTA: 0,5 mM, pH 7,5; $T = 10\,°C$. Les niveaux initiaux ont été déplacés arbitrairement au 'stop'. (b) Variation de l'amplitude de la phase rapide avec la concentration de GTP.

(a) Si la GDH est incubée initialement avec le couple effecteur NADP + GTP avant l'addition du L-glutamate, on voit sur la Figure 3 que la réaction est très lente, pratiquement inobservable à saturation en GTP, tandis que la réaction présente la première phase rapide correspondant à la formation du complexe II ($k = 80\ s^{-1}$) pour le témoin sans GTP.

A concentration plus faible de GTP, la réaction est biphasique: il y a une phase rapide de vitesse similaire au témoin sans GTP suivie d'une phase lente. La phase rapide représente très probablement la phase préstationnaire des molécules d'enzyme qui ne sont pas combinées au GTP tandis que la phase lente correspondrait aux phases postérieures de ces mêmes molécules. L'amplitude de la phase rapide varie avec la concentration de GTP (Figure 3b) de façon approximativement hyperbolique. Les faits suggèrent que la constante de dissociation du complexe E–GTP en présence des autres ligands dans les conditions expérimentales est de l'ordre de 100 μM.

(b) Si la GDH est incubée initialement sans l'ensemble du couple effecteur, c'est-à-dire soit avec GTP seul, soit avec NADP seul, le complément se faisant au moment du déclanchement réactionnel, la première phase rapide préstationnaire (I et II) ne semble pas affectée (Figure 4). Une série d'expériences, dans des conditions particulières n'a montré de modification de cette phase ni en vitesse, ni en amplitude [16]. Cependant, l'évolution ultérieure est pratiquement bloquée (98% d'inhibition). Le transfert d'hydrogène se fait donc normalement au sein du complexe quaternaire. L'évolution comparée des spectres d'absorption et de la fluorescence permet d'identifier le produit de blocage comme le complexe II [16].

Si l'on essaie maintenant d'avoir des informations sur une dépolymérisation éventuelle pendant ces mêmes phases réactionnelles, en observant la cinétique de variation de la lumière diffusée dans des expériences parallèles, on observe (Figure 4) que le processus de dépolymérisation prend place effectivement, avec une vitesse $k \sim 100\ s^{-1}$ à 10°C.

La différence considérable observée dans la cinétique réactionnelle suivant que l'enzyme a été, ou non, préincubé avec le couple effecteur, prouve que celui-ci entraîne une transition lente, plus lente que les processus réactionnels préstationnaires, qui aboutit à une inhibition quasi-totale des phases préstationnaires (et stationnaires?). On peut écrire le schéma suivant:

$$(\text{E–GTP–NADP})_{\text{actif}} \rightarrow (\text{E–GTP–NADP})_{\text{inactif}}.$$

D'après les vitesses de fixation déterminées pour NADP (en absence de GTP), et pour GTP (en présence de NADPH), il semble que la réaction ne puisse être limitée par les fixations aboutissant au complexe comprenant enzyme et le couple effecteur.

L'un des faits importants est que dans les conditions ou GTP est ajouté en dernier lieu, au moment du déclanchement réactionnel, la dépolymérisation prend place sans qu'il y ait d'inhibition de la phase préstationnaire alors que si le couple effecteur est préincubé avec l'enzyme, cette phase préstationnaire est totalement inhibée. Il n'y a donc pas simultanéité entre dépolymérisation et inhibition. D'autre part si les deux

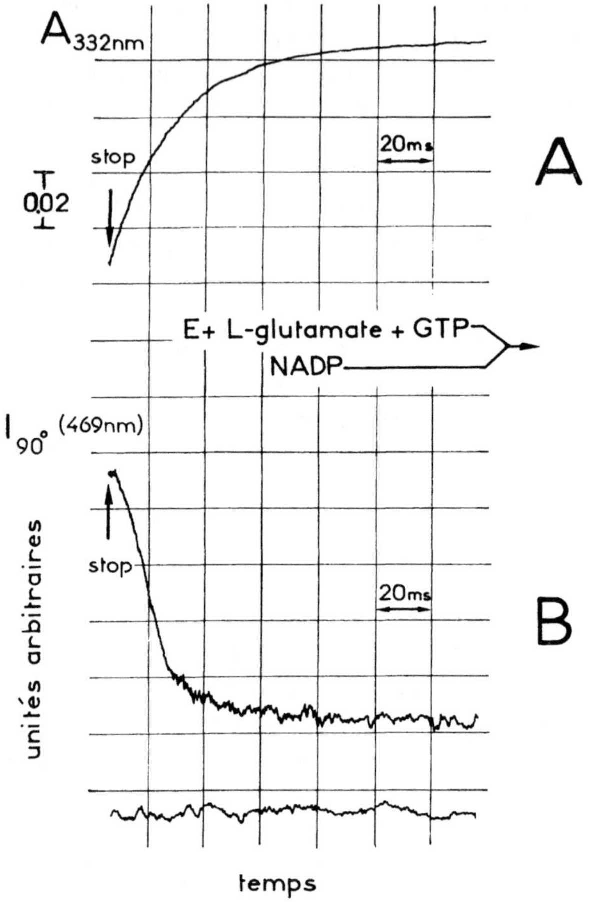

Fig. 4. Réaction catalytique et changement de poids moléculaire lorsque le complexe E–NADP–GTP n'est pas préformé. Cinétique de variation de l'absorption à 332 nm (A). Cinétique de variation de diffusion de la lumière (B), lors de la réaction catalytique en présence de GTP. Conditions: E: 15 μM; NADP: 1,1 mM; L-glutamate: 50 mM; GTP: 250 μM; tampon phosphate: 0,1 M, EDTA: 0,5 mM, pH 7,8; $T = 10\,°C$.

processus ne sont pas indépendants, la dépolymérisation précède la transition correspondant à l'inhibition.

Nous suggérons que l'inhibition correspond à une très forte réduction de la vitesse de fixation du glutamate (voir paragraphe 6).

B. PROCESSUS TRANSITOIRES DANS LE SENS DE L'OXYDATION DU COENZYME RÉDUIT

Lorsque la réaction catalytique est observée dans le sens de l'oxydation du NADPH par le couple des deux substrats (2-oxoglutarate plus NH_4^+), il est très difficile de détecter un phénomène transitoire, car la réaction atteint très rapidement son régime stationnaire [16]. Cependant, si l'on étudie la transformation d'un complexe intermédiaire préformé, enzyme–NADPH–2-oxoglutarate, à la suite de l'addition d'ammoniaque, on observe deux phases (Figure 5a), une extrêmement rapide ($k_a \gg 200\ s^{-1}$) et une plus lente qui correspond à la phase stationnaire. En présence du couple effecteur GTP+NADPH préincubé à l'enzyme, à concentration saturante, on n'observe que peu d'oxydation du NADPH, la réaction étant très lente.

Si cependant l'enzyme n'est pas incubé initialement avec le 'couple effecteur' NADPH+GTP, par exemple, si le complexe E–NADPH réagit avec les substrats (2-oxoglutarate et NH_4^+) et avec le GTP (Figures 5b, 6a), la réaction est ralentie

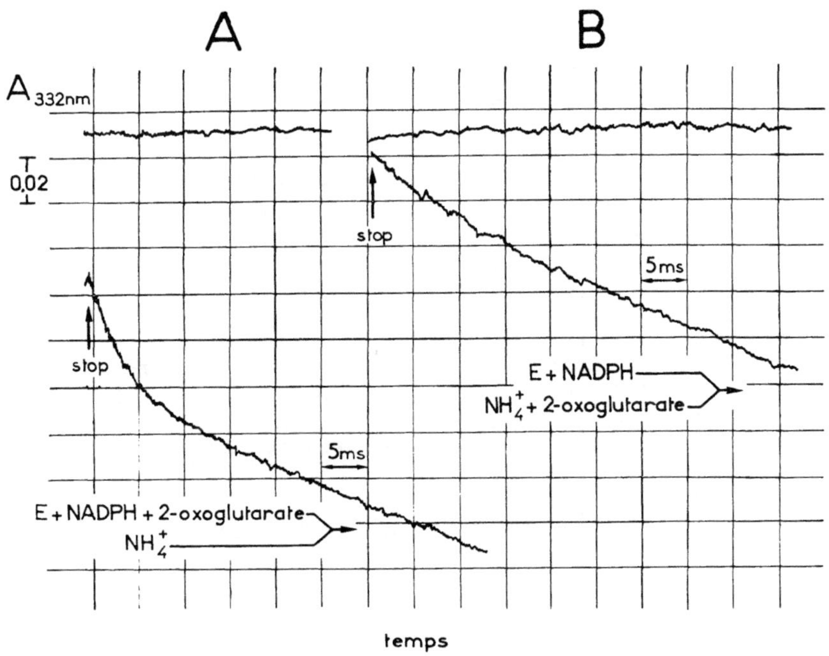

Fig. 5. Cinétique transitoire de l'oxydation de NADPH par les substrats 2-oxoglutarate et NH_4^+ pour différents ordres d'addition (A et B). Absorption à 332 nm. Les niveaux initiaux sont indiqués. Conditions: E: 23 μM; NADPH: 95 μM; 2-oxoglutarate: 4,9 mM; NH_4Cl: 0,1 M; tampon tris-HCl: 0,1 M, phosphate: 10 mM, EDTA: 0,5 mM, pH 7,5; $T = 10\,°C$.

progressivement. Elle présente une phase stationnaire très inhibée (98%). Une analyse soignée de l'évolution réactionnelle (par la méthode de Guggenheim) montre que l'inhibition s'installe progressivement, avec une constante de vitesse de l'ordre de $20\ s^{-1}$.

Pendant qu'évoluent ces réactions, la dépolymérisation, observée par l'intensité de lumière diffusée, se produit aussi, avec une constante apparente de $100\ s^{-1}$, donc plus rapide que l'inhibition.

La même succession d'événements, (1) dépolymérisation, (2) inhibition se produit donc que l'on observe la réaction dans le sens direct ou dans le sens inverse.

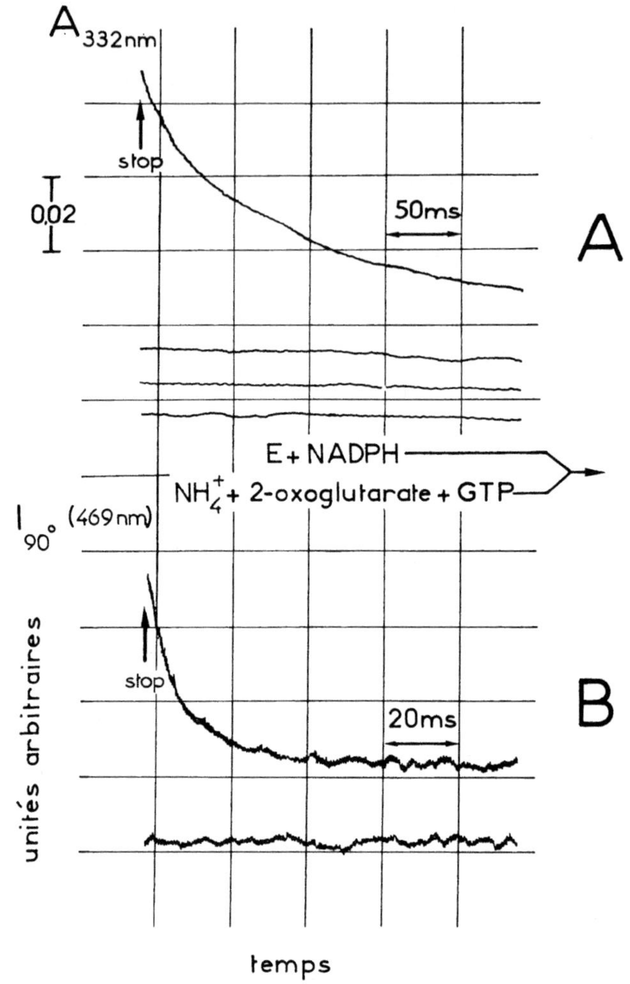

Fig. 6. Remaniements structuraux aboutissant à l'inhibition par le GTP lors de l'oxydation catalytique du NADPH. Cinétique de variation d'absorption à 332 nm (A). Cinétique de variation de diffusion de la lumière (B), lors de la réaction catalytique en présence de GTP. Conditions: E: 20 μM: NADPH: 90 μM; 2-oxoglutarate: 10 mM; NH_4Cl: 0,1 M; tampon tris-HCl: 0,1 M, phosphafe- 10 mM, EDTA: 0,5 mM, pH 7,5; $T = 10$ °C.

6. Effet de GTP, en présence du coenzyme, sur les paramètres cinétiques élémentaires de fixation des autres ligands spécifiques

En suivant soit l'extinction de la fluorescence soit les changements d'absorption ou de fluorescence du coenzyme réduit fixé à l'enzyme, nous avons pu déterminer les constantes cinétiques de fixation des autres ligands spécifiques: coenzymes et substrats [17]. La Figure 7 montre les graphiques aboutissant à des déterminations de ce type, en présence de GTP, concernant les substrats L-glutamate et 2-oxoglutarate. Les constantes de vitesse tirées d'expériences analogues ou d'expériences de déplacement du

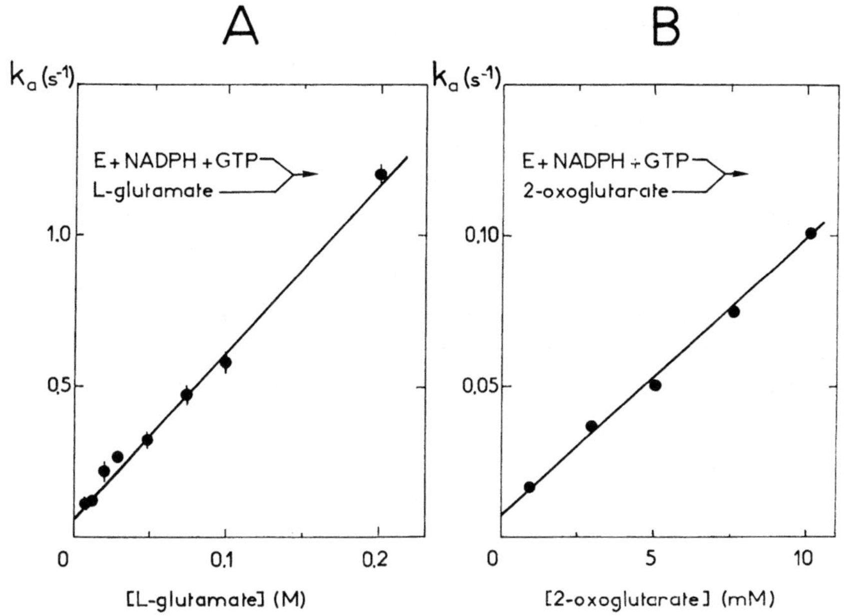

Fig. 7. Paramètres cinétiques de fixation des substrats au complexe E–NADPH en présence de GTP. Variations de la constante apparente de diminution de fluorescence (excitée à 350 nm) avec la concentration de L-glutamate (A) et de 2-oxoglutarate (B). Conditions: E: 25 μM; NADPH: 24 μM; GTP: 470 μM; tampon tris-HCl: 0,1 M, phosphate: 10 mM, EDTA: 0,5 mM, pH 7,5; $T = 10$ °C.

TABLEAU I

Modifications des paramètres cinétiques de fixation des substrats sur le complexe E–NADPH sous l'action du GTP ($T = 10$ °C, voir Figure 6)

	sans GTP		avec GTP	
	k_+ (M^{-1} s^{-1})	k_- (s^{-1})	k_+ (M^{-1} s^{-1})	k_- (s^{-1})
E + NADPH	2×10^6	30	6×10^5	3
E + 2-oxoglu + NADPH	8×10^5	0,30	8×10^5	0.2
E–L-glu + NADPH	$1{,}5 \times 10^6$		10^6	
E–NADPH + 2-oxoglu	3×10^4	2	9	0,007
E–NADPH + L-glu	1×10^4	8	6	0,2

coenzyme réduit par le coenzyme oxydé ou d'un substrat par un de ses analogues sont rassemblés dans le Tableau I. La vitesse de fixation du NADPH à l'enzyme est peu affectée par la présence de GTP mais sa dissociation paraît très ralentie (environ 10 fois). Le phénomène le plus remarquable est l'effet du GTP sur la fixation de chacun des substrats au complexe enzyme–coenzyme. Leur dissociation est très ralentie (50 à 100 fois). Leur fixation est encore davantage ralentie (> 1000 fois). La structure 'lente' doit être celle appelée [E–NADP(H)–GTP]$_{inactif}$ dans laquelle la porte d'entrée des substrats est quasiment verrouillée.

7. Étude de la rigidité autour de marqueurs de spin; variation en présence de GTP

Les marqueurs de spin reflètent (dans leurs spectres de résonance paramagnétique électronique) la mobilité de leur site de fixation. Nous avons recherché un marquage spécifique de la glutamate déshydrogénase.

Les travaux de nombreux auteurs ont montré qu'une lysine (Lys 126) était essentielle soit pour la fixation des substrats, soit pour la catalyse elle-même [13, 18, 19]. Nous avons pu former une base de Schiff entre la GDH et une cétone-nitroxyde. Cette modification semble spécifique puisque le substrat 2-oxoglutarate protège l'enzyme. Le spectre RPE de l'enzyme marqué est considérablement immobilisé (par rapport au marqueur libre) (Figure 8). En plus, nous avons observé que le couple effecteur

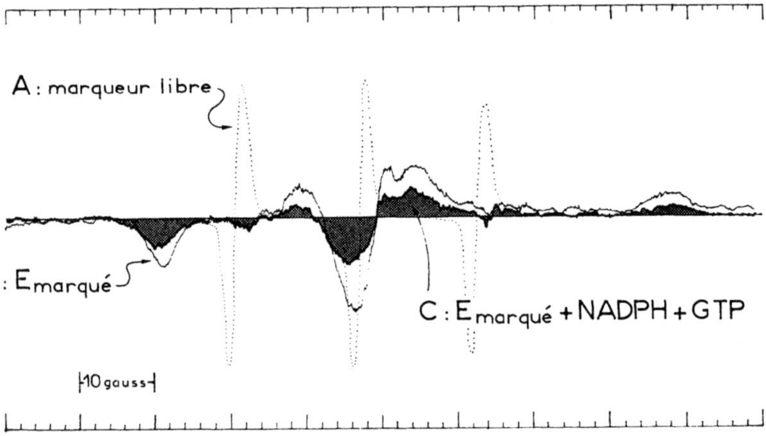

Fig. 8. Spectres de résonance paramagnétique électronique du marqueur de spin 2,2,4,4-tétramethyl piperidone 1-oxyde. Marqueur de spin A: libre en solution; B: fixé à l'enzyme; C: fixé à l'enzyme en présence de NADPH et GTP. Conditions: E: 90 μM; NADPH: 200 μM; GTP: 500 μM; tampon tris-HCl: 0,1 M, EDTA: 0,5 mM, pH 7,5; température ambiante.

NADPH + GTP provoquait une immobilisation supplémentaire du marqueur de spin. Malheureusement, il n'a pas été possible d'obtenir un marquage supérieur à 15% des chaines.

Avec deux autres marqueurs de spin, dérivés du maléimide et du PCMB, qui, eux, réagissent avec un résidu par chaîne de la protéine [20], nous observons les mêmes types d'immobilisation. Ces données apportent une preuve supplémentaire de la rigidification du site des ligands actifs provoquée par le GTP.

8. Conclusion

Nous présentons ici la description cinétique de la dépolymérisation de la glutamate déshydrogénase induite par divers couples de ligands. Il apparaît que les actions concertées de deux ligands au moins, de deux classes différentes (substrats, coenzymes, effecteur GTP) sont nécessaires: cette action concertée était déjà connue [1] pour le couple coenzyme réduit NADPH et l'effecteur GTP, qui sont particulièrement efficaces et provoquent une dépolymérisation rapide et totale. Cependant, ce processus est lui-même modulé par la présence d'un autre ligand: l'addition supplémentaire d'un substrat accélère la dépolymérisation. On se trouve donc devant une interaction concertée étrange entre les trois classes d'effecteurs qui se répercute sur les aires d'association entre hexamères, qui implique donc des changements de conformation secondaire-tertiaire se propageant à partir des sites de fixation.

Le problème du rapport entre l'activité catalytique et l'état conformationnel de l'enzyme au niveau de la structure quaternaire (dépolymérisation) ou seulement secondaire-ternaire (transitions sur l'unité hexamère) a été très débattu depuis une quinzaine d'années. On a d'abord pensé que seul l'état polyhexamérique était actif mais les travaux de plusieurs auteurs ont clairement démontré que l'activité spécifique était indépendante de l'état de polymérisation [16, 23]. Cependant, les travaux de Tomkins [24, 25], basées seulement sur des études cinétiques à l'état stationnaire, ont mis en évidence un phénomène curieux en rapport avec la spécificité large de l'enzyme: alors que l'activité glutamate déshydrogénasique était inhibée par le GTP et activée par l'ADP, l'activité (très faible) avec les acides aminés monocarboxyliques (comme l'alanine ou la norvaline) était accélérée par le GTP et inhibée par l'ADP. Pour interpréter ces résultats (établis avec la GDH de bœuf), Tomkins a présenté successivement deux schémas dont le plus récent faisait intervenir un équilibre entre deux formes dépolymérisées de l'enzyme en équilibre allostérique, l'une active sur le glutamate, l'autre sur l'alanine. Les effecteurs GTP et ADP agissent en déplaçant en sens inverse cet équilibre.

Cependant, en abordant l'étude de ce système en cinétique rapide, au 'stopped-flow', avec la GDH de porc, Iwatsubo et Pantaloni [14] ont montré que la réaction évoluait avec une phase préstationnaire (jet initial) et que le GTP et l'ADP n'agissaient pas de la même manière sur cette phase (inhibée par ADP) et sur la phase stationnaire (activée par ADP). Du fait qu'en cinétique stationnaire le processus limitant est la fixation du substrat pour l'alanine et la libération des produits pour le glutamate, ils concluaient que les arguments expérimentaux sur lesquels s'appuyaient l'hypothèse de la modification de spécificité entre les deux conformations de l'enzyme n'était pas fondée.

En ce qui concerne les rapports entre activité et état conformationnel, nous apportons ici des informations:

(1) sur l'action du couple GTP–coenzyme en tant que dépolymérisant, avec la détermination des constantes de vitesse en l'absence d'autres réactifs et en présence des substrats au cours de la réaction catalytique dans les deux sens de la réaction (réduction ou oxydation du coenzyme);

(2) sur l'existence d'une transconformation lente (relativement) prenant place apparemment sur la forme dépolymérisée, et du type unité hexamérique active → unité hexamérique inactive.

L'unité hexamérique inactive peut être comparée au polyhexamère actif: Le complexe inactif possède un certain nombre de caractéristiques spectrales largement documentées avec le NADPH: spectre d'absorption centré aux environs de 350 nm, ellipticité positive (légèrement supérieure à celle de E–NADPH), large rendement quantique de fluorescence, taux de polarisation de fluorescence élevé. Les remaniements de la protéine elle-même se manifestent par les changements de polarité de l'environnement d'un marqueur hydrophobe l'ANS et l'immobilisation de marqueurs de spin au site actif. Mais le fait le plus remarquable est la modification très importante des paramètres thermodynamiques et cinétiques de l'interaction avec les substrats. Au sein de ce complexe, les substrats se dissocient moins vite. Mais surtout leur vitesse de fixation est ralentie plus de mille fois. Ceci permet d'expliquer que le complexe (E–NADPH–GTP) ne fixe que très difficilement le glutamate. et le 2-oxoglutarate. Si ces complexes ne sont pas préformés, l'inhibition apparaît dès que l'ensemble des remaniements structuraux décrits dans le schéma s'est effectuée.

L'intermédiaire 'unité hexamérique active' ne peut être détecté que dans les premières phases préstationnaires de la réaction observée au 'stopped flow' lorsque l'enzyme n'est pas préincubé avec l'ensemble du couple effecteur. En effet, quand on procède à une telle préincubation, l'enzyme est déja dépolymérisé et inhibé avant le début de la réaction.

La conclusion que ce n'est pas la transition de dépolymérisation en elle-même qui donne la forme inhibée, mais une seconde transition plus lente de la forme depolymérisée est très clairement établie dans les expériences présentées ici. La relation séquentielle entre dépolymérisation et transition 'actif→inactif' n'est pas claire; il peut s'agir de deux phénomènes indépendants mais simultanés avec des vitesses différentes ou de deux phénomènes séquencés.

Une compréhension profonde des relations entre changements de structure quaternaire et modulation de l'activité catalytique pour la glutamate déshydrogénase semble d'autant plus nécessaire que le même type de régulation est probablement très général, se rencontre dans un grand nombre de protéines enzymatiques [26], mais n'a jamais été étudié de façon détaillée.

Remerciements

Nous tenons à remercier le Professeur P. Slonimski pour l'intérêt constant et pour les moyens matériels qu'il a mis à notre disposition, et Madame F. Labeyrie, Directeur de notre groupe de recherches, pour des discussions fructueuses, les Drs. Douzou et F. Leterrier pour nous avoir permis l'accès à leur équipement RPE. D'autre part nous reconnaissons l'aide particulièrement efficace de Mesdames T. Iwatsubo et Y. Risler pour la partie expérimentale et de Monsieur A. Bertin pour la mise au point de l'appareillage.

Ce travail n'a été possible que grâce à la généreuse contribution financière de la Délégation Générale à la Recherche Scientifique et Technique (Convention no. 727 0500 'Interactions Moléculaires en Biologie').

Références

[1] Frieden, C.: dans *The Enzymes*, Vol. 7 (Boyer), p. 3.
[2] Eisenberg, H. et Tomkins, G. M.: 1968, *J. Mol. Biol.* **31**, 37.
[3] Sund, H. et Burchard, W.: 1968, *Eur. J. Biochem.* **6**, 202.
[4] Smith, E. L., Landon, M., Pishzkiewicz, D., Bratten, W. J., Langeley, T. J., et Melamed, M. D.: 1969, *Proc. Nat. Acad. Sci. US* **64**, 338.
[5] Huang, C. Y et Frieden, C.: 1969, *Proc. Nat. Acad. Sci. US* **64**, 338.
[6] Huang, C. Y. et Frieden, C.: 1972, *J. Biol. Chem.* **247**, 3638.
[7] Jallon, J. M., Di Franco, A., et Iwatsubo, M.: 1970, *Eur. J. Biochem.* **13**, 428.
[8] Dessen, P.: communication personnelle.
[9] Jallon, J. M. et Iwatsubo, M.: 1971, *Biochem. Biophys. Res. Commun.* **45**, 964.
[10] Kubo, H., Iwatsubo, M., Watari, H., et Soyama, T.: 1959, *J. Biochem. Japan* **46**, 1.
[11] Iwatsubo, M. et Capeillère, C.: 1967, *Biochim. Biophys. Acta* **146**, 349.
[12] Colman, R. F. et Frieden, C.: 1966, *J. Biol. Chem.* **241**, 3652.
[13] Anderson, B. M., Anderson, C. D., et Churchich, J. E.: 1966, *Biochemistry* **5**, 2893.
[14] Iwatsubo, M. et Pantaloni, D.: 1967, *Bull. Soc. Chim. Biol.* **49**, 1563.
[15] Di Franco, A. et Iwatsubo, M.: 1971, *Biochimie* **53**, 153.
[16] Di Franco, A.: 1971, Thèse de Doctorat d'État, Université de Paris Sud.
[17] Di Franco, A. et Iwatsubo, M.: 1972, *Eur. J. Biochem.* **30**, 517.
[18] Pishzkiewicz, D., Landon, M., et Smith, E. L.: 1970, *J. Biol. Chem.* **245**, 2622.
[19] Holbrook, J. J. et Jeckel, R.: 1969, *Biochem. J.* **111**, 689.
[20] Jallon, J. M. et Piette, L. H.: 1972, résultats non publiés.
[21] Brocklehurst, J. R., Dodd, G. H., Freedman, R. B., Malcolm, A. D. B., Price, N. C., et Radda, G. K.: 1970, dans *Pyridine Nucleotide-Dependent Dehydrogenases* (H. Sund, éd.).
[22] Jallon, J. M.: 1969, Thèse 3ème cycle, Université de Paris Sud.
[23] Fisher, A. F., Cross, D. G., et McGregor, L. L.: 1962, *Nature* **196**, 895.
[24] Bitensky, M., Yielding, K., et Tomkins, G. M.: 1965, *J. Biol. Chem.* **240**, 1077.
[25] Tomkins, G. M., Yielding, K., Curran, J., Summer, M., et Bitensky, M.: 1965, *J. Biol. Chem.* **240**, 3793.
[26] Frieden, C.: 1971, *Ann. Rev. Biochem.* **40**, 653.

DISCUSSION

Pantaloni: (1) Avez-vous pu déterminer la vitesse de dissociation du GTP du complexe E–NADPH–GTP et E–NADPH–glu–GTP?

(2) Peut-on supposer que la diminution des vitesses de fixation et de libération du glutamate en présence du GTP et de la coenzyme puisse être dûe à la vitesse lente de libération du GTP des complexes E–NADPH–GTP et E–NADPH–glu–GTP, cette libération de GTP pouvant être nécessaire pour permettre la fixation ou la libération du substrat?

Iwatsubo: (1) La constante de vitesse de fixation de GTP, k_+, déterminée par Mlle Ann di Franco est de 1×10^6 M^{-1} s^{-1}. Si l'on combine cette valeur à la constante d'équilibre de dissociation $K_{\text{diss}} \simeq$ $\simeq 0{,}5 \times 10^{-6}$ M, on peut estimer la valeur de la constante de vitesse, k_-, égale à 0,5 s^{-1}.

(2) Je pense que ce n'est pas le cas. D'après la courbe expérimentale (Figure 7) la constante apparente de vitesse de fixation k_a augmente linéairement avec la concentration de substrat dans une gamme comprise entre 10 mM et 200 mM. Or, si la fixation de substrat était limitée par la dissociation de GTP, on devrait avoir une courbe de saturation au niveau très faible de concentration en substrat ($\sim$ 50 μM pour le L-glutamate).

H. Buc: Il serait intéressant de savoir si le pourcentage d'inhibition de la première étape est proportionnel au taux de fixation du GTP, ce qui serait compatible avec une action intraprotomérique ou en diffère.

Iwatsubo: En effet, cette relation entre la partie inactive de l'enzyme et les concentrations de GTP ajouté (20–700 μM) nous a surpris, car la valeur connue de Kequil de GTP pour l'enzyme seul est de l'ordre de 1×10^{-5} M. Nous avons remarqué que la valeur de K (demi-effet) de GTP augmente lorsque la concentration de NADP ajouté est plus élevée. Il faut donc examiner la fixation de GTP à l'enzyme en présence de NADP par la méthode d'équilibre de dialyse mise au point chez vous.

MOUVEMENTS MOLÉCULAIRES CONSÉCUTIFS À LA FIXATION DE MÉTAUX DANS LA β-GALACTOSIDASE

JEANNINE M. YON et JEAN-PIERRE TENU

Laboratoire d'Enzymologie Physico-chimique et Moléculaire, G.R. C.N.R.S., associé à l'Université de Paris 11, 91 Orsay, France

Résumé. La β-galactosidase de *E. coli* est activée par les ions Na^{++} et Mg^{++}. Le mécanisme de l'activation de l'enzyme par Mg^{++} a été étudié en détail. D'autres métaux Mn^{++} et Co^{++} ont également un effet activateur important. L'activation par les métaux, de même que la réaction inverse correspondent à des phénomènes lents. L'analyse cinétique montre que l'activation est un processus en deux étapes, une première fixation rapide du métal sur l'enzyme libre, suivie d'une étape plus lente qui s'accompagne de l'augmentation d'activité impliquant probablement un changement dans la conformation de l'enzyme.

Par suite de l'existence de cette étape lente, l'affinité apparente de l'enzyme pour le Mg^{++} est relativement forte ($1{,}5 \times 10^6$ M^{-1}). Ces mouvements moléculaires lents peuvent être associés à la taille et à la complexité de l'enzyme (tétramère de poids moléculaire 540000). Les résultats obtenus avec cet enzyme et la grande spécificité apparente pour l'activateur suggèrent de poser le problème de la spécificité biologique en terme de spécificité dynamique. Des systèmes plus complexes sont probablement le siège de mouvements moléculaires plus lents encore, ce qui se manifeste par des constantes d'affinité apparente encore plus élevées, exemple les hormones-récepteurs ...).

Summary. β-Galactosidase from *E. coli* is activated by Mg^{++} ions. However, a residual activity is found in the absence of this cation. The mechanism of the activation has been considered in detail for the hydrolysis of phenylgalactoside substrate.

Both activation and deactivation are slow processes, the rate of which depends on the Mg^{++} concentration. For the activation process, a lag period preceding the steady state is observed. The increase of the catalytic rate constant (k_{cat}) at the steady state as a function of Mg^{++} concentration undergoes a Michaelian behavior; the corresponding Hill plot exhibits a slope equal to 1, indicating the absence of cooperativity. The apparent dissociation constant for the enzyme–Mg^{++} complex is very small at pH 7.0 (0.6×10^{-6} M).

The kinetics of activation and deactivation has been analysed. A scheme consistent with the experimental data is proposed. It involves a two-step reaction for Mg^{++} binding. The first step is a rapid binding of the metal to the enzyme without enhancement of the activity. The second step probably consists of a slow rearrangement of the complex with an increase of the enzymatic activity.

The occurrence of the slow step explains the high apparent affinity of the Mg^{++} for the enzyme. On the basis of this simple model, some suggestions are presented to explain the very high biological specificity, as found, for instance, in the membrane systems or for hormone-receptor interactions. The concept of 'dynamic specificity' is proposed.

1. Introduction

La β-galactosidase de *E. coli* est un tétramère de poids moléculaire 540000, formée de quatre sous-unités identiques [1–3]. Cet enzyme catalyse l'hydrolyse des β-D-galactosides. Son activité requiert la présence de Na^+ et de Mg^{++} [4]. L'enzyme possède cependant une faible activité catalytique en l'absence de ce dernier cation; les caractéristiques cinétiques de l'enzyme dépourvu de Mg^{++} ont été déterminées [5].

Des études antérieures utilisant la méthode de compétition nucléophile [6, 7] ont montré que la réaction procède via la formation d'un galactosyl enzyme consécutif au

C. Sadron (ed.), Dynamic Aspects of Conformation Changes in Biological Macromolecules. 447–458. *All Rights Reserved.*

complexe de Michaelis. Le schéma réactionnel minimum peut donc s'écrire:

$$E + S \rightleftharpoons ES \rightarrow ES' \rightarrow E + P_2, \quad + P_1$$

P_1 étant la partie aglycone du substrat et P_2 la partie glycosidique.

La cinétique de la réaction a été étudiée avec divers substrats dans des conditions de saturation en Mg^{++} (10^{-3} M) au pH optimum de la réaction (pH 7,0), les différentes constantes cinétiques, K_s, k_2 et k_3 déterminées [6–8].

Pour un certain nombre de substrats (*o*-nitrophényl β-D-fucoside; méthyl cinnamyl, *o*-aminophényl, phényl, *p*-nitrophényl β-D-galactosides), k_2 est limitant [9]. Pour les dinitrophényl (2,4 et 3,5) galactosides, k_3 est limitant [9]. Pour l'hydrolyse des *o*- et *m*-nitrophénylgalactosides, il n'y a pas d'étape limitante; k_2 et k_3 sont du même ordre de grandeur.

L'étude des paramètres cinétiques en fonction du pH a également été effectuée. L'activité de l'enzyme est sous la dépendance d'un groupe qui s'ionise aux pH acides, vraisemblablement un carboxylate; aux pH alcalins, elle dépend de l'ionisation d'un groupe de pK 8,5.

La β-galactosidase est le siège d'un certain nombre de phénomènes lents. L'activation par les ions Mg^{++} en est un exemple qui a fait l'objet d'une étude particulière.

2. Activation lente de l'enzyme par les ions Mg^{++}

L'enzyme est très sensible à l'influence de divers cations. Les ions Na^+ sont indispensables à l'activité. Certains cations divalents en particulier Mg^{++} et Mn^{++} sont activateurs [4]; d'autres au contraire sont inhibiteurs (comme Be^{++} et Ca^{++} [10]).

L'étude quantitative de l'activation par le Mg^{++} a été étudiée au pH optimum de la réaction, dans des conditions très contrôlées, en particulier en gardant constantes la concentration de Na^+ (0,145 M) et la force ionique ($0{,}17 \pm 0{,}02$). Le phénylgalactoside a été choisi comme substrat, la première étape chimique étant l'étape limitante, et dans des conditions de concentrations grandes devant K_m, la constante de Michaelis; ces conditions facilitent l'interprétation. La Figure 1 montre l'activation lente pour diverses concentrations de Mg^{++}. Lorsqu'on ajoute le Mg^{++} à l'enzyme libre de tout cation bivalent au temps zéro de la réaction, on observe une phase de latence initiale. Au bout d'un temps variable selon la concentration de Mg^{++}, la réaction atteint un régime stationnaire qui se traduit par la phase linéaire de la cinétique; la vitesse devient constante. Lorsque l'enzyme est préalablement incubé, pendant 10 mn en présence de Mg^{++}, on n'observe plus de phase de latence. Pour les fortes concentrations de Mg^{++}, la phase de latence devient trop rapide pour être observée par les méthodes classiques. On ne peut la suivre que par les techniques de 'stopped flow' [17].

L'activation lente par le Mg^{++} est indépendente de la nature du substrat; avec d'autres substrats, en particulier l'*o*-nitrophénylgalactoside, le phénomène est analogue.

Le processus inverse, la désactivation, est aussi un phénomène lent. Si, au temps

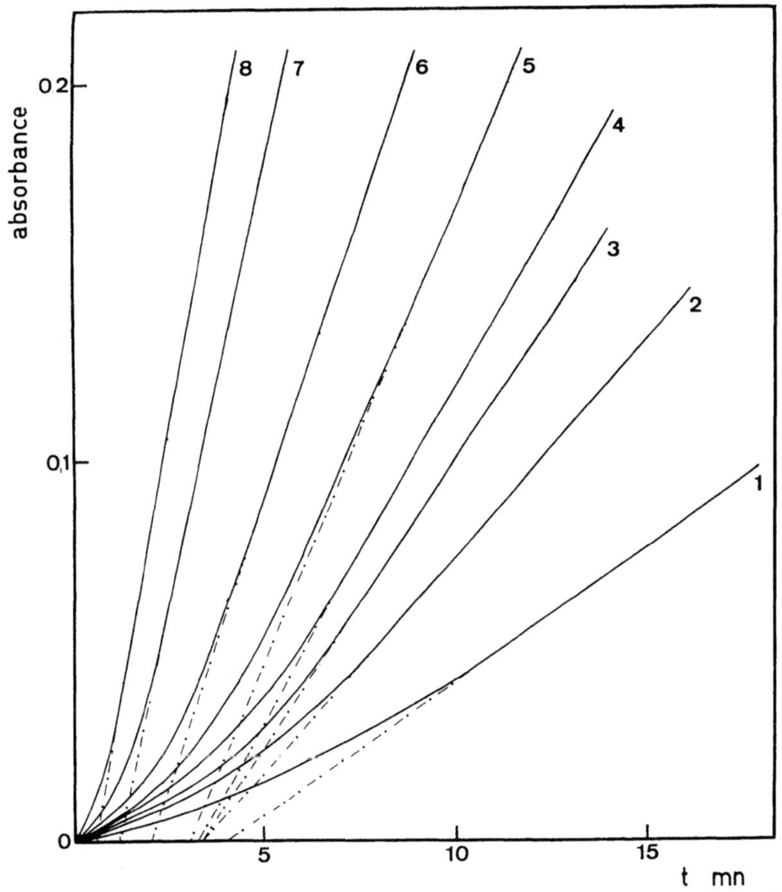

Fig. 1. Activation par le Mg^{++} de l'hydrolyse enzymatique du phényl-D-galactoside. Concentration d'enzyme $3,25 \times 10^{-8}$ M. Concentration de substrat 10^{-2} M; pH 7,0; $T = 25°$. Concentration de Mg^{++}: (1) $7,8 \times 10^{-8}$ M; (2) $2,02 \times 10^{-7}$ M; (3) $3,17 \times 10^{-7}$ M; (4) $4,2 \times 10^{-7}$ M; (5) $6,7 \times 10^{-7}$ M; (6) $1,23 \times 10^{-6}$ M; (7) $7,03 \times 10^{-6}$ M; (8) 5×10^{-5} M.

zéro de la réaction, on ajoute un agent complexant tel l'EDTA ou l'hydroxyquinoléine, on observe une lente désactivation de l'enzyme.

Il est donc possible de suivre la cinétique d'activation par les ions Mg^{++} ainsi que la réaction réverse (Figure 2). Avant d'aborder l'aspect cinétique de l'activation et de la désactivation, l'étude de la phase stationnaire sera rapidement décrite.

3. Étude de la phase stationnaire

L'activation par les ions Mg^{++} étant indépendante de la nature du substrat, seule la fixation de Mg^{++} sur l'enzyme peut être prise en considération. En étudiant les variations de la constante catalytique en fonction de la concentration de Mg^{++}, on peut déterminer une constante apparente de dissociation du complexe enzyme–Mg^{++}, par

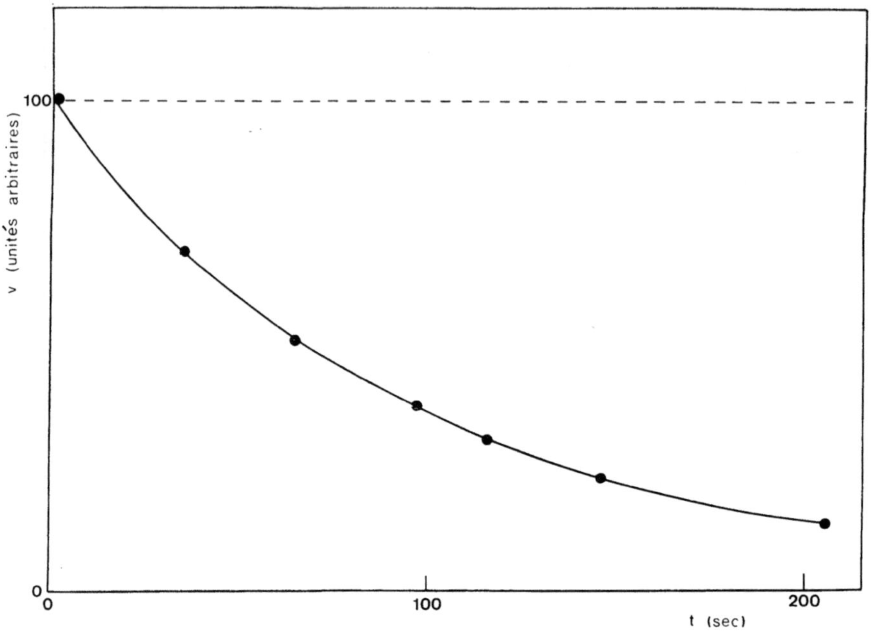

Fig. 2. Désactivation consécutive à l'addition d'EDTA à pH 7,0 et 25°. Substrat 10^{-2} M.

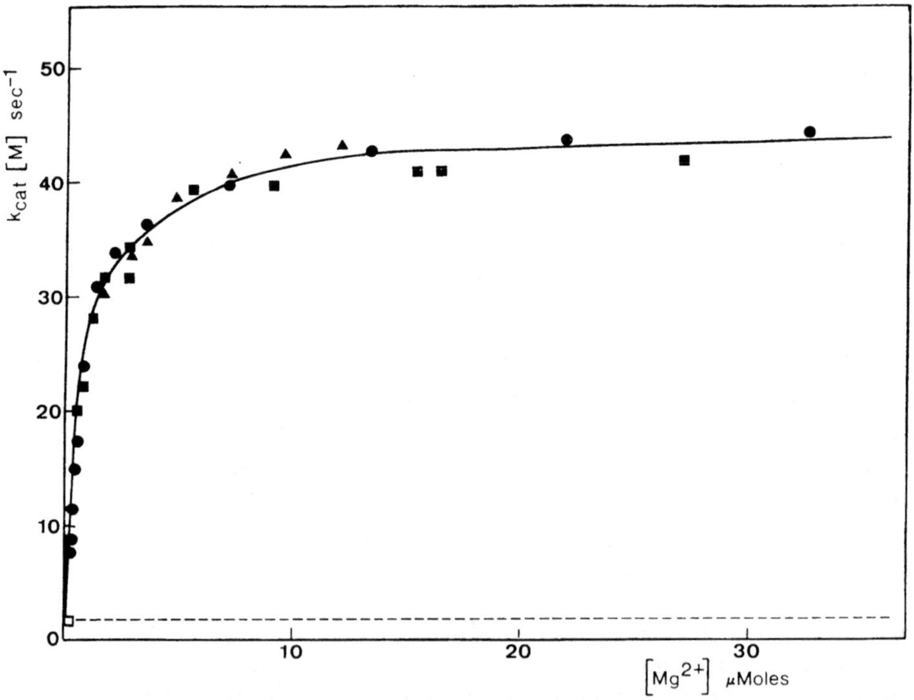

Fig. 3. Variation de k_{cat} en fonction de la concentration de Mg^{++} (substrat 10^{-2} M).

une représentation du type:

$$\Delta k_{cat} = (\Delta k_{cat})_{max} \frac{(M)}{(M) + K}.$$

Δk_{cat} représente la différence des valeurs de k_{cat} (la constante catalytique) pour une concentration donnée (M) de Mg^{++} et pour une valeur de Mg^{++} nulle (k_{cat} étant différent de zéro dans ces conditions). $(\Delta k_{cat})_{max}$ est la différence correspondante pour Mg^{++} saturant. K est la constante de dissociation apparente du complexe enzyme–Mg^{++}. La Figure 3 montre que l'on obtient une courbe de saturation michaélienne.

En utilisant la représentation de Hill (Figure 4), il apparaît plus clairement encore que le phénomène d'activation par le Mg^{++} ne présente aucune coopérativité. La constante K ainsi déterminée a une valeur de $(0{,}65 \pm 0{,}05)\ 10^{-6}$ M, ce qui correspond à une affinité apparente relativement forte de l'enzyme pour le métal. Des mesures directes de fixation du Mg^{++} sur l'enzyme ont permis de montrer en outre qu'un seul Mg^{++} par protomère se fixe sur l'enzyme.

Une étude analogue a été effectuée à pH 5,16. Dans ces conditions la constante de

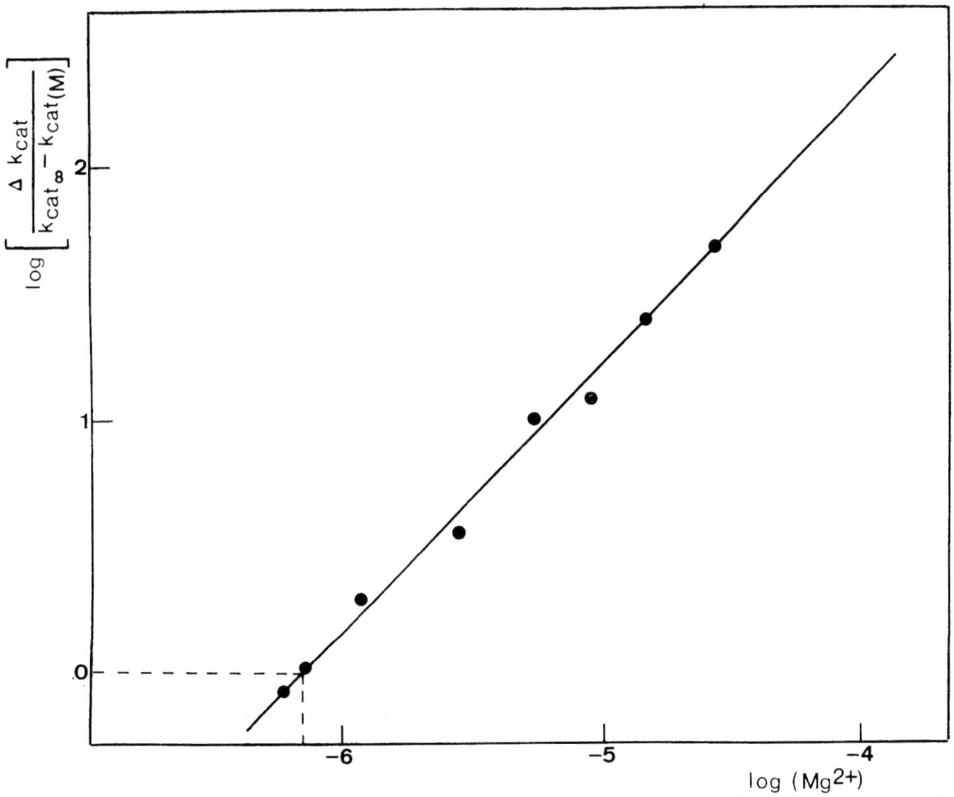

Fig. 4. Représentation de Hill d'après les données de la Figure 3.

dissociation est sensiblement différente: $6{,}85 \times 10^{-3}$ M. Une simple compétition entre protons et Mg^{++} pour le site de fixation ne peut suffire à expliquer cette différence.

4. Étude cinétique des processus d'activation et de désactivation

Une analyse de la période de latence, pour différentes concentrations de magnésium, permet de suivre la cinétique de l'activation qui est décrite d'une manière satisfaisante par une relation exponentielle simple:

$$v_t = v_0 + (v_\infty - v_0)\,[1 - \exp(-\,a_{(M)}t)]\,.$$

v_0 est la vitesse au temps zéro de la réaction, v_∞ la vitesse au temps infini, c'est-à-dire lorsque le complexe enzyme–substrat–magnésium a atteint sa concentration d'équilibre. $a_{(M)}$ est un paramètre complexe, qui dépend de la concentration de Mg^{++} et du mécanisme de l'activation.

Le paramètre $a_{(M)}$ peut être déterminé par une représentation logarithmique dérivée de cette expression (Figure 5). On peut l'obtenir plus précisément à partir des courbes données sur la Figure 1, d'après la relation:

$$P_t = v_\infty \left\{t - \frac{1}{a_{(M)}}\left(1 - \frac{v_0}{v_\infty}\right)[1 - \exp(-\,a_{(M)}t)]\right\}.$$

Le calcul est alors effectué au moyen d'une méthode de régression linéaire d'après Cleland [11] adaptée pour une calculatrice Wang.

Les valeurs obtenues montrent que pour une concentration donnée de Mg^{++}, le paramètre $a_{(M)}$ varie avec la concentration du substrat. Pour la désactivation, le para-

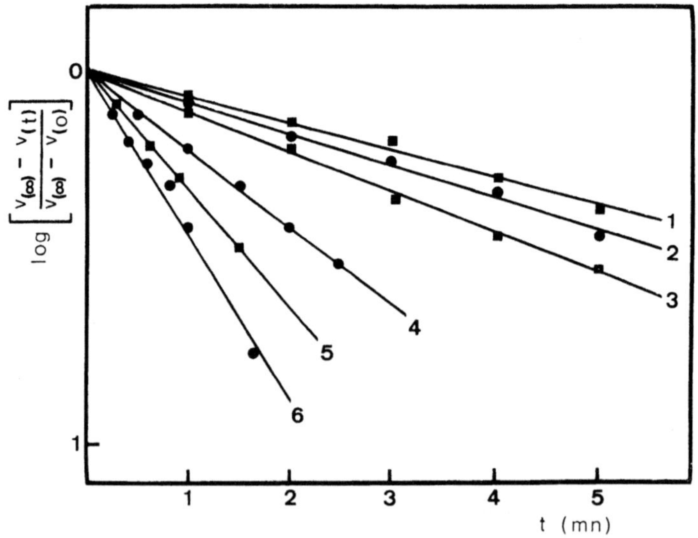

Fig. 5. Détermination du paramètre $a_{(M)}$ (voir texte) d'après une représentation logarithmique. Les conditions sont les mêmes que pour la Figure 1.

mètre a_0 correspondant varie aussi avec la concentration de substrat. Pour rendre compte de ces résultats nous avons été conduits à admettre une fixation préférentielle du cation sur l'enzyme libre; la fixation de Mg^{++} sur les complexes intermédiaires étant pratiquement négligeable.

Le schéma le plus simple compatible avec les résultats expérimentaux est le suivant:

$$\begin{array}{cccccccc}
 & E & +S \overset{K_s}{\rightleftharpoons} & ES & \overset{k_2}{\rightarrow} & ES' & \overset{k_3}{\rightarrow} & E + P_2 \\
M & \downarrow\uparrow & & & & +P_1 & M & \downarrow\uparrow \\
 & EM & +S \overset{K_s}{\rightleftharpoons} & EMS & \overset{k_2}{\rightarrow} & EMS' & \overset{k_3}{\rightarrow} & EM + P_2 \\
k_{23} & & k_{32} & & & +P_1 & & \downarrow\uparrow \\
 & EM^* & +S \overset{K'_s}{\rightleftharpoons} & EM^*S & \overset{\alpha k_2}{\rightarrow} & EM^*S' & \overset{\beta k_3}{\rightarrow} & EM^* + P_2 . \\
 & & & & & +P_1 & &
\end{array} \qquad \text{(I)}$$

Il implique que la fixation du Mg^{++} sur l'enzyme s'effectue selon un processus à deux étapes, la seconde étape s'accompagnant de l'activation de l'enzyme. La résolution de ce schéma permet de montrer [5] que le paramètre $a_{(M)}$ est de la forme:

$$a_{(M)} = \frac{k_{23}K_{Mg}(M)}{(1 + s/K_s)(1 + K_{Mg}(M))} + \frac{k_{32}}{1 + s/K'_s}.$$

Cette expression traduit une relation hyperbolique entre $a_{(M)}$ et la concentration de magnésium, pour une concentration donnée de substrat. La Figure 6 montre que les résultats expérimentaux sont en bon accord avec le schéma proposé.

Un schéma réactionnel impliquant une seule étape de fixation du Mg^{++} ne saurait rendre compte des résultats expérimentaux:

$$\begin{array}{cccccccc}
 & E & +S \rightleftharpoons & ES & \rightarrow & ES' & \rightarrow & E + P_2 \\
M & \downarrow\uparrow & & & & +P_1 & & \\
 & EM^* & +S \rightleftharpoons & EM^*S & \rightarrow & EM^*S' & \rightarrow & EM^* + P_2 . \\
 & & & & & +P_1 & &
\end{array} \qquad \text{(II)}$$

Il conduit à une variation linéaire du paramètre $a_{(M)}$ en fonction de la concentration de Mg^{++}:

$$a_{(M)} = \frac{k_{12}(M)}{1 + s/K_s} + \frac{k_{21}}{1 + s/K'_s}.$$

Les résultats donnés par la Figure 6 permettent une interprétation sans ambiguité. L'association du Mg^{++} à l'enzyme libre correspond bien à un processus à deux étapes. La première consiste en une fixation rapide du Mg^{++} à l'enzyme. On sait, d'après les résultats de Eigen [12] que lors de la fixation de Mg^{++} à un ligand en milieu aqueux, l'étape limitante est le départ de molécules d'eau qui a une constante de vitesse de $10^5\ sec^{-1}$. Il s'agit d'un processus beaucoup trop rapide pour rendre compte des phénomènes lents observés ici. La formation du premier complexe ne s'accompagne pas d'augmentation de l'activité enzymatique. Celle-ci se produit lors de la seconde

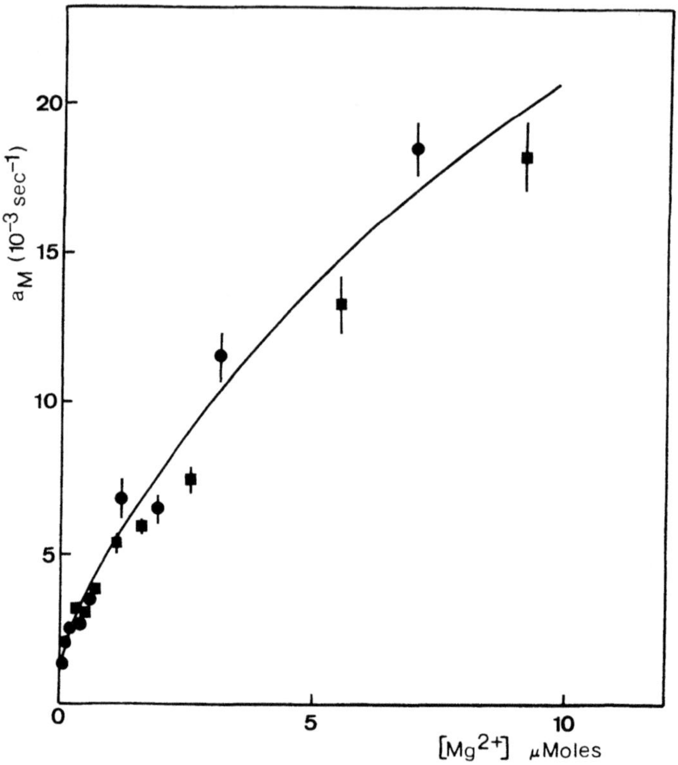

Fig. 6. Variation de $a_{(M)}$ en fonction de la concentration de Mg^{++}. Les conditions sont les mêmes que pour la Figure 1.

étape, beaucoup plus lente et résulte vraisemblablement d'un réarrangement de la molécule ou d'une partie de la molécule de *β*-galactosidase. Les variations conformationnelles induites par la fixation du Mg^{++} sont vraisemblablement très localisées. En effet, jusqu'à maintenant, aucune différence significative n'a pu être décelée par les méthodes classiques d'analyse conformationnelle: O.R.D., spectrophotométrie différentielle.

5. Caractéristiques de l'activation par le magnésium

La présence de Mg^{++} entraîne une modification des paramètres catalytiques. Avec le phénylgalactoside, k_2 est la vitesse limitante de la réaction avec et sans magnésium; sa valeur de 1,9 sec^{-1} en l'absence de métal atteint 45 sec^{-1} pour Mg^{++} 10^{-3} M. Dans les mêmes conditions K_m varie de 0,8 à 0,1 10^{-3} M. Avec un autre substrat, l'*o*-nitrophényl-*β*-D-galactoside, la présence de Mg^{++} entraîne un changement de l'étape limitante de la réaction. En l'absence de métal, k_2 est la vitesse limitante, alors qu'en présence de Mg^{++} 10^{-3} M, k_2 et k_3 sont du même ordre de grandeur. En fait l'activation entraînée par la fixation du métal à l'enzyme est beaucoup plus importante

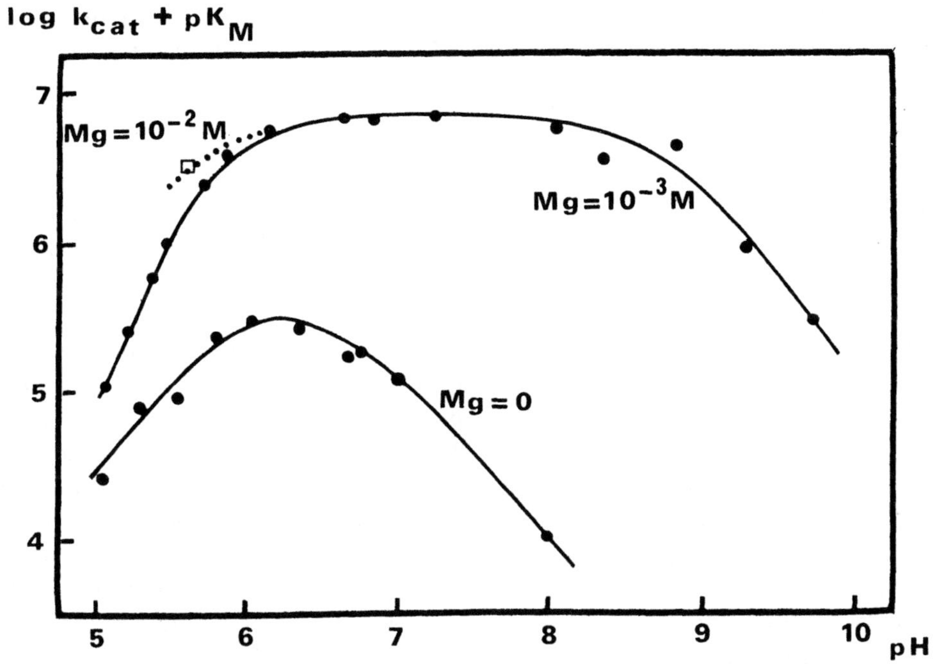

Fig. 7. Variation de k_{Fat}/K_m en fonction du pH. Substrat: *o*-nitrophényl-β-D galactoside, I en absence Mg^{++}, II en présence de Mg^{++} 10^{-3} M.

sur l'étape de formation du galactosylenzyme (k_2 est multiplié par un facteur 40) que sur sa décomposition (k_3 est multiplié par un facteur 10 au maximum).

Les profils de pH, c'est-à-dire l'allure des variations des paramètres catalytiques en fonction du pH, sont modifiés par la présence de Mg^{++}. En particulier, le pK du groupe qui contrôle l'activité de l'enzyme aux pH alcalins (évalué à partir des variations du rapport k_{cat}/K_m), est affecté par la présence de métal (Figure 7). De 6,5 pour l'enzyme sans métal le pK augmente jusqu'à 8,4 pour l'enzyme en présence de Mg^{++} 10^{-3} M. Il est d'ailleurs intéressant de noter que la fixation du substrat sur l'enzyme sans métal se traduit par un effet analogue, ce qui suggère que le groupe impliqué pourrait avoir un rôle conformationnel [8]. Des travaux récents de Yariv et coll. [14] vont dans le sens de notre hypothèse.

Parmi les cations divalents le Mg^{++} n'est pas le seul effecteur de l'enzyme; Mn^{++} et Co^{++} sont aussi activateurs. Avec ces cations on note également une phase de latence traduisant un réarrangement lent de la molécule enzymatique. L'affinité apparente de ces cations pour l'enzyme est supérieure à celle du Mg^{++} (pour Mn^{++}, $K \sim 5 \times 10^{-9}$ M, pH 7).

6. Signification de la constante apparente de dissociation du complexe enzyme–métal. Spécificité dynamique

La constante de dissociation du complexe enzyme–Mg^{++} déterminée par la représentation donnée sur la Figure 3, ou par une représentation linéaire correspondante est une constante complexe. En effet le schéma I conduit à l'expression suivante:

$$K = \frac{1}{K_{Mg}} \frac{(k_{32}/k_{23})(1 + s/K_s)}{(1 + s/K_s') + (k_{32}/k_{23})(1 + s/K_s)}.$$

Pour une concentration donnée de substrat, K dépend donc de la constante vraie d'association enzyme métal, mais aussi des constantes de vitesse de la réaction consécutive. Plus lente sera l'étape de vitesse k_{32}, plus grande sera l'affinité apparente du métal pour l'enzyme. Sa valeur importante résulte donc des mouvements moléculaires lents de la protéine consécutifs à la fixation du métal.

A partir de cet exemple, on peut se demander si la spécificité biologique d'une manière plus générale ne résulte pas de mouvements moléculaires lents. L'association d'un certain nombre d'effecteurs à leur récepteur spécifique correspond à de très fortes affinités apparentes; des constantes allant de 10^6 et atteignant parfois des valeurs de 10^{10} et même 10^{14} se rencontrent fréquemment dans les systèmes biologiques, bien que les forces mises en jeu soient généralement non covalentes. Dans les systèmes membranaires par exemple, on note des affinités assez grandes. Le Tableau I donne quelques valeurs de constantes de dissociation de complexes entre certains métabolites et leur protéines de transport dans la membrane de *E. coli.* Les valeurs relevées représentent des ordres de grandeur beaucoup plus faibles que pour les associations enzyme–substrat, par exemple, où la constante de dissociation dans la plupart des cas est de l'ordre de 10^{-3}, 10^{-4} M.

TABLEAU I

Valeurs des constantes de dissociation des complexes spécifiques pour quelques protéines de transport dans les membranes [15]

Source	Substrat transporté	K_{diss}(M)
E. coli	L-Leucine	3×10^{-7}
E. coli	D-Galactose	10^{-6}
E. coli	L-Arabinose	2×10^{-6}
Duodenum de poulet	Ca^{2+}	4×10^{-6}

Des affinités encore plus fortes peuvent être relevées pour les interactions hormones–récepteurs. D'après Baulieu *et al.* [16], l'association de l'oestradiol à une protéine réceptrice de l'utérus s'effectue avec une constante de 5×10^{10} M^{-1} d'après les études effectuées à l'équilibre. Les mesures cinétiques indiquent une très faible vitesse de dissociation expliquant la grande stabilité du complexe ($k_{-1} = 3{,}6 \times 10^{-7}$ s^{-1}). Cependant, d'après les auteurs, les résultats semblent indiquer que le phénomène est plus complexe qu'un équilibre simple. Bien qu'ils n'en donnent aucune explication, on est en droit de se demander si, dans ce cas comme celui du magnésium et de la β-galactosidase, on ne se trouve pas en présence d'un phénomène en plusieurs étapes impliquant des réarrangements lents de la protéine réceptrice.

Quoiqu'il en soit, il semble que la stéréospécificité des récepteurs ne suffise pas à expliquer les affinités importantes observées en biologie. La spécificité biologique est vraisemblablement une 'spécificité dynamique' impliquant des relaxations lentes des molécules ou des ensembles moléculaires concernés.

Références

[1] Craven, G., Anfinsen, C. B., et Steers, E.: 1965, *J. Biol. Chem.* **240**, 2468.
[2] Zipser, D.: 1963, *J. Mol. Biol.* **7**, 113.
[3] Ullmann, A., Jacob, F., et Monod, J.: 1968, *J. Mol. Biol.* **32**, 1.
[4] Cohn, M. et Monod, J.: 1951, *Biochim. Biophys. Acta* **7**, 153.
[5] Tenu, J. P., Viratelle, O. M., et Yon, J. M.: 1972, *Eur. J. Biochem.* **26**, 112.
[6] Viratelle, O. M., Tenu, J. P., Garnier, J., et Yon, J. M.: 1969, *Biochem. Biophys. Res. Commun.* **37**, 1036.
[7] Viratelle, O. M. et Yon, J. M.: 1972, *Eur. J. Biochem.* **26**, 112.
[8] Tenu, J. P., Viratelle, O. M., Garnier, J., et Yon, J. M.: 1971, *Eur. J. Biochem.* **20**, 363.
[9] Sinnott, M. et Viratelle, O. M.: 1973, *Biochem. J.* **133**, 81.
[10] Kuby, S. A. et Lardy, H. A.: 1953, *J. Am. Chem. Soc.* **75**, 890.
[11] Cleland, W. W.: 1967, *Advan. Enzymol.* **29**, 1.
[12] Eigen, M.: 1963, *Pure Appl. Chem.* **6**, 97.
[13] Eigen, M.: 1951, dans *Proc.* Roberts A. Welsh Foundation, Conference on Chemical Research, Houston.
[14] Yariv, J.: communication personnelle.
[15] Kotyk, A. et Janacek, K.: 1970, dans *Cell Membrane Transport, Principles and Techniques*, Plenum Press, New York.
[16] Baulieu, E. E., Alberga, A., Jung, I., Lebeau, M. C., Mercier-Bodard, C., Milgrom, E., Raynaud, J. P., Raynaud-Jammet, C., Rochefort, H., Truong, H., et Robel, P.: 1971, *Recent Progr. Hormone Res.* **27**, 351.
[17] Strom, R., Attardi, D. G., Forsen, S., Turini, P., Celada, F., et Antonini, E.: 1971, *Eur. J. Biochem.* **23**, 118.

DISCUSSION

M. Lazdunski: (1) Peut-on étudier l'état préstationnaire avec les dinitrophénylgalactosides pour lesquels k_3 est déterminant? (2) La présence de Na^+ est-elle indispensable dans les expériences que vous décrivez? Si oui, quel est l'effet Na^+ ou l'effet Mg^{++}?

Mlle Yon: (1) Non, cela n'est pas possible par suite de la grande valeur de k_3. Bien que limitant, k_3 est de 1200 sec^{-1} à pH 7,0; cette étape est trop rapide pour être suivie par 'stopped flow' et observer les 'burst'. Même en jouant sur le pH, il ne nous a pas été possible avec les dinitrophénylgalactosides de trouver les conditions pour observer le burst. Pour cela il faudra trouver des substrats plus adéquats. (2) La présence de Na^+ est indispensable. Cependant, nous n'avons pas étudié l'effet Na^+. Nous avons simplement travaillé à concentration constante en Na^+ pour toutes nos expériences (Na^+: 0,145 M) et à force ionique constante ($\mu = 0,17 \pm 0,02$). La présence de Na^+ affecte la vitesse l'activation par Mg^+(17).

Ptak: (1) Le départ de molécules d'eau dont vous avez parlé accompagnant la fixation de Mg^{++}, correspond-il bien à la désolvatation des ions? Ne doit-il pas y avoir également une désolvatation des sites ioniques de la protéine fixant les métaux? (2) Que pouvez-vous dire sur la stoechiométrie de la réaction de fixation du Mg^{++}?

Mlle Yon: (1) Le départ des molécules d'eau qui s'effectue avec une vitesse de 10^5 M^{-1} sec^{-1}, correspond bien à la désolvatation de l'ion. La désolvatation de la protéine au site de fixation de l'ion doit bien entendu se produire également. Toutefois, ce phénomène me semble trop rapide pour expliquer les phénomènes lents observés. (2) J'ai en effet omis de dire qu'il se fixe un Mg^{++} par protomère d'enzyme, c'est-à-dire quatre par molécule. Nous avons pu le montrer par des expériences directes.

Goldberg: (1) L'augmentation par des facteurs 40 et 10 des constantes k_2 et k_3 par Mg^{++} est-elle la même pour les différents substrats? (2) L'augmentation de la constante apparente d'association de certains 'récepteurs biologiques' par un phénomène 'séquentiel' tel que vous le proposez, peut-elle rendre compte du fait que la vitesse d'échange d'un ligand sur de tels récepteurs est parfois considérablement plus grande que la vitesse de dissociation?

Mlle Yon: (1) L'augmentation par un facteur 40 de k_2 et par un facteur 10 de k_3 est indépendante de la nature du substrat. (2) J'avoue n'y avoir pas réfléchi. Ceci me semble a priori possible. Lorsque l'équilibre est atteint entre récepteur R et substrat S, on a:

rapide lent
| R + S ⇌ RS ⇌ RS* |
→ S′ ⇅ à l'équilibre
RS′
⇅
RS′*

une certaine quantité de R libre et le RS; on observe peut-être d'abord la réversion de l'étape rapide, mais au bout d'un certain temps on devrait observer la réversion de l'équilibre lent. On devrait observer une cinétique biphasique.

MOUVEMENTS INTRAMOLÉCULAIRES DANS LA CHYMOTRYPSINE

B. LABOUESSE

Institut de Biochimie, Orsay, France

Résumé. La chymotrypsine, malgré son faible poids moléculaire (25000), sa structure globulaire, et la présence de cinq liens covalents (disulfure) entre différents points de sa chaîne peptidique, oscille en permanence entre deux états conformationnels, l'un catalytiquement actif, l'autre inactif. La conformation active est favorisée par la présence d'une liaison électrostatique précise à l'intérieur de la molécule et est seule à pouvoir assurer la fixation des substrats. Le passage d'une conformation à l'autre concerne un nombre restreint d'acides aminés de la chaîne.

Summary. Chymotrypsin is known to exist in more than one conformational state. At 15°, KCl 0,15 M, one state (E) predominates at neutral pH, another (E′) state at alkaline pH. From its alkaline pH conformation the protein can resume its E form upon either binding a competitive inhibitor such as indole or making an acylenzyme with a pseudosubstrate such as DFP. The same conformation E appears to exist whether it is hold by binding of a specific ligand at the active site or by binding of a proton to a unique ionizing group, the α-amino group of the N-terminal isoleucine 16. This apparent structural identity is evidenced by the fact that there are no significant changes when the protein in its neutral form E binds an inhibitor or a substrate as seen from absorbancy, fluorescence, optical rotation or circular dichroism measurements. The conversion from one form to the other is a slow process (as compared to any of the catalytic steps of the enzyme action), in which the fastest isomerization path has a rate constant of the order of sec^{-1}. This sluggishness shows up in a lag phase when a specific substrate such as N-acetyl tyrosine ethyl ester is added to the E′ form of the enzyme: this form presents no catalytic activity towards such a substrate and has to isomerize into the E state before exhibiting any activity. Several characteristic features appear to differ between states E and E′: (1) When its N-terminal isoleucine 16 residue is blocked by alkylation or removed by leucine aminopeptidase, chymotrypsin is forced into its E′ state; it becomes inactive towards specific substrates and presents spectroscopic properties close to those of chymotrypsinogen. (2) Reaction of isoleucine 16 is prevented when the protein is in the E state (acylenzyme such as DIP-chymotrypsin). (3) In the zymogen as well as in the E′ form of the enzyme, methionine 192 is chemically nonreactive. In the E form of the enzyme this residue can be alkylated or oxidized. (4) In the acidic pH range a conformational transition depending on one carboxylic group occurs at the same rate and with the same fluorescence changes as the one controlled by the ionization of isoleucine 16.

From these kinetic, physicochemical and chemical data it appears that chymotrypsin oscillates slowly and continuously between two structures, one being catalytically inactive, the other active. In the active structural state E the N-terminal isoleucine residue of the protein lies inside the molecule, where it interacts with a carboxylic group, and its methionine 192 residue is out in the solvent; in the inactive structural state E′ the isoleucine 16 residue has moved out at the protein surface and the methionine residue has slipped into the protein interior. This behaviour is identical to what would happen if the shift E→E′ were identical to the structural change observed between chymotrypsin and chymotrypsinogen as shown by Blow's and Kraut's cristallographic data. The amplitude of the movements of methionine 192 and isoleucine 16 (half the radius of the protein) would then explain the high enthalpy and entropy of activation needed by the different paths of the isomerization process. Therefore, it is suggested that the essential features of the intramolecular structural oscillation of chymotrypsin are those that are seen in the cristallographic models of chymotrypsin and chymotrypsinogen, and that the enzyme, in its native state, is able to exist alternatively in both these structures.

1. Introduction

Le concept de protéine rigide, au creux de laquelle vient s'insérer un substrat de structure complémentaire ('la serrure et la clef' d'Émile Fischer), a été progressivement

C. Sadron (ed.), Dynamic Aspects of Conformation Changes in Biological Macromolecules. 459–476.

abandonné pour tenir compte de la souplesse de spécificité de beaucoup d'enzymes ('induced-fit') et surtout des phénomènes de régulation enzymatique (régulation allostérique) qui ne sont compatibles qu'avec une certaine mobilité conformationnelle de ces molécules. A titre d'exemple les deux états structuraux distincts présentés respectivement par l'oxyhémoglobine et la désoxyhémoglobine [1], illustrent à quel point la fixation d'un très petit ligand peut retentir sur l'organisation architecturale d'une protéine. Cependant, entre de tels états cristallins limites on ne sait pas très bien ce qui se passe et les cristaux étant par nature des structures relativement figées on ne peut évaluer l'importance statistique de telle ou telle conformation, l'existence d'états intermédiaires, et la désynchronisation éventuelle de certains des mouvements intramoléculaires qui permettent de passer de l'un à l'autre; plus encore on ne peut déceler l'existence d'états conformationnels non cristallisables, si l'on ne s'en tient qu'à la cristallographie comme technique d'étude structurale. Le cas de l'hémoglobine, où une protéine a pu être cristallisée et 'cristallographiée' dans deux situations architecturales distinctes, est pour l'instant relativement exceptionnel et l'on peut se demander si de nombreux exemples similaires pourront être examinés et si par conséquent d'autres approches expérimentales ne doivent pas être nécessairement utilisées pour analyser ces mouvements ou recouper les données cristallographiques partielles qui peuvent exister.

Bien qu'en principe un équilibre apparaissant entre deux états conformationnels puisse être déplacé par un agent physique (saut de température, par ex.), c'est dans la plupart des cas à la fixation de ligandes plus ou moins spécifiques que sont liés les changements de conformation et c'est à la suite de telles fixations qu'on est venu à les mettre en évidence. Réciproquement, sous l'angle expérimental, il faut évidemment provoquer un tel changement de conformation pour pouvoir en déceler l'existence et en analyser la nature, et l'addition d'un ligand, que celui-ci soit substrat, régulateur ou simplement un ion, est la phase initiale, l'amorce d'une telle analyse. S'il y a changement structural, les variations des propriétés physiques, chimiques, catalytiques de la protéine permettront d'en pénétrer la nature. Par les propriétés physiques, en particulier spectrales de la protéine, il est possible de situer les types de groupements perturbés (chromophores, squelette peptidique, métal, etc.), d'accéder aux vitesses du phénomène, et de là, aux forces qui le contrôlent si l'on peut opérer expérimentalement à plusieurs températures. Par les propriétés chimiques on peut mettre en évidence au moins certains éléments structuraux concernés par le remaniement observé. Par les propriétés catalytiques ou l'affinité, on peut évaluer à quel point les sites 'actif' ou 'régulateur' sont concernés par les variations provoquées et détectées. C'est l'ensemble des données ainsi confrontées qui permettra de remplacer partiellement ou de compléter les données cristallographiques, qui demeurent évidemment essentielles à toute analyse et toute compréhension approfondies.

La chymotrypsine offre un assez bon exemple de ce que peut apporter une étude convergente effectuée par voie cinétique, physique ou chimique des mouvements intramoléculaires d'une protéine, non seulement parce qu'étant étudiée depuis trente ans beaucoup de données ont été accumulées à son sujet mais également parce que les informations cristallographiques que l'on possède sur elle, pour remarquablement fines qu'elles soient n'avaient précisément pas fait émerger l'idée que cette protéine puisse présenter une quelconque mobilité structurale et auraient même pu conduire à en faire un exemple de protéine 'rigide'. Bien des propriétés laissent en effet penser qu'elle est relativement figée: sa forme presque sphérique jointe à sa petite dimension (PM 25000), la superstructuration que lui apportent cinq ponts disulfure, le fait qu'entre enzyme libre et acylenzyme pratiquement aucune différence architecturale ne soit apparue aux cristallographes, malgré les différences discrètes qui avaient pu se manifester lors de l'étude en solution de cette protéine [2].

L'analyse présentée ici a donc pour but de résumer l'évolution des idées relatives à la rigidité de cette protéine et de situer la méthodologie qui a pu être mise en œuvre, aussi bien pour parvenir à la conclusion qu'elle présente une réelle mobilité architecturale, que pour évaluer ce en quoi consiste ses mouvements intramoléculaires.

2. Existence de plusieurs états conformationnels dans la chymotrypsine

L'une des premières observations suggérant que la chymotrypsine pouvait présenter des changements de structure dans les conditions usuelles de température et de force ionique (20–25°; KCl 0,1–0,15 M) a été celle qui montrait que la rotation optique de cette protéine variait avec le pH [3]. Le fait que le pouvoir rotatoire d'une protéine présente un tel changement peut être en effet dû soit à la simple apparition d'une charge au voisinage d'un chromophore asymétrique soit à une modification structurale. L'hypothèse d'un changement de conformation ne pouvant être étayée par la mise en évidence de la seule variation d'un paramètre physique de la molécule, il fallait rechercher si d'autres éléments changeaient dans les mêmes conditions expérimentales. Ce peut être des propriétés spectrales intrinsèques de la protéine (absorption, fluorescence, dichroisme, etc.) ou extrinsèques telles que celles que pourrait apporter un réactif couplé de façon covalente; ce peut être également les propriétés d'affinité pour les substrats ou les inhibiteurs, une variation d'affinité suggérant un changement de conformation.

De telles variations ont pu être observées pour les paramètres a_0 et b_0 de la rotation optique [4], pour l'absorption dans l'ultraviolet au niveau des bandes caractéristiques de la tyrosine et du tryptophane (279, 288, 292 nm) [5], pour l'émission de fluorescence du tryptophane (340 nm) [6], pour le dichroisme circulaire (230 nm) [7], pour l'affinité spécifique pour les ions H^+ [8] et pour la résonance de spin d'un marqueur attaché de façon covalente [9]; d'autre part ces changements correspondent à ceux que l'on observe en mesurant la vitesse de fixation d'un pseudosubstrat [10] et en mesurant la fluorescence d'un inhibiteur compétitif (proflavine) [11]. Ainsi le nombre d'éléments architecturaux concernés par ces variations (squelette peptidique, résidus de tryptophane ou de tyrosine, résidus périphériques fixant les ions H^+ ou marqués par covalence, site actif de l'enzyme) sont trop nombreux et trop différents pour que l'hypothèse d'un changement de conformation ne s'impose pas.

Si un changement de conformation, et en particulier un changement au niveau du site actif, a lieu sous l'influence d'une modification de pH, c'est qu'un ou plusieurs

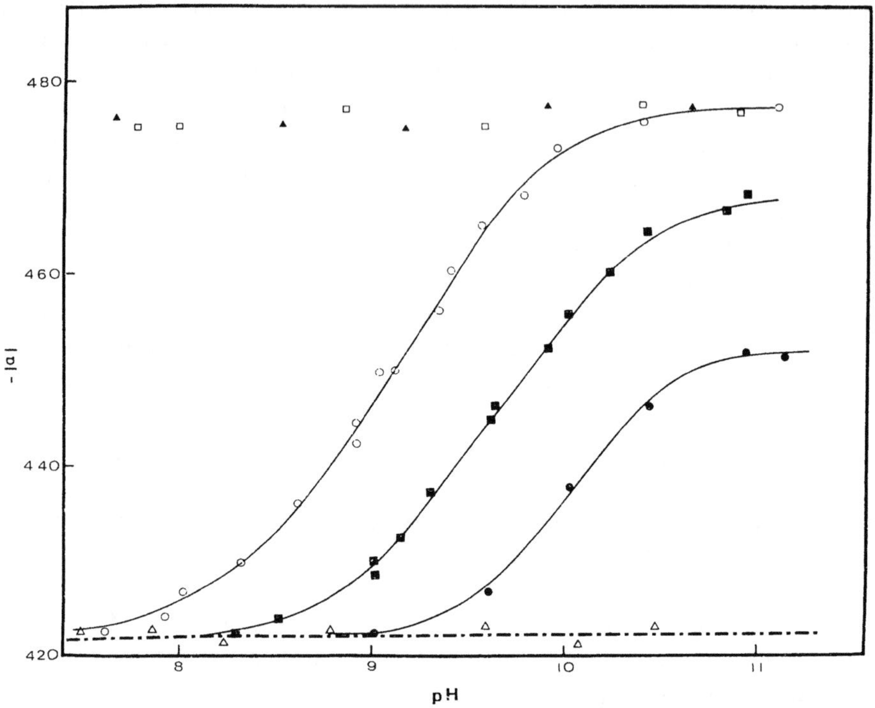

Fig. 1. Variation du pouvoir rotatoire $|\alpha|_{320}$ à 15°, KCl 0,15 M, pour le N(ε)acétyl-chymotrypsinogène (□), pour la N(ε)acétyl-DIP-chymotrypsine (acylenzyme) (△), pour la N(ε)acétyl-chymotrypsine acétylée sur l'isoleucine 16 (▲), et pour la N(ε)acétyl-chymotrypsine en absence d'indole (–○–) et en présence d'indole 2×10^{-3} M (–■–) et 9×10^{-3} M (–●–). La ligne pointillée (·–·–·) représente la variation de $|\alpha|_{320}$ de la N(ε)acétyl-chymotrypsine pour une concentration en indole extrapolée à l'infini [27, 28].

groupements ionisables de la protéine changent de position et par là d'interaction avec leur voisinage lors de l'acquisition ou de la perte d'une charge. Réciproquement si le changement de conformation a lieu indépendamment d'un changement de charge, lors de la fixation d'un substrat par exemple, les interactions de ces groupements ionisables doivent se trouver modifiées et leur pK sera déplacé [12]. La Figure 1 montre que c'est bien ce que l'on observe: le même changement de conformation peut être imposé par fixation d'indole (inhibiteur compétitif) au site actif de l'enzyme ou par fixation de protons. L'analyse de ces courbes montre que la fixation d'une molécule d'indole ou celle d'un proton, par molécule d'enzyme, provoque le même changement de conformation; l'affinité de la protéine pour l'indole change lorsque sa structure est modifiée par fixation d'ions H^+ et réciproquement son affinité pour les ions H^+ change lorsque sa structure est modifiée par fixation d'indole: le pK du groupe critique contrôlant la structure (pK de 9) tend vers une valeur plus élevée lorsque la concentration en indole croît (pK de 10,5 à saturation en indole); de même l'affinité de la protéine pour l'indole ($K_i = 5 \times 10^{-4}$ M à pH neutre) décroît lorsque la concentration en ions H^+ décroît ($K_i = 1{,}5 \times 10^{-2}$ M à pH 11). La chymotrypsine présente donc au moins deux états structuraux dont l'un est apparemment lié à la présence d'un inhibiteur au site actif ou d'un proton en un site spécifique, et dont l'autre est prédominant en absence d'inhibiteur, à pH élevé. La Figure 1 montre en outre que le premier état structural semble être maintenu tant par fixation non covalente d'un inhibiteur compétitif que par fixation covalente d'un pseudosubstrat (DIP), tandis que le deuxième état est proche de celui du zymogène.

3. Vitesse d'interconversion des conformations l'une en l'autre

Si la protéine est susceptible d'occuper plusieurs états conformationnels et que l'un d'eux est favorisé par la présence d'un ligand spécifique, l'addition d'un tel ligand à la protéine augmentera la proportion de conformation favorisée. Si le ligand spécifique est un proton, il est en outre possible de diminuer, aussi bien que d'augmenter, sa

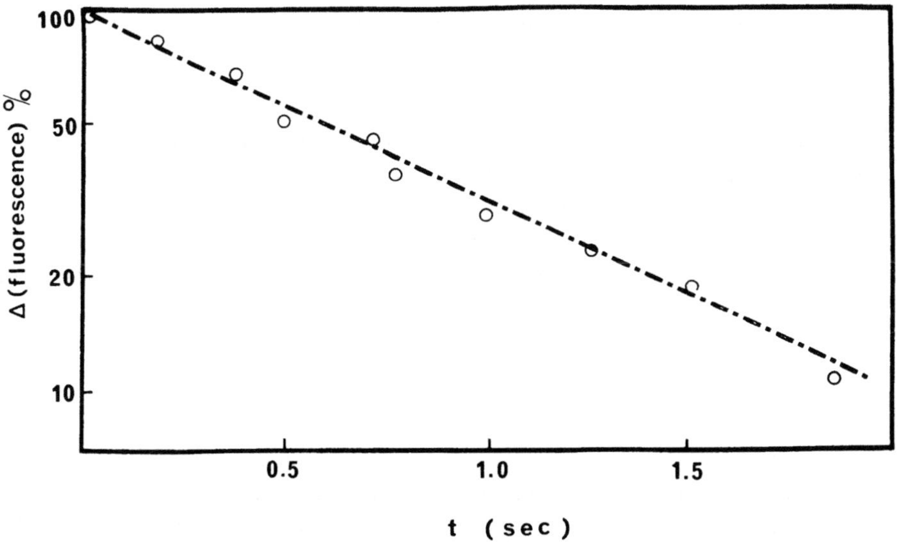

Fig. 2. Variation de la fluorescence de la N(ε)acétyl-chymotrypsine (15°, KCl 0,15 M) après un saut de pH (de pH 7,5 à pH 10,2) [5].

concentration. Par changement de concentration de ce seul ligand on peut donc amener la protéine à l'une ou l'autre des deux (ou des *n*) structures que l'on sait imposer. La Figure 2 montre la variation de la fluorescence de la chymotrypsine après un saut de pH, en absence de tout substrat en inhibiteur.

On peut constater:

(1) que le changement obéit à une cinétique de premier ordre; il est monophasique et ne permet de mettre en évidence ni phénomène plus rapide ni phénomène plus lent.

(2) que l'amplitude du changement mesuré pendant la phase exponentielle rend

compte de la totalité de la variation observable à l'équilibre; aucun changement très rapide n'a précédé le processus observé.

Ces constatations permettent de suggérer qu'après un saut de pH (positif ou négatif) la protéine passe d'un état structural à un autre sans intermédiaire décelable par fluorescence. Le même résultat est obtenu lorsque l'on mesure la variation d'absorbance, à plusieurs longueurs d'onde; en outre, il est retrouvé lorsque l'on suit la variation de rotation optique de la protéine après changement de pH ou après addition d'indole: quelle que soit la technique d'analyse utilisée le processus observé demeure monophasique, réversible, sans phase rapide préalable et sans phase lente ultérieure. Ces données sont essentiellement compatibles avec l'existence de deux états structuraux discrets [12] où aucun intermédiaire de durée de vie significative n'est mis en évidence. Ces deux états sont réversiblement convertibles l'un en l'autre, et l'on ne peut distinguer, par aucune des techniques physiques utilisées, l'état structural où la protéine a fixé une molécule d'indole (ou une molécule de pseudosubstrat) au site actif, de celui où elle a fixé un proton (en un site qui sera caractérisé ultérieurement) ou les deux simultanément.

Il est donc possible de ramener cette interconversion structurale à un équilibre tel que:

$$\mathrm{EL} \rightleftharpoons \mathrm{E'} + \mathrm{L}, \tag{1}$$

où E et E′ sont les deux états conformationnels de la protéine et L est l'un des ligands entraînant l'interconversion. Cet équilibre ne peut évidemment être aussi simple puisque l'association $\mathrm{E'+L \rightarrow EL}$ doit nécessairement se faire en deux étapes: $\mathrm{E'+L \rightarrow E'L \rightarrow EL}$ ou bien $\mathrm{E'+L \rightarrow E+L \rightarrow EL}$ ou les deux, tout comme la dissociation, et le schéma minimum rendant compte de la formation du complexe et de l'isomérisation est:

$$\begin{array}{ccc} \mathrm{EL} & \rightleftharpoons & \mathrm{E + L} \\ \upharpoonleft\downharpoonright & & \upharpoonleft\downharpoonright \\ \mathrm{E'L} & \rightleftharpoons & \mathrm{E' + L}. \end{array} \tag{2}$$

Si L est un proton, H^+, il sera possible d'élever ou d'abaisser réversiblement sa concentration et d'examiner cinétiquement l'évolution du système lorsque l'on part de l'état EH^+ (à forte concentration en H^+) ou de l'état E′ (à faible concentration en H^+). En milieu suffisamment tamponné la formation du complexe EH^+ ou $E'H^+$ ne sera pas limitée par la concentration en protons, même à pH élevé [13], et la formation de ces complexes (protonation d'un groupe ionisable) est de plusieurs ordres de grandeur plus rapide que les processus d'isomérisation observés.

Le schéma précédent, dans le cas de la simple fixation d'un proton peut s'écrire*:

$$\begin{array}{ccc} \mathrm{EH} & \overset{K}{\rightleftharpoons} & \mathrm{E + H} \\ k_1 \Big\uparrow\Big\downarrow k_{-1} & \searrow K_a & k_2 \Big\uparrow\Big\downarrow k_{-2} \\ \mathrm{E'H^+} & \underset{K'}{\rightleftharpoons} & \mathrm{E' + H}. \end{array}$$

* Dans le schéma (3) et l'Équation (4) le proton ou sa concentration sont représentés par la simple notation H.

Dans ce schéma, K_a est la constante de dissociation apparente du proton dans le processus d'isomérisation $EH \rightleftharpoons E'+H$, K et K' sont les constantes de dissociation individuelles pour chacun des deux états $\bar{E}$ (c'est-à-dire E+EH) et $\bar{E}'$ (c'est-à-dire, E'+E'H), k_1, k_{-1}, k_2, k_{-2}, sont les constantes de vitesse de chaque étape d'isomérisation de la protéine.

La constante de vitesse d'isomérisation apparente, k, après un changement de pH, est liée aux six constantes précédentes par la relation:

$$k = \frac{k_1 H + k_2 K}{H + K} + \frac{k_{-1} H + k_{-2} K'}{H + K'}, \tag{4}$$

a concentration initiale en ions H n'intervenant pas dans la vitesse du processus d'interconversion. En outre le schéma carré précédent demande que l'on ait $K_a = (k_1/k_{-1})\, K' = (k_2/k_{-2})\, K$.

Dans les conditions favorables,* la variation de k avec le pH final du milieu permet d'accéder aux deux constantes d'équilibre et aux quatre constantes de vitesse, et connaissant ces vitesses, aux équilibres $EH \rightleftharpoons E'H$ et $E \rightleftharpoons E'$.

Une telle étude a permis de montrer:

(1) que le groupe auquel se fixe spécifiquement un proton pour provoquer le changement de structure (un groupe α aminé) possède un pK à peu près normal dans la conformation $\bar{E}'$ (pK'=7,5) et qu'il est extrêmement anormal dans la conformation $\bar{E}$ (pK=10,5);

(2) que les constantes de vitesse de cette isomérisation sont faibles: de l'ordre de la sec^{-1} ou moins, ce qui laisse penser à un remaniement structural impliquant beaucoup plus que des oscillations de chaîne latérale d'acides aminés (Tableau I);

TABLEAU I

Constantes de vitesse et d'équilibre du schéma (3), à 15°, KCl 0,15 M

k_1	k_{-1}	k_2	k_{-2}	K	K'	K_a
0,005 s^{-1}	0,23 s^{-1}	3,3 s^{-1}	0,07 s^{-1}	3×10^{-11} mole/l	6×10^{-8} mole/l	10^{-9} mole/l

(3) que l'équilibre entre les deux conformations est fortement déplacé en faveur de l'état $\bar{E}$ à pH neutre (~98% de $\bar{E}$, 2% de $\bar{E}'$ à pH 7), et de l'état $\bar{E}'$ à pH alcalin (~98% de $\bar{E}'$, 2% de $\bar{E}$ à pH 11), mais qu'en absence d'autres ligand que H^+ ces deux états structuraux coexistent en permanence, quelle que soit la concentration en proton. Dans les conditions expérimentales ou l'activité enzymatique est généralement déter-

* Il faut en particulier que K, K' et K_a soient suffisamment distincts (un ordre de grandeur par exemple) pour permettre l'analyse avec quelque confiance.

minée (pH 8 à 8,5), 10 à 20% de la protéine sont sous la forme structurale $\bar{E}'$, et 80–90% sous forme $\bar{E}$, à 15°.

Des conclusions similaires ont été également obtenues en examinant la formation du complexe ternaire protéine-proton-inhibiteur, l'inhibiteur (proflavine) servant de signal (fluorescence) à la place de la protéine [14].

(4) que les vitesses et les équilibres précédents sont fortement dépendants de la température (Tableaux II et III); on voit en particulier que la constante d'équilibre apparente K_a croît rapidement avec la température (pK_a de l'ordre de 8 à 30°: à cette température les molécules de chymotrypsine sont réparties moitié-moitié entre les états $\bar{E}$ et $\bar{E}'$ au pH où l'activité est habituellement mesurée). On voit d'autre part que le processus d'isomérisation de la protéine libre vers la forme de faible affinité pour les ligands ($\bar{E} \rightarrow \bar{E}'$) présente une enthalpie d'activation négative: cette isomérisation est d'autant plus lente que la température est plus élevée.

TABLEAU II

Valeurs des paramètres thermodynamiques relatifs aux équilibres impliqués dans la transition alcaline de l'acétylchymotrypsine[a]

Processus	Constante	ΔH (kcal/mole)	ΔS (u.e./mole)	ΔG à 15° (kcal/mole)
$EH \leftrightarrow E' + H$	K_a	28 ± 3	55 ± 5	$11{,}9 \pm 0{,}1$
$EH \leftrightarrow E + H$	K	-20 ± 3	-117 ± 10	$13{,}9 \pm 0{,}2$
$E'H \leftrightarrow E' + H$	K'	$8 \pm$	-6 ± 3	$9{,}5 \pm 0{,}2$
$E \leftrightarrow E'$	L	43 ± 4	157 ± 12	$-2{,}1 \pm 0{,}2$
$EH \leftrightarrow E'H$	L'	16 ± 2	46 ± 6	$2{,}3 \pm 0{,}2$

[a] Acétylée sur les groupes ε aminés, sur le groupe α aminé de la cystine N-terminale et sur les deuxtyrosines accessibles. [29]

TABLEAU III

Valeurs des paramètres d'activation relatifs aux isomérisations de l'acétylchymotrypsine[a]

Réaction	Constante	$\Delta H^{\neq}$ (kcal/mole)	$\Delta S^{\neq}$ (u.e./mole)	$\Delta G^{\neq}$ à 15° (kcal/mole)
$EH \rightarrow E'H$	k_1	37 ± 3	69 ± 8	$17{,}4 \pm 0{,}4$
$E'H \rightarrow EH$	k_{-1}	22 ± 2	24 ± 5	$15{,}2 \pm 0{,}4$
$E \rightarrow E'$	k_2	30 ± 2	58 ± 6	$13{,}6 \pm 0{,}4$
$E' \rightarrow E$	k_{-2}	-12 ± 3	-96 ± 10	$15{,}8 \pm 0{,}4$

[a] Cf. Tableau II.

4. Conséquence de l'équilibre conformationnel sur l'activité catalytique de la chymotrypsine

La forme 'alcaline', $\bar{E}'$, de la chymotrypsine a une moins bonne affinité pour l'indole que la forme 'neutre', Ē, ainsi que l'a montré la Figure 1. Elle est moins active sur les pseudosubstrats non spécifiques tels que le diisopropylfluorophosphate [10] ou le diphénylcarbamate de nitrocarboxyphénol [4]. Pourtant l'activité sur les substrats spécifiques de l'enzyme ne varie pas avec le pH [15] et semble identique pour Ē et Ē′, ce qui apparaît contradictoire. En fait la fixation non covalente d'un inhibiteur, ou la formation covalente de l'acylenzyme, même avec un pseudosubstrat, entraînant la

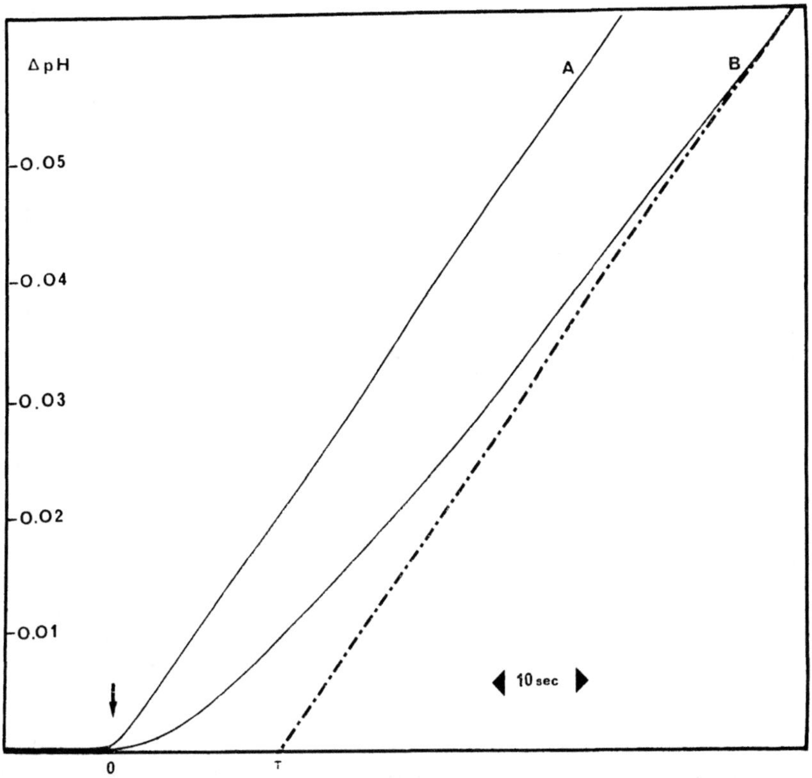

Fig. 3. Activité estérasique de la chymotrypsine à pH 8,0 (15°, KCl 0,15 M) vis-à-vis du N-acétyl tyrosine éthyl ester. L'activité est mesurée par la variation de pH pendant l'hydrolyse de l'ester. (A) pH initial de l'enzyme 7,0. (B) pH initial de l'enzyme 11,0. L'extrapolation (·–·–·) de l'asymptote à la courbe B sur l'axe des temps correspond au temps de latence τ d'apparition de l'activité catalytique de la protéine préincubée à pH 11 (Garel, J. R., résultats non publiés).

conversion $\bar{E}' \rightarrow \bar{E}$, on peut se demander si l'activité de la protéine en milieu alcalin n'est pas consécutive à une isomérisation 'imposée' par le substrat (déplaçant l'équilibre $\bar{E}' \rightleftharpoons \bar{E}$ vers Ē), et si par conséquent elle ne doit pas attendre la conversion spontanée $\bar{E}' \rightarrow \bar{E}$ pour apparaître.

La Figure 3 montre que l'activité catalytique au pH optimum de l'enzyme, vis-à-vis d'un substrat spécifique, est immédiatement atteinte si la protéine est initialement dans l'état Ē (pH 7 à 15°) et n'est en effet que progressivement atteinte lorsqu'elle est initialement dans l'état $\bar{E}'$ (c'est-à-dire à pH alcalin): une phase de latence précède alors l'établissement de l'état stationnaire et le temps de relaxation τ de cette phase de latence est identique à celui que l'on observe au même pH final pour l'isomérisation de la protéine en absence de substrat. Si le pH final auquel est ajouté le substrat aussi bien que le pH initial où a été maintenu l'enzyme, est alcalin, c'est-à-dire si la protéine est en principe sous forme $\bar{E}'$, son activité catalytique n'apparaît que progressivement: faible ou nulle au moment de l'addition du substrat, identique à ce qu'elle est à pH neutre à la fin de la phase de latence. Mesurée à l'état stationnaire, cette activité semble ainsi la même quel que soit le pH; en fait, elle n'apparaît qu'après un processus lent

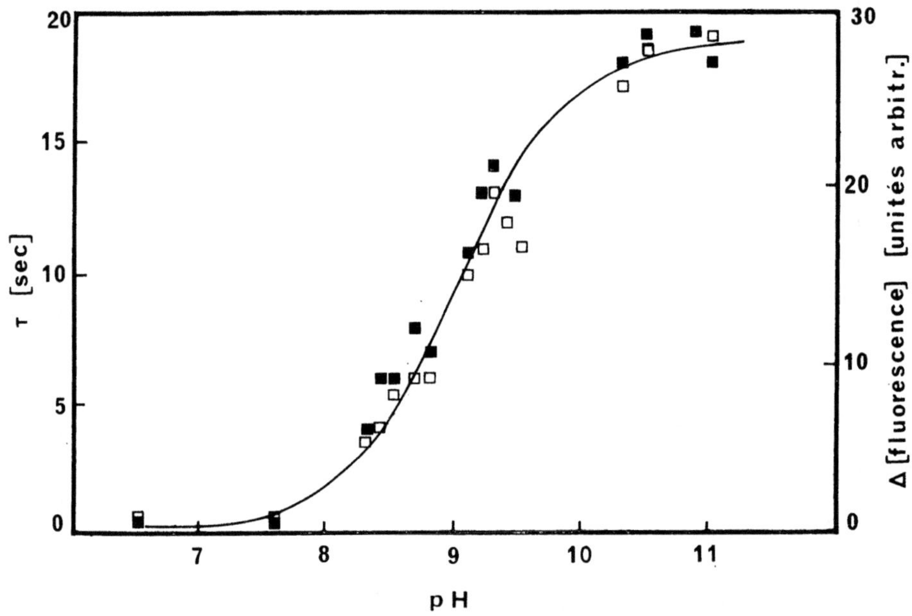

Fig. 4. Variation du temps de latence τ (–□–□–) (Figure 3) en fonction du pH initial auquel est préincubée la chymotrypsine (Garel, J. R., résultats non publiés). Cette variation est comparable à celle de la fluorescence de la protéine (–■–■–).

et la durée de ce processus, à saturation en substrat, est la même que celle de l'isomérisation de l'enzyme libre. Cette phase de latence est supprimée par la présence d'un inhibiteur compétitif tel que l'indole (à concentration saturante); l'apparition de l'activité catalytique résulte donc d'une isomérisation de la protéine et les événements moléculaires liés à cette isomérisation sont beaucoup plus lents ($\sim 0{,}1\ s^{-1}$) que ceux au cours desquels le substrat est converti en produit ($100\ s^{-1}$ pour l'hydrolyse de l'acétyltyrosine éthyl ester).

Si la forme $\bar{E}'$ est inactive, le temps de relaxation de la phase de latence doit être proportionnel à la concentration relative en forme $\bar{E}'$: nul si toutes les molécules sont déjà sous forme $\bar{E}$ (c'est-à-dire à pH neutre ou en présence de quantités suffisantes d'un ligand spécifique ayant entraîne l'isomérisation $\bar{E}' \rightarrow \bar{E}$), maximum en l'absence de tout ligand: c'est-à-dire qu'il doit varier avec le pH initial où est préincubée la protéine de la même façon que la quantité relative de forme $\bar{E}'$.

La Figure 4 montre que τ dépend en effet d'un groupe de pK 9, tout comme le fait la structure de la protéine (Figure 1). Quelle que soit la concentration en substrat, il n'est pas possible de mettre en évidence une activité catalytique significative vis-à-vis d'un substrat spécifique, au temps zéro (Figure 3), qui soit supérieure à celle dont puisse rendre compte la quantité d'ezyme déjà présent sous forme $\bar{E}$.

Ainsi est-on amené à conclure que vis-à-vis de substrats spécifiques aussi bien que non spécifiques, l'une des deux formes structurales de la chymotrypsine est active, l'autre apparemment inactive, que le passage de l'une à l'autre est spontané et ininterrompu, et que seule la proportion relative des deux formes, c'est-à-dire l'équilibre entre leur concentration, est commandée par le milieu qui entoure la protéine. Une telle conclusion est en tout point comparable à celle à laquelle parvenait le modèle allostérique [16], les formes appelées ici $\bar{E}$ et $\bar{E}'$ étant respectivement les formes R et T de ce modèle.

5. Analyse et nature des changements structuraux

L'existence d'une oscillation permanente entre deux états conformationnels étant établie, il reste à tenter d'évaluer en quoi ces états diffèrent sous l'angle architectural: s'agit-il de mouvements restreints engageant la chaîne latérale de quelques résidus ou plutôt d'un remaniement en profondeur de l'assemblage polypeptidique comme le suggère la lenteur du phénomène?

L'organisation moléculaire de la chymotrypsine a été analysée par le groupe de Blow [17], à une résolution de 2 Å, dans des conditions de milieu où il est permis de penser que la conformation occupée par les molécules dans le cristal est celle qui est dénotée ici par $\bar{E}$. De fait, après addition de pseudosubstrats, aucune modification structurale notable n'apparaît dans les cristaux [2, 18], de la même façon qu'aucune modification significative des propriétés spectroscopiques n'apparaît en solution quand on ajoute un substrat ou un inhibiteur à $\bar{E}$ (Figure 1). Cependant, en l'absence de données cristallographiques obtenues dans les conditions où la protéine est sous sa deuxième forme, $\bar{E}'$, une approche non cristallographique doit être utilisée pour déceler les différences qui peuvent exister entre les deux structures; cette approche ne peut être que chimique c'est-à-dire basée sur un marquage de la protéine en un ou plusieurs points dont le comportement diffère dans les conformations $\bar{E}$ et $\bar{E}'$, si l'on veut pouvoir localiser au niveau de points précis le comportement de la molécule. Il faut bien, néanmoins, réaliser que de telles approches ne pourront avoir la précision topologique de la cristallographie, sauf si l'on peut s'appuyer au maximum sur celle-ci.

Le principe de l'analyse repose sur la réactivité différentielle des groupes chimiquement réactifs qui participent au mouvement d'isomérisation. On doit s'attendre en effet à ce qu'au cours du remaniement structural, cette réactivité croîsse ou décroîsse en raison du changement de l'entourage de ces groupes. La même réaction chimique effectuée sur une catégorie de groupements réactifs donnés (fonction amine, carboxyl, thiol, phénol, etc.) dans des conditions où la protéine est dans l'une ou l'autre structure, aura lieu plus vite si le groupe est plus réactif, moins vite ou pas du tout si le groupe est peu ou pas réactif tandis que le marquage des groupes dont l'entourage ne change pas demeurera identique quel que soit l'état structural de la protéine. Il est cependant évident que si un groupe est marqué à la même vitesse dans l'une et l'autre conformation, aucune information ne pourra être acquise par une telle technique quel que soit le mouvement que subit ce groupe. Le marquage différentiel consistera donc idéalement à effectuer la même réaction en deux fois: la première fois dans des conditions où le groupe dont on recherche la participation réagit le moins possible tandis que tous les autres réagissent; la seconde fois dans les conditions où ce groupe peut être couplé, lui seul l'étant puisque les autres ont déjà réagi. La réaction chimique utilisée la seconde fois doit naturellement faire appel à un composé différent, par exemple radioactif ou chromophore, pour que son identification soit réalisable. Si l'isomérisation n'entraîne pas une variation très profonde de réactivité (tout ou rien) on peut néanmoins examiner cinétiquement la réaction (vitesse de marquage des différents groupes et identification des groupes marqués à différentes vitesses [19].

Le premier groupe dont la participation au mouvement intramoléculaire $\bar{E} \rightarrow \bar{E}'$ était connue et dont pouvait rechercher l'identification par voie chimique, est celui dont la protonation entraîne le basculement de $\bar{E}'$ à $\bar{E}$ (Figure 1), non seulement du

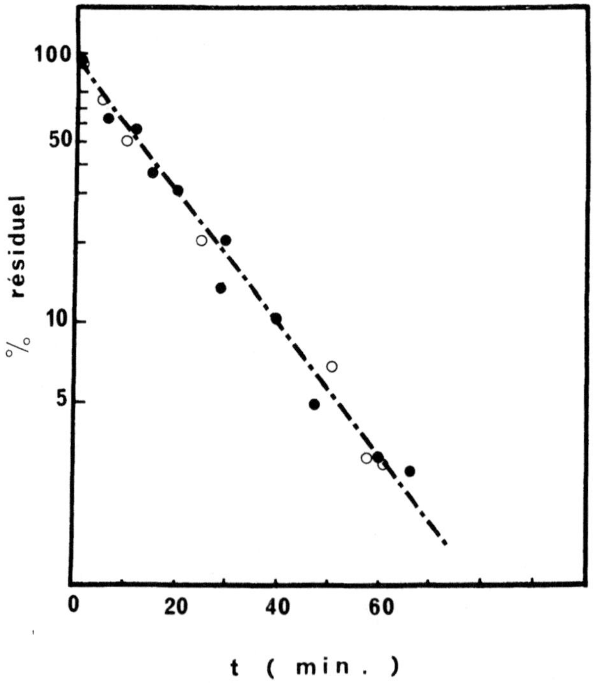

Fig. 5. Variation de l'activité hydrolytique de la N(ε)acétyl-chymotrypsine vis-à-vis du N-acétyl tyrosine éthyl ester pendant la réacétylation de la protéine (acétylation spécifique sur l'isoleucine N-terminale). –●– activité résiduelle; –○– isoleucine N-terminale libre résiduelle [21].

fait que parmi les résidus présents dans une chaîne protéique les groupes ionisables sont particulièrement réactifs, mais parce que celui qui est impliqué dans le remaniement structural de la protéine doit vraisemblablement avoir une réactivité différente dans les deux états $\bar{E}$ et $\bar{E}'$ (son pK change): réactivité vraisemblablement normale dans la structure $\bar{E}'$ où son pK est normal, et soit très élevée, soit très basse dans $\bar{E}$ où son pK est déplacé de 3 unités pH. De plus, on sait que lors du processus d'activation du chymotrypsinogène en chymotrypsine un groupe α aminé libre se trouve apparaître, celui de l'extrémité N terminale de la chaîne B de l'enzyme (isoleucine 16 du zymogène) [20].

L'identification de ce groupe ionisable critique a pu en effet être réalisée par alkylation spécifique: si tous les résidus nucléophiles accessibles du chymotrypsinogène sont acétylés ou maléylés, la protéine demeure transformable et une chymotrypsine (alkylée) qui présente toutes les particularités décrites jusqu'ici. Si par contre le groupe α aminé qui apparaît lors de l'activation du chymotrypsinogène en chymotrypsine, est à son tour bloqué (au cours d'une seconde alkylation effectuée après l'activation), l'activité catalytique de l'enzyme disparaît en même temps que ce groupe aminé (Figure 5) et ses propriétés physiques changent [21]. La rotation optique, la fluorescence, le dichroisme circulaire, deviennent ceux de l'enzyme sous forme $\bar{E}'$ c'est-à-dire aussi

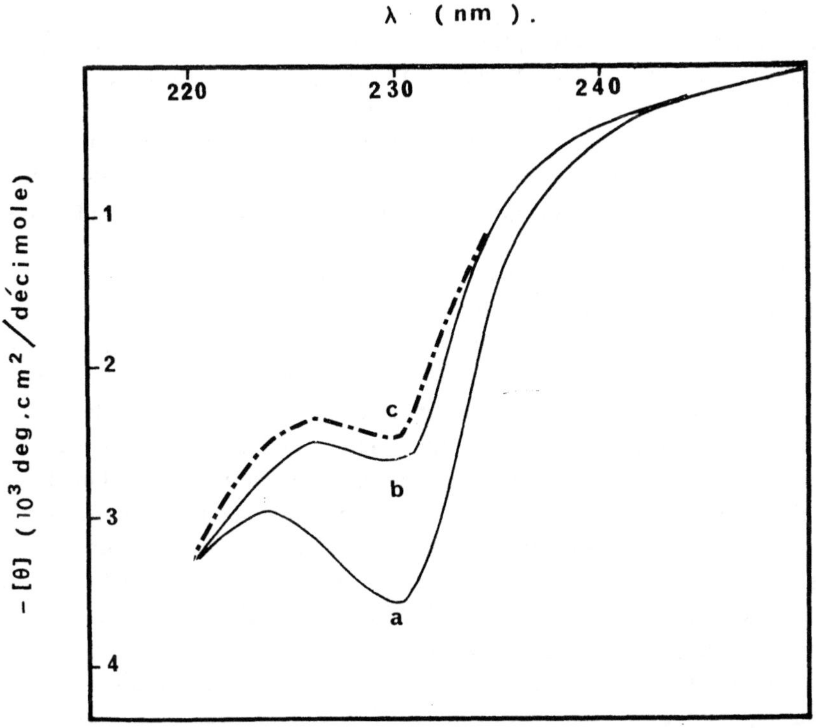

Fig. 6. Dichroïsme circulaire de la N(ε)acétyl-chymotrypsine à pH 7 (courbe (a)), de la N(ε)acétyl-chymotrypsine à pH 11, et du N(ε)acétyl-chymotrypsinogène à pH 7 et pH 11 (courbe (b)) et de la N(ε)acétyl-chymotrypsine acétylée sur l'isoleucine N-terminale, à pH 7 et pH 11 (courbe (c)).

ceux du zymogène (Figures 1 et 6). Cette perte d'activité et le changement de propriétés spectrales sont réversibles par désalkylation spécifique: si tous les groupes alkylables du chymotrypsinogène ont été alkylés en un dérivé stable (acétyl) et si celui de l'isoleucine 16 est marqué par un composé instable (maléyl), il est possible de libérer ce dernier par une désalkylation sélective, l'activité catalytique de l'enzyme et ses propriétés physiques initiales sont alors retrouvées. Enfin ce résidu N-terminal peut être enlevé par action de leucine aminopeptides; il y a alors inactivation irréversible.

Ces observations amènent à deux conclusions:

(1) Si le groupe α aminé critique de l'isoleucine N terminale de la chaîne B de l'enzyme est chimiquement réactif, ou si ce résidu N terminal est éliminable, c'est que celui-ci est obligatoirement accessible au solvant, et plus encore, à une protéine extérieure (la leucine aminopeptidase), donc à l'extérieur de la molécule. Or on le trouve à l'intérieur de la protéine dans le modèle cristallographique, à une dizaine d'angströms de la surface.

(2) Si les propriétés physiques de l'enzyme spécifiquement alkylé sur cette isoleucine N terminale ressemblent à celles du zymogène et non à celles de l'enzyme, c'est qu'une plus grande parenté existe entre ces deux protéines qu'entre le zymogène et l'enzyme actif. Dans le zymogène l'isoleucine 16 est effectivement à l'extérieur de la molécule, puisque c'est par hydrolyse trypsique de la liaison arginine 15 – isoleucine 16 qu'à lieu l'activation.

On est ainsi amené à suggérer que l'enzyme dont l'extrémité N-terminale d'isoleucine 16 est bloquée est structuralement proche du zymogène, et puisque ces deux protéines ont les mêmes propriétés physiques et la même absence d'activité catalytique que la forme $\bar{E}'$ de l'enzyme, que cette dernière est également proche du zymogène. Si cette hypothèse est fondée, on peut alors s'aider des données cristallographiques comparées entre chymotrypsine et chymotrypsinogène pour rechercher d'autres analogies ou oppositions entre les formes $\bar{E}$ et $\bar{E}'$ de l'enzyme.

Deux différences importantes existent entre l'enzyme et son précurseur [22]:

(1) Dans l'enzyme à l'état cristallin une interaction électrostatique doit exister au sein de la protéine entre le groupe aminé de l'isoleucine N-terminale et le carboxyle d'un résidu aspartate (no. 194) en raison de la proximité de ces deux groupes ionisés [23]. Elle doit se traduire par une forte perturbation dans l'ionisation des deux participations de cette interaction. C'est ce qui a lieu pour l'isoleucine N-terminale: dans l'une des conformations, le pK en est à peu près normal (pK 7,5), dans l'autre, très anormal (pK 10,5). Cette dernière valeur est celle trouvée pour la conformation $\bar{E}$, l'autre pour $\bar{E}'$, et cette opposition est en accord avec l'hypothèse d'un passage de ce groupe de l'extérieur à l'intérieur de la molécule: pK normal à l'extérieur, pK anormalement élevé à l'intérieur en raison de l'interaction électrostatique dans le milieu peu polaire de la protéine. A ce passage dedans-dehors, c'est-à-dire à la perte de la charge portée par le groupe aminé de l'isoleucine 16, sont associées toutes les variations physiques notées dans la transition précédemment décrite (Figure 1): fluorescence, absorbance, ellipticité, etc.: on doit donc s'attendre à retrouver la même transition si le partenaire négatif du groupe aminé perd sa charge. C'est en effet ce qui a lieu,

dans la zone d'ionisation des carboxyles (Figure 7): une transition dépendant d'un groupe de pK apparent de 3, présentant un pK 'normal' de 4 dans la conformation $\bar{E}'$ et un pK très anormal de 1,7 dans la conformation $\bar{E}$ est observée. Bien entendu, une telle transition ne démontre pas que le groupe qui en est responsable est l'aspartate 194 trouvé par cristallographie; elle est cependant conforme à ce que l'on attend; on sait de plus que seuls deux carboxyles sont anormaux par leur réactivité, l'aspartate 194 et l'aspartate 102 [24, 25]. (Le rôle et la position de ce dernier ont été assignés par cristallographie à celui d'un relais de charge avec les résidus du 'site actif' de l'enzyme (aspartate 102 - histidine 57 - sérine 195) [26].)

(2) Dans l'enzyme à l'état cristallin, le résidu de méthionine 192 apparaît à la surface de la molécule tandis que dans le zymogène il est enfoui dans la poche du site actif [22]. Si ce comportement est conservé en solution, la chaîne latérale de ce résidu de méthionine doit être chimiquement plus réactive dans l'enzyme que dans son zymogène. C'est ce que l'on observe (Figure 8): cette méthionine n'est pas alkylable dans le zymogène alors qu'elle l'est dans l'enzyme.

On a suggéré précédemment que le blocage chimique de l'isoleucine 16 contraint l'enzyme à conserver une structure proche de celle du zymogène. Si cette hypothèse

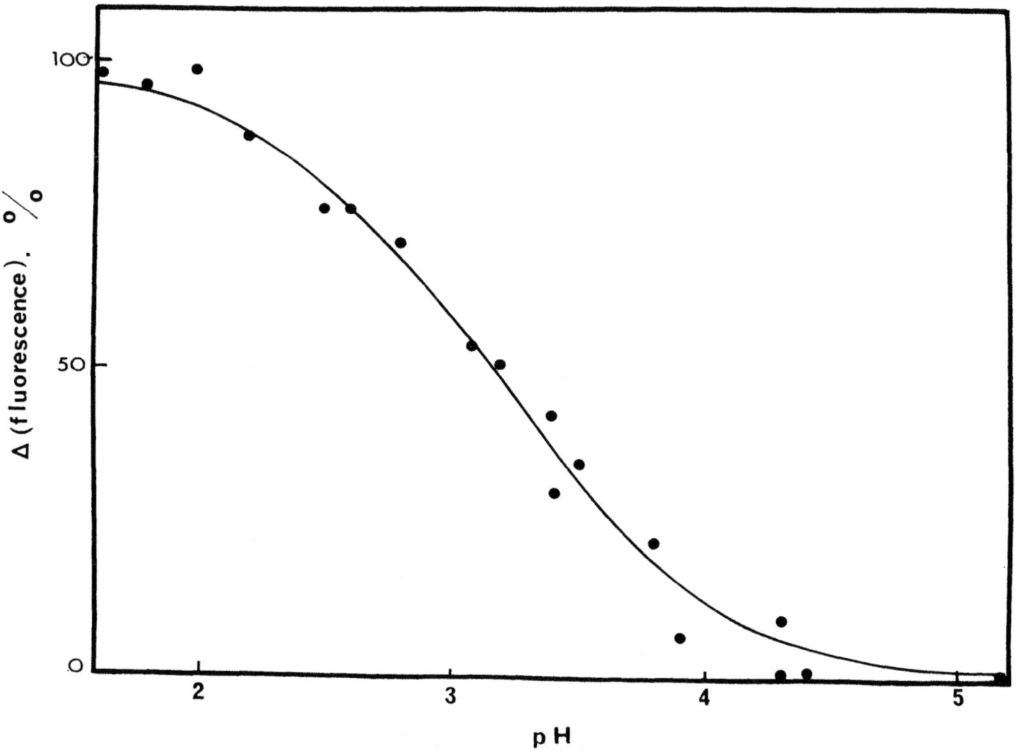

Fig. 7. Amplitude relative du changement de fluorescence (Δ [fl]) de la chymotrypsine après un saut de pH, à 15°, KCl 0,15 M. Le pH porté en abscisse est le pH final de la protéine. Le changement de fluorescence présente une cinétique monophasique (cf. Figure 2) (Epely et Garel, résultats non publiés).

est correcte, on doit avoir la même absence de réactivité de la méthionine 192 chez le zymogène et chez l'enzyme dans lequel l'isoleucine 16 est maintenue hors de la molécule par blocage chimique: c'est ce que montre la Figure 8.

Enfin si la conformation Ē′ est elle aussi proche de celle du zymogène la méthionine 192 n'y sera pas réactive. Elle acquerra lentement la réactivité observée dans la structure Ē si le mouvement qu'elle effectue est synchrone ou dépendant de celui que l'on suit par variation de fluorescence, absorption ou rotation. Par conséquent cette réactivité au temps zéro devra varier avec le pH comme la concentration relative en forme Ē (Figure 1) ou comme le temps de latence du passage de Ē′ à Ē (Figure 4): c'est ce que l'on constate (Figure 9); la réactivité initiale de la méthionine 192 est nulle ou très faible lorsque la protéine est préincubée à pH alcalin et croît avec le pH comme la population de molécules sous forme Ē. Cette coincidence et les données précédentes permettent donc de suggérer que la forme Ē′ de la chymotrypsine présente une grande parenté structurale avec le chymotrypsinogène, et si cette parenté structurale est exacte, permettent d'examiner comparativement les deux modèles cristallographiques (celui de l'enzyme et celui du zymogène), afin de savoir ce que sont les différences présentées par les deux molécules. On constate que ces différences sont assez peu nombreuses

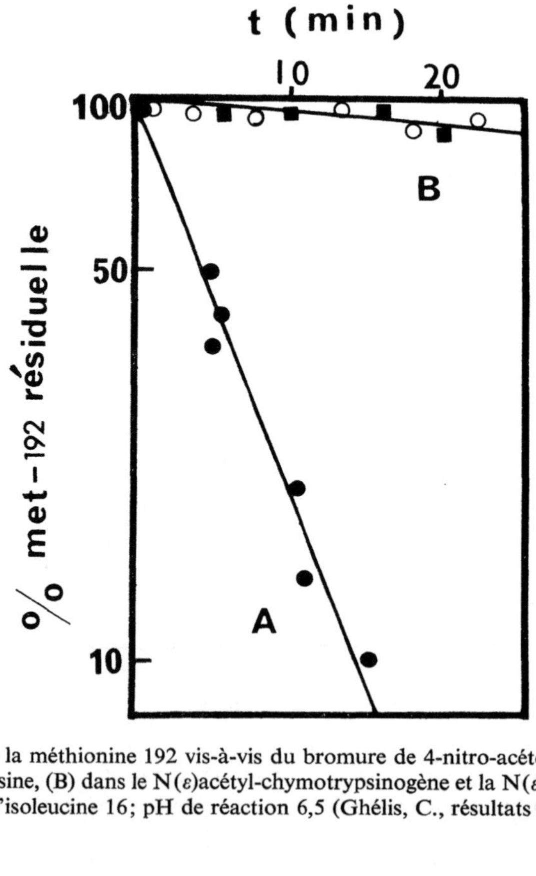

Fig. 8. Réactivité de la méthionine 192 vis-à-vis du bromure de 4-nitro-acétophénone, (A) dans la N(ε)acétyl-chymotrypsine, (B) dans le N(ε)acétyl-chymotrypsinogène et la N(ε)acétyl-chymotrypsine acétylée sur l'isoleucine 16; pH de réaction 6,5 (Ghélis, C., résultats non publiés).

[22]: les aspects les plus saillants du passage zymogène → enzyme consistent effectivement en une 'entrée' de l'isoleucine 16, qui se déplace de 11 Å et vient se fixer à l'intérieur de la protéine face à l'aspartate 194 (qui la retient), et en une 'sortie' de la méthionine 192, qui se déplace de 8 Å et libère l'accès du site actif, qui préexiste dans le zymogène. D'autres mouvements sont observés mais leur amplitude est plus faible; ils accompagnent le remaniement du squelette peptidique auquel est associé le déplacement de l'isoleucine 16 et de la méthionine 192.

En conclusion les résultats qui viennent d'être résumés montrent qu'une protéine apparemment rigide comme la chymotrypsine a une structure oscillant continuellement entre deux états fonctionnellement très distincts. L'état 'natif' n'en est donc pas unique, et c'est là un concept que les données cristallographiques ne permettaient pas de proposer. Bien que le nombre de résidus de la chaîne peptidique que ce remaniement concerne ne semble pas élevé, les mouvements de certains d'entre eux sont importants et paraissent nécessiter une forte énergie d'activation, où enthalpie et entropie se compensent partiellement. La vitesse de ce remaniement est donc très faible par rapport aux mouvements que présentent normalement des groupes oscillant librement en solution [13].

Sous l'angle méthodologique ces résultats montrent en outre que l'on peut non

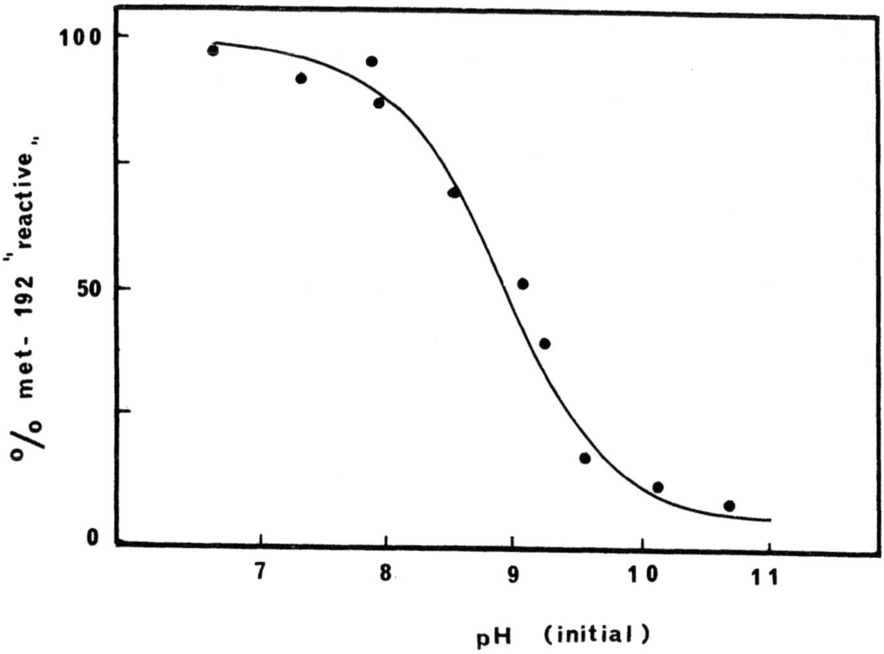

Fig. 9. Réactivité initiale de la méthionine 192 de la chymotrypsine vis-à-vis du bromure de 4-nitroacétophénone, à pH 6,5, suivant le pH auquel la protéine a été préincubée (préincubation à 15°, KCl 0,15 M) (Ghélis, C., résultats non publiés).

seulement mettre en évidence l'existence de mouvements intramoléculaires dans une protéine, mais également localiser avec précision certains des résidus qui sont entraînés dans ces mouvements, en se basant sur les corrélations qu'apportent les informations cinétiques, physicochimiques et chimiques. Cependant, ils prouvent au moins autant que sans les données structurales apportées par les cristallographes de tels mouvements intramoléculaires ne s'exprimeraient guère pour leur plus grande part qu'en termes de sec^{-1}.

Références

[1] Perutz, M. F.: 1970, *Nature* **228**, 726.
[2] Sigler, P. B., Jeffery, B. A., Matthews, B. W., et Blow, D. M.: 1966, *J. Mol. Biol.* **15**, 175.
[3] Parker, H. et Lumry, R.: 1963, *J. Am. Chem. Soc.* **85**, 483.
[4] Karibian, D., Laurent, C., Labouesse, J., et Labouesse, B.: 1968, *Eur. J. Biochem.* **5**, 260.
[5] Garel, J. R. et Labouesse, B.: 1971, *Biochimie* **53**, 9.
[6] Hess, G. P., McConn, J., Ku, E., et McConkey, G.: 1970, *Phil. Trans. Roy. Soc. London* **257**, 89.
[7] Ghélis, C.: 1971, Thèse, Orsay, France.
[8] Glick, D. M.: 1968, *Biochemistry* **7**, 3391.
[9] Kosman, D. J. et Piette, L. H.: 1972, *Arch. Biochem. Biophys.* **149**, 452.
[10] Moon, A. Y., Sturtevant, J. M., et Hess, G. P.: 1965, *J. Biol. Chem.* **240**, 4204.
[11] Fersht, A. R. et Requena, Y.: 1971, *J. Mol. Biol.* **60**, 279.
[12] Tanford, C.: 1968, *Adv. Protein Chem.* **23**, 121.
[13] Hammes, G. G.: 1968, *Adv. Protein Chem.* **23**, 1.
[14] Fersht, A. R.: 1972, *J. Mol. Biol.* **64**, 497.
[15] Bender, M. L., Clement, G. E., Kezdy, F. J., et d'A. Heck, H.: 1964, *J. Am. Chem. Soc.* **86**, 3680.
[16] Monod, J., Wyman, J., et Changeux, J. P.: 1965, *J. Mol. Biol.* **12**, 88.
[17] Matthews, B. W., Sigler, P. B., Henderson, R., et Blow, D. M.: 1967, *Nature* **214**, 652.
[18] Henderson, R.: 1970, *J. Mol. Biol.* **54**, 341.
[19] Koshland, D. E. et Neet, K. E.: 1968, *Ann. Rev. Biochem.* **37**, 359.
[20] Desnuelle, P.: 1960, dans *The Enzymes* (P. Boyer, H. Lardy, et K. Myrbäck, éds.), Vol. 4, Academic Press, New York.
[21] Ghélis, C., Garel, J. R., et Labouesse, J.: 1970, *Biochemistry* **9**, 3902.
[22] Freer, S. T., Kraut, J., Robertus, J. D., Wright, H. T., et Xuong, N. H.: 1970, *Biochemistry* **9** 1997.
[23] Sigler, P. B., Blow, D. M., Matthews, B. W., et Henderson, R.: 1968, *J. Mol. Biol.* **35**, 143.
[24] Carraway, K. L., Spoerl, P., et Koshland, D. E.: 1969, *J. Mol. Biol.* **42**, 133.
[25] Abita, J. P. et Lazdunski, M.: 1969, *Biochem. Biophys. Res. Commun.* **35**, 707.
[26] Blow, D. M., Birktoft, J. J., et Hartley, B. S.: 1969, *Nature* **221**, 337.
[27] Oppenheimer, H. L., Labouesse, B., et Hess, G. P.: 1966, *J. Biol. Chem.* **242**, 2720.
[28] Garel, J. R. et Labouesse, B.: 1970, *J. Mol. Biol.* **47**, 41.
[29] Garel, J. R.: 1971, Thèse, Orsay, France.

THE DIFFUSION OF PHOSPHOLIPIDS IN MEMBRANES

ROGER D. KORNBERG*

Stauffer Laboratories for Physical Chemistry, Stanford University, Stanford, Calif. 94305, U.S.A.

Summary. Two kinds of self-diffusion are possible for phospholipids in a bilayer membrane: phospholipids may move parallel to the surface of the membrane, remaining within one monolayer (lateral diffusion), and they may move perpendicular to the surface of the membrane, passing from one monolayer to the other (flip-flop). We have determined the rates of both processes by magnetic resonance measurements on spin-labeled phosphatidylcholine bilayers. We find that lateral diffusion is greater than eight orders of magnitude more rapid than flip-flop. Flip-flop is so slow that it is possible for a membrane to be asymmetric in the distribution of phospholipids between its two sides. We have investigated transmembrane electrical potential differences as possible sources of membrane asymmetry. For this purpose we developed a spectroscopic method of measuring transmembrane electrical potential differences in microscopic phospholipid vesicles.

Résumé. Deux types de 'self-diffusion' sont possibles pour les phospholipides dans une membrane à double couche: ils peuvent se déplacer parallèlement à la membrane en restant à l'intérieur d'une monocouche (diffusion latérale) et ils peuvent se déplacer perpendiculairement à la surface de la membrane, passant d'une monocouche à une autre (flip-flop). Nous avons déterminé les vitesses de ces deux processus par des mesures de resonance magnetique sur des bicouches de phosphatidylcholine à spin marqué. La diffusion latérale est plus rapide que le flip-flop de plus de huit ordres de grandeur. Le flip-flop est si lent qu'on peut avoir une distribution asymétrique des phospholipides entre les deux faces de la membrane. Nous avons étudié les différences de potentiel électrique à travers la membrane comme sources possibles d'asymétrie de la membrane. A cette fin nous avons mis au point une méthode spectroscopique qui permet la mesure des différences de potentiel électrique dans la membrane de vésicules microscopiques de phospholipides.

Most of you know that membranes are practically unique among complex structures in biology in their high content of small molecules. Roughly half the volume of most membranes is occupied by phospholipids, which are molecules of about 1000 molecular weight. And I think one must wonder why. Why are the lipids in membranes so numerous and so small? Why aren't there a few large lipids instead of many small ones? I think the answer a chemist would give is that the components of a membrane may be mobile with respect to one another if many of them are small. That is, a collection of large molecules is necessarily solid, whereas a collection of small molecules is a liquid if the molecules are small and chemically independent enough. And that is really what we are led to ask about phospholipids – are phospholipids small and also chemically independent enough for a collection of them to be liquid? That is, are the members of an array of phospholipids essentially fixed in place or do they move about?

Now I would like to describe an experiment which answers this question for lecithin bilayers. The experiment is really quite simple. What we did was to measure the effect of a paramagnetic analog of lecithin on the nuclear resonance spectrum of an otherwise pure lecithin bilayer. Our paramagnetic analog of lecithin, called spin-labeled lecithin, is shown in Figure 1. The effect of spin-labeled lecithin on the proton reso-

* Present address: MRC Laboratory of Molecular Biology, Hills Road, Cambridge CB2 2QH, Great Britain.

C. Sadron (ed.), Dynamic Aspects of Conformation Changes in Biological Macromolecules. 477–486.

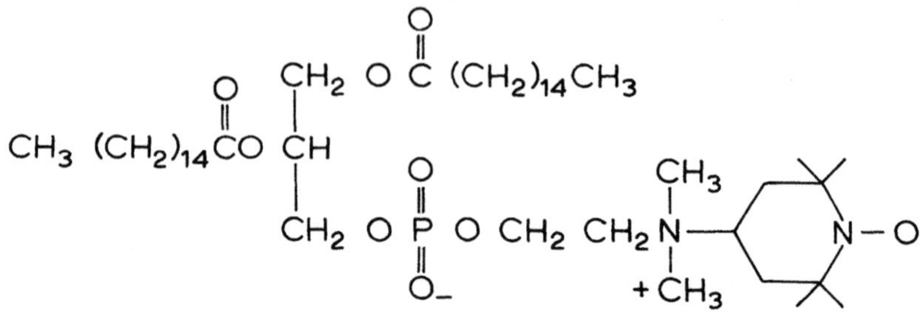

Fig. 1. Spin-labeled lecithin.

nance spectrum of a lecithin bilayer is shown in Figure 2. The effect is a broadening of the lecithin N-methyl proton line.

In order to interpret this broadening you must know that the magnetic interaction between a spin label and a proton which gives rise to the broadening has a range of about 30 Å under the conditions of our experiment. Now suppose on the one hand that the lecithin molecules in a bilayer are not mobile with respect to one another. Then, based on a distance between nearest neighboring lecithin molecules of about

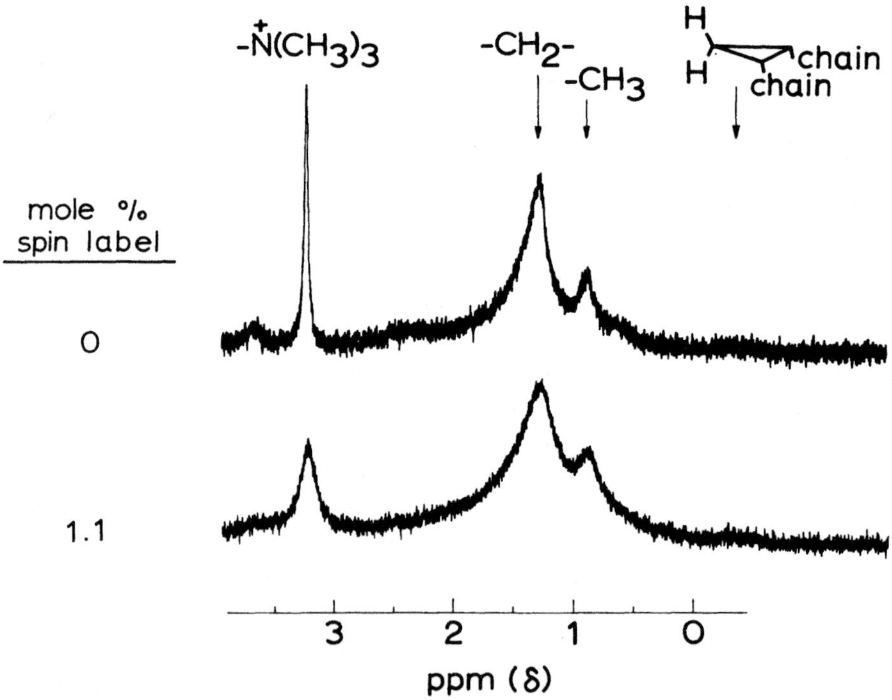

Fig. 2. Effect of spin-labeled lecithin on the proton resonance spectrum of lecithin vesicles at 100 MHz, 35°. The assignments of peaks to N-methyl, hydrocarbon chain methylene, hydrocarbon chain terminal methyl, and cyclopropane methylene protons are indicated.

10 Å and 1 molecule in 100 of spin-labeled lecithin, one calculates that 29% of the lecithin molecules are always within 30 Å of a spin-labeled lecithin molecule and give rise to a broad proton resonance line, whereas 71% of the lecithin molecules are always more than 30 Å distant from a spin-labeled lecithin molecule and give rise to a narrow unperturbed proton resonance line. The proton resonance line we should expect to observe is the sum of these broad and narrow lines and will therefore have at least 71% the amplitude of the resonance line of a pure lecithin bilayer. Now suppose on the other hand that the lecithin molecules in a bilayer are rapidly mobile with respect to one another. Then every lecithin molecule comes occasionally within 30 Å of a spin-labeled lecithin molecule so every lecithin molecule gives rise to a broad proton resonance line. You can see intuitively that the more frequent the encounters of lecithin with spin-labeled lecithin molecules the broader is the proton resonance line. So if the lecithin molecules in a bilayer are sufficiently rapidly mobile with respect to one another then the proton resonance line will be very broad and have, in particular, much less than 71% the amplitude of the resonance line of a pure lecithin bilayer.

Looking again at Figure 2, we see that the N-methyl proton line of a lecithin bilayer containing 1 molecule in 100 of spin-labeled lecithin has plainly much less than 71% the amplitude of the resonance line of a pure lecithin bilayer. We conclude that the lecithin molecules in a bilayer are rapidly mobile with respect to one another. Of course, this reasoning is only semiquantitative. To establish a completely quantitative distinction between what we should expect for a bilayer of lecithin molecules which are not mobile with respect to one another and what we observe, we have calculated the shapes of the broad and narrow resonance lines we should expect and made measurements on lecithin bilayers containing various amounts of spin-labeled lecithin. The results of the comparison are shown in Figure 3. The results make us confident of our conclusion that the lecithin molecules in a bilayer are rapidly mobile with respect to one another.

From the absolute magnitude of the broadening we measure we can calculate the rate of the movement of lecithin molecules. Our result, expressed in terms of the frequency of pairwise exchanges of a lecithin molecule with its nearest neighbors, is a frequency of much greater than 3×10^3 sec^{-1}.

Of course this nuclear resonance experiment does not distinguish between the two possible ways in which lecithin molecules in a bilayer may be mobile with respect to one another. Lecithin molecules may pass from one monolayer to the other and they may move about within each monolayer. Now I would like to describe a paramagnetic resonance experiment in which we measure specifically the rate of lecithin molecules passing from one monolayer to the other. We have measured this rate for spin-labeled lecithin molecules in the bilayer membrane of a lecithin vesicle. Since a vesicle is a closed structure we speak of the passing of molecules from one monolayer to the other as inside–outside transitions.

We measure rates of inside–outside transitions by means of the effect on spin labels of chemical reducing agents such as ascorbic acid. Whereas spin labels are paramagnetic and give rise to paramagnetic resonance spectra, the product of reduction of a

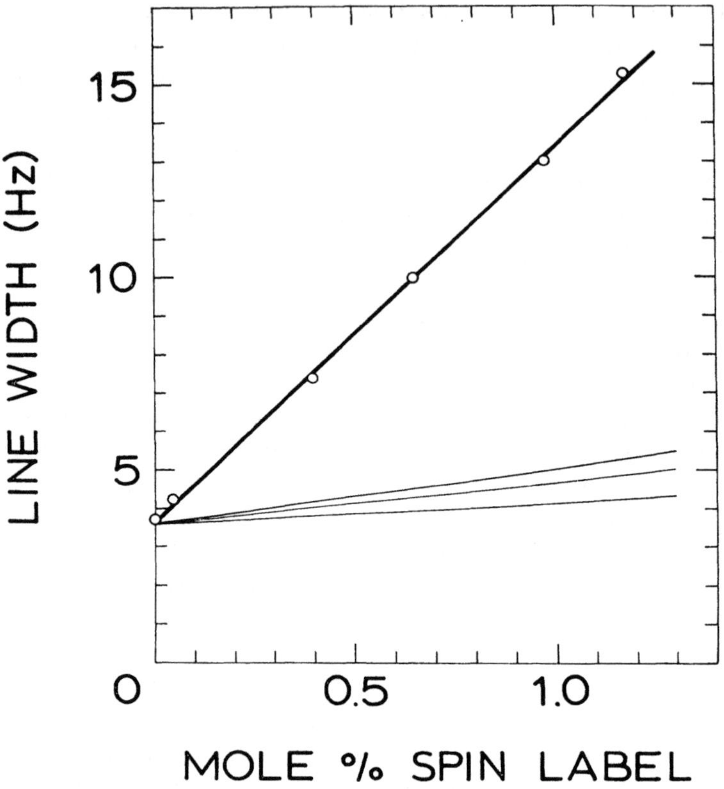

Fig. 3. Dependence of the N-methyl proton line width on the spin label content of spin-labeled lecithin vesicles. Circles are data on pure and spin-labeled lecithin vesicles. The heavy line is a straight line with slope to fit the data. The thin lines are calculated curves for a bilayer of lecithin molecules which are not mobile with respect to one another, with three values of the range of the interaction between N-methyl protons and spin label.

spin label by ascorbic acid is nonparamagnetic and does not give rise to a paramagnetic resonance spectrum. We made use of this at first by simply mixing ascorbic acid with spin-labeled lecithin vesicles at room temperature and measuring the time course of the amplitude of the paramagnetic resonance spectrum. We observed the rapid disappearance of part of the paramagnetic resonance amplitude and the slow disappearance, with a half-time of about 10 minutes, of the rest. This we attributed to the reduction by ascorbic acid of first the spin labels in the outer monolayer of the vesicle membrane and then the spin labels in the inner monolayer. We could not say from this experiment whether the rate of reduction of the spin labels in the inner monolayer was determined by the rate of inside to outside transitions of spin labels or by the rate of inward penetration of ascorbic acid. Unfortunately we found, in an independent experiment, that it was the rate of inward penetration of ascorbic acid. But we also found, in an independent experiment, that at 0° there was no inward penetration of ascorbic acid at all. So then we mixed ascorbic acid with spin-labeled lecithin vesicles at 0° and measured the time course of the amplitude of the paramagnetic resonance spectrum. We observed the rapid disappearance of 65% of the paramagnetic resonance amplitude and no change at all in the residual 35%. This coincides in a rather striking way with the geometries of our lecithin vesicles. In these experiments our vesicles were uniformly about 250 Å in outside diameter, which is so small compared with a bilayer thickness of 40 Å that the surface area of the bilayer is significantly greater on the outside of a vesicle than on the inside. Assuming the surface area per lecithin molecule to be the same on both sides, one calculates that 68% of the lecithin molecules in a vesicle are on the outside and 32% on the inside. So you can see why we conclude that ascorbic acid at 0° reduces all the spin labels in the outer monolayers of the vesicle membranes and none of the spin labels in the inner monolayers.

We make use of this result to measure the rate of inside–outside transitions of spin-labeled lecithin at temperatures such as 30°, temperatures at which the inward penetration of ascorbic acid occurs. Our procedure is as follows. First, we add ascorbic acid to spin-labeled lecithin vesicles at 0°. Second, we remove the ascorbic acid by gel filtration, still at 0°. Third, we keep the gel filtrate for some time at a temperature such as 30°. And finally, we add ascorbic acid at 0° and record the paramagnetic resonance. I think you can see that if any inside to outside transitions occur during the time at 30° then some spin labels which were not reduced by ascorbic acid in the first step are reduced by ascorbic acid in the last step. The results of this procedure are shown in Figure 4. Evidently, inside–outside transitions do occur, with a half-time of 6.5 hours at 30°. This corresponds to a frequency of inside–outside transitions of 2×10^{-5} sec^{-1}.

The contrast with the result of our nuclear resonance experiment (a frequency of

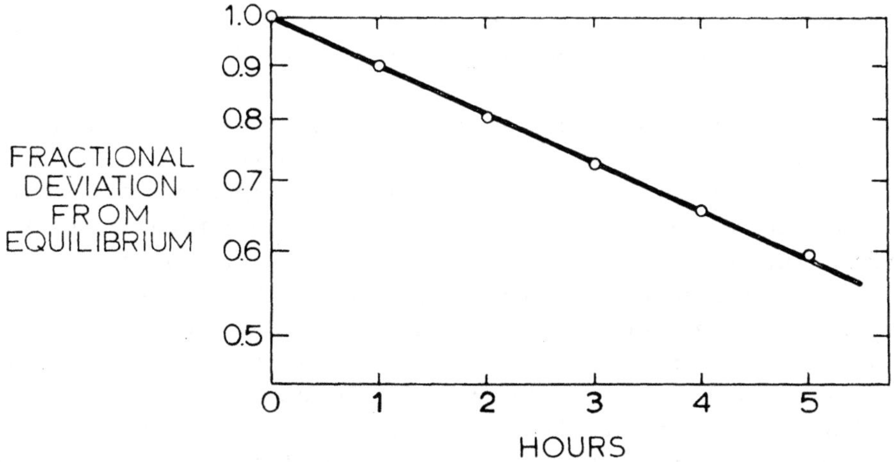

Fig. 4. Time course of inside–outside spin label distribution in ascorbic acid-treated spin-labeled lecithin vesicles at 30°. On the abscissa is plotted how long the vesicles were kept at 30° in the third step of the experiment. On the ordinate is plotted the fractional deviation of the paramagnetic resonance amplitude in the final step of the experiment from its equilibrium value.

pairwise exchanges of a lecithin molecule with its nearest neighbors of much greater than 3×10^3 sec^{-1}) is striking. Clearly in the nuclear resonance experiment we detected only motions of lecithin molecules within each monolayer of the bilayer. We call that process lateral diffusion. I might best summarize the results of our nuclear and paramagnetic resonance experiments by saying we have found that lateral diffusion in a lecithin bilayer is at least eight orders of magnitude more rapid than inside–outside transitions.

One implication of the slow rate of inside–outside transitions is that the membrane of a cell can be asymmetric in the distribution of phospholipids between its two sides. That is, it may be possible for a cell to keep different molecular species of phospholipids on the two sides of its membrane with very little expenditure of energy. Of course one must wonder how such an asymmetry would arise. One obvious source of asymmetry in cells is a transmembrane electrical potential difference. If, for example, the inside of a cell were electrically positive then there must be an excess of negatively charged phospholipids on the inner surface of its membrane at equilibrium. The main difficulty in thinking along these lines is that there are no methods of measuring membrane potentials in most cells. Microelectrodes are too large for use in most cells. So we have developed an alternative to the use of microelectrodes which is applicable to objects as small as 250 Å lecithin vesicles.

Our approach to measuring membrane potentials is quite straightforward. All we do is allow a spin-labeled ion to equilibrate between the inside and outside of a membrane-bounded object and calculate the membrane potential from the inside–outside distribution of spin label by means of the Boltzmann relation,

$$C_o/C_i = \exp(-zV/26).$$

Here C_o and C_i are the outside and inside concentrations of spin-labeled ion, z is the charge on the ion, and V is the membrane potential in millivolts. We measure C_o and C_i by measuring the paramagnetic resonance amplitude before and after adding ascorbic acid at 0°.

Our main difficulty in this approach was finding a spin-labeled ion which equilibrates between the inside and outside of membrane-bounded objects. We got around this difficulty by preparing a spin-labeled ion which penetrates membranes in an uncharged form. This spin label is shown in Figure 5. It is, as you can see, a derivative of tartaric acid. The uncharged form which penetrates membranes is the protonated acid. The evidence for this is shown in Figure 6. It is that the rate of equilibration between the inside and outside of lecithin vesicles is proportional to the hydrogen ion concentration.

There is one very great difficulty in using such a spin label to measure membrane potentials. The difficulty is that a molecule cannot reflect a membrane potential in its distribution between the two sides of a membrane unless it traverses the membrane in a charged form. You can see the way around this difficulty by considering the dissociation of spin-labeled acid on both sides of a membrane,

$$(H^+)_i\,(T^-)_i/(HT)_i = K, \qquad (H^+)_o\,(T^-)_o/(HT)_o = K.$$

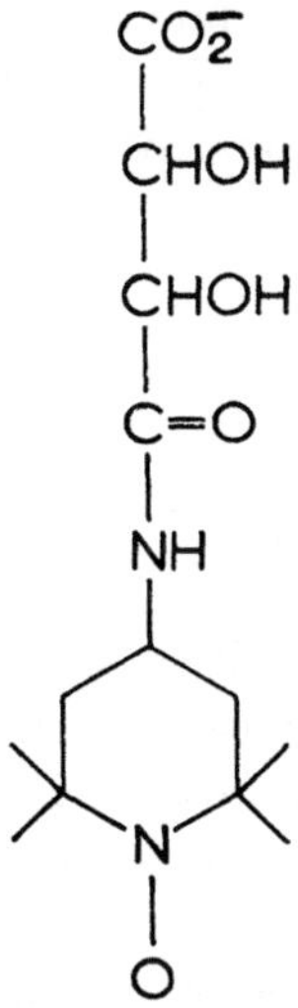

Fig. 5. Spin label derivative of tartaric acid.

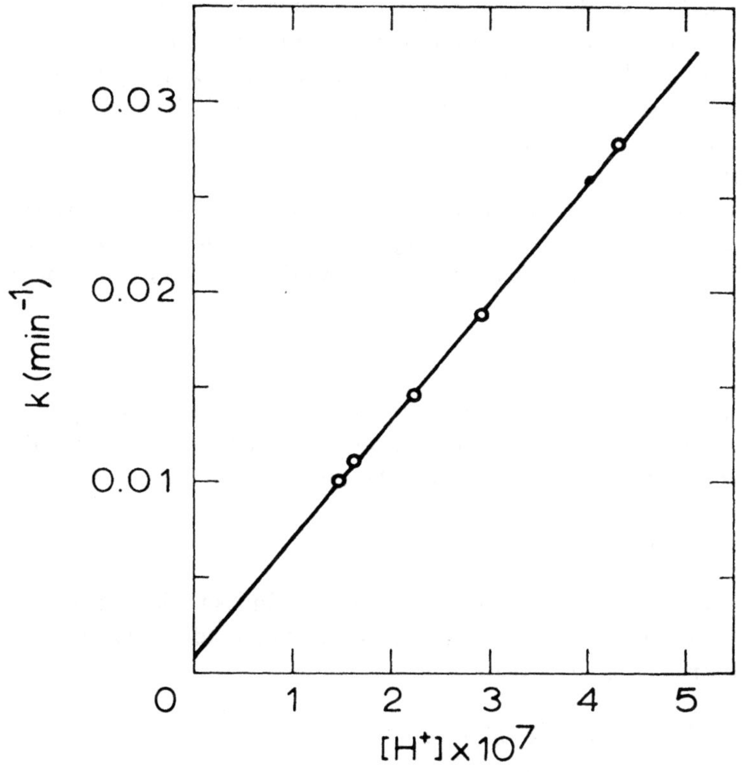

Fig. 6. The rate of approach to equilibrium of the inside–outside distribution of the spin label derivative of tartar:c acid in lecithin vesicles is proportional to the H^+ concentration. The values of k are first-order ıate constants for the approach to equilibrium.

Here the subscripts i and o denote inside and outside, HT denotes the protonated form of the spin label derivative of tartaric acid, and K is its dissociation constant. Since membranes are permeable to the protonated acid, its equilibrium concentration must be the same on both sides,

$$(\mathrm{HT})_i = (\mathrm{HT})_o .$$

It follows from these equations that

$$(\mathrm{H}^+)_o/(\mathrm{H}^+)_i = (\mathrm{T}^-)_i/(\mathrm{T}^-)_o . \tag{1}$$

In other words, the inside–outside distribution of spin label measures the transmembrane pH difference. So all we must do to measure transmembrane electrical potential differences is couple them to transmembrane pH differences. And fortunately that is easy to do. Many molecules are known which make membranes permeable to hydrogen ions.

Now I would like to give an example of an experiment in which we have used this spin label method to measure membrane potentials in lecithin vesicles. We prepared the vesicles from solutions containing high concentrations of Na^+ and low concentrations of K^+, and passed the vesicles by gel filtration into solutions containing the same sum of Na^+ and K^+ concentrations but with their ratio reversed. We made the vesicles permeable to K^+ and H^+ by adding the carrier molecule valinomycin. And finally we added a small amount of the tartaric acid spin label and allowed its inside–outside distribution to come to equilibrium.

What should we have expected to find? First we would expect K^+ to diffuse down its concentration gradient, making the insides of the vesicles electrically positive. This process should continue until the driving force of the concentration gradient is just balanced by the membrane potential, which condition is expressed by the Boltzmann relation

$$(\mathrm{K}^+)_o/(\mathrm{K}^+)_i = \exp(-V/26) . \tag{2}$$

Next, since the interiors of the vesicles are electrically positive, we expect an efflux of H^+ which continues until the H^+ concentration gradient is in accord with the membrane potential through the Boltzmann relation

$$(\mathrm{H}^+)_o/(\mathrm{H}^+)_i = \exp(-V/26) . \tag{3}$$

Finally we expect an influx of spin label which continues until the spin label concentration gradient is in accord with the H^+ concentration gradient through Equation (1). From Equations (1)–(3) we expect

$$(\mathrm{K}^+)_o/(\mathrm{K}^+)_i = (\mathrm{T}^-)_i/(\mathrm{T}^-)_o . \tag{4}$$

That is, we expect the equilibrium inside–outside distributions of K^+ and spin label to be reciprocals of one another.

TABLE I

$(K^+)_o/(K^+)_i$	$(T^-)_i/(T^-)_o$
6.6	6.2
12.0	9.8

The results of such experiments are shown in the Table I. The results conform to Equation (4), so our method evidently works. It remains now for the method to be put to use in investigating such matters as membrane asymmetry.

Acknowledgements

This work is the result of my collaboration with Drs. Harden M. McConnell and Mark G. McNamee. Details of the work are given in the References.

References

Kornberg, R. D. and McConnell, H. M.: 1971, *Proc. Nat. Acad. Sci. US* **68**, 2564.
Kornberg, R. D. and McConnell, H. M.: 1971, *Biochemistry* **10**, 1111.
Kornberg, R. D., McNamee, M. G., and McConnell, H. M.: 1972, *Proc. Nat. Acad. Sci. US* **69**, 1508

DISCUSSION

Shulman: Is there a dependence of the inside–outside transfer rate upon vesicle size?

Kornberg: We have only measured the rate for vesicles of 250 Å outside diameter, so we don't know.

Fromageot: You mentioned the use of a valinomycin/vesicle ratio of 0.2 to allow K^+ permeability. Is that ratio high enough?

Kornberg: The results are not affected by the use of a higher valinomycin-to-vesicle ratio. We conclude that valinomycin is exchanged between vesicles. This confirms the same finding by Johnson and Bangham (*Biochim. Biophys. Acta* **193** (1969) 82).

Changeux: Serait-il possible d'effectuer des mesures de potentiel de membrane avec des indicateurs fluorescents au lieu de marqueurs de spin?

Kornberg: There are many conceivable ways of measuring membrane potentials with fluorescent probes, but I don't know of anyone who has succeeded in doing it.

Leach: In your studies of inside–outside displacement of the spin-labeled species it is assumed that its flip-flop diffusion rate would be typical of unlabeled molecules. Can this be demonstrated by independent means?

Kornberg: At the present time, no.

Grivet: In the measurement of trans-membrane potentials, how does one ascertain that thermodynamic equilibrium is indeed established?

Kornberg: There is no change in the results between 3.5 and 5 hours of equilibration.

Karplus: (1) Does the ascorbic-acid permeability show a sharp or gradual change as a function of temperature and does this tell you something interesting about the membrane?

(2) Is it possible that dislocations introduced by the presence of the spin label could contribute significantly to the observed diffusion rate?

Kornberg: (1) We have only measured the ascorbic-acid permeability at 0° and room temperature, so we can't say.

(2) In the case of lateral diffusion we think not, since the spin label is in the part of the molecule.

which is solvated by water and should not interfere with the packing of nearby lecithin molecules. In the case of inside–outside transitions we think the rate for spin-labeled lecithin is probably greater than the rate for lecithin, because spin-labeled lecithin is less polar than lecithin in thin-layer chromatography.

Mme Alfsen: In your model, which should be the place for a protein molecule involved in the membrane structure?

Kornberg: Protein molecules may be bound to the surface, partly penetrate or span the phospholipid bilayer.

ÉTUDE PAR MARQUAGE DE SPIN DE LA DIFFUSION LATÉRALE DANS LES MEMBRANES BIOLOGIQUES

PHILIPPE DEVAUX*, CARL J. SCANDELLA, et H. M. McCONNELL
Stauffer Laboratories for Physical Chemistry, Stanford University, Stanford, Cal. 94305, U.S.A.

Résumé. Les constantes de diffusion latérale de phospholipides marqués avec des radicaux paramagnétiques ont été déterminées par deux méthodes différentes dans des multicouches de lécithine de jaune d'œuf et dans le réticulum sarcoplasmique des muscles dorsaux du lapin. Une constante de diffusion de 6×10^{-8} cm²/s a été déterminée à 37 °C dans ce dernier système; des valeurs légèrement supérieures sont obtenues à cette température dans les bicouches de lécithine. Ceci implique qu'une molécule peut diffuser sur une distance de plusieurs microns en 1 s.

Summary. The lateral diffusion constants of spin-labeled phospholipids have been determined in egg lecithin multilayers and in the sarcoplasmic reticulum from rabbit skeletal muscle. The diffusion constant is 6×10^{-8} cm²/s at 37 °C in this latter system; slightly higher values at 37 °C are found in lecithin multilayers. This implies that a molecule can diffuse a distance of the order of several microns in 1 s.

A la suite des expériences de Kornberg [1] mettant en évidence une diffusion latérale rapide des phospholipides dans les vésicules de phosphatidylcholine, nous avons réalisé des mesures effectives de la constante de diffusion latérale dans des membranes modèles ainsi que dans les vésicules du réticulum sarcoplasmique des muscles du lapin. Ces expériences ont été réalisées avec des phospholipides contenant un radical paramagnétique stable greffé, soit la chaîne β (I(m, n), avec en général $m = 1$ et $n = 14$), soit sur la tête polaire (II).

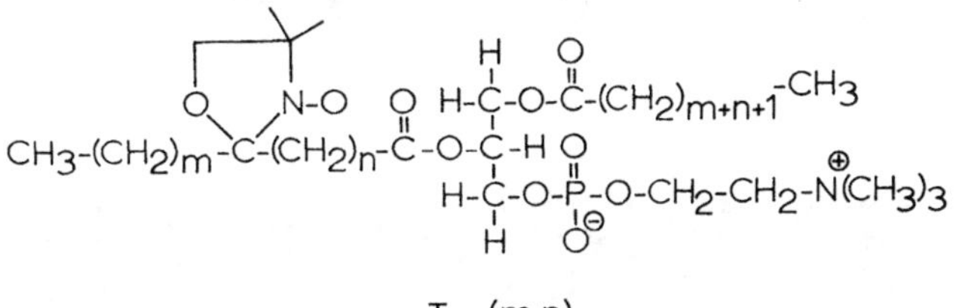

I (m,n)

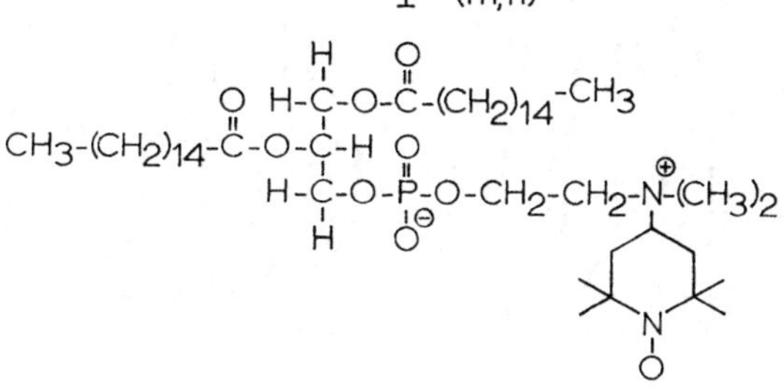

II

* Adresse permanente: Groupe de Physique des Solides de l'École Normale Supérieure, Paris, France.

C. Sadron (ed.), Dynamic Aspects of Conformation Changes in Biological Macromolecules. 487–492.

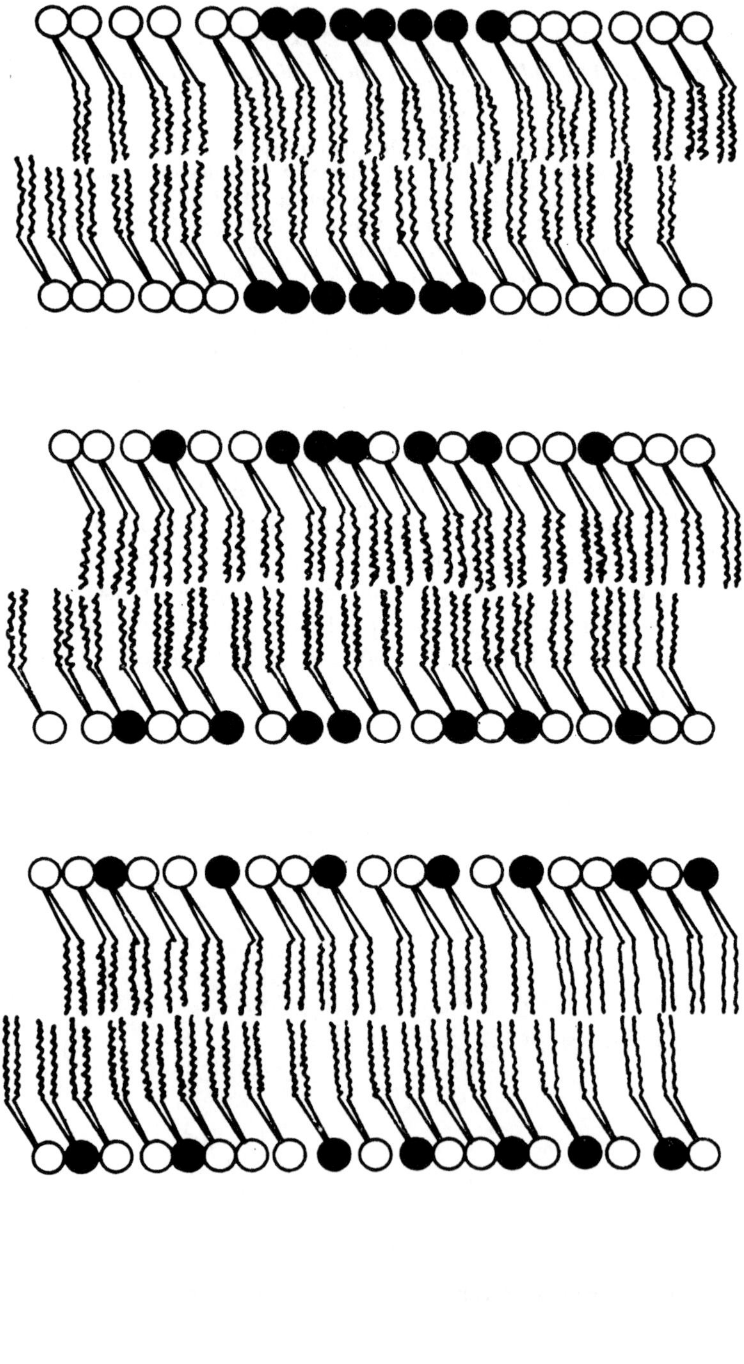

Fig. 1. Représentation schématique d'une expérience de dilution de phospholipides marqués (représentés avec des têtes polaires noires) dans une bicouche lipidique.

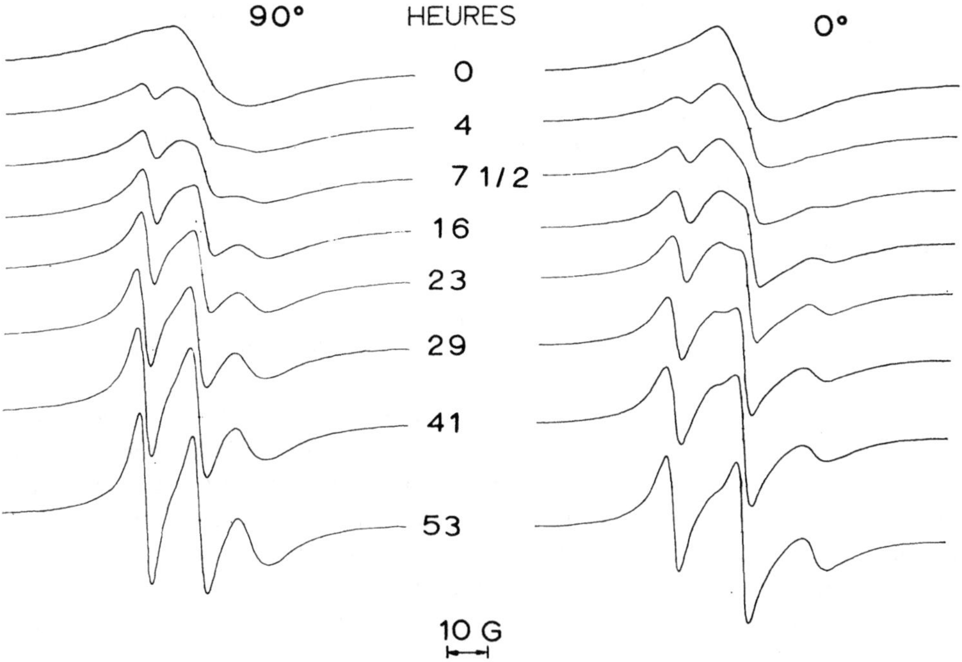

Fig. 2. Evolution du spectre de résonance paramagnétique électronique de II diffusant dans des bicouches orientées de lécithine (0° et 90° correspondent à deux directions du champ magnétique statique).

Des perturbations dans les spectres de résonance paramagnétique électronique apparaissent lorsque les distances entre molécules marquées sont faibles ($\leqslant$20 Å). Nous avons analysé ce type de perturbation (spin–spin) pour en déduire les constantes de diffusion des lipides.

Le premier type d'expérience a été réalisé sur des bicouches orientées de lécithine de jaune d'œuf contenant au début de l'expérience une (ou plusieurs) région de forte concentration de phospholipides marqués. L'expérience est schématisée sur la Figure 1. Le détail des conditions expérimentales est décrit ailleurs (voir [2]).

Au début de l'expérience, le spectre RPE comprend une raie large et unique faiblement anisotrope. Lorsque les molécules marquées se trouvent diluées par diffusion latérale dans les bicouches, le spectre RPE se transforme progressivement pour donner les trois raies caractéristiques du nitroxyde (Figure 2). L'anisotropie du spectre permet de vérifier que le déplacement se fait bien dans un plan. Typiquement une tache initiale d'environ 0,8 mm de diamètre sera diluée en environ 50 heures. L'étude cinétique de ce système permet d'estimer la constante de diffusion à environ 2×10^{-8} cm^2/s. Ceci correspond à des fréquences de sauts de l'ordre de 10^7 s^{-1} (résultats consistants avec les résultats antérieurs de Kornberg [1]). Cette diffusion est donc rapide puisque cela signifie qu'en 1 sec une molécule peut diffuser de plusieurs microns (distance correspondant par exemple à la taille d'une bactérie).

Ces résultats ont été confirmés par un 2e type d'expérience fondé sur l'analyse complète de la forme du spectre RPE obtenu à des concentrations élevées et *homogènes* de phospholipides marqués (C=0,1 à 0,001). Un tel type d'expérience a pu être fait avec des membranes biologiques dans lesquelles des phospholipides marqués ont été introduits. La méthode d'incorporation est décrite dans [3]. Elle consiste essentiellement à incuber les vésicules du réticulum avec des vésicules soniqués de phospholipides marqués. L'incorporation dans la membrane est indiquée par la transformation du spectre RPE du mélange. Le spectre initial ne comprend qu'une raie large; cette dernière est progressivement remplacée par trois raies fines signifiant une dilution des marqueurs (sans perte de spin). Les vésicules du reticulum sont lavés par centrifugation. Les spectres finaux correspondent alors à une répartition *homogène* et sont très semblables à ceux que l'on peut obtenir en incorporant des phospholipides marqués dans des vésicules de phosphatidylcholine.

Nous avons pu montrer [3, 4] qu'à *haute température* ($\gtrsim$35°C) la forme de ces spectres, où une interaction spin–spin existe, est déterminée principalement par la fréquence de collisions entre molécules marquées. L'étude détaillée du spectre permet donc d'obtenir la fréquence de collisions pour une concentration donnée de molécules marquées. De là, on peut déduire la fréquence d'échange entre deux molécules voisines (marquées ou non) et finalement la constante de diffusion latérale.

Les valeurs obtenues à haute température sont indiquées sur la Figure 3. L'extrapo-

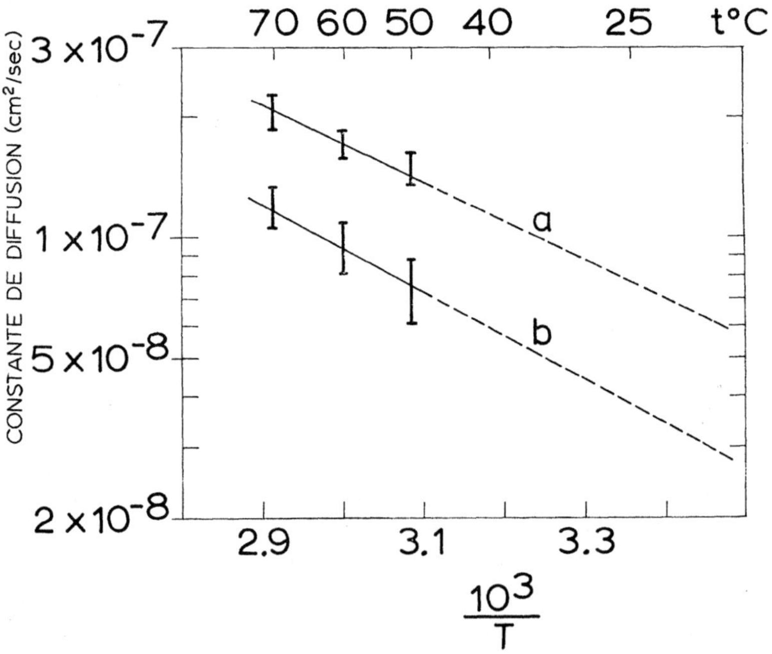

Fig. 3. Constantes de diffusion latérale déduites de l'analyse de la forme des spectres de résonance à haute température avec des échantillons isotropes contenant une distribution *homogène* de I(1,14): (a) lécithine/cholestérol (4: 1), liposomes non soniqués; (b) vésicules du réticulum sarcoplasmique.

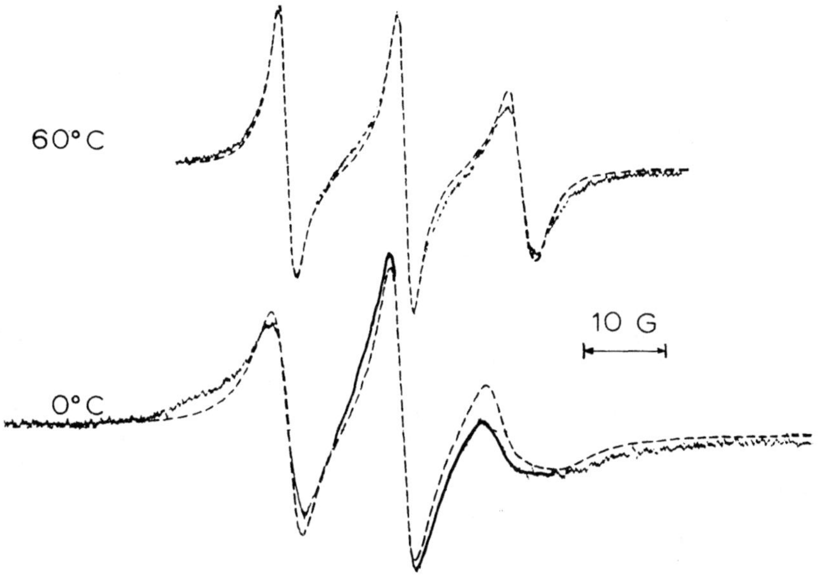

Fig. 4. Comparaison des spectres de résonance à basse et à haute température de I(1,14) incorporés respectivement dans des liposomes formés avec les lipides extraits du reticulum (----) et dans des vésicules du réticulum sarcoplasmique entier (——).

lation à 25° C donne des résultats tout à fait comparables à ceux précédemment obtenus.

Il est à noter que nous avons pu observer dans la membrane du reticulum une séparation de phase très nette à 0° C, ce qui indique que les extrapolations à basse température doivent être faites prudemment.

La Figure 4 montre la différence entre les spectres RPE de deux systèmes voisins. A haute température (60° C) les spectres des deux systèmes se différencient à peine. A basse température (0° C), le spectre RPE des phospholipides marqués incorporés dans le reticulum laisse apparaître un début de ségrégation des phospholipides marqués (un pic très large est superposé aux trois raies attendues). Ce phénomène est reversible, ce qui est une preuve supplémentaire qualitative de la diffusion des lipides.

Notons donc que cette ségrégation n'apparait pas en l'absence de protéines, ce qui implique un effet possible des protéines dans l'organisation des lipides.

La constante de diffusion ne doit pas en principe dépendre de façon dramatique de la taille des molécules. En conséquence, si une protéine se trouve libre de se mouvoir dans une membrane lipidique, on peut s'attendre à trouver des constantes de diffusion d'un ordre de grandeur seulement plus faible. De récents travaux de Edidin et Fambrough sur la diffusion d'antigènes fluorescents sur la surface de fibres musculaires [5] corroborent cette hypothèse.

Remerciements

Ce travail a été soutenu par N.S.F. grant NP 26456 et N.I.H. NB-08050-04.

Références

[1] Kornberg, R. D. et McConnell, H. M.: 1971, *Proc. Nat. Acad. Sci. US* **68**, 2564.
[2] Devaux, P. et McConnell, H. M.: 1972, *J. Am. Chem. Soc.* **94**, 4475.
[3] Scandella, C. J., Devaux, P., et McConnell, H. M.: 1972, *Proc. Nat. Acad. Sci. US* **69**, 2056
[4] Devaux, P., Scandella, C. J., et McConnell, H. M.: 1973, *J. Magn. Res.* **9**, 474.
[5] Edidin, M. et Fambrough, D.: 1973, *J. Cell. Biol.*, sous presse.

DISCUSSION

Ptak: (1) Vous avez mentionné un procédé de reconstitution des spectres. Pouvez-vous préciser ce procédé et le modèle de mouvement choisi? (2) Que pouvez-vous dire sur l'effet du contenu en eau sur la vitesse de diffusion des molécules de phospholipide dans les 'multilayers' et aussi sur l'effet des ions bivalents?

Devaux: (1) On commence par déterminer les paramètres des spectres RPE des échantillons faiblement dopés (c'est-à-dire sans interaction spin–spin). On introduit des formes de raie lorentziennes par l'intermédiaire de systèmes d'équations de Bloch (un système pour chaque orientation). Dans le cas d'interactions d'échange, on introduit en plus une fréquence d'échange dans des systèmes dits 'd'équations de Bloch modifiés''. (2) Les échantillons orientés sont saturés en eau et les échantillons isotropes sont en solution.

Bourgarit: (1) Quelle est la valeur de l'énergie d'activation de la diffusion que vous avez utilisée pour extrapoler vos résultats en fonction de la température? (2) La technique RPE permet-elle de mettre en évidence des étapes intermédiaires (interstitielles) dans le processus de diffusion?

Devaux: (1) Environ 3 kcal/mole, valeur très approximative. (2) On ne peut pas déterminer quel est le mouvement exact dans la bicouche mais simplement que les molécules se déplacent effectivement dans un plan.

Changeux: Pouvez-vous distinguer entre des mouvements de translation locaux (10–100 Å) et des mouvements sur des distances beaucoup plus grandes ($\sim \mu$)?

Devaux: Dans les expériences avec des bicouches orientées, les molécules se déplacent effectivement sur de longues distances. Le fait que l'on retrouve à peu près les mêmes constantes de diffusion dans le réticulum indique seulement que les domaines (s'ils existent) ont au moins 30 à 40 Å de rayon. Par contre, si nous arrivons à prouver que le mécanisme d'incorporation des marqueurs dans la membrane est bien une fusion des vésicules de lipides marqués avec les microsacs, nous aurons prouvé que les phospholipides peuvent se déplacer aussi sur de grandes distances dans le réticulum (puisque l'on obtient une complète dilution).

Buc: Avez-vous une idée des valeurs relatives des deux constantes de diffusion latérales d'une protéine et d'un lipide inclus dans le même système membraneux?

Devaux: Si l'on utilise l'équation de Stokes-Einstein, la diffusion varie en $1/R$ (R étant le rayon de la molécule supposée sphérique). Ceci permet d'estimer les constantes de diffusion de protéines à environ un ordre de grandeur plus faible. Les mesures d'Edidin sur les antigènes donneraient plutôt deux ordres de grandeur.

Lopez-Delgado: Dans votre cliché, on constate une parfaite corrélation entre votre modèle donnant la distribution des molécules marquées et les spectres RPE prévus d'une part et les spectres expérimentaux d'autre part. Maintenant, comment 'préparez-vous' expérimentalement le 'moment initial' (où toutes les molécules marquées sont ensemble)? Ensuite, quelle est, à votre avis, la structure de votre échantillon 'vue' par le spectromètre?

Devaux: Les détails de cette expérience ont été publiés dans *J. Am. Chem. Soc.* **94** (1972) 4475.

Goldberg: De quels arguments disposez-vous pour distinguer entre une migration 'intra-vésiculaire' du marqueur de spin et un échange intervésiculaire?

Devaux: Les molécules sont extrêmement peu solubles; d'ailleurs dans les deux types d'expérience présentés, on verrait par RPE s'il y avait une fraction notable de molécules en solution. Ceci suggère donc des mouvements intravésiculaires (les molécules se déplaçant en restant dans une bicouche).

ANGIOTENSINE II ET ANALOGUES: ÉTUDES CONFORMATIONNELLES COMPARATIVES

SERGE FERMANDJIAN, DANIEL GREFF, et PIERRE FROMAGEOT
Centre d'Études Nucléaires, Saclay, B.P.2, 91190, *Gif-sur-Yvette, France*

Résumé. L'étude par dichroïsme circulaire dans l'ultraviolet et par résonance magnétique nucléaire de l'angiotensine II et de plusieurs de ses analogues structuraux, a d'une part confirmé le modèle 'cross β' proposé pour l'hormone et d'autre part montré qu'il existe une assez bonne relation entre la conformation adoptée par ces peptides en solution et leur activité biologique.

Summary. The characteristics of angiotensin II and some of its analogs have been investigated by circular dichroism and by NMR. AII data collected indicate the tendency of the molecule of angiotensin II to fold on itself making a first turn at the level of valine in position 3 and of tyrosine in position 4, and a second turn taking place at the His–Pro linkage. It is interesting to note that a rather good correlation exists between the conformation of the analogs obtained from one amino acid substitution and the associated biological response. For instance, the loss of 50 % of biological activity for Pro^3[AGII] fits very well with the greater rigidity introduced in the first turn which leads to a new orientation of the biologically very important tyrosine ring.

1. Généralités – introduction

L'angiotensine II est un octapeptide naturel [1] qui intervient dans la régulation de plusieurs fonctions physiologiques importantes, dont les principales sont:

(a) la régulation de la pression artérielle [2], d'où son nom (hypertensine);

(b) la secrétion de l'aldostérone;

(c) une intervention dans le mécanisme de la soif au niveau du système nerveux central;

(d) des effets antidiurétiques (action au niveau du rein).

L'[AGII] fait partie des 'kinin hormones' qui sont des polypeptides biologiquement actifs 'relachés' à partir de précurseurs inactifs dans le plasma sanguin (angiotensinogène, bradykinogène, kallidinogène). Rocha e Silva [3, 4] a appliqué le terme 'kinin' à toutes les hormones tissulaires de nature peptidique qui agissent sur les muscles lisses et qui possèdent une activité hypotensive ou hypertensive. Cette définition, qui exclut les hormones neurohypophysaires, couvre cependant les angiotensines.

Le système rénine–angiotensine met en jeu un certain nombre de réactions enzymatiques [5, 6] qui conduisent à l'activation de l'hormone ([AGII] = octapeptide) et à sa dégradation (fragments peptidiques).

Le diagramme suivant décrit les principales étapes de ces évènements.

L'[AGII] est rapidement détruite dans l'organisme animal. Des mesures récentes [9] ont montré que la période biologique du peptide dans le plasma est de 15 sec.

L'effet hypertenseur de l'hormone, déterminé par la méthode de Peart (rat binéphrectomisé) [10], est décelable à des doses variant entre 0,5 et 1×10^{-9} g. Sur des organes isolés (aorte, utérus, colon), l'activité hormonale est décelable à des doses plus élevées.

C. Sadron (ed.), Dynamic Aspects of Conformation Changes in Biological Macromolecules. 493–509.

ANGIOTENSINOGÈNE
↓ ←- - - - - - - - - - - - - - - - - RÉNINE (action protéolytique) [7, 8]
ANGIOTENSINE I

Pro – phe + HIS – LEU – OH
blocage de l'activité durant le transport

Cl⁻ activation
↓ ←- - - - - - - - - - - - - - - - - enzyme de conversion (peptidase)

Substance active = ANGIOTENSINE II

↓ ←- - - - - - - - - - - - - - - - - Angiotensinases (peptidases)
HEPTAPEPTIDE

Des $(Asn)^1$ – Val^5 [AGII] actif à 50% = heptapeptide central: séquence active; transport de l'activité au récepteur

↓ ←- - - - - - - - - - - - - - - - - Aminopeptidase A
HEXAPEPTIDE
↓ ←- - - - - - - - - - - - - - - - - Peptidases
←- - - - - - - - - - - - - - - - - Peptidases
FRAGMENTS PEPTIDIQUES
biologiquement et immunologiquement inactifs

Des progrès récents ont montré que l'[AGII] agit par l'intermédiaire de l'AMP cyclique [11] et produit dans les cellules isolées des muscles lisses la relâche (ou fuite) des ions Ca^{++} [12] dont on connaît le rôle dans la contraction musculaire. La caractérisation des sites de liaisons spécifiques, à l'aide de l'[AGII] tritiée [13], et la mise en évidence de plusieurs classes de récepteurs ont été réalisées par l'équipe du Pr. P. Meyer de Broussais [14, 15]. La synthèse récente d'un analogue non compétitif [16] et l'utilisation de l'hormone tritiée vont permettre sûrement dans un proche avenir d'isoler le récepteur membranaire de l'[AGII].

Au niveau moléculaire, le problème le plus intéressant est celui des relations entre la structure chimique, la conformation (déterminée par la structure primaire) et l'activité biologique. Chacune des possibilités conformationnelles est caractérisée par une 'topochimie' unique de la molécule. La relation entre l'activité biologique et la conformation moléculaire a été illustrée dans le cas des hormones par le travail de Manning et du Vigneaud [17] et par Eisler *et al.* [18] sur l'ocytocine et certains de ses analogues.

Parmi la centaine d'analogues de l'angiotensine II synthétisée jusqu'à ce jour, beaucoup se sont montrés fructueux dans l'établissement d'une relation activité biologique–structure chimique tout en dégageant le rôle prédominant de certains aminoacides de la chaîne (Tableau I). Les principales remarques sont les suivantes:

(a) L'inégale importance des acides aminés des extrémités N et C terminales de la chaîne; en effet, si après délétion du résidu aspartyl, l'heptapeptide 2-8 possède encore 50% d'activité biologique, par contre celui en 1-7, sans phénylalanine n'est pratiquement plus actif. Si le remplacement dans une large mesure du résidu N terminal par d'autres aminoacides est réalisable sans trop de dommages [19], celui du C terminal provoque des effets importants sur l'activité biologique. Ces résultats peuvent permettre de distinguer entre le rôle purement structural ou de protection joué par l'acide aspartique et celui plus fondamental (structural et d'intervention directe dans le mécanisme de déclenchement du signal biologique) de la phénylalanine. L'importance de ce dernier résidu est soulignée par le fait qu'il ne peut être déplacé dans la chaine et que son remplacement par l'alanine [20, 21] ou l'isoleucine conduit à deux inhibiteurs de type compétitif [22].

(b) La valine en position 3, ainsi que la valine ou l'isoleucine en position 5, apparraissent jouer un rôle stérique. Ainsi le remplacement de la valine en position 3 par l'alanine [23] ou bien par la proline [24] provoque une perte d'environ 50% de l'activité biologique. Une plus grande spécificité semble être demandée en position 5 car Leu^5[AGII] et Ala^5[AGII] ne possèdent plus qu'environ 25% et 5% d'activité biologique [25].

Le fait que Pro^3[AGII] soit encore actif à 50% montre bien que cet analogue possède une conformation peu différente de celle de l'[AGII].

(c) Les résidus aromatiques $tyrosyl^4$ et $histidyl^6$ présentent un grand intérêt car on connaît le rôle important qu'ils tiennent dans les mécanismes de catalyse ou d'interaction. Dans le cas de la tyrosine, le cycle aromatique est essentiel puisque Phe^4[AGII] possède encore 10% d'activité biologique [26], alors que l'introduction de tout autre acide aminé en position 4 entraîne la perte totale de l'activité. D'autre part, pour les

TABLEAU I – Activités biologiques de quelques analogues de l'angiotensine II comparées à celles de Val^5 angiotensine II Asp^1 β-amide

Modifications au niveau des acides aminés	Analogues	Séquence des acides aminés 1	2	3	4	5	6	7	8	Activités biologiques[a]
–	Val^5[AGII]	Asp–	Arg–	Val–	Tyr–	Val- (Ileu)	His–	Pro–	Phe	100
–	Val^5[AGII]Asp$^1\beta$-amide	Asp (CONH$_2$)								100
1	p-Nitrobenzoyl Val^5[AGII]Asp$^1\beta$-amide	R-Asp								40–50
1	D-Asp^{1}Val^5[AGII]Asp$^1\beta$-amide	D-Asp								100
1	Gly^{1}Val^5[AGII]	Gly								50
1	Heptapeptide 2–8	–								30–50
1–2	Hexapeptide 3–8	–	–							0
2	Orn^{2}Val^5[AGII]Asp$^1\beta$-amide		Orn							20
2	D-Arg^{2}Val^5[AGII]Asp$^1\beta$-amide		D-Arg							3
2	(Im-Orn)^{2}Val^5[AGII]Asp$^1\beta$-amide[b]		Im-Orn							80
2	Val^{2}Val^5[AGII]Asp$^1\beta$-amide		Val							5
3	Pro^3[AGII]			Pro						40
3	Leu^{3}Val^5[AGII]Asp$^1\beta$-amide			Leu						100
3	Ala^3[AGII]			Ala						30–68
1–2–3	Pentapeptide 4–8	–	–	–						0
4	Phe^{4}Val^5[AGII]Asp$^1\beta$-amide				Phe					10
4	Ala^4[AGII]				Ala					0,3–3
4	D-Tyr^{4}Val^5[AGII]Asp$^1\beta$-amide				D-Tyr					0
5	Leu^5[AGII]Asp$^1\beta$-amide					Leu				25
5	Ala^5[AGII]					Ala				5
6	Lys^{6}Val^5[AGII]Asp$^1\beta$-amide						Lys			0
6	D-HisVal^5[AGII]Asp$^1\beta$-amide						D-His			1
7	Ala^7[AGII]							Ala		0
7	HO–Pro^7[AGII]							HO–Pro		7
8	Val^5[AGII]Asp$^1\beta$-amide Phe^8 méthylester								Phe COOMe	
8	Tyr^8[AGII]								Tyr	10–30
8	Ala^8[AGII]								Ala	inh. comp.
8	Ileu^8[AGII]								Ileu	inh. comp.

[a] Les données pour les activités biologiques (pression artérielle chez le rat binéphrectomisé) ont été extraites des livres de Schröder et Lübke (1966) et de

monoiodo et diiodotyrosyl[AGII], il persiste encore 30% d'activité pour le premier composé, alors que le second en est dépourvu.

L'histidine est importante par les particularités du cycle imidazole qui se résument par les possibilités de tautomérisme, d'échanges et de résonance. Elles expliquent sa grande réactivité et peuvent faire comprendre pourquoi la molécule d'[AGII] (et bien d'autres) n'admet aucun remplacement de l'histidine par un autre acide aminé, même aromatique (Lys^6[AGII] et Phe^6[AGII] sont dénuées de toute activité biologique).

(d) Le rôle de la proline est souligné par l'absence d'activité biologique de l'analogue Ala^7[AGII], et sa haute spécificité est mise en évidence par la faible activité biologique que possède HO–Pro^7[AGII].

(e) L'arginine, en position 2 dans la molécule d'[AGII], joue également un rôle important. Si pour l'heptapeptide 2–8 une activité biologique de 30 à 50% est mesurée, l'hexapeptide 3–8 est inactif. Son remplacement par l'ornithine ou la valine provoque une chute considérable de l'activité biologique.

Les résultats relatifs à la conformation, pour la plupart discordants il y a encore peu de temps [27, 28] aboutissent peu à peu à la conclusion que l'angiotensinamide II en solution adopte préférentiellement une structure du type 'cross *β*' [29–31]. Cette structure accorde un grand rôle conformationnel aux résidus valyls en position 3 et 5 ainsi qu'au résidu prolyl en position 7 de la chaîne. Leur présence induit le repliement de la chaine en deux boucles, l'une autour du résidu tyrosyl en position 4, l'autre au niveau de la proline projetant le cycle phényl du résidu C-terminal vers l'intérieur de la molécule. Dans cette situation une zone hydrophobe, comprenant les chaînes latérales des résidus valyls et phénylalanyl est délimitée d'un côté de la molécule d'angiotensine II. La présente étude a pour objet d'examiner d'une part la conformation de l'angiotensinamide II et de plusieurs de ses analogues (Tableau II) dans l'HFIP par la technique du dichroïsme circulaire. L'ensemble des résultats concernant aussi bien

TABLEAU II

Analogues de l'angiotensine II utilisés pour les études de dichroïsme circulaire

Analogues	Séquence des acides aminés	PM
	1 2 3 4 5 6 7 8	
Angiotensine II	H_2N–Asp–Arg–Val–Tyr–Ileu–His–Pro–Phe–COOH	
Val^5[AGII]Asp^1*β*-amide	←—Asp————Val————→ (CONH₂)	1030
Ala^8[AGII]	←————————————Ala————→	969
$Ileu^8$[AGII]	←————————————Ileu————→	1011
Pro^3[AGII]	←————Pro————————→	1043
Phe^4Val^5[AGII]Asp^1-*β*-amide	←—Asp————Phe–Val————→ (CONH₂)	1014
D-His^6Val^5[AGII]Asp^1-*β*-amide	←—Asp————Val–D-His————→ (CONH₂)	1030
D-Arg^2Val^5[AGII]Asp^1-*β*-amide	←— Asp–D-Arg————Val————→ (CONH₂)	1030
(Im-Orn)$^2Val^5$[AGII]-Asp^1*β*-amide	←—Asp–Im-Orn————Val————→ (CONH₂)	1124

l'activité biologique des peptides examinés que leur conformation est discuté dans le but de dégager l'existence possible d'une relation activité biologique – conformation moléculaire. D'autre part, quelques résultats de résonance magnétique nucléaire sont présentés.

2. Matériel et méthodes

L'angiotensinamide II, ainsi que les analogues D-Arg^2[AGII], D-His^6[AGII] et Phe^4[AGII] sont un don du Dr. Riniker (Ciba-Geigy, Bâle). Les analogues Ileu^8[AGII], Ala^8[AGII] ainsi que Pro^3[AGII] ont été synthétisés dans le laboratoire du Pr. Bumpus (Cleveland Clinic, Cleveland, Ohio). Le dérivé CHD-[AGII], ((imidazolidinone-ornithine)^2–Val^5[AGII]) a été préparé dans notre laboratoire. l'HFIP (hexafluoroisopropanol) est un produit Mann-Research. Le d_6-DMSO est un produit C.E.A-Saclay. La cyclohexanedione provient des laboratoires Koch-Light.

L'enregistrement des spectres de dichroïsme circulaire (entre 320 et 185 nm) a été réalisé avec un dichrographe II (Jouan Roussel) équipé d'une source xénon. Les expériences de RMN du proton ont été effectuées sur un spectromètre TSN 250 MHz (Cameca). Celles de RMN du ^{13}C ont été menées sur un appareil JEOL du type PS 100 F.T. En dichroïsme circulaire la gamme de concentration des peptides varie de 10^{-3} à 2×10^{-3} M, pour des trajets optiques de 0,1 cm à 0,01 cm. Pour les expériences de RMN du proton, des concentrations de 0,05 M et des tubes de 5 mm sont utilisés. En résonance du ^{13}C les concentrations sont plus élevées (10^{-1} M) les tubes possédant un diamètre de 8 ou 10 mm.

3. Résultats – discussion

Les spectres de dichroïsme circulaire dans la région aromatique de l'angiotensinamide II et de ses analogues (Tableau II) sont reportés sur la Figure 1. Entre 250 et 320 nm, comme le montrent les spectres des acides aminés libres, apparaissent les bandes de dichroïsme dues aux cycles aromatiques de la tyrosine et de la phénylalanine [32]. Le spectre de l'angiotensinamide II dans cette région est caractérisé par la présence de trois bandes situées respectivement à 260, 268 et 277 nm dont les ellipticités sont voisines de -200. Les deux premières bandes sont dues à la contribution de la chaîne latérale de la phénylalanine. Alors que la dernière (277 nm) attribuée au cycle tyrosyl est absente dans le spectre de l'analogue (Phe^4)[AGII] les deux premières (260 et 268 nm) apparaissent renforcées. La position de la bande due à la contribution de la tyrosine et son intensité laissent supposer qu'aucune interaction ou blocage contrarient la libre rotation du cycle aromatique. La titration du groupement hydroxyle de la tyrosine par spectrophotométrie UV, par DCM, ainsi que les mesures par RMN des déplacements chimiques des protons en ortho du OH confirment ce résultat. Toutefois des variations importantes du signal dû au cycle tyrosyl apparaissent à l'examen des spectres de Val-tyr-val (286 nm, $\theta = +235$ nm) (Figure 2), de D-His^6[AGII] (283 et 273 nm, $\theta = +100$), de Pro^3[AGII] (274 nm, $\theta = +500$; 281 nm, $\theta = +460$), de Ala^8[AGII] (283 nm, $\theta = +50$; 278 nm, $\theta = 0$), et enfin de Ileu^8[AGII] (288 nm,

$\theta = -30$; 284 nm, $\theta = -5$; 279 nm, $\theta = -30$). De telles modifications au niveau des spectres peuvent s'expliquer par des orientations et des environnements particuliers du cycle tyrosyl dans chacun des peptides qui d'ailleurs peuvent être présents sous forme de plusieurs conformères en solution et c'est sûrement le cas pour les deux derniers analogues cités. Une place importante doit être accordée à l'analogue Pro^3[AGII] dont le signal dédoublé et très intense correspondant au cycle tyrosyl montre bien que le remplacement de la valine en position 3 par la proline crée dans cette région de la molécule une réorganisation spatiale dont une des conséquences est le blocage du cycle tyrosyl. Enfin, la contribution du cycle tyrosyl dans D-Arg^2[AGII] (276,5 nm, $\theta = -240$) est voisine de celle dans l'[AGII], alors que dans le spectre de

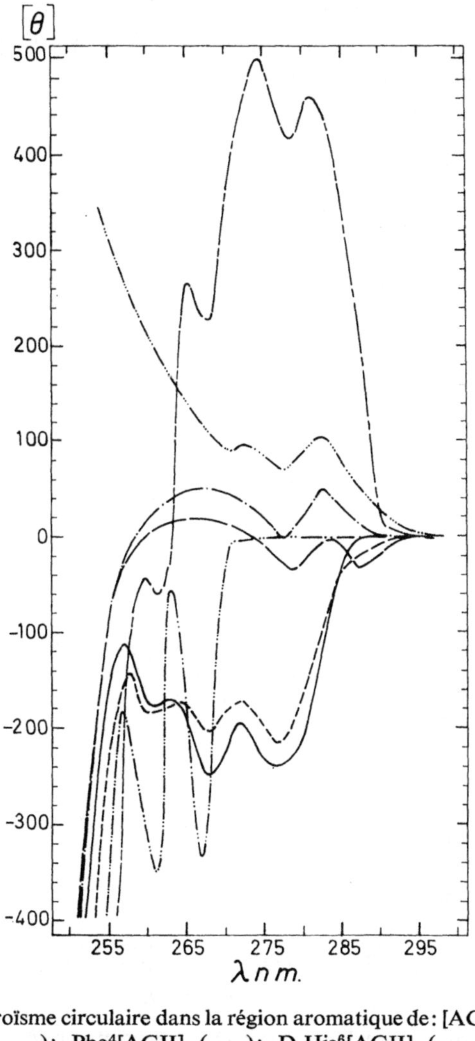

Fig. 1. Spectres de dichroïsme circulaire dans la région aromatique de: [AGII] (------); D-Arg^2[AGII] (——); Pro^3[AGII] (— — —); Phe^4[AGII] (-··-); D-His^6[AGII] (-···-); Ala^8[AGII] (— · —); Ileu^8[AGII] (- - -); en solution dans l'hexafluoroisopropanol.

CHD[AGII] elle est nettement positive (275 nm, $\theta = +220$). Notons que le spectre de cet analogue possède une large bande négative centrée vers 355 nm qui caractérise le chromophore imidazolidinone.

L'examen de la région peptidique (250 à 185 nm) (Figure 3) du spectre de dichroïsme circulaire de l'angiotensinamide II montre la présence de trois bandes: une négative intense à 188 nm, une positive à 200 nm, et une troisième négative à 223 nm possédant un épaulement vers 235 nm. La bande à 223 nm est affectée par la bande positive due à la tyrosine et bien mise en évidence dans le spectre de Val–tyr–val (Figure 4). Si la bande située à 188 nm reflète en grande partie la contribution des chaînes latérales des résidus aromatiques tyrosyl, phénylalanyl et histidyl [33], l'épaulement à 235 nm est le résultat des contributions à ce niveau du résidu prolyl et des formes désordonnées. L'[AGII] en solution dans l'HFIP possède donc en majorité la conformation β antiparallèle, comme le montre la bande caractéristique positive à 200 nm. L'intensité de celle-ci est toutefois diminuée par celle négative à cette même longueur d'onde correspondant aux formes désordonnées. Ainsi il apparait que dans l'hexafluoroisopro-

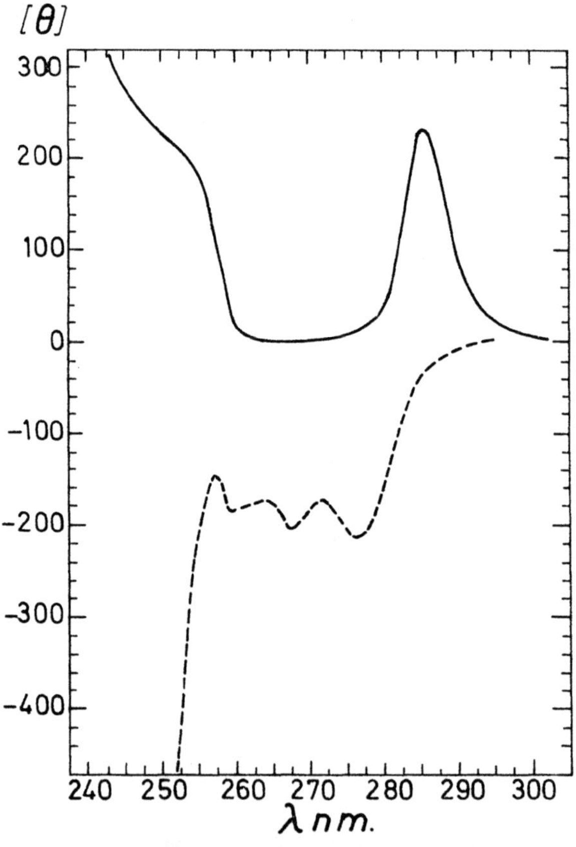

Fig. 2. Spectres de dichroïsme circulaire dans la région aromatique de: Val–tyr–val (—) et de AG II (----) en solution dans l'hexafluoroisopropanol.

panol la molécule d'angiotensinamide II est beaucoup plus stabilisée que dans l'eau mais moins que dans le trifluoroéthanol [29] où seule la structure β antiparallèle est décelable.

Les spectres de l'Ileu[8][AGII] et de l'Ala[8][AGII] présentent des similitudes de forme bien que des ellipticités (négatives) légèrement différentes caractérisent leurs bandes communes localisées respectivement aux alentours de 216 et 200 nm. Celles-ci ainsi que la bande située à 195 nm décrivent les structures hélicoïdales. La constatation d'une telle différence conformationnelle entre l'hormone et ses deux inhibiteurs compétitifs, où la seule variable est le résidu C-terminal, met en relief l'importance tant structurale que biologique de l'acide aminé en position 8. Ainsi est souligné non seulement le rôle joué par le cycle phényl mais également la nécessité pour la molécule d'angiotensine II d'avoir une conformation bien définie.

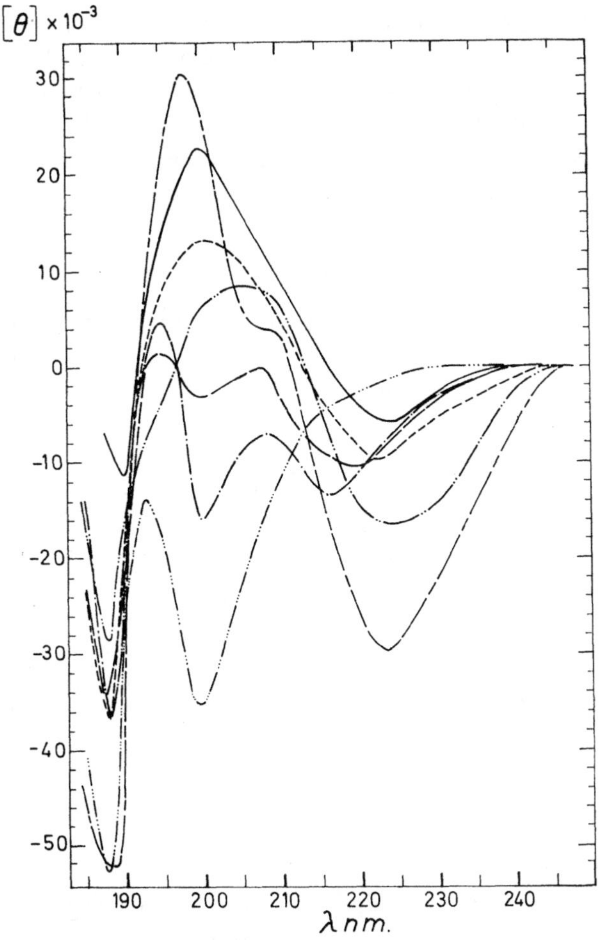

Fig. 3. Spectres de dichroïsme circulaire dans la région peptidique de: [AGII] (------); D-Arg[2][AGII] (——); Pro[3][AGII] (———); Phe[4][AGII] (-..-); D-His[6]][AGII] (-...-); Ala[8][AGII] (—.—); Ileu[8] [AGII] (- - -); en solution dans l'hexafluoroisopropanol.

L'analogue Phe[4][AGII] comme le montre les bandes situées à 205 nm et à environ 225 nm (l'intensité de cette dernière différente de celle relevée dans le spectre de l'angiotensine II est dûe à l'absence de la contribution positive de la tyrosine remplacée par la phénylalanine dans cet analogue) possède une structure du type β comme l'angiotensine II. Ainsi le remplacement du groupe phénolique par un groupe phényl dans la chaine peptidique ne modifie pas sa conformation. L'analogue Phe[4][AGII] ne possédant plus que 10% d'activité biologique il faut admettre que l'hydroxyle de la tyrosine de l'angiotensine n'entre dans aucune interaction susceptible de modifier la conformation du peptide. Les différentes mesures titrimétriques, spectrophotométriques, de RMN ... accordent à ce groupe un pK normal (10,1). Le rôle de l'hydroxyle se limiterait donc à une intervention dans l'activité biologique.

Peu de différences sont relevées lorsqu'on compare le spectre de la D-Arg[2][AGII] à celui de l'angiotensine II. Là encore, beaucoup plus que des modifications conformationnelles, c'est la mauvaise orientation de la chaîne latérale de la D-arginine qui doit provoquer la chute presque totale de l'activité biologique (97% de perte).

Comparé au spectre de l'angiotensinamide II, celui de l'analogue D-His[6][AGII] montre des différences dont la principale est l'exaltation de la bande négative à 200 nm qui caractérise l'apparition d'une large quantité de forme désordonnée. La perturbation provoquée au niveau de la chaîne par le remplacement d'un acide aminé L- par

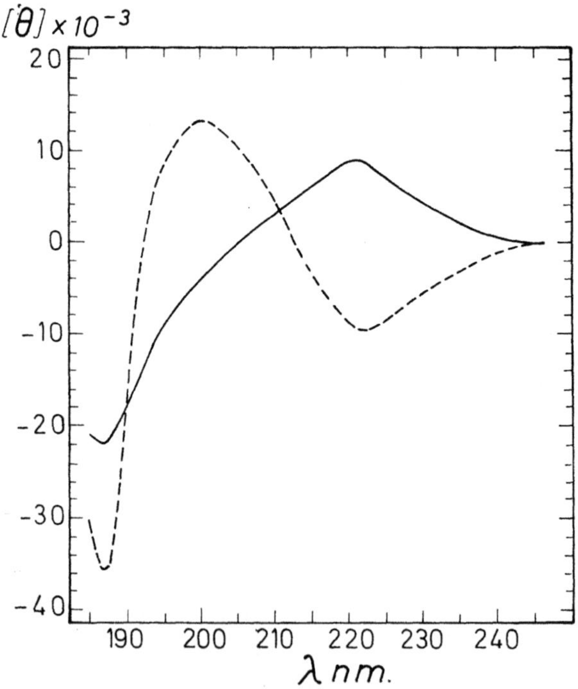

Fig. 4. Spectres de dichroïsme circulaire dans la région peptidique de: Val–tyr–val (——) et de [AGII] (----) en solution dans l'hexafluoroisopropanol.

un acide aminé D- a pour effet d'inverser les positions spatiales de la chaine latérale et du proton portés par le carbone α.

La conséquence importante serait d'empêcher pour des raisons stériques (le cycle imidazole se retrouvant vers l'intérieur de la molécule) la création de la boucle stabilisée par une liaison hydrogène, bien établie dans le modèle proposé pour l'angiotensine II [29–31], entre le groupe carbonyle de l'histidine et le NH de la phénylalanine. Cette déstabilisation de la chaine associée à une orientation inadéquate du cycle imidazole peut expliquer la chute de l'activité biologique observée.

Le spectre de l'analogue CHD[AGII] est très proche de celui de l'angiotensine II (Figure 5). La présence d'un groupement cyclique dans lequel toutefois les principales caractéristiques du système guanidilique se retrouvent n'affecte donc que peu la conformation et l'activité biologique (20% de perte).

Les plus grandes modifications sont notées dans le spectre de Pro^3[AGII] qui présente une large bande négative à 223 nm, un épaulement à 208 nm, une bande intense positive à 198 nm et enfin une bande négative très intense à 188 nm. L'allure générale

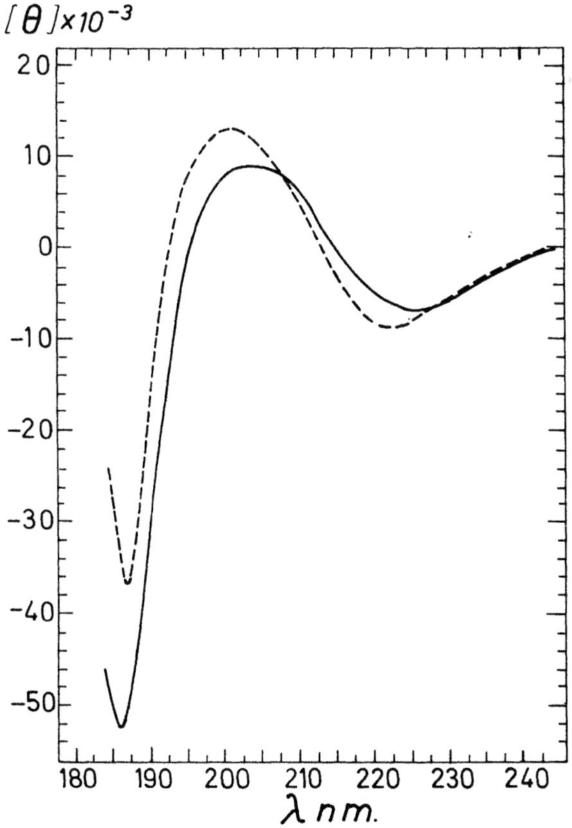

Fig. 5. Spectres de dichroïsme circulaire dans la région peptidique de: CHD-[AGII] (——) et de [AGII] (----) en solution dans l'hexafluoroisopropanol.

du spectre est en faveur de structure β antiparallèle grandement stabilisée par rapport à l'angiotensine II. Le remplacement de la valine en position 3 par la proline aurait donc pour effet de renforcer la boucle située à ce niveau dans le modèle proposé de l'angiotensine II. La perte d'activité biologique de 50% pourrait être la conséquence d'une mauvaise orientation du cycle phénolique de la tyrosine vicinale qui apparait bloqué dans le spectre de ce même composé, enregistré dans la région aromatique.

4. Études de resonance magnétique nucléaire

Rappelons que la résonance magnétique nucléaire a permis de mettre en évidence par des mesures de variation de température l'existence d'au moins une liaison hydrogène

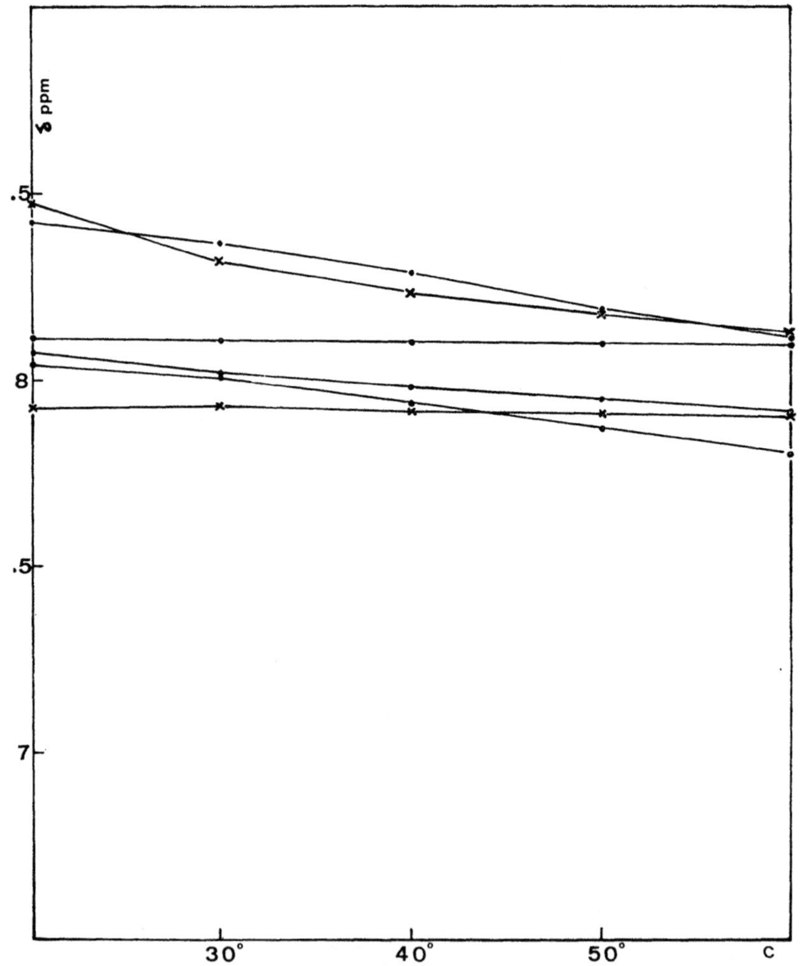

Fig. 6. Courbes des déplacements chimiques en fonction de la température des protons amides du tétrapeptide (5→8) — × — et de l'hexapeptide (3→8) —·— dans le d_6-DMSO à 250 Mz.

NH---O=C impliquant la phénylalanine et l'histidine [29–31]. Cette liaison hydrogène est retrouvée dans le tétrapeptide (5 → 8) ainsi que dans l'hexapeptide (3 → 8) dérivant de l'hormone comme le montrent les courbes obtenues en fonction de la température de la Figure 6. Ainsi, c'est entre les résidus de part et d'autre de la proline qu'apparait une liaison hydrogène, la création de celle-ci ne nécessitant pas l'intégralité de la molécule d'hormone. En fait, cette interaction pourrait bien stabiliser la boucle induite par la proline impliquant les trois acides aminés C-terminaux de l'angiotensine II. Pour mieux comprendre le rôle de ce résidu qui peut adopter dans les peptides la

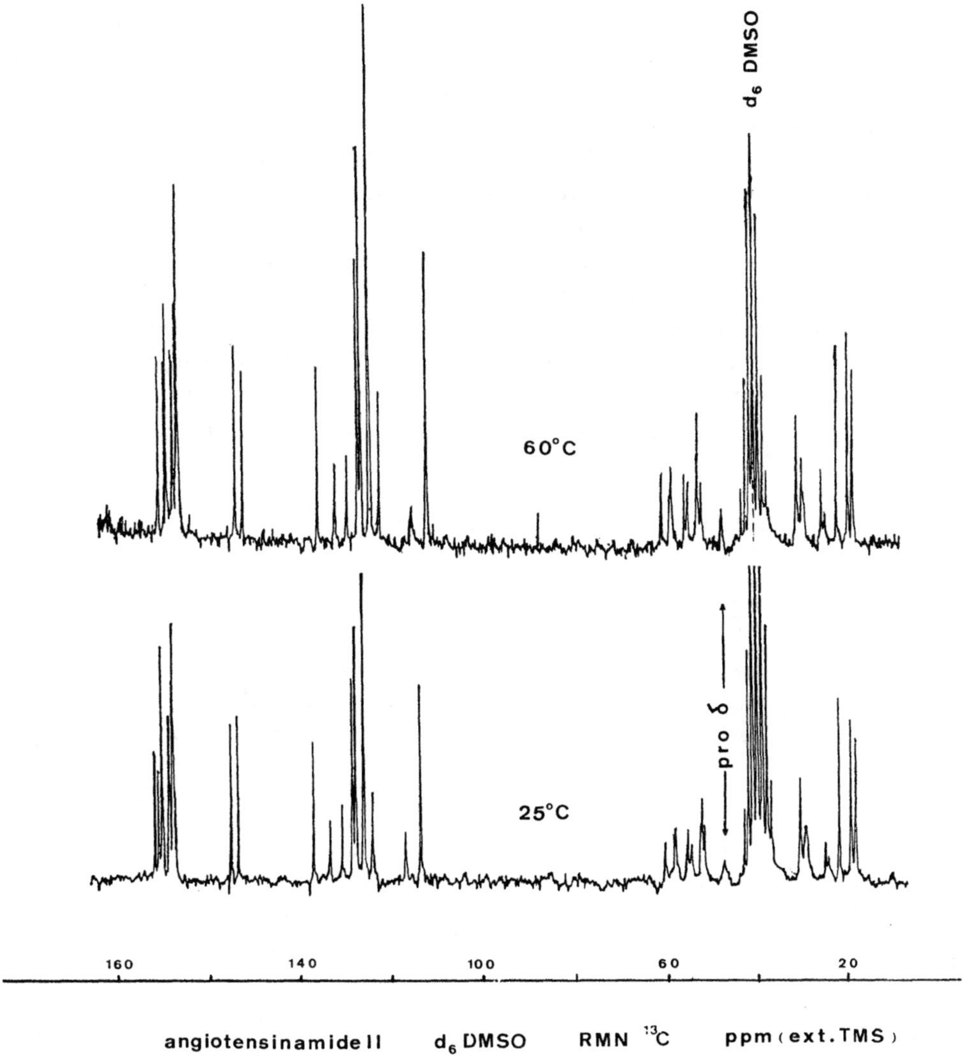

Fig. 7. Spectres de RM ^{13}C de l'angiotensinamide II dans d_6-DMSO à 25 et 60 °C. TMS utilisé comme référence externe 25, 15 Mz.

configuration cis ou trans [33–35], les spectres de résonance magnétique du ^{13}C de l'angiotensine II à 25 et 60° C, dans le d_6-DMSO, ont été obtenus (Figure 7). On constate que le signal du Cδ de la proline s'affine en fonction de l'élévation de la température. Ce résultat peut traduire, soit la mutarotation de la proline à l'intérieur du peptide soit son déblocage à partir d'une position figée (enveloppe) ce qui redonnerait au cycle prolyl une plus grande souplesse. Il est donc difficile de choisir entre l'une ou l'autre des possibilités bien que les résultats du dichroïsme circulaire soient en faveur de la mutarotation [29].

Les spectres de résonance magnétique nucléaire du proton, de l'angiotensine II et de son analogue Pro^3[AGII] sont représentés sur la Figure 8. C'est par double résonance homonucléaire (proton–proton) et par comparaison avec les spectres des peptides constitutifs de l'hormone à plusieurs températures que l'attribution des différents protons a été réalisée. Toutefois, si pour l'angiotensine II la corrélation CαH–NH a été rendue possible par une bonne résolution des protons amides elle est difficile à établir dans le cas de Pro^3[AGII]. C'est pourquoi dans ce qui suit seront comparées d'une part les régions CαH et d'autre part les régions aromatiques des spectres des deux peptides. Ainsi les CαH des Val^3 et Val^5 de l'angiotensinamide II donnent un signal unique apparaissant dans un pic (3 protons) centré à 4,17 ppm, alors que celui de Pro^7 résonne à environ 4,30 ppm.

Dans le cas de Pro^3–$Ileu^5$–[AGII] il n'apparaît à 4,15 ppm plus que le signal de l'$Ileu^5$ alors que celui de Pro^3 (résidu qui remplace Val^3 dans l'analogue) apparait à 4,26 ppm superposé à celui de Pro^7. Les déplacements chimiques des autres CαH ne semblent pas affectés et en particulier les signaux de l'histidine et surtout de la tyrosine

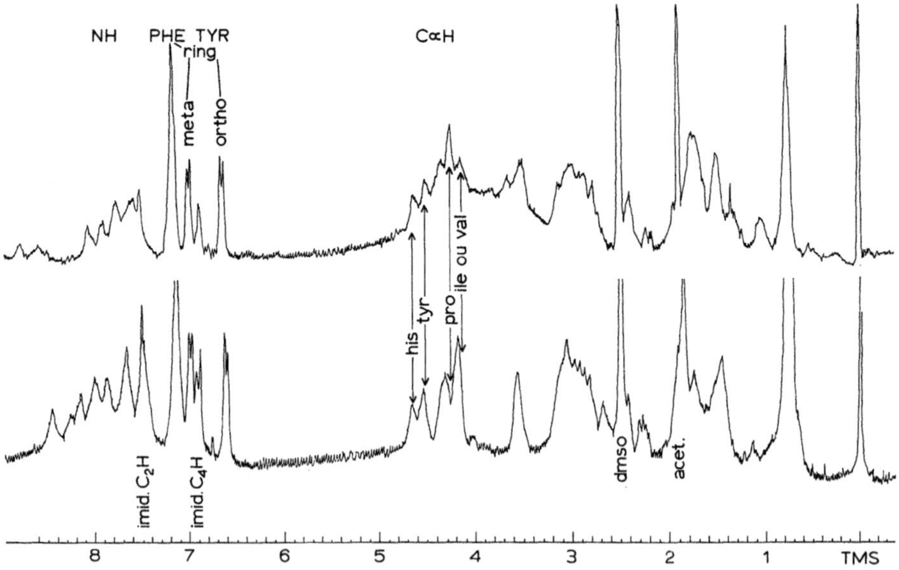

Fig. 8. Spectres de RMN ^1H de l'angiotensinamide II et de son analogue Pro^3[AGII] dans d_6-DMSO à 20 °C. TMS utilisé comme référence interne 250 Mz.

apparaissent à 4,65 et 4,52 ppm dans chacun des spectres. Des déplacements chimiques identiques caractérisant les CαH des deux prolines de l'analogue il est probable que la configuration de ces résidus soit voisine.

D'autre part, la bonne superposition des signaux du CαH de la tyrosine, adjacente soit à la Val^3 dans l'angiotensinamide II soit à la Pro^3 dans le cas de l'analogue, indique bien que la conformation de ces peptides est voisine. Cette observation parait intéressante car la substitution d'un acide aminé par un autre dont la géométrie et les propriétés sont différentes devrait se traduire dans le cas de modification conformationnelle, par la perturbation des déplacements chimiques des CαH des acides aminés adjacents. Plus loin dans la chaine peptidique le résidu histidyl en position 6 n'est pas affecté, ce qui montre bien que la conformation de l'angiotensine n'est pas modifiée même à longue distance. Le deuxième point important découle de l'examen de la région aromatique. En effet, une légère différence de déplacement chimique existe entre les protons ortho du cycle tyrosyl de l'angiotensinamide II (6,62 ppm) et ceux de l'analogue (6,85 ppm.) Bien que cet écart soit faible il pourrait refléter, et ceci confirmant les résultats du dichroïsme circulaire dans la région aromatique, l'existence d'une contrainte stérique au niveau du cycle tyrosyl. Le blocage de la chaîne latérale du résidu aromatique devrait alors être dû à l'encombrement important qu'apporte le cycle prolyl vicinal, cette situation n'affectant pas ou peu la conformation du squelette peptidique proprement dite.

Concernant le cycle imidazole de l'histidine aucune différence n'est notée entre les deux composés.

5. Conclusion

L'examen par dichroïsme circulaire des analogues de l'angiotensine a permis de mettre en évidence trois catégories d'acides aminés. D'une part, ceux qui jouent un rôle conformationnel, leur substitution par un autre acide aminé provoquant des modifications spectrales et consécutivement une perte de l'activité biologique. Ceci souligne la nécessité pour la molécule d'angiotensine II de posséder une conformation moléculaire précise pour être biologiquement active.

On a vu l'exemple de la valine, en position 3 de la chaine peptidique, qui substituée par la proline conduit à l'analogue Pro^3[AGII] dont la structure β est stabilisée par rapport à celle de l'angiotensine II. Il s'ensuit une perte d'activité biologique d'environ 50%. On sait de par la littérature (Tableau I) que l'analogue Ala^3[AGII]* possède une activité biologique modifiée dans les mêmes proportions que Pro^3[AGII]. La seconde categorie rassemble les acides aminés qui possèdent un rôle davantage fonctionnel et dont la substitution ne semble pas modifier la conformation moléculaire mais perturbe l'activité biologique. Ces acides aminés sont en particulier la tyrosine, l'histidine, et l'arginine. En fait pour ceux-ci c'est la nature et l'orientation spatiale de leur chaine latérale qui jouent un rôle déterminant.

* Il serait intéressant de comparer les spectres de Pro^3[AGII] et d'Ala^3[AGII]. Malheureusement nous ne possédons pas ce dernier analogue.

Ainsi l'analogue D-Arg[AGII] qui possède un spectre voisin de celui de l'angiotensine II n'est pratiquement plus actif, et enfin l'ornithine^2[AGII] ne possède plus que 20% d'activité biologique (voir Tableau I). Ces deux observations soulignent d'une part qu'une bonne orientation de la chaine latérale est nécessaire et d'autre part l'importance du groupe guanidilique (bien que la spécificité ne soit pas totale) pour l'activité biologique.

La phénylalanine en position C-terminale de l'hormone entre dans la troisième catégorie. Les différences notées entre l'angiotensinamide II et ses deux inhibiteurs compétitifs Ala^8[AGII] et Ileu^8[AGII] qui possèdent une structure à tendance hélicoïdale traduisent bien l'importance du résidu C-terminal. En fait l'acide aminé dans cette position de la chaîne possède un double rôle: il module d'une part la conformation générale de la molécule et d'autre part la nature de sa chaine latérale définit les interactions qui prennent naissance au niveau de la membrane cellulaire.

Le dernier point important découle des études de résonance magnétique du proton et en particulier de la comparaison des spectres de l'angiotensinamide II et de Pro^3[AGII]. Comme le montre la bonne superposition des résonances dans la région des CαH peu de différence semble exister entre les conformations de l'hormone et de son analogue. Ces deux peptides si on se réfère aux spectres de dichroïsme circulaire dans la région peptidique, possèdent tous les deux des structures du type β et seulement des différences d'ellipticités les distinguent accordant à l'analogue Pro^3[AGII] une plus grande stabilité. Il faut donc admettre que la boucle qui prend naissance dans l'hormone au niveau de Val^3 possède la même forme mais est plus stable à cause de la rigidité introduite par le cycle prolyl dans l'analogue Pro^3[AGII]. La perte d'activité biologique observée pour cet analogue provient d'une mauvaise orientation du cycle tyrosyl comme semblent l'indiquer, d'une part le spectre de dichroïsme circulaire et d'autre part le spectre de résonance magnétique nucléaire tous les deux examinés dans la région aromatique.

Remerciements

Nous remercions le Docteur Riniker (Ciba-Geigy, Bâle) qui nous a gracieusement donné l'angiotensinamide II ainsi que les Professeurs Bumpus et Smeby qui ont synthétisé les analogues utilisés. Deux d'entre nous, Daniel Greff, et Serge Fermandjian, sous contrat RCP 220, remercient le C.N.R.S.

Références

[1] Page, I. H. et Bumpus, F. M.: 1961, *Physiol. Rev.* **4**, 331.
[2] Skeggs, L. T., Kahn, J. R., et Shuway, N. P.: 1956, *J. Exp. Med.* **103**, 301.
[3] Rocha e Silva, M.: 1961, *Biochem. Pharmacol.* **8**, 177.
[4] Rocha e Silva, M.: 1963, *Ann. N.Y. Acad. Sci.* **104**, 190.
[5] Skeggs Jr., L. T., Kahn, J. R., Lentz, K., et Shuway, N. P.: 1957, *J. Exptl. Med.* **106**, 439.
[6] Poulsen, K.: 1970, dans '*In Vitro* Procedures with Radio-isotopes in Medicine', Proceedings of a Symposium, Vienna 8–12 Sept. 1969.
[7] Braun-Menendez, E., Fasciolo, J. C., Leloir, L. F., et Munoz, J. M.: 1939, *Rev. Soc. Arg. Biol.* **15**, 420.

[8] Katz, Y. L. et Goldblatt, H.: 1943, *J. Exptl. Med.* **78**, 67.
[9] Meyer, P.: publication en cours.
[10] Peart, W. S.: 1955, *Biochem. J.* **59**, 300.
[11] Anglese d'Auriac, G. et Meyer, P.: *Proc. Congr. Cyclic AMP*, Milan 1971.
[12] Baudouin, M., Meyer, P., Fermandjian, S., et Morgat, J. L.: 1972, *Nature* **336**, 235.
[13] Morgat, J. L., Lam Thanh Hung, et Fromageot, P.: 1970, *Biochim. Biophys. Acta* **207**, 374.
[14] Baudouin, M., Meyer, P., et Worcel, M.: 1971, *Biochem. Biophys. Res. Commun.* **42**, 434.
[15] Meyer, P., Papadimitriou, A., et Worcel, M.: 1970, *Brit. J. Pharmacol.* **4**, 541.
[16] Paiva, T. B., Paiva, A. C. M., Freer, R. J., et Stewart, J. M.: 1972, *J. Med. Chem.* **15**, 1, 6.
[17] Manning, M. et Du Vigneaud, J.: 1965, *J. Am. Chem. Soc.* **87**, 3978.
[18] Eisler, K., Rudinger, J., et Sorm, F.: 1966, *Coll. Czech Chem. Commun.* **31**, 4563.
[19] Arakawa, K., Smeby, R. R., et Bumpus, F. M.: 1962, *J. Am. Chem. Soc.* **84**, 1424.
[20] Park, W. K., Smeby, R. R., et Bumpus, F. M.: 1968, *Biochemistry* **6**, 3458.
[21] Khairallah, P. A., Toth, A., et Bumpus, F. M.: 1970, *J. Med. Chem.* **13**, 181.
[22] Bumpus, F. M.: 1971 dans *Protein and Polypeptide Hormones*, Part 1, Liège, 1971.
[23] Khosla, M. C., Smeby, R. R., et Bumpus, F. M.: 1967, *Biochemistry* **6**, 754.
[24] Khosla, M. C., Chaturvedi, N. C., Smeby, R. R., et Bumpus, F. M.: 1968, *Biochemistry* **7**, 3417.
[25] Jorgensen, E. C., Rapaka, S. R., Windridge, G. C., et Lee, T. C.: 1971, *J. Med. Chem.* **14**, 10, 899.
[26] Riniker, B. et Schwyzer, R.: 1961, *Helv. Chim. Acta* **44**, 677.
[27] Smeby, R. R., Arakawa, K., Bumpus, F. M., et Marsh, M. M.: 1962, *Biochim. Biophys. Acta* **58**, 550.
[28] Paiva, T. B., Paiva, A. C. M., et Scheraga, H. A.: 1963, *Biochemistry* **2**, 1327.
[29] Fermandjian, S., Morgat, J. L., et Fromageot, P.: 1971, *Eur. J. Biochem.* **24**, 252.
Fermandjian, S., Greff, D., et Fromageot, P.: 1972, dans *Chemistry and Biology of Peptides.*
[30] Weinkam, R. J. et Jorgensen, E. C.: 1971, *J. Am. Chem. Soc.* **93**, 25, 7038.
[31] Printz, M. P., Williams, H. P., et Craig, L. C.: 1972, *Proc.Mat. Acad. Sci. US* **69**, 2, 378.
[32] Fasman, G. D., Bodenheimer, E., et Lindblow, G.: 1964, *Biochemistry* **3**, 1665.
[33] Legrand, M. et Viennet, A.: 1965, *Bull. Soc. Chim. France* **679**.
[34] Bovey, F. A. et Hood, F. P.: 1967, *Biopolymers* **5**, 325.
[35] Deslauriers, R., Walter, R., et Smith, I. C. P.: 1972, *Biochem. Biophys. Res. Commun.* **48**, 834.

DISCUSSION

De Coen: Vous avez montré que l'angiotensine dissoute dans le TFE donnait un spectre de dichroïsme circulaire du type β. Avez-vous fait des études en fonction de la concentration susceptibles de distinguer entre la forme β-étendue et la forme β-croisée?

Fermandjian: Deux types d'étude ont été menés. L'une toujours par dichroïsme circulaire et l'autre par diffusion Rayleigh. Les deux méthodes montrent que dans le TFE, et ceci même à des concentrations élevées, aucune agrégation n'apparaît, ce qui confirme bien que l'angiotensine II se trouve sous la forme β-croisée.

CARACTÉRISATION, ISOLEMENT ET PURIFICATION DE LA PROTÉINE RÉCEPTRICE DE L'ACÉTYLCHOLINE À PARTIR DE L'ORGANE ÉLECTRIQUE DU GYMNOTE

JEAN-PIERRE CHANGEUX
Institut Pasteur, 75015 Paris, France

Summary. With the electric organ of *Electrophorus electricus* the electrogenic action of cholinergic agonists can be studied at three levels: (1) the *cellular level*: electric parameters, potential and conductance, are recorded with the isolated electroplax by electrophysiological methods, dose-response curves obtained and 'apparent' dissociation constants determined; (2) the *membrane level*: membrane fragments purified from homogenates of electric tissue make closed bags or microsacs which respond to the binding of agonists by an increase of permeability to Na^+, K^+ and Ca^{++}. Specific binding of ^{14}C-decamethonium to the microsacs is measured and the binding curve obtained compared to the dose-response curve; (3) the *molecular level*: the receptor protein solubilized by detergent treatment of membrane fragments binds reversibly cholinergic agonists and antagonists with binding constants close to those inferred from the dose-response curves obtained with electroplax and microsacs.

A pure snake venom α-toxin from *Naja nigricollis*, either cold or tritium-labelled (14 Ci/nmole), displaces cholinergic agonists or antagonists from their binding site, is displaced by them from the same site and thus constitutes, like α-bungarotoxin, a highly specific reagent of the cholinergic receptor. It is used to assay the receptor protein bound to the membrane or in solution.

The physical properties: $S_{20,w}$, Stokes radius and molecular weight of the free receptor protein in the presence of deoxycholate and Triton X-100 are measured and evidence for binding of detergents to the protein given. The receptor protein is purified by affinity chromatography.

Reintegration of the soluble receptor protein into membrane fragments after removal of detergents is demonstrated.

1. Introduction

L'ensemble du travail que nous avons effectué au cours des dernières années est fondé sur l'idée que la régulation de la perméabilité ionique d'une membrane excitable par l'acétylcholine présente des analogies profondes avec le contrôle d'un enzyme régulateur par son signal métabolique spécifique [1–3]. De fait, il est bien établi que l'accroissement du courant transmembranaire provoqué par l'acétylcholine ne résulte pas du transport de l'acétylcholine, ni de sa transformation chimique [4]. De plus, aucune analogie de structure n'a jamais été rencontrée entre l'ion transporté et l'acétylcholine quelle que soit la préparation utilisée.

Il était alors légitime de postuler que l'acétylcholine agit comme un ligand régulateur et que l'élément de structure membranaire minimum qui se trouve sous son contrôle, comprenne deux éléments au moins partiellement distincts: la protéine réceptrice qui porte le site récepteur et un ionophore impliqué dans la translocation sélective des cations à travers la membrane [5]. Le couplage entre acétylcholine et ion perméant, c'est-à-dire entre protéine réceptrice et ionophore, est considéré comme une interaction indirecte et donc allostérique s'effectuant par l'intermédiaire d'une transition de structure [6]. En d'autres termes, récepteur et ionophore joueraient un rôle semblable à celui des sous-unités régulatrices et catalytiques d'un enzyme régulateur tel que l'aspartate transcarbamylase [7].

C. Sadron (ed.), Dynamic Aspects of Conformation Changes in Biological Macromolecules. 511–519.

Nous allons résumer ici les principaux résultats obtenus avec la protéine réceptrice de l'acétylcholine et des ligands cholinergiques. Le travail biochimique a été effectué à partir d'un matériel biologique particulièrement riche en synapses cholinergiques qui, par surcroit, sont d'un type très homogène: l'organe électrique du Gymnote (*Electrophorus electricus*) et de la Torpille (*Torpedo marmorata*) [8–9]. Un organe électrique de Gymnote pèse plusieurs centaines de grammes et peut contenir jusqu'à 10^{10} synapses identiques. Les cellules électrogènes ou électroplaques peuvent être isolées et disséquées à partir de l'organe électrique ce qui permet la réalisation d'expériences d'électrophysiologie sur le même matériel que celui employé pour les expériences biochimiques. Il devient donc possible de comparer les données *in vivo* avec des résultats obtenus *in vitro*, ce qui est nécessaire pour la réduction d'un mécanisme physiologique du niveau cellulaire au niveau moléculaire.

Les expériences ont été effectuées à trois niveaux d'organisation différents: (1) L'électroplaque isolée: la réponse aux agonistes et antagonistes cholinergiques est suivie en mesurant le potentiel de membrane de régime obtenu après l'application en bain du ligand pharmacologique [10–12]. (2) Les microsacs excitables préparés à partir d'un homogénat de tissu électrique: il est possible d'isoler des fragments de membrane ou microsacs qui se referment sur eux-mêmes en vésicules closes et à partir desquelles des flux d'ions sont mesurés. La liaison des ligands cholinergiques radioactifs est suivie facilement avec la même préparation et la corrélation entre liaison et réponse de perméabilité établie [13, 14]. (3) La protéine réceptrice: les courbes de liaison à l'équilibre de plusieurs agonistes et antagonistes cholinergiques sont effectuées avec la protéine extraite des microsacs excitables par des détergents non dénaturant (cholate de sodium, Triton X-100)[15, 16]. Cette protéine est purifiée par chromatographie d'affinité [41].

2. Caractérisation du récepteur physiologique de l'acétylcholine à l'aide de ligands cholinergiques et de toxines α de venins de serpent

La caractérisation du site membranaire qui est à l'origine de l'effet électrogénique de l'acétylcholine se trouve affrontée à un difficile problème de spécificité. La plupart des ligands cholinergiques employés jusqu'à ce jour ne possèdent pas une spécificité suffisante et présentent souvent une affinité élevée pour des sites distincts du site récepteur cholinergique existant dans la membrane excitable: par exemple les sites catalytiques et allostériques de l'acétylcholinestérase, d'éventuels sites de transport présynaptiques de l'acétylcholine etc. Il devenait donc nécessaire d'avoir à sa disposition un test de spécificité complémentaire et totalement indépendant. La solution a été apportée par des protéines toxiques de petit poids moléculaire présentes dans le venin de divers serpents. Bien qu'elles ne présentent aucune analogie de structure évidente avec les effecteurs cholinergiques, ces protéines agissent comme des agents curarisants très efficaces au niveau de la jonction neuromusculaire des vertébrés. Très souvent l'effet obtenu est irréversible, et la D-tubocurarine protège efficacement contre celui-ci. Les plus étudiées de ces toxines sont les α toxines de *Bungarus multicinctus* [17, 18], *Naja n. siamensis* [19–21], *Naja n. naja* [22], *Naja nigricollis* [23–26], *Laticauda semifasciata* [27,28], etc.

L'α-bungarotoxine et l'α-toxine de *Naja nigricollis* bloquent d'une manière très peu réversible, dans nos conditions expérimentales, la réponse électrique de l'électroplaque à la carbamylcholine et au décaméthonium [16, 29]. Elles bloquent aussi la réponse de perméabilité mesurée *in vitro* avec des microsacs excitables [14, 29]. De plus, la D-tubocurarine protège efficacement contre l'effet de ces toxines. *In vitro* ces toxines déplacent une fraction importante du décaméthonium ^{14}C, un agoniste cholinergique puissant, lié d'une manière réversible à des fragments de membrane excitable ou à un extrait soluble de ces fragments [29]. La quantité de décaméthonium déplacée par la toxine α est la même que celle déplacée par des concentrations élevées de deux antagonistes réversible: la D-tubocurarine ou la gallamine. La liaison résiduelle qui est résistante à la toxine α comme à celle des antagonistes en question est déplacée par des ligands cholinergiques inhibiteurs de l'acétylcholinestérase [16]. Cette dernière classe de sites résistants à la toxine est éliminée par traitement thermique dans des conditions qui inactivent l'acétylcholinestérase sans inactiver la protéine réceptrice. Il s'agit vraisemblablement du site catalytique de l'acétylcholinestérase [30].

Le parallélisme observé entre l'effet des toxines de venin de serpent et celui d'antagonistes cholinergiques typiques suggère que la fraction de décaméthonium déplacée par ces toxines est spécifiquement liée au site récepteur cholinergique. L'accord obtenu entre les constantes de liaison *intrinsèques* mesurées en suivant la liaison du décaméthonium sur des fragments de membrane excitable et les constantes de dissociation *apparente* mesurées *in vivo* ou *in vitro* en suivant les flux de sodium renforce cette conclusion [14].

L'utilisation des toxines α révèle donc plusieurs classes de sites de liaison du décaméthonium ^{14}C dans une membrane excitable, mais la même question se pose pour la toxine: est-ce qu'il existe plusieurs classes de sites de liaison des α toxines dans nos fragments de membrane? Afin de répondre à cette question qui devint essentielle lorsque furent utilisées les toxines α pour marquer la protéine réceptrice, la toxine α de *N. nigricollis* a été tritiée. La tritiation a été effectuée en deux étapes: l'iodation d'une histidine accessible de la toxine, suivie par la dehalogenation catalytique de la toxine iodée en présence de gaz tritium [31]. La toxine tritiée est ensuite purifiée par chromatographie. Elle possède la toxicité et les propriétés pharmacologiques de la toxine native avec une radioactivité comprise entre 10 et 15 curies par millimole. A peu près 90% du matériel marqué était constitué par la toxine active et approximativement 40% de la toxine était marqué.

Utilisant une technique de filtration sur Millipore, la cinétique de fixation de la toxine sur des fragments de membrane excitable se mesure facilement [32]. En première approximation les données cinétiques sont en accord avec l'hypothèse d'un mécanisme bimoléculaire de liaison de la toxine sur une classe homogène de sites membranaires. A la température ordinaire, en l'absence de sucrose, mais en présence de la solution physiologique de Ringer, la constante de vitesse d'association $\vec{k}$ est $(2{,}5 \pm \pm 0{,}5) \times 10^7$ M^{-1} min^{-1}, et la constante de vitesse de dissociation mesurée en présence d'un excès de toxine froide $(2 \pm 1) \times 10^{-4}$ min^{-1}. A partir de ces données cinétiques on calcule une constante de dissociation à l'équilibre de la toxine pour son site récepteur $K = \overleftarrow{k}/\overrightarrow{k} \# 10^{-11}$ M.

Dans une première série d'expériences nous avons étudié l'effet de concentrations physiologiques d'agonistes et d'antagonistes cholinergiques sur les vitesses initiales de fixation de la toxine α tritiée de *Naja nigricollis*. En accord avec les observations de Lee [18], Changeux *et al.* [29], Miledi *et al.* [33], Meunier *et al.* [34], les agonistes et antagonistes cholinergiques réduisent d'une manière très efficace les vitesses initiales de liaison des toxines. A concentrations saturantes d'effecteur, la vitesse initiale, dans nos conditions expérimentales, devient négligeable. Si l'on fait l'hypothèse que la toxine et l'effecteur cholinergique se lient d'une manière exclusive pour la même classe de site, et qu'un équilibre rapide est obtenu entre l'effecteur cholinergique et son site de liaison, alors il devient possible de proposer une formulation mathématique de l'effet protecteur du ligand cholinergique. Ce traitement montre que la vitesse initiale mesurée en présence d'effecteur cholinergique est proportionnelle à la fraction de site récepteur libre [32]. Jusqu'à 80% de protection environ, la courbe obtenue avec le décaméthonium et la D-tubocurarine suit l'isotherme de Langmuir. De plus, la courbe de protection se superpose à la courbe de liaison obtenue avec les mêmes ligands cholinergiques et les microsacs excitables, ainsi qu'avec les courbes dose-réponse obtenues *in vivo* et *in vitro*. En première approximation, la courbe de protection peut donc être considérée comme une courbe de liaison. Utilisant cette méthode, Weber *et al.* [32] ont montré que agonistes et antagonistes ne diffèrent pas dans leur manière de diminuer les vitesses initiales de liaison de toxine tritiée. Leur efficacité de protection est toujours en relation directe avec leur affinité réelle ou apparente. Les constantes de dissociation mesurées par cette méthode sont toujours en accord avec les constantes de dissociation apparentes [35], même dans le cas de la nicotine avec laquelle Eldefrawi *et al.* [36] ont mesuré des constantes d'affinité environ trois ordres de grandeur plus grandes que celles mesurées *in vivo*. Ces observations ainsi que le résultat qu'une complète protection a lieu à saturation d'effecteur, montrent clairement que dans les conditions expérimentales présentes, la toxine tritiée se lie avec une haute sélectivité sur le site récepteur de l'acétylcholine.

Les anesthésiques locaux comme la tétracaïne, la procaïne etc. bloquent la réponse électrique aux agents cholinergiques d'une manière non compétitive et diffèrent en cela des antagonistes compétitifs classiques tels que D-tubocurarine ou gallamine. Il est remarquable de noter qu'avec les anesthésiques locaux, les courbes de protection contre la toxine ne se superposent pas aux courbes dose-réponse. Ces composés agissent comme anesthésiques à des concentrations beaucoup plus faibles, souvent de 1 à 2 ordres de grandeur, que celles à laquelle ils protègent contre la liaison de toxine. L'interprétation la plus immédiate de ce résultat est que les anesthésiques locaux agissent en tant qu'anesthésiques à des sites qui sont distincts du site sur lequel les agonistes et les antagonistes se lient sur la protéine réceptrice ou son proche environnement.

Dans une autre série d'expériences, a été étudié l'effet de divers ligands cholinergiques sur l'équilibre obtenu après incubation du mélange toxine–fragments de membrane pendant un à trois jours à la température du laboratoire. Avec le décaméthonium, dans la mesure où la réaction a été déclanchée en ajoutant la toxine après le récepteur cholinergique, les valeurs obtenues au plateau sont en accord avec les données cinétiques. D'autres effets très intéressants des agents cholinergiques ont été obtenus en suivant la dissociation de la toxine déjà fixée au récepteur cholinergique. A des concentrations *très élevées* de décaméthonium ou D-tubocurarine, environ 1000 fois leur constante de dissociation à l'équilibre pour les sites récepteurs libres, un accroissement remarquable de la constante de vitesse de dissociation est observé: de cinq fois avec la D-tubocurarine 10^{-3} M, de 18 fois avec du décaméthonium 10^{-3} M. Dans ces conditions, un déplacement complet de la toxine a lieu en quelques heures. L'interprétation la plus simple de ce résultat est qu'il existe des sites pour les ligands cholinergiques sur la protéine réceptrice une fois que la toxine est liée. La liaison des ligands cholinergiques sur ces sites accroît la vitesse de dissociation de la toxine. Il n'est pas encore clair si ces sites secondaires font partie de sites cholinergiques dont certaines aires resteraient accessibles aux ligands cholinergiques une fois le complexe toxine–récepteur établi, ou s'il s'agit de sites antérieurement distincts.

3. Propriétés physiques de la protéine réceptrice en solution aqueuse de détergent

Les premières expériences signalées plus haut avaient montré que la molécule qui porte le site récepteur de l'acétylcholine peut être extraite sous une forme soluble par des détergents chargés négativement tel que le cholate ou le déoxycholate de sodium, ou par des détergents non chargés tel que le Triton X-100. La molécule réceptrice une fois solubilisée fixe encore les agonistes et les antagonistes cholinergiques ainsi que les toxines de venin de serpent. Cette molécule ne traverse pas les membranes Diaflo qui retiennent des molécules sphériques de masse moléculaire supérieure à 50000. De plus elle perd son activité ligante après chauffage à 100°, exposition à pH 4,75 et digestion par la pronase. La molécule réceptrice est donc une protéine ou contient au moins une partie protéique. Cette protéine se sépare de l'acétylcholinestérase sur une colonne de billes de Sépharose à laquelle la toxine α de *Naja nigricollis* a été attachée d'une manière covalente: la protéine réceptrice reste adsorbée sur la colonne tandis que dans ces conditions l'acétylcholinestérase n'est pas retenue [30].

La toxine α de *Naja nigricollis* a été employée pour suivre les propriétés hydrodynamiques de la protéine réceptrice en solution aqueuse. La protéine réceptrice était marquée, soit avant solubilisation, soit après solubilisation, la protéine en solution étant titrée par exemple par précipitation sélective du complexe récepteur–toxine par le sulfate d'ammonium. Le complexe récepteur–toxine ainsi que la protéine réceptrice sédimentent sous forme d'un pic unique en gradient de saccharose contenant les détergents mentionnés précédemment. Il est élué sous forme d'un pic unique d'une colonne de Sépharose 6B. Dans les conditions expérimentales présentes, et suivant les critères employés, la molécule protéique sur laquelle la toxine se fixe apparait homogène. Elle précipite en l'absence de détergent [34, 37].

Un paradoxe intéressant se trouve posé par la comparaison de données de sédimentation et de filtration sur gel d'acrylamide. En présence de Triton X-100 1%, le complexe récepteur–toxine sédimente légèrement plus lentement que la catalase de foie de bœuf (11,4S, masse moléculaire 240000) mais est exclu d'une colonne de Sépharose 6B légèrement avant la β-galactosidase du colibacille (16,0S, masse moléculaire 540000). En d'autres termes, le complexe récepteur–toxine, et cela est vrai aussi avec la protéine réceptrice libre, sédimente plus lentement que ce qui est prévisible de la mesure de son rayon de Stokes, faisant l'hypothèse que la protéine réceptrice est une sphère de densité voisine de celle des protéines globulaires classiques. Deux hypothèses peuvent rendre compte de ce comportement hydrodynamique inhabituel: ou bien une forme non globulaire, ou bien une densité différente de celle des protéines solubles employées comme standards.

L'ultracentrifugation du complexe récepteur–toxine dans des gradients de saccharose où la densité est variée en remplaçant l'eau légère par de l'eau lourde, a montré que le complexe récepteur–toxine possède une densité inhabituelle [38]. L'accroissement de densité du milieu de centrifugation s'accompagne d'une diminution de vitesse de sédimentation qui est beaucoup plus prononcée avec le complexe récepteur–toxine qu'avec des protéines globulaires employées comme standards. En présence de Triton X-100 1%, la densité apparente de la protéine réceptrice est plus petite ou le volume partiel spécifique apparent: $\bar{v}$ plus grand que celui d'une protéine globulaire habituelle $\bar{v}=0{,}78$ au lieu de 0,73. Sur le Tableau I sont rassemblées les données obtenues avec le Triton X-100 et les masses moléculaires apparentes, et le coefficient de friction (f/f_0) calculé à partir de ces valeurs.

On peut rendre compte de l'anomalie de densité observée en présence de Triton X-100 en supposant que la protéine fixe jusqu'à 20% de sa masse totale de Triton X-100 ou de lipides, ce calcul étant fondé sur l'hypothèse que la densité de la partie protéique est voisine de celle d'une protéine classique. Dans ces conditions, la masse moléculaire de la partie protéique sans Triton serait de l'ordre de 360000.

Le traitement d'un extrait au déoxycholate par du dodécylsulfate de sodium donne une sous-unité dont la masse moléculaire est voisine de 50000 sur gel d'acrylamide.

TABLEAU I

Quelques propriétés physiques de la protéine réceptrice de l'acétylcholine extraite de l'organe d'*Electrophorus electricus*

Solubilité	solvent aqueux + Triton X-100 1% ou Na cholate 1% ou Na deoxycholate 1%
Coefficient de sédimentation	$S_{20,w} = 12{,}5 \pm 0.01$
Rayon de Stokes	a (Å) $= 73 \pm 3$
Volume spécifique partiel	$\bar{v} = 0{,}78 \pm 0.01$
Masse moléculaire apparente en Triton X-100	470000 ± 30000
Triton X-100 lié (calculé)	21% de la masse totale
Masse moléculaire *apparente* de la partie protéique	360000 ± 25000
Masse moléculaire de la 'sous-unité'	45000 ± 5.000

Cette sous-unité est la plus petite observée avant la dissociation du complexe récepteur–toxine. Il est remarquable qu'une valeur semblable a été trouvée indépendamment par Karlin *et al.* [39] pour la protéine marquée par un réactif d'affinité, et par Raftery [40] et nous-mêmes avec une préparation purifiée de récepteur de Torpille. Il apparait donc plausible que la forme moléculaire qui est solubilisée par le déoxycholate est un polymère bien défini de sous-unités plus petites. Pour l'instant nous ne savons pas si ce polymère est constitué exclusivement par de telles unités et s'il correspond à l'état natif de la protéine réceptrice intégrée dans la membrane.

4. Purification de la protéine réceptrice

Puisque la protéine réceptrice peut être extraite sans perte de sa capacité de liaison par des détergents non dénaturants et puisque cette protéine est stable pendant plusieurs jours en présence de ces détergents, il devrait être possible de la purifier. Les extraits habituels d'organe électrique d'*Electrophorus* contiennent environ 10–20 nmoles de

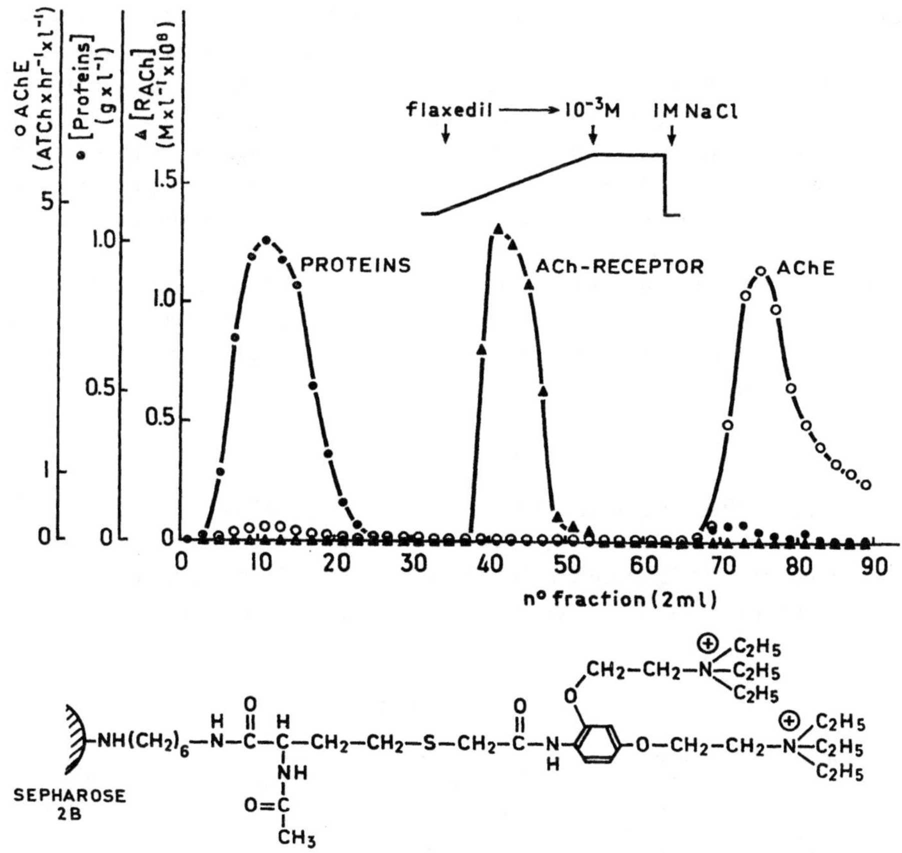

Fig. 1. Purification du récepteur cholinergique par chromatographie d'affinité (d'après Olsen *et al.* [41]).

récepteur par gramme de protéine. Si on fait l'hypothèse (non vérifiée) que la protéine réceptrice est constituée par une seule chaîne de masse moléculaire 50000, l'activité spécifique de la protéine réceptrice pure devrait être de 20000 nmoles par gramme de protéine. Nous avons utilisé la chromatographie d'affinité pour purifier cette protéine. Les premières colonnes que nous avons utilisées étaient constituées par des granules de Sépharose couplés à la toxine α de *Naja nigricollis* [30]. L'adsorption de la protéine réceptrice a lieu, mais les quantités de récepteur libéré de la colonne étaient toujours très faibles. Cela sans doute à cause de l'affinité très élevée de la toxine pour le récepteur. Les meilleurs résultats ont été obtenus avec une colonne constituée par un analogue du flaxedil, le CT 5263 (voir Figure 1), couplé à du Sépharose 2B [41]. La figure montre la purification de la protéine réceptrice obtenue avec une colonne de ce type en présence de Triton X-100 1%. Toute la protéine réceptrice s'adsorbe tandis que 90% des protéines ne s'adsorbe pas. La même figure montre que la protéine réceptrice est relarguée de la colonne en présence de flaxedil avec une très petite quantité de protéine et des quantités encore plus faibles d'acétylcholinestérase. Cette dernière protéine reste sur la colonne mais est libérée par du chlorure de sodium 1 M. Le rendement de la colonne était proche de 50% et le facteur de purification de 150 fois environ en une seule étape. Le matériel obtenu contenait moins d'un site catalytique de l'acétylcholine (en faisant l'hypothèse qu'un site pèse 65000 g) pour 100 sites récepteurs cholinergiques, son activité spécifique était de 2000 ± 500 nmoles de site récepteur cholinergique par gramme de protéine en utilisant notre méthode de dosage du récepteur par filtration sur Millipore. Les activités spécifiques les plus élevées atteignent 5500 nmoles par gramme de protéine. Ce matériel très purifié donne plusieurs bandes par électrophorèse en gel en présence de SDS.

La protéine réceptrice purifiée fixe encore les agonistes et antagonistes cholinergiques. Le décaméthonium est lié à cette protéine et complètement déplacé par la toxine α de *Naja nigricollis* ainsi que par le flaxedil: un résultat en accord total avec nos premiers travaux [15].

Remerciements

Les recherches effectuées dans l'Unité de Neurobiologie Moléculaire à l'Institut Pasteur bénéficient de l'aide du Collège de France, du Centre National de la Recherche Scientifique, de la Délégation Générale à la Recherche Scientifique et Technique, des 'National Institutes of Health' et du Commissariat à l'Énergie Atomique.

Références

[1] Changeux, J. P.: 1966, *Mol. Pharmacol.* **2**, 369.
[2] Changeux, J. P., Thiéry, J., Tung, Y., et Kittel, C.: 1967, *Proc. Nat. Acad. Sci. US* **57**, 335.
[3] Changeux, J. P. et Thiéry, J.: 1968, dans *Regulatory Functions of Biological Membrane* (ed. by J. Järnefelt), Elsevier, Amsterdam, p. 116.
[4] Fatt, P. et Katz, B.: 1951, *J. Physiol.* **115**, 320.
[5] Changeux, J. P., Podleski, T. R., et Meunier, J. C.: 1969, *J. Gen. Physiol.* **54**, 224 S.
[6] Monod, J., Changeux, J. P., et Jacob, F.: 1963, *J. Mol. Biol.* **6**, 306.
[7] Gerhart, J. C. et Schachman, H. K.: 1965, *Biochemistry* **4**, 1054.

[8] Nachmansohn, D.: 1959, *Chemical and Molecular Basis of Nerve Activity*, Academic Press, New York.
[9] Nachmansohn, D.: 1971, dans *Handbook of Sensory Physiology*, Vol. I (ed. by W. R. Lowenstein). Springer-Verlag.
[10] Higman, H., Podleski, T. R., et Bartels, E.: 1963, *Biochim. Biophys. Acta* **75**, 187.
[11] Higman, H., Podleski, T. R., et Bartels, E.: 1964, *Biochim. Biophys. Acta* **79**, 138.
[12] Podleski, T. et Changeux, J. P.: 1969, *Nature* **221**, 541.
]13] Kasai, M. et Changeux, J. P.: 1970, *C.R. Acad. Sci. Paris* **270**, 1400, D.
[14] Kasai, M. et Changeux, J. P.: 1971, *J. Membrane Biol.* **6**, 1.
[15] Changeux, J. P., Kasai, M., Huchet, M., et Meunier, J. C.: 1970a, *C.R. Acad. Sci. Paris* **270**, 2864, D.
[16] Changeux, J. P., Meunier, J. C., et Huchet, M.: 1971, *Mol. Pharmacol.* **7**, 538.
[17] Lee, C. Y. et Chang, C. C.: 1966, *Mem. Inst. Butantan Sim. Intern.* **33**, 555.
[18] Lee, C. Y.: 1970, *Clin. Toxicol.* **3**, 457.
[19] Karlsson, E., Arnberg, H., et Eaker, D.: 1971, *Eur. J. Biochem.* **21**, 1.
[20] Lester, H.: 1970, *Nature* **227**, 727.
[21] Lester, H.: 1971, Thèse Docteur ès Sciences, Université Rockefeller.
[22] Karlsson, E., Eaker, D., et Ponterius, G.: 1972, *Biochim. Biophys. Acta* **257**, 235.
[23] Boquet, P., Izard, Y., Jouannet, M., et Meaume, J.: 1966, *C.R. Acad. Sci. Paris* **262**, 1134, D.
[24] Boquet, P., Izard, Y., Jouannet, M., et Meaume, J., 1967, dans *Animal Toxins*, Pergamon Press, Oxford et New York, p. 293.
[25] Boquet, P., Izard, Y., et Ronseray, A. M.: 1970, *C.R. Acad. Sci. Paris* **271**, 1456, D.
[26] Karlsson, E., Eaker, D., et Porath, J.: 1966, *Biochim. Biophys. Acta* **127**, 505.
[27] Arai, H., Tamiya, N., Toshioka, S., Shinonaga, S., et Kano, R.: 1964, *J. Biochem. (Tokyo)* **56**, 568.
[28] Sato, S. et Tamiya, N.: 1971, *Biochem. J.* **122**, 453.
[29] Changeux, J. P., Kasai, M., et Lee, C. Y.: 1970b, *Proc. Nat. Acad. Sci. US* **67**, 1241.
[30] Meunier, J. C., Huchet, M., Boquet, P., et Changeux, J. P.: 1971a, *C.R. Acad. Sci. Paris* **272**, 117, D.
[31] Menez, A., Morgat, J. L., Fromageot, P., Ronseray, A. M., Boquet, P., et Changeux, J. P.: 1971, *FEBS Letters* **17**, 333.
[32] Weber, M., Menez, A., Fromageot, P., Boquet, P., et Changeux, J. P.: 1972, *C.R. Acad. Sci. Paris* **274**, 1575, D.
[33] Miledi, R., Molinoff, P., et Potter, L.: 1971, *Nature* **229**, 554.
[34] Meunier, J. C., Olsen, R., Menez, A., Fromageot, P., Boquet, P., et Changeux, J. P.: 1972a, *Biochemistry* **11**, 1200.
[35] Bartels, E. et Podleski, T. R.: 1964, *Biochim. Biophys. Acta* **79**, 511.
[36] Eldefrawi, M. E., Eldefrawi, A. T., et O'Brien, R. D.: 1971, *Proc. Nat. Acad. Sci. US* **68**, 1047.
[37] Meunier, J. C., Olsen, R., Menez, A., Morgat, J. L., Fromageot, P., Ronseray, A. M., Boquet, P., et Changeux, J. P.: 1971b, *C.R. Acad. Sci. Paris* **273**, 595, D.
[38] Meunier, J. C., Olsen, R., et Changeux, J. P.: 1972b, *FEBS Letters* **24**, 63.
[39] Karlin, A., Prives, J., Deal, W., et Winnik, M.: 1971, *J. Mol. Biol.* **61**, 175.
[40] Raftery, M.: 1972, dans *Neuroscience Research Program, Work Session on Receptor Biochemistry and Biophysics*, mars 1972.
[41] Olsen, R. W., Meunier, J. C., et Changeux, J.-P.: 1972, *FEBS Letters* **28**, 96.

GPSR Compliance

The European Union's (EU) General Product Safety Regulation (GPSR) is a set of rules that requires consumer products to be safe and our obligations to ensure this.

If you have any concerns about our products, you can contact us on ProductSafety@springernature.com

In case Publisher is established outside the EU, the EU authorized representative is:

Springer Nature Customer Service Center GmbH
Europaplatz 3
69115 Heidelberg, Germany

Batch number: 10815974

Printed by Printforce, the Netherlands